Wireless Sensor Networks

From Theory to Applications

OTHER TELECOMMUNICATIONS BOOKS FROM AUERBACH

Ad Hoc Mobile Wireless Networks: Principles, Protocols, and Applications, Second Edition
Subir Kumar Sarkar, T.G. Basavaraju, and C. Puttamadappa
ISBN 978-1-4665-1446-1

Building Next-Generation Converged Networks: Theory and Practice
Al-Sakib Khan Pathan, Mostafa Monowar, and Zubair Md. Fadlullah (Editors)
ISBN 978-1-4665-0761-6

Game Theory in Communication Networks: Cooperative Resolution of Interactive Networking Scenarios
Josephina Antoniou and Andreas Pitsillides
ISBN 978-1-4398-4808-1

Green Communications and Networking
F. Richard Yu, Xi Zhang, and Victor C.M. Leung (Editors)
ISBN 978-1-4398-9913-7

Handbook on Mobile and Ubiquitous Computing: Status and Perspective
Laurence T. Yang, Evi Syukur, and Seng W. Loke (Editors)
ISBN 978-1-4398-4811-1

Identification and Management of Distributed Data: NGN, Content-Centric Networks and the Web
Giovanni Bartolomeo and Tatiana Kováčiková
ISBN 978-1-4398-7907-8

Intelligent Sensor Networks: The Integration of Sensor Networks, Signal Processing and Machine Learning
Fei Hu (Editor)
ISBN 978-1-4398-9281-7

The Internet of Things in the Cloud: A Middleware Perspective
Honbo Zhou
ISBN 978-1-4398-9299-2

Linear Programming and Algorithms for Communication Networks: A Practical Guide to Network Design, Control, and Management
Eiji Oki
ISBN 978-1-4665-5263-0

Multihomed Communication with SCTP (Stream Control Transmission Protocol)
Victor C.M. Leung, Eduardo Parente Ribeiro, Alan Wagner, and Janardhan Iyengar
ISBN 978-1-4665-6698-9

Next-Generation Mobile Broadcasting
David Gómez-Barquero
ISBN 978-1-4398-9866-6

Optimal Resource Allocation for Distributed Video Communication
Yifeng He and Ling Guan
ISBN 978-1-4398-7514-8

SC-FDMA for Mobile Communications
Fathi E. Abd El-Samie, Faisal S. Al-kamali, Azzam Y. Al-nahary, and Moawad I. Dessouky
ISBN 978-1-4665-1071-5

Security and Privacy in Smart Grids
Yang Xiao (Editor)
ISBN 978-1-4398-7783-8

Security for Wireless Sensor Networks using Identity-Based Cryptography
Harsh Kupwade Patil and Stephen A. Szygenda
ISBN 978-1-4398-6901-7

Transmission Techniques for 4G Systems
Mário Marques da Silva, Americo Correia, Nuno Souto, and Joao Carlos Silva
ISBN 978-1-4665-1233-7

Wireless Sensor Networks: Current Status and Future Trends
Shafiullah Khan, Nabil Ali Alrajeh, and Al-Sakib Khan Pathan (Editors)
ISBN 978-1-4665-0606-0

Wireless Sensor Networks

From Theory to Applications

Edited by
Ibrahiem M. M. El Emary
S. Ramakrishnan

CRC Press is an imprint of the
Taylor & Francis Group, an **informa** business

CRC Press
Taylor & Francis Group
6000 Broken Sound Parkway NW, Suite 300
Boca Raton, FL 33487-2742

Version Date: 20130701

International Standard Book Number-13: 978-1-4665-1810-0 (Hardback)

Library of Congress Cataloging-in-Publication Data

Wireless sensor networks : from theory to applications / edited by Ibrahiem M.M. El Emary and S. Ramakrishnan.
pages cm
Summary: "Supplying comprehensive coverage of Wireless Sensor Networks (WSNs), this book covers the latest advances in WSN technologies. It considers some of theoretical problems in WSN, including issues with monitoring, routing, and power control, and details methodologies that can provide solutions to these problems. It examines applications of WSN across a range of fields, including health, defense military, transportation, and mining. Addressing the main challenges in applying WSNs across all phases of our life, it explains how WSNs can assist in community development"-- Provided by publisher.
Includes bibliographical references and index.
ISBN 978-1-4665-1810-0 (hardback)
1. Wireless sensor networks. I. El Emary, Ibrahiem M. M.

TK7872.D48W583 2013
004.6--dc23 2013014364

Visit the Taylor & Francis Web site at
http://www.taylorandfrancis.com

and the CRC Press Web site at
http://www.crcpress.com

Contents

Preface

A wireless sensor network (WSN) is a simple low-cost approach that can be used in a distributed environment. A WSN is a group of distributed devices that could be employed to monitor physical and environmental conditions in real time. It is also used in the control of instruments, with the major benefit being that it provides efficient reliable communications via a wireless network. WSNs support the formation of connectivity independently in addressing and routing structures without much help from human beings. WSNs have certain traits of their own. The limitations of WSNs include power, low battery life, redundant data acquisition, low duty cycle, and many-to-one flows. New design methodologies could be adapted to overcome these limitations. Despite these problems, many WSN solutions are developed based on notions regarding wireless communication and environment. Moreover, it requires an enormous breadth of knowledge from a variety of disciplines such as electromechanical computers and communications.

There are many books on WSNs, mainly focused on beginners, which are already available in the market. A few books are available for advanced readers but they fail to provide comprehensive coverage. Hence, we thought of writing a book to fill this gap. However, because of the diverse richness and rapid development of the subject, we believe that only teamwork will yield good material for this hot topic. Consequently, we collected chapters from domain experts working on various subfields of WSNs around the globe and edited these chapters to develop this book. We approached CRC Press with our proposal on editing a book on the said topic during August 2011. CRC Press accepted our proposal and subsequently we announced the call for book chapters to various academic and industrial experts working on WSNs worldwide.

Forty-three expert teams submitted contributions from countries such as the United States, the United Kingdom, Canada, Mexico, Taiwan, Jordan, India, Iran, Italy, France, Egypt, Malaysia, Japan, Algeria, Saudi Arabia, Greece, Nigeria, Korea, and China. The materials presented in this book were reviewed by 36 reviewers and finally we accepted only 25 out of the 43 submissions based on the recommendations and comments from these reviewers.

This book is an edited volume prepared mainly for senior undergraduate and postgraduate students, researchers, scholars and academics, industrial researchers, and practicing engineers working in the field of WSNs and aiming to develop some genuine solutions for WSNs. We assume that the readers have some prior knowledge of computer networks, wireless communication, and basic electronics. The reader is provided with a concise list of references at the end of each chapter.

Organization of the Chapters

These 25 chapters are divided into seven parts, namely, Part I: Data Collection; Part II: Physical Layer and Interfacing; Part III: Routing and Transport Protocols; Part IV: Energy Saving Approaches; Part V: Mobile and Multimedia WSN; Part VI: Data Storage and Monitoring; and Part VII: Applications.

Data collection is a fundamental function provided by WSNs. How to efficiently collect sensing data from all sensor nodes, how to aggregate the collected data, and how to increase the coverage area are all critical to the performance of sensor networks. Part I addresses these issues and contains three chapters—one each on data collection, aggregation, and spatial coverage. Chapter 1 discusses data collection from different types of sensor networks. It starts with network models, communicative models, and related work. Later, this chapter focuses on data collection in random sensor networks. This chapter also elaborates data collection in arbitrary sensor networks under different categories. Chapter 2 gives a clear picture about data aggregation and data gathering in WSNs. This chapter deals with energy-efficient data aggregation protocols for a heterogeneous WSN. Various algorithms and protocols for data gathering and data aggregation are also discussed. Chapter 3 provides a survey of the various methods proposed for the estimation and optimization of spatial coverage of sensor networks. In addition, the chapter also concentrates on the state-of-the-art sensor networks and their related issues in terms of phenomena type of the environment and sensor as well as issues such as coverage communication and energy-saving problems.

Part II deals with physical layer and interfacing and also comprises three chapters. Chapter 4 deals with IEEE 802.15.4 WSNs. The architecture and functionalities of the ZigBee sensor networks are discussed. The IEEE 802.15.4 standard and the characteristics of 3-D terrains in WSNs are introduced. The performances of IEEE 802.15.4 WSNs in five artificial 3-D terrains are evaluated. A brief presentation on 3-D terrain visualization software is also available in this chapter. Chapter 5 is about multi-interface model. The idea is to exploit the heterogeneity of the interfaces available in modern devices in order to reduce energy consumption and prolong network lifetime. Several well-known combinatorial optimization problems are then reconsidered with respect to this new feature. Chapter 6 deals with the sensor bus architecture for a real-time WSN, and this applicability has been tested with three different sensor networks, which includes a real-world temperature sensor network from Texas Instruments. It has been integrated with the sensor bus to show its practical application in real-time temperature monitoring applications. Sun SPOT sensor networks, which generate random values using a Solarium emulator and a simulated sensor network, are also discussed.

Part III has five chapters on routing and transport protocols. Chapter 7 addresses the development of a specific framework for the computation of the network topology. Link layer performance is evaluated and the link level throughput and delay are defined. Routing performance is analyzed for IEEE 802.15.6. Chapter 8 focuses on the routing protocols that cover low energy adaptive clustering hierarchy (LEACH), C-LEACH, V-LEACH, N-LEACH, and PEGASIS, which are modified versions of the LEACH protocol. In addition to discussions on multihop routing in wireless ad hoc and sensor networks, a detailed survey on various connected dominating set construction techniques for ad hoc sensor networks is given in Chapter 9. This chapter also provides details on various network models used for virtual backbone construction. In Chapter 10, the authors discuss the different types of transport protocols in WSNs, and their guidelines performance metrics and congestion control mechanisms are shown. The existing transport protocols for WSNs are briefly reviewed and several problems in the existing protocols are also listed. Chapter 11 discusses energy-efficient MAC protocols for WSNs. It comprises radio energy consumption models, MAC layer issues for WSNs design goals, energy trade-offs, and metrics. It also elaborates about the emerging MAC layer protocols such as Y-MAC, EM-MAC, and PIP.

Part IV has four chapters exclusively focusing on energy efficient methods in WSNs. Chapter 12 focuses on event monitoring and energy saving in the WSNs. Energy saving is accomplished using ultralight pulse switching protocol for resource-constrained sensors in event monitoring and target tracking applications. The joint MAC routing architectures for pulse switching with the hop angular and cellular event localization are also presented in this chapter. Chapter 13 explains about energy handling with the help of previous research findings and points out the major issues in energy conservation congestion control and avoidance in WSNs. Chapter 14 starts with the concept of cooperative multiple-input/multiple-output (MIMO) in WSNs and different forms of MIMO and its configurations. Then, the chapter focuses on energy consumption techniques, the energy model, and a complete study of the parameters that affect the system in various situations. The clustering algorithm, which is a kind of key technique used to reduce energy consumption in WSNs, is dealt with in Chapter 15. It details clustering requirement formation maintenance and the factors affecting clustering network architecture, etc. The focus is given to discussing the application of WSNs in watering for irrigation and effective usage of water in agriculture.

Part V is on mobile and multimedia WSNs and has three chapters. Chapter 16 focuses on GPS-free and anchor-free indoor localization schemes. This chapter also comprises information on works related to mobile sensor networks along with the proposed network model. A cognitive approach in mobile WSNs is addressed in Chapter 17. The chapter also describes the various challenges and design issues faced by MWSNs in addition to the major focus on cognitive radio-based WSNs. Chapter 18 introduces the acorrelation-based communication framework which leverages the spatial correlation of visual information in communication protocols for wireless multimedia sensor networks. A novel analytical spatial correlation model based on the projection geometry of camera sensors is provided in this chapter. It also provides a method for predicting the compression efficiency of correlated camera sensors, which is used to designed clustering and routing algorithms for multimedia sensor networks.

Data storage and monitoring are the themes of four chapters in Part VI. Distributed data storage and retrieval schemes using IPv6 in WSNs are discussed in Chapter 19. The state-of-the-art on distributed data storage techniques and current standard Internet of Things protocols are discussed. Then, a distributed data storage scheme called the low-complexity greedy mechanism is presented and analyzed. Chapter 20 deals with the monitoring mechanism for WSNs. This chapter introduces various monitoring mechanisms and their challenges. This chapter also provides solutions for those problems using a wireless distributed intrusion detection system. Building and orchestrating centralized remote management procedures for WSNs using the TinyOS platform and OpenRSM is dealt with in Chapter 21. It details the application development in TinyOS. This chapter also provides details for installing TinyOS remotely using web-based technologies such as HTML5, AJAX, and middleware, which provide platform-independent operations. Chapter 22 addresses the challenges for QoS support in WSNs. This chapter describes QoS performance metrics in WSNs at different levels. This chapter provides details on different types of services in WSNs and various mechanisms to achieve QoS in WSNs.

Most of the chapters that have been discussed thus far are oriented toward applications of WSNs in addition to their theoretical discussions. However, the last three chapters of this book in Part VI exclusively focus on the applications of WSNs. The ways and means to use the wireless body area networks (WBANs) in artificial eye vision is discussed in Chapter 23. Details about WBANs, which includes the system prototype and the data rates, are given. This is followed by discussions on the types of sensor benefits and challenges of WBANs, such as how WBANs can be used in an artificial retina. Chapter 24 considers WSNs from a medical perspective. It presents an overview of the effect of WSNs on medical health care as well as highlighting the state-of-the-art in the applications of WSNs for tele–health care. Also, this chapter describes the challenges introduced by health care applications to WSNs. Chapter 25 deals with applying WSNs for monitoring various phenomena in different environmental fields. This chapter describes the challenges introduced by smart environmental monitoring applications of WSNs.

Acknowledgments

We are directly and indirectly indebted to a number of individuals who assisted in the preparation of the book. In particular, we are grateful to the 36 referees who performed the task of reviewing chapters and providing valuable suggestions, comments, and criticisms in improving the quality of the work. Our special thanks and congratulations are due to the authors of the 25 chapters for successfully submitting their work according to the recommendations of the reviewers. The support of CRC Press is really tremendous and is to be appreciated greatly. We are pleased to thank our colleagues for their cooperation in sharing our academic load during this work. We cannot find the right words to express our thankfulness to our family members for their love, prayers, and backing.

We are open to comments and criticisms from the readers in improving the quality of this book for future editions.

Ibrahiem M. M. El Emary
Information Technology Deanship
King Abdulaziz University Jeddah
Kingdom of Saudi Arabia

S. Ramakrishnan
Professor and Head
Department of Information Technology
Dr. Mahalingam College of Engineering & Technology
Pollachi, India

Editors

Ibrahiem M. M. El Emary received his doctor of engineering degree in 1998 from the Electronic and Communication Department, Faculty of Engineering, Ain Shams University, Egypt. From 1998 to 2002, he was an assistant professor of computer sciences in different faculties and academic institutions in Egypt. From 2002 to 2010, he worked as visiting assistant and associate professor of computer science and engineering in two universities in Jordan. Currently, he is a professor of computer science and engineering at King Abdulaziz University, Jeddah, Kingdom of Saudi Arabia. His research interests cover various analytic and discrete event simulation techniques, performance evaluation of communication networks, application of intelligent techniques in managing computer communication networks, and performing comparative studies between various policies and strategies of routing congestion control subnetting of computer communication networks. He has published more than 150 articles in various refereed international journals and conferences covering computer networks, artificial intelligent expert systems, software agents, information retrieval, e-learning, case-based reasoning, image processing, and pattern recognition wireless sensor networks, cloud computing, and robotic engineering. Also, he has participated in publishing seven book chapters in three international books (published by Springer-Verlag, IGI Global, and Nova Science Publishers) as well as editor of two books from international publishers (LAP Lampert-Germany). He has been included in the 2013 *Marquis Who's Who in the World*.

S. Ramakrishnan received his bachelor of engineering degree in electronics and communication engineering in 1998 from the Bharathidasan University Trichy, and his mechanical engineering degree in communication systems in 2000 from the Madurai Kamaraj University, Madurai. He received his Ph.D. degree in information and communication engineering from Anna University, Chennai in 2007.

He has 12 years of teaching experience and 1 year of industry experience. He is a professor and the head of the Department of Information Technology, Dr. Mahalingam College of Engineering and Technology, Pollachi, India.

Dr. Ramakrishnan is a reviewer of 14 international journals such as *IEEE Transactions on Image Processing*, *IET Journals* (formerly IEE), *ACM Computing Reviews*, *International Journal of Vibration and Control*, *IET Generation Transmission and Distribution*, etc. He is on the editorial board of six international journals. He is a guest editor of special issues in three international journals including *Telecommunication Systems* (Springer). He has published 103 papers in international/national journals and conference proceedings. Dr. S. Ramakrishnan has published three books on computational techniques and speech processing. He has also reviewed three books for McGraw-Hill International Editions and four books for *ACM Computing Reviews*. He is a convenor of the IT board at Anna University of Technology-Coimbatore Board of Studies (BoS). He is guiding 10 PhD research scholars. His biography has been included in the 2011 *Marquis Whos's Who in the World*. His areas of research include digital image processing, soft computing, human–computer interaction, wireless sensor networks, and cognitive radio.

Contributors

Abraham Lamesgin Addisie
Department of Electrical and Computer Engineering
Addis Ababa University
Addis Ababa, Ethiopia

Kemal Akkaya
Department of Computer Science
Southern Illinois University
Carbondale, Illinois

Meysam Argany
Center for Research in Geomatics
Laval University
Quebec, Canada

Lalit Kumar Awasthi
CSED
National Institute of Technology
Hamirpur, India

Soumya Banerjee
Birla Institute of Technology
Mesra, India

and

Scientific Research Group in Egypt (SRGE)
Cairo University
Cairo, Egypt

Ali Bicak
Department of Information Technology and Management Science
Marymount University
Arlington, Virginia

Subir Biswas
Electrical and Computer Engineering
Michigan State University
East Lansing, Michigan

Siddhartha Chauhan
CSED
National Institute of Technology
Hamirpur, India

Lu Chen
School of Software
Dalian University of Technology
Dalian, China

Yuanfang Chen
School of Software
Dalian University of Technology
Dalian, China

Suchismita Chinara
Department of Computer Science and Engineering
National Institute of Technology Rourkela
Odisha, India

Rui Dai
Department of Computer Science
North Dakota State University
Fargo, North Dakota

Gianlorenzo D'Angelo
MASCOTTE Project
INRIA/I3S (CNRS/UNSA)
Sophia Antipolis, France

Ariyam Das
Yahoo! Research & Development
India

Ph. De Doncker
OPERA Department
Université Libre de Bruxelles
Brussels, Belgium

Vivek S. Deshpande
Department of Information Technology
MIT College of Engineering
Pune, India

Gabriele Di Stefano
Dipartimento di Ingegneria e Scienze dell'Informazione e Matematica
Università degli Studi dell'Aquila
L'Aquila, Italy

Bo Dong
Electrical and Computer Engineering
Michigan State University
East Lansing, Michigan

J.-M. Dricot
OPERA Department
Université Libre de Bruxelles
Brussels, Belgium

Nashwa El-Bendary
Arab Academy for Science, Technology, and Maritime Transport
and
Scientific Research Group in Egypt
Cairo University
Cairo, Egypt

Shaimaa Ahmed Elsaid
Electronics and Communications Department
Zagazig University
Sharkeya, Egypt

Gianluigi Ferrari
Department of Information Engineering
University of Parma
Parma, Italy

Mohamed Mostafa M. Fouad
Arab Academy for Science, Technology, and Maritime Transport
Cairo, Egypt
and
Scientific Research Group in Egypt
Cairo University
Cairo, Egypt

Pietro Gonizzi
Department of Information Engineering
University of Parma
Parma, Italy

Aboul Ella Hassanien
Information Technology Department
Cairo University
Cairo, Egypt

Qiong Huo
Electrical and Computer Engineering
Michigan State University
East Lansing, Michigan

Manas Ranjan Kabat
Department of Computer Science and Engineering
VSS University of Technology
Burla Sambalpur, India

Michael N. Kalochristianakis
Department of Applied Science
Technological Educational Institute of Crete
Heraklion, Greece

Farid Karimipour
Department of Surveying and Geomatics
University of Tehran
Tehran, Iran

Aristotelis Kretsis
Department of Computer Engineering and Informatics
University of Patras
Patras, Greece

Sumit Kumar
Communication Research Center
IIIT Hyderabad
Andhra Pradesh, India

Jérémie Leguay
Advanced Studies Department
THALES Communications and Security
Colombes, France

Jenq-Shiou Leu
Department of Electronic Engineering
National Taiwan University of Science and Technology
Taipei, Taiwan

Kuen-Han Li
Department of Electronic Engineering
National Taiwan University of Science and Technology
Taipei, Taiwan

Mu-Sheng Lin
Department of Electronic Engineering
National Taiwan University of Science and Technology
Taipei, Taiwan

Chittaranjan Mandal
IIT Kharagpur
Kharagpur, West Bengal, India

Nikhil Marriwala
Electronics and Communication Engineering Department
Kurukshetra University
Kurukshetra, India

Paolo Medagliani
Advanced Studies Department
THALES Communications and Security
Colombes, France

Prabhudutta Mohanty
Department of Computer Science and Engineering
VSS University of Technology
Burla Sambalpur, India

Mir Abolfazl Mostafavi
Center for Research in Geomatics
Laval University
Quebec, Canada

Tamer Nadeem
Department of Computer Science
Old Dominion University
Norfolk, Virginia

Alfredo Navarra
Dipartimento di Matematica e Informatica
Università degli Studi di Perugia
Perugia, Italy

A. Nonclercq
OPERA Department
Université Libre de Bruxelles
Brussels, Belgium

Evangelia Psilidou
Department of Computer Engineering and Informatics
University of Patras
Patras, Greece

Abderrezak Rachedi
Gaspard Monge Computer Science
Université Paris-Est Marne-la-Vallée
Marne-la-Vallée, France

Kumudha Raimond
Department of Computer Science and Engineering
Karunya University
Coimbatore, India

Rabie A. Ramadan
Computer Engineering Department
Cairo University
Cairo, Egypt

G. Ramamurthy
Communication Research Center
IIIT Hyderabad
Andhra Pradesh, India

Chris Reade
Kingston University
London, United Kingdom

M. Rinaudo
OPERA Department
Université Libre de Bruxelles
Brussels, Belgium

and

Department of Information Engineering
University of Parma
Parma, Italy

Rawya Yehia Rizk
Electrical Engineering Department
Port Said University
Port Said, Egypt

Lei Shu
Guangdong Petrochemical Equipment Fault Diagnosis Key Laboratory
Guangdong University of Petrochemical Technology
Maoming, China

M. Sujeethnanda
Communication Research Center
IIIT Hyderabad
Andhra Pradesh, India

Asis Kumar Tripathy
Department of Computer Science and Engineering
National Institute of Technology Rourkela
Odisha, India

S. Van Roy
OPERA Department
Université Libre de Bruxelles
Brussels, Belgium

Emmanouel (Manos) Varvarigos
Department of Computer Engineering and Informatics
University of Patras
Patras, Greece

Pu Wang
School of Electrical and Computer Engineering
Georgia Institute of Technology
Atlanta, Georgia

Yu Wang
Department of Computer Science
University of North Carolina at Charlotte
Charlotte, North Carolina

Jean-Lien C. Wu
Department of Computer and Communication Engineering
St. John's University
Taipei, Taiwan

Mohamed Younis
Department of Computer Science and Electrical Engineering
University of Maryland, Baltimore County
Baltimore, Maryland

DATA COLLECTION I

Chapter 1

Data Collection in Wireless Sensor Networks: A Theoretical Perspective

Yu Wang

University of North Carolina at Charlotte

Contents

1.1 Introduction

A wireless sensor network (WSN) consists of a set of sensor devices that are spread over a geographical area [1]. These sensors are able to perform processing as well as sensing and are additionally capable of communicating with each other. Due to the wide range of its potential applications, such as in the battlefield, emergency relief, environment monitoring, and so on, sensor networking has recently emerged as a premier research topic. For WSNs, the ultimate goal is often to collect sensing data from all sensors to certain sink nodes and then perform further analyses at these sink nodes. Thus, data collection is one of the most common services used in sensor network applications. Figure 1.1 shows an example of the data collection process in a WSN, in which a single sink node s at the center collects sensing values from every sensor using a collection tree.

The performance of data collection in sensor networks can be characterized by the rate at which sensing data can be collected and transmitted to sink nodes. In particular, theoretical measures that capture the possibilities and limitations of collection processing in sensor networks are the *delay* and *capacity* for many-to-one data collection. The delay of data collection is the time to transmit one single snapshot to sinks from its generation at sensors. Considering the size of data in the snapshot, we can define *delay rate* as the ratio between the data size and the delay. Clearly, a large delay rate is desired. When multiple snapshots from sensors are generated continuously, data transport can be pipelined in the sense that further snapshots may begin to transport before sinks receive the prior snapshot. The maximum data rate at the sinks to continuously receive snapshot data from sensors is defined as the capacity of data collection. Note that the capacity is always larger than or equal to the delay rate. Both delay rate and capacity reflect how fast the sinks can collect sensing data from all sensors. It is critical to understand the limitations of many-to-one information flows and devise efficient data collection algorithms to maximize the performance of WSNs. In this chapter, we are particularly interested in how the delay rate and capacity of data

Figure 1.1 Data collection in a WSN with a single sink s.

collection vary in theory as the number of sensors increases. We will study some fundamental capacity problems arising from different types of data collection scenarios in WSNs. For each problem, we will introduce the asymptotic upper bound of transport capacity and present some efficient algorithms to achieve or approximate the upper bound.

1.1.1 Network Model

We focus on the theoretical capacity bound of data collection in WSNs. We consider a static sensor network, which includes n wireless sensor nodes $V = \{v_1, v_2, \ldots, v_n\}$ and k sink nodes $S = \{s_1, s_2, \ldots, s_k\}$ (when $k = 1$, we use s to denote the single sink). We assume that both sensor nodes and sink nodes are deployed in a two-dimensional square area. Two types of networks will be considered in this chapter: random networks and arbitrary networks (Figure 1.2 shows an example for each case). In random networks, sensor nodes are uniformly and randomly deployed in the area. Usually, under this model, the number of sensor nodes in the network is assumed to be very large. Such an assumption is useful to simplify the analysis and derive nice theoretical limits. Thus, the random network model has been widely used in the community for analyzing network performance. On the other hand, random networks may be invalid in many practical sensor applications in which the number of sensors is limited and the distribution of sensors is uneven inside the deployment region. In these cases, the arbitrary network model can be used. In arbitrary networks, sensors are deployed in any distribution and can form any network topology. Obviously, this model is more general and the random network model is just a special case of it.

Throughout this chapter, we assume each sensor node transmits at a fixed transmission power P. Then, a fixed transmission range r can be defined such that a node v_i can successfully receive the signal sent by node v_j only if $||v_i - v_j|| \leq r$. Here, $||v_i - v_j||$ is the Euclidean distance between v_i and v_j. We can further define a communication graph $G = (V, E)$, where V is the set of all nodes (including the sink) and E is the set of all possible communication links. This graph model is called a disk graph model. We assume the communication graph G is connected.

At regular time intervals, each sensor node measures the field value at its position and transmits the value to one of the sink nodes. We assume that the channel bandwidth for all wireless links is W bits per second. We also assume that all packets have the unit size of b bits. Time is divided into slots with $t = b/W$ seconds. Accordingly, only one packet can be transmitted in each time slot between two neighboring nodes. Time division multiple access (TDMA) scheduling is used at the media access control (MAC) layer.

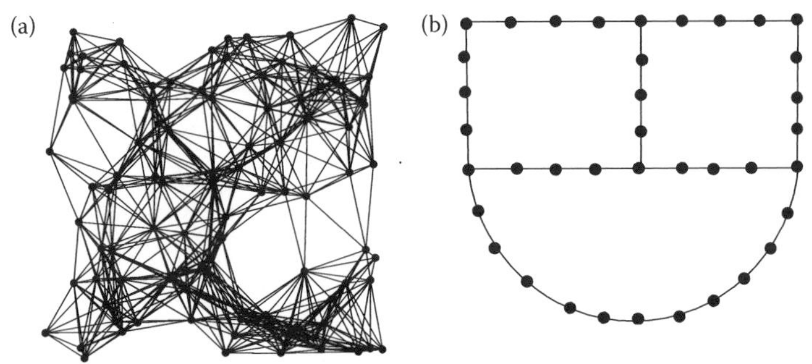

Figure 1.2 A random network (a) versus an arbitrary network (b).

1.1.2 Communication Model

Due to spatial separation, several sensors can successfully transmit at the same time if these transmissions do not cause any destructive wireless interference. There are three widely used communication models [2,3] to capture such interference constraints: the protocol model, the physical model, and the generalized physical model.

In the protocol model (also called protocol interference model), all nodes are assumed to have a uniform interference range R. When node v_i transmits to node v_j, node v_j can receive the signal successfully if no other node within a distance R of v_j is transmitting simultaneously. Figure 1.3 illustrates an example of the protocol model in which the transmission between v_k and v_q will cause interference at v_j. Usually, R/r is assumed as a constant α larger than 1. The protocol model is the simplest communication model considering the interference among nodes, and has been widely used in the literature. However, it is sometimes too simple to capture the complexity of interference.

In the physical model (also called physical interference model), node v_j can correctly receive the signal from the sender v_i if and only if, given a constant $\eta > 0$, the signal-to-noise ratio (SINR)

$$\frac{P \cdot l(v_i, v_j)}{N_0 + \sum_{k \in I} P \cdot l(v_k, v_j)} \geq \eta.$$

Here, $l(v_i, v_j)$ is the transmission loss between v_i and v_j, $N_0 > 0$ is the background Gaussian noise, I is the set of actively transmitting nodes when node v_i is transmitting, and P is the fixed transmission power. In this chapter, we consider the attenuation function $l(v_i, v_j) = \min\{1, \|v_i - v_j\|^{-\beta}\}$ where $\beta > 2$ is the path loss exponent and $\|v_i - v_j\|$ is the Euclidean distance between v_i and v_j. Hereafter, we assume that all P, N_0, β, and η are fixed constants. Notice that the values of P, N_0, η, and transmission range r should satisfy $\frac{P \cdot r^{-\beta}}{N_0} \geq \eta$. Thus, $r \leq \left(\frac{P}{N_0 \cdot \eta}\right)^{1/\beta}$.

For both the protocol model and the physical model, as long as the value of a given conditional expression (such as transmission distance or SINR value) reaches some threshold, the sender can

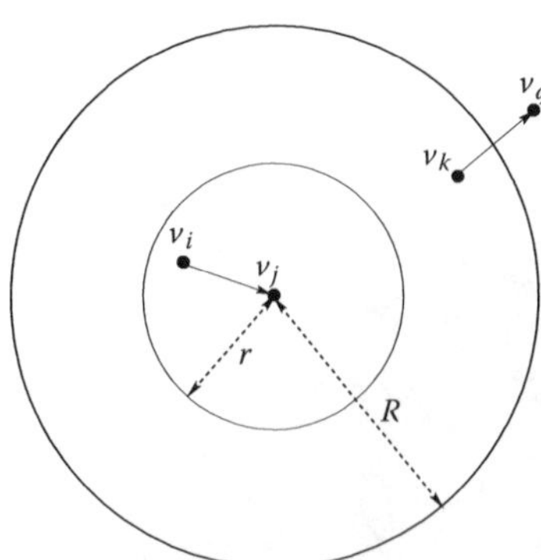

Figure 1.3 In the protocol model, node v_j can receive the signal successfully from v_i if v_i is within v_j's transmission range r, and no other transmitting node within a distance R of v_j. In this example, v_k's transmission will cause interference at node v_j.

send data successfully to a receiver at a specific constant rate W due to the fixed rate channel model. However, the fixed rate channel model may not capture the feature of wireless communication well. As a result, a more realistic model, the generalized physical model (also called Gaussian channel model) is introduced. Such a model determines the rate under which the sender can send its data to the receiver reliably, based on a continuous function of the receiver's SINR. Any two nodes v_i and v_j can establish a direct communication link v_iv_j over a channel of bandwidth W, of rate

$$W_{ij} = W \log_2\left(1 + \frac{P \cdot l(v_i, v_j)}{N_0 + \sum_{k \in I} P \cdot l(v_k, v_j)}\right).$$

This model assigns a more realistic transmission rate at a larger distance than the fixed rate channel models (protocol model and physical model). In this chapter, we will cover the data collection in WSNs under all these communication models. Without specific notification, we use the protocol model as the default model in our analysis.

1.1.3 Capacity and Delay in Data Collection

We now formally define delay and capacity of data collection in WSNs. Recall that each sensor, at regular time intervals, generates an independent field value with b bits and wants to transport it to one of the sink nodes. The union of all sensing values from n sensors at a particular sampling time is called a *snapshot* of the sensing data. The task of data collection is to collect these snapshots from all sensors to sinks as quickly as possible.

Definition 1

The delay of data collection Δ is the time transpired between the time a snapshot is taken by the sensors and the time the sinks have all data of this snapshot. ■

Definition 2

The delay rate of data collection Γ is the ratio between the data size of one snapshot nb and the delay Δ. ■

It is clear that we prefer smaller delay and larger delay rate so that the sink can get each snapshot more quickly. On the other hand, the data transport can be pipelined in the sense that further snapshots may begin to transport before the sinks receive prior snapshots completely. Therefore, we need to define a new data rate of data collection under pipelining (sometimes called continuous data collection).

Definition 3

The *usage rate* of data collection U is the number of time slots needed at sinks between completely receiving one snapshot and completely receiving the next snapshot. ■

Thus, the time used by sinks to successfully receive a snapshot is $T = U \times t$. Notice that due to pipelining, T is always smaller than or equal to Δ. Clearly, small usage rate and T are desired.

Definition 4

The capacity of data collection C is the ratio between the size of data in one snapshot and the time to receive such a snapshot (i.e., nb/T) at the sinks. ■

Thus, the capacity C is the maximum data rate at the sinks to continuously receive the snapshot data from sensors. Clearly, C is at least as large as the delay rate Γ, and is usually substantially larger.

In this chapter, we analyze the delay rate and capacity for data collection in both random and arbitrary WSNs under various communication models. Notice that in our definitions, we require data from every sensor to reach the sink in the same rate; thus, fairness among all sensors is guaranteed.

1.1.4 Related Works

Gupta and Kumar [4] initiated research on the capacity of wireless ad hoc networks by studying fundamental capacity limits in their seminal article under both protocol and physical models. The following articles studied capacity under different communication scenarios in wireless networks: unicast [5,6], multicast [7–9], and broadcast [10,11] capacities. In this chapter, we focus on the capacity of data collection, which is an all-to-one communication scenario different from the unicast, multicast, and broadcast capacities.

The capacity of data collection in random WSNs has been previously studied [12–25]. In previous work by Duarte-Melo et al. [12,13], they first studied the many-to-one transport capacity in random sensor networks under the protocol model and gave the results of overall capacity of data collection as $\Theta(W)$. They also showed that compressing data is inefficient in improving capacity when the density of the sensor network increases to infinity [13]. El Gamal [14] studied data collection capacity subject to a total average transmitting power constraint. The assumption that every node can only receive a packet from one source node at a time was relaxed, and it was shown that the capacity of random networks scales as $\Theta((\log n)\ W)$ when n goes to infinity and the total average power remains fixed. Their methods used antenna sharing and channel coding. Barton and Zheng [15,16] also investigated data collection capacity under more complex physical models [a noncooperative SINR model and a cooperative time reversal (CTR) communication model]. They first demonstrated that $\Theta((\log n)\ W)$ is optimal and achievable by using CTR for a regular grid network [15], then showed that the capacities of $\Theta((\log n)\ W)$ and $\Theta(W)$ are optimal and achievable by CTR when operating in fading environments with power path-loss exponents that satisfy $2 < \beta < 4$ and $\beta \geq 4$ for random networks [16]. Liu et al. [17] recently introduced the capacity of a more general some-to-some communication paradigm in random networks where there are $s(n)$ randomly selected sources and $d(n)$ randomly selected destinations. They derived the upper and lower bounds for such a problem. Note that data collection is a special case for their problem when $s(n) = n$ and $d(n) = 1$. Most recently, Ji et al. [18–21] also studied data collection methods in random WSNs under different communication models, such as dual-radio multichannel networks [18], asynchronous WSNs [20], or probabilistic network models [21].

This chapter mainly covers the recent results from the author and his colleagues [22–27] on data collection capacity in both random and arbitrary WSNs. However, readers are encouraged to further read the references listed above to get the complete picture of capacity research in wireless networks.

The rest of the chapter is organized as follows. We first discuss the study on data collection capacity of random WSNs under various communication models and network scenarios in Section 1.2. We then consider the data collection capacity for an arbitrary WSN under different models. Finally, we conclude this chapter with a short summary in Section 1.4.

1.2 Data Collection in Random Sensor Networks

In this section, we focus on data collection in a large-scale random WSN and study how fast it can be performed under the existence of interference among sensors. We consider both cases with a single sink or multiple sinks under the protocol model, and also discuss the cases under physical models.

1.2.1 Preliminaries

We consider a random WSN in which n sensor nodes are randomly and uniformly deployed in a square area with side length l. Two types of random network models [28] can be defined: random dense network and random extended network. In the random dense network, sensors are uniformly deployed in a unit square area ($l = 1$). Thus, its node density is n. In the random extended network, sensors are uniformly deployed in a square region with $l = \sqrt{n}$, thus its density is 1. In most of this chapter (except for Section 1.2.4), we use the random dense network.

We now introduce a classic grid-partition method that is essential for the proposed data collection methods and theoretical analysis. As shown in Figure 1.4, the network (e.g., the unit square) is divided into a^2 micro cells of the size $d \times d$. Here, $a = 1/d$. We assign each cell a coordinate (i,j), where i and j are between 1 and a, indicating its position at the jth row and ith column.

The following lemma gives a guidance of the cell size.

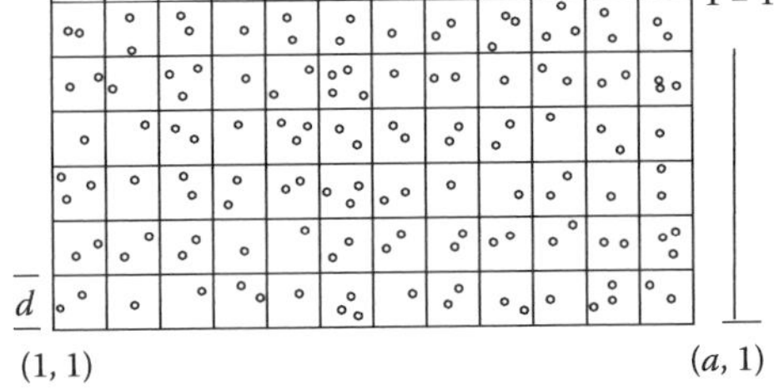

Figure 1.4 Grid partition of the WSN: a^2 cells with a cell size of $d \times d$.

Lemma 1

Given n random nodes in a unit square [29], dividing the square into micro cells of the size $\sqrt{3\frac{\log n}{n}} \times \sqrt{3\frac{\log n}{n}}$, every micro cell is occupied with a probability of at least $1-\frac{1}{n^2}$. ■

Therefore, if $d=\sqrt{3\frac{\log n}{n}}$ (i.e., $a=\sqrt{\frac{n}{3\log n}}$), every micro cell has at least one node with a high probability (the probability converges to one as $n\to\infty$).

We can also derive the upper bound of the number of nodes inside a single cell.

Lemma 2

Given n random nodes in a unit square [22,23], dividing the unit square into micro cells of the size $\sqrt{3\frac{\log n}{n}} \times \sqrt{3\frac{\log n}{n}}$, the maximum number of nodes in any cell is $O(\log n)$ with a probability of at least $1-\frac{3\log n}{n}$. ■

The proof is straightforward from the following lemma when the number of balls $\gamma = n$ and the number of bins $\delta = a^2 = \frac{n}{3\log n}$. Lemma 2 indicates the number of nodes inside any cell is bounded from above by $O(\log n)$ with high probability.

Lemma 3

Randomly putting γ balls into δ bins [30], with a probability of at least $1-\frac{1}{\delta}$, the maximum number of balls in any bin is $O\left(\frac{\gamma}{\delta}+\log\delta\right)$. ■

To make the whole network connected, the transmission range r needs to be equal to or larger than $\sqrt{5}d$ so that any two nodes from two neighboring cells are inside each other's transmission range. Hereafter, we set $r=\sqrt{5}d=\sqrt{15\frac{\log n}{n}}$.

1.2.2 Data Collection with a Single Sink

In this subsection, we consider the simplest situation: data collection under the protocol model in a sensor network in which a single sink s is located in the upper right corner of the deployment region (i.e., cell (a,a) as shown in Figure 1.5). Notice that if the sink is located at the center of the region or anywhere in the region, it only adds a constant in the analysis. We first construct a data

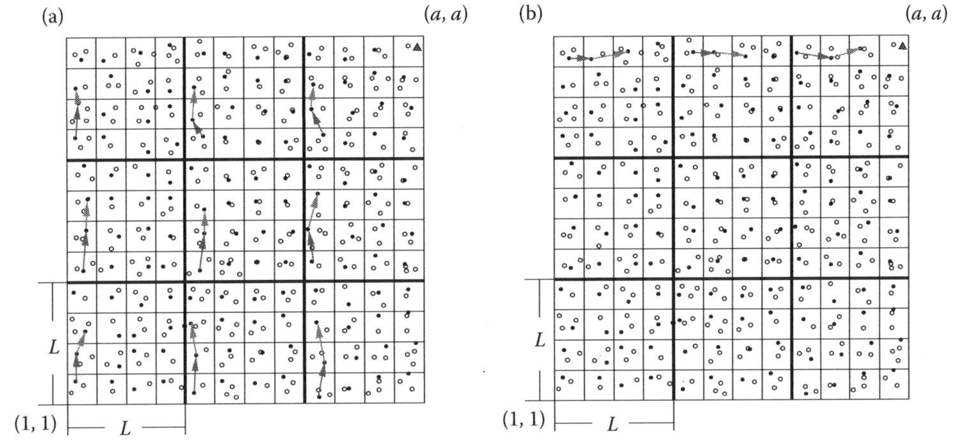

Figure 1.5 Our collection method: (a) every node sends its data to the upper cell in Phase I; (b) then each node in the top row sends its data to the cell to its right in Phase II.

collection scheme whose delay and delay rate are $O(nt)$ and $\Omega(W)$, respectively, and then prove that these values are order-optimal.

Our collection algorithm has two phases. In the first phase (phase I), every sensor sends its data up to the highest cell in its column (in the ath row) as shown in Figure 1.5a, and in the second phase (phase II), all data is sent via cells in the ath row to the sink as shown in Figure 1.5b. We define the time needed for these two phases as T_1 and T_2, respectively.

By Lemma 2, the number of nodes in each cell is at most $O(\log n)$. Every node needs one time slot t to send one packet to its neighbor in the next cell. However, due to wireless interference, when node v_i transmits a packet to v_j, nodes within R distance from v_j cannot transmit any packets in the same time slot. Let $L = \left(\frac{R}{d} + 2\right)$. Thus, every $L \times L$ cell (we call it an *interference block* hereafter) can only have one node send a packet to its upper neighbor in every time slot t during phase I. In Figure 1.5, bold lines show interference blocks. Remember that $\frac{R}{r} = \alpha$ and $\frac{r}{d} = \sqrt{5}$, so $\frac{R}{d}$ and L are also constants, and a packet in the lowest row (i.e., cell $(0,k)$) has to walk a cells to reach nodes in the highest cell in the rectangle. Hence,

$$T_1 \leq L \times L \times t \times O(\log n) \times a = O(t \log n)a = O(t \log n)\sqrt{\frac{n}{3\log n}} = O\left(t\sqrt{n \log n}\right).$$

In the beginning of phase II, all data are already at the cells of the top row. The sink s lies in the same row with these cells. We now estimate the time T_2 needed for sending all data to s. Each cell in the top row has at most $a \times O(\log n)$ nodes' data and the interference block is now $1 \times L$. Similarly, we can get

$$T_2 \leq L \times t \times a \times O(\log n) \times a = O(t \log n)a^2 = \frac{n}{3\log n} O(t \log n) = O(nt).$$

Therefore, the total time needed to collect b bits information from every sensor to the sink is $T_1 + T_2 = O(nt)$. That is, the total delay Δ for the sink to receive a complete snapshot is at most $O(nt)$. Consequently, the total delay rate of this collection scheme is

$$\Gamma = \frac{nb}{\Delta} = \Omega\left(\frac{nb}{nt}\right) = \Omega(W).$$

It has been proved that the upper bound of delay rate or capacity of data collection is W [12,13]. It is obvious that the sink cannot receive at a rate faster than W because W is the fixed transmission rate of an individual link. Therefore, the delay rate of our collection scheme achieves the order of the upper bound, and the delay rate of data collection is $\Theta(W)$. Notice that even for individual sensors, the lowest achievable delay rate of our method was $\Theta(W/n)$, which also meets the upper bound. In other words, our approach can achieve the order-optimal capacity for each individual sensor too.

Next, we consider the situation with pipelining. It is clear that the upper bound of capacity is still W. Because our scheme above already reaches the upper bound, the pipelining operation can only improve the capacity within a constant factor.

With pipelining, in phase I, the sensor can begin to transfer the data to its up-cell from next snapshot after sensors in its interference block finish their transmissions of previous snapshot. Whenever the cells in the top row receive $a \times b$ data (every cell in the top row receives a data from its lower cell), phase II can begin at the top row. We consider the improvements of pipelining on both phases. With the pipelining, the time T_1' for the highest cell to receive a new set of $a \times b$ data in phase I is

$$T_1' \le L \times L \times t \times O(\log n) = O(t \log n)$$

And the time T_2' for the sink to receive a new set of $a \times b$ data in phase II is

$$T_2' \le \max\{nt, L \times t \times a\} = O(nt).$$

Therefore, the total time for sink to receive $a \times b$ data is $T_1' + T_2' = O(nt)$. Thus, the capacity of our method with pipelining is still

$$C = \frac{a \cdot b}{T_1' + T_2'} = \Omega(W).$$

This also meets the upper bound W in order.

In summary, we have the following theorem:

Theorem 1

Under the protocol model [22,23], the delay rate Γ and the capacity C of data collection in random sensor networks with a single sink are both $\Theta(W)$. ■

Notice that the scheduling algorithm presented here is order-optimal but the constant behind the big Θ could be large. There are different methods (such as those used by Ji et al. [19]) that could further improve the achieved capacity by constant times.

1.2.3 Data Collection with Multiple Sinks

Now we consider networks with multiple sinks (e.g., k sinks). With more sinks, the collection task can be divided into small subtasks (i.e., collections in subareas) and each subtask can be assigned to a single sink. Multiple sinks can collect data from their areas simultaneously if they are not interfering with each other. This can increase the capacity and decrease the delay of data collection. We will derive the bounds of data collection for multiple sinks using the results in the case with a single sink. Because the delay rate and the capacity are always of the same order in both cases, we will not distinguish between them and instead use only the term of capacity. Two scenarios are considered in the following subsections: sinks are regularly deployed on a grid or randomly deployed in the field.

1.2.3.1 Regularly Deployed Multiple Sinks

When sinks are displayed regularly on a $\sqrt{k}\times\sqrt{k}$ grid, the capacity of collection depends on the number of k sinks. Here, we divide the unit area into k subareas, which are $\frac{1}{\sqrt{k}}\times\frac{1}{\sqrt{k}}$ squares. There are two cases: $k<\frac{n}{15(\alpha+1)^2\log n}$ or $k\geq\frac{n}{15(\alpha+1)^2\log n}$.

Case 1: When $k<\frac{n}{15(\alpha+1)^2\log n}$, $k<\frac{1}{(R+r)^2}$ because $R=\alpha r$ and $r=\sqrt{15\frac{\log n}{n}}$. Thus, each subarea assigned to a sink is larger than or equal to $(R+r)^2$. Therefore, we can perform the data collection in each subarea without interfering with the neighboring subareas. Because we have k subareas, the total delay rate and the total capacity of the whole area is at most $k\cdot\Theta(W)=\Theta(kW)$.

Case 2: When $k\geq\frac{n}{15(\alpha+1)^2\log n}$, $k\geq\frac{1}{(R+r)^2}$. Thus, the area of each subarea is smaller than $(R+r)^2$, which indicates that there will be interference between neighboring subareas. Therefore, the total delay rate or capacity is bounded by $\frac{1}{(R+r)^2}\cdot\Theta(W)=\Theta\left(\frac{n}{\log n}W\right)$ from above, due to interference.

To achieve these upper bounds, the collection method for a single sink case can be used. When $k<\frac{n}{15(\alpha+1)^2\log n}$, we partition the field into k subareas with the size of $\frac{1}{\sqrt{k}}\times\frac{1}{\sqrt{k}}$ and every sink performs the collection method to collect their subareas. When $k\geq\frac{n}{15(\alpha+1)^2\log n}$, we partition the field into $\frac{1}{(R+r)^2}$ subareas with a size of $(R+r)\times(R+r)$ as shown in Figure 1.6. Then, $\frac{1}{(R+r)^2}$ sinks can be selected to perform the collection method. Note that one selected sink may still cause interference with other selected sinks in an adjacent block. However, the number of such adjacent selected sinks is bounded by eight. Thus, a simple scheduling can avoid the interference and the capacity of data collection is still in the order of the theoretical bound. Figure 1.6 shows

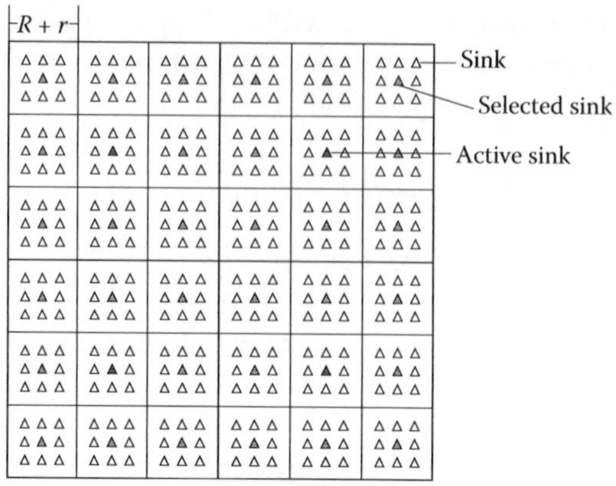

Figure 1.6 When k is large, we partition the field into $\frac{1}{(R+r)^2}$ subareas. Each subarea selects one sink as its selected sink (shown as a gray triangle). Only one selected sink inside nine subareas is active for data collection (shown as a black triangle). It will collect data from the surrounding nine subareas using the single sink method. Notice that the adjacent nine subareas will not interfere with each other when applying the collection method.

a possible scheduling in which only one of nine selected sinks collects data from its surrounding blocks, and thus, we have our second theorem.

Theorem 2

Under the protocol model [22,23], the delay rate Γ and the capacity C of data collection in random sensor networks with k regularly deployed sinks are

$$
\begin{cases}
\Theta(kW) & \text{when } k < \dfrac{n}{15(\alpha+1)^2 \log n}, \\[2ex]
\Theta\left(\dfrac{n}{\log n} W\right) & \text{when } k \geq \dfrac{n}{15(\alpha+1)^2 \log n}.
\end{cases}
$$

■

Because when $k = \Theta\left(\frac{n}{\log n}\right)$, the capacity (or delay rate) of two cases are all equal to $\Theta(kW) = \Theta\left(\frac{n}{\log n} W\right)$. Therefore, the above equations can also be written as follows:

$$
\begin{cases}
\Theta(kW) & \text{when } k = O\left(\dfrac{n}{\log n}\right), \\[2ex]
\Theta\left(\dfrac{n}{\log n} W\right) & \text{when } k = \Omega\left(\dfrac{n}{\log n}\right).
\end{cases}
$$

1.2.3.2 Randomly Deployed Multiple Sinks

Consider the scenario when k sinks are randomly distributed in the network. It is clear that if k is very large, the capacity is still bounded by the interference area. However, when the k is very small, the achievable capacity of collection may not reach the upper bound of $\Theta(kW)$ because the distribution of k sinks could be unbalanced in the field. In that case, even though the two neighboring sinks may not interfere with each other, they cannot fully operate over the whole period because some of them may finish their collection earlier and have no data to collect.

We first derive the upper bound of data collection capacity. Because the interference range is $R = \alpha r = \alpha \cdot \sqrt{15\frac{\log n}{n}}$, we partition the whole area into interference blocks with a size of $(R + r) \times (R + r)$. Thus, there are $B = \frac{n}{15(1+\alpha)^2 \log n}$ interference blocks. We then consider three cases when we randomly put k sinks into B interference blocks:

Case 1: When $k = o\left(\frac{n}{\log n}\right)$. For this case, the capacity of data collection is bounded by $\Theta(kW)$ from above because the collection rate of each sink is bounded by W. Notice that data collection with a single sink is a special case when $k = 1$.

Case 2: When $k = \Theta\left(\frac{n}{\log n}\right)$. We calculate the probability that an arbitrary interference block has at least one sink.

$$\Pr\text{ (an interference block has at least one sink)} = 1-\left(1-\frac{1}{B}\right)^k = 1-\left[1-\frac{1}{\Theta\left(\frac{n}{\log n}\right)}\right]^k$$

$$= 1-\left[1-\frac{1}{\Theta\left(\frac{n}{\log n}\right)}\right]^{\Theta\left(\frac{n}{\log n}\right)}.$$

When $n \to \infty$, this probability is equal to $1-\frac{1}{e}$. Let Pr be this probability. Then, we define the number of interference blocks occupied by at least one sink as a random variable X. The expectation and variance of X are $E[X] = \Pr \times B = \left(1-\frac{1}{e}\right)\frac{n}{60\alpha^2 \log n}$ and $\sigma^2 = \Pr \times (1-\Pr) \times B = \frac{1}{e}\left(1-\frac{1}{e}\right)\frac{n}{60\alpha^2 \log n}$. Based on the Chebyshev inequality, we have the following:

$$\Pr\left(|X - E[X]| \geq \varsigma\sigma\right) \leq \frac{1}{\varsigma^2}.$$

Let $\varsigma = \frac{1}{2}\sqrt{\frac{\left(1-\frac{1}{e}\right)\frac{n}{60\alpha^2 \log n}}{\frac{1}{e}}}$, we have

$$\Pr\left(|X - E[X]| \geq \frac{1}{2}E[X]\right) \leq \frac{4 \cdot \frac{1}{e}}{\left(1-\frac{1}{e}\right)\frac{n}{60\alpha^2 \log n}},$$

which goes to 0 when $n \to \infty$. This means that $\frac{1}{2}E[X] \leq X \leq \frac{3}{2}E[X]$ with a high probability. In other words, the number of occupied interference blocks is $\Theta\left(\frac{n}{\log n}\right)$. Therefore, the capacity of data collection is bounded by $\Theta\left(\frac{n}{\log n}W\right)$, which is also $\Theta(kW)$.

Case 3: When $k = \omega\left(\frac{n}{\log n}\right)$. We also consider the probability that an arbitrary interference block has at least one sink.

$$\Pr(\text{an interference block has at least one sink}) = 1 - \left[1 - \frac{1}{\Theta\left(\frac{n}{\log n}\right)}\right]^k$$

$$= 1 - \left[1 - \frac{1}{\Theta\left(\frac{n}{\log n}\right)}\right]^{\Theta\left(\frac{n}{\log n}\right) \cdot \frac{k}{\Theta\left(\frac{n}{\log n}\right)}}$$

$$= 1 - \left[\cdot 1 - \frac{1}{\Theta\left(\frac{n}{\log n}\right)}\right]^{\Theta\left(\frac{n}{\log n}\right)\frac{\Omega\left(\frac{n}{\log n}\right)}{\Theta\left(\frac{n}{\log n}\right)}}$$

When $n \to \infty$, this probability goes to 1. In other words, every interference block has at least one sink with high probability. Thus, we can select only one sink in each block to collect data at the same time. Then, the capacity of data collection is bounded by $\Theta\left(\frac{n}{\log n}W\right)$ from above.

From the previous analysis, we find that the capacity upper bounds for the randomly distributed case are the same with the ones for the regularly distributed case. Next, we present the lower bounds of data collection capacity by giving our data collection methods.

When $k = O\left(\frac{n}{\log n}\right)$, we first partition the network into interference blocks with the size $\sqrt{3\frac{\log k}{k}} \times \sqrt{3\frac{\log k}{k}}$. From Lemma 1, we know that each of the blocks is occupied by at least one sink with a high probability. Because $k = O\left(\frac{n}{\log n}\right)$, the size of a block is $\sqrt{3\frac{\log k}{k}} > R + r$. Thus, we select one sink for each block, and use the same technique for grid-deployed sinks (Section 1.2.3.1) to schedule a subset of selected sinks to collect data from its surrounding area. The capacity achieved is $\Theta\left(\frac{k}{\log k}W\right)$ because the number of selected sinks is $\Theta\left(\frac{k}{\log k}\right)$. Notice that there is a gap between this lower bound and the upper bound $\Theta(kW)$. This is due to the possibly uneven distribution of k sinks in this case, thus each sink may not have the same number of sensors (or areas) to perform the collection to achieve $\Theta(kW)$ capacity in total.

When $k = \omega\left(\frac{n}{\log n}\right)$, we first partition the network into interference blocks with size $(R + r) \times (R + r)$. As shown in Case 3, with high probability, each block has at least one sink. Using the same collection method, the achievable capacity is $\Theta\left(\frac{n}{\log n}W\right)$, which meets the upper bound perfectly.

Theorem 3

Under the protocol model [22,23], the delay rate Γ and the capacity C of data collection in random sensor networks with k randomly deployed sinks are

$$\begin{cases} \Theta\left(\frac{k}{\log k}W\right) \le C \le \Theta(kW) & \text{when } k = O\left(\frac{n}{\log n}\right), \\ C = \Theta\left(\frac{n}{\log n}W\right) & \text{when } k = \omega\left(\frac{n}{\log n}\right). \end{cases}$$

■

In summary, with multiple sinks (either grid or random deployment of k sinks), the capacity of data collection increases from that of the single sink case. When the capacity is constrained by the number of sinks $\left(\text{i.e., } k = O\left(\frac{n}{\log n}\right)\right)$, it is beneficial to add more sinks. However, when the capacity is constrained by the interference among sinks $\left(\text{i.e., } k = \omega\left(\frac{n}{\log n}\right)\right)$, adding more sinks has no substantial capacity improvement. Similar observations were made by Liu et al. [17] for many-to-many capacity.

1.2.4 Data Collection under the Physical and Generalized Physical Models

Thus far, we only consider the protocol model, which is ideal but unrealistic in WSNs, in which the interference is modeled as a localized phenomenon. However, a receiver can be interfered with by a group of actively transmitting sensors even if its location is extremely far away from the group of sensors. Thus, we now consider more accurate models to reflect the influence of interference: the physical model and the generalized physical model. Please refer to Section 1.1.2 for their definitions.

Again, we consider a random sensor network with n sensor nodes and a single sink. We now use the random extended network model [28], in which all sensor nodes are uniformly deployed in a square region with side length $l = \sqrt{n}$, by use of Poisson distribution with density 1. The grid partition method we introduced in Section 1.2.1 is the same except for the size length of every cell and the transmission range of each sensor are $\sqrt{n}$ times larger than those in Section 1.2.1.

1.2.4.1 Data Collection under the Physical Model

For the case of data collection under the physical model, our collection scheme and analysis are almost the same with the one under the protocol model (Section 1.2.2). The only difference is that a new size of interference block is used.

We first divide the field into big blocks with size $L \times L$ as shown in Figure 1.7. We call these blocks *interference blocks* and L *interference distance*. Thus, the number of interference blocks is $\frac{l^2}{L^2}$. We label each block with (i,j) where i and j are the indexes of the block as in Figure 1.7. In our collection scheme, we schedule data transmission in parallel at all blocks but make sure that there is only one sensor in each interference block transferring at any time. To avoid interference from senders in other interference blocks, we need an interference distance L that is larger than a certain value.

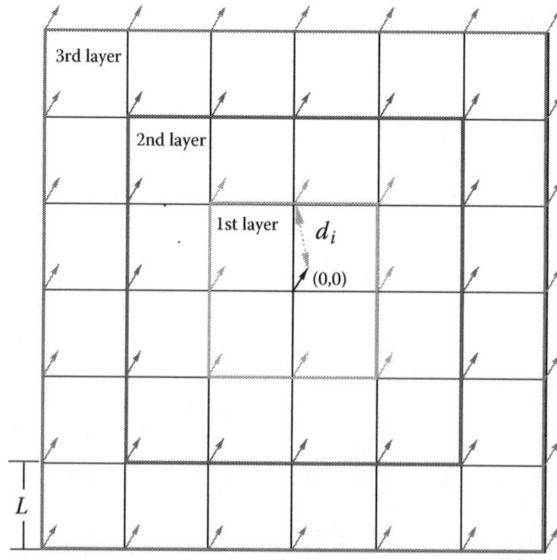

Figure 1.7 Grid partition of interference blocks with size of $L \times L$ and simultaneous transmissions around the center block (0,0) by layers.

Next, we derive the lower bound of interference distance such that all simultaneous transmissions, as shown in Figure 1.7, can be successfully received. Here, we consider the SINR at the receiver in interference block (0,0), which is in the center of the field, because it has the minimum SINR among all receivers. Similar to the technique used by Franceschetti et al. [31], we now label all simultaneous transmissions by layers from block (0,0), as shown in Figure 1.7.

Based on the physical interference model, its SINR is at least

$$\frac{P \cdot r^{-\beta}}{N_0 + \sum_{\text{all layers } i \geq 1} c_i P \cdot (d_i)^{-\beta}}.$$

Here, d_i is the minimum distance from a transmitter on the ith layer to the receiver in block (0,0) and c_i is the number of transmitters on the ith layer. Therefore, we need to derive L such that SINR $\geq \eta$, that is,

$$\sum_{\text{all layers } i \geq 1} c_i (d_i)^{-\beta} \leq \frac{r^{-\beta}}{\eta} - \frac{N_0}{P}.$$

Notice that $d_i \geq iL - 2d$ and $c_i = 8i$. For example, there are 8 transmitters at the first layer with distance at least $L - 2d$ and 16 transmitters at the second layer with a distance of at least $2L - 2d$, and so on. Thus,

$$\sum_{i \geq 1} c_i (d_i)^{-\beta} \leq \sum_{i \geq 1} 8i(iL - 2d)^{-\beta} \leq \sum_{i \geq 1} 8i(iL - 2id)^{-\beta} = 8(L - 2d)^{-\beta} \sum_{i \geq 1} i^{-(\beta-1)}.$$

Because $\beta > 2$, $\sum_{i \geq 1} i^{-(\beta-1)}$ converges to a constant, let it be denoted by ϕ. Then, we only need

$$8\phi(L - 2d)^{-\beta} \leq \frac{r^{-\beta}}{\eta} - \frac{N_0}{P},$$

to guarantee that the SINR at the receiver in the center is at least η. This can be satisfied by setting

$$L \geq \left[\frac{1}{8\phi} \cdot \left(\frac{r^{-\beta}}{\eta} - \frac{N_0}{P}\right)\right]^{-\frac{1}{\beta}} + 2d.$$

Remember that $r \leq \left(\frac{P}{N_0 \cdot \eta}\right)^{1/\beta}$, this makes sure we can find such suitable L. We can further select $L = \left[\frac{1}{8\phi} \cdot \left(\frac{r^{-\beta}}{\eta} - \frac{N_0}{P}\right)\right]^{-\frac{1}{\beta}} + 2d$. Because $r = \sqrt{5}d$,

$$\frac{L}{d}=\left[\frac{1}{8\phi}\cdot\left(\frac{(\sqrt{5}d)^{-\beta}}{\eta d^{-\beta}}-\frac{N_0}{Pd^{-\beta}}\right)\right]^{-\frac{1}{\beta}}+2=\left[\frac{1}{8\phi}\cdot\left(\frac{5^{-\beta/2}}{\eta}-\frac{N_0 d^{\beta}}{P}\right)\right]^{-\frac{1}{\beta}}+2.$$

When $n \rightarrow \infty$, this ratio goes to a constant, denoted by α.

Notice that now the physical model can be treated as a protocol model because the ratio between the sizes of interference block and micro cell $\frac{L}{d}$ is bounded by the constant α. By having interference blocks, we can simply use the data collection scheme for protocol model, which has already been presented in Section 1.2.2, to perform the data collection. It is easy to show that the same capacity can be achieved compared with the protocol model. In summary, we have the following theorem for the physical model:

Theorem 4

Under the physical model [24,25], the delay rate Γ and the capacity C of data collection in random sensor networks with a single sink are both $\Theta(W)$. ■

1.2.4.2 Data Collection under the Generalized Physical Model

The physical model assumes a threshold-based channel in which the signal can be decoded at a fixed constant rate of W bits per second only if the SINR is greater than a certain threshold. If the SINR is below this threshold, no throughput is received at all. However, in practice, the throughput is usually a function of the SINR at the receiver. Thus, the generalized physical model is a more realistic communication model than the protocol or physical models, especially under random extended networks [28]. Therefore, we also study the theoretical bounds of data collection capacity under the generalized physical model. Notice that because the data rate is now related to SINR and interference, the capacity analysis becomes much more complex and challenging.

First, we give a lemma to derive an upper bound of data collection capacity under the generalized physical model.

Lemma 4

Under the generalized physical model [25], the capacity of data collection in random sensor networks is at most $O[(\log n)\ W]$. ■

Proof

We first order all the incoming links of sink s according to their length as follows: $||v_1 - s|| \leq ||v_2 - s|| \leq \ldots \leq ||v_{n'} - s||$. Here, n' is the number of incoming links at sink s, which transmits simultaneously to s; clearly, $n' \leq n$. Next, we try to bound the SINR of the sink node s. For any link $v_i s$ $(i \neq 1)$, its SINR

$$\mathrm{SINR}_{is} \le \frac{P \cdot l(v_i,s)}{N_0 + \sum_{k=1}^{i-1} P \cdot l(v_k,s)} \le \frac{P \cdot l(v_i,s)}{N_0 + \sum_{k=1}^{i-1} P \cdot l(v_i,s)} < \frac{1}{i-1}.$$

■

Therefore, for $i \neq 1$,

$$W_{is} = W \log_2(1 + \mathrm{SINR}_{is}) < W \log_2\left(\frac{i}{i-1}\right).$$

So the maximum rate at sink s is at most

$$W_{1s} + \sum_{i=2}^{n'} W \log_2\left(\frac{i}{i-1}\right) = W_{1s} + \log_2\left(\prod_{i=2}^{n'} \frac{i}{i-1}\right)$$

$$\le \max_i(W_{is}) + W \cdot \log_2 n' \le \max_i(W_{is}) + W \cdot \log_2 n.$$

The first part of this upper bound depends on the rate of the shortest incoming link at the sink, whereas the second part depends on the total number of nodes. Notice that $\max_i(W_{is}) \le W \cdot \log_2\left(1 + \frac{P}{N_0}\right)$. Thus, which part of the bound that is playing an important role depends on the relationship between n and $1 + \frac{P}{N_0}$. If P and N_0 are constants as we assumed, $\max_i(W_{is}) \le O(W)$. Then, the upper bound of capacity can be written as $O((\log n)\ W)$.

We now can introduce our data collection algorithm, which uses the same partition method and scheduling algorithm as that of the physical model. The only difference is the size of the interference block.

We now divide the field into big interference blocks of a certain size $L(d) \times L(d)$ as shown in Figure 1.7. Thus, the number of interference blocks is $\frac{l^2}{L(d)^2}$. In our collection scheme, we will schedule data transmission in parallel at all blocks but make sure that there is only one sensor in each interference block transferring at any time.

We now prove that the transmission rate of each transmitting sensor node in such data collection scheme is at least $\Omega\left((\log n)^{-\frac{\beta}{2}} W\right)$, if $L(d) = \kappa d$ and $\kappa > 2$ is a constant.

Lemma 5

In each interference block with size of $\kappa d \times \kappa d$ [25], there exists a node that can transmit at rate $\Omega\left((\log n)^{-\frac{\beta}{2}} W\right)$ to any destination in its adjacent cell. ■

Proof

Let us focus on one given sensor node v_i, which transmits to a destination v_j in v_i's adjacent cell. Its transmission rate is:

$$W_{ij} = W \log_2\left(1 + \frac{P \cdot l(v_i, v_j)}{N_0 + \sum_{k \in I} P \cdot l(v_k, v_j)}\right).$$

■

Because the distance between v_i and v_j is at most $\sqrt{5}d$, $P \cdot l(v_i, v_j) \geq P \cdot (\sqrt{5}d)^{-\beta} = \Omega(d^{-\beta})$.

We then need to find the upper bound of the interference at the receiver v_j from simultaneous transmitters. Using the same technique in Section 1.2.4.1, we consider layers of simultaneous transmissions in surrounding interference blocks as shown in Figure 1.7. Once again, assume that $d_i \geq iL - 2d$ is the minimum distance from an ith layer transmitter to v_j and $c_i = 8i$ is the number of transmitters on the ith layer. Therefore,

$$\sum_{k \in I} P \cdot l(v_k, v_j) \leq \sum_{i=1}^{\infty} 8iP[iL(d) - 2d]^{-\beta}$$

$$\leq \sum_{i=1}^{\infty} 8iP(i\kappa - 2)^{-\beta} d^{-\beta} \leq 8Pd^{-\beta} \cdot \sum_{i=1}^{\infty} i(i\kappa - 2)^{-\beta}.$$

Because $\beta > 2$, the summation $\sum_{i=1}^{\infty} i(i\kappa - 2)^{-\beta}$ converges to a constant ρ. Therefore,

$$\sum_{k \in I} P \cdot l(v_k, v_j) \leq 8P\rho \cdot d^{-\beta} = O(d^{-\beta}).$$

When $n \to \infty$, $d \to \infty$; hence, the SINR

$$\frac{P \cdot l(v_i, v_j)}{N_0 + \sum_{k \in I} P \cdot l(v_k, v_j)} = \Omega(d^{-\beta}).$$

Therefore, the transmission rate from v_i to v_j

$$W_{ij} = \Omega(d^{-\beta} W).$$

We use the same data collection scheme in Section 1.2.2. The total time we need to collect all the n packets is

$$T \leq \left[\left(\frac{\kappa d}{d}\right)^2 O(\log n)m + \left(\frac{\kappa d}{d}\right) O(\log n)m^2\right] \cdot \frac{b}{\Omega(Wd^{-\beta})}$$

$$\leq O((\log n)m^2)\cdot\frac{b}{\Omega(Wd^{-\beta})}=\frac{O(n)}{\Omega(d^{-\beta})}\cdot t\leq O\left(n(\log n)^{\frac{\beta}{2}}\right)\cdot t.$$

Thus, the achieved capacity of data collection under the generalized physical model is $\Omega\left((\log n)^{-\frac{\beta}{2}}W\right)$.

In summary, the bounds of data collection capacity can be summarized as the following:

Theorem 5

Under the generalized physical model [25], the capacity of data collection in random sensor networks is between $\Omega\left((\log n)^{-\frac{\beta}{2}}W\right)$ and $O((\log n)\ W)$. ■

1.3 Data Collection in Arbitrary Sensor Networks

We have studied the capacity of data collection on large-scale random WSNs. However, all results are based on a strong assumption that sensors are deployed randomly in an environment and the number of nodes n must be extremely large. Such an assumption is useful to simplify the analysis and derive nice theoretical limits, but may be invalid in many practical sensor applications. In most of the practical sensor applications, the sensor network is not uniformly deployed and the number of sensors may not be as huge as in theory. Therefore, it is necessary to study the capacity of data collection in an arbitrary network. In this section, we consider an arbitrary WSN in which n sensors and a single sink s are arbitrarily deployed in a finite geographical region. Figure 1.2 illustrates the difference between a random network and an arbitrary network.

1.3.1 Data Collection under the Protocol Model

Recall that the upper bound of data collection capacity in random networks is W. Obviously, this upper bound also holds for any arbitrary networks because sink s cannot receive at a rate faster than W due to the fixed transmission rate at each link. Therefore, we now introduce a simple breadth first search (BFS) tree-based data collection scheme to achieve capacity in the same order of the upper bound, that is, $\Theta(W)$. The data collection method includes two steps: data collection tree formation and data collection scheduling.

1.3.1.1 Data Collection Tree: BFS Tree

The data collection tree used in our method is a classic BFS tree rooted at the sink s. The time complexity to construct such a BFS tree is $O(|V| + |E|)$. Let T be the BFS tree and $v_1^l, \ldots, v_c^l$ be all leaves in T. For each leaf v_i^l, there is a path P_i from itself to the root s. Let $\delta^{P_i}(v_j)$ be the number of nodes on path P_i that are inside the interference range of v_j (including v_j itself). Assume the maximum interference number Δ_i on each path P_i is $\max\{\delta^{P_i}(v_j)\}$ for all $v_j \in P_i$. Hereafter, we call Δ_i path interference of path P_i. Then, we can prove that T has a nice property that the path interference of each branch is bounded by a constant.

Lemma 6

Given a BFS tree T under the protocol model [26,27], the maximum interference number Δ_i on each path P_i is bounded by a constant $8\alpha^2$, that is, $\Delta_i \leq 8\alpha^2$. ■

Proof

We prove by contradiction with a simple area argument. Assume that there is a v_j on P_i whose $\Delta_i > 8\alpha^2$. In other words, more than $8\alpha^2$ nodes on P_i are located in the interference region of v_j. Because the area of interference region is πR^2, we consider the number of interference nodes inside a small disk with a radius of $\frac{r}{2}$ (see Figure 1.8 for illustration). The number of such small disks is at most $\frac{\pi R^2}{\pi\left(\frac{r}{2}\right)^2} = 4\alpha^2$ inside πR^2. By the pigeonhole principle, there must be more than $\frac{8\alpha^2}{4\alpha^2} = 2$ nodes inside a single small disk with radius $\frac{r}{2}$. In other words, three nodes v_x, v_y, and v_z on the path P_i are connected to each other as shown in Figure 1.8. This is a contradiction with the construction of the BFS tree. As shown in Figure 1.8, if v_x and v_z are connected in G, then v_z should be visited by v_x not v_y during the construction of the BFS tree. This finishes our proof. ■

1.3.1.2 Branch Scheduling Algorithm

We now illustrate how to collect one snapshot from all sensors. Given the collection tree T, our scheduling algorithm basically collects data from each path P_i in T one by one.

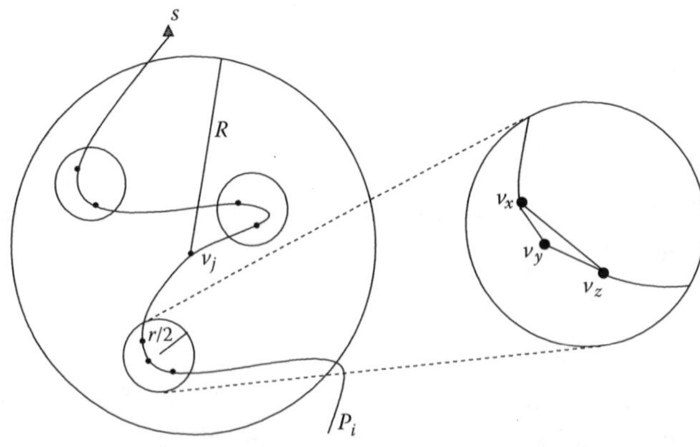

Figure 1.8 Proof of Lemma 6: on a path P_i in BFS tree T, the interference nodes for a node v_j is bounded by a constant.

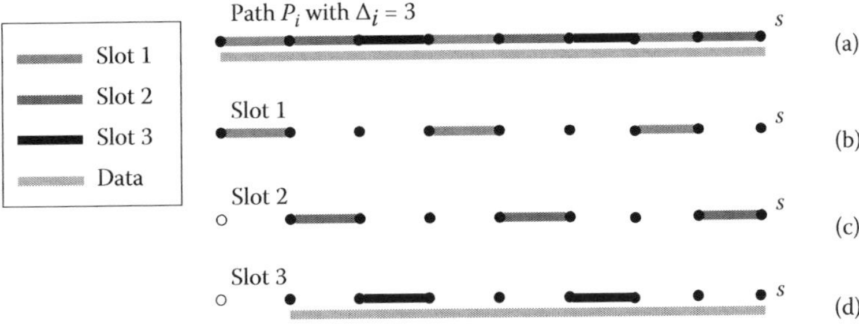

Figure 1.9 Scheduling on a path: after Δ_i slots, the sink obtains one set of data.

First, we explain how to schedule the collection on a single path. For a given path P_i, we can use Δ_i slots to collect one unit of data in the snapshot at the sink (see Figure 1.9 for illustration). In this figure, we assume that $R = r$, that is, only adjacent nodes interfere with each other. Thus, $\Delta_i = 3$. Then, we color the path using three shades as in Figure 1.9a. Notice that each node on the path has unit data to transfer. Links with the same color are active in the same slot. After three slots (Figure 1.9d), the leaf node has no data in this snapshot and the sink received one unit of data from its child. Therefore, to receive all data on the path, at most $\Delta_i \times |P_i|$ time slots are needed. We call this scheduling method *path scheduling*.

Now, we describe our scheduling algorithm on the collection tree T. Remember that T has c leaves, which define c paths from P_1 to P_c. Our algorithm collects data from path P_1 to P_c in order. We define the ith branch B_i as part of P_i from v_i^l to the intersection node with P_{i+1} for $i = (1, c - 1)$ and cth branch $B_c = P_c$. For example, in Figure 1.10b, there are four branches in T: B_1 is from v_1^l to v_a, B_2 is from v_2^l to s, B_3 is from v_3^l to v_b, and B_4 is from v_4^l to s. Notice that the union of all branches is the whole tree T. Algorithm 1 (in Figure 1.11) shows the detailed branch scheduling algorithm. Figure 1.10c through j gives an example of scheduling on T. In the first step (Figure 1.10c), all nodes on P_1 participate in the collection using the scheduling method for a single path (for every Δ_1 slot, sink s receives one unit of data). Such a collection stops until there is no data in this snapshot on branch B_1 (as shown in Figure 1.10d). Then step 2 collects data on path P_2. This procedure is repeated until all data in this snapshot reaches s (Figure 1.10j).

1.3.1.3 Capacity Analysis

We now analyze the achievable capacity of our data collection method by counting how many time slots the sink needs to receive all data in one snapshot.

Theorem 6

The data collection method based on path-scheduling in the BFS tree can achieve a data collection capacity of $\Theta(W)$ at the sink [26,27].

Proof

In Algorithm 1, the sink collects data from all c paths in T. In each step (lines 3 and 4), data are transferred on path P_i and it takes at most $\Delta_i \times |B_i|$ time slots. Recall that path scheduling

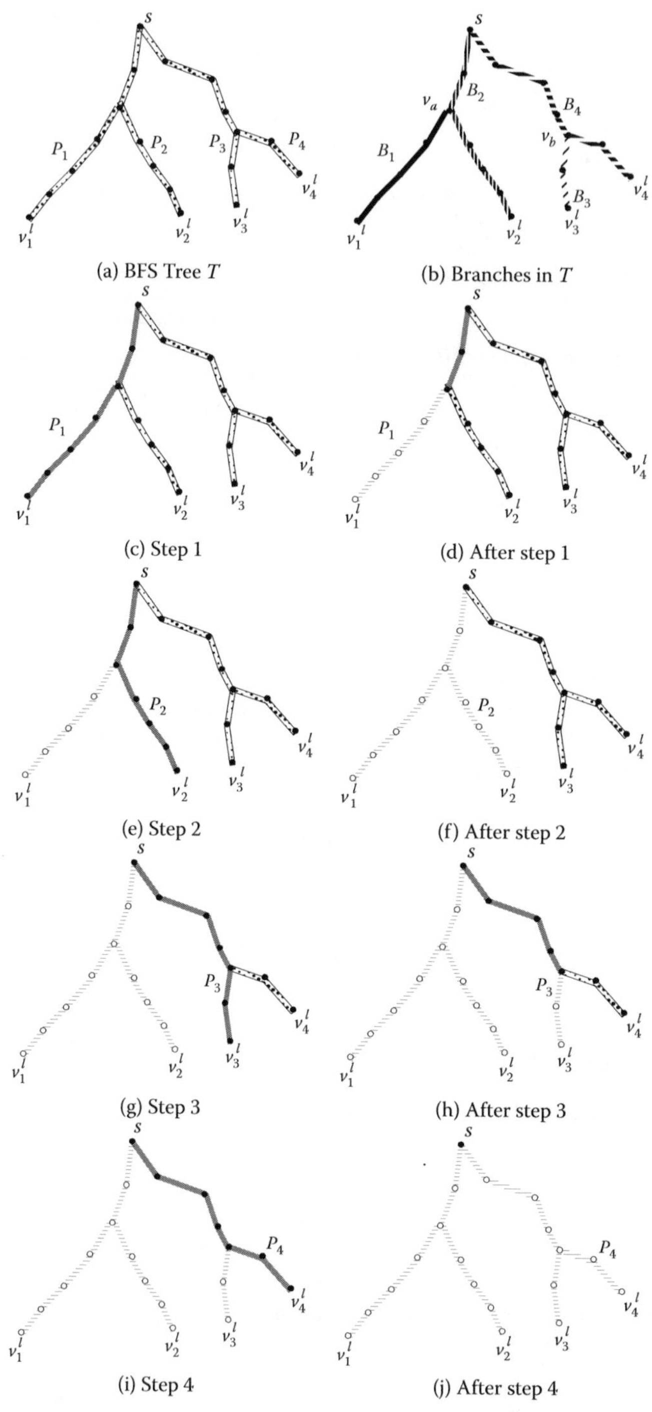

Figure 1.10 Illustrations of our scheduling on the data collection tree *T*.

Algorithm 1 Branch Scheduling on BFS Tree

Input: BFS tree T rooted at s.

```
for each snapshot do
    for t = 1 to c do
        Collect data on path P_i. All nodes on P_i transmit data towards the sink s using path scheduling.
        The collection terminates when nodes on branch B_i do not have data for this snapshot. The total slots used are at most Δ_i · |B_i|, |B_i| is the hop length of B_i.
    end for
end for
```

Figure 1.11 Branch-scheduling algorithm on a BFS tree.

needs at most $\Delta_i \times k$ time slots to collect k packets from path P_i. Therefore, the total number of time slots needed for Algorithm 1, denoted by τ, is at most $\sum_{i=1}^{c}(\Delta_i \times |B_i|)$. Because the union of all branches is the whole tree T, that is, $\sum_{i=1}^{c}|B_i| = n$, $\tau \le \sum_{i=1}^{c}(\Delta_i \times |B_i|) \le \sum_{i=1}^{c}(\tilde{\Delta} \times |B_i|) = \tilde{\Delta} n$.

Here, $\tilde{\Delta} = \max\{\Delta_1, \cdots, \Delta_c\}$. Then, the delay of data collection $D = \tau t \le \tilde{\Delta} n t$. The capacity $C = \frac{nb}{D} \ge \frac{nb}{\tilde{\Delta} n t} \le \frac{W}{\tilde{\Delta}}$. From Lemma 6, we know that $\tilde{\Delta}$ is bounded by a constant. Therefore, the data collection capacity is $\Theta(W)$.

Remember that the upper bound of data collection capacity is W; thus, our data collection algorithm is order-optimal. Consequently, we have the following theorem.

Theorem 7

Under the protocol and disk graph models [26,27], data collection capacity for arbitrary WSNs is $\Theta(W)$. ■

1.3.2 Data Collection under the General Graph Model

In previous parts of this chapter, our collecting algorithm and analysis was based on a disk graph model in which two nodes could communicate if and only if their distance was less than or equal to the transmission range r. However, a disk graph model is idealistic because, in practice, two nearby nodes may be unable to communicate due to various reasons such as barriers and path fading. Therefore, in this subsection, we consider a more general graph model $G = (V, E)$ in which V is the set of sensors and E is the set of possible communication links. Every sensor still has a fixed transmission range r such that the necessary condition for v_j to correctly receive the signal from v_i is $\|v_i - v_j\| \le r$. However, $\|v_i - v_j\| \le r$ is not the sufficient condition for an edge $v_iv_j \in E$. Some links do not belong to G because of physical barriers or the selection of routing protocols. Thus, G is a subgraph of a disk graph. Under this model, the network topology G can be any general graph (for example, setting $r = \infty$ and putting a barrier between any two nodes v_i and v_j if $v_iv_j \notin G$).

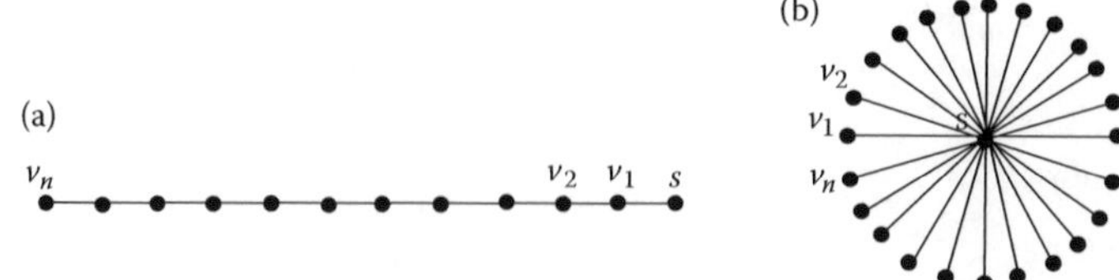

Figure 1.12 Two extreme cases in general graph models: (a) straight-line topology and (b) star topology.

In the general graph model, the capacity of data collection could be $\frac{W}{n}$ in the worst case. We consider a simple straight-line network topology with n sensors as shown in Figure 1.12a. Assume that the sink s is located at the end of the network and the interference range is large enough to cover every node in the network. Because the transmission on one link will interfere with all the other nodes, the only possible scheduling is transferring data along the straight-line via all links. The total time slots needed are $\frac{n(n+1)}{2}$, thus the capacity is at most $\frac{nb}{\frac{n(n+1)}{2}t} = \Theta\left(\frac{W}{n}\right)$. Notice that in this example, the maximum interference number Δ of graph G is n. It seems the upper bound of data collection capacity could be $\frac{W}{\Delta}$. We now show an example whose capacity can be much larger than $\frac{W}{\Delta}$. Again, we assume all n nodes with the sink interfering with each other. The network topology is a star with the sink s in the center, as shown in Figure 1.12b. Clearly, a scheduling that lets every node transfer data in order can lead to a capacity W, which is much larger than $\frac{W}{\Delta} = \frac{W}{n}$. From these two examples, we find that the capacity problem for the general graph model is more complex. Next, we analyze the upper and lower bounds of the collection capacity under the protocol model for the general graph model.

1.3.2.1 Upper Bound of Collection Capacity

We first present a tighter upper bound of data collection capacity for the general graph model than the natural one W. Consider all packets from one snapshot, we use p_i to represent the packet generated by sensor v_i. For any v_i, let $l(v_i)$ be its level in the BFS tree rooted at the sink s (which is the minimum number of hops required for packet p_i or a packet at v_i to reach s). We use $D(s,l)$ to represent a virtual disk centered at the sink node s with a radius of hop distance l. The *critical level* (or the *critical radius*) l^* is the greatest level l such that no two nodes within l level from the sink node s can receive a message in the same time slot, that is, $l^* = \max\{l \mid \forall v_i, v_j \in D(s,l)$ cannot receive packets at the same time$\}$. The region defined by $D(s,l^*)$ is called the *critical region* (see Figure 1.13 for illustration). For any packet p_i originating at node v_i, we define

$$\lambda_i^* = \begin{cases} l(v_i), & \text{if } v_i \in< D(s,l^*), \\ l^*+1, & \text{otherwise.} \end{cases}$$

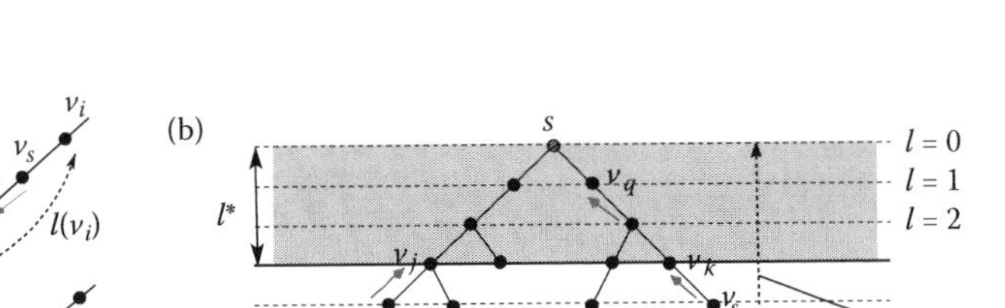

Figure 1.13 Illustration of the definition of the critical region, that is, l^*. The gray area is the critical region, where no two nodes can receive a message in the same time slot due to interference around s. Critical region around sink s (a) and a tree view of the critical region (b).

Here, λ_i^* gives the minimum number of hops needed to reach the sink s after packet p_i reaches the critical region around s. Let $\lambda^* = \max_i\{\lambda_i^*\}$. Then, we can prove the following lemma on the lower bound of delay for data collection.

Lemma 7

For all packets from one snapshot [27], the delay to collect them at sink s

$$D \geq t\sum_i \lambda_i^*.$$

■

Proof

It is clear that the critical region around sink s is a bottleneck for the delay. Any packet inside the critical region can only move one step at each time slot. First, the total delay must be larger than the delay, which is needed for the case in which all packets originating from outside the critical region are just one hop away from the critical region. In other words, assume that we can move all packets originating from outside the critical region to the surrounding area without spending any time. Then, each packet p_i needs λ_i^* time slots to reach the sink. By the definition of the critical region, no simultaneous transmissions around the critical region (one hop away from it) can be scheduled in the same slot. Therefore, the delay is at least the summation of λ_i^*. ■

Let $\Delta^* = \frac{\sum_i \lambda_i^*}{n}$, we have a new upper bound of data collection capacity, $C \leq \frac{W}{\Delta^*} \leq W$. Notice that $\Delta^* \geq 1$ and it represents the limit of scheduling due to interference around the sink (and its critical region).

1.3.2.2 Lower Bound of Collection Capacity

The data collection algorithm based on branch scheduling in the BFS tree can still achieve the capacity of $\frac{W}{\tilde{\Delta}}$. However, in the general graph model, $\tilde{\Delta}$ is no longer bounded by a constant, and it could be $O(1)$ or $O(n)$. Thus, there is a gap between our lower bound of data collection $\frac{W}{\tilde{\Delta}}$ and the natural upper bound W. Considering both examples shown in Figure 1.12 of the article, the BFS tree-based method method matches their tight upper bounds $\frac{W}{n}$ and W. For the star topology, even though the sink has the maximal interference $\Delta = n$, each individual path has the path interference $\Delta_i = 1$, which leads to a capacity of W. For the straight-line topology, the path interference of the single path $\Delta_i = n$, thus the capacity is $\frac{W}{n}$. In both cases, $\frac{W}{\tilde{\Delta}}$ matches the optimal capacity. However, similar to $\frac{W}{\Delta}$, $\frac{W}{\tilde{\Delta}}$ is still not a tight bound. We will show such an example in Figure 1.14. In this subsection, we will provide two new tighter lower bounds for data collection in the general graph: one based on the branch scheduling method and the other based on a greedy scheduling method.

We first look at the branch scheduling–based method (Algorithm 1). We modify the basic path scheduling of the BFS tree-based method to achieve better collection capacity. Recall that in Section 1.3.1.2, we claim that the path scheduling for a path P_i can be done in $\Delta_i \times |P_i|$ time slots. However, we can perform path scheduling in the following way to save more time slots. Assume that path $P_i = s, v_1, v_2, \ldots v_{|P_i|}$ includes $|P_i|$ hops. Let $\delta_k^{P_i} = \max\{\delta^{P_i}(v_1), \ldots, \delta^{P_i}(v_k)\}$, that is, $\delta^{P_i}(v_k)$ is the maximum interference number among the first k nodes v_1 to v_k in path P_i. Clearly, $\delta^{P_i}(v_k) \le \delta^{P_i}(v_{k+1})$. In the first step, using $\delta_{|P_i|}^{P_i}$ slots, every node on the path transfers its data to its parent in the BFS tree. After the first step, the leaf $v_{|P_i|}$ already finishes its task in this round and has no data from the current snapshot. In the second step, using $\delta_{|P_i|-1}^{P_i}$ slots, the current snapshot data will move up one more level along the path in a BFS tree. Repeat these steps until all data along this path reaches the sink. It is easy to show that the total number of time slots used by the above procedure is $\sum_{k=1}^{|P_i|} \delta_k^{P_i}$. Because $\delta_k^{P_i} \le \Delta_i$, $\sum_{k=1}^{|P_i|} \delta_k^{P_i} \le \Delta_i \times |P_i|$.

Figure 1.14 shows an example in which $\sum_{k=1}^{|P_i|} \delta_k^{P_i}$ is much smaller than $\Delta_i \times |P_i|$. Again, we have n sensors and the sink distributed on a line P as shown in the figure. Assume that $R = r$. On the left side, there are $\log n$ nodes close to each other, thus their $\delta(v_i) = \log n$ except for $\delta(v_{n-\log n+1}) = \log n + 1$. On the right side, every node has $\delta(v_i) = 3$. Thus, $\Delta = \Delta_i = \log n + 1$ and $\Delta_i \times |P_i| = \Theta(n \log n)$. In addition, $\delta_k^P = \log n + 1$ for $k = n - \log n + 1, \ldots, n$ and $\delta_k^P = 3$ for $k = 3, \ldots, n - \log n$,

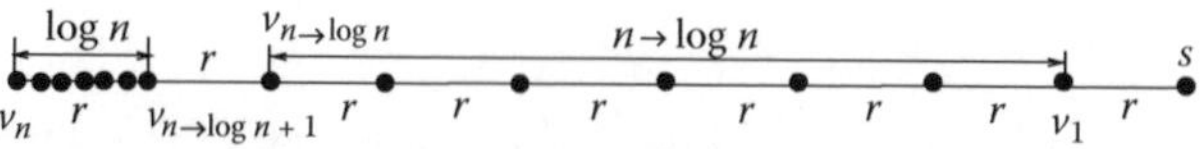

Figure 1.14 Illustration of the advantage of a new path scheduling. Here, *R* = *r*.

$\delta_2^P = 2$, and $\delta_1^P = 1$. Therefore, $\sum_{k=1}^{|P|} \delta_k^P = (\log n + 1)\log n + 3(n - \log n) - 3 = \Theta(n)$. It is obvious that $\sum_{k=1}^{|P|} \delta_k^P = \Theta(n)$ is smaller than $\Delta_i \times |P_i| = \Theta(n \log n)$ in order.

Using the new path scheduling analysis described above, we now derive a tight lower bound for our BFS tree-based method. Recall that our method transfers data based on branches in the BFS tree T. Given T, there are c paths P_i and c branches B_i as shown in Figure 1.10a and b. Then, the total number of time slots used by Algorithm 1 with the new path scheduling is at most

$$\sum_{i=1}^{c} \sum_{k=|P_i|-|B_i|+1}^{|P_i|} \delta_k^{P_i}.$$

It is clear that this number is much smaller than $\sum_{i=1}^{c} (\Delta_i \times |B_i|)$ from a previous analysis. Notice that for path P_i our algorithm (lines 3 and 4 in Algorithm 1) will terminate the transmission until branch B_i does not have data for the current snapshot and switches to the next path P_{i+1}. Thus, the index of k is only from $|P_i|$ to $|P_i| - |B_i| + 1$. Therefore, the capacity achieved by our algorithm is at least

$$\frac{W}{\frac{\sum_{i=1}^{c} \sum_{k=|P_i|-|B_i|+1}^{P_i} \delta_k^{P_i}}{n}}.$$

Let $\Delta^{**} = \frac{\sum_{i=1}^{c} \sum_{k=|P_i|-|B_i|+1}^{|P_i|} \delta_k^{P_i}}{n}$, which can be derived given the BFS tree. We now have a new lower bound of collection capacity as $\frac{W}{\Delta^{**}}$ [26,27]. Here, Δ^{**} is a kind of weighted average of the maximum interference among paths P_i and branches B_i in the BFS tree. We then have the following relationship:

$$n \geq \Delta \geq \tilde{\Delta} \geq \Delta^{**} \geq 1,$$

among the maximum interference number Δ in the whole graph, the maximum interference number $\tilde{\Delta}$ in the paths/branches of the BFS tree, and the "average" maximum interference Δ^{**} in the paths/branches of the BFS tree. These three interference numbers can be different from each other in order.

Now we introduce a new greedy-based scheduling algorithm inspired by Bonifaci et al. [32] and show that it can achieve a nice approximation ratio and lead to another tighter lower bound of collection capacity. The scheduling algorithm still uses the BFS tree as the collection tree. All messages will be sent along the branch toward the sink s. For n messages from one snapshot, it works as follows. In every time slot, it sends each message along the BFS tree from the current node to its parent, without creating interference with any higher-priority message. The priority ρ_i of each packet p_i is defined as $\frac{1}{l(v_i)}$. It is clear that packets originating from the children of the sink have the highest priority $\rho_i = 1$, whereas packets originating from other nodes have lower priority $\rho_i < 1$. For two packets with the same priority (on the same level in the BFS tree), ties can be broken arbitrarily. Given a schedule, let v_j^{τ} be the node of packet p_j in the end of time slot τ. The detailed greedy algorithm (Algorithm 2) is given in Figure 1.15.

Now we analyze the capacity achieved by this greedy data collection method. Before presenting the analysis, we first introduce some new notations. For two nodes v_i and v_j, $h(v_i, v_j)$ denotes the shortest hop number from v_i and v_j in graph G. The delay of packet p_j is defined as the time until it reaches the sink s, that is, $D_j = t \cdot \min\{\tau : v_j^{\tau} = s\}$.

Let λ_i be the minimal number of hops that a packet needs to be forwarded from node v_i before a new packet at v_i can be safely forwarded along the BFS tree. So $\lambda_i = \max\{l | \exists v_j, h(v_i, v_j) = l$ and transmission from v_i to par(v_i) interferes with transmission from v_j to par$(v_j)\} + 1$. Here, par(v_i) is the parent of v_i in T (see Figure 1.16 for illustration). Here, $\lambda_i = 4$ for v_i. We define $\lambda = \max_i\{\lambda_i\}$. Both λ and λ_i are integers (hop counts). In addition, we can prove $\lambda \geq \lambda^*$ as follows.

Algorithm 2 Greedy Scheduling on BFS Tree

Input: BFS tree T rooted at s.

```
Compute the priority ρ_i = 1/l(v_i) of each message p_i.
for each snapshot do
    while ∃p_j such that v_j^τ ≠ s do
        for all such p_i in decreasing order of priority ρ_i do
            if sending p_i from node v_j^τ will not create interference with any higher-priority messages that are
            already scheduled for this slot then
                node v_j^τ sends p_i to its parent par(v_j^τ) in T.
            end if
        end for
        τ = τ + 1.
    end while
end for
```

Figure 1.15 Greedy scheduling on a BFS tree.

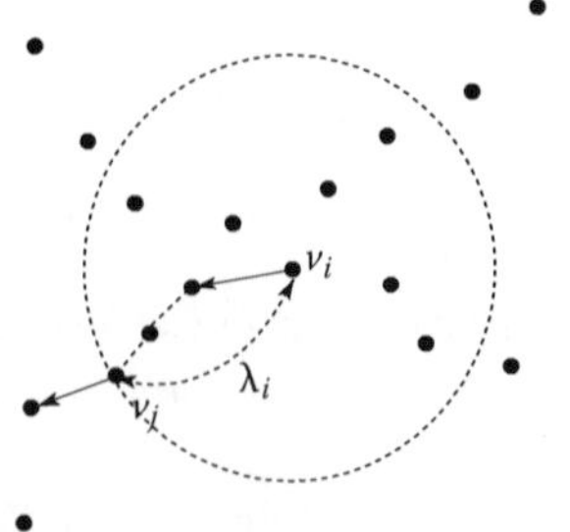

Figure 1.16 Illustration of the definitions of λ_i.

Lemma 8

In the general graph model, $\lambda \geq \lambda^*$ [27]. ■

Proof

Let v_k be the node inside critical region with the largest level. We now consider two cases.

Case 1: If there is a node outside the critical region, as shown in Figure 1.13a, the transmission from v_s to v_k should interfere with the transmission from v_q to s. Thus, in view of v_s, its $\lambda_s \geq l^* + 1 = \lambda^*$. Therefore, $\lambda \geq \lambda^*$.

Case 2: If all nodes are inside the critical region, again consider the v_k with the largest level. Then, $\lambda = \lambda_k = l(v_k) + 1 > l(v_k) = \lambda^*$.

Consequently, we conclude that $\lambda \geq \lambda^*$. ■

Packet p_j is said to be blocked in time slot τ if, in time slot τ, p_j is not sent out. We define the following blocking relation in our greedy algorithm schedule: $p_k \prec p_j$ if in the last time slot in which p_j is blocked by the transmission of higher priority packets in that time slot, p_k is the one closest to p_j in terms of hops among these packets (ties broken arbitrarily). The blocking relation induces a directed blocking tree T_D in which nodes are all message p_i and edge (p_k, p_j) representing $p_k \prec p_j$. The root p_r of the tree T_D is a message with the highest priority (originating from a child of s), which is never blocked. Let $P(j)$ the path in T_D from p_r to p_j and $h(j)$ be the hop count of $P(j)$. We then derive an upper bound on the delay D_j of packet p_j in the greedy algorithm.

Lemma 9

For each packet p_j in the snapshot [27], its delay

$$D_j \leq t \cdot \sum_{p_i \in P(j)} \min\{l(v_i), \lambda\}.$$

■

Proof

We prove this lemma by induction on $h(j)$. For any packet p_j, if $h(j) = 0$, which means p_j is the root p_r of T_D, it will not be blocked. So, $D_j \leq t \cdot l(v_j)$. Then, consider the right side of the inequation $t \cdot \sum_{p_i \in P(j)} \min\{l(v_i), \lambda\} = t \cdot \min\{l(v_j), \lambda\}$. Because p_j is the packet with the highest priority, $l(v_j) = 1$ and $l(v_j) \leq \lambda$. Thus, $t \cdot \sum_{p_i \in P(j)} \min\{l(v_i), \lambda\} = t \cdot l(v_j)$ and the claim in this lemma holds for the case in which $h(j) = 0$.

If $h(j) > 0$, that is, $p_j \neq p_r$, let τ be the last time slot in which p_j is blocked by packet p_k, that is, $p_k \prec p_j$. Notice that $t \cdot h(v_k^\tau, s) \leq D_k - t \cdot \tau$, otherwise p_k would not reach s by time D_k. Also, $h(v_j^\tau, v_k^\tau) \leq \lambda - 1$ because after p_k moves one hop, p_j is safe to move. From time slot $\tau + 1$, p_j may be forwarded toward s over one hop in each time slot, and reach s at the earliest time slot,

$$D_j \leq t \cdot \left[\tau + 1 + h(v_j^\tau, s)\right] \leq t \cdot \left[\tau + 1 + h(v_k^\tau, s) + h(v_j^\tau, v_k^\tau)\right]$$

$$\leq t \cdot (\tau + 1) + D_k - t \cdot \tau + t \cdot (\lambda - 1) = D_k + t \cdot \lambda.$$

On the other hand, $D_j \leq D_k + t \cdot l(v_j)$ because after p_k reaches the sink s, p_j needs at most $l(v_j)$ to reach the sink. Consequently, $D_j \leq D_k + t \cdot \min\{l(v_j), \lambda\}$. This completes our proof. ■

Lemma 10

The data collection capacity of our greedy algorithm [27] is at least $\frac{\lambda^*}{\lambda}\frac{W}{\Delta^*}$. ■

Proof

Let p_j be the packet having a maximum of D_j. By Lemma 9 and Lemma 8 ($\lambda \geq \lambda^*$),

$$D_j \leq t \sum_{p_i \in P(j)} \min\{l(v_i), \lambda\} \leq \frac{\lambda}{\lambda^*} t \sum_{p_i \in P(j)} \min\{l(v_i), \lambda^*\}$$

$$\leq \frac{\lambda}{\lambda^*} t \left[\sum_{v_i \in D(s,l^*)} l(v_i) + \sum_{v_i \notin D(s,l^*)} (l^* + 1)\right] = \frac{\lambda}{\lambda^*} t \sum_i \lambda_i^* = \frac{\lambda}{\lambda^*} nt\Delta^*$$

Thus, the capacity achieved by our greedy algorithm is at least $\frac{nb}{D_j} = \frac{\lambda^*}{\lambda}\frac{W}{\Delta^*}$. ■

In summary, we show that under the protocol and general graph models, the data collection capacity for arbitrary sensor networks has the following bounds:

Theorem 7

Under the protocol and general graph models [27], the data collection capacity for arbitrary sensor networks is at least $\frac{\lambda^*}{\lambda}\frac{W}{\Delta^*}$ and at most $\frac{W}{\Delta^*}$. ■

Here, λ^* describes the interference around the sink s, whereas λ describes the interference around a node v_i. Because $\lambda \geq \lambda^*$, $\frac{\lambda^*}{\lambda} \geq 1$. For the disk graph model, $\frac{\lambda^*}{\lambda}$ is a constant. However, for the general graph model, it may not be. Thus, there is still a gap between the lower and upper bounds (such an example is given in Figure 1.14). We leave finding tighter bounds to close the gap for future works. For two examples in Figure 1.12, the greedy method matches the optimal solutions in order. For the straight-line topology in Figure 1.12a, $\lambda = \lambda^* = n$ and $\Delta^* = \Theta(n)$. Thus, the capacity $\frac{\lambda^*}{\lambda}\frac{W}{\Delta^*} = \Theta\left(\frac{W}{n}\right)$ matches the upper bound. For the star topology in Figure 1.12b, $\lambda = \lambda^* = 1$ and $\Delta^* = 1$. In this case, $\frac{\lambda^*}{\lambda}\frac{W}{\Delta^*} = \Theta(n)$ also matches the upper bound. Compared with the branch scheduling method, the greedy method can achieve much better capacity in practice because it allows packet transmissions among multiple branches of the BFS tree in the same time slot.

Compared with the lower bound of $\frac{\lambda^*}{\lambda}\frac{W}{\Delta^*}$, which we derive from greedy scheduling on the BFS tree, this lower bound of $\frac{W}{\Delta^{**}}$, which we derive from branch scheduling on the BFS tree, may be smaller in some cases. Consider the example in Figure 1.14, $\frac{\lambda^*}{\lambda}\frac{W}{\Delta^*} = \Theta\left(\frac{W}{\log n}\right)$, whereas $\frac{W}{\Delta^{**}} = \frac{W}{\Delta^{**}} = \Theta(W)$. However, the reason is mainly due to the rough relaxation in our capacity analysis of greedy scheduling.

Finally, the bounds of collection capacity could be revised as the following:

Theorem 8

Under the protocol and general graph models [27], data collection capacity for arbitrary sensor networks is at least $\min\left\{\frac{\lambda^*}{\lambda}\frac{W}{\Delta^*}, \frac{W}{\Delta^{**}}\right\}$ and at most $\frac{W}{\Delta^*}$. ■

1.3.3 Data Collection under the Physical and Generalized Physical Models

Similar to the random network part, we can also consider data collection under a physical model or a generalized physical model instead of a protocol model for arbitrary networks.

1.3.3.1 Data Collection under the Physical Model

Chen et al. [27] proved the following theorem for data collection in arbitrary WSNs under the physical model.

Theorem 9

Under the physical and disk graph models [27], the data collection capacity for arbitrary WSNs is $\Theta(W)$. ■

The basic idea of their proof is as follows. To give an upper bound on the capacity of data collection, an artificial transmission range r_0 and an artificial interference range R_0 are defined, such that (1) the receiving node v_j of a sender v_i is within distance r_0, and (2) a transmitting node v_k

will cause interference at node v_j within distance R_0. That is, if there is any interference among the nodes in the protocol model with these artificial ranges, there is also interference among them in the physical model. By artificially setting r_0 and R_0 (which are both constants), we convert the physical model into a protocol model. Using previous proofs in protocol model, it is straightforward to show that the upper bound on the capacity under the disk graph model is bounded by $\Theta(W)$. Similarly, to give a lower bound on the capacity of data collection, an artificial transmission range r_1 and an artificial interference range R_1 are defined, such that, when all simultaneously transmitting nodes are separated by a distance R_1, and the receiving nodes of a transmitting node is within r_1, the SINR of every receiving node is at least η. In other words, if there is no interference among nodes in the protocol model with artificial ranges r_1 and R_1, there is no interference among the nodes in the physical model as well. Thus, we can convert the physical model into a protocol model. Using previous collection algorithms for the protocol model, it can be shown that the lower bound $\Theta(W)$ on the capacity of data collection under the disk graph model is achievable.

1.3.3.2 Data Collection under the Generalized Physical Model

For the capacity of data collection under a generalized physical model, we can derive an upper bound by considering the congestion near the sink node. In particular, we can prove that whatever scheduling scheme is implemented, the total transmission rate of all the incoming links at the sink node is upper bounded by some value. As a bottleneck, the capacity of the whole network is always bounded by that value, as stated in the following theorem.

Theorem 10

Under the generalized physical and general graph models [27], data collection capacity for arbitrary sensor networks is at most

$$\max_i(W_{is}) + W \cdot \log_2 n.$$

■

The first part of this upper bound depends on the rate of the shortest incoming link at the sink, whereas the second part depends on the total number of nodes. Notice that $\max_i(W_{is}) \leq W \cdot \log_2\left(1+\frac{P}{N_0}\right)$. Thus, which part of the bound plays an important role depends on the relationship between n and $1+\frac{P}{N_0}$. When the network is a regular grid or a random homogeneous topology, we have $\max_i(W_{is}) + O(W \log n)$. Therefore, the total rate of all incoming links at sink node s is at most $O((\log n)W)$. The detailed proof of this theorem is similar to the one for Lemma 4 in Section 1.2.4.2, and it is true for any general graph. A lower bound of data collection capacity in this model is still open.

1.4 Conclusion

In this chapter, we investigate the theoretical limitations of data collection in terms of capacity for both random and arbitrary WSNs under different communication models. Table 1.1 briefly summarizes all completed work.

Table 1.1 Summary of Capacity Limits on Data Collection in WSNs

Network Model	*Graph Model*	*Communication Model*	*Sink no. k*	*Capacity C*
Random net	Disk graph	Protocol	1	$C = \Theta(W)$
Random net	Disk graph	Protocol	k (regularly deployed)	$C = \Theta(kW)$ if $k = O\left(\frac{n}{\log n}\right)$ $C = \Theta\left(\frac{n}{\log n}W\right)$ if $k = \Omega\left(\frac{n}{\log n}\right)$
Random net	Disk graph	Protocol	k (regularly deployed)	$\Theta\left(\frac{k}{\log k}W\right) \le C \le C = \Theta(kW)$ if $k = O\left(\frac{n}{\log n}\right)$ $C = \Theta\left(\frac{n}{\log n}W\right)$ if $k = \omega\left(\frac{n}{\log n}\right)$
Random net	Disk graph	Physical	1	$C = \Theta(W)$
Random net	Disk graph	Generalized physical	1	$\Omega\left((\log n)^{-\frac{\beta}{2}}\right) \le C \le O((\log n)W)$
Arbitrary net	Disk graph	Protocol	1	$C = \Theta(W)$
Arbitrary net	General graph	Protocol	1	$\min\left\{\Theta\left(\frac{\lambda^*}{\lambda}\frac{W}{\Delta^*}\right), \Theta\left(\frac{W}{\Delta^{**}}\right)\right\} \le C \le \Theta\left(\frac{W}{\Delta^*}\right)$
Arbitrary net	Disk graph	Physical	1	$C = \Theta(W)$
Arbitrary net	General graph	Generalized physical	1	$C \le \max_i(W_{i0}) + W \cdot \log_2(n)$

There are other advanced techniques, which can be applied in the data collection process to further improve capacity, such as using multiradios to reduce interference [18], using data aggregation to merge data packets [22,23,33], or using compressive data gathering to compress data packets [18,19]. For example, if each sensor can aggregate its received data (multiple packets) into a single packet, the following theorem can be proved, showing improved data rate and capacity for random networks over Theorem 1.

Theorem 11

Under the protocol model [22,23], the delay rate Γ and the capacity C of data aggregation in random sensor networks with a single sink are $\Theta\left(\sqrt{n \log n} W\right)$ and $\Theta\left(\frac{n}{\log n} W\right)$, respectively. ■

Notice that for data collection, the delay rate and the capacity are in the same order (Theorem 1), that is, pipelining can improve only a constant factor of the data rate. However, for data aggregation, it is very interesting to see that pipelining can increase the data rate in order.

Finally, all results presented in this chapter focus on how fast the data collection can be performed under the existence of interferences among sensors. However, in practice, there are also other metrics that should be considered for data collection in WSNs, such as total energy consumption [33], message complexity [33], load balancing among sensors, or possible retransmissions [21]. Readers are encouraged to check relevant references in the literature.

Acknowledgments

The author is grateful to his PhD students (Siyuan Chen and Minsu Huang) and colleagues (Xiang-Yang Li, Shaojie Tang, and Xinghua Shi) for working together on this research topic. This work was supported in part by the U.S. National Science Foundation under grants CNS-0915331 and CNS-1050398.

References

1. I. Akyildiz, W. Su, Y. Sankarasubramaniam, and E. Cayirci. A survey on sensor networks. *IEEE Communications Magazine* 40:102–114, 2002.
2. C. Wang, C. Jiang, X.-Y. Li, S. Tang, Y. He, X. Mao, and Y. Liu. Scaling laws of multicast capacity for power-constrained wireless networks under Gaussian channel model. *IEEE Transactions on Computers* 61(5):713–725, 2012.
3. A. Agarwal and P. R. Kumar. Capacity bounds for ad hoc and hybrid wireless networks. *ACM SIGCOMM Computer Communication Review* 34(3):71–81, 2004.
4. P. Gupta and P. Kumar. The capacity of wireless networks. *IEEE Transactions on Information Theory* 46(2):388–404, 2000.
5. M. Grossglauser and D. Tse. Mobility increases the capacity of ad hoc wireless networks. In *Proceedings of IEEE INFOCOM*, 2001.

6. B. Liu, P. Thiran, and D. Towsley. Capacity of a wireless ad hoc network with infrastructure. In *Proceedings of ACM MobiHoc '07*, 2007.
7. X.-Y. Li, S. Tang, and O. Frieder. Multicast capacity for large scale wireless ad hoc networks. In *Proceedings of ACM MobiCom*, 2007.
8. X. Mao, X.-Y. Li, and S. Tang. Multicast capacity for hybrid wireless networks. In *Proceedings of ACM MobiHoc '08*, 2008.
9. S. Shakkottai, X. Liu, and R. Srikant. The multicast capacity of large multihop wireless networks. In *Proceedings of ACM MobiHoc '07*, 2007.
10. A. Keshavarz-Haddad, V. Ribeiro, and R. Riedi. Broadcast capacity in multihop wireless networks. In *Proceedings of ACM MobiCom*, 2006.
11. B. Tavli. Broadcast capacity of wireless networks. *IEEE Communications Letters* 10:68–69, 2006.
12. E. J. Duarte-Melo and M. Liu. Data-gathering wireless sensor networks: Organization and capacity. *Computer Networks* 43:519–537, 2003.
13. M. Liu, D. L. Neuhoff, D. Marco, and E. J. Duarte-Melo. On the many-to-one transport capacity of a dense wireless sensor network and the compressibility of its data. In *Proceedings of ACM IPSN*, 2003.
14. H. El Gamal. On the scaling laws of dense wireless sensor networks: The data gathering channel. *IEEE Transactions on Information Theory* 51(3):1229–1234, 2005.
15. R. J. Barton and R. Zheng. Order-optimal data aggregation in wireless sensor networks using cooperative time-reversal communication. In *Proceedings of the Annual Conference on Information Sciences and Systems*, 2006.
16. R. Zheng and R. J. Barton. Toward optimal data aggregation in random wireless sensor networks. In *Proceedings of IEEE INFOCOM*, 2007.
17. B. Liu, D. Towsley, and A. Swami. Data gathering capacity of large scale multihop wireless networks. In *Proceedings of IEEE MASS*, 2008.
18. S. Ji, Y. Li, and X. Jia. Capacity of dual-radio multi-channel wireless sensor networks for continuous data collection. In *Proceedings of IEEE INFOCOM*, 2011.
19. S. Ji, R. Beyah, and Y. Li. Continuous data collection capacity of wireless sensor networks under physical interference model. In *Proceedings of IEEE MASS 2011*, 2011.
20. S. Ji and Z. Cai. Distributed data collection and its capacity in asynchronous wireless sensor networks. In *Proceedings of IEEE INFOCOM*, 2012.
21. S. Ji, R. Beyah, and Z. Cai. Snapshot/continuous data collection capacity for large-scale probabilistic wireless sensor networks. In *Proceedings of IEEE INFOCOM*, 2012.
22. S. Chen, Y. Wang, X.-Y. Li, and X. Shi. Order-optimal data collection in wireless sensor networks: Delay and capacity. In *Proceedings of 6th Annual IEEE Communications Society Conference on Sensor, Mesh, and Ad Hoc Communications and Networks (SECON 2009)*, 2009.
23. S. Chen, Y. Wang, X.-Y. Li, and X. Shi. Capacity of data collection in randomly-deployed wireless sensor networks. *ACM Springer Wireless Networks* 17(2):305–318, 2011.
24. S. Chen, Y. Wang, X.-Y. Li, and X. Shi. Data collection capacity of random-deployed wireless sensor networks. In *Proceedings of IEEE Global Telecommunications Conference (Globecom 2009)*, 2009.
25. S. Chen and Y. Wang. Data collection capacity of random-deployed wireless sensor networks under physical models. *Tsinghua Science and Technology* 17(5):485–498, 2012.
26. S. Chen, S. Tang, M. Huang, and Y. Wang. Capacity of data collection in arbitrary wireless sensor networks. In *Proceedings of IEEE 29th Conference on Computer Communications (INFOCOM 2010), Mini-Conference*, 2010.
27. S. Chen, M. Huang, S. Tang, and Y. Wang. Capacity of data collection in arbitrary wireless sensor networks. *IEEE Transactions on Parallel and Distributed Systems* 23(1):52–60, 2012.
28. C. Wang, X.-Y. Li, C. Jiang, and S. Tang. General capacity scaling of wireless networks. In *Proceedings of IEEE INFOCOM*, 2011.
29. S. R. Kulkarni and P. Viswanath. A deterministic approach to throughput scaling in wireless networks. *IEEE Transactions on Information Theory* 50(6):1041–1049, 2004.
30. S. Rao. The *m* balls and *n* bins problem. Lecture Note for Lecture 11, CS270, University of California, Berkeley, 2003.

31. M. Franceschetti, O. Dousse, D. Tse, and P. Thiran. Closing the gap in the capacity of random wireless networks via percolation theory. *IEEE Transactions on Information Theory* 53(4):1009–1018, 2007.
32. V. Bonifaci, P. Korteweg, A. Marchetti-Spaccamela, and L. Stougie. An approximation algorithm for the wireless gathering problem. *Operations Research Letters* 36:605–608, 2008.
33. X.-Y. Li, Y. Wang, and Y. Wang. Complexity of data collection, aggregation, and selection for wireless sensor networks. *IEEE Transactions on Computers* 60(3):386–399, 2011.

Chapter 2

Data Aggregation and Data Gathering

Lalit Kumar Awasthi and Siddhartha Chauhan
National Institute of Technology

Contents

2.1 Introduction

A wireless sensor network (WSN) is a self-organizing network that does not need user intervention for configuration or setting up of routing paths. Due to their advantageous characteristics (low cost, multifunctionality, small size, and mobility), WSNs can be widely used in agriculture, industry, transportation, health care, and everyday life. WSNs can be used in virtually any environment, even in tough terrain or where the physical placement is difficult. WSNs can combine various readings and computations over a very large area of observation to impart aggregated values in different formats and with different observed parameters. Thus, they make it possible to monitor real-world events to an unprecedented level of granularity. However, distinctive features, such as limited energy and memory or unreliable communication, bring up many problems for scientists and engineers working in this field. The ideal WSN should be scalable, low cost, less power-consuming, efficient in data gathering, reliable and accurate, and above all, maintenance-free.

Sensor nodes (SNs), which are deployed over a sensor field, organize themselves into a network, sense real-world phenomena, and forward observed measurements back to base stations or sinks. As shown in Figure 2.1, the SNs sense an event and forward their data toward the sink through other SNs. Data delivery to a sink may include a large number of wireless hops among the networked set of small, resource-limited SNs. Data gathered at the sink from various SNs has to be pieced together into meaningful information; therefore, it is important that all the SNs are well synchronized. This makes the problem of data gathering and dissemination in WSNs a real challenge for researchers. SNs are subject to frequent failures because of the operational environment as well as energy and memory constraints. It is important for a large-scale deployment of WSNs that the network provide reliable, robust, and accurate measurements. WSNs should be self-healing so that critical information is delivered promptly despite node failures.

SNs can have dense and large-scaled deployment due to their small size and low cost. Consequently, monitoring can be done more efficiently and at a higher sensing resolution compared

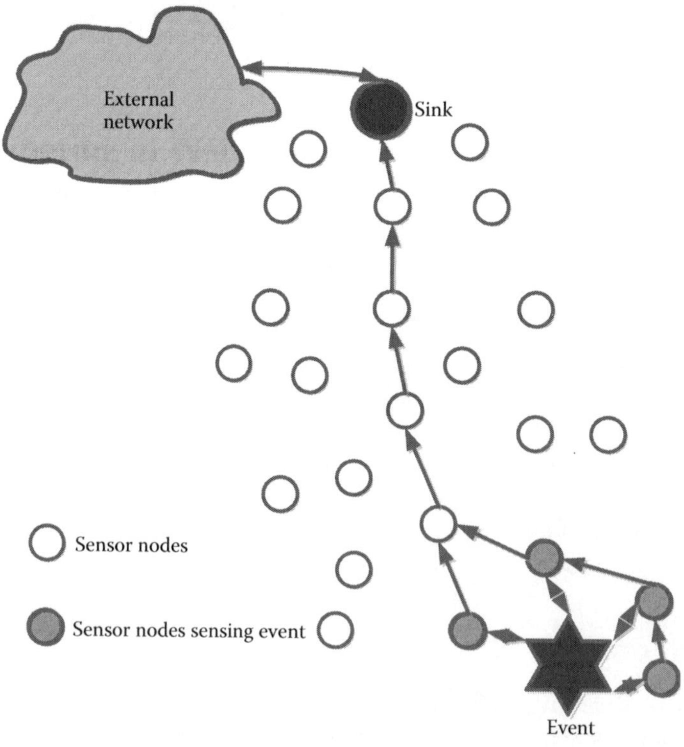

Figure 2.1 Scenario of WSN deployment.

with traditional sensor systems. The previously unobservable becomes observable. Abstracting simple sensor readings to interrelated events allows for the guarding of complex processes. Due to intelligence and collaboration between the individual nodes, the network (by itself) can carry out complex tasks related to observing.

The dense and large deployment of sensor nodes creates a large amount of redundant sensor data. The transmission of such redundant data consumes precious resources from the network and nodes. Data aggregation is an effective technique in this context because it reduces the number of packets to be sent to the sinks by aggregating similar packets. Data aggregation not only saves energy but also increases network life. In this chapter, we present some data aggregation and data-gathering algorithms proposed by several researchers and academicians.

2.2 Data Aggregation

Data aggregation is defined as the process of aggregating the data from multiple sensors to eliminate redundant transmission and provide fused information to the base station. Data latency and accuracy are important in many applications such as environmental monitoring, in which the freshness of data is also an important factor. It is critical to develop energy-efficient and fast data-aggregation algorithms. Aggregation can be done using two basic approaches. In the first approach, every sensor node sends the sensed values to the sink. After receiving all messages, the sink computes the aggregated value. In the second approach, all the nodes send the sensed values to their neighbor/parent. The parent node has to wait for its children before computing the aggregate to be forwarded to the sink. The amount of transmitted data depends on the type of aggregate function.

2.3 Data Aggregation and Data Gathering in WSNs

Various protocols proposed for aggregation, data gathering, and routing are interdependent and some authors have used the terms interchangeably. Routing in WSNs takes into consideration data aggregation at some nodes and accordingly decides packet routing. Similarly, for data aggregation, the routing protocol used underneath plays a vital role. Data generated by different sensors can be jointly processed while being forwarded toward the sink. This data generated by different nodes can be fused together (data fusion), processed locally, and any redundancy can be removed before transmission. It is important that for data fusion or other data aggregation operations, the WSNs should be time-synchronized. Data aggregation techniques are closely related to the way data is gathered at SNs as well as to how packets are routed through the network. Data aggregation has a significant effect on energy consumption and overall network energy consumption.

Data aggregation is the simplest in-network processing activity in which data from different nodes is combined into a single entity. In-network aggregation is the global process of gathering and routing information through a multihop network, processing data at intermediate nodes with the objective of reducing resource consumption (in particular energy), thereby increasing network lifetime [1]. There are two ways in which data from different sources can be combined. The first method combines data packets from different sources into one data packet that is smaller compared with the combined size of the data packets from all sources. The reduction in size of the data packet transmitted into the network is energy-conserving because smaller-sized aggregated packets are transmitted. However, this type of aggregation is lossy aggregation and sometimes results in the loss of granularity of the sensed observations. The second method combines data packets from different sources into one bigger packet. The packets' overheads in this case are reduced and the granularity of data is preserved.

Data-gathering protocols are formulated for configuring the network and collecting information from the desired environment [2]. In each round of the data-gathering protocol, data from the nodes needs to be collected and transmitted to the sink [3], where the end user can access the data. A simple way of doing this is by aggregating (sum, average, min, max, count) the data originating from different nodes [4]. A more elegant solution is data fusion, which can be defined as a combination of several unreliable data measurements to produce a more accurate signal by enhancing the common signal and reducing the uncorrelated noise. Sensor nodes use different data aggregation techniques to achieve energy efficiency. The aim is the efficient transmission of all data to the base station so that the lifetime of the network is maximized. Existing data-gathering protocols [5] can be classified into different categories based on the network topology and on routing protocols [3,6], which are aimed at power saving and prolonging network lifetime [7,8].

2.4 Protocols for Data Aggregation and Data Gathering

In the literature, there are different heuristics to a Steiner tree problem, which is a well-known NP hard problem. A network represented by graph; $G = (V,E)$, where $V = \{v_1, v_2, \ldots, v_n\}$ is a set of SNs, and E are the edges that represent the connections among them with a cost associated with each edge. The problem is how to build a minimal cost tree that connects all source nodes $S = \{s_1, s_2, \ldots, s_m\}$, $S \subseteq V$, and the sink node. The cost of the Steiner tree is the sum of the costs of its edges. Some proposals [9,10] that have tackled this problem require large amounts of messages to set up a routing tree and consequently lead to high energy consumption in the SNs. There are distributed approaches to solve this problem, such as the shortest path tree (SPT) [11], center-at-nearest source tree (CNS) [12], greedy incremental tree (GIT) [13], and information fusion–based role assignment (InFRA) [14]. However, none of these approaches were designed with fault-tolerance in mind. SPT is the simplest strategy to build a routing tree in a distributed fashion. In this approach, every node that detects an event reports its collected information by using the shortest path to the sink. Information fusion occurs whenever paths overlap (opportunistic information fusion).

In the CNS algorithm [12], every node that detects an event sends its information to a specific node, called an aggregator, by using the shortest path. The aggregator is the closest node to the sink (in hops) that detects an event. In the GIT strategy [13], when the first event is detected, the nodes send their information similar to the SPT and, for every new event, the information is routed using the shortest path to the current tree. There is a new aggregation point every time a new branch is created. Some practical issues make GIT unaffordable for WSNs. For example, each node needs to identify the shortest path to all nodes in the network. The communication cost to create this infrastructure is $O(n^2)$. Furthermore, the space needed to store this information at each node is $O(Dn)$, where D is the number of hops in the shortest path connecting the farthest node $v \in V$ to the sink node (network diameter). After the initial phase, the algorithm needs $O(m\ n)$ messages to build the routing tree.

The InFRA algorithm [14] builds a cluster for each event; including only those nodes that were able to detect it. Then, cluster heads merge the data within the cluster and send the results to the sink node. The InFRA algorithm aims to build a SPT that maximizes this information fusion. Thus, once clusters are formed, cluster heads choose the shortest path (to the sink node) that maximizes information fusion by using the aggregated coordinators' distance. The disadvantage of the InFRA algorithm is that at each new event that arises in the network, the information about the event must be flooded throughout the network to inform other nodes about the occurrence of the new event.

Pattem et al. [15] proposed a model to describe the spatial correlation in terms of joint entropy. They analyzed a symmetric line network with different degrees of correlation among neighboring nodes. The authors showed that, for uncorrelated data, the best routing strategy was to forward packets along the shortest paths. In case of correlated information, the data is aggregated as soon as possible and a single aggregated packet is sent to the sink along the shortest path. In a study by Zhu et al. [16], the effect of data correlation on energy expenditure of data distribution protocols was studied. The focus was on various energy-aware data aggregation trees under different network conditions such as node density, source density, source distribution, and data-aggregation degree.

Chen et al. [17] proposed an energy-efficient protocol for aggregator selection (EPAS) for single-level aggregation. The optimal number of aggregators with generalized compression and power consumption models were derived and fully distributed algorithm hierarchical EPAS, an extension of EPAS, were presented.

A tree-based aggregation algorithm that exploits data correlation based on a shallow length tree (SLT), which unifies the properties of the minimum Steiner tree (MST) and the SPT, was presented by von Rickenbach and Wattenhofer [18]. In an SLT, the total cost of the tree is only a constant factor larger than that of the MST, whereas the distances (and delays thereof) between any node and the sink are only a constant factor larger than the shortest paths. Cristescu et al. [19] analyzed the aggregation properties of a tree structure that is based on an SPT of nodes close to the sink node, whereas nodes that are further away are connected to the leaves of the SPT via paths found by an approximation algorithm for the traveling salesman problem. In studies by Albert et al. [20], Dasgupta et al. [21], and Ding et al. [22], the ways through which the sink organizes routing paths to evenly and optimally distribute energy consumption while favoring the aggregation of data at the intermediate nodes were investigated. Dasgupta et al. [21] used linear programming to compute aggregation topologies by taking into account the residual energy of each node.

Hong and Kim [23] have suggested an integrated gateway node control protocol (IGCP). An algorithm, namely, the integrated gateway node (IGN) algorithm that compensates for the vulnerability of both hierarchical and flat structures in the network has been proposed. The study suggests a mixed algorithm with virtual gateway nodes, which include both the advantages of existing hierarchical structure algorithms and flat structure algorithms in WSN. The algorithm forms virtual gateway nodes consisting of several nodes such as the cluster of a hierarchical structure routing protocol and allows flat structure routing protocols between virtual nodes. The suggested algorithm communicates with the application of flat structure–type protocols after bundling up several nodes like the hierarchical structure cluster and making them work as one node. Virtual gateway nodes allow efficient energy management because it not only increases energy efficiency but also creates virtual nodes. It also makes up for the disadvantages of existing hierarchical structure routing protocols by allowing even energy use of each node as well as performing data-aggregation and in-network processing that are characterized in existing WSN routing protocols.

Nodes in the network have been exploited as aggregation points for optimal performance. Al-Karaki et al. [24] present exact and approximate algorithms to find the minimum number of aggregation points to maximize the network lifetime. Algorithms use a fixed virtual wireless backbone that is built on top of the physical topology. The tradeoffs between energy savings and the potential delay involved in the data-aggregation process have also been studied. Studies by Hartl and Li [25] and Solis and Obraczka [26] focused on the nodes that should be entrusted with the transmission of the sensed values, whereas in a study by Erramilli et al. [27] the emphasis was put on the proper scheduling of sleeping/active periods. Optimal paths are calculated in a centralized manner at the sink by exploiting different assumptions on the data correlation and selecting the best aggregation points using cost functions [28].

An energy-efficient data-gathering algorithm for prolonging the lifetime of WSNs has been proposed by Zhu et al. [29]. The authors have proposed a routing algorithm, called energy-efficient routing algorithm to prolong lifetime (ERAPL), that is able to dramatically prolong network lifetime while efficiently spending energy. In ERAPL, a data-gathering sequence (DGS) to avoid mutual transmission and loop transmission among nodes is first constructed. Each node proportionally transmits traffic to the links confined in the DGS.

A novel data-gathering scheme called data-aggregating ring (DAR) has been proposed by Bi et al. [30]. In the DAR scheme, all sensor nodes are classified according to the number of hop counts to the sink (hop grades). The nodes from different hop grades, ordered in a certain sequence, would spend different amounts of time taking charge of gathering the data packets from the nodes in other hop grades and transmitting them to the sink directly by one hop, respectively. The sequence is elaborated according to the traffic characteristic and routing strategy of the given network to balance the workloads between the nodes in different hop grades. As a result, the nodes at a distance of only one hop to the sink tend to consume an equal amount of energy as those with more hops upon delivering data. Therefore, the DAR scheme can nearly balance the energy consumption over a whole network range, increase energy efficiency and extending network lifetime notably.

Chuan-Ming Liu et al. [31] argue that a cluster-based architecture is an effective architecture for data-gathering in WSNs. However, in a mobile environment, the dynamic topology poses a challenge to designing an energy-efficient data-gathering protocol. In this study, a cluster-based architecture was considered. Two distributed clustering algorithms for mobile sensor nodes, which minimize the energy dissipation for data-gathering in a wireless mobile SN, have been proposed. There are two steps in the clustering algorithm: cluster-head election step and cluster formation step. Two distributed algorithms proposed for cluster-head election are the algorithm of cluster-head election by counting (ACE-C) and the algorithm of cluster-head election with location (ACE-L). Considering the effect of node mobility, a mechanism has been proposed (i.e., clusters with mobility or CM) for sensor nodes to select a proper cluster-head to join the cluster formation. The CM mechanism is shown to achieve better performance in terms of energy consumption and system lifetime when the sensor nodes are capable of mobility. Proposed clustering algorithms achieve the following three objectives: (1) there is at least one cluster-head elected, (2) the number of cluster-heads generated is uniform, and (3) all the generated clusters have the same cluster size.

Steiner points grid routing [32] reduces the total energy consumption for data transmission between the source node and the sink node. A virtual grid structure is constructed based on the square Steiner trees [33]. Once the sensor nodes are deployed in the sensor field, the sink node starts to construct the grid structure. The sink divides the plane into a grid of cells. Cross-points of the grid are the dissemination points (DPs). The size of the cells, denoted as α, is determined by the sink such that DPs are not within direct transmission range (the sink is the first DP). Recognizing its own position and the size of each cell, the sink is able to send a data request (in the form of a data announcement message) to each adjacent DP in the grid. Any node that is within the target region of a received QUERY message stores the appropriate routing information and starts to send the sensed data (in the form of a DATA message) to the sink. The routing information contains the appropriate upstream dissemination nodes (DNs) through which DATA messages will be forwarded. DN will find the appropriate path to transmit the DATA message depending on which DP it belongs to.

Tree-based schemes for real-time or time-constrained applications have been proposed [34,35]. An approach that relies on the construction of connected dominating sets has been proposed by Gupta et al. [36]. These consist of a small subset of nodes that form a connected backbone and whose positions are such that they can collect data from any point in the network. Nodes that do not belong to these sets are allowed to sleep when they do not have data to send. Some rotation of the nodes in the dominating set is recommended for energy balancing. More algorithms on data aggregation and data gathering have presented in other studies [23–27,37,38].

The advances in sensor node architectures, such as the inclusion of multiple sensing units and other components with variable power mode capability, have made data gathering more challenging. A sensor node with multiple sensing units is usually unable to simultaneously process the data generated by multiple sensing units, thereby resulting in missed events.

The multiple sensing unit scheduling (MSUS) [39] algorithm is the first of its kind in dealing with the tasks of multiple sensing units of the same sensor node. It schedules the tasks of different sensing units based on their priority and according to the timing constraints imposed by the application, and the existing as well as predicted future tasks for all the sensing units of a sensor node. MSUS treats task timing constraints as hard requirements whereas minimizing energy consumption and missed events. This work addresses the problem of scheduling in multiple sensing units of a sensor node in which the arrival of events and the corresponding tasks are not known in advance. Therefore, MSUS first predicts the time and the type of the next task that will soon arrive, then determines the best power state for the sensor node by considering the power and timing constraints of the current and future tasks. In MSUS, the prediction of future tasks for the sensing units is based on the past history of the occurrence of events for a particular environment.

Ozgur Sanli et al. [40] present a collaborative task scheduling algorithm (CTAS), to minimize event misses and energy consumption by exploiting power modes and overlapping sensing areas of sensor nodes. CTAS enables sensor nodes to keep only a subset of their sensing units' active at any time even though each sensor node has multiple sensing units. Although some of the sensing tasks are scheduled to neighboring nodes, the degree of coverage in the network is still maintained at a specific level for each event type. The novel idea of CTAS is that it employs a two-level scheduling approach to the execution of tasks collaboratively at group and individual levels among neighboring sensor nodes. CTAS first implements coarse grain scheduling at the group level to schedule the event types to be detected by each group member. Then, CTAS performs fine-grain scheduling to schedule the tasks corresponding to the assigned event types. The coarse grain scheduling of CTAS is based on a new algorithm that determines the degree of overlapping among neighboring sensor nodes.

2.5 Energy-Efficient Clustering and Data Aggregation Protocol for Heterogeneous WSNs

A novel energy-efficient clustering and data aggregation (EECDA) [41] protocol for heterogeneous WSNs combines the ideas of energy-efficient cluster-based routing and data aggregation to achieve a better performance in terms of lifetime and stability. The EECDA protocol includes a novel cluster-head election technique and a path would be selected with the maximum sum of energy residues for data transmission instead of the path with minimum energy consumption. In their work, the authors have shown that EECDA balances the energy consumption and prolongs the network lifetime by a factor of 51%, 35%, and 10% when compared with low-energy adaptive clustering hierarchy (LEACH), energy-efficient hierarchical clustering algorithm (EEHCA), and effective data-gathering algorithm (EDGA), respectively.

The main goal of the EECDA protocol is to efficiently maintain the energy consumption of sensor nodes by involving them in a single-hop communication within a cluster. The data aggregation and fusion technique is used to reduce the number of transmitted messages to the base station to save the energy and prevent the congestion. The authors have adopted a few reasonable assumptions for implementing the protocol as follows: (i) sensor nodes are uniformly dispersed within a square field, (ii) all sensor nodes and the base station are stationary after deployment, (iii) the WSN consists of heterogeneous nodes in terms of node energy, (iv) cluster heads (CHs) perform data aggregation, and (v) the base station is not energy limited in comparison with the energy of other nodes in the network.

A novel cluster-head election technique and a path with the maximum sum of energy residual for data transmission can maintain the balance of energy consumption in the network.

2.6 A Noble Data Aggregation Algorithm for Low Latency in Wireless SNs

In their work, Tianbo Wang et al. have optimized the energy consumption and latency of data transmission. A bilayer-based data aggregation scheme [42] has been adopted in which the wireless network is divided into two layers and each layer has a different number of cluster heads optimized. The members (nodes or cluster heads) in the region of detection of each layer send data to the related head it belongs to and the head aggregates the data. First, the number of nodes in a certain cluster is calculated, and then nodes of the network are divided into two layers. Employing these strategies, as well as the new cluster head selection method in which the aggregation is done at the cluster heads, the authors have shown that energy consumption and latency can be reduced compared with the LEACH process.

The number of nodes in a certain cluster is calculated with the cover rate and overlap rate in every round. In WSN, the cluster heads selected by the LEACH process broadcast in a radius (R) to form the measuring area. The nodes in the measuring area would return the signal to the respective cluster head, so the cluster head would acquire the number of nodes in the measuring area. The broadcasting radius of the measuring area is affected by the following two factors. The broadcasting radius R designed is long enough to measure the nearby nodes' distribution area. The total measuring area of the cluster head would reach a certain cover rate of the network area so the node distribution situation in the measuring area could reflect the node distribution situation in the clustering area. As the R extends, the overlap rates of the measuring areas of the cluster heads increase accordingly. Furthermore, the energy consumption in the measuring process would also increase with the length of R. According to the two factors previously mentioned, the length analysis of the broadcasting radius has to be optimized. The bilayer-based algorithm has been designed to improve on the LEACH process. The bilayer structure is constructed according to the following algorithms:

1. Select the group heads under probability of randomness. The selected nodes have a flag that cannot be selected in the future N/Kk rounds, and it will be stored in an array in the meantime (N is the number of nodes, K is the number of group heads, and k is the number of cluster heads).
2. The normal nodes select the nodes with the shortest distance to the group head.
3. The group selects the cluster heads distribution. The selected heads compute the distance to the upper heads and the energy consumption.
4. The normal nodes select the nodes in the same group as the cluster head according to the distance and the selected heads will be stored in another array as the cluster head. Similar to the group heads, the cluster heads will assign a flag marking that will not be selected in future N/Kk rounds. The normal nodes transmit the collected data to the related head. After network construction, each layer has the optimal heads and each node has the related head with least energy consumption.

The WSN is thus divided into several clusters in which the cluster head aggregates the sensed data from the nodes. The data sensed by the nodes in a zone is aggregated and relayed to members in the upper layers (aggregators or sensors) by the CH in the respective zone. A set of clusters that monitor the same phenomenon form a group. Each group has only one group head responsible for collecting and aggregating the data from its zone.

2.7 A Scalable and Dynamic Data Aggregation Aware Routing Protocol for WSNs

Villas et al. [43] have considered the problem of constructing a dynamic and scalable structure for data aggregation in WSN that addresses the load balancing problem. A protocol called dynamic and scalable tree (DST) reduces the number of messages required for setting up a routing tree, maximizes the number of overlapping routes, and selects routes with the highest aggregation rate. The DST is a routing tree with the shortest routes (in distance) that connects all source nodes to the sink node. The routing tree created by DST does not depend on the order of events and is not held fixed along the occurrence of events. DST considers the following roles for the creation of routing infrastructure:

- Collaborator: a node that detects events
- Coordinator: a node that gathers events from collaborator nodes, aggregates data, and notifies them
- Aggregator: a node that forwards aggregated data from two or more source nodes
- Sink: a node interested in receiving data from a set of coordinator and collaborator nodes
- Relay: a node that forwards data toward the sink

The DST has four phases. In phase 1, the sensor nodes store the sink's position and the neighbor's position. This is done by the sink, which floods a configuration message from the neighbors' position (CMNP). Phase 2 consists of cluster formation and the election of a coordinator among the nodes that detected the occurrence of a new event in the network. When an event is detected by one or more nodes, the leader election algorithm is started with the nodes running for leadership (group coordinator). For this election, all nodes are eligible; however, the group leader is the node that is closest to the sink. In phase 3, when an event occurs, the coordinator sends a package to the sink node reporting its position. The sink then notifies all other coordinators of the new coordinators' position. The sink also notifies the new coordinator of the positions of the previous coordinators. The node chosen as the event coordinator in phase 2 gathers the information collected by the collaborators. Based on its position and the sink's position, the coordinator creates a straight line segment that connects itself to the sink. The sensor nodes closest to this straight line segment and to the sink are chosen to notify of the occurrence of a new event and send data to the sink node. Finally, phase 4 is responsible for creating the routing tree connecting all coordinators to the sink node and sending the collected data to the sink node.

2.8 A Hierarchical Multiparent Cluster-Based Data Aggregation Framework for WSNs

Alemu and colleagues [44] have proposed an efficient fault-tolerant data aggregation framework based on hierarchical clustering. The proposed scheme is not only an energy-efficient aggregation scheme but is also able to overcome faulty readings and detect faulty nodes. Each cluster has two heads—primary and secondary—to aggregate sensed data. Sensor nodes in each cluster have two parents wherein the secondary parent decreases the packet drop rate by overhearing those packets that fail to reach the primary head. This maximizes the communication efficiency within the cluster. Both the spatial and temporal characteristics of nodes within and across clusters have been exploited to substitute missed values based on regression analysis.

The framework considers a network with the aggregation tree formed at the initialization stage. Cluster heads (selected based on LEACH) in each cluster elect the secondary parent based on the average distance of the nodes. The two parents synchronize with each other using identical time slots so that data packets that fail to reach the primary parent can be heard and transmitted by the secondary parent without the need for any transmission from the source node.

The aggregation process has three phases. The first phase is for building an aggregation tree. It follows a top-down hierarchical flow of messaging from the sink to the leaf nodes for constructing the basic clustering structure of the system. The data collection phase is the bottom-up process of sending the sensed data back to the sink node. Based on the maximum allowable round-trip time, each node in each cluster will send their data to their cluster head; at the same time, the secondary parent will overhear it. In this process, transient error is highly minimized as the reading is being heard by either of the parents. In addition, the secondary parent can perform simple aggregation functions like MIN/MAX, and send a single value to the primary head. The third phase is an error recovery phase. One characteristic of sensor readings is their correlation within the neighboring nodes (spatial) and a high probability of repeated values within itself (temporal). Hence, at the intracluster level, data value is lost only if it is not received by both parents. In this case, the primary parent estimates the lost data from the spatially related neighboring nodes using multiple regression analysis. If data value is heard by the secondary parent, the primary parent will get the reading before performing a regression analysis.

2.9 Cluster Tree–Based Data Gathering in WSN

Network lifetime, scalability, and load balancing are important requirements for many data-gathering SN applications. The proposed scheme is an improved version that uses both cluster- and tree-based protocols to improve the performance. The proposed protocol improves the power consumption. Chhabra and Sharma [45] have proposed the combination of cluster-based and tree-based protocols. The protocol has been designed with the following assumptions:

1. Each node or sink has the ability to transmit messages to any other node and sink directly.
2. Each sensor node has a radio-powered control node that can tune the magnitude according to the transmission distance.
3. Each sensor node has the same initial power in WSNs.
4. Each sensor node has location information.
5. Every sensor node is fixed after they are deployed.
6. WSNs would not be maintained by humans.
7. Every sensor node has the same process and communication ability in WSNs, and they play the same role.
8. Wireless sensor nodes are deployed densely and randomly in a sensor field.

The setup phase is composed of cluster formation and cluster head selection. The cluster, once formed, will not be changed much but the selected cluster head may be different in each round.

During the first round, the base station first splits the network into two subclusters, and proceeds further by splitting the subclusters into smaller clusters. The whole process is repeated until the desired number of clusters is formed. At the end of this entire process, a cluster head for each cluster will be selected by the base station. After the first round, the primary cluster topology is formed; the task of cluster formation is shifted from the base station to the sensor nodes. The decision to choose a new cluster head is made locally within each cluster based on the node's weight value.

The next phase is for constructing a cluster-based tree in which the minimum spanning tree is used to compute the tree path after labeling the cluster head. After the routing mechanism has been established, every tip node transmits gathered data to nodes in the upper level. Then, the upper level nodes will fuse received data and sensed data by itself, and send these data to next upper level nodes. This proposed method has several advantages in WSNs for data gathering. It reduces power consumption by avoiding direct communication between sink and sensor nodes. Use of the threshold mechanism increases the network lifetime. The threshold mechanism protects the death of the parent node death slowly, as each node has the chance to be a parent (Figure 2.2).

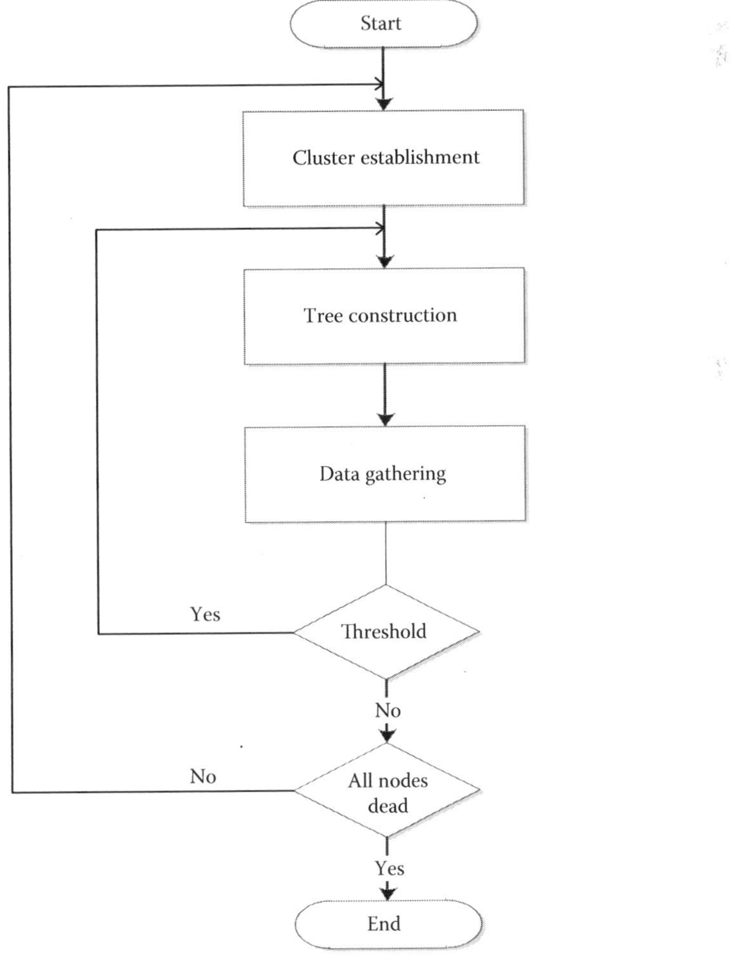

Figure 2.2 Flow chart for the cluster tree–based data-gathering protocol.

2.10 Compressive Data Gathering for Large-Scale WSNs

Luo and colleagues [46] presented a complete design to apply compressive sampling theory to sensor data gathering for large-scale WSNs. The proposed compressive data gathering is able to reduce global scale communication costs without introducing intensive computation or complicated transmission controls. The scheme used load balancing, which extends the lifetime of the network. The proposed scheme compresses sensor readings to reduce data traffic and distributes energy consumption evenly thus improving the network lifetime. In compressive data gathering (CDG), higher efficiency can be achieved by transmitting correlated sensor readings jointly.

The data-gathering process of CDG is illustrated in Figure 2.3, which is a detailed view of a small fraction of a routing tree. Leaf nodes initiate the transmission process when all nodes have acquired their readings. S2 generates a random number φ_{i2}, computes $\varphi_{i2}d_2$, and transmits the value to S1. The index i denotes the ith weighted sum ranging from 1 to M. Similarly, $s4$, $s5$, and $s6$ transmit $\varphi_{i4}d_4$, $\varphi_{i5}d_5$, and $\varphi_{i6}d_6$ to S3, respectively. Once S3 receives the three values, it computes $\varphi_{i3}d_3$, adds it to the sum of relayed values, and transmits $\sum_{j=3}^{6}\varphi_{ij}d_j$ to S1. Then, S1 computes $\varphi_{i1}d_1$ and transmits $\sum_{j=1}^{8}\varphi_{1j}d_j$. Finally, the message containing the weighted sum of all readings in a subtree is forwarded to the sink.

2.11 An In-Network Approximate Data-Gathering Algorithm Exploiting Spatial Correlation in WSNs

Several schemes have been proposed that utilize the spatial correlation of sensor readings to achieve energy savings, but most of the proposed schemes experienced high control overhead or did not fully exploit spatial correlation. To overcome these shortcomings, an in-network approximate data-gathering algorithm exploiting spatial correlation has been proposed by Huangy and colleagues [47]. The proposed algorithm consists of two phases: an in-network clustering phase and a reading streaming phase. In the in-network clustering phase, the first clusters are initialized; then,

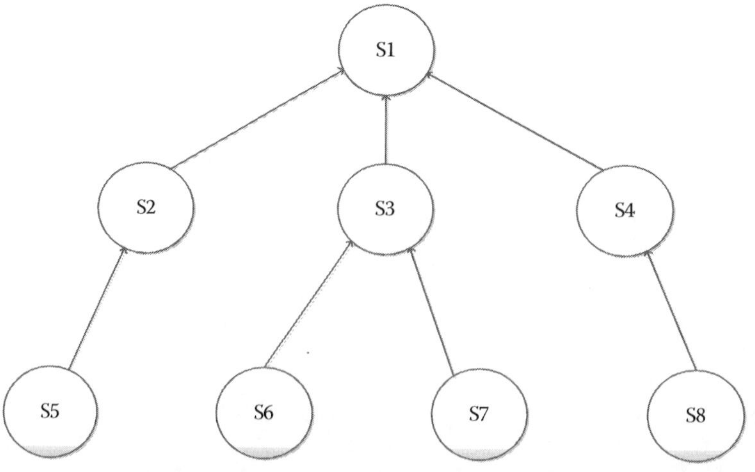

Figure 2.3 The data-gathering process.

cluster heads collect the readings from the nodes. In this phase, the in-network clustering scheme exploits the spatial correlations of sensor readings as well as cluster readings to further reduce the number of representative nodes. On the other hand, the reading streaming phase employs an adaptive cluster maintenance scheme which ensures that the user will obtain the reading answers of the desired quality despite changing sensor readings.

2.11.1 Cluster Initialization

When a user wants to monitor a phenomenon of interest p_i (e.g., temperature) and tolerates an error threshold of α, the user submits a query Q = (Query ID; p_i; α) to the sink. The sink floods the query Q into the SN and executes cluster initialization. In cluster initialization, the spatial correlation of sensor readings to group sensor nodes into disjoint clusters is exploited and representative cluster heads are determined. A data-gathering tree is created at the same time. Because the focus is not on initial clustering, the clustered aggregation technique (CAG) algorithm for initial clustering (due to its simple and distributed nature) was adopted. When the query flooding process terminates, the SN is partitioned into disjoint clusters and organized into a data-gathering tree. The cluster heads are called the initial cluster heads.

2.11.2 Cluster Reading Collection

After cluster initialization, the cluster head invokes the cluster reading collection mechanism to obtain the readings of cluster members in the same cluster, and the most representative cluster member is represented as the new cluster head. After a sensor node joins a cluster and rebroadcasts the received query, the node sends the current reading to the cluster head. With the readings of cluster members received, the cluster head can calculate the introduced reading range and tolerance range of the cluster.

2.11.3 Cluster Merging Problem

The reading of a cluster head can represent the cluster members in the same cluster or reading of a cluster head is likely to represent other nearby cluster heads. The number of representative cluster heads is further reduced by grouping multiple neighboring clusters into a larger cluster. To preserve spatial correlation, only the clusters in a merging candidate set are considered to be merged together.

2.11.4 In-Network Cluster Merging

To enable in-network cluster merging, when a cluster head i reports a reading to the sink, it attaches additional information to the report. A report is formatted as (CID_i; R_i; $MaxDiff_i$; $State_i$). CID_i is the ID of the cluster head i. This is used to identify the merging candidate clusters. An intermediate node receiving a report modifies the value of the state according to its current value.

2.11.5 Reading Streaming Phase

The clusters must be dynamically readjusted due to the changes of sensor readings because the reported readings are bounded by the user-tolerable error threshold. In this phase, an adaptive cluster maintenance scheme to merge or split clusters, which offers a threshold guarantee while minimizing the number of representative cluster heads, is used. According to changes in the sensor readings, the cluster maintenance scheme readjusts the clusters in the following three cases:

Case 1—Cluster merging: a merging check node maintains a merging table to record CIDs, R_s, $MaxDiff_s$, as well as the reading and the tolerance range of received reports. On the basis of the reports, the clusters are merged.

Case 2—Cluster creation: when a cluster member *A* realizes that its new reading lies outside the tolerance range of the cluster head, it creates a new cluster because the cluster head cannot be representative of it. Node *A* first elects itself as a new cluster head. $MaxDiff_A$ is set to 0 because no other cluster member is in the cluster. Then, new cluster head *A* notifies the original cluster head of its departure by sending a leave message which upon receiving the leave message decrements the number of the cluster members by 1. Although a new cluster head causes an increase in the number of report transmissions, it is highly likely that the new cluster will soon be merged with other nearby clusters.

Case 3—Cluster splitting: when a slave cluster head *A* refreshes $MaxDiff_A$ and it is not dominated by the master cluster, it simply resumes to report the readings to the sink at regular intervals.

EXERCISES

1. What is the difference between data fusion and data gathering?
2. Many authors have used these words interchangeably, that is, data aggregation and data gathering. What is the major difference between data aggregation and data gathering?
3. Assume that there are 4 clusters with 10 cluster nodes per cluster. One node among each cluster is chosen as the cluster head. Each cluster head is capable of transmitting the packets directly to the sink. Assuming that the transmission and per bit for 1 m is 50 nJ. The reception and processing energy per bit is 50 and 0.5 nJ, respectively. The distance between the sink and cluster heads is 4 m. The length of the aggregated data packet is 64 bytes. Calculate the energy consumed by the cluster head in 5 min if four packets of 36 bytes in size are received in 1 s from all its child nodes.
4. Assume the scenario of question 3. Calculate the lifetime (number of rounds, transmission rounds) and number of total packets the cluster head will transmit with an initial energy of 2 J.
5. Considering the energy consumption of sensor nodes given in question 3, suppose the network topology is as given in the figure in which each cluster has five sensor nodes:

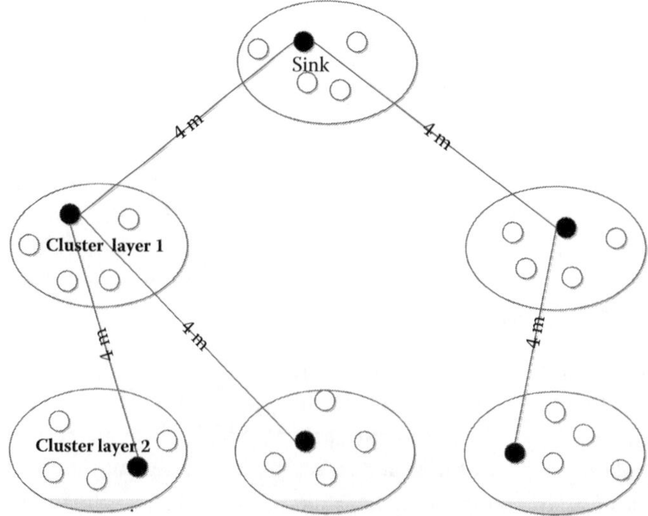

i. Calculate the energy consumed by clusters of layers 2 and 3 if the sensing event is continuous for 10 min.
ii. If all the cluster heads can transmit directly to the sink and 64-byte packets are received from cluster nodes by each cluster head. The cluster head combines these packets to form one bigger packet in which the aggregated packet size is 256 bytes. Which is more energy-consuming in a network?
   a. To transmit directly to sink by each cluster head, where layer 2 cluster heads are at a distance of 6 m from the sink.
   b. Layer 2 cluster heads transmit their packet to layer 1 cluster heads, which aggregate the packets received (aggregated packet size is 256 bytes).

References

1. E. Fasolo, M. Rossi, J. Widmer, and M. Zorzi. In-network aggregation techniques for wireless sensor networks: a survey. *IEEE Wireless Communications*, Vol. 14, No. 2, pp. 70–87, April 2007.
2. P. Mohanty, S. Panigrahi, N. Sarma, and S. S. Satapathy. Security issues in wireless sensor network data gathering protocols: a survey. *Journal of Theoretical and Applied Information Technology*, pp. 14–27, 2010.
3. J. Norman, J. P. Joseph, and P. P. Roja. A faster routing scheme for stationary wireless sensor networks—A hybrid approach. *International Journal of Ad Hoc, Sensor and Ubiquitous Computing*, Vol. 1, Issue 1, pp. 1–10, 2010.
4. M. R. E. Jebarani and T. Jayanthy. An analysis of various parameters in wireless sensor networks using adaptive FEC technique. *International Journal of Ad Hoc, Sensor and Ubiquitous Computing*, Vol. 1, Issue 3, pp. 33–43, 2010.
5. P. Samundiswary, D. Sathian, and P. Dananjayan. Secured greedy perimeter stateless routing for wireless sensor networks. *International Journal of Ad Hoc, Sensor and Ubiquitous Computing*, Vol. 1, Issue 2, pp. 9–20, 2010.
6. M. P. Singh and Md. Z. Hussain. A top-down hierarchical multi-hop secure routing protocol for wireless sensor networks. *International Journal of Ad Hoc, Sensor and Ubiquitous Computing*, Vol. 1, Issue 2, pp. 33–52, 2010.
7. J. Liang, J. Wang, and J. Chen. A delay-constrained and maximum lifetime data gathering algorithm for wireless sensor networks. *5th International Conference on Mobile Ad-Hoc and Sensor Networks (MSN '09)*, pp. 148–155, December 2009.
8. S. K. Narang, G. Shen, and A. Ortega. Unidirectional graph-based wavelet transforms for efficient data gathering in sensor networks. *IEEE International Conference on Acoustics Speech and Signal Processing (ICASSP 2010)*, pp. 2902–2905, March 2010.
9. G. Robins and A. Zelikovsky. Improved Steiner tree approximation in graphs. *Proceedings of the 11th Annual ACM-SIAM Symposium on Discrete Algorithms (SODA 2000)*, Philadelphia, PA, USA, pp. 770–779, 2000.
10. S. Hougardy and H. J. Prömel. A 1.598 approximation algorithm for the Steiner problem in graphs. In *Proceedings of the 10th Annual ACM-SIAM Symposium on Discrete Algorithms (SODA '99)*, Philadelphia, PA, USA, pp. 448–453, 1999.
11. B. Krishnamachari, D. Estrin, and S. B. Wicker. The impact of data aggregation in wireless sensor networks. *Proceedings of the 22nd International Conference on Distributed Computing Systems (ICDCSW '02)*, Washington, DC, USA, pp. 575–578, 2002.
12. A. Roumy and D. Gesbert. Optimal matching in wireless sensor networks. *IEEE Journal on Selected Topics in Signal Processing*, Vol. 1, No. 4, December 2007.
13. F. Bauer and A. Varma. Distributed algorithms for multicast path setup in data networks. *IEEE/ACM Transaction on Networking*, Vol. 4, pp. 181–191, 1996.
14. E. F. Nakamura, H. A. B. F. de Oliveira, L. F. Pontello, and A. A. F. Loureiro. On demand role assignment for event-detection in sensor networks. *Proceedings of the 11th IEEE Symposium on Computers and Communications (ISCC '06)*, Washington, DC, USA, pp. 941–947, 2006.

15. S. Pattem, B. Krishnamachari, and R. Govindan. The impact of spatial correlation on routing with compression in wireless sensor networks. *ACM Transactions on Sensor Networks (TOSN)*, Vol. 4, No. 4, April 2008.
16. Y. Zhu, K. Sundaresan, and R. Sivakumar. Practical limits on achievable energy improvements and useable delay tolerance in correlation aware data gathering in wireless sensor networks. *Proceedings of the IEEE International Conference on Sensor and Ad Hoc Communications and Networks (SECON 2005)*, pp. 328–329, September 2005.
17. Y. P. Chen, A. L. Liestman, and J. Liu. A hierarchical energy efficient framework for data aggregation in wireless sensor networks. *IEEE Transactions on Vehicular Technology*, Vol. 55, Issue 3, pp. 789–796, May 2006.
18. P. von Rickenbach and R. Wattenhofer. Gathering correlated data in sensor networks. *Proceedings of the ACM Joint Workshop on Foundations of Mobile Computing (DIALM-POMC 2004)*, pp. 60–66, October 2004.
19. R. Cristescu, B. Beferull-Lozano, M. Vetterli, and R. Wattenhofer. Network correlated data gathering with explicit communication: NP-completeness and algorithms. *IEEE/ACM Transactions on Networking (TON)*, Vol. 14, No. 1, pp. 41–54, February 2006.
20. H. Albert, R. Kravets, and I. Gupta. Building trees based on aggregation efficiency in sensor networks. *Elsevier Ad Hoc Networks*, Vol. 5, No. 8, pp. 1317–1328, 2007.
21. K. Dasgupta, K. Kalpakis, and P. Namjoshi. An efficient clustering based heuristic for data gathering and aggregation in sensor networks. *IEEE Wireless Communications and Networking (WCNC 2003)*, Vol. 3, pp. 1948–1953, March 2003.
22. M. Ding, X. Cheng, and G. Xue. Aggregation tree construction in sensor networks. *Proceedings of the 58th IEEE Vehicular Technology Conference (VTC 2003)*, Vol. 4, pp. 2168–2172, October 2003.
23. S.-H. Hong and B-K. Kim. An efficient data gathering routing protocol in sensor networks using the integrated gateway node. *IEEE Transactions on Consumer Electronics*, Vol. 56, Issue 2, pp. 627–632, May 2010.
24. J. N. Al-Karaki, R. Ul-Mustafa, and A. E. Kamal. Data aggregation in wireless sensor networks: Exact and approximate algorithms. *Proceedings of the IEEE Workshop on High Performance Switching and Routing (HPSR 2004)*, pp. 241–245, April 2004.
25. G. Hartl and B. Li. Infer: A Bayesian approach towards energy efficient data collection in dense sensor networks. *Proceedings of the 25th IEEE International Conference on Distributed Computing Systems (ICDCS 2005)*, pp. 371–380, June 2005.
26. I. Solis and K. Obraczka. Isolines: Energy-efficient mapping in sensor networks. *Proceedings of the 10th IEEE Symposium on Computers and Communications (ISCC 2005)*, pp. 379–385, June 2005.
27. V. Erramilli, I. Matta, and A. Bestavros. On the interaction between data aggregation and topology control in wireless sensor networks. *Proceedings of the 1st IEEE Communications Society Conference on Sensor and Ad Hoc Communications and Networks (SECON 2004)*, pp. 557–565, October 2004.
28. H. O. Tan and I. Korpeoglu. Power efficient data gathering and aggregation in wireless sensor networks. *ACM SIGMOD Record*, Vol. 32, No. 4, pp. 66–71, December 2003.
29. Y. Zhu, W. Wu, J. Pan, and Y. Tang. An energy-efficient data gathering algorithm to prolong lifetime of wireless sensor networks. *Computer Communications*, Vol. 33, pp. 639–647, 2010.
30. Y. Bi, N. Li, and L. Sun. DAR: An energy-balanced data-gathering scheme for wireless sensor networks. *Computer Communications*, pp. 2812–2825, 2007.
31. C.-M. Liu, C.-H. Lee, and L.-C. Wang. Distributed clustering algorithms for data-gathering in wireless mobile sensor networks. *Journal of Parallel and Distributed Computing*, Vol. 67, Issue 11, pp. 1187–1200, 2007.
32. C.-K. Liang, J.-D. Lin, and C.-S. Li. Steiner points routing protocol for wireless sensor networks. *5th International Conference on Future Information Technology (FutureTech)*, pp. 1–5, May 2010.
33. C. Schurgers, V. Tsiatsis, S. Ganeriwal, and M.B. Srivastava. Topology management for sensor networks: exploiting latency and density. *Proceedings of the 3rd ACM International Symposium on Mobile Ad Hoc Networking and Computing (MobiHoc)*, pp. 135–145, 2002.

34. H. Cheng, Q. Liu, and X. Jia. Heuristic algorithms for real-time data aggregation in wireless sensor networks. *Proceedings of the ACM International Conference on Communications and Mobile Computing*, pp. 1123–1128, July 2006.
35. J. Choi, J. W. Lee, K. Lee, S. Choi, W. H. Kwon, and H. S. Park. Aggregation time control algorithm for time constrained data delivery in wireless sensor networks. *Proceedings of the 63rd IEEE Vehicular Technology Conference*, Vol. 2, pp. 563–567, May 2006.
36. H. Gupta, V. Navda, S. R. Das, and V. Chowdhary. Efficient gathering of correlated data in sensor networks. *ACM Transactions on Sensor Networks (TOSN)*, Vol. 4, No. 1, January 2008.
37. C. Hua and T. P. Yung. Optimal routing and data aggregation for maximizing lifetime of wireless sensor networks. *IEEE/ACM Transactions on Networking*, Vol. 16, Issue 4, pp. 892–903, August 2008.
38. L. A. Villas, A. Boukerche, R. B. Araujo, and A. A. F. Loureiro. A reliable and data aggregation aware routing protocol for wireless sensor networks. *Proceedings of the 12th ACM International Conference on Modeling, Analysis and Simulation of Wireless and Mobile Systems (MSWIM'09)*, pp. 245–252, October 2009.
39. R. Poornachandran, H. Ahmad, and H. Cam. Energy-efficient task scheduling for wireless sensor nodes with multiple sensing units. *Proceedings of IWSEEASN'05*, April 2005.
40. H. Ozgur Sanli, R. Poornachandran, and H. Çam. Collaborative two-level task scheduling for wireless sensor nodes with multiple sensing units. *Proceedings of the 2nd Annual IEEE Communications Society Conference on Sensor and Ad Hoc Communications and Networks (SECON)*, pp. 350–361, September 2005.
41. D. Kumar, T. C. Aseri, and R. B. Patel. EECDA: Energy efficient clustering and data aggregation protocol for heterogeneous wireless sensor networks. *International Journal of Computers, Communications and Control, ISSN 1841-9836, E-ISSN 1841-9844*, Vol. VI, No. 1 (March), pp. 113–124, 2011.
42. T. Wang, C. Wu, P. Ji, and J. Zhang. A noble data aggregation algorithm for low latency in wireless sensor network. *Proceedings of the 2010 IEEE International Conference on Mechatronics and Automation*, Xi'an, China, August 4–7, 2010.
43. L. A. Villas, D. L. Guidoni, R. B. Araujo, A. Boukerche, and A. A. F. Loureiro. A scalable and dynamic data aggregation aware routing protocol for wireless sensor networks. *MSWiM'10*, Bodrum, Turkey, October 17–21, 2010.
44. Y. Alemu, J.-B. Koh, and D.-K. Kim. A hierarchical multi-parent cluster-based data aggregation framework for WSNs. *2010 International Conference on Complex, Intelligent and Software Intensive Systems*. DOI 10.1109/CISIS.2010.110.
45. G. S. Chhabra and D. Sharma. Cluster-tree based data gathering in wireless sensor network. *International Journal of Soft Computing and Engineering (IJSCE) ISSN: 2231-2307*, Vol. 1, Issue 1, March 2011.
46. C. Luo, F. Wu, J. Sun, and C. W. Chen. Compressive data gathering for large-scale wireless sensor networks. *MobiCom'09, September 20–25, 2009*, Beijing, China. ACM 978-1-60558-702-8/09/09.
47. C.-C. Huangy, J.-L. Huangy, J.-A. Yany, and L.-Y. Yehz. An in-network approximate data gathering algorithm exploiting spatial correlation in wireless sensor networks. *SAC'12 March 25–29, 2012*, Riva del Garda, Italy. ACM 978-1-4503-0857-1/12/03.
48. A. Anand, S. Sachan, K. Kapoor, and S. Nandi. QDMAC: An energy efficient low latency MAC protocol for query based wireless sensor networks. *Proceedings of the 10th International Conference on Distributed Computing and Networking (ICDCN)*, pp. 306–317, January 2009.
49. A. Dobra, M. Garofalakis, J. Gehrke, and R. Rastogi. Processing complex aggregate queries over data streams. *International Conference on Data Management (SIGMOD)*, pp. 61–72, 2002.
50. A. Ephremedis and T. Truong. Scheduling broadcasts in multihop radio networks. *IEEE Transactions on Communications*, Vol. 38, No. 4, pp. 456–460, 1990.

Chapter 3

Spatial Coverage Estimation and Optimization in Geosensor Networks Deployment

Farid Karimipour
University of Tehran
Meysam Argany and Mir Abolfazl Mostafavi
Laval University

Contents

3.1 Introduction

Recent advances in electrical, mechanical, and communications systems have led to development of efficient low cost and multifunction sensors that are capable of sensing the environment, performing data processing, and communicating with each other. The efficiency of sensors, in terms of data collection and communication, is constrained by limitations of sensing range, battery power, connection ability, memory, and computation capabilities. As a result, an individual sensor can sense only a small region. However, a group of sensors collaborating with each other can overcome this limitation and cover the whole region of interest. They are arranged in a wireless network, each monitoring and collecting physical and environmental data such as motion, temperature, humidity, pollutants, and traffic flow for a certain area. The data is then communicated to a processing center where it is aggregated and analyzed to produce the desired information for different applications. Sensors could be deployed either randomly or based on a predefined distribution over the region of interest. They may be spread with various densities, from 10 m apart to as high as 20 nodes per square meter, depending on the application as well as the type and quality of the desired information.

The efficient deployment of sensors in a wireless (geo)sensor network is an important issue that affects the coverage as well as the communication between sensors. Nodes use their sensing modules to detect the events occurring in the region of interest. Each sensor is assumed to have a sensing range, which may be constrained by the phenomenon being sensed, obstacles, the environment, etc. In a network of sensors, these constraints affect the coverage and may result in making holes in the sensing area. Communication between the nodes is equally important. Information collected from the region covered by a sensor should be transferred to a processing center, directly or via its adjacent sensors. In the latter case, each sensor must be aware of the position of other adjacent sensors in its proximity. Any failure in communication between sensors may result in holes in the aggregated information. Several optimization methods (i.e., global or local, deterministic, or stochastic) have been proposed to detect and eliminate holes and hence increase the coverage of sensor networks. Furthermore, some methods use general optimization techniques, whereas some others consider the problem as a geometric issue and use the structures and tools existing in computational geometry.

This chapter provides a survey on the main methods proposed for the estimation and optimization of spatial coverage of sensor networks, with a special focus on geometrical methods. Section 3.2 presents the state of the art on sensor networks and their related issues in terms of phenomena, type of environment and sensor, as well as issues such as coverage, communication, and energy-saving problems. Section 3.3 focuses on geometrical issues in the deployment of sensor networks. It describes the concepts of sensing and communication models as well as sensor network topologies. Some preliminary geometric definitions are also provided at the end of this section. In Section 3.4, the spatial coverage problem is discussed in more detail and a general review on global and local methods for sensor deployment optimization is presented. Section 3.5 explains the main methods proposed in the literature for sensor deployment to optimize spatial coverage. In Section 3.6, the current challenging issues in the sensor coverage problem, which are open for more research, are discussed. Finally, Section 3.7 concludes the chapter and introduces some research perspectives in the field.

3.2 Wireless Geosensor Networks: An Overview

Business Week (1999) proclaimed that sensor networks are one of the most important technologies for the twenty-first century. These networks are usually composed of a set of small, smart, and low-cost sensors with limited onboard processing capabilities, storage, and short-range wireless communication links based on radio technology. Previously, sensor networks consisted of a small number of sensor nodes that were usually wired to a central processing station. However, nowadays, the focus is more on wireless, distributed, sensing nodes (Worboys and Duckham 2006). A sensor node is characterized by its sensing field, memory, and battery power as well as its computation and communication capabilities. Due to restrictions in sensing and communications range, a sensor can only cover a small area. However, collaboration among a group of sensors can cover a more significant sensing field and hence accomplish much larger tasks. Each element of a group of sensors can sense and collect data from the environment, apply local processing, communicate it to other sensors, and perform aggregations on the observed information (Sharifzadeh and Shahabi 2004). These tiny and ingenious devices are usually deployed in a wireless network for accessing remote and inaccessible areas without a wired communication, and often, without even power lines. Deploying sensor networks allows covering inaccessible areas by minimizing the sensing costs compared with the use of separate sensors. The size reduction of computing and storage platforms has led to low power consumption and provided computational platforms that run on battery power for longer periods. In addition, advances in real-time data input and output have improved the data collection and information preparation devices, and the data collected by sensor networks could be available directly on the Web. From a computation capability point of view, onboard computing advances (including local data analysis, data filtering, and sampling) have adapted the occurring events and reduced the data transmission and battery consumption.

Sensor networks are also referred to as geosensor networks because they are intensively used to acquire spatial information (Nittel 2009) by being deployed on the ground, in the air, under water, on bodies, in vehicles, and inside buildings. Hereafter, we will use both of the terms *sensors* and *geosensors*, interchangeably.

A broad classification of geosensor network applications is monitoring continuous phenomena (e.g., to assess plant health and growth circumstances, to observe and measure geophysical processes), detecting real-time events (e.g., floods and volcanoes), and tracking objects (e.g., animal monitoring; Szewczyk et al. 2004; Worboys and Duckham 2006; Nittel 2009). Sensor networks have several applications including environmental monitoring, change detection, traffic monitoring, border security, public security, etc. They are used for collecting the information needed by smart environments, quickly and easily, whether in buildings, utilities, industries, home, shipboard, transportation systems automation, and elsewhere. Sensor networks are useful in vehicle traffic monitoring and control. Most traffic intersections have either overhead or buried sensors to detect vehicles and control traffic lights. Furthermore, video cameras are frequently used to monitor road segments with heavy traffic, through the videos sent to human operators at central locations (Chong and Kumar 2003). Sensor networks can be used for infrastructure security in critical buildings and facilities, such as power plants and communication centers. Networks of video, acoustic, and other sensors provide early detection of possible threats (Soro and Heinzelman 2005). Commercial industries have long been interested in sensing as a means of lowering cost and improving machine (and perhaps user) performance and maintainability.

Sensor networks have some limitations when it comes to the modeling, monitoring, and detecting of environmental processes. Monitoring and analyzing dynamic objects in real-time is also difficult. Examples of such processes include observations of dynamic phenomena (e.g., air pollution) or monitoring of mobile objects (e.g., animals in a habitat). It is necessary to know how to use this technology to detect and monitor those phenomena, appropriately and efficiently. For this purpose, one needs to identify the relevant mix of hardware platforms for the phenomena type, accessibility or inaccessibility of the observation area, hazardous environmental conditions, and power availability, etc. Due to battery constraints, today's wireless sensor network technology can be more effective at detecting and monitoring time-limited events (e.g., earthquake tremors) instead of continuous sampling in remote areas (Nittel 2009). Data acquisition and distribution network are two aspects of complexity of the wireless sensor network. Thus, choosing the components of such systems is difficult due to the abundance of available technologies as well as the design of a consistent, reliable, robust overall system. The study of wireless sensor networks is a challenging task because it requires an enormous breadth of knowledge from a great variety of disciplines.

3.3 Geometrical Issues in Deployment of Geosensor Wireless Networks

A simple description of the coverage problem in geosensor networks considers the sensors as a set of points (nodes) in Euclidean space, each is assigned a sensing region; the problem is to place them such that the space is fully covered with the union of the sensing regions providing that the sensors can communicate. The key points of this definition are *sensing* and *communication*, whose modeling has a direct effect on the sensor deployment. This section presents the geometrical issues related to sensing and communication modeling. Two geometric structures, that is, Delaunay triangulation and Voronoi diagrams, which are frequently used in the geometrical geosensor deployment strategies, are introduced at the end of this section.

3.3.1 Sensing Models

The simplest model of sensing is the *binary disk model*, which confines the sensibility of a sensor within a certain disk. It considers a sensing range, that is, a circular disk of radius R_s for each sensor. The points lying within the sensing range of a sensor are fully covered by that sensor and the points beyond it are not covered at all (Figure 3.1a):

$$S(s_i, P) = \begin{cases} 1 & d(s_i, P) \leq R_{s_i} \\ 0 & d(s_i, P) > R_{s_i} \end{cases}$$

where S is the sensitivity of the sensor s_i, R_{s_i} is the sensitivity disk of the sensor s_i, and $d(s_i, P)$ is the Euclidean distance between sensor s_i and the point P. This model assumes that there are no obstacles in the environment and ignores the decrease in the strength of the signal. The coverage

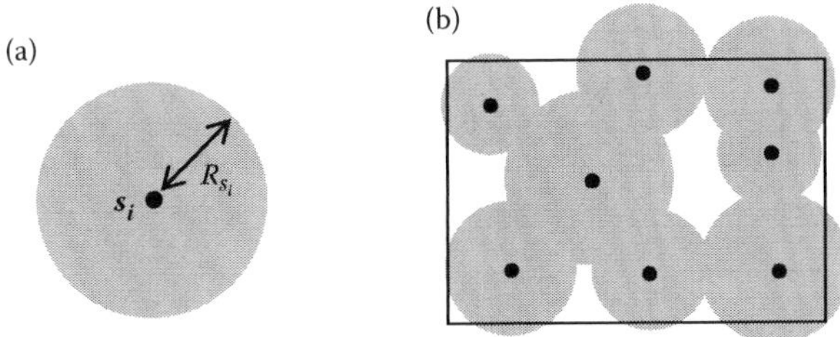

Figure 3.1 Binary disk model of sensing: (a) only the gray region is covered by the sensor s_i; (b) modeling the coverage problem of a sensor network using the binary disk model.

problem of a sensor network is simply modeled using the binary disk model through computing the union of the (not necessarily equal) sensing disks (Figure 3.1b). To achieve more realistic extensions of this model, the following variants are applied:

- The sensibility is not binary, but varies with distance to the sensor node
- The sensing region is not circular

3.3.1.1 *Variable Sensibility*

In practice, the sensing capability is not binary, but gradually attenuates with increasing distance (Figure 3.2), that is,

$$S(s_i, P) = f\left(\frac{1}{d(s_i, P)}\right)$$

This concept is used in a *probabilistic sensing model* to model the sensitivity of devices such as infrared and ultrasound sensors (Zou and Chakrabarty 2004; Hossain et al. 2008). In this model, two quantities R_1 and R_{max} are defined, which is the beginning of uncertainty in sensor detection and the maximum sensing range of the sensor, respectively. The points with a distance less than R_1 to s_i are surely covered; the points with a distance greater than R_{max} to s_i are not

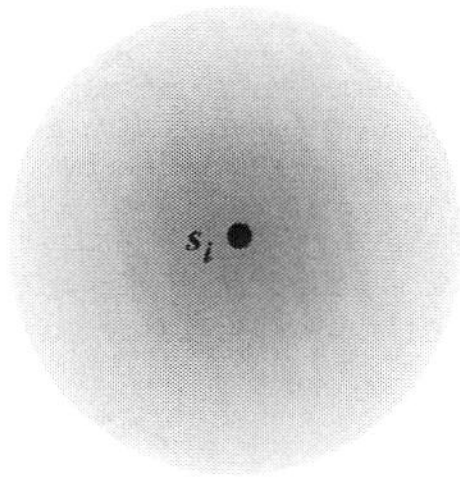

Figure 3.2 Attenuating the sensitivity with increasing distance to the sensor s_i.

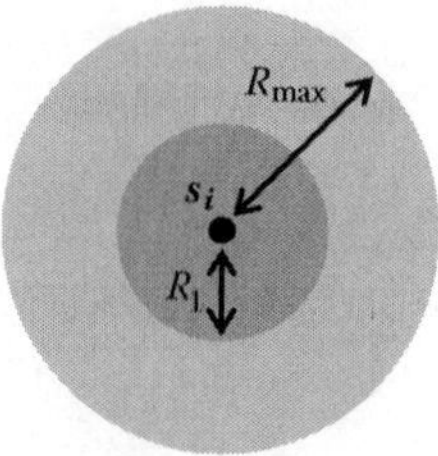

Figure 3.3 Probability sensing model: the dark gray disk is covered; the coverage of the light gray nut is a probability function of distance to s_i. It is not covered elsewhere.

covered, the coverage of the points between the two above disks is a probability function of distance (Figure 3.3):

$$S(s_i,P)=\begin{cases}1 & d(s_i,P)\le R_1\\ f(d) & R_1<d(s_i,P)<R_{max}\\ 0 & d(s_i,P)\ge R_{max}\end{cases}$$

3.3.1.2 Noncircular Sensibility

If the sensor emits signals to all directions and the environment is homogeneous, the circular sensing region fairly models the reality. However, there are cases where these assumptions are not true:

- Directional sensors, for example, cameras, whose covered region is restricted to a certain direction (Figure 3.4; Wang and Cao 2011a,b).
- In the presence of obstacles in the environment, or if the environment is not homogeneous, the sensing ability of the sensor is not uniform in all directions (Hossain et al. 2008).

Several modeling strategies and strategies for coverage estimation of such cases have been proposed in the literature (Hwang et al. 2007; Wu and Chung 2009; Yi and Guohong 2011).

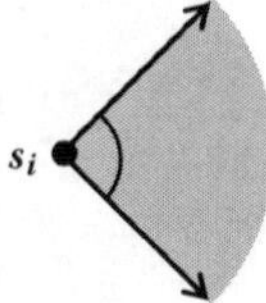

Figure 3.4 A directional sensor.

3.3.2 Communication Models

Wireless sensors are small, low-power sensors with limited storage and short-range wireless communication links based on radio technology. Communication between the sensors is very power-consuming; thus, optimizing the communications between sensors is crucial for prolonging the lifetime of wireless sensor networks (Luo and Hubaux 2005; Madan et al. 2005; Wang et al. 2005). On the other hand, radio links between sensors are extremely unreliable, and thus making realistic modeling of radio communications is very challenging (Ghosh and Das 2008).

Like modeling the sensor coverage, the simplest model of sensor communication is the binary disk model, which assumes a communication radius R_{c_i} for each node s_i. This means that s_i is capable of communicating to sensors located up to a distance of R_c from it (Figure 3.5a). However, empirical measurements have challenged this model of communication (Zuniga and Krishnamachari 2004), because in reality, the strength of radio signals emitted from the sensors attenuates with increasing the distance. Furthermore, "the signal undergoes several disruptive physical phenomena, such as interference, scattering, diffraction, and reflection due to the presence of other transmissions and obstacles along its path" (Ghosh and Das 2008).

Based on the binary disk model, two nodes s_i and s_j can communicate with each other if the minimum of their communication radii is greater than their Euclidean distance, that is, $\min\{R_{c_i}, R_{c_i}\} > d(s_i, s_j)$. It means that the sensor with a smaller communication range falls in the communication range of the other sensor (Figure 3.5b).

Two sensors are called *one-hop neighbors* if they can directly communicate with each other. On the other hand, two sensors may not directly communicate (i.e., at least one of them does not fall in the communication range of another), but they could communicate through a sequence of intermediate sensors. Such sensors are called *multihop neighbors*. This idea leads to different topologies (i.e., communication strategies) in wireless sensor networks to maximize the life-time and communication reliability of the whole network (Salhieh et al. 2001; Deb et al. 2002; Yu et al. 2005; Cao et al. 2006; Muthukumar et al. 2010). Figure 3.6 illustrates the basic sensor network topologies.

The communication in wireless sensor networks is perfectly modeled using graphs: each sensor is a node in the graph. Two nodes are connected through an edge if their Euclidean distance is less

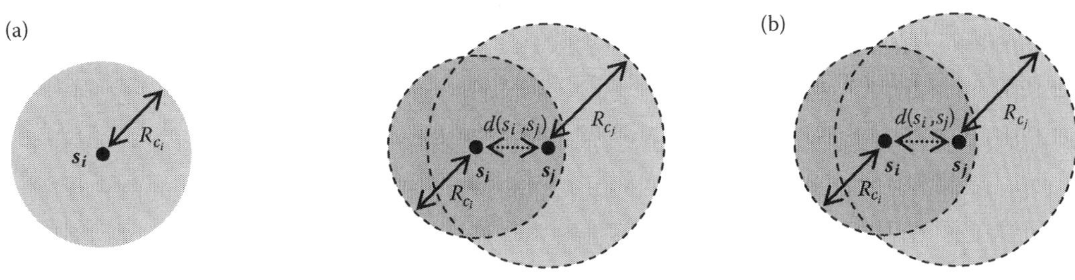

Figure 3.5 Binary disk model of communication: (a) the sensor s_i is able to communicate with the sensors located in the gray region; (b) the sensors s_i and s_j can communicate with each other because the one with the smaller communication range falls in the communication range of the other one.

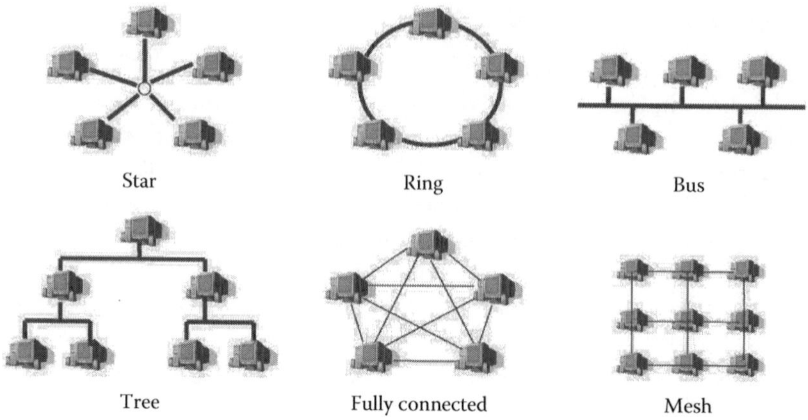

Figure 3.6 Basic sensor network topologies.

than the minimum of their communication radii. Several parameters could be extracted from the induced communication graph (Ghosh and Das 2008):

- The number of one-hop neighbors of a sensor is the degree of its corresponding node in the graph.
- All sensors in the network can communicate (i.e., there is no isolated sensor) if the induced communication graph is connected. It means that there is a path between every pair of sensors. In other words, there is an edge between them (i.e., they are one-hop neighbors) or they are connected through a sequence of edges (i.e., they are multihop neighbors).
- The sensor network is k node–connected, if for every pair of nodes there are at least k node–disjoint paths connecting them. This parameter is an indicator of the reliability of the network.

3.3.3 Preliminary Geometric Structures

This subsection introduces Delaunay triangulation and Voronoi diagrams—two geometric structures that are frequently used later in geosensor network deployment strategies.

Given a point set P in the plane, the Delaunay triangulation is a unique triangulation of the points in P, which satisfies the empty circumcircle property: the circumcircle of each triangle does not contain any other point $p \in P$ (Figure 3.7).

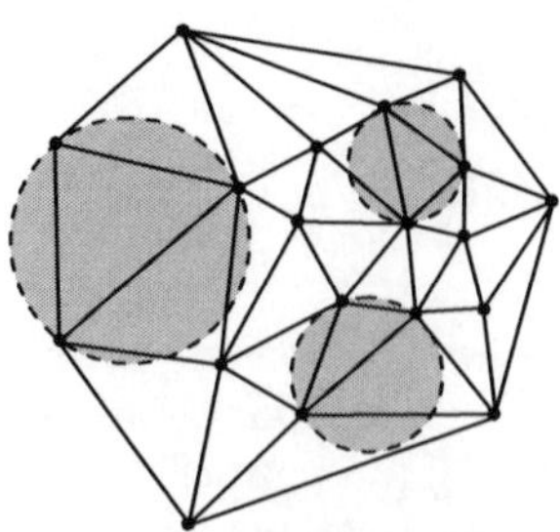

Figure 3.7 Two-dimensional Delaunay triangulations; three of the circumcircles are shown.

Delaunay triangulation is the dual structure of the Voronoi diagram. The Voronoi diagram (VD) of a set of points is defined as follows: let P be a set of points in an n-dimensional Euclidean space R^n. The Voronoi cell of a point $p \in P$, noted $V_p(P)$, is the set of points $x \in R^n$ that are closer to p than to any other point in P:

$$V_p(P) = \{x \in R^n \mid \|x - p\| \leq \|x - q\|, q \in P, q \neq p\}$$

The union of the Voronoi cells of all points $p \in P$ form the Voronoi diagram of P, noted as $VD(P)$:

$$VD(P) = \cup V_p(P), p \in P$$

Figure 3.8 shows an example of the Voronoi diagrams of a set of two-dimensional (2-D) points.

Delaunay triangulation and Voronoi diagram are dual structures. This means that each node in Delaunay triangulation corresponds to a Voronoi cell, each Delaunay edge corresponds to a Voronoi edge and each Delaunay triangle corresponds to a Voronoi vertex. The centers of circumcircles of Delaunay triangulation are the Voronoi vertexes; and joining the adjacent generator points in a Voronoi diagram yields their Delaunay triangulation (Figure 3.9). This duality is very useful because construction, manipulation, and storage of the Voronoi diagram are more difficult than Delaunay triangulation, so all the operations can be performed on Delaunay triangulation, and the Voronoi diagram is only extracted on demand.

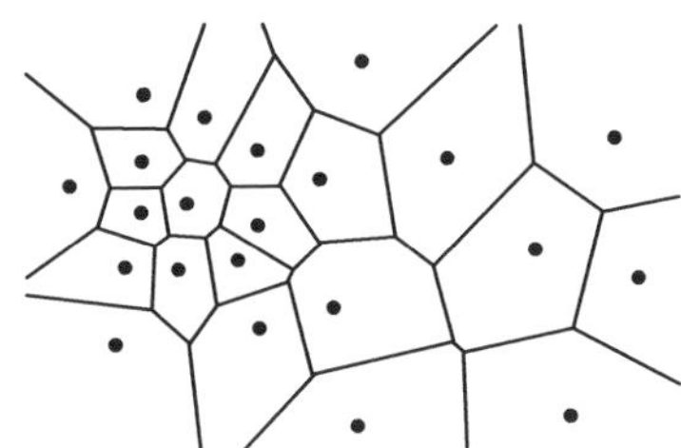

Figure 3.8 Voronoi diagram of a set of points in the plane.

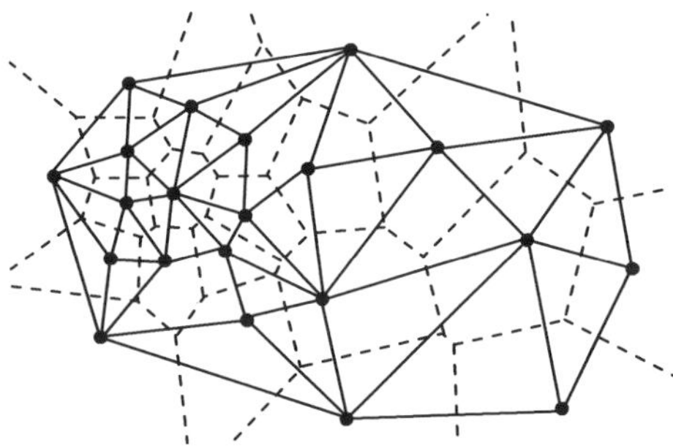

Figure 3.9 Duality of Delaunay triangulation (solid lines) and Voronoi diagram (dashed lines).

3.4 Spatial Coverage in Geosensor Networks

An important issue in deploying a sensor network is finding the best sensor location to cover the region of interest. The definition of *coverage* differs from one application to another. The so-called art gallery problem, for example, aims to determine the minimum number of required sensors, that is, cameras, to cover an art gallery room such that every point is seen by at least one sensor (De Berg et al. 2000). Hence, in this example, coverage is defined based on a direct visibility between the camera and the target point. In sensor networks, however, the coverage of a point means that the point is located in the sensing range of a sensor node. As stated in Section 3.3, a uniform sensing range is represented by a disk around the sensor. Failing this condition for some points in the region of interest will result in coverage holes (Figure 3.10).

Regarding the above definition of coverage in sensor networks, the coverage problem basically is placing the minimum number of nodes in an environment, such that every point in the sensing field is optimally covered (Ghosh and Das 2008; Aziz et al. 2009). Nodes can either be placed manually at predetermined locations or dropped randomly in the environment. It is difficult to find a random scattering solution that satisfies all the coverage and connectivity conditions. Thus, the term *area coverage* plays an important role in sensor networks and their connectivity.

The Voronoi diagram elegantly models the sensor coverage problem (Argany et al. 2010, 2011). In a Voronoi diagram, all the points within a Voronoi cell are closest to the generating node that lies within this cell. Thus, having constructed the Voronoi diagram of the sensor nodes and overlaid the sensing regions on it (Figure 3.11), if a point of a Voronoi cell is not covered by its

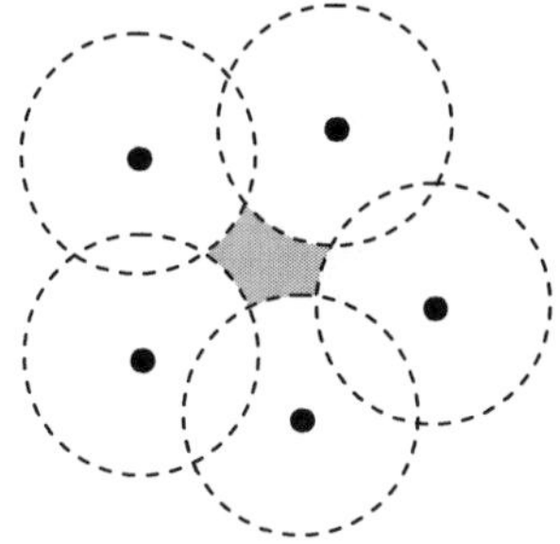

Figure 3.10 Coverage hole (shaded region) in a sensor network with disk model sensing range.

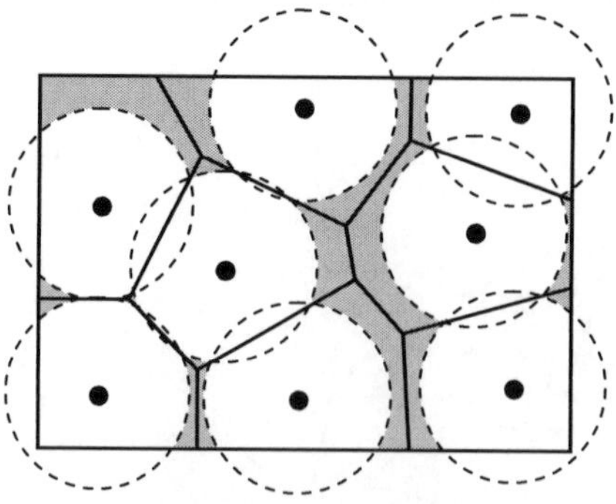

Figure 3.11 Using the Voronoi diagram to detect the coverage holes (shaded regions) in a sensor network.

generating node, this point is not covered by any other sensors (Wang et al. 2003, 2009; Ghosh 2004; Ahmed et al. 2005). Although computing the area of a Voronoi cell is straightforward, computing the area of the uncovered region in a Voronoi cell is a complicated task because the sensing regions may protrude into the Voronoi cells and overlay each other. Strategies for this computation have been described by Wang et al. (2003) and Ghosh (2004).

3.4.1 Coverage Estimation Based on the Concept of Exposure

The estimation of coverage can be defined as a measure of the ability to detect objects within a sensor field. The notion of *exposure* can represent such a measurement. It is described as the expected average ability of observing a target moving in a sensor field. It is related to the concept of coverage in the sense that "it is an integral measure of how well the sensor network can observe an object (exists in the field or) moving on an arbitrary path, over a period of time" (Meguerdichian et al. 2001a,b,c).

A very simple, but nontrivial, example of an exposure problem is illustrated in Figure 3.12. An object moves from point *A* to point *B* and there is only one sensor node *S* in the field. Obviously, path 2 has the maximum exposure because it is the shortest path from *A* to *B* and it passes through the sensor node *S*. Thus, an object moving along this path is certainly tracked by the sensor *S*. However, finding the path with the minimum exposure is tricky: although path 1 is the farthest path from the sensor node *S* and so intuitively seems to have the lowest exposure, it is also the longest path. Therefore, travelling along this path takes longer and the sensor has a longer time to track the moving object. It is shown that the minimum exposure path is 3, which is a trade-off between distance from the sensor and travelling time (Huang and Tseng 2005).

The so-called *worst-case* and *best-case* coverage are examples of methods for exposure evaluation (Meguerdichian et al. 2001a,b,c; Megerian et al. 2005). Worst-case coverage involves the regions of lower observability from sensor nodes, so objects moving along this path have the minimum probability of being detected. Best-case coverage, however, involves the regions of higher observability from sensors, thus the probability of detecting an object moving along this path is at a maximum (Ghosh and Das 2008). Together, these two parameters give an insight of the coverage quality of the network and can help decide if additional sensors must be deployed. Different approaches have been proposed in the literature for the worst-case and best-case coverage problems (Meguerdichian et al. 2001a,b,c; Huang 2003; Veltri et al. 2003).

A Voronoi approach based on the notion of *exposure* to evaluate the coverage of a sensor network has been proposed (Meguerdichian et al. 2001a,b,c; Megerian et al. 2005). To solve the

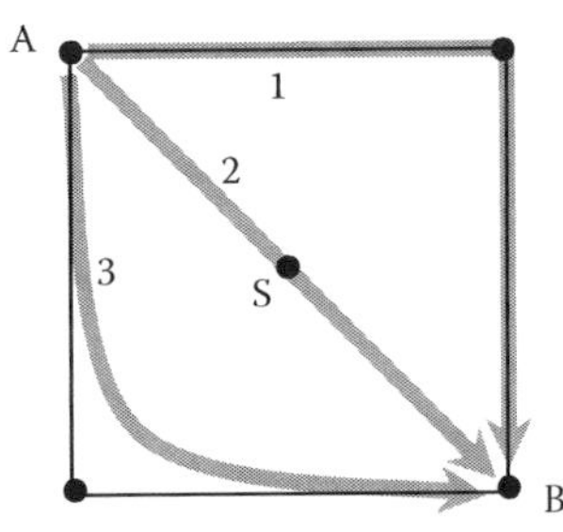

Figure 3.12 Minimum and maximum exposure paths in a simple sensor network. (From Huang, C. F. and Y. C. Tseng. *J. Internet Technol.* 6(1): 1–8, 2005.)

worst-case coverage problem, a very similar concept, that is, the *maximal breach path* is used. It is the path through a sensing field between two points such that the distance from any point on the path to the closest sensor is maximized. Because the line segments of the Voronoi diagram have the maximum distance from the closest sites, the maximal breach path must lie on the line segments of the Voronoi diagram corresponding to the sensor nodes (Figure 3.13). The Voronoi diagram of the sensor nodes is first constructed. This diagram is then considered as a weighted graph in which the weight of each edge is the minimum distance from the closest sensor. Finally, an algorithm uses breadth first and binary searches to find the maximal breach path.

The best-case coverage problem is solved through the similar concept of a *maximal support path*. This is the path through a sensing field between two points for which the distance from any point on it to the closest sensor is minimized. Intuitively, this is traveling along straight lines connecting sensor nodes. The Delaunay triangulation produces triangles that have minimal edge lengths among all possible triangulations. Thus, the maximal support path must lie along the lines of the Delaunay triangulation of the sensors (Figure 3.14). Delaunay triangulation of the sensor nodes is constructed and is considered as a weighted graph, where the weight of each edge is the length of that edge. The maximal breach path is found through an algorithm that uses breath first and binary searches.

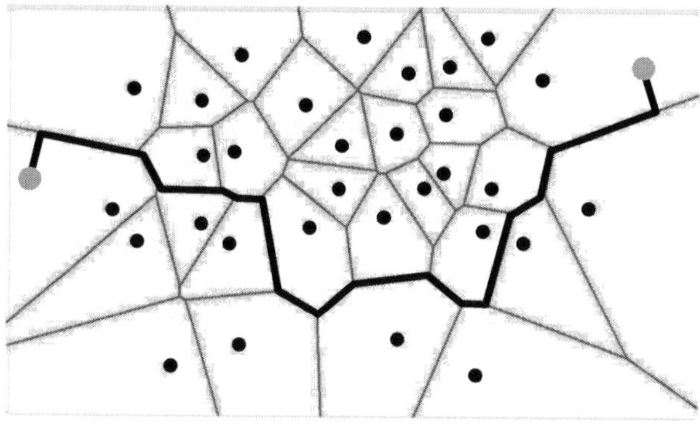

Figure 3.13 Maximal breach path in a sensor network and its connection to Voronoi diagram.

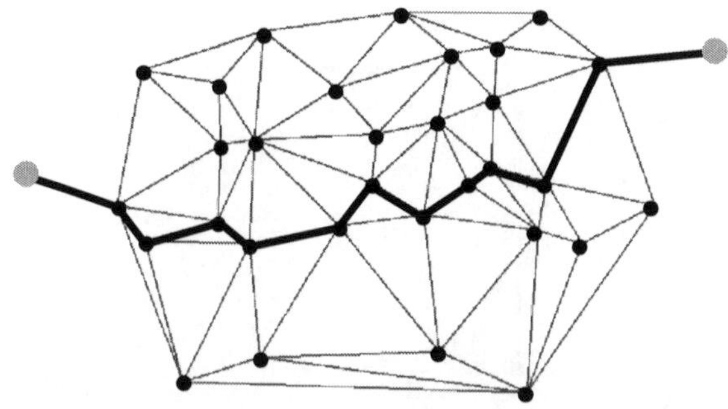

Figure 3.14 Maximal support path in a sensor network and its connection to Delaunay triangulation.

3.5 Optimization Algorithms in Geosensor Networks Deployment

Having detected the coverage holes, the sensors must be relocated to heal the holes. In a broad view, the existing approaches to optimally place the sensors are classified into local and global methods.

3.5.1 Global Approaches

Global optimization approaches are used to find the global maximum or minimum of a function. These approaches usually deal with the entire region of interest and look for the optimum of a function inside the whole part of the searching area. Therefore, having a general knowledge about the searching area, its characteristics, and its reaction to the optimization algorithm are necessary to get the desirable results. Especially in iterative algorithms, the entire area of interest must be considered in each iteration.

Several global optimization methods have been proposed for sensor network deployment in the literature. Niewiadomska-Szynkiewicz and Marks (2009) used a classic version of simulated annealing to solve the deployment problem in sensor networks and it was implemented as a computer simulation of a stochastic process. Simulated annealing is a stochastic search algorithm based on the concept of "annealing." The annealing process increases the temperature of a solid to a point where its atoms can freely move, and then lowering the temperature, forcing the atoms to rearrange themselves into a lower energy state (i.e., a crystallization process). During this process, the free energy of the solid is minimized (the crystalline state is the state of minimum energy of a system). The cooling schedule is crucial: if the solid is cooled too quickly, or if the initial temperature of the system is too low, it is not able to become a crystal and instead, the solid arrives at an amorphous state with higher energy. In this case, the system reaches a local minimum (a higher energy state) instead of the global minimum, that is, the minimal energy state. The algorithmic analogue of this process begins with a random guess of the cost function variable values. Heating means randomly modifying the variables, and higher heat implies greater random fluctuations. The cost function returns the output associated with a set of variables.

Akbarzadeh et al. (2011) applied covariance matrix adaptation evolutionary strategy (CMA-ES) method in a topography aware sensor deployment. They encoded the position and orientation of the sensors inside the individuals, and then the population of individuals was evolved through generations. Finally, the individual with the best coverage is chosen as the best solution. CMA-ES is categorized as evolutionary strategy that is a stochastic optimization method for nonlinear or nonconvex problems. In this method, new candidate solutions are sampled according to a multivariate normal distribution. Pairwise dependencies between the variables in this distribution are described by a covariance matrix. The CMA is a method to update the covariance matrix of this distribution.

Genetic algorithms are widely utilized as global optimizing intelligent techniques. The genetic algorithm is a heuristic optimization method that has been inspired by the process of natural evolution in four steps: initialization, selection, reproduction, and termination. In the sensor deployment problem, they place the nodes such that both of optimal coverage and energy consumption are achieved in the entire sensor networks. Romoozi and Ebrahimpour-komleh (2010) have proposed a genetic algorithm method to create energy efficient node positioning in wireless sensor networks and showed that intelligent algorithms can extend the network lifetime by finding the optimum position of nodes. Ferentinos and Tsiligiridis (2007) applied genetic algorithm for self-organizing and adaptive wireless sensor network design. They showed that optimal sensor network designs constructed by the genetic algorithm satisfy all application-specific requirements, fulfill the connectivity constraints and manage energy consumption to guarantee the maximum nodes' lifetime. Finally, Jourdan and Weck (2004) proposed a framework that served to benchmark a multi objective genetic algorithm for sensor deployment to reach the optimum coverage and network lifetime.

3.5.2 Local Approaches

Unlike global approaches, local optimization methods find the local optima among a number of candidate solutions. These methods move from solution to solution in the searching space until the optimal solution is reached or a time bound is elapsed. A local optimization algorithm starts from a candidate or initial value or solution, and then iteratively moves to neighbor values or solutions. Typically, every candidate solution has more than one neighbor solution. Therefore, choosing the next solution or value depends on the neighborhood information of the current solution as well as the previous minimum or maximum founded value regarding to the optimization objective (this is why they are called "local search" or "local optimization"). Normally, local optimization methods are applied for solving computationally hard optimization problems that can be formulated as finding a solution maximizing a criterion among a number of candidates.

Cortes et al. (2004) proposed the gradient descent algorithm for coverage control and optimal sensing policies in mobile sensor networks. Gradient descent is an optimization algorithm that takes steps proportional to the negative of the gradient of the function at the current point to find a local minimum.

Many of the local optimization approaches use the concept of mobility, which exploit moving properties of nodes to get better coverage conditions and try to relocate sensor nodes to optimal locations that serve maximum coverage. For sensor deployment approaches, where there is no information available about the terrain surface and its morphology, random sensor deployment is used. This method does not guarantee the optimized coverage of the sensing region. Thus, some deployment strategies take advantage of mobility options and try to relocate sensors from their initial places to optimize the network coverage. Potential field–based, virtual force–based, and incremental self-deployment methods are examples of such approaches.

The idea of potential field is that every node is exposed to two forces: (i) a repulsive force that causes the nodes to repel each other, and (ii) the attractive force that makes nodes move toward each other when they are on the verge of being disconnected (Howard et al. 2002a,b). These forces have an inverse proportion with the square of distance between nodes. Each node repels all its neighbors. This action decreases the repulsive force but, at the same time, it stimulates the attractive force. Eventually, it ends up in an arrangement wherein all the nodes reach an equilibrium situation and uniformly cover the sensing field.

The virtual force–based method is very similar to the potential-based methods but, here, each node is exposed to three types of forces: (i) a repulsive force exerted by obstacles, (ii) an attractive force exerted by areas in which a high degree of coverage is required, and (iii) an attractive or repulsive force by another point based on its location and orientation (Zou and Chakrabarty 2003, 2004).

In incremental self-deployment algorithms, each node finds its optimal location through previous deployed nodes information in four steps: (i) *initialization*, which classifies the nodes into three groups: waiting, active, and deployed; (ii) *goal selection*, which selects the best destination for the node to be deployed based on previous node deployment; (iii) *goal resolution*, which assigns this new location to a waiting node, and specifies the plan for moving to this location; and (iv) finally, *execution*, which deploys the active nodes in their place (Howard and Mataric 2002; Howard et al. 2002a,b; Heo and Varshney 2003).

As it is illustrated in the above algorithms, spatial coverage of sensor networks is significantly related to the spatial distribution of the sensors in the environment. In other words, the described algorithms try to distribute the sensors in the field so that as much coverage as possible is obtained. Voronoi diagrams and Delaunay triangulation have been used in many of the mobility-based methods because they directly satisfy the required distribution. We classify the Voronoi-based solutions based on the sensor types used in the network as (1) static sensor networks, (2) mobile sensor networks, and (3) hybrid sensor networks, in which a combination of static and mobile sensors is deployed. For static sensor networks, new sensors are added. For mobile and hybrid networks, however, existing sensors move to heal the holes.

3.5.2.1 Voronoi-Based Solutions for Static Sensor Networks

To the best of our knowledge, there are two suggestions to deploy an additional sensor to heal the holes in a static sensor network. Ghosh (2004) proposes that for each Voronoi vertex, one node should be added to heal the coverage hole around this Voronoi vertex. As Figure 3.15 shows, to heal the hole around the Voronoi vertex v_2, the target location p_1 lies on the bisector of the angle $v_1v_2v_3$ and $d(s, p_1) = \min\{2R, d(s,v_2)\}$, where d is the Euclidean distance and R is the sensing radius of the sensors. Wang et al. (2003), however, deploy only one mobile node to heal the coverage hole of a Voronoi cell. As illustrated in Figure 3.15, the target location p_2 lies on the line connecting the sensor node and its furthest Voronoi vertex (v_4 here) and $d(s, p_2) = \max\left\{\sqrt{3}R, d(s,v_4)\right\}$.

3.5.2.2 Voronoi-Based Solutions for Mobile Sensor Networks

In mobile sensor networks, all sensors have the ability to move and heal the holes. Wang et al. (2004) proposes three Voronoi-based strategies for this movement: vector-based (VEC), Voronoi-based (VOR), and Minimax. They all are iterative approaches and gradually improve the coverage of the sensor network.

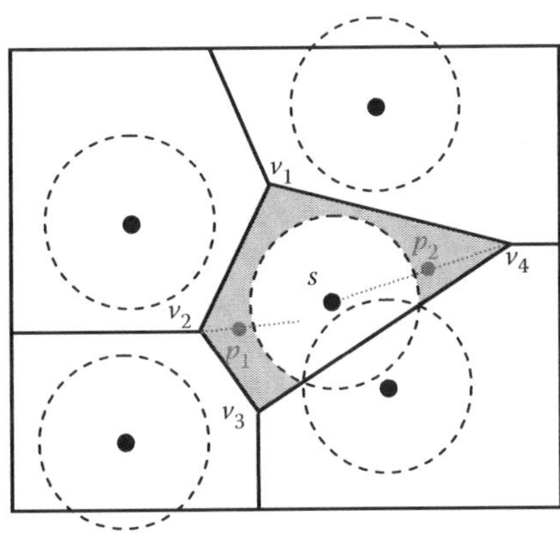

Figure 3.15 Deploying an additional sensor to heal the hole in a static sensor network.

3.5.2.2.1 Vector-Based Algorithm

VEC pushes sensors away from a densely covered area. It imitates the electromagnetic force that exists between two particles: if two sensors are too close to each other, they exert a repulsive force. By knowing the target area and the number of sensors, an average distance between the sensors, d_{avg}, can be calculated beforehand. If the distance between two sensors s_i and s_j is smaller than d_{avg} and none of their Voronoi cells is completely covered, the virtual force pushes them to move $[d_{avg} - d(s_i, s_j)]/2$ away from each other. However, if one of the sensors completely covers its Voronoi cell, and so it should not move, then the other sensor pushes $[d_{avg} - d(s_i, s_j)]$ away.

In addition to the repulsive forces between sensors, the boundaries also exert forces to push sensors that are too close to the boundary inside. If the distance of the sensor i, that is, $d_b(s_i)$, from its closest boundary is smaller than $d_{avg}/2$, then it moves $[d_{avg}/2 - d_b(s_i)]$ toward the inside of the network.

Note that movements of the sensors change the shape of the Voronoi cells, which may result in decreasing the coverage in the new configuration. Thus, the sensors move to the target position only if their movement increases the local coverage within their Voronoi cell. Otherwise, they take the midpoint position between its current and target positions as the new target position, and again check the improvement, and so on (this process is called movement adjustment). Figure 3.16 shows an example of the VEC algorithm in use.

3.5.2.2.2 Voronoi-Based Algorithm

Unlike the VEC algorithm, VOR is a pulling strategy so that sensors cover their local maximum coverage holes. In this algorithm, each sensor moves toward its furthest Voronoi vertex until this vertex is covered (Figure 3.17). The movement adjustment mentioned for VEC is also applied here. Furthermore, VOR is a greedy algorithm that heals the largest hole. However, after moving a sensor, a new hole may be created that is healed by a reverse movement in the next iteration, so it results in an oscillation moving. An *oscillation control* is added to overcome this problem. This control does not allow sensors to move backward immediately: before a sensor moves, it first checks if the direction of this movement is opposite to that in the previous round. If so, it stops for one round to see if the hole is healed by the movement of a neighboring sensor. Figure 3.18 shows an example that moves the sensors based on the VOR algorithm.

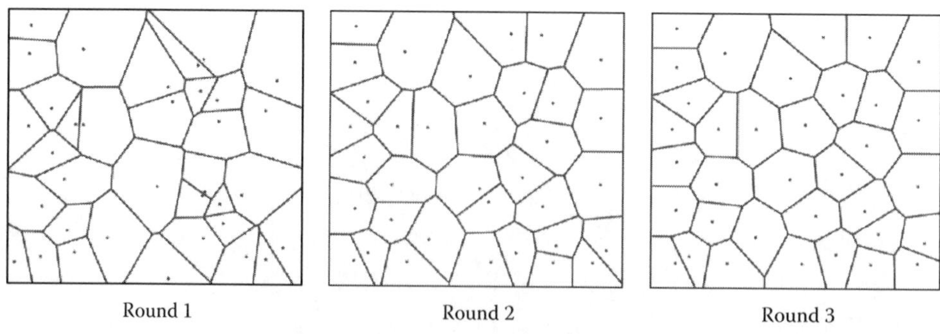

Figure 3.16 An example using the VEC algorithm to move the sensors. (From Wang, G. et al. Movement-assisted sensor deployment. *IEEE INFOCOM '04*, 2004.)

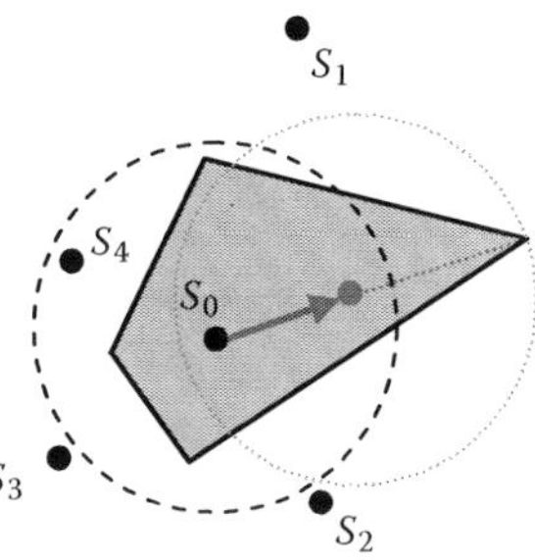

Figure 3.17 Movement of a sensor in the VOR algorithm.

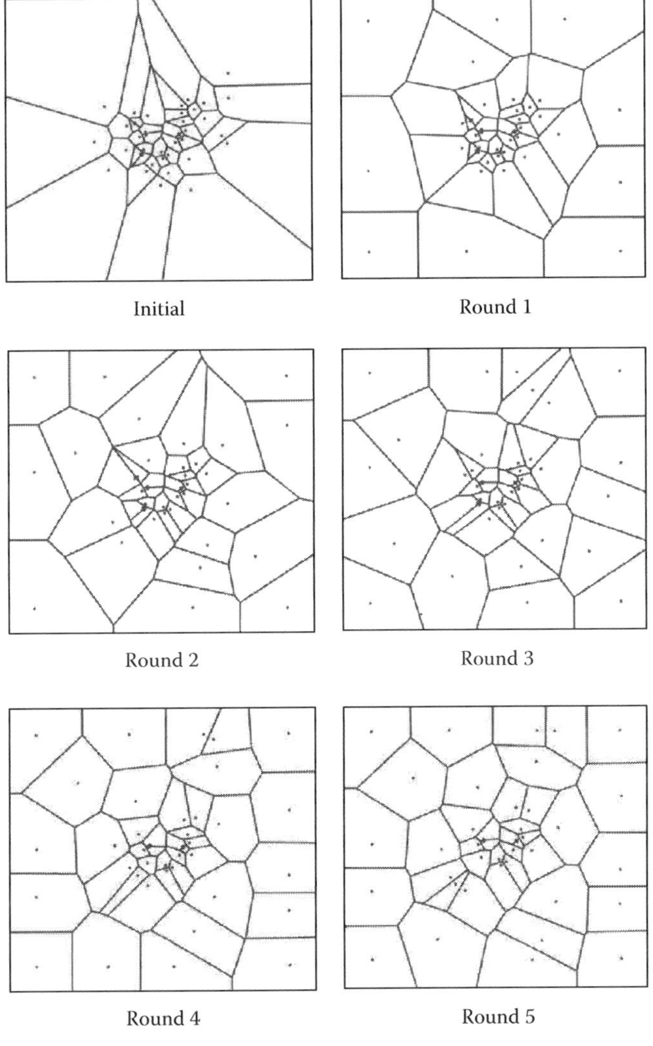

Figure 3.18 An example using the VOR algorithm to move the sensors. (From Wang, G. et al. Movement-assisted sensor deployment. *IEEE INFOCOM '04*, 2004.)

3.5.2.2.3 Minimax Algorithm

This algorithm is based on the fact that when the sensors are evenly distributed, a sensor should not be too far away from any of its Voronoi vertices. In other words, the disadvantage of the VOR algorithm is that it may result in a case in which a vertex that was originally close becomes the new farthest vertex. The Minimax algorithm solves this by choosing the target location as the point inside the Voronoi cell whose distance to the farthest Voronoi vertex is minimized. This point, which is called the *Minimax point*, is the center of the smallest enclosing circle of the Voronoi vertices and can be calculated by the algorithms described by Megiddo (1982), Skyum (1991), and Welzl (1991). The Minimax algorithm has some advantages. First, it reduces the variance of the distances to the Voronoi vertices, resulting in more regularly shaped Voronoi cells, which better utilizes the sensor's sensing circle. Second, Minimax considers more information than VOR, and it is more conservative. Third, Minimax is more "reactive" than VEC, that is, it heals the holes more directly by moving toward the farthest Voronoi vertex.

3.5.2.3 Voronoi-Based Solutions for Hybrid Sensor Networks

In a hybrid sensor network, having detected a hole around a static sensor, a mobile sensor moves to heal this hole. The location to which the mobile sensor should move is computed similar to the solutions proposed for the static networks. Then, the static sensor requests the neighboring mobile sensors to move to the calculated destination. Each of the mobile sensors that have received this request calculates the coverage holes formed at its original location due to its movement. It decides to move if the new hole is smaller than the hole size of the requesting static sensor. Because movements of the mobile sensors may create new (but smaller) holes, this solution is an iterative procedure. More discussion on this movement and its technical considerations (e.g., bidding protocols) can be found in studies by Wang et al. (2003) and Ghosh (2004).

3.5.3 Node Scheduling

As mentioned previously, energy is an important issue in sensor networks. Thus, strategies to save energy are of high interest in this regard. A relevant case to save energy is temporarily turning some sensor nodes to sleep mode in the multicovered areas. This is also important to avoid other problems (e.g., the intersection of sensing area, redundant data, and communication interference) in areas with a high density of sensor nodes (Vieira et al. 2003). Different methods have been proposed for this problem (Tian and Georganas 2002; Ruiz et al. 2003).

As an example, Vieira et al. (2003) proposed a Voronoi-based algorithm to find the nodes to be turned on or off. The Voronoi diagram of the sensor nodes is constructed. Each Voronoi cell represents the area for which the corresponding node is responsible. The sensors whose responsible areas are smaller than a predefined threshold are turned off. By updating the Voronoi diagram, the neighbors of that sensor, which have been turned off, become responsible for that area. This process continues until there is no node responsible for an area smaller than the given threshold.

3.6 Open Problems and Research Challenges

This section introduces more challenging issues in sensor coverage problem, which are open for more research.

3.6.1 K-Coverage Sensor Networks

In some applications, such as military or security control, it is required that each point of the region is covered by at least k ($k > 1$) sensors for the sake of higher reliability. Among the different solutions proposed in the literature (Zhou et al. 2004), So and Ye (2005) have developed an algorithm based on the concept of Voronoi regions. Suppose that $P = \{p_1, p_2,..., p_n\}$ is a set of n points in $\mathbf{R}^n$. For any subset U of P, the Voronoi region of U is the set of points in $\mathbf{R}^n$ closer to all points in U than to any point in $P - U$. The proposed algorithm checks the k-coverage for the area, but developing the algorithms to heal the holes is still an open question.

3.6.2 Sensor Networks with Various Sensing Ranges

Thus far, we have assumed that all sensors are identical. In reality, however, a sensor network could be composed of multiple types of sensors with different specifications, including their sensing range and sensing model, for example, circular, ellipsoidal, or irregular (Ahmed et al. 2005; So and Ye 2005). As illustrated in Figure 3.19, a weighted Voronoi diagram is the solution in such cases to examine the coverage quality of the network (So and Ye 2005). However, to the best of our knowledge, the movement strategies have not been researched deeply for such heterogeneous sensor networks.

3.6.3 Directional Sensor Networks

Coverage determination for directional sensor networks (i.e., networks composed of sensors with limited field of views like cameras) is a practical area of research. Adriaens et al. (2006) have extended the research done by Meguerdichian et al. (2001a,b,c) and Megerian et al. (2005) and developed a Voronoi-based algorithm to detect the worst-case coverage (maximal breach path) in such networks.

3.6.4 Sensor Networks in a Three-Dimensional Environment

The approaches mentioned in this chapter assume that a sensor network is deployed in a 2-D flat environment (i.e., a 2-D Euclidean plane). However, this assumption oversimplifies the sensor network reality. The real-world environment is mostly a three-dimensional (3-D) heterogeneous field containing obstacles (Figure 3.20). Hence, 3-D sensor networks have considerable interest in diverse

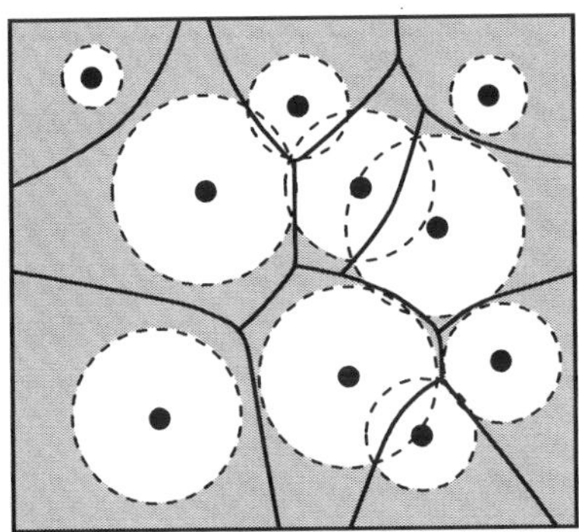

Figure 3.19 Using a weighted Voronoi diagram to examine the coverage quality of a sensor network with various sensing ranges.

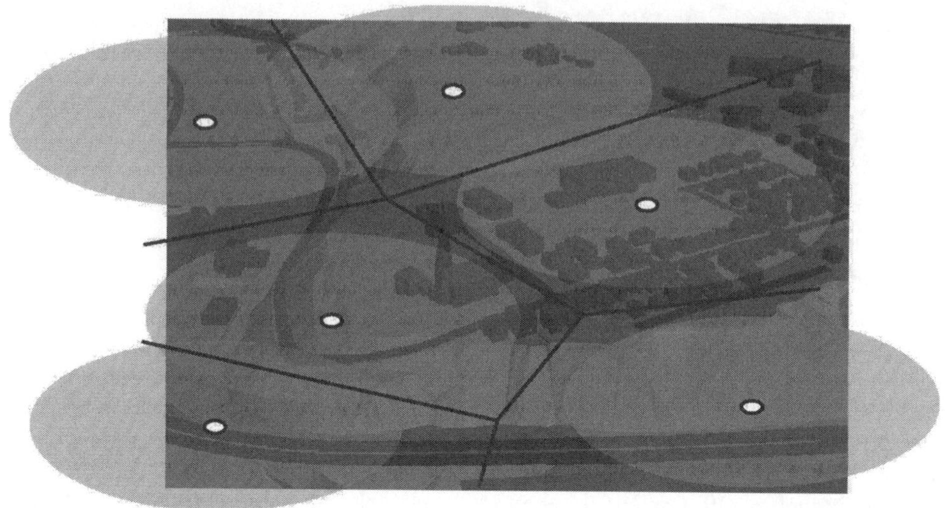

Figure 3.20 A sensor network in a 3-D environment with various obstacles. The superimposed 2-D Voronoi diagram cannot determine the network coverage.

applications including structural monitoring networks and underwater networks (Huang and Tseng 2005). In addition to the form and the relief of the sensor network area, various obstacles may prevent the sensors from covering an invisible region or communicating data between each other.

Several algorithms have been proposed for the coverage problem of 3-D sensor networks (Chong and Kumar 2003; Huang and Tseng 2005; Bahramgiri et al. 2006). The algorithms presented here can be extended to use the 3-D Delaunay triangulation and the Voronoi diagram for coverage determination and optimization of such sensor networks (Ghosh et al. 2007; Lei et al. 2007). There are also suggestions to use Delaunay triangulation and Voronoi diagram when the environment contains obstacles (Wu et al. 2007). Although these extensions are interesting in some applications, they may have deficiencies for the geographical fields because they consider 3-D Euclidean field and man-made obstacles, for example, walls. The real-world environment, however, is a 3-D heterogeneous field full of man-made and natural obstacles. Even, the terrain could play the role of an obstacle in this case. Using capabilities of geographical information systems (GIS) seems a promising solution in this regard, which has not been investigated. It can provide the information (e.g., digital terrain models) or spatial analyses (e.g., visibility analysis) required to evaluate and optimize the sensor networks installed in the natural environment. Hence, 3-D Delaunay triangulation and Voronoi diagrams present interesting solutions for the sensor network modeling and optimization in a 3-D environment. However, their application is not straightforward and several challenging conceptual and implementation problems should be addressed.

3.6.5 Integration of Spatial Information in Sensor Deployment Procedure

A major deficiency of the existing sensor deployment methods is that they are not adequately adopted to consider the reality of the terrain and the environment. They do not consider the form and the topography of the area covered by the sensor network as well as various existing obstacles that may prevent the sensors from covering the entire area or allowing data communication between sensors. On the other hand, the region of interest may change over the sensing experiments, which may significantly affect the sensor network coverage. For instance, in a battlefield, all parameters of the study area may rapidly change. In urban areas, new constructions may be created, urban facilities may be added or removed, and land cover and land use information may change. Furthermore, the characteristics of sensor platforms may change during the sensing steps. For example, fluctuation of the battery power for each platform decreases the sensing range of the nodes, thus the network arrangement must be modified to maintain the network performance. These changes must be considered by the network and the development methods must be adopted to deal with them.

Argany et al. (2011) proposed a realistic sensor network approach that integrates the terrain information with the Voronoi-based optimization methods discussed in Section 3.5. For this method, an optimization process is coupled to a GIS for integrating spatial information, including man-made (buildings, bridges, etc.) or natural objects. Moreover, the functions and capabilities available in the GIS (e.g., visibility, line of sight, and viewshed analysis) serve more facilities in sensor network deployment. Finally, a dynamic geometric data structure based on a Voronoi diagram is deployed to consider the topology of the sensor network and its dynamics, that is, insertion, deletion, and movement. Figure 3.21 presents the results of using the proposed approach for

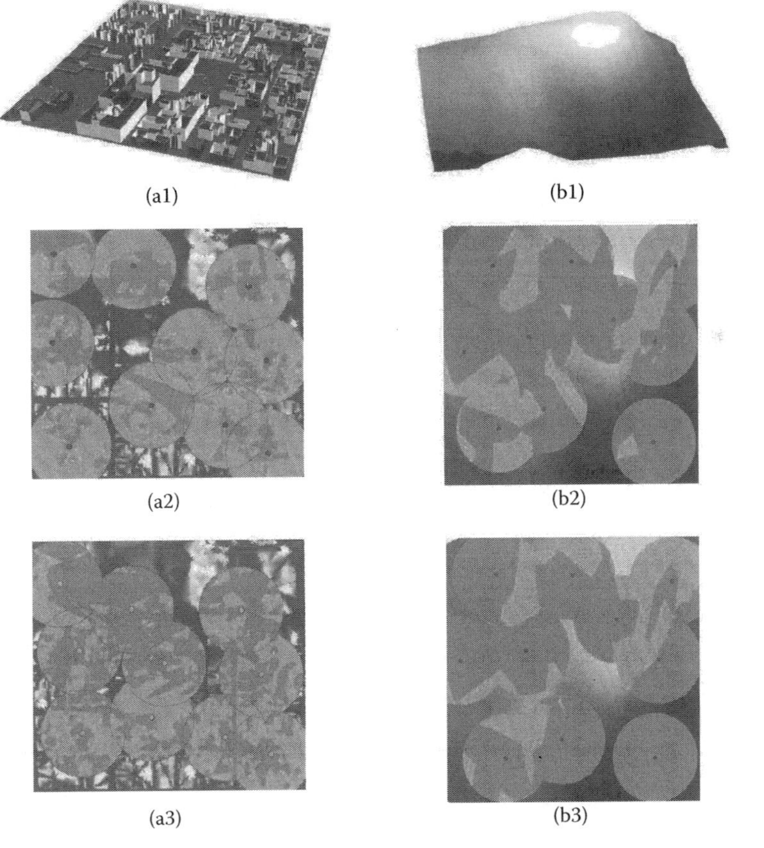

Figure 3.21 Results of the realistic sensor network for the (a1) urban and (b1) natural areas, the covered regions in the sensing field of each sensor node in the initial deployment for the (a2) urban and (b2) natural areas, and the covered regions in the sensing field of each sensor node in the second deployment for the (a3) urban and (b3) natural areas. Green and pink regions are the covered and uncovered regions, respectively. (From Argany, M. et al., *Trans. Comput. Sci.* XIV, 151–172, 2011.)

an urban area and a natural area from two case studies. In both cases, the study areas are initially covered by 10 randomly distributed sensors with sensing range of 50 m, which could rotate −90° to 90° vertically and 0° to 360° horizontally. Figure 3.21a2 and b2 show the area covered by the initial position of the sensors on the Digital Terrain Model (DTM) (which was 23% and 66% for the urban and natural areas, respectively). An extension of the VOR algorithm is used, which allows better consideration of the topography of the terrain and the presence of various obstacles in the sensing area. For this, the sensors are moved toward the farthest Voronoi vertex, but with the restriction that the sensor stops if it reaches a position with a higher elevation than its current position. As illustrated in Figure 3.21a2 and b2, this consideration significantly improves the spatial coverage of the sensor network in both cases (14% and 5% for urban and natural areas, respectively).

3.7 Conclusions and Perspectives

This chapter introduces the coverage problem in geosensor networks and surveys the methods proposed in the literature for its resolution. In particular, algorithms that use the Voronoi diagram and Delaunay triangulation were intensively investigated. As discussed in the chapter, most of the existing methods oversimplify the coverage problem and do not consider the characteristics of the environment in which they are deployed. Spatial coverage of sensor networks is significantly related to the spatial distribution of the sensors in the environment. The coverage determination algorithms try to distribute the sensors in the field so that maximum coverage is obtained.

The Voronoi diagram and Delaunay triangulation are well adapted for abstraction and modeling of sensor networks and spatial data structures. However, their application is still limited when it comes to the determination and optimization of spatial coverage of more complex sensor networks (e.g., sensor networks with the presence of obstacles).

To overcome the limitation of these methods, a novel approach based on the Voronoi diagram has been proposed, which considers spatial information in senor network deployment and coverage optimization. To evaluate the proposed method, two case studies, which presented and compared the sensor network deployment and its spatial coverage in urban and natural areas, were discussed. The preliminary results obtained from these experiments are very promising. We observed a considerable improvement in the spatial coverage of the geosensor networks in both cases. Improving the quality and the performance of the proposed method is an interesting direction for future work.

Acknowledgments

The authors thank the GEOID Networks of Centres of Excellence and the Natural Sciences and Engineering Research Council of Canada for funding of this research work.

References

21 Ideas for the 21st Century (1999). *Business Week* 78–167.

Adriaens, J., S. Megerian et al. (2006). Optimal worst-case coverage of directional field-of-view sensor networks. *Sensor and Ad Hoc Communications and Networks (SECON '06), IEEE* 336–345.

Ahmed, N., S. S. Kanhere et al. (2005). The holes problem in wireless sensor networks: a survey. *ACM SIGMOBILE Mobile Computing and Communications Review* 9(2): 4–18.

Akbarzadeh, V., A. Ko et al. (2011). Topography-aware sensor deployment optimization with CMA-ES. *Parallel Problem Solving from Nature—PPSN XI* 11: 141–150.

Argany, M., M. A. Mostafavi et al. (2010). Voronoi-based approaches for geosensor networks coverage determination and optimisation: a survey. *International Symposium on Voronoi Diagrams in Science and Engineering 2010, IEEE* 115–123.

Argany, M., M. Mostafavi et al. (2011). A GIS based wireless sensor network coverage estimation and optimization: a Voronoi approach. *Transactions on Computational Science XIV* 14: 151–172.

Aziz, N. A. A., K. A. Aziz et al. (2009). Coverage strategies for wireless sensor networks. *World Academy of Science, Engineering and Technology* 50: 145–150.

Bahramgiri, M., M. Hajiaghayi et al. (2006). Fault-tolerant and 3-dimensional distributed topology control algorithms in wireless multi-hop networks. *Wireless Networks* 12(2): 179–188.

Cao, Q., T. He et al. (2006). Efficiency centric communication model for wireless sensor networks. *Proceedings of INFOCOM 2006.*

Chong, C. Y. and S. P. Kumar (2003). Sensor networks: evolution, opportunities, and challenges. *Proceedings of the IEEE* 91(8): 1247–1256.

Cortes, J., S. Martinez et al. (2004). Coverage control for mobile sensing networks. *IEEE Transactions on Robotics and Automation* 20(2): 243–255.

De Berg, M., M. van Kreveld et al. (2000). *Computational Geometry: Algorithms and Applications*. New York.

Deb, B., S. Bhatnagar et al. (2002). A topology discovery algorithm for sensor networks with applications to network management. Short paper. Department of Computer Science, Rutgers University DCS-TR-441.

Ferentinos, K. P. and T. A. Tsiligiridis (2007). Adaptive design optimization of wireless sensor networks using genetic algorithms. *Computer Networks* 51(4): 1031–1051.

Ghosh, A. (2004). Estimating coverage holes and enhancing coverage in mixed sensor networks. *29th Annual IEEE Conference on Local Computer Networks, IEEE, Tampa, Florida, USA.*

Ghosh, A. and S. K. Das (2008). Coverage and connectivity issues in wireless sensor networks: a survey. *Pervasive and Mobile Computing* 4(3): 303–334.

Ghosh, A., Y. Wang et al. (2007). Efficient distributed topology control in 3-dimensional wireless networks. *4th Annual IEEE Communications Society Conference on Sensor and Ad Hoc Communications and Networks (SECON 2007), IEEE, San Diego, CA, USA.*

Heo, N. and P. K. Varshney (2003). A distributed self spreading algorithm for mobile wireless sensor networks. *IEEE Wireless Communications and Networking Conference (WCNC'03), New Orleans, USA, IEEE.*

Hossain, A., P. Biswas et al. (2008). Sensing models and its impact on network coverage in wireless sensor network. *Proceedings of the 10th Colloquium and the 3rd ICIIS, IEEE, Kharagpur, India.*

Howard, A. and M. J. Mataric (2002). Cover me! A self-deployment algorithm for mobile sensor networks. *IEEE International Conference on Robotics and Automation (ICRA'02), Washington DC, USA.*

Howard, A., M. J. Matarić et al. (2002a). An incremental self-deployment algorithm for mobile sensor networks. *Autonomous Robots* 13(2): 113–126.

Howard, A., M. J. Matarić et al. (2002b). Mobile sensor network deployment using potential fields: A distributed, scalable solution to the area coverage problem. *6th International Symposium on Distributed Autonomous Robotic Systems (DARS'02), Fukuoka, Japan.*

Huang, Q. (2003). Solving an open sensor exposure problem using variational calculus. Tech. Rep. WUCS-03-1. Department of Computer Science, Washington University, St. Louis, MO.

Huang, C. F. and Y. C. Tseng (2005). A survey of solutions to the coverage problems in wireless sensor networks. *Journal of Internet Technology* 6(1): 1–8.

Hwang, J., Y. Gu et al. (2007). Realistic sensing area modeling. *Proceedings of INFOCOM, IEEE, Anchorage, Alaska, USA.*

Jourdan, D. B. and O. L. d. Weck (2004). Layout optimization for a wireless sensor network using a multi-objective genetic algorithm. *IEEE Semiannual Vehicular Technology Conference. Milan, Italy.*

Lei, R., L. Wenyu et al. (2007). A coverage algorithm for three-dimensional large-scale sensor network. *Intelligent Signal Processing and Communication Systems, IEEE, Xiamen, China.*

Luo, J. and J. P. Hubaux (2005). Joint mobility and routing for lifetime elongation in wireless sensor networks. *Proceedings of INFOCOM, IEEE, Miami, Florida, USA.*

Madan, R., S. Cui, R. Madan, C. Shuguang, S. Lall, A. Goldsmith (2005). Cross-layer design for lifetime maximization in interference-limited wireless sensor networks. *Proceedings of INFOCOM, IEEE.*

Megerian, S., F. Koushanfar et al. (2005). Worst and best-case coverage in sensor networks. *IEEE Transactions on Mobile Computing* 4(1): 84–92.

Megiddo, N. (1982). Linear-time algorithms for linear programming in R3 and related problems. *SIAM Journal on Computing* 12: 329–338.

Meguerdichian, S., F. Koushanfar et al. (2001a). Coverage problems in wireless ad-hoc sensor networks. *Proceedings of INFOCOM, IEEE, Anchorage, Alaska, USA.*

Meguerdichian, S., F. Koushanfar et al. (2001b). Exposure in wireless ad-hoc sensor networks. *Proceedings of MOBICOM'01, IEEE, Rome, Italy.*

Meguerdichian, S., S. Slijepcevic et al. (2001c). Localized algorithms in wireless ad-hoc networks: location discovery and sensor exposure. *ACM International Symposium on Mobile Ad Hoc Networking and Computing (MobiHOC), ACM, Long Beach, California, USA.*

Muthukumar, M., N. Sureshkumar et al. (2010). A wireless sensor network communication model for automation of electric power distribution. *The Technology Interface Journal* 10(3).

Niewiadomska-Szynkiewicz, E. and M. Marks (2009). Optimization schemes for wireless sensor network localization. *International Journal of Applied Mathematics and Computer Science* 19(2): 291–302.

Nittel, S. (2009). A survey of geosensor networks: advances in dynamic environmental monitoring. *Sensors* 9(7): 5664–5678.

Romoozi, M. and H. Ebrahimpour-komleh (2010). A positioning method in wireless sensor networks using genetic algorithms. *International Journal of Digital Content Technology and Its Applications* 4(9): 174–179.

Ruiz, L. B., J. M. Nogueira et al. (2003). A management architecture for wireless sensor networks. *Communications Magazine, IEEE* 41(2): 116–125.

Salhieh, A., J. Weinmann et al. (2001). Power efficient topologies for wireless sensor networks. *Proceedings of the International Conference on Parallel Processing, IEEE, Valencia, Spain.*

Sharifzadeh, M. and C. Shahabi (2004). Supporting spatial aggregation in sensor network databases. *12th Annual ACM International Workshop on Geographic Information Systems, ACM, Washington, DC, USA.*

Skyum, S. (1991). A simple algorithm for computing the smallest enclosing circle. *Information Processing Letters* 37(3): 121–125.

So, A. and Y. Ye (2005). On solving coverage problems in a wireless sensor network using Voronoi diagrams. *Workshop on Internet and Network Economics (WINE'05), Hong Kong, China* 584–593.

Soro, S. and W. B. Heinzelman (2005). On the coverage problem in video-based wireless sensor networks. *2nd Workshop on Broadband Advanced Sensor Networks (BaseNets'05), IEEE, Boston, MA, USA* 932–939.

Szewczyk, R., E. Osterweil et al. (2004). Habitat monitoring with sensor networks. *Communications of the ACM* 47(6): 34–40.

Tian, D. and N. D. Georganas (2002). A coverage-preserving node scheduling scheme for large wireless sensor networks. *1st ACM International Workshop on Wireless Sensor Networks and Applications, ACM, Atlanta, GA, USA.*

Veltri, G., Q. Huang et al. (2003). Minimal and maximal exposure path algorithms for wireless embedded sensor networks. *ACM International Conference on Embedded Networked Sensor Systems (SenSys), ACM, Los Angeles, California, USA.*

Vieira, M. A. M., L. M. Vieira et al. (2003). Scheduling nodes in wireless sensor networks: a Voronoi approach. *28th IEEE Conference on Local Computer Networks (LCN2003), Bonn, Germany, IEEE.*

Wang, Y. and G. Cao (2011a). Barrier coverage in camera sensor networks. *12th ACM International Symposium on Mobile Ad Hoc Networking and Computing, ACM, Paris, France.*

Wang, Y. and G. Cao (2011b). On full-view coverage in camera sensor networks. *Proceedings of INFOCOM 2011, IEEE, Shanghai, China.*

Wang, G., G. Cao et al. (2003). A bidding protocol for deploying mobile sensors. *11th IEEE International Conference on Network Protocols (ICNP'03), IEEE, Atlanta, Georgia, USA.*

Wang, G., G. Cao et al. (2004). Movement-assisted sensor deployment. *IEEE Infocom (INFOCOM'04), IEEE, Hong Kong, China.*

Wang, W., V. Srinivasan et al. (2005). Using mobile relays to prolong the lifetime of wireless sensor networks. *Proceedings of ACM MobiCom, ACM, Cologne, Germany.*

Wang, B., H. B. Lim et al. (2009). A survey of movement strategies for improving network coverage in wireless sensor networks. *Computer Communications* 32(13): 1427–1436.

Welzl, E. (1991). Smallest enclosing disks (balls and ellipsoids). *New Results and New Trends in Computer Science* 359–370.

Worboys, M. and M. Duckham (2006). Monitoring qualitative spatiotemporal change for geosensor networks. *International Journal of Geographical Information Science* 20(10): 1087–1108.

Wu, C. H. and Y. C. Chung (2009). A polygon model for wireless sensor network deployment with directional sensing areas. *Sensors* 9(12): 9998–10022.

Wu, C. H., K. C. Lee et al. (2007). A Delaunay triangulation based method for wireless sensor network deployment. *Computer Communications* 30(14–15): 2744–2752.

Yi, W. and C. Guohong (2011). Barrier coverage in camera sensor networks. *12th ACM International Symposium on Mobile Ad Hoc Networking and Computing. Paris, France.*

Yu, Y., V. K. Prasanna et al. (2005). Communication models for algorithm design in networked sensor systems. *Proceedings of the 19th IEEE International Parallel and Distributed Processing Symposium, IEEE, Denver, Colorado, USA.*

Zou, Y. and K. Chakrabarty (2003). Sensor deployment and target localization based on virtual forces. *IEEE Infocom (INFOCOM'03), San Francisco, USA, IEEE.*

Zou, Y. and K. Chakrabarty (2004). Sensor deployment and target localization in distributed sensor networks. *ACM Transactions on Embedded Computing Systems* 3(1): 61–91.

Zhou, Z., S. Das et al. (2004). Connected K-coverage problem in sensor networks. *ICCCN 2004, IEEE, Chicago, IL, USA.*

Zuniga, M. and B. Krishnamachari (2004). Analyzing the transitional region in low power wireless links. *Proceedings of the First IEEE International Conference on Sensor and Ad Hoc Communications and Networks, Santa Clara, CA, USA, IEEE.*

PHYSICAL LAYER AND INTERFACING

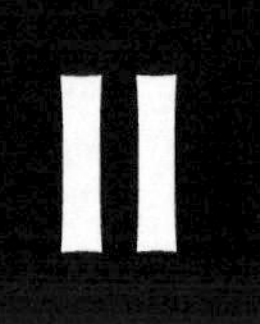

Chapter 4

Overview of IEEE 802.15.4 Wireless Sensor Networks in Three-Dimensional Terrains

Mu-Sheng Lin, Jenq-Shiou Leu, and Kuen-Han Li

National Taiwan University of Science and Technology

Jean-Lien C. Wu

St. John's University

Contents

Mobile ad hoc networks (MANET) and wireless sensor networks (WSN) have become popular research topics in recent years because of the rapid development of microelectromechanical systems (MEMS) technology. Because tiny wireless sensors use battery power, they can sense environmental conditions, and are capable of wireless communications and processing information. They are not only capable of sensing and detecting changes in the target but can also manage and process the collected data and transmit them to a collection center. Based on the features of sensors, several studies have focused on the enhancement of network performance, reliability, and power-saving mechanisms. Even though many articles have been proposed to evaluate the performance of IEEE 802.15.4 WSNs, they still lack the performance evaluation under different three-dimensional (3-D) terrains. In this chapter, we first describe the architecture and functionalities of the ZigBee sensor networks. Second, we introduce the IEEE 802.15.4 standard. Furthermore, the characteristics of 3-D terrains in WSNs are introduced. Finally, we evaluate the performance of IEEE 802.15.4 WSNs in five artificial 3-D terrains.

4.1 Introduction

ZigBee [1] is a specification for a suite of high-level communication protocols using small, low-power digital MEMS and builds on the physical layer and medium access control (MAC) defined in IEEE standard 802.15.4 for low-rate wireless personal area networks (LR-WPANs). The IEEE 802.15.4 standard [2] defines the protocol and interconnection of devices via radio communication in a WPAN. It provides reliable transmissions among devices. The specification goes on to complete the standard by adding four main components: network layer, application layer, ZigBee device objects (ZDOs), and manufacturer-defined application objects, which allow for customization and favor total integration. Two types of devices called the full-function device (FFD) and the reduced-function device (RFD) are used in a LR-WPAN. The FFD is a fully functional device that could be a PAN coordinator, a coordinator, or just a device. The RFD is a device with reduced functionality that can only function as an end device; it cannot communicate with any other device in addition to the coordinator.

The PAN coordinator, which acts as a coordinator for the entire WPAN, is authorized to provide synchronization services in an established network. Applications include home entertainment and control, security alarms, industrial monitoring and control, personal mobile health care and tele-assist, wireless light switches, and other consumer and industrial equipment that require short-range wireless transfer of data at relatively low rates.

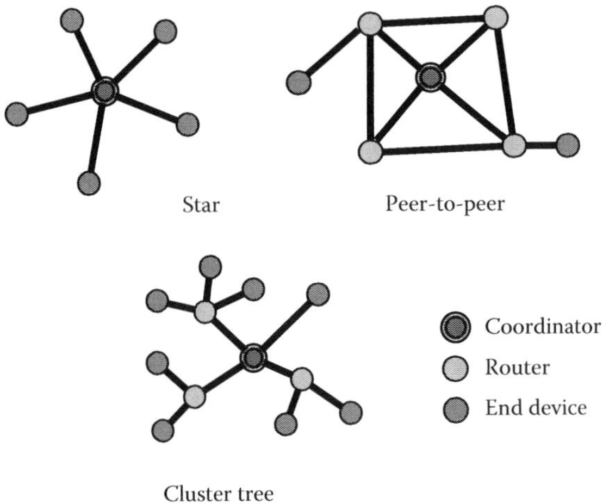

Figure 4.1 ZigBee network topologies.

4.1.1 ZigBee Topology

Figure 4.1 shows the three topologies that are proposed in the IEEE 802.15.4 protocol standard: star topology, peer-to-peer topology, and cluster-tree topology. The coordinator and routers are FFDs and the end devices are the RFDs. In the star topology, communication is established between devices and a PAN coordinator. In peer-to-peer topology, a device can communicate with other devices as long as they are in range of one another. A peer-to-peer network can be ad hoc, self-organizing, or self-healing. In cluster-tree topology, most devices are FFD and an RFD may connect to a cluster-tree network as leave nodes at the end of a branch [1].

4.1.2 IEEE 802.15.4 MAC Superframe Structure

The IEEE 802.15.4 MAC protocol supports beacon-enabled and non–beacon-enabled modes. Figure 4.2 shows the IEEE 802.15.4 MAC superframe structure [2]. In beacon-enabled mode, the access to the channel is managed through a superframe, starting with the beacon packet transmitted by the PAN coordinator. The superframe is subdivided into a contention access period (CAP),

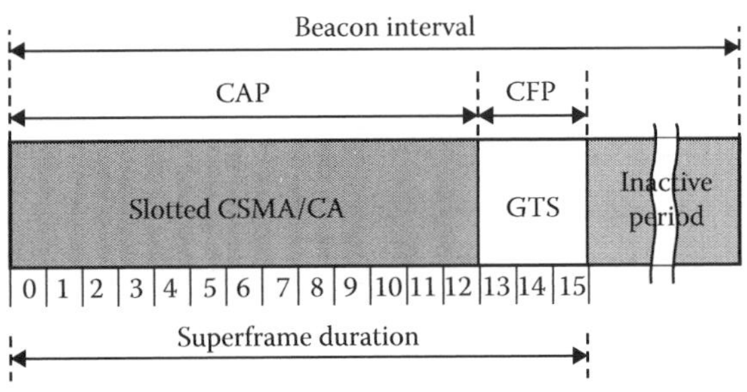

Figure 4.2 IEEE 802.15.4 MAC superframe structure.

contention-free period, and an inactive period. Nodes in the CAP use a slotted carrier sense multiple access with collision avoidance (CSMA/CA) to contention for channel and CAP containing a number of guaranteed time slots that can be allocated by the PAN coordinator to specific nodes. In the non–beacon-enabled mode, there are no regular beacons, but the coordinator may unicast beacons to a soliciting device. Communication among devices in the non–beacon-enabled mode uses unslotted CSMA for decentralized access.

The structure of this superframe is described by the values of macBeaconOrder (BO) and macSuperframeOrder (SO). The MAC PAN information based (PIB) attribute (BO) describes the interval at which the coordinator shall transmit its beacon frames. The superframe order is the variable that is used to determine the length of the superframe duration, which is divided into 16 time slots. Similarly, the beacon interval (BI) is determined by the variable BO.

4.1.2.1 Beacon-Enabled Mode

Because the time of the superframe duration cannot exceed the time of a BI, the condition for both parameters is $0 \le SO \le BO \le 14$. When BO is greater than SO, this indicates that there is an inactive portion present in the superframe. Also, for SO that is equivalent to BO, the BI is the same as the superframe duration indicating that there is no inactive portion.

The values of BO and the BI are related as follows [2]:

$BI = aBaseSuperframeDuration \times 2^{BO}$

$SD = aBaseSuperframeDuration \times 2^{SO}$

aBaseSuperframeDuration = 960 symbols = 15.36 ms, each time slot has a duration of 15.36/16 = 0.96 ms

4.1.2.2 Non–Beacon-Enabled Mode

For BO = SO = 15, it operates in a non–beacon-enabled mode. If BO = 15, the coordinator will not transmit beacon frames except when requested to do so, such as upon receipt of a beacon request command. The value of SO shall be ignored if BO = 15.

In CSMA/CA algorithms, the nodes apply the randomized binary exponential backoff process if the collision of transmissions, which start at almost the same time, occurs. This process works after it initializes the backoff exponent (BE) equivalent to the minimum backoff exponent (macMinBE), which is 3, and the number of backoffs (NB) equivalent to 0:

1. It waits for a backoff delay time, which is a random integer number between 0 and $2^{BE} - 1$ (i.e., 7)
2. If it still has collision or the channel is busy, both BE and NB are increased by 1 and wait for a backoff delay time between 0 and 15 ($2^{BE} - 1 = 2^4 - 1 = 15$)
3. If NB exceeds macMaxCSMABackoffs, which is 4, the protocol completes with a channel access failure, or repeats step 2 if it still has collision or if the channel is busy

Figure 4.3 shows data transmission from a device to the coordinator in a star topology, and Table 4.1 shows part of an NS-2 trace file in the beacon-enabled mode. Node 0 is the coordinator and node 1 is a FFD that sends constant bit rate (CBR) traffic to node 0. We set the values of BO and SO to 3, and so the value of BI is 15.36 ms × 2^3 = 122.88 ms. From Table 4.1, we can observe node 0 send BCN (beacon) every 122.88 ms (1.775232000 – 1.652352000).

Figure 4.4 shows data transmission from the coordinator to the device in a tree topology, and Table 4.2 shows node 1 sending an association request (CM1) to node 0. After node 0 receives the request and sends back an ACK, a connection is established. Then, node 1 sends a data request (CM4) and node 0 sends an ACK. Node 0 sends an association response (CM2) and node 1 sends back an ACK.

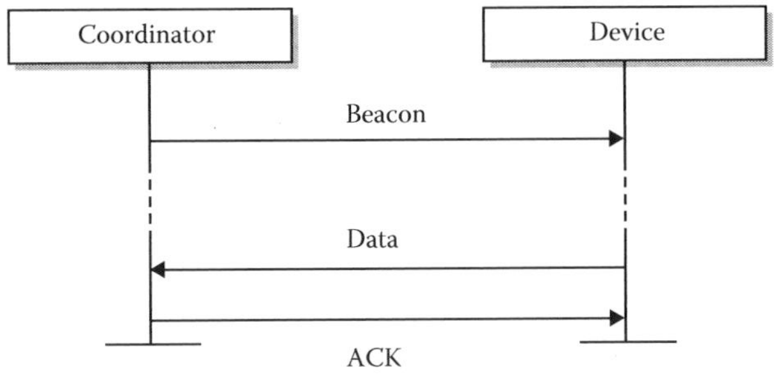

Figure 4.3 Data transmission from the device to the coordinator.

Table 4.1 Part of NS-2 Trace File (BO = SO = 3)

s 1.652352000 _0_ MAC— - 0 BCN 12 [0 ffffffff 0 0]
s 1.775232000 _0_ MAC— - 0 BCN 12 [0 ffffffff 0 0]
s 1.789120000 _1_ MAC— - 0 CM1 17 [0 0 1 0]
r 1.789856033 _0_ MAC— - 0 CM1 17 [0 0 1 0]
s 1.790272000 _0_ MAC— - 0 ACK 5 [0 1 0 0]
r 1.790624033 _1_ MAC— - 0 ACK 5 [0 1 0 0]
s 1.898112000 _0_ MAC— - 0 BCN 20 [0 ffffffff 0 0]

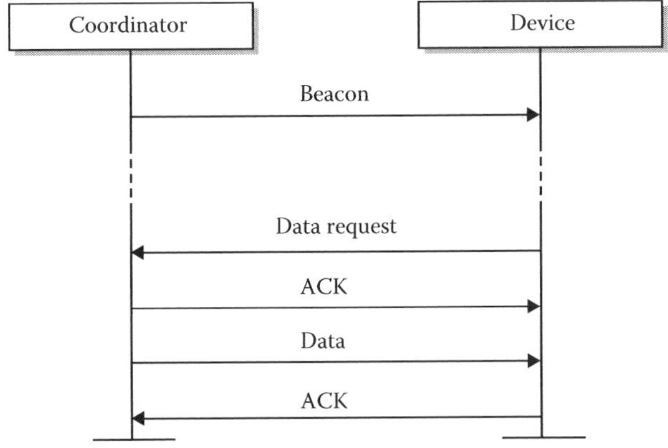

Figure 4.4 Data transmission from the coordinator to the device.

Table 4.2 Part of NS-2 Trace File (Node 1 Sends Data Request to Coordinator)

s 1.789120000 _1_ MAC— - 0 CM1 17 [0 0 1 0]
r 1.789856033 _0_ MAC— - 0 CM1 17 [0 0 1 0]
s 1.790272000 _0_ MAC— - 0 ACK 5 [0 1 0 0]
r 1.790624033 _1_ MAC— - 0 ACK 5 [0 1 0 0]
....
s 2.282144033 _1_ MAC— - 0 CM4 16 [0 0 1 0]
r 2.282848067 _0_ MAC— - 0 CM4 16 [0 0 1 0]
s 2.283072000 _0_ MAC— - 0 ACK 5 [0 1 0 0]
r 2.283424033 _1_ MAC— - 0 ACK 5 [0 1 0 0]
s 2.283712000 _0_ MAC— - 0 CM2 25 [0 1 0 0]
r 2.284704033 _1_ MAC— - 0 CM2 25 [0 1 0 0]
s 2.284896033 _1_ MAC— - 0 ACK 5 [0 0 1 0]
r 2.285248067 _0_ MAC— - 0 ACK 5 [0 0 1 0]

4.1.3 IEEE 802.15.4 Association Procedure

In the IEEE 802.15.4 standard, any sensor node has to scan the listed channels before it joins the network. Once the sensor node finds the network ID and the coordinator, then it would send an association request to the coordinator. If the association is successful, the sensor node can join the network via the coordinator. Otherwise, the sensor node has to rescan the listed channels, sending a new association request to find a new coordinator if the association procedure fails.

The following steps are the synchronization procedure. Briefly, the sensor node starts to track the beacon frames and then competes for the access channel in the superframe duration if it can receive the beacon frame. If a sensor node succeeds in completing the synchronization procedure, it can transmit data to the coordinator. The synchronization procedure fails when the sensor node cannot receive the beacon frame from its coordinator, and the sensor node would lose its connectivity. To regain connectivity, the sensor node must first check the number of lost beacon frames. If the number is lower than the threshold value "aMaxLostBeacon" (default is 4), the sensor node sends the synchronization request to its coordinator for regaining the connectivity. Otherwise, the sensor node would perform the reassociation procedure or start a new association. Figure 4.5 shows the IEEE 802.15.4 association procedure [2].

4.1.4 Effect of 3-D Terrains in WSNs

Several traditional routing protocols, such as AODV, DSDV, DSR, OSLR, and TORA were previously studied, and their performances were compared for two-dimensional (2-D) wireless networks. However, nodes in the real world are deployed in a 3-D space. The unrealistic 2-D terrain assumption, in which communication and sensing is not blocked because of neglected topographic formations, results in optimistic and unrealistic WSN performance. Figure 4.6 shows the data transmission from source node to the sink node in a 3-D realistic terrain. The source node

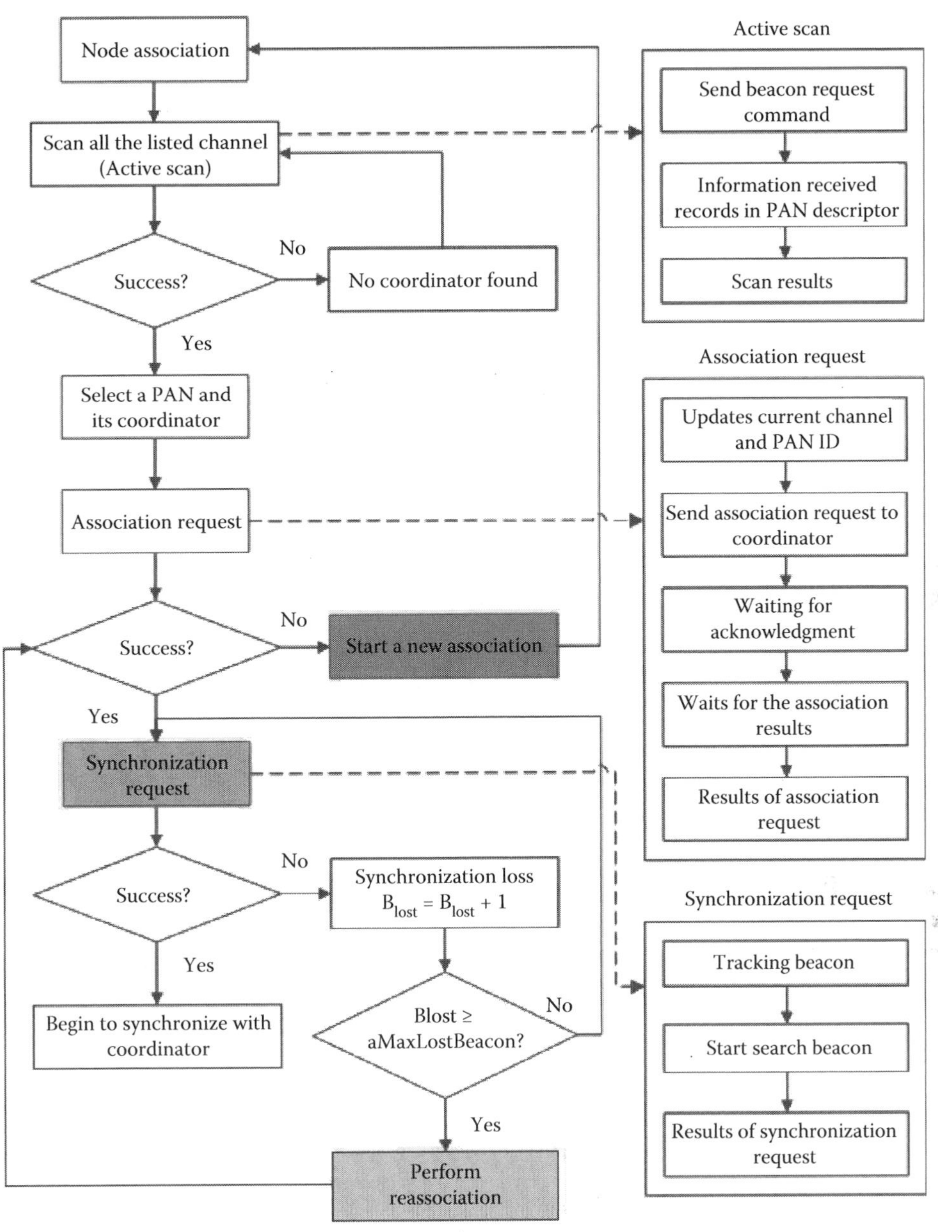

Figure 4.5 IEEE 802.15.4 association procedure [2].

might not transmit data within line of sight (LOS) because the transmission could be obstructed by hills, trees, or mountains.

Figure 4.7 shows an example of nodes displayed in a realistic terrain. Most currently proposed routing protocols in the literature assume a distance-based sensing and 2-D free space communications for a randomly deployed WSN scenario. However, randomly deployed sensors occur in 3-D terrains in a realistic world. The nodes' connections can be obstructed by the variety of terrain.

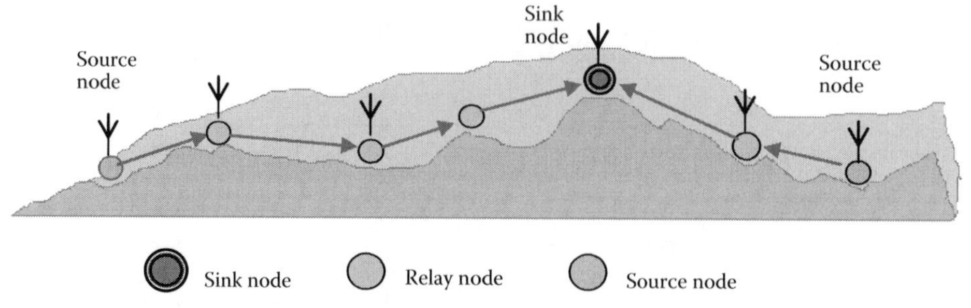

Figure 4.6 Data transmission on a 3-D terrain.

Figure 4.7 An example of connections on a 3-D terrain.

4.1.5 ZigBee Nodes for Internet of Things

The intrinsic characters of the Internet of things (IoT) are the Internet of everything, the Internet of services, and the Internet of networks. It can identify, trace, and control trillions of objects over networks. The IoT is expected to provide a resource fabric interfacing the physical world with a ubiquitously deployed substrate of embedded networked devices. The resources provided by the IoT include sensors, actuators, radiofrequency identification (RFID) tags and readers, near-field

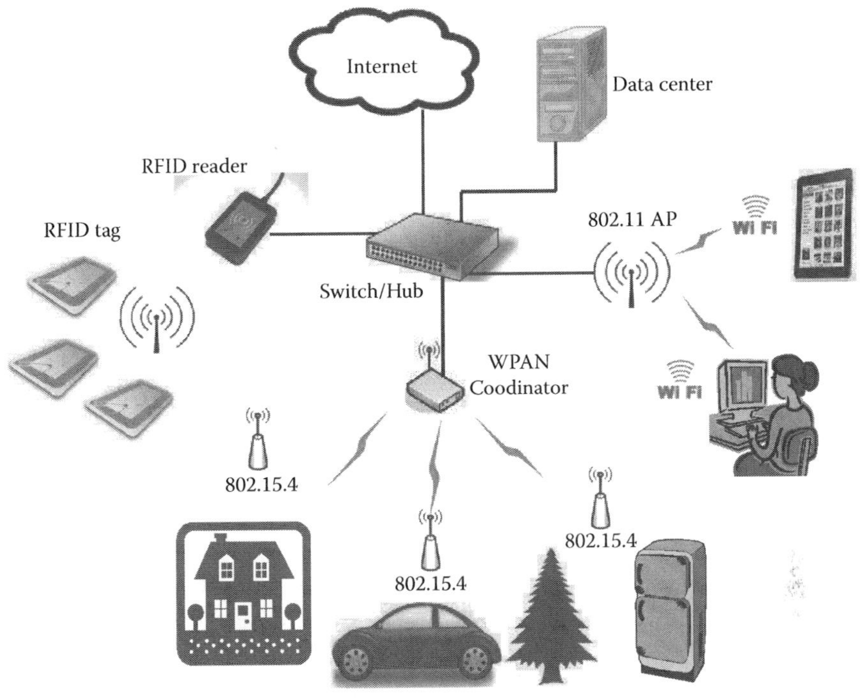

Figure 4.8 Structure of the IoT.

communication (NFC) enabled devices, etc. Some of these devices will have the ability to interact with or record information/events concerning the real world and the entities contained within. Future IoT is highly unified with networks, services, and the real world. Figure 4.8 shows the structure of the IoT. Everything is connected or sensed by ubiquitous sensors. WSNs represent a reasonably cheap sensory extension to Internet-connected devices; moreover, their computational capabilities allow for further flexibility of use and functional expansion [3,4].

ZigBee is the latest and most advanced wireless technology being built into millions of home automation and smart energy devices worldwide. Lights, thermostats, alarms, fridges, doors, appliances, utility meters—all are being ZigBee enabled, and this is why the technology is often referred to as "the Internet of things" [5].

4.2 Related Work

Many performance analyses or evaluations of the IEEE 802.15.4 MAC for LR-WPAN have been proposed. Buratti et al. [6,7] performed analyses of IEEE 802.15.4 non–beacon-enabled and beacon-enabled modes. The results showed how the distribution of traffic, goodput, and probability of success changes when different loads are offered. Pollin et al. [8] proposed a performance analysis of slotted carrier sense IEEE 802.15.4 medium access layer in a star topology network. They provided a detailed analytical evaluation of its performance for uplink and acknowledged uplink traffic. Koubaa et al. [9] analyzed the performance limits of the slotted CSMA/CA mechanism of IEEE 802.15.4 in the beacon-enabled mode for broadcast transmissions in WSNs. The performance was evaluated and analyzed for different network settings to understand the effect of protocol attributes on network performance.

Most of the current literature assumes a distance-based sensing and 2-D free space communication for a randomly deployed WSN scenario. However, in the real world, randomly deployed sensors occur in 3-D terrains. Ortiz et al. [10] investigated the study of 3-D networks and their behavior in crucial aspects, such as cover, connectivity, and energy consumption. Jain et al. [11] proposed an energy-aware beaconless geographic routing approach for 3-D WSNs. It considers the energy budget of nodes, excluding the distance, in the next hop forward in the selection process. Babaei et al. [12] proposed a new signal obstruction submodel used in realistic mobility models. It considers all obstacles in the environment in 3-D and simulates the obstruction of signals by 3-D obstacles. Vieira et al. [13] presented analytical results for link probability, node degree, and network coverage for 3-D MANETs assuming random uniform distribution. The approach of increasing duty cycle reduces energy consumption and increases network lifetime. Song et al. [14] proposed hybrid position-based 3-D routing algorithms for routing in 3-D environments. The proposed local hybrid algorithm combined greedy routing with adaptive least-squares projective (ALSP) face routing on projection planes. They demonstrated that this hybrid ALSP greedy-face-greedy (GFG) routing algorithm on static 3-D ad hoc networks could achieve nearly guaranteed delivery whereas discovering routes that are considerably closer in length to the shortest paths. Filiposka and Trajanov [15–17] presented Durkin's propagation model extension for the NS-2 network simulator to calculate the distance between nodes, which depends on all three coordinates. They revealed that this amount was more than tolerable when considering that the use of Durkin's propagation model was substantially closer to a more realistic environment. Lee et al. [18,19] presented the integration of immerse 3-D graphics contents as the representative of guidance information in real-time indoor environments. The designed system enables the tracking of nearby targets by using the same system within the WSN sensing range.

4.3 Background Knowledge

4.3.1 Durkin's Propagation Model

Radio channels are more complicated to model compared with wired channels. Their characteristics may change rapidly because of the variety in their environment (obstacle, terrain, etc.). The crucial components of a wireless network simulator include the mobility model, the signal propagation model, and the routing protocol. Most current wireless network simulators, for example, NS-2, OMNeT++, TOSSIM, and OPNET, use simplified propagation models such as the free space model, the two-ray ground model, the shadowing model, and the Ricean and Rayleigh fading models. Three propagation models are supported by NS-2: free space, two-ray ground, and shadowing [20]. The free space propagation model assumes the ideal propagation condition that there is only one clear LOS path between the transmitter and the receiver. It is used to predict received signal strength when the transmitter and receiver have a clear, unobstructed LOS path between them. The two-ray ground reflection model considers both the direct path and a ground reflection path between the transmitter and the receiver. This model has been found to provide a more accurate prediction at long distances compared with the free space model. The received power at distance is predicted by

$$P_r(d) = \frac{P_t G_t G_r h_t^2 h_r^2}{d^4 L}$$

where

P_t, the transmitted power
P_r, the received power
G_t, the antenna gains of the transmitter
G_r, the antenna gains of the receiver
d, the distance between transmitter and receiver
h_t, the heights of the transmit antennas
h_r, the heights of the receive antennas
L, the system loss, the original equation assumes $L = 1$

The algorithm then selects the smaller value as the appropriate received power for the terrain profile. If the profile is within the LOS of the inadequate first Fresnel zone clearance, there is an additional diffraction loss that is added (in decibels) to the appropriate received power according to the approximate solution [21].

Despite many simulators supporting various propagation models, not all of them are suitable for the 3-D space. However, radio transmission in a mobile communications system often takes place over irregular terrain. Therefore, the terrain profile of a particular area needs to be taken into account when estimating the path loss because the transmission path between the transmitter and the receiver can vary from simple LOS to one that is severely obstructed by hills, trees, or mountains. In 1969, Durkin [22,23] proposed a computer simulator for predicting field strength contours over irregular terrain. It assumes that the receiving antenna receives all of its energy along that radial, without considering multipath propagation. Durkin's model is based on one of the basic mechanisms of radiopropagation, that is, diffraction. Diffraction allows radio signals to propagate around the curved surface of the earth, and to propagate behind obstructions. It uses LOS and diffraction from obstacles along the radial, and excludes reflections from other surrounding objects and local scatterers. Path loss estimation consists of two parts: the first part of the algorithm addresses a topographic digital elevation model (DEM) file turned into a topographical database and reconstructs the ground profile information along the path between transmitter and receiver. The second part of the algorithm calculates the expected loss along that path.

A function of the Fresnel–Kirchoff diffraction parameter ν, is defined as

$$\nu = h\sqrt{\frac{2(d_1 + d_2)}{\lambda d_1 d_2}}$$

where

h, the relative height of the obstruction
d_1, distanced from the obstacle to the transmitter
d_2, distanced from the obstacle to the receiver
λ, the radio signal wavelength

If the terrain profile failed the first Fresnel zone test, then there are two possibilities: non-LOS and LOS, but with inadequate first Fresnel zone clearance. Fresnel zones represent successive regions that have the effect of alternately providing constructive and destructive interference to the total received signal. It is critical if an unobstructed transmission path is to be approximated by at least 60% of the zone radius [21]. Figure 4.9 shows the Fresnel zone.

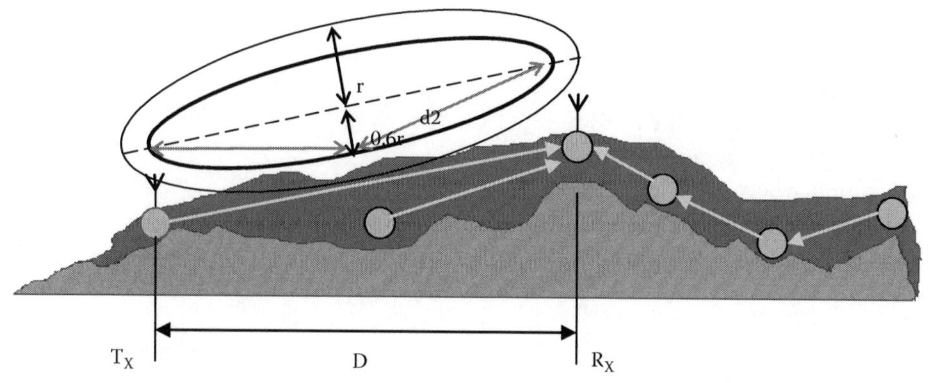

Figure 4.9 Fresnel zones.

$$Fn = \sqrt{\frac{n\lambda d_1 d_2}{d_1 + d_2}}$$

Fn, *n*th Fresnel zone
λ, the wavelength of the transmitted signal in meters
n, Fresnel zone number
d_1, the distance of *P* from one end in meters
d_2, the distance of *P* from the other end in meters

We use Durkin's propagation model for the NS-2 simulator to investigate more realistic simulations and analyze the effect of terrain profiles on the MANET and WSN performances. The antenna height (in 3-D space), the carrier frequency of the radio signals, and node position in 3-D space are input parameters to the propagation model. The overall node elevation is the sum of terrain elevation and antenna height. The authors also modified the CSMA/CA channel in NS-2 to determine the presence of the terrain. The current design of the NS-2 wireless channel considers the square shape of the physical carrier sense range of the transmitting node. Hence, the transmission and physical carrier sensing ranges of the signal propagation of the nodes are not ideal circles, but depend on the terrain configuration. Figure 4.10 shows the distance between two nodes in a 3-D space.

Assign the coordinates of two nodes N_1 and N_2, (x_1, y_1, z_1) and (x_2, y_2, z_2), and make the distance *d*, the actual distance used as the transmission and sensing ranges to calculate the RSSI value.

$$d = \sqrt{(x_2 - x_1)^2 + (y_2 - y_1)^2 + (z_2 - z_1)^2}$$

Although many algorithms and models have been proposed to improve the routing in 3-D space, these models still lack performance evaluation for IEEE 802.15.4 under different 3-D terrains. Our target was to investigate the evaluations combining the Durkin's propagation model and IEEE 802.15.4 MAC. This dissertation is based on Durkin's propagation model extension for the NS-2 network simulator, which was presented by Filiposka and Trajanov.

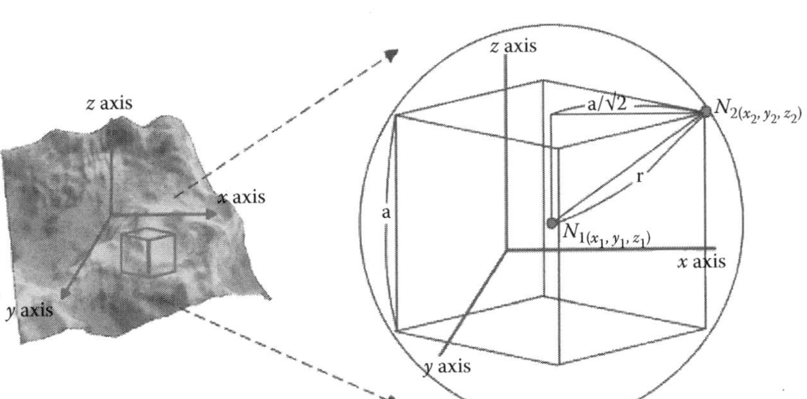

Figure 4.10 Distance between two nodes in a 3-D terrain.

4.3.2 *The Digital Elevation Model*

A DEM is a digital file consisting of terrain elevations for ground positions at regularly spaced horizontal intervals. It represents only height information without any further definition about the surface [24,25]. The DEM files can be used to determine the morphology of the ground, slopes, and other aspects of the terrain. The DEM format is an ASCII flat file that is organized into three parts (records), as follows: A, B, and C. The A record defines DEM characteristics, including name, borders, units, minimal and maximal levels, projection parameters, and number of B records. The B record contains the elevation data. Each elevation is an integer. The DEM file is read from west to east; however, profile elevations are from south to north. The C record contains the root-mean-squared error (RMSE) quality control statistical data about the accuracy of the DEM file. Each terrain data set consists of a set of vertices and edges. The vertices are represented by three Cartesian coordinates (x, y, z), with the x and y coordinates evenly distributed at unit intervals apart. Although the x and y coordinates are distributed in an even grid, the z coordinate (altitude) can assume any value, either positive or negative. We used terrain data as the simulation terrains through USGS 7.5-min DEM files, which is supported by Filiposka and Trajanov. It enabled us to place and move the ad hoc network nodes in an irregular 3-D terrain, defined by a DEM file that held the digitalized elevation values for a specified terrain [26,27].

4.3.3 *3DEM Software*

3DEM is a terrain visualization software [28]. It offers the ability to merge multiple DEMs to provide 3-D terrain images and flyby animations. It has the capacity to produce realistic terrain by inputting various topographic data files containing rows and columns of elevation data or random numbers, and outputting high-resolution overhead maps and 3-D projections of large areas. It can render 3-D terrain scenes and MPEG flyby animations from the following data sources:

- USGS DEM files
- USGS Spatial Data Transfer Standard (SDTS DEM) files
- USGS National Elevation Dataset (NED) files
- LIDAR Point Cloud (LAS) files

- XYZ Point Cloud files (ASCII or floating point binary)
- USGS Global 30 Arc Second Elevation Data Set (GTOPO30 DEM) files
- NOAA Global Land One-km Base Elevation (GLOBE) files
- GEOTIFF DEMs
- NASA Mars Orbiter Laser Altimeter (MOLA) files
- NASA Mars Orbiter Laser Altimeter (MOLA POLAR) files
- Any topographic data file organized by rows and columns of elevation data

3DEM uses the SGI/Microsoft OpenGL libraries for high-speed 3-D rendering. 3DEM will render 24-bit color 3-D projections, red–blue projections requiring red–blue 3-D glasses for viewing, or color side-by-side stereo 3-D projections. 3DEM scenes can be saved in the following formats:

- Windows bitmap (*.bmp, *.dib)
- Joint Photographic Experts Group (*.jpg, *.jpeg)
- Tagged image file (*.tif, *.tiff)
- Flyby animation AVI (*.avi)
- Flyby animation MPEG (*.mpg, *.mpeg)
- USGS ASCII digital elevation model (*.dem)
- GEOTIFF digital elevation model (*.tif)
- Binary terrain matrix (*.bin)
- VRML world (*.wrl)
- Terragen terrain (*.ter)

Figure 4.11 presents the 3DEM software. Figure 4.11a shows user can select the file type that corresponds to the terrain data. Figure 4.12 illustrates the variety of height in different terrains. For testing purposes, Filiposka and Trajanov provided the following artificial terrains: flats, hillsides, hills, ravines, and pyramids. Their dimensions were 1000 × 1000 m, with the highest relative point being 200 m.

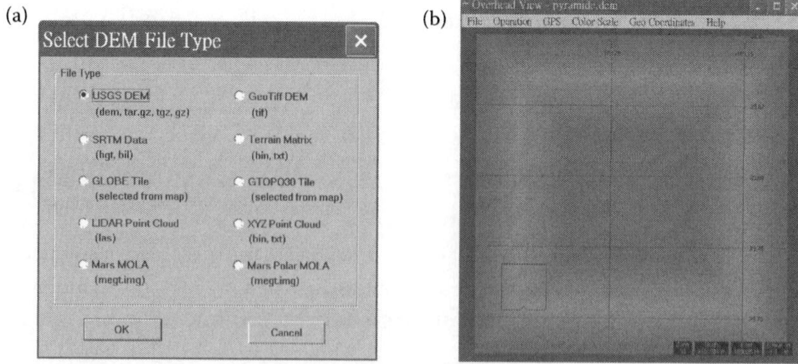

Figure 4.11 3DEM software. (a) Various file types can be loaded in 3DEM. (b) Loading "pyramid" terrain DEM file into 3DEM.

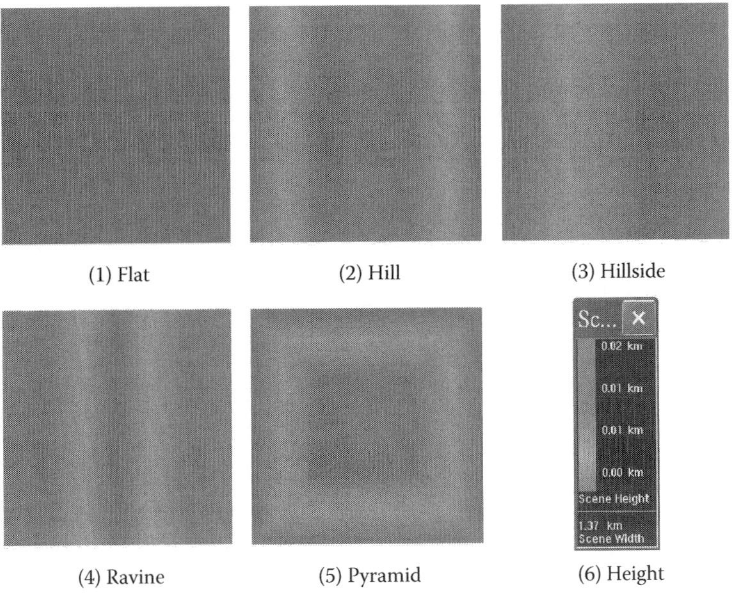

Figure 4.12 Five artificial terrains and heights.

4.4 Simulation Model for 3-D ZigBee Sensor Networks

The crucial components of a wireless network simulator include the mobility model, the signal propagation model, and the routing protocol. In this chapter, we introduce Durkin's propagation model and the DEM used for 3-D terrain in ZigBee sensor networks. This chapter also describes the proposed simulation model and presents many comprehensive simulation results. We investigate the performance of WSNs on different 3-D terrains.

4.4.1 Simulation Model in a 3-D Terrain Model

Our target was to investigate the performance of networks combining Durkin's propagation model and the IEEE 802.15.4 MAC adapted scheme. Figure 4.13 shows a model simulating ZigBee sensor networks in different 3-D terrains (the gears represent the utility tools used in this model). Although many algorithms and models have been proposed to improve the routing in 3-D space, they still lack the performance measurement of IEEE 802.15.4 MAC on different 3-D terrains. The following steps have been adopted:

1. The "*cbrgen.tcl*" connection pattern file generator, which is contained in an NS-2 independent utility, is used for generating CBR connections. The following parameters are shown:

   ```
   $ns cbrgen.tcl [-type cbr|tcp] [-nn nodes] [-seed seed] [-mc
   connections] [-rate rate]
   ```

   For example,

   ```
   $ns cbrgen.tcl -type cbr -nn 100 -seed 1 -mc 1 -rate 10.0 >
   connections.tcl
   ```

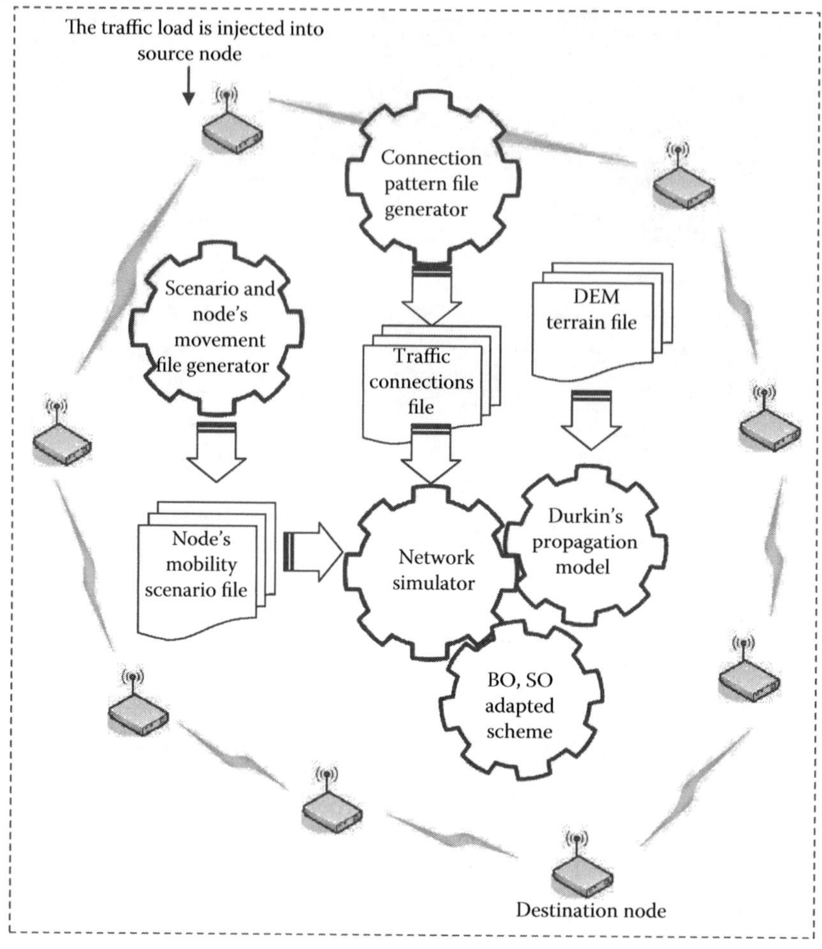

Figure 4.13 Simulation model in a 3-D terrain.

2. We used "*setdest.cc*" scenario and node-movement file generator to generate mobility scenario file.

```
$./setdest [-n nodes] [-p pause time] [-M maximum speed] [-x x
dimension of space] [-y y dimension of space] > mobilityScen.tcl
```

   For example,

```
$./setdest -n 100 -p 0.0 -M 10.0 -x 1000 -y 1000 > mobilityScen.tcl
```

3. The DEM terrain file containing the shape model of a 3-D terrain is imported into NS-2 TCL script file. Durkin's propagation model calculates the communication channel between nodes and cooperates with NS-2 simulator. The following shows how to set the radiopropagation model as Durkin's in NS-2 TCL script.

```
set val(prop) Propagation/Durkins ;# radio-propagation model
Propagation/Durkins set percent_ 60
Propagation/Durkins set nodes_ $val(nn) ;# number of nodes
```

4. Running NS-2 simulator to generate the trace file. NS-2 has two options to generate traces with different formats. We use $ns use-newtrace to generate the new trace format in this dissertation.
5. Analysis and statistics of the results. After generating a trace file from running an NS TCL script, we are able to analyze it without the use of any tools. We use a set of AWK scripts to get node-to-node statistics including packet delivery ratio (PDR), goodput, end-to-end delay, and energy consumption.

4.4.2 BO and SO Adapted Scheme

We can adapt the BO and SO values to change the active period in IEEE 802.15.4 MAC according to the intrinsic characters of different terrains. How to correctly decide the BO and SO values according to the different kinds of traffic loads and varying terrains is the target in this study.

We describe the BO and SO adaptation state of a coordinator node as follows:

State 0: Packet transmission begins.

State 1: The coordinator initializes the BO and SO values. Then, the coordinator begins to broadcast beacons. User can set the initial BO and SO value according to different terrains.

State 2: If the traffic load is light, the system is in a stable state. If the traffic load becomes heavy as the mobile node moves over a variable terrain, the system moves to State 3. If the transmission is finished, the system moves to State 6.

State 3: In this state, BO and SO values are increased by one to enlarge the BI and superframe duration. Then, it moves to State 4 and decreases BO and SO values. If the traffic load is getting light, the system moves to State 2. The system tries to find the optimal BO and SO values to adapt the variety of terrains.

State 4: In this state, the BO and SO values are decreased by one to shorten the BI and superframe duration. If the traffic load is still light and the SO value is still greater than zero, the SO value is decreased by one continually. BO and SO are adapted to find the optimal values for successfully transmitting the data over the variable terrain. If the traffic load is stable, the system moves to State 2. If the traffic load is getting heavy, the system moves to State 3. If BO and SO values reach zero, the system returns to State 5.

State 5: In this state, BO and SO values are set to BO_max and SO_max values. If the traffic is light, it moves to State 4. If the traffic load is stable, the system moves to State 2. The default BO_max and SO_max values are 15 in IEEE Std. P802.15.4a/D5. The user can set the BO_max and SO_max values according to different terrains. Longer BO_max values mean longer BIs. Similarly, the longer SO_max values mean longer superframe durations. BO_max and SO_max must be less than or equal to 14 in the beacon-enabled mode. When the BO_max and SO_max values are set to 15, the system will operate in a non–beacon-enabled mode.

State 6: Transmission is finished in this state.

4.5 Performance Evaluation

The performance metrics measured in this chapter are the following: PDR, network goodput, average end-to-end delay, and residual energy of the sink node. The sink nodes' residual energy affect the system's lifetime. We investigate which combination of BO and SO values can maximize the longest sink nodes' lifetime.

(1) Packet Delivery Ratio

Packet loss may occur at any stage of a network transmission, mainly due to link failures and CSMA/CA channel access mechanisms. It is an important metric that can be used as an indicator of a congested network. We use PDR to denote the performance. The PDR is only considered for data packets in this study.

$$\text{Packet delivery ratio} = \frac{\text{(number of received packets)}}{\text{(number of transmitted packets)}} \times 100$$

(2) Network Goodput

Network goodput is the average rate of successful data delivery over a communication channel. In our simulation, this metric only measures the total data goodput over the network, ignoring all other overhead. The goodput of a node is measured by first counting the total number of data packets successfully received at the node and computing the number of bits received, which is finally divided by the total simulation runtime.

$$\text{Goodput of a node} = \frac{\text{(total data bits received)}}{\text{(simulation runtime)}}$$

The goodput of the network is finally defined as the average of the goodput of all nodes involved in data transmission.

$$\text{Network goodput} = \frac{\text{(sum of goodput of nodes involved in data transmission)}}{\text{(number of nodes)}}$$

(3) Average End-to-End Delay

Average end-to-end delay is one of the most important metrics of emergent events. In WSNs, the end-to-end delay is the total time delay to deliver a packet from source to sink node. It is the sum of delays at all links within the end-to-end path. The delay at an intermediate node usually includes the following components: processing delay, queuing delay, transmission delay, propagation delay, and retransmission delay. We mainly consider the average end-to-end delay for all source traffic along a multihop path to sink node. By decreasing the packet retransmission time, we can decrease the average end-to-end delay.

$$\text{Packet delay} = T_r(d) - T_t(s)$$

where

$T_r(d)$, the received time at destination node
$T_t(s)$, the transmitted time at source node

$$\text{Average delay} = \frac{\text{(sum of all packet delays)}}{\text{(total no. of received packets)}}$$

(4) Energy Consumption

The coordinator consumes energy when it transmits the beacon and ACK packets, receives data packets, and listens to the channel. Where rxPower is power consumption in receiving a packet, txPower is power consumption in transmitting a packet, sleepPower is power consumption in sleep state, and idlePower is power consumption in idle state. The initial energy and the final energy left in the node at the end of the simulation run are measured. To measure the energy consumption in our scenarios, we use the energy model in NS-2.

We measure the energy consumption of all nodes in beacon-enabled/non–beacon-enabled mode with two metrics listed as follows. The percentage of energy consumed by a node is calculated as the energy consumed to the initial energy.

(1) *Network Residual Energy Ratio.* This ratio is measured by the total residual energy of all nodes divided by the total initial energy of all nodes at the end of the simulation time. We can use this metric to compare the energy consumption in different terrains.

$$\text{Network residual energy ratio} = \frac{\sum_{i-1}^{\text{nn}} E_{\text{r}}(i)}{\sum_{i=1}^{\text{nn}} E_{\text{i}}(i)}$$

where

nn, the total node's number

$E_{\text{r}}(i)$, the residual power of node i

$E_{\text{i}}(i)$, the initial power of node i

(2) *Active Path Energy Ratio.* We defined the active path energy ratio as being equal to the total nodes' residual energy in the path from the source node to the sink divided by the total nodes' initial energy in the path. We can use this metric to compare the energy consumption of all flows in different terrains in the paths.

$$\text{Active path energy ratio} = \frac{\sum_{i \in s} E_{\text{r}}(i)}{\sum_{i \in s} E_{\text{i}}(i)}$$

where

S, the set of all nodes in the path

$E_{\text{r}}(i)$, the residual power of node i

$E_{\text{i}}(i)$, the initial power of node i

The coordinator consumes energy when it transmits beacons and ACK packets, receives data packets, and listens to the channel. The initial energy and the final energy left at the end of the simulation are measured. To measure the energy consumption in our scenarios, we use the default energy model in NS-2.

4.5.1 Network Configuration and Assumptions

The solution performance evaluation is carried out under the NS-2 simulator. We use the NS-2 module developed by Zheng and Lee [29] for IEEE 802.15.4 (NS-2.31) and Durkin's propagation model developed by Filiposka and Trajanov in our simulation. For the simulation setup, a random waypoint (RWP) mobility model combined with Durkin's propagation model was used.

Figure 4.14 illustrates the network topology with 100 FFD nodes and one PAN coordinator (node 100). Each sensor node is set 50 m away from the other nodes. The sink node is a coordinator that is placed at the center of this scenario (we assume that the nodes do not affect each other). Maximum node transmission range is 200 m. However, in the presence of obstructions, the actual transmission range of each individual node is likely to be limited.

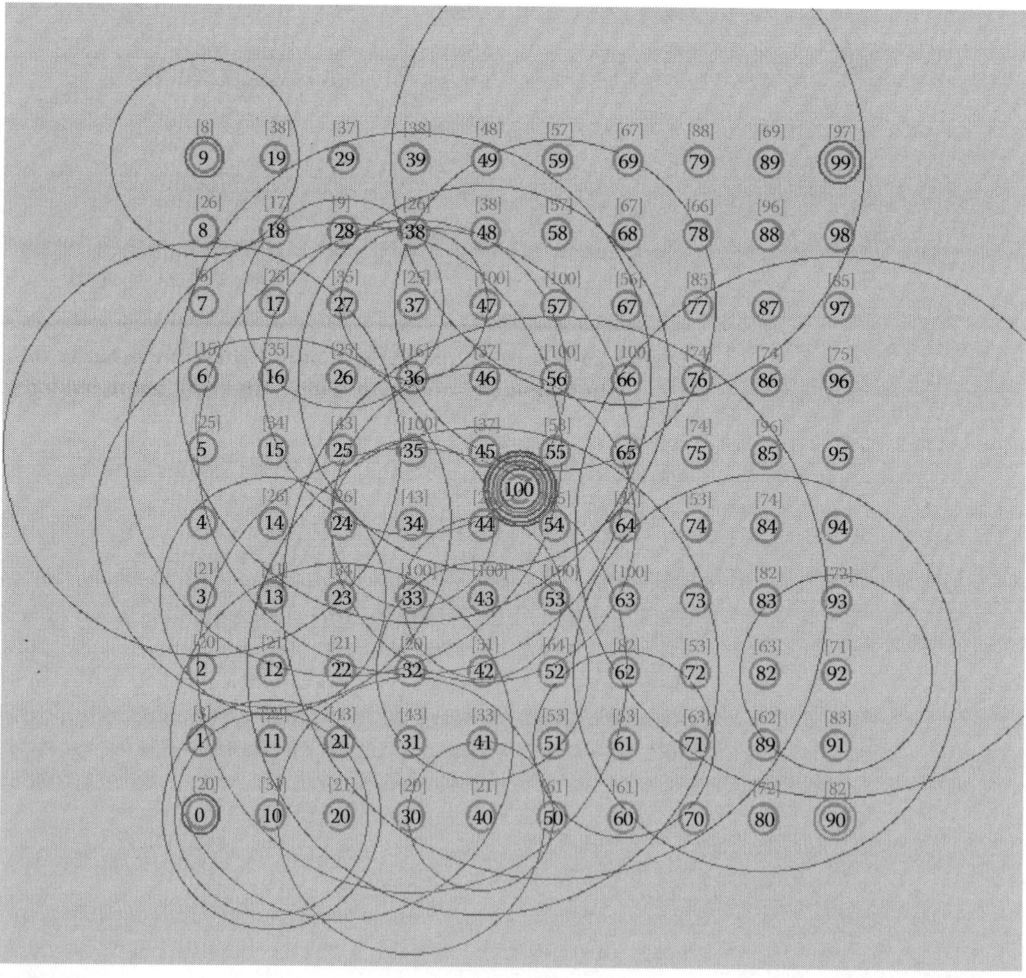

Figure 4.14 Simulation topology.

Four CBR traffic flows (CBR0–CBR3) start from four vertexes to sink node (node 100). Node 0 sends CBR0 traffic from 25 to 100 s. Node 9 sends CBR1 traffic from 30 to 105 s. Node 90 sends CBR2 traffic from 35 to 110 s. Node 99 sends CBR3 traffic from 40 to 115 s. Each node's queue size is set to 50 packets. All nodes set the same BO, and thus values are from 0 to 9 in simulation 1 whereas the fixed BO and SO values are set to 3 in simulation 2. To investigate the performance of IEEE 802.15.4 WSNs, we studied various metrics including operation modes, terrain types, and traffic loads. Simulation duration was set to 125 s, the first 25 s of which allowed the nodes to associate with the PAN coordinator, and the remaining time was used for sending application traffic.

We conclude these assumptions as follows:

- The physical layer consists of IEEE 802.15.4 compliant radio transmitter (tx) and receiver (rx), which operate in the ISM band at 2.4 GHz, with raw data rate at 250 kbps. The modulation technique is called quadrature phase shift keying (QPSK).
- Ignoring the hidden node and exposing node problems.
- Ignoring the interference problem.
- The MAC sublayer implements the slotted/unslotted CSMA/CA.
- The application layer includes four CBR traffic sources with different data rates and one sink in this article.
- The antenna height is set to 2 m.
- The proposed values for the frequencies and antenna height equalize the conditions for the simulations with the Fresnel propagation model compared with the statistical propagation models.

The simulation parameters are summarized in Table 4.3.

4.5.2 Simulation Results

4.5.2.1 Simulation 1

In this simulation, we investigate the effects of different BI and superframe duration in the IEEE 802.15.4 MAC superframe in a 3-D pyramid terrain. From the IEEE 802.15.4 MAC superframe structure, the BO and SO values determine the time between two successive beacons and the length of the superframe, respectively. Many simulations are run by fixing SO and adapting BO or by fixing BO and adapting SO. When the BO value is small, the BI is short. It results in the probability of collision increase. When the BO is fixed, higher SO values have lower average end-to-end delay because the inactive period decreases. However, from repeated experiments, we found that some simulation results were not useful because of the variable simulation conditions such as terrain, traffic loads, and BI. We set the BO to an equivalent SO value to keep the duty cycle at 1. Sensor nodes did not change into sleep mode when BO was equivalent to SO.

4.5.2.1.1 Effect of Various BO and SO Value Settings

Table 4.4 shows a comparison of default BO and SO values with pyramid terrain in the PDR, goodput, average end-to-end delay, and energy consumption metrics (the simulation parameters are listed in Table 4.3). In this simulation, two data flows are sent from the source to the sink. Traffic load is one packet per second. We observed that when setting BO = SO = 4, the maximal

Table 4.3 Simulation Parameters

Name	*Setting Values*
Simulation tool	NS-2.31
MAC layer protocol	802.15.4
Network dimensions	1000 m ⊆ 1000 m ⊆ 200 m
Number of nodes	Simulation 1, simulation 2: 1 static coordinator, 100 static nodes Simulation 3: 1 static coordinator, 100 mobile nodes
Propagation model	Durkin's model
Terrain	Flat, hillside, hill, ravine, pyramid
Traffic type	CBR
BO and SO values	Simulation 1: BO, SO is setting from 0 to 9 Simulation 2: BO = 3, SO = 3 Simulation 3: BO = 5, 4, SO = 4, 3
Traffic loads	Simulation 1: 1, 20 packets/s Simulation 2: 20, 40, 60, 80, 100 packets/s Simulation 3: 2 packets/s
Data bit rate	250 kbps
Node's moving speed	1–5 m/s
Packet size	100 bytes
Routing protocol	AODV
Transmission range	200 m
Simulation time	125 s
Node's initial energy	10 J
T_X, R_X, idle, sleep power	56.4 mW, 15.138 mW, 1.278 mW, 0.045 mW

PDR value of 98.67% was retrieved. When BO is set to 4, it means BI is 15.36 ⊆ 24 = 245.76 ms. This period is divided by 16 time slots. Therefore, each time slot has a duration of 245.76/16 = 15.36 ms. If the BO is too small, the data packet cannot be transmitted in a time slot, resulting in a loss of PDR. If the BO is too large, nodes that want to transmit data must wait for a long time, and this also results in the loss of PDR.

The residual energy of an active path is equal to the total nodes' residual energy in the path from the source node to the sink divided by the total nodes' initial energy in the path. The residual energy rate of the active path was 97.4938%. Meanwhile, we observed that the residual energy of

Table 4.4 Performance of Various BO and SO Settings (Two Flows, Traffic Load is 1 Packet/s)

BO = SO	*PDR (%)*	*Goodput (kbps)*	*Average Delay(s)*	*Network Residual Energy Ratio (%)*	*Active Path Energy Ratio (%)*	*Residual Energy (J)*
0	50.34	0.581573	0.046024	91.4851	92.585	9.256968
1	48.99	0.555442	0.007316	93.4207	95.3636	9.549977
2	51.28	0.620207	0.040316	95.709	96.8331	9.689052
3	92.31	0.800635	1.256612	96.8387	96.7689	9.74593
4	98.67	0.928719	0.212432	97.6154	97.4938	9.779229
5	79.08	0.757318	0.756995	97.9869	97.9282	9.801046
6	52.03	0.580196	0.006147	98.234	98.2388	9.818893
7	50.67	0.576309	0.00601	98.2897	98.2829	9.822346
8	34.44	0.419918	6.585134	98.3656	98.3357	9.828471
9	0	0	—	98.4147	98.4455	9.844393

node 100 was 9.779229 J when BO and SO are set to 4. It is better than BO and SO values from 0 to 3. In addition to the residual energy of node 100, total nodes' residual energy rate and residual energy rate of the active path are also better than the others. Because the transmitted and received power are set very low, the residual energy of node 100 is still large when the simulation ends. When BO and SO values are greater than 5, it seems the total nodes' residual energy rate, residual energy rate of active path, and residual energy of node 100 are greater than BO = SO = 4. Despite their high residual energy, they retrieve low PDR values. The reason is that data packets cannot be transmitted successfully from the source node to the sink node, thus less transmission power and received power are consumed.

4.5.2.1.2 Effect of Various Data Flows

In this experiment, we increase the number of data flows from 2 to 4. Table 4.5 shows that when the data flow increases, more connections intend to content the time slot. The number of collisions increase resulting in PDR value decreases. We observed that when setting BO = SO = 4, the maximal PDR value is 98.35%. It still retrieves the highest PDR value compared with other BO and SO settings. The higher the PDR value is, the better the goodput.

4.5.2.1.3 Effect on Lifetime of Sink Node

For accelerating the power consumption to observe the lifetime of a sink node, we amplify the transmission power and receiving power by 100 times and increase the traffic load to 20 packets/s. From Table 4.6, we can observe that the residual energy of sink node 100 has an optimal lifetime. It runs out at 106.099392 s when setting BO and SO values to 3. Choosing larger BO and SO

Table 4.5 Performance of Various BO and SO Settings (Four Flows, Traffic Load is 1 Packet/s)

BO = SO	*PDR (%)*	*Goodput (kbps)*	*Average Delay(s)*	*Network Residual Energy Ratio (%)*	*Active Path Energy Ratio (%)*	*Residual Energy (J)*
0	24.5	0.566509	0.0464	90.8282	93.7077	9.268391
1	25.5	0.58502	0.007361	93.393	95.681	9.54864
2	72.04	1.501975	1.5283	95.3	94.8575	9.653263
3	75.91	1.446006	0.406976	96.7509	96.5539	9.723979
4	98.35	1.884508	0.149885	97.4402	97.28	9.75272
5	76.71	1.409687	0.916082	97.8422	97.7328	9.782332
6	32.3	0.670916	0.626245	98.1854	98.1219	9.814575
7	25.08	0.572651	0.006027	98.1798	98.2741	9.820714
8	17.11	0.414775	6.613428	98.3009	98.3387	9.827996
9	0	0	—	98.3928	98.4335	9.843446

values does not increase the lifetime. Overall, the average end-to-end delays are higher than the two data flows because more data connection content in the channel results in an increase in collisions. An increase in the number of data packet retransmissions results in long average end-to-end delays. Similarly, the PDR decreases rapidly when traffic load increases. An optimal PDR is retrieved from setting BO and SO to 4. Despite their high residual energy, they retrieve low PDR. The reason behind the long lifetime of node 100 (such as BO = SO = 1, BO = SO = 3) is that data packets cannot transmit successfully from source node to sink node; therefore, less transmission power and received power are consumed.

4.5.2.2 Simulation 2

In this simulation, we evaluate the performance in the beacon-enabled and non–beacon-enabled modes.

4.5.2.2.1 Effect of Various Terrains on PDR

Simulation time is 125 s and queue size is 50 packets. For each configuration, we vary the interarrival times of the flows in the source node to obtain different offered loads, assuming a constant packet size. Figure 4.15 shows the PDR using different terrains with different traffic loads varying from 20 to 100 packet/s. Generally speaking, PDR is not high and, with the traffic load, increases will follow the decline in PDR. Although the PDR in the beacon-enabled mode is lower than in the non–beacon mode, the PDR is more stable than in the non–beacon mode because

Table 4.6 Performance of Various BO and SO Settings (Four Flows, Traffic Load is 20 Packet/s)

BO = SO	*PDR (%)*	*Goodput (kbps)*	*Average Delay(s)*	*Network Residual Energy Ratio (%)*	*Active Path Energy Ratio (%)*	*Residual Energy (J)*
0	0	0	—	31.5408	66.4959	0 J (run out at 81.277632 s)
1	3.63	11.015486	0.006938	0.4043	4.5266	0 J (run out at 108.126912 s)
2	16.15	18.357373	0.920007	0	0	0 J (run out at 46.441152 s)
3	7.72	12.256902	0.715166	1.8286	2.435	0 J (run out at 106.099392 s)
4	52.28	14.666157	0.522085	10.1415	6.3753	0 J (run out at 45.273792 s)
5	36.41	11.575479	0.566132	39.6867	17.6214	0 J (run out at 51.943929 s)
6	33.91	11.302758	0.005917	73.4605	58.7317	0 J (run out at 55.509369 s)
7	33.1	11.261729	0.07582	73.9644	58.4981	0 J (run out at 56.410489 s)
8	3.72	1.655324	3.350338	85.4476	85.197	7.035674 J
9	0	0	—	86.7999	91.804	9.174113 J

of synchronization. Every device in the network should compete for the channel when it is ready to transmit data. Because the beacon mode and non–beacon mode are all adapted CDMA/CA mechanisms, the packets will be sent to a time slot. On the contrary, even though there are fewer collisions in the beacon mode, it still has higher PDR values.

Figure 4.15a shows that the PDR values are not stable with varying terrain in the non–beacon-enabled node. In Figure 4.15b, the PDR value is 18.45% in the flat terrain and 10.92% in the hilly terrain when traffic load is 20 packets/s. When traffic load increases to 100 packets/s, the PDR value decreases to 2.93% in the flat terrain and to 1.1% in the hilly terrain. The results show that PDR values in the flat terrain will always be higher than in the hilly terrain because data transmission on flat terrains has no obstacles.

4.5.2.2.2 Effect of Various Terrains on Goodput

Figure 4.16 shows the goodput using different terrains. It shows that goodput differs violently with varying terrain, except for the flat terrain in the non–beacon-enabled node. In Figure 4.16, the goodput is quite stable in the flat terrain in the beacon-enabled node, but the others are volatile

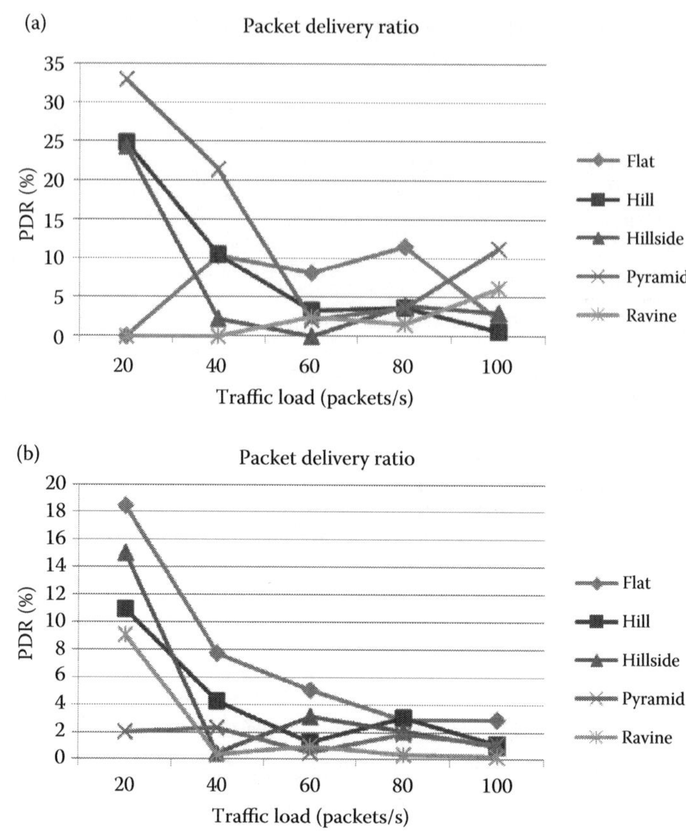

Figure 4.15 PDR versus traffic loads (from 20 to 100 packets/s). (a) Non–beacon-enabled mode. (b) Beacon-enabled mode.

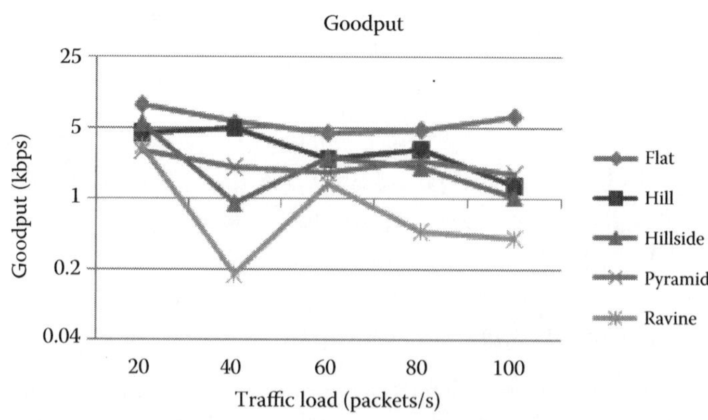

Figure 4.16 Goodput versus traffic loads in beacon-enabled mode (from 20 to 100 packets/s).

in other terrains. Especially in the ravine terrain, the goodput is very low in the beacon-enabled mode. For higher offered loads, the total goodput is decreased for each terrain type. Because the traffic load becomes heavier, it results in decreased PDR values.

4.5.2.2.3 Effect of Various Terrains on Average End-to-End Delay

Figure 4.17 shows the average end-to-end delay using different terrains. This measurement includes the latency for selecting routes. As the traffic load gets heavier, the delay increases in the non–beacon-enabled mode. However, in the beacon-enabled mode, the change is not obvious. We denote the value in the *y* axis using a logarithm to the base 5. The value cannot be correctly represented in hillside terrains because it uses the logarithm. In Figure 4.17, the average end-to-end delay increases from 0.633376 s to 0.899271 s when traffic load increases from 20 to 40 packets/s in flat terrain because of the failures in packet delivery. When traffic load increases to 100 packets/s, the latency is 0.67221 s in flat terrain and 1.473401 s in hilly terrain. Moreover, we observed that the latency in the flat terrain is more stable and less than in hilly and other terrains.

4.5.2.2.4 Effect of Various Terrains on Energy Consumption

Figure 4.18 shows the residual energy ratio of a network using different terrains. The CSMA/CA algorithm reduces energy costs due to idle listening in the backoff period but increases the number of collisions at a higher rate and with a larger number of sources. We observe that the residual energy ratio increases when traffic load becomes heavier because the transmission failure increases. Data packets cannot successfully transmit results as the residual energy increases.

4.5.2.2.5 Effect of Various Terrains on Energy Consumption of Active Path

Figure 4.19 shows the residual energy ratio of an active path using different terrains. When traffic load becomes heavier, contention is also increased. This results in the stimulation of residual energy in the active path because fewer packets are being transmitted.

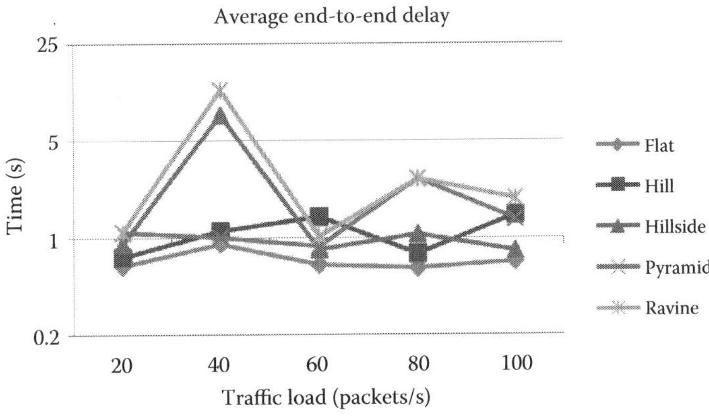

Figure 4.17 Average end-to-end delay versus traffic loads in beacon-enabled mode.

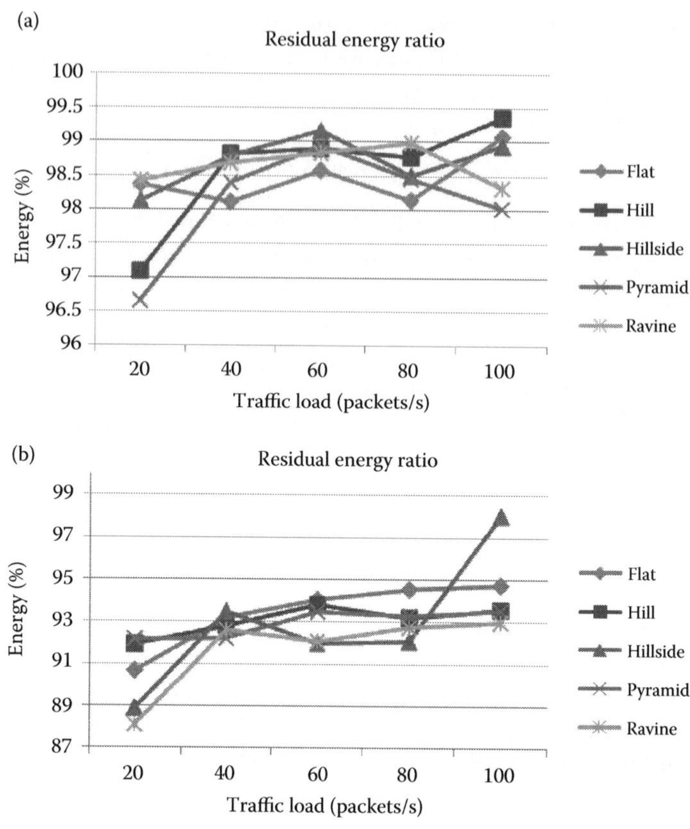

Figure 4.18 Residual energy ratio of network versus traffic loads (from 20 to 100 packets/s). (a) Static nodes in non–beacon-enabled mode. (b) Static nodes in beacon-enabled mode.

4.5.2.3 Simulation 3

In this simulation, we investigate the effects of different moving speeds in beacon/non–beacon-enabled modes. Figure 4.20 illustrates the network topology with 100 mobile FFD nodes and one static PAN coordinator (node 100). Sensor nodes are randomly deployed on a 1000 × 1000 m terrain. The sink node is a coordinator that is placed at the center of this scenario. Also, four CBR traffic flows (CBR0–CBR3) start from the source nodes (node 0, 9, 90, 99) to the sink node (node 100). The BO and SO values are set to 4 and 3, respectively. The traffic load is 2 packets/s. The other parameters are shown in Table 4.3.

4.5.2.3.1 Effect of Various Nodes' Moving Speed on PDR

Figure 4.21 shows the mobile nodes' PDR on different terrains when moving speeds vary from 1 to 5 m/s. The PDR in the beacon-enabled mode still follows the trend in which the moving speed increases as the PDR decreases.

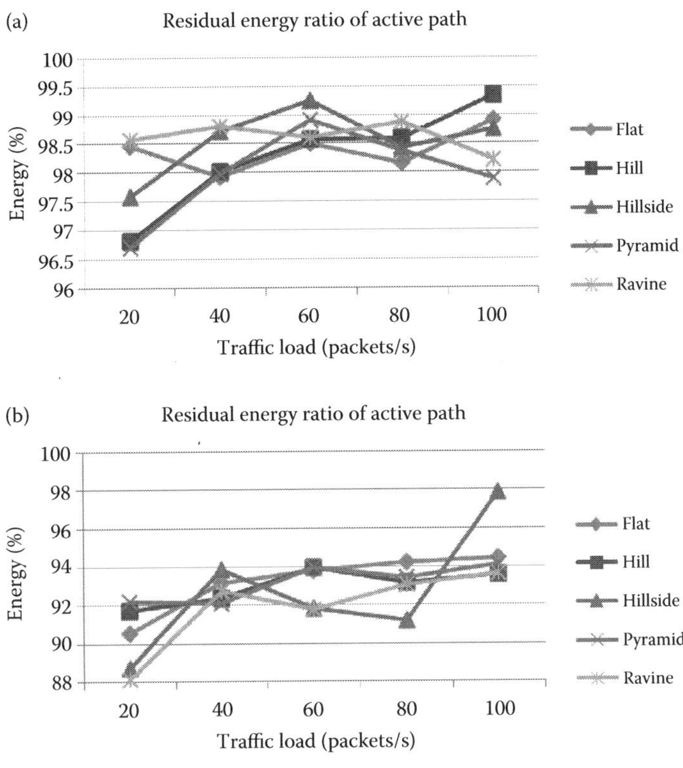

Figure 4.19 Residual energy ratio of active path versus traffic loads (from 20 to 100 packets/s). (a) Static nodes in non–beacon-enabled mode. (b) Static nodes in beacon-enabled mode.

4.5.2.3.2 Effect of Various Nodes' Moving Speed on Goodput

Figure 4.22 shows the mobile nodes' goodput on different terrains. The goodput in the beacon-enabled mode follows the trend in which the moving speed increases as the goodput decreases.

4.6 Summary

In this chapter, a comprehensive performance evaluation of the IEEE 802.15.4 protocol in 3-D ZigBee WSNs is proposed. We present a simulation model to investigate the performance of 3-D WSNs on different terrains. We adapt the BI and superframe duration to change the active period in the IEEE 802.15.4 superframe structure according to the intrinsic characters of different terrains and traffic loads. We evaluated the effects of the following parameters on the performance of slotted CSMA/CA: (1) different BO and SO values, (2) different terrains, (3) different traffic loads, (4) different nodes' moving speeds, and (5) beacon-enabled and non–beacon-enabled modes. Simulation results show that considering obstacles and pathways in 3-D can affect performance evaluation metrics of WSNs, such as data packet delivery, average end-to-end delay, goodput, and energy consumption.

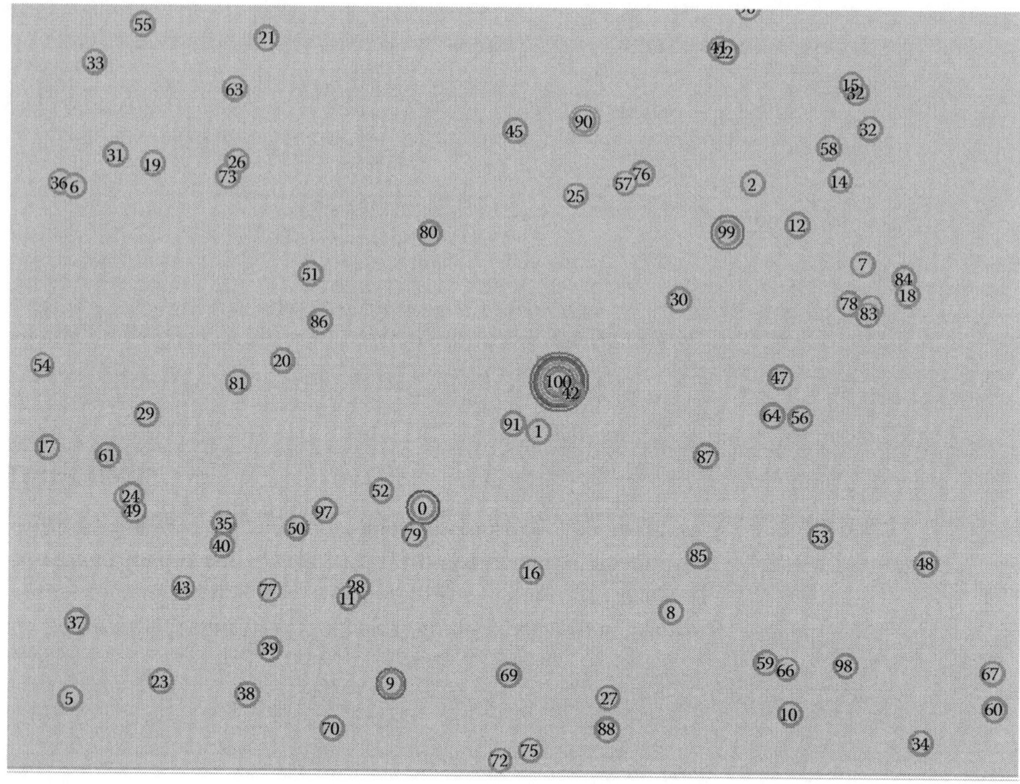

Figure 4.20 Network topology with one static sink node and 100 mobile nodes.

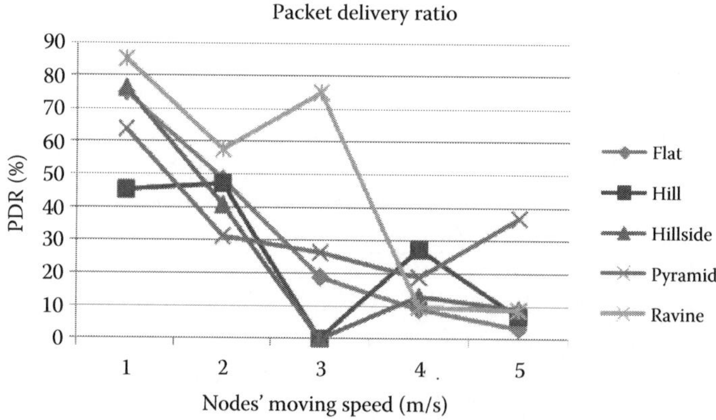

Figure 4.21 PDR versus nodes' moving speed (from 1 to 5 m/s).

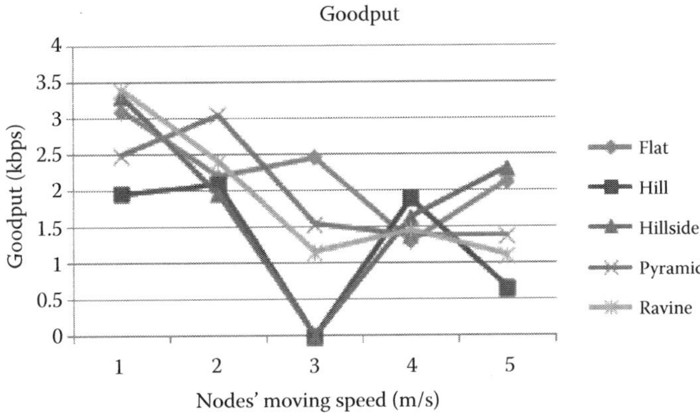

Figure 4.22 Goodput versus nodes' moving speed (from 1 to 5 m/s).

In future work, we will observe mobile sink and cluster heads as possibilities in cluster-tree topology. How to cluster the nodes in a 3D IEEE 802.15.4 wireless sensor networks is the direction. Also, the BO and SO selecting strategy on different terrains should also be considered as a study topic.

References

1. ZigBee, http://www.zigbee.org/, April 28, 2013.
2. PART 15.4: wireless medium access control (MAC) and physical layer (PHY) specifications for low-rate wireless personal area networks (LR-WPANs). IEEE Std. P802.15.4a/D5, 2006.
3. N. Huansheng and W. Ziou. Future internet of things architecture: like mankind neural system or social organization framework? *IEEE Communications Letters* 15(4), 2011, 461–463.
4. M. Zorzi, A. Gluhak, S. Lange, and A. Bassi. From today's INTRAnet of things to a future INTERnet of things: a wireless- and mobility-related view. *IEEE Wireless Communications* 17(6), 2010, 44–51.
5. ZigBee—The Internet of Things. http://www.vesternet.com/zigbee-c-276.html, April 28, 2013.
6. C. Buratti and R. Verdone. Performance analysis of IEEE 802.15.4 non beacon-enabled mode. *IEEE Transactions on Vehicular Technology* 58(7), 2000, 3480–3493.
7. C. Buratti. Performance analysis of IEEE 802.15.4 beacon-enabled mode. *IEEE Transactions on Vehicular Technology* 59(4), 2010, 2031–2045.
8. S. Pollin, M. Ergen, S. Ergen, B. Bougard, L. Der Perre, I. Moerman, A. Bahai, P. Varaiya, and F. Catthoor. Performance analysis of slotted carrier sense IEEE 802.15.4 medium access layer. *IEEE Transactions on Wireless Communications* 7(9), 2008, 3359–3371.
9. A. Koubaa, M. Alves, and E. Tovar. A comprehensive simulation study of slotted CSMA/CA for IEEE 802.15.4 wireless sensor networks. *2006 IEEE International Workshop on Factory Communication Systems* 2006, 183–192.
10. C. D. Ortiz, J. M. Puig, C. E. Palau, and M. Esteve. 3D wireless sensor network modeling and simulation. *Sensor Technologies and Applications, 2007. International Conference on SensorComm 2007*, 14–20, 2007, 307–312.
11. M. Jain, M. K. Mishra, and M. M. Gore. Energy aware beaconless geographical routing in three dimensional wireless sensor networks. *Advanced Computing, 2009. First International Conference on ICAC 2009*, 13–15, 2009, 122–128.

12. G. Babaei, M. Romoozi, and M. Rostmai. A 3D signal obstruction model for realistic mobility models in mobile ad hoc networks. *IJCSI International Journal of Computer Science Issues* 8, issue 4(2), 2011, 168–175.
13. L. F. M. Vieira, M. G. Almiron, and A. A. F. Loureiro. 3D MANETs: Link probability, node degree, network coverage and applications. *IEEE Wireless Communications and Networking Conference (WCNC), 2011* 28–31 March 2011, 2042–2047.
14. L. Song, T. Fevens, and A. E. Abdallah. Hybrid position-based routing algorithms for 3D mobile ad hoc networks. *Mobile Ad-Hoc and Sensor Networks, 2008. The 4th International Conference on MSN 2008* 10–12 December 2008, 177–186.
15. S. Filiposka and D. Trajanov. Terrain-aware three-dimensional radio-propagation model extension for NS-2. *Simulation* 87(1–2), 2011, 7–23.
16. S. Filiposka and D. Trajanov. 3D simulations for wireless ad hoc networks in grid environment. *Proceedings of Small Systems Simulation Symposium 2010, Niš, Serbia,* February 2010, 12–14.
17. A. Dzambaski, D. Trajanov, S. Filiposka, and A. Grnarov. Ad hoc networks simulations with real 3D terrains. *15th Telecommunications forum (TELFOR) 2007, Serbia, Belgrade,* November 20–22, 2007, 95–98.
18. B. G. Lee, K. H. Do, and W. Y. Chung. WSN based 3D mobile indoor multiple user tracking. *IEEE SENSORS 2009 Conference,* 2009, 1598–1603.
19. B. G. Lee, Y. S. Lee, and W. Y. Chung. 3D navigation real time RSSI-based indoor tracking application. *Journal of Ubiquitous Convergence Technology* 2(2), 2008, 67–77.
20. The network simulator—NS-2. http://www.isi.edu/nsnam/ns/, April 28, 2013.
21. Fresnel zone. http://www.vias.org/wirelessnetw/wndw_04_08b.html, April 28, 2013.
22. Durkin's propagation model. http://e-tnc.com/etnc/Home/tabid/57/Default.aspx.
23. M. Vuckovik, D. Trajanov, and Sonja Filiposka. Durkin's propagation model based on triangular irregular network terrain. *ICT Innovations 2010, CCIS 83,* 333–341. Berlin, Heidelberg: Springer.
24. Digital elevation model. http://tahoe.usgs.gov/DEM.html, April 28, 2013.
25. Digital elevation model. http://rmmcweb.cr.usgs.gov/elevation/dpi_dem.html.
26. U.S. Geological Survey National Mapping Division: Part 1. General. Standards for digital elevation models, U.S. Department of the Interior, Washington, DC.
27. U.S. Geological Survey National Mapping Division: Part 2. Specifications. Standards for digital elevation models, U.S. Department of the Interior, Washington, DC.
28. 3DEM. http://www.hangsim.com/3dem/, April 28, 2013.
29. J. Zheng and M. J. Lee. Will IEEE 802.15.4 make ubiquitous networking a reality? A discussion on a potential low power, low bit rate standard. *IEEE Communication Magazine* 42(6), 2004, 140–146.

Chapter 5

Multi-Interface Wireless Networks: Complexity and Algorithms

Alfredo Navarra
Università degli Studi di Perugia

Gianlorenzo D'Angelo
INRIA/I3S (CNRS/UNSA)

Gabriele Di Stefano
Università degli Studi dell'Aquila

Contents

5.1 Introduction

Sensors networks and, more broadly speaking, wireless networks have received significant attention during recent years. As technology advances and hardware costs are reduced, very powerful devices are available for a wide range of applications. Moreover, heterogeneous devices may communicate with each other using different protocols and interfaces. The connection among heterogeneous devices might result as a fundamental means for communication among different local area networks that altogether form a wide area network, or the Internet.

In this chapter, we are interested in networks composed of heterogeneous devices that exploit different communication interfaces to establish the desired connections. It is very common to find devices equipped with Bluetooth, Wi-Fi, and GPRS interfaces, but very few applications usually take advantage of such available heterogeneity. Selecting the best radio interfaces for specific connections depends on several factors. Namely, the choice might depend on the availability of a specific interface on some devices, the required communication bandwidth, the cost (in terms of energy consumption) for maintaining an active interface, the neighborhood, and so forth. For instance, in an article by Friedman et al. [1], the authors provide an experimental study on the choice of using either Bluetooth or Wi-Fi for the transmission of data packets among various smartphones. The choice has been based on the energy consumed to accomplish the required communications by varying the size of the packets. Because devices are usually portable or mobile, a lot of effort must be devoted to energy consumption issues to prolong the network lifetime. In fact, the failure of a device due to drained batteries is something that could be delayed if suitable solutions are provided when building the connections of a desired network in accordance with specific requirements.

This introduces challenging and natural optimization problems that must take care of different parameters at the same time. Some problems that will be presented in this chapter are specific for the considered multi-interface networks, whereas some other problems come directly from the algorithm theory and are adapted to the considered model.

In general, a network of devices will be described by a graph $G = (V,E)$, where V represents the set of devices and E is the set of possible connections defined according to the distance between devices and the available interfaces that they share. Each $v \in V$ is associated with a set of available interfaces $W(v)$. The set of all the possible available interfaces in the network is then determined by $\cup_{v \in V} W(v)$; we denote the cardinality of this set by k. We say that a connection is established when the end points of the corresponding edge share at least one active interface. If an interface x is activated at some node u, then u consumes some energy $c(x)$ for maintaining x as active, and it provides a maximum communication bandwidth $b(x)$ with all its neighbors that share interface x.

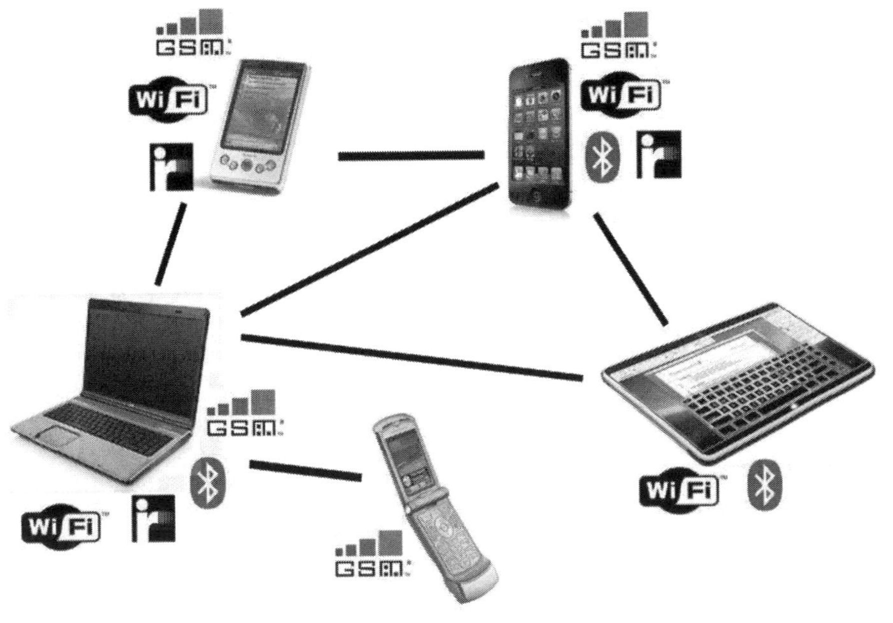

Figure 5.1 Example of multi-interface network composed of heterogeneous devices connected by different interfaces such as IrDA, Bluetooth, GSM, and Wi-Fi.

An example of a network instance is shown in Figure 5.1, in which mobile phones, smartphones, tablets, and laptops can communicate using different interfaces and protocols such as infrared data association (IrDA), Bluetooth, Wi-Fi, global system for mobile communications (GSM), Edge, universal mobile telecommunications system (UMTS), and satellite. All the possible connections can be established by at least one interface. Note that some devices are not directly connected even though they share some interfaces. This can be caused by many factors such as obstacles, distances, or protocols used. To provide a full instance, one should provide (if necessary) the cost for each device to switch-on a specific interface, and the corresponding bandwidth that can be handled. Although bandwidths usually concern the interfaces, the energy spent by each device to switch-on a specific interface may vary substantially. To simplify the model, the costs might be referred to as the percentage of battery consumed by each device, and hence it might be considered the same for each device with respect to a specific interface among the whole network. Nevertheless, different assumptions may lead to completely different problems that point out the specific peculiarities of the composed networks. Our interests concern algorithmic approaches to the optimization problem arising from the classic algorithmic graph theory [2], but extend to the multi-interface networks. The problems we are going to consider within such networks are summarized below.

5.1.1 Coverage

The coverage problem consists of finding the cheapest way to establish all the connections defined by an input graph G, no matter the interface used to accomplish each connection, no bandwidth requirements are provided. The problem only asks to ensure that for each edge of G, there is a common active interface at its end points. The objective is to minimize the overall cost of activation in the network. Another interesting objective function that has been studied concerns the minimization of the maximum cost on a single node. The coverage problem does not have a counterpart in the classic algorithmic graph theory but this is the only exception.

5.1.2 Connectivity

The connectivity problem consists of finding the cheapest way to ensure the connectivity of the entire network. In other words, it aims to find, at each node, a subset of the available interfaces that must be activated to guarantee a path between every pair of nodes in G while minimizing the overall cost of the interfaces switched-on among the whole network. As for the case of coverage, another objective function studied is that of minimizing the maximum cost paid on a single node. The connectivity problem corresponds to a generalization of the well-known minimum spanning tree problem. In fact, costs are not on the edges but on the interfaces held by the nodes. The same interface may be used by a node to establish several connections, hence saving energy. This property of multi-interface networks reveals the advantage and higher complexity of the problems being studied.

5.1.3 Cheapest Path

The cheapest path problem has the goal of finding a minimum cost subset of available interfaces that can be activated in some nodes to guarantee a connecting path between two specified nodes of the network. This problem corresponds to a generalization of the well-known shortest path problem between two nodes in standard networks. The cheapest path and maximum flow problems, among those studied, are actually the only ones that maintain their computational complexity close to the classic problems, and hence they can be solved efficiently.

5.1.4 Maximum Matching

Similar to its classic version, the problem asks for the maximum subset of connections that can be established concurrently without sharing any common nodes. That is, a solution is provided by a set of disjointed edges of the input graph, and hence each node appears in the solution (at most) once. The maximum matching problem in the field of multi-interface networks reveals another interesting peculiarity of this kind of network. An instance of the problem is simply provided by an input graph G without specifying costs and bandwidths. In fact, the problem becomes difficult to solve because one must consider that when a solution includes two edges established using the same interface, they cannot be directly connected by another edge; otherwise, the connection represented by this third edge would be established as well because both its end points share the same active interface, hence making the current solution invalid.

5.1.5 Maximum Flow

This problem as well as the next one aims to guarantee a connection between two given nodes, taking into account bandwidth constraints. The maximum flow problem consists of finding the maximum possible bandwidth between two selected nodes. In particular, we consider all the interfaces of the network as active, so that all the allowed connections are established. Then, we look for a suitable flow function that guarantees the maximum communication bandwidth between the two given nodes. This problem corresponds to a generalization of the maximum flow problem in standard networks.

5.1.6 Minimum Cost Flow

The minimum cost flow problem consists of establishing a communication subnetwork between two given nodes of minimum cost in terms of energy consumption while guaranteeing a minimum communication bandwidth B. In other words, we look for the minimum cost set of active interfaces among the input network in such a way that a certain node s is guaranteed to exchange data with another specified node t with a bandwidth of at least B. In general, the solution is not a path between s and t, but a more complex graph consisting of nodes with active interfaces might be required according to the topology and the available interfaces. This problem corresponds to a generalization of the minimum cost flow problem in standard networks.

Besides the relevance of the problems that will be addressed within this chapter, the educational perspective that this survey can constitute with respect to optimization, approximation, and hardness studies is also very interesting. Indeed, many basic problems such as set cover, vertex cover, minimum spanning tree, shortest path, Hamiltonian path, minimum hitting set, and minimization knapsack (borrowed from the classic algorithmic graph theory) are exploited to obtain hardness results as well as an approximation or exact algorithm for the problems mentioned.

5.2 Preliminaries

In this section, we provide the basic definitions and the necessary notation for a full understanding and for the formal presentation of the problems mentioned.

A network of devices is represented by a graph $G = (V,E)$, where V represents a set of devices, whereas E is the set of connections that can be established by at least one interface. It follows that if there exists an edge $e = (u,v)$, then both u and v share at least one interface. G is assumed to be simple (i.e., without multiple edges), undirected and connected. Moreover, the cardinality of the sets V and E are denoted by n and m, respectively. The degree of node $v \in V$ is denoted by Δ_v and the set of its neighbors by $N(v)$. The minimum node degree of graph G is denoted by δ, and its maximum node degree by Δ.

To provide a global characterization concerning the interfaces that each device holds, we use an interface assignment function W, according to the following definition.

Definition 1

A function $W: V \rightarrow 2^{\{1,\ldots,k\}}$ is said to cover graph $G = (V, E)$ if for each $\{u, v\} \in E$ the set $W(u) \cap W(v) \neq \emptyset$. ■

The cost of activating an interface for a node is assumed to be identical for all nodes and is given by cost function c: $\{1,\ldots, k\} \rightarrow R^{+}$. The cost of interface i is denoted by $c(i)$. As described above, it represents the percentage of energy consumed by a device to switch-on interface i.

Sometimes costs are considered to be the same for all the interfaces. In such a case, the corresponding problem is referred to as its *uniform cost* form. This variant of a problem might be very useful for the study of complexity results because it may reveal implicit peculiarities.

Another important variant to some of the problems faced within multi-interface networks concerns the number k of interfaces appearing in the whole network. If such a number is provided as an input constant, the corresponding problem is referred to as in its *bounded* form; otherwise, it is said in the *unbounded* form. The latter case is again very important for analytical results whereas the former is more representative of practical cases.

As will be shown in Section 5.6 with problems referring to flow, each interface x may provide a maximum communication bandwidth $b(x)$. This means that a node holding interface x is potentially able to manage $b(x)$ units of flow (usually expressed in terms of Mbps, that is, megabits per second) with all its neighbors holding the same interface x. Clearly, for each interface, there might be a subset of neighbors with which the node can transfer at a predetermined amount of bandwidth. The bandwidth provided by an interface for a node is assumed to be identical for all nodes and is given by bandwidth function b: $\{1,\ldots, k\} \rightarrow N$. The bandwidth of interface x is denoted by $b(x)$. As for the costs, the amount of bandwidth is associated with the interfaces, and hence with the nodes rather than with the edges as it is in classic flow problems.

Bandwidth constraints represent a very important means of dealing with quality of service issues because they refer to the amount of data that can be transferred between the nodes of specified connections.

5.3 Coverage and Connectivity

This section addresses both the coverage and the connectivity problems because they share many properties and have already been treated together in the literature from different perspectives. Although coverage consists of finding the cheapest way to establish all the connections defined by an input graph G, connectivity requires the establishment of just a subset of edges that ensures connectivity among all the nodes of G.

5.3.1 Coverage

The coverage problem has been addressed in various forms [3–6]. The acronym k-CMI, standing for cost minimization in multi-interface networks, refers to the coverage problem when the number of k interfaces is given as an input parameter, that is, the number of available interfaces among the whole network is bounded by a given constant k. This is one of the most application-oriented assumptions because it is reasonable to not have an infinite number of interfaces or a number of interfaces proportional to the number of devices appearing. However, as we are going to see later in this section, the unbounded case also provides interesting results, that is, when k is not an input parameter and deserves particular attention for analytical studies. Formally, k-CMI can be stated as follows:

k-CMI: Cost Minimization in Multi-Interface Networks

Input	A graph $G = (V, E)$, an allocation of available interfaces $W: V \rightarrow 2^{\{1,\ldots,k\}}$ covering graph G, an interface cost function $c:\{1,\ldots, k\} \rightarrow R^+$
Solution	An allocation of active interfaces $W_A: V \rightarrow 2^{\{1,\ldots,k\}}$ covering G such that $W_A(v) \subseteq W(v)$ for all $v \in V$
Goal	Minimize the total cost of the active interfaces, $c(W_A) = \sum_{v \in V} \sum_{l \in W_A(v)} c(l)$

As an example, we consider the graph provided in Figure 5.1. To define a full input instance of k-CMI, we need to define a cost function on the ordered sequence of $k = 4$ interfaces (IrDA, Bluetooth, GSM, Wi-Fi). Let $c(1) = 0.2$; $c(2) = 0.6$; $c(3) = 1$; $c(4) = 1.2$; an optimal solution is then given by the activation shown in Figure 5.2. The total cost is given by $3c(1) + 3c(2) + 2c(3)$. Interestingly, the connection between the laptop and the iPhone is established twice, with both IrDA and Bluetooth. It

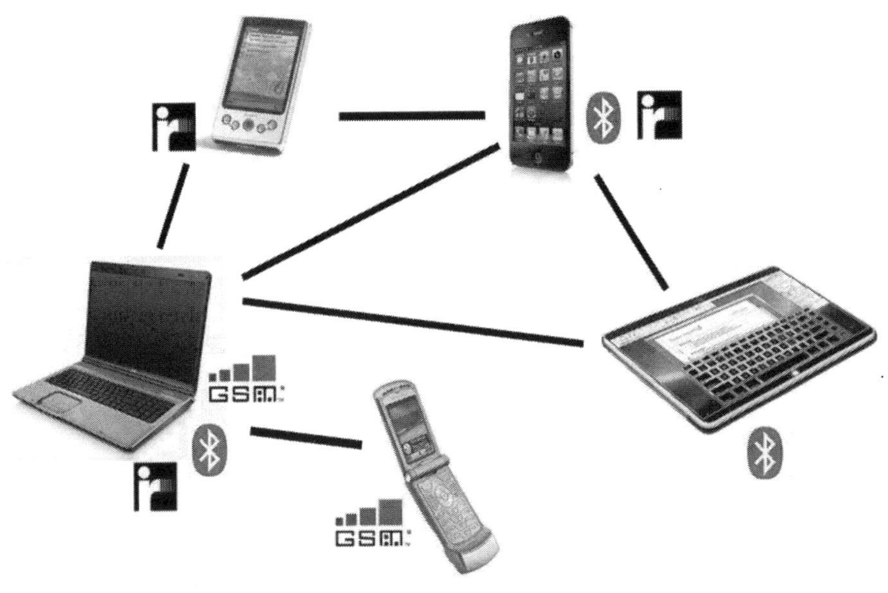

Figure 5.2 Optimal solution for the coverage problem when the input instance is provided by the network of Figure 5.1 and the cost function is defined on the ordered sequence of $k = 4$ interfaces (IrDA, Bluetooth, GSM, and Wi-Fi) with $c(1) = 0.2$; $c(2) = 0.6$; $c(3) = 1$; $c(4) = 1.2$.

can be verified that any other solution would be more expensive. This reveals an interesting property of multi-interface networks because the cheapest solution does not always correspond to a solution that minimizes the number of connections established. These problems were originally from the work of Caporuscio et al. [3], in which a slightly different model of k-CMI was introduced. That model also considered the possibility of having mutually exclusive interfaces, that is, interfaces that, if activated, preclude the activation of some other interfaces. The motivation is quite technical; for instance, the Wi-Fi interface can operate in different modalities (infrastructure and ad hoc). If a device activates Wi-Fi in the infrastructure modality, it cannot satisfy connections that require ad hoc modality and vice versa. In the formulation of k-CMI, this further constraint of mutual exclusion as well as bandwidth constraints and other peculiarities presented in the original model have not been addressed here. The problem is, in fact, already of practical relevance and not always so easy to approach.

A summary of the obtained results is shown in Table 5.1. Very different results have been obtained by varying the graph's topology, the number of available interfaces k, and considering both the uniform and nonuniform cost cases. For some graph topologies, such as complete graphs and trees, the problem becomes polynomially solvable as well as for the case of $k = 2$. Although it is enough to have $k = 3$ to substantially increase the complexity of the problem. In fact, k-CMI becomes APX-hard even for graphs of bounded degree Δ, whereas it is NP-hard but admitting a polynomial time approximation scheme (PTAS) in the case of planar graphs. Between the uniform and nonuniform cases, results are substantially the same. Only some better approximation algorithms can be devised for the uniform case. This means that the complexity of k-CMI is, in general, not dependent on the costs but it comes from the nature of the problem.

In the remainder of this section, we give more details on the results obtained for k-CMI, emphasizing the interesting results and techniques used.

Table 5.1 Hardness and Approximation Results for k-CMI

Graph Class	*Interfaces*	*Nonuniform Costs*	*Uniform Costs*
General graphs	$k = 2$	$O(n^3)$	$O(nm)$
	$k \geq 3$	$(k-1)$-approx, APX-hard	$\min\left\{\frac{k+1}{2}, \frac{2m}{n}\right\}$-approx, APX-hard
Graphs of bounded degree Δ	$k \geq 3$	Δ-approx, APX-hard for $\Delta \geq 5$	$\frac{\Delta+1}{2}$ – approx, APX-hard for $\Delta \geq 5$
Planar graphs	$k \geq 3$	NP-hard, PTAS	NP-hard, PTAS
Trees	any k	$O(n)$	$O(n)$
Complete graphs	any k	$O(n^2)$	$O(n^2)$

5.3.1.1 Computational Complexity for k-CMI

In this section, we describe, in more detail, the most general results concerning the complexity of k-CMI that have been obtained for $k \geq 3$ on graphs of bounded degree $\Delta \geq 5$. In fact, for such a case, the problem becomes APX-hard and the technique used deserves particular attention. The proof proceeds by a polynomial transformation from the well-known vertex cover problem on subcubic graphs to k-CMI. The vertex cover problem can be stated as follows:

Vertex Cover

Input	A graph $G = (V, E)$, a positive integer $K \leq \|V\|$
Question	Is there a subset V' of V with $\|V'\| \leq K$ s.t. for each edge $\{u,v\} \in E$ at least one of u and v belongs to V'?

On subcubic graphs, vertex cover is known to be APX-hard [7]. Given a subcubic graph $G = (V, E)$, that is, a graph with $\Delta \leq 3$, it is known that in general its chromatic number is at most three [8] (unless the input graph is a clique of four nodes).

That is, three colors are always sufficient to obtain a solution to the coloring problem on such graphs and such a solution can be found in polynomial time. The transformation proceeds by partitioning the nodes of G into three subsets V_1, V_2, and V_3 according to an optimal coloring (Figure 5.3). Clearly, $V_1 \cup V_2 \cup V_3 \equiv V$ and, for each edge $e = \{u,v\} \in E$, u, and v do not belong to the same

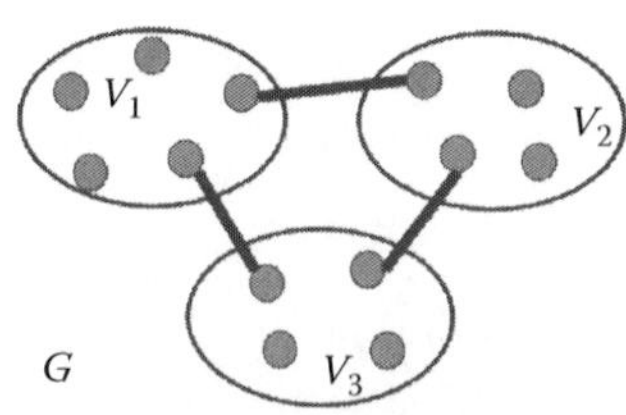

Figure 5.3 A subplanar graph *G* divided into three sets according to a 3-coloring problem. The three edges represent the only possible connections among the nodes.

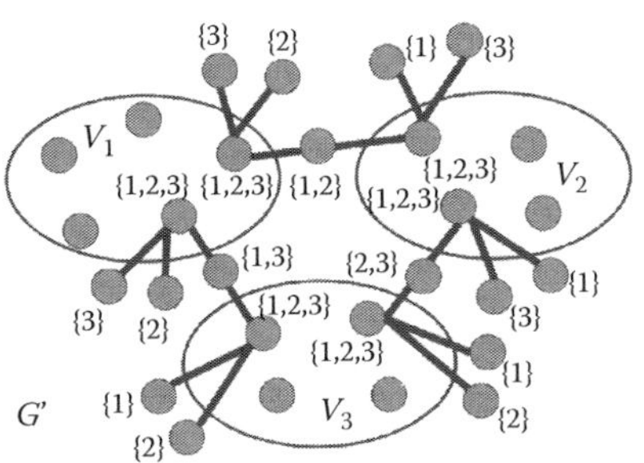

Figure 5.4 The modified graph G' obtained from G along with the associated interfaces.

subset V_i for every $i = 1, 2$, or 3. To construct an instance for k-CMI, with each node $v \in V$, three interfaces are associated, namely, 1, 2, and 3 (Figure 5.4).

The uniform case is considered where all the interfaces have the same cost (i.e., function c is constant). Moreover, to each $v \in V$, there are two new nodes connected. Those new nodes have only one interface: one has interface 2 and the other has interface 3 (1 and 3 or 1 and 2, respectively), if $v \in V_1$($v \in V_2$ or $v \in V_3$, respectively). Each edge of G is replaced by a path of two edges, that is, one further node is added. With such a node, two interfaces are associated. If the considered edge connects V_1 and V_2(V_1 and V_3 or V_2 and V_3, respectively) then interfaces 1 and 2 (1 and 3 or 2 and 3, respectively) are associated. Considering a generic edge $e = \{u,v\}$ such that $u \in V_1$ and $v \in V_2$, to solve k-CMI on the new graph of maximum degree of 5 built from G, a solution necessarily has to activate interfaces 2 and 3 in u, and 1 and 3 in v. For both u and v to be able to communicate with the new intermediate node, such a node must either activate both its interfaces or one among u and v has to activate its third available interface. Both the solutions are locally equivalent, on the other hand, activating the third interface for either u or v may lead to a decrease in the number of activated interfaces in the global solution. This is implied by the fact that the neighborhood of the added intermediate node between u and v is constituted by only u and v, whereas both u and v may have many other connections. This implies that one can look for solutions in which, for each edge of the original graph, at least one end point has all its three interfaces activated. Note that this reflects exactly the requirement of vertex cover.

5.3.1.2 Approximation Results for k-CMI

About the approximation algorithms designed for k-CMI in its uniform and nonuniform variants, we provide two sample cases. For the general case of nonuniform costs, one of the achieved approximations leads to a factor of $k - 1$. The algorithm uses a greedy technique. It starts activating the cheapest interface 1 among each node that has a neighbor holding that interface. Let $V_1 \subseteq V$ be the set of nodes in which the algorithm activated interface 1 and let $E(V_1)$ be the corresponding set of covered edges. Note that the optimal solution restricted to $E(V_1)$ [i.e., the set of activated interfaces of an optimal solution at the end points of the edges belonging to $E(V_1)$] clearly costs at least as much as the cost of this greedy algorithm. In the second step, the same is done for the next cheapest interface 2 among the remaining connections $E/E(V_1)$. Again, the cost of the optimal solution restricted to $E(V_2)$ is at least the cost paid by the algorithm. This is implied by the fact that any connection belonging to $E(V_2)$ cannot be covered by interface 1; otherwise, the algorithm would have covered it in the previous step. This process is continued for all the interfaces in a nondecreasing cost order but for the last two interfaces. When the two most expensive interfaces remain, in fact, the optimal algorithm for the case of $k = 2$ can be applied. As shown in Table 5.1, in fact, k-CMI is optimally solvable for $k = 2$. Because each step costs at most as much as the optimal solution, the $(k - 1)$ approximation holds by observing that the whole process requires $k - 1$ steps.

For the uniform case, a very simple algorithm, still greedy, obtains a $\frac{2m}{n}$-approximation. The algorithm simply chooses one interface for each edge to establish the connection. This means that for each edge, one interface (at most) in each end point is activated. It follows that for m edges, it activates at most $2m$ interfaces for n nodes.

5.3.1.3 Solvable Cases for k-CMI

The most important optimal result for k-CMI is certainly provided for the case of $k = 2$. Because the technique used for devising the algorithm is not so easy to describe without providing all the details, we point the interested reader to the full proof contained in the study by Klasing et al. [5]. Very roughly, the problem is transformed into a partitioning problem that has to respect some properties. Such a partition will divide the set of nodes V into three subsets, one for the nodes that will activate only interface 1 in the final solution, one subset for the nodes that will activate only interface 2, and the third subset for all the nodes that are required to activate both interfaces 1 and 2.

One algorithm that can be described in this contest concerns the case of complete graphs. For this class of graphs, it is interesting to note how fundamental the property of k is to be a given constant number rather than depending on the input instance. In fact, the set of nodes is divided into classes according to the available interfaces they have. In this way, there will be at most 2^k classes (i.e., a constant number of different classes). Because the graph is complete, symmetry implies that every node belonging to the same class has the same subset of interfaces activated in some optimal solution. The maximum number of interfaces that a node can activate is of course k. Hence, by trying all the possible configurations (at most 2^k) for the available interfaces in each class, the optimal solution can be computed in $(2^k)^{2^k}$ steps by checking each time if all the edges are covered. In practice, the algorithm applies an exhaustive search among all the possible solutions, of which there are a finite number, once its formation has been observed.

5.3.1.4 Unbounded CMI

In this section, we provide a description of the results obtained thus far for the coverage problem in case the number of interfaces k is not an input parameter but depends on the devices composing the input graph. The problem is referred to in the literature as *unbounded CMI*. As already pointed out, the relevance of this problem is more theoretical with respect to k-CMI. Moreover, all the negative results holding for k-CMI are clearly inherited in this variant. Formally, CMI can be stated as follows:

CMI: Cost Minimization in Unbounded Multi-Interface Networks

Input	A graph $G = (V, E)$, a positive integer k, an allocation of available interfaces $W: V \rightarrow 2^{\{1,\ldots k\}}$ covering graph G, an interface cost function $c:\{1,\ldots, k\} \rightarrow R^+$
Solution	An allocation of active interfaces $W_A: V \rightarrow 2^{\{1,\ldots k\}}$ covering graph G such that $W_A(v) \subseteq W(v)$ for all $v \in V$
Goal	Minimize the total cost of the active interfaces, $c(W_A) = \sum_{v \in V} \sum_{l \in W_A(v)} c(l)$

As mentioned previously, the only difference with k-CMI is that parameter k is part of the input. A summary of the obtained results for CMI is shown in Table 5.2.

As shown in Table 5.2, adding the parameter k from the input of the coverage problem results in increasing its hardness. Apart from the results marked by (*) that are directly inherited from k-CMI, the obtained problem becomes APX-hard even for trees. On planar graphs, CMI is APX-hard as well, whereas its counterpart k-CMI is NP-hard and admits a PTAS. Very surprisingly, on complete graphs in which k-CMI has been shown to be polynomially solvable, CMI provides the worst result in terms of complexity. In fact, it is not approximable within $O(\log k)$ even for the uniform costs case.

The proof proceeds by reduction to the minimum hitting set problem [9].

Minimum Hitting Set

Input	Collection of nonempty subsets $C_1, C_2, \ldots, C_l \subseteq \{1, 2, \ldots, k\}$.
Solution	Set $S \subseteq \{1, 2, \ldots, k\}$ s.t. $\forall_{1 \le i \le l} C_i \cap S \neq \emptyset$
Goal	Minimize the cardinality of S

The transformation can be sketched as follows. Let graph G be a complete bipartite graph, in which the set of vertices $V = X \cup Y$, and all vertices from set $X = \{x_1, x_2, \ldots, x_l\}$ are connected with all vertices from set $Y = \{y_1, y_2, \ldots, y_p\}$. For all vertices from set Y, allocate the entire set of available interfaces $W(y_i) = \{1, 2, \ldots, k\}$, whereas for all vertices from set X, allocate fixed subsets of interfaces to each vertex $W(x_i) = C_i$. Let W_A be an arbitrary activation function that solves CMI for the provided instance. For any fixed j, $1 \le j \le p$, $\forall_{1 \le i \le l} W_A(x_i) \cap W(x_i) \neq \emptyset$, and because $W_A(x_i) \subseteq W(x_i) = C_i$, obviously $\forall_{1 \le i \le l} W_A(x_i) \cap C_i \neq \emptyset$. It follows that $W_A(y_j)$ is a hitting set for the collection of sets $\{C_i\}$.

Thus, given a solution to the considered CMI instance with a total cost of c, using the above transformation, one can immediately show a hitting set S for $\{C_i\}$ such that $|S| \le \frac{c-l}{p}$. Conversely, given a hitting set S for $\{C_i\}$, it is easy to construct a valid solution for the CMI instance. In doing

Table 5.2 Hardness and Approximation Results for CMI

Graph Class	*Nonuniform Costs*	*Uniform Costs*
General graphs	$(k-1)$-approx (*), $[\sqrt{n}(1+\ln n)]$-approx, not approx within $O(\log k)$	$\min\left\{\frac{2m}{n}\right\}$-approx(*), not approx within $O(\log k)$
Graphs of bounded degree Δ	Δ-approx (*), APX-hard for $\Delta \ge 5$, $k \ge 3$ (*)	$\frac{\Delta+1}{2}$-approx (*), APX-hard for $\Delta \ge 5$, $k \ge 3$ (*)
Planar graphs	6-approx, APX-hard	
Trees	2-approx, APX-hard	
Complete graphs	Not approx within $O(\log k)$	

Note: Results marked with (*) come directly from k-CMI.

so, it follows that any polynomial time a-approximation algorithm for CMI leads to a polynomial time a-approximation algorithm for the minimum hitting set problem. The claim then follows from the stated hardness result for the approximation of minimum hitting set that is as hard as the minimum set cover problem [10] and, consequently, is hard to approximate within a factor of $O(\log k)$.

5.3.2 Connectivity

As for the coverage, the connectivity problem has been addressed in various forms (see studies by Kosowski et al. [11,12]) by considering either uniform or nonuniform costs, and varying the number of available interfaces k that might be bounded or unbounded. Formally, connectivity can be stated as follows:

Connectivity in Multi-Interface Networks

Input	A graph $G = (V, E)$, an allocation of available interfaces $W: V \rightarrow 2^{\{1,\ldots,k\}}$ covering graph G, an interface cost function $c:\{1,\ldots, k\} \rightarrow R^+$
Solution	An allocation of active interfaces $W_A: V \rightarrow 2^{\{1,\ldots,k\}}$ covering a connected spanning graph $G' = (V,E')$ *of* G such that $W_A(v) \subseteq W(v)$ for all $v \in V$, and $E' \subseteq E$
Goal	Minimize the total cost of the active interfaces, $c(W_A) = \sum_{v \in V} \sum_{l \in W_A(v)} c_l$

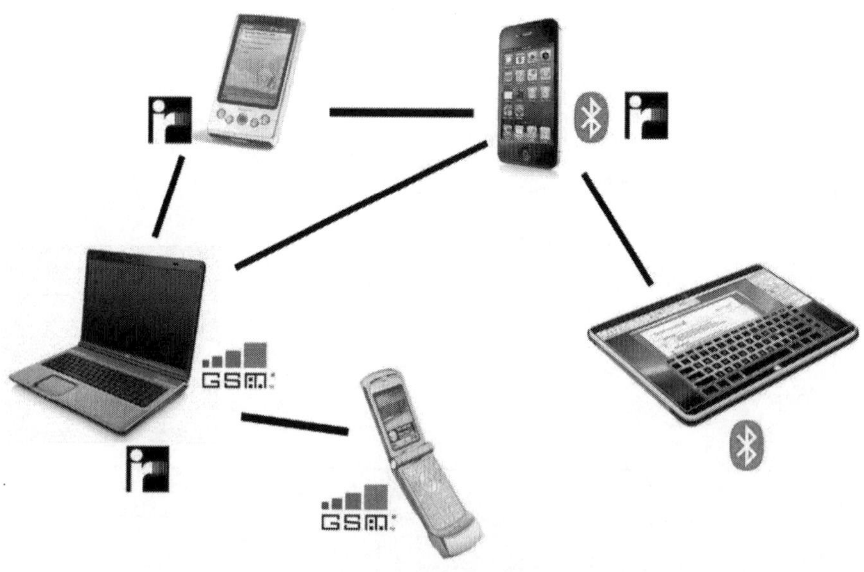

Figure 5.5 Optimal solution for the connectivity problem when the input instance is provided by the network of Figure 5.1 and the cost function is defined similar to that of Figure 5.2.

The difference with coverage is that the problem asks to establish the minimum cost set of connections to guarantee the connectivity among all the nodes of the network. In this respect, connectivity can be considered as a counterpart in the context of multi-interface networks of the well-known minimum spanning tree problem studied in the classic algorithmic graph theory.

It follows that connectivity coincides with coverage when the input graph is a tree. Again considering the example of Figure 5.1, the optimal solution for connectivity on that instance is shown in Figure 5.5.

Minimum Spanning Tree

Input	A graph $G = (V, E)$, a cost function $W: E \rightarrow R^+$
Solution	A subset S of $\|V\| - 1$ edges from E s.t. for every pair of nodes u, v in V, there exists a sequence of edges from S connecting u to v
Goal	Minimize $\sum_{e_i \in S} W(e_i)$

It follows that connectivity coincides with coverage when the input graph is a tree. Again considering the example of Figure 5.1, the optimal solution for connectivity on that instance is shown in Figure 5.3. It costs $3c(1) + 2c(2) + 2c(3)$. It is worth noting that the provided solution is not a tree, but a spanner graph of the input graph G. This is one of the most important peculiarities that one has to face when dealing with problems on multi-interface networks. That is, some connections are simply implied by the activation of necessary interfaces even though they are not required. This property is not a problem in the context of connectivity, but for other problems (like the matching problem in Section 5.5) this drawback must be opportunely managed. Although the minimum spanning tree problem is polynomially solvable [2]. Connectivity turns out to be APX-hard, in general, even for the uniform costs case with bounded $\Delta \geq 4$ and unbounded $k \geq 2$.

Table 5.3 Hardness and Approximation Results for Connectivity

Graph Class	*Nonuniform Costs*	*Uniform Costs*
General graphs	APX-hard, $k \geq 2$ $\left(\frac{3}{2}+\varepsilon\right)$-approx, $k \geq 2$	APX-hard, $k \geq 2$ $\frac{4}{3}$-approx, $k = 2$ $\left(\frac{3}{2}+\varepsilon\right)$-approx, $k > 2$
Graphs of bounded degree Δ	NP-hard, $\Delta \geq 3$, $k \geq 2$ APX-hard, $\Delta \geq 4$, $k \geq 2$	
Planar graphs	NP-hard, $\Delta \geq 6$, $k \geq 10$	
Trees	Polynomial (bounded k)	APX-hard (unbounded k)
Complete graphs	$\min\left\{\frac{3}{2}+\varepsilon, C\left(1+\frac{k}{n}\right)\right\}$-approx $O(n^2)$ (bounded k)	$\min\left\{*, 1+\frac{k}{n}\right\}$-approx APX-hard (unbounded k)

5.3.2.1 Computational Complexity for Connectivity

In this section, we describe (in more detail) the most general results concerning the complexity of connectivity that have been obtained for $k \geq 2$ on graphs of bounded degree $\Delta \geq 4$ in the uniform costs case. In fact, in such a setting, the problem becomes APX-hard. The proof proceeds by a polynomial transformation from the well-known minimum 3-set cover problem with bounded occurrences.

X3C: Minimum 3-Set Cover with at Most Three Occurrences

Input	Collection *C* of subsets, each of cardinality at most 3, of a finite set *S* s.t. each element occurs in at most three subsets of *C*
Solution	A set cover for *S*, i.e., a subset $C' \subseteq C$ s.t. every element of *S* belongs to at least one member of C'
Goal	Minimize the cardinality of C'

Problem X3C is known to be APX-hard [13]. Given an instance (C,S) of X3C, it can be transformed into an instance (G,W) of connectivity with $k = 2$ interfaces of uniform costs. Each element of *S* becomes node of the input graph of connectivity holding only interface 1. For each subset belonging to the collection *C*, a gadget of nine nodes is introduced to the connectivity input instance as shown in Figure 5.6. To complete the instance transformation in the reduction, it is necessary to explain the meaning of the ground symbol in the figure. This is a connection to an auxiliary graph. For the moment, it can be assumed that all the ground connections lead to the same additional node (called the root) shared by all the gadgets. Such a node holds both interfaces 1 and 2, and it is connected to two further auxiliary nodes holding only interface 1 and interface 2, respectively. In the modified graph, it is easy to see that to connect all the nodes of a single gadget to the root, it is enough to activate nine interfaces. This is done by activating interface 1 at nodes *d*, *h*, and *i*, and interface 2 at all the remaining six nodes. In this way, all the nodes of the gadget admit a path to at least one ground connection. Note that this is the only possible activation of nine interfaces having the property that each node of the gadget admits a connection path to the root. Moreover, this activation cannot satisfy any connection (dotted lines of Figure 5.6) from the gadget to the elements of the represented subset because, by construction, the

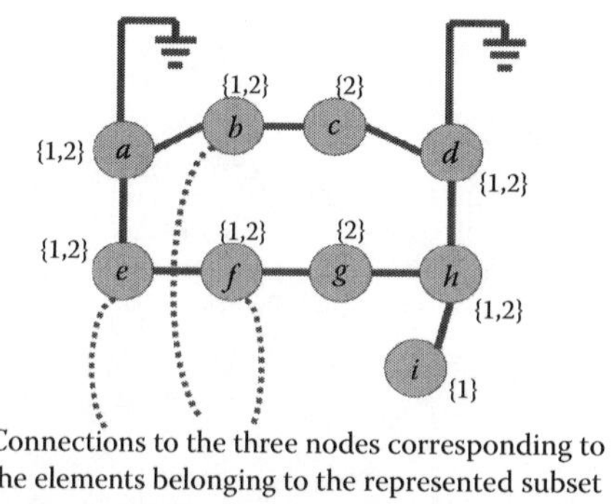

Figure 5.6 Auxiliary gadget for the reduction from X3C to connectivity.

corresponding nodes hold only interface 1. Although by allowing a cost bigger than 9 for a gadget, then there exists an interface allocation of cost 10, which permits a connection between the gadget to both the root and the three external nodes corresponding to the represented subset. This can be realized by activating interface 1 at node i, and both interfaces 1 and 2 at node h, then interface 1 at nodes a, b, e, and f, and interface 2 at the remaining nodes c, d, and g. This configuration connects the gadget to all the nodes of the represented subset to the root. Finally, regardless of the constraints for the other nodes, interface 1 must be activated at each node representing elements of S, both interfaces at the root, and the one available interface for the two auxiliary neighbors of the root. In doing so, one obtains that there exists a solution to the original X3C instance with a cardinality of $|C'|$ if and only if there exists a solution to connectivity in the modified graph with a cardinality of $9|C| + |C'| + |S| + 4$. As X3C is known to be APX-hard even with the additional restriction that $|C| \leq |S|$ (see article by Garey and Johnson [13]), and obviously $|C'| \geq |S|/3$; clearly, connectivity is APX-hard as well.

Because the root has high degree, the construction can be opportunely modified using a tree structure to obtain $\Delta \leq 4$.

5.3.2.2 Approximation Results for Connectivity

A simple algorithm that provides a 2-approximation works as follows. To each edge of the input graph G, it associates a weight equal to two times the cost of the cheapest interface that can be used to establish the corresponding connection. Then, it chooses a minimum spanning tree T_G of the input graph G. Once T_G is chosen, the algorithm simply activates the cheapest interface for covering each edge of T_G. This means that for each covered edge, one interface (at most) at both end points is activated. Moreover, according to the considered cost function, the cost of T_G, that is, the sum of the costs of its edges, is minimum. In fact, the set of edges required for connectivity purposes by an optimal solution cannot have a cost smaller than the cost of T_G. Indeed, consider any optimal interface activation function W_{opt} for connectivity in G. It induces some connected spanning subgraph $G' \subseteq G$ that contains a spanning tree $T_{G'}$ of G'. An edge e is said to cost a, which is denoted by cost(e) = a, if the cheapest common interface available at both its end points has a cost of $a/2$. For each edge $e \in E(T_{G'})$, there must exist an interface of cost not less than cost(e), which is activated in W_{opt} by both the end points of e. Because the edges of $T_{G'}$ can be mapped into distinct vertices of V, and by the minimality of spanning tree T_G, it follows that $c(W_{\text{opt}}) \geq \sum_{e \in E(T_{G'})} \text{cost}(e) \geq \sum_{e \in E(T_G)} \text{cost}(e)$.

On the other hand, for each edge $e \in E(T_{G'})$, the proposed algorithm activates the interface of cost(e) for both its end points, leading to an activation W_A such that $c(W_A) \leq 2 \sum_{e \in E(T_G)} \text{cost}(e)$. By combining the two inequalities, one obtains $c(W_A) \leq 2c(W_{\text{opt}})$.

Actually, as shown in Table 5.3, there exists an algorithm providing a $\left(\frac{3}{2} + \varepsilon\right)$ approximation factor. The interested reader can find all the details in an article by Athanassopoulos et al. [14]. The algorithm is based on a challenging technique that makes use of an "almost" minimum spanning tree in an appropriately defined hypergraph and transforms it to an efficient solution for connectivity. In the same article, another variant for the connectivity problem is considered. In fact, the case wherein different devices have different costs for the same interface is introduced along with some inapproximability results. Clearly, the same model could be applied to other problems as well.

5.3.2.3 Solvable Cases for Connectivity

Similar to the coverage problem, connectivity is solvable when the input graph is a tree or a complete graph, and the case of k bounded is considered. For trees, dynamic programming techniques can be used, whereas for complete graphs, it is possible to exploit the finite number of feasible solutions so that a simple exhaustive search would solve the problem.

5.3.3 Min–Max Version

In this section, both the coverage and the connectivity problems are reconsidered with respect to a different objective function. That is, instead of finding the solutions that minimize the overall cost due to the activation of the interfaces along the whole network, more local objective functions are investigated. In particular, the functions that minimize the maximum cost spent by a single node of the network have been addressed in studies by D'Angelo et al. [15–17]. This is very important for practical purposes because it is the first step toward the study of multi-interface networks in distributed settings. Maintaining a balanced energy consumption in the single nodes helps in prolonging the lifetime of the network. Formally, the coverage problem with respect to the new objective function, denoted by MMCov, can be stated as follows:

MMCov: Minimum–Maximum Cost Coverage in Multi-Interface Networks

Input	A graph $G = (V, E)$, an allocation of available interfaces W: $V \rightarrow 2^{\{1,\ldots,k\}}$ covering graph G, an interface cost function c:$\{1,\ldots, k\} \rightarrow R^+$
Solution	An allocation of active interfaces W_A: $V \rightarrow 2^{\{1,\ldots,k\}}$ covering G such that $W_A(v) \subseteq W(v)$ for all $v \in V$
Goal	Minimize the maximum cost of the active interfaces among all the nodes, i.e., $\min_{W_A} \max_{v \in V} \sum_{l \in W_A(v)} c(l)$

With the new objective function, coverage may change substantially. It is enough to see the example from Figure 5.1. In such a case, MMCov would provide the solution shown in Figure 5.7.

Note that the overall cost of the solution is 5.8 according to the input parameters. However, the results are cheaper than those shown in Figure 5.2 because the maximum spent by one single node is now 1.6 instead of 1.8.

Similarly, the connectivity problem with respect to the new objective function, denoted by MMCon, can be stated as follows:

MMCon: Minimum–Maximum Cost Connectivity in Multi-Interface Networks

Input	A graph $G = (V, E)$, an allocation of available interfaces W: $V \rightarrow 2^{\{1,\ldots,k\}}$ covering graph G, an interface cost function c:$\{1,\ldots, k\} \rightarrow R^+$

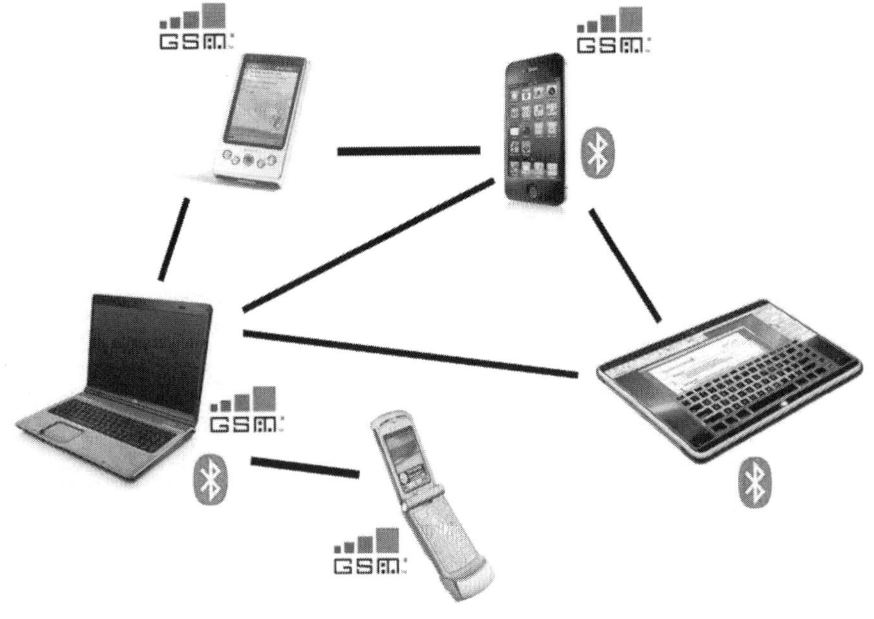

Figure 5.7 Optimal solution for the MMCov when the input instance is provided by the network of Figure 5.1 and the cost function is defined as for Figure 5.2.

Solution	An allocation of active interfaces $W_A: V \rightarrow 2^{\{1,\ldots,k\}}$ covering a connected subgraph $G' = (V, E')$ of G such that $W_A(v) \subseteq W(v)$ for all $v \in V$, and $E' \subseteq E$
Goal	Minimize the maximum cost of the active interfaces among all the nodes, i.e., $\min_{W_A} \max_{v \in V} \sum_{l \in W_A(v)} c(l)$

As expected, moving toward a distributed setting increases the hardness of the corresponding problems. For both problems, in fact, it can be shown that there cannot exist an approximation algorithm with ratio guarantee within $\eta \cdot \ln \Delta$, for a certain constant η, unless $P = NP$. The result comes from a transformation of any instance of the well-known minimum set cover problem into an instance of coverage in the uniform costs case.

SC: Minimum Set Cover

Input	A set U with n elements and a collection $S = \{S_1, S_2, \ldots, S_q\}$ of subsets of U
Solution	A cover for U, i.e., a subset $S' \subseteq S$ such that every element of U belongs to at least one member of S
Goal	Minimize $\|S'\|$

Because the proof exploits a star graph to build the corresponding instance of MMCov, the same result also holds for MMCon because the two problems coincide on tree networks. A summary of the obtained results for MMCov is shown in Table 5.4.

Actually, the general approximation ratio can be refined with respect to different graph classes to $(\ln \Delta + 1 + b \cdot \min\{c_{max}, \ln \Delta + 1\})$ approximation, with parameter b, which may vary according to the considered case. As shown in Table 5.4, this is equivalent to $O\left(\frac{m}{n}\right)$ in the general case. Other values with respect to different graph classes are shown in Table 5.5.

As a sampling result for the MMCov problem, we describe the simple algorithm that provides an optimal solution for the case of $k \le 3$ in the uniform costs case. Be reminded that the objective function aims at minimizing the maximum cost spent among the nodes. The algorithm is based on the next analysis. When $k = 2$, either there exists one common interface for all the nodes or the optimal solution costs 2, which implies that all the available interfaces should be activated. When $k = 3$, if there exists a solution that costs 1, again it is easily verifiable by checking whether all the nodes hold one common interface. If not, to check whether there exists a solution that costs 2, the algorithm proceeds as follows. If a node holds less than three interfaces, all such interfaces are activated. For each node v holding three interfaces, it is possible to check whether at most two of the interfaces among the available three are enough to connect v to all its neighbors holding less than three interfaces. If not, then the optimal solution costs 3 and all the nodes can activate all their interfaces to accomplish the coverage task. If yes, then v activates the two interfaces induced by its neighborhood. In this way, all the edges connecting two nodes holding at most two interfaces and all the edges connecting a node holding three interfaces with a node holding at most two interfaces are established. To conclude the algorithm, it is necessary to ensure the establishment of all the edges between nodes holding three interfaces are from the designed activation function. Indeed, because each node holding three interfaces activates two interfaces, every two such neighbors must share at least one common interface, and hence the algorithm is correct.

This simple algorithm cannot be extended to MMCon. Indeed, similar arguments can be applied only for instances with $k \le 2$, in which either all the nodes share one common interface or at least one node has to switch-on two interfaces. In the latter case, according to the defined

Table 5.4 Hardness and Approximation Results for MMCov

Graph Class	*Nonuniform Costs*	*Uniform Costs*
General graphs	Polynomial for $\Delta \le 2$ $\min\{1+(k-2),\Delta\}\frac{c_{max}}{2c_{min}}$-approx	Polynomial for $\Delta \le 2$ or $k \le 3$ $\min\left\{\frac{k}{2},\frac{\Delta}{2}\right\}$-approx
	NP-hard, $\Delta \ge 5$, $k \ge 16$ Not approx within $\eta \ln \Delta$ $\left(\ln\Delta+1+O\left(\frac{m}{n}\right)\min\{c_{max},\ln\Delta+1\}\right)$-approx	
Trees	Not approx within $\eta \ln \Delta$ (both Δ and k unbounded)	Polynomial (bounded k or bounded Δ)

Table 5.5 Parameter b with Respect to Different Graph Classes

Graph Class	*Parameter b*
Planar graphs	$b \le 3$
Graphs with genus g	$b = O(1+\sqrt{g})$
Graphs with arboricity a	$b \le a$
Graphs of bounded degree Δ	$b \le \Delta$
Graphs with page number p	$b \le p$
Graphs with treewidth t	$b \le t$

cost function, the solution where all the nodes switch-on all the available interfaces will have the same cost of the solution in which only one node has activated two interfaces. A summary of the obtained results for MMCon is shown in Table 5.6.

Just to mention one further problem from the classic graph theory used in the context of multi-interface networks to obtain hardness results, we point out that the proof for the NP-hardness holding even for the uniform costs case with fixed parameters $\Delta \ge 3$ and $k \ge 10$ makes use of a polynomial time transformation from the well-known Hamiltonian path problem:

HP: Hamiltonian Path

Input	A graph $G = (V, E)$
Question	Does G contain a Hamiltonian path, i.e., a path that visits each node exactly once?

In this respect, it is also worth mentioning that in the unit cost case, if the input graph is a polynomially recognizable Hamiltonian graph (i.e., it admits Hamiltonian path), MMCon can be optimally solved in $O(k^{\Delta}n)$ time.

Table 5.6 Hardness and Approximation Results for MMCon

Graph Class	*Nonuniform Costs*	*Uniform Costs*
General graphs	$\min\{1+(k-2),\Delta\}\frac{c_{max}}{2c_{min}}$-approx	$\min\left\{\frac{k}{2},\frac{\Delta}{2}\right\}$-approx
	Polynomial for $k \le 2$ Not approx within $\eta \ln \Delta$	
Graphs of bounded degree Δ	NP-hard for $\Delta \ge 3$ and $k \ge 10$ polynomial for $\Delta \le 2$	
Trees	$(\ln \Delta + 1 + \min\{c_{max} \ln \Delta + 1\})$-approx	$(\ln \Delta + 2)$-approx
	Optimally solvable if k or Δ are bounded	

5.4 Cheapest Paths

The cheapest path problem was first addressed by Barsi et al. [18] and Kosowski et al. [12]. This corresponds to the well-known shortest path problem but in the context of multi-interface networks in which costs are associated to the interfaces (and hence to the nodes) rather than to the edges of the input graph G. A path P in G from a given source node s to a target node v is denoted by a sequence of couples: for each node $v_j \in P$, besides node v_j itself, the interface i_j used to reach v_j is given. Namely, $i_j \in W(v_j)$. For example, the sequence $P = (s \equiv v_0), (v_1, i_1), \ldots, (v \equiv v_t, i_t)$ denotes a path P from s to v that moves on the nodes $s, v_1, \ldots, v_{t-1}, v$ and that reaches node v_j via interface i_j, for $1 \leq j \leq t$. Interface 0 is used to denote "no interface" because the source is not reached by any other node in P. Conventionally, $c(0) = +\infty$. However, the source needs to activate interface i_1 to reach v_1, hence the cost of activating the edge (s, v_1) is $2c(1)$. In general, the cost for activating the path P is

$$d_P(v) = \sum_{j=1}^{t} \text{cost}\big((v_{j-1}, i_{j-1}), (v_j, i_j)\big)$$

where

$$\text{cost}\Big(\big(v_{j-1}, i_{j-1}\big), (v_j, i_j)\Big) = \begin{cases} 2c(i_j) \text{ if } i_{j-1} \neq i_j \\ c(i_j) \text{ otherwise} \end{cases}$$

Let $\delta(v)$ be the minimum cost to activate a path from the source node s to node v, that is, $\delta(v) = \min\{d_P(v)\colon P$ is any path from s to $v\}$. In addition, let the cheapest path CP_v from the source s to v be any path P from s to v such that $d_P(v) = d(v)$. An i-path P from s to v is a path from s to v that reaches v via interface i. Let $d_P(v,i)$ denote the cost of the i-path P, whereas $d(v,i)$ denotes the minimum cost among all the i-paths from s to v. Furthermore, let the cheapest i-path $CP_{v,i}$ from the source node s to node v be any i-path P such that $d_P(v, i) = d(v, i)$. Clearly, $\delta(v) = \min\{\delta(v,i)\colon i \in W(v)\}$. Whenever clear by the context, we remove P from the notation d_P. We study the so-called shortest path problem but in the context of multi-interface networks. Actually, in these networks, dealing with shortest paths is neither of practical nor of theoretical interest. In fact, as shown above, the cost of an edge is not set up a priori as in the standard problem, but depends on the activated interfaces at its end point. The cheapest path (CP) problem can be formulated as follows:

CP: Cheapest Path

Input	A graph $G = (V, E)$, an allocation of available interfaces $W\colon V \rightarrow 2^{\{1,\ldots,k\}}$ covering graph G, an interface cost function $c\colon\{1,\ldots, k\} \rightarrow R^+$, and a source node $s \in V$
Solution	For each node $v \in V/\{s\}$, a path P from s to v must be specified by a sequence of couples of the form (v_j, i_j), with $v_0 = s$, $i_0 = 0$, $v_t = v$ and $v_j \in V$, $i_j \in W(v_j)$ for $1 \leq j \leq t$ with the meaning that node v_j is reached with interface i_j
Goal	For each node $v \in V$, find $\delta(v)$ along with CP_v

Considering the example of Figure 5.1, but with the cost function specified by $c(1) = 0.6$; $c(2) = 0.8$; $c(3) = 1$; $c(4) = 1.2$, then the optimal solution for the cheapest path when assuming the PDA on the top left of the figure as the source s requires the activation of three GSM interfaces on the nodes from s to the cellular phone on the bottom of the figure, whereas it requires two IrDA interfaces to communicate s with the laptop. Therefore, it turns out that the main property of the standard shortest path problem does not hold. That is, a subpath of the cheapest path is not necessarily the cheapest path itself. Although the suboptimality property does not hold in general, it holds when a subpath of the cheapest path is characterized not only by its final node but also by the interface used in its last hop. In fact, it can be proved that given a graph $G = (V, E)$ and a source node $s \in V$, let CP_v be the cheapest path from s to v that passes through node u and reaches u via interface i. Then, the i subpath of CP_v from s to u is the cheapest i path. As a consequence, the cost of the cheapest path from s to v can be easily determined when the end points of the final edge of such a path and the interface used to reach v are given.

Thus, the cheapest path problem on a graph $G = (V, E)$ can be solved by using the standard Dijkstra algorithm [2] on a slightly modified input instance of G'. The directed weighted graph $G' = (V',E')$ with $n' = |V'|$ and $m' = |E'|$ is obtained from G as follows. Each node $v \in V$ is replaced by $|W(v)| + 1$ nodes: one node for each interface available at node v, $\{v_i: i \in W(v)\}$, and one extra node v'. These nodes are connected as a star with center v' as follows: for all $i \in W(v)$, edges (v_i, v') have weight 0, whereas edges (v', v_i) have weight $c(i)$. Moreover, each edge $\{u, v\} \in E$ is replaced by $|W(u) \cap W(v)|$ pairs of edges: for each $i \in W(u) \cap W(v)$, a new pair of arcs (u_i, v_i) and (v_i, u_i) is defined, each with a weight of $c(i)$. By considering the instance of Figure 5.1, an example of the transformation is

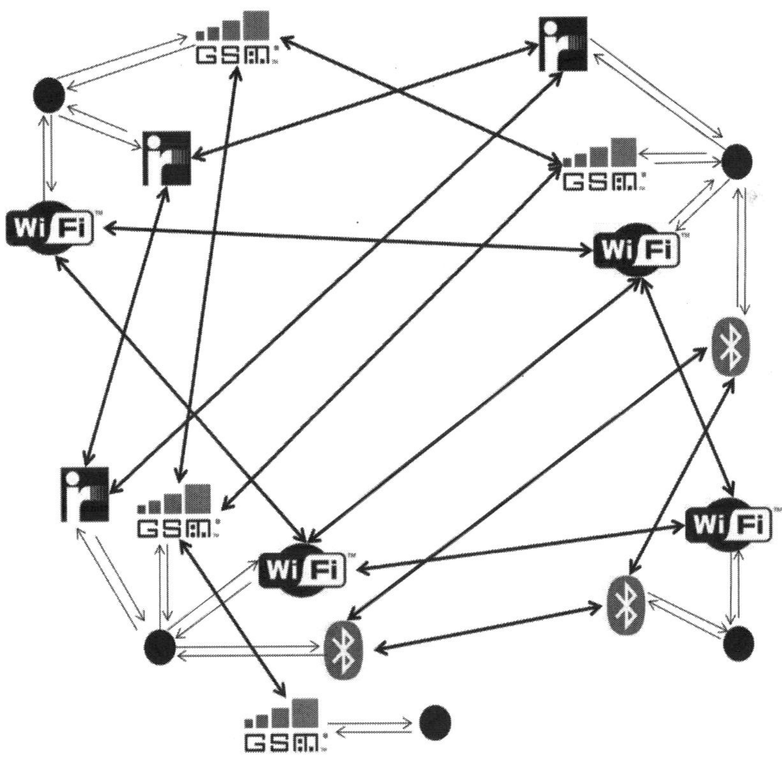

Figure 5.8 The directed graph G' obtained from the graph G of Figure 5.1. Costs can now be associated directly on edges hence allowing Dijkstra's algorithm to solve the cheapest paths problem.

shown in Figure 5.8. Solving the cheapest path problem with source s on G is equivalent to solving the shortest path tree problem on G' with source s', and the cost of the path from the source to a generic node $v \in V$ in G is equal to $d_P(v')$ in G'. Hence, Dijkstra's algorithm [2] can be applied to G', achieving a time complexity of $O(m' + n' \log n') = O[km + kn \log (kn)]$. For the uniform cost case, each of the arcs of G' has a weight of either 0 or 1, and thus the time complexity is reduced to $O(m' + n') = O[k(m + n)]$.

5.5 Maximum Matching

A natural continuation to investigating multi-interface networks is certainly to consider another basic combinatorial problem such as maximum matching. From the standard algorithmic graph theory, maximum matching can be formally stated as follows:

Maximum Matching

Input	Graph $G = (V,E)$
Solution	Set $M \subseteq E$ of pairwise disjointed edges
Goal	Maximize the cardinality of M

That is, no two edges in the solution M share a common node. The maximum matching problem is well-known and it is polynomially solvable [19]. For a positive integer p, the distance $- p$ maximum matching problem asks for a maximum size set of edges whose pairwise distance is at least p in G. The distance between two edges is defined as the length (number of edges) of the shortest path connecting them. The distance $- p$ maximum matching is, in general, NP-hard [20,21]. In particular, the special case of $p = 2$ is referred to in the literature as the *maximum induced matching* (MIM) problem and it has been extensively studied. For instance, a solution for MIM on the underlying graph shown in Figure 5.1 would be given by only one edge. MIM has been shown to be NP-hard for various graph classes like bipartite graphs [21], line graphs (and hence including claw-free graphs) [22], and trapezoid graphs [23]. Moreover, it is APX-complete in d regular graphs, $d \geq 3$ [24], and a $(d - 1)$-approximation algorithm has been provided for d regular graphs, $d \geq 3$. The size of MIM on twinless graphs and planar graphs has been studied by Kanj et al. [25], and for subcubic planar graphs by Kang et al. [26]. Polynomial time algorithms have been devised for chordal graphs [21], trees[27], and other special graphs. It is worth mentioning that, when $p = 3$, the problem is NP-hard [20] even for strongly chordal graphs.

In Kosowski et al. [28], the maximum matching problem has been studied in the context of multi-interface networks. The aim is to maximize the number of parallel connections without incurring interferences. Two communications/edges are assumed to interfere if they share a node or if they are activated by the same interface and connected by one edge. Note that, in the latter case, when two edges at one distance are activated by the same interface, the edge in the middle is also activated by the same interface because its end points share a common active interface.

An example of such a behavior is shown in Figure 5.9. The two black full edges represent a possible solution for maximum matching in the context of multi-interface networks if the black edge on top is activated by the Wi-Fi interface and the one on the bottom by the GSM interface. If both edges are activated by the GSM interface, then it will not represent a feasible solution because the connection between the laptop and the PDA on its top will be activated as well. It follows that the

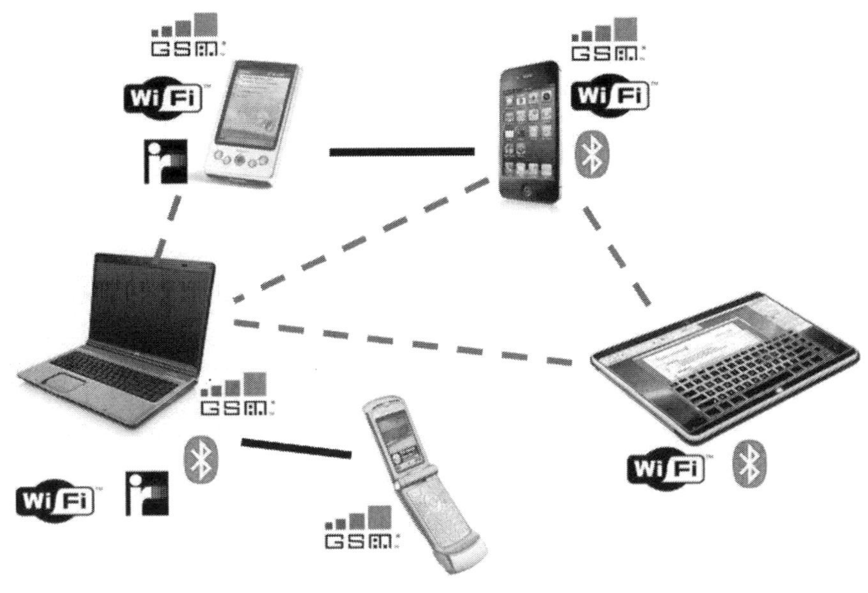

Figure 5.9 A possible solution for the maximum matching in the context of multi-interface networks from the instance provided in Figure 5.1 composed of the black full edges that must be covered by different interfaces.

main difference with the standard matching problem is that when an edge is selected, it is associated with an interface by which the connection is established. Hence, if an edge $e \in E$ is selected for matching along with a communication interface i, then any other edge $e' \in E$ selected for the matching at distance 1 from e in G must be selected with interface $j \neq i$. This problem is referred to as the *maximum matching in multi-interface* (3MI) networks and it can be formulated as follows:

3MI: Maximum Matching in Multi-Interface Networks

Input	A graph $G = (V, E)$, an allocation of available interfaces W: $V \rightarrow 2^{\{1,\ldots,k\}}$ covering graph G
Solution	An allocation of active interfaces W_A: $V \rightarrow [\emptyset, \{1\}, \{2\}, \ldots, \{k\}]$, $W_A(v) \subseteq W(v)$ for all $v \in V$ such that if $W_A(v) \neq \emptyset$ then there is exactly one neighbor of v, say w, with $W_A(w) = W_A(v)$
Goal	Maximize the number of edges $(v,w) \in E$ such that $W_A(v) = W_A(w) \neq \emptyset$

3MI has been considered in both the bounded and unbounded cases. Moreover, when $k = 1$, that is, a single interface is available, 3MI coincides with the well-known induced matching. When each edge could be covered by an interface different from any other, 3MI coincides with the standard maximum matching problem. It follows that the size of a solution to an instance G of 3MI cannot be bigger than that of the solution provided by the standard maximum matching, and cannot be smaller than that of a solution provided by its induced version, that is, MIM $(G) \leq$ 3MI $(G) \leq$ matching (G).

Table 5.7 Complexity Results Achieved for 3MI

Graph Classes	*k*	*Complexity*
Complete	Unbounded	Optimally solvable in $O(n^{3/2}k^{3/2})$
Complete bipartite	Unbounded	Optimally solvable in $O(n^{3/2}k^{3/2})$
Bounded treewidth	Unbounded	Optimally solvable in polynomial time
Interval	Bounded	Optimally solvable in polynomial time
Unit interval	Unbounded	NP-hard
Bipartite, $\Delta > 3$	Bounded	NP-hard (from Cameron [21])
Claw-free	Bounded	NP-hard (from Kobler and Rotics [22])

In particular, when $k = 1$, 3MI coincides with MIM. Hence, all the hardness results concerning MIM also hold for 3MI. However, it is worth mentioning that MIM has been proved to be polynomially solvable on chordal graphs whereas the distance-3 maximum matching, whose size is clearly smaller than MIM, is NP-hard for chordal graphs but polynomially solvable on strongly chordal graphs. In an article by Kosowski et al. [28], it is proved that 3MI is NP-hard on proper interval graphs (which are chordal graphs), and some polynomial time algorithms are presented for complete graphs, complete bipartite graphs, bounded treewidth graphs, and even for interval graphs but only in the bounded case.

5.5.1 Computational Complexity for 3MI

In this section, the most general result concerning the complexity of 3MI is described. It has been obtained for unit interval graphs in the unbounded case. In graph theory, an interval graph is the intersection graph of a multiset of intervals on the real line. It has one vertex for each interval in the set, and an edge between every pair of vertices corresponding to intervals that intersect. A unit interval graph is an interval graph in which all the intervals have the same length. The proof of NP-hardness for 3MI proceeds by a polynomial transformation from the well-known three-satisfiability problem [13] to the underlying decisional problem $3MI_D$ obtained from 3MI by adding a bound B and asking whether there exists a solution for 3MI whose size is not less than B. The 3-satisfiability problem can be stated as follows:

3SAT: 3-Satisfiability

Input	A set U of variables, a collection C clauses over U, such that each $c \in C$ satisfies $\lvert c \rvert = 3$
Question	Is there a satisfying truth assignment for C?

Given an instance of 3SAT, an instance of $3MI_D$ can be obtained with a proper interval graph G in which one has to decide which interfaces to switch-on at each node (Figure 5.10). For each variable $v \in U$, there is a chain of unit intervals (as shown in Figure 5.10 for the four variables x, y, z, and w), each interval is associated with interfaces v, $-v$, and one further interface denoted as 1. Actually, all the nodes of G contain interface 1. Each clause $c \in C$, composed of three literals, corresponds to G in 10 unit intervals vertically disposed as shown in the figure. There are two

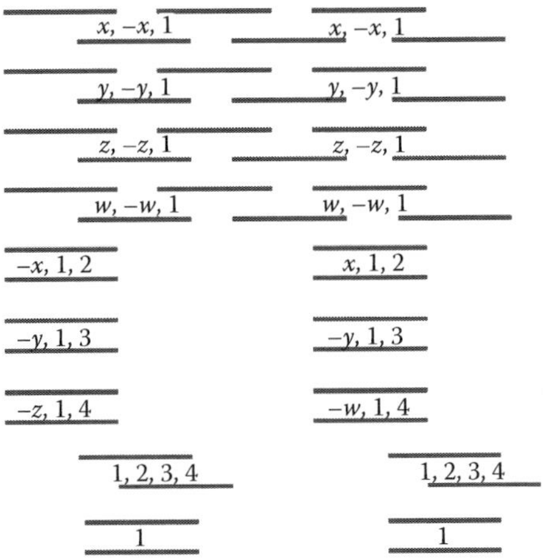

Figure 5.10 The unit interval graph G obtained from the transformation of clauses *x, y, z* and *−x, y, w* to $3MI_D$. Interfaces held by the intervals are specified between the lines.

intervals associated for each literal l with interfaces $-l$, 1, and one further interface among the set {2, 3, 4} (a different interface per pair). Plus, there are two intervals associated with interfaces 1, 2, 3, and 4, and two other intervals with only interface 1. The bound B is set to $(5 + 2|U|)|C|$.

When 3SAT admits a positive answer, that is, there exists a satisfying truth assignment for C, then $3MI_D$ also has a positive answer, that is, there exists an activation of the available interfaces such that the number of edges in the obtained matching is at least B, and vice versa. In fact, for each variable v, if v is set to true in the solution of 3SAT, then, along the chain of unit intervals corresponding to v, interface v is switched-on at the first two intervals, interface $-v$ at the second two intervals, again interface v at the third two intervals, and so forth, always alternating between v and $-v$. If v is set to false in the solution of 3SAT, then the activations of interfaces v and $-v$ in the chain corresponding to the variable v would still be alternated but starting with $-v$. This provides a contribution of $2|U||C|$ established edges to the final solution. Note that, in this way, all the intervals corresponding to the clauses will be located below intervals that have activated interface v in the first case, and interface $-v$ in the second case. In the first case (in the second case, respectively), it follows that all the clauses containing the literal v ($-v$, respectively) have two intervals in their representation of G containing interface $-v$ (v, respectively). Hence, such intervals are allowed to switch-on the interface corresponding to the negation of the apparent literal because this is not in conflict with the activation in the chain above. Because 3SAT is supposed to have been solved, then there will always be an available interface for each clause that determines the connection of one pair of intervals among the 10 associated with the clause, as described previously. This will contribute $|C|$ edges to the final solution. Then, interface 1 is switched-on at all the bottom edges that hold only such an interface, for a contribution of other $|C|$-activated edges. Finally, the remaining three pairs of intervals corresponding to each clause can be connected by interfaces 2, 3, and 4, hence contributing other $3|C|$ edges. In total, the obtained solution for $3MI_D$ was composed of $(5 + 2|U|)|C|$ edges.

On the contrary, when $3MI_D$ is assumed to have a positive answer, as graph G is composed of $2(5 + 2|U|)|C|$ nodes, the solution for $3MI_D$ must be composed of exactly $(5 + 2|U|)|C|$ edges, hence all the nodes participate to the matching. This means that nodes holding only interface 1 must be coupled among themselves. Any other node cannot use interface 1. For each generic clause

$c = l_1 l_2 l_3$, the corresponding nodes holding only interfaces 1, 2, 3, and 4, must be connected among themselves by using one interface among the set {2, 3, 4} because one of them does not share any available interface with other intervals. This means that among the other six intervals representing c, at least two of them must be connected among themselves by interface $-l_1$, or $-l_2$, or $-l_3$. Without loss of generality, let $-l_1$ be the interface used. It follows that on top of graph G, in the chain corresponding to variable $|l_1|$, the two interval neighbors to the ones connected by interface $-l_1$ must be connected by interface l_1 because no other options are available. As a "chain effect," the subsequent two intervals on the same chain (if any), and the previous two intervals on the same chain (if any) must use interface $-l_1$. The same arguments can be applied to each chain, and the corresponding truth assignment for the underlying 3SAT problem is given by the interface used at the first two intervals of each chain. In fact, from the assumption that the solution for $3MI_D$ has size $(5 + 2|U|)|C|$, it follows that all the intervals representing a clause have used at least one interface corresponding to the set of variables appearing in the instance of 3SAT. That is, the assignment provided by the solution of $3MI_D$ is compatible with the satisfaction of all the clauses of 3SAT.

It is worth noting that 3MI can be optimally solved in the bounded case for interval graphs. According to the result shown above, this is one further hint on the drastic change in the complexity when moving from the unbounded to the bounded case of the considered problems, as it was for the coverage problem on some graph topologies, for instance. The algorithm to solve 3MI in the bounded case on interval graphs makes use of standard dynamic programming techniques that lead to an overall runtime bounded by $O(n^{8k+2})$.

5.6 Maximum Flow and Minimum Cost Flow

This section focuses on the issues arising from considering bandwidth capacities for the interfaces (see articles by Bertossi et al. [29] and D'Angelo et al. [30,31]). In real-world networks, it is unavoidable to consider bandwidth constraints as they correspond to technological bounds that are always present in real devices. Thus, this can be considered as a step forward toward the implementation of multi-interface algorithms in real networks. In more detail, this section surveys the results of two fundamental optimization problems in the context of multi-interface networks that arise when bandwidth bounds are given on the interfaces, that is, each interface is associated with a maximum bandwidth capacity that it can guarantee. The first problem, called *maximum flow in multi-interface* (MFMI) networks, was introduced by D'Angelo et al. [31] and aims to find the maximal communication bandwidth that can be guaranteed between two given nodes. This is similar to the classic problem of finding the maximum flow between two nodes in a network. The main difference resides in the fact that, in MFMI, the bandwidth capacities are associated with the interfaces instead of the edges. Therefore, a node v can communicate with many other nodes using a single interface i but, if v uses the whole bandwidth of i to transmit to (receive from, respectively) with a neighbor u, it cannot use i to to transmit to (receive from, respectively) with another neighbor w, even if i belongs to both v and w. The second problem was introduced in an article by Ahuja et al. [32], and aims to establish the cheapest way of communication between two given nodes while guaranteeing a minimum bandwidth of communication. Such a problem, called *minimum-cost flow in multi-interface* (MCFMI) networks is similar to the better known minimum cost flow problem [30]. Again, we do not consider costs and capacities for the edges but we have to cope with interfaces at the nodes that require some costs and can manage some maximum bandwidths.

The MCFMI and MFMI problems have been introduced and theoretically studied by Ahuja et al. [32] and D'Angelo et al. [31], respectively. Moreover, the MCFMI problem has been experimentally studied by D'Angelo et al. [31]. A unified study of both problems is given in an article by D'Angelo et al. [33]. In the following, we survey the results given in these articles after having formally defined the two problems. Actually, in the formalization, the main difference with the multi-interface model presented thus far deserves some attention. In fact, when dealing with flow problems, the model has been refined to cope with the case in which two devices are not guaranteed to communicate through all common interfaces shared if they are close enough, but only by a subset of such interfaces. This reflects more practical cases in which the communication range of different interfaces may vary. Given a set of nodes V, it follows that in the input parameters of both MFMI and MCFMI, the usual covering function W is "substituted" by the so called *sharing* function X: $V \times V \rightarrow 2^{\{1, 2, \ldots, k\}}$, which defines the interfaces that each pair of nodes can use to communicate. Function X must also satisfy two properties: for each $u \in V$, $X(u,u) = \emptyset$; for each $u,v \in V$, $X(u,v) = X(v,u)$. In doing so, function X induces a global assignment of interfaces to the nodes in V given in terms of an appropriate interface assignment function $W: V \rightarrow 2^{\{1, 2, \ldots, k\}}$ defined as $W(v) = \bigcup_{u \in V} X(u,v)$. If an interface i is in $X(u, v)$ for some nodes u and v, then $i \in W(u)$, $i \in W(v)$, and u and v are close enough to communicate via interface i. It follows that, for each u, $v \in V$, $X(u, v) \subseteq W(u) \cap W(v)$. The use of those functions represents a generalization of the model with respect to earlier works on the subject. Note that, the above definitions of V and X induce a graph $G = (V,E)$ where $\{u, v\} \in E$ if and only if $X(u, v) \neq \emptyset$.

MFMI: Maximum Flow in Multi-Interface Networks

Input	A set of nodes V, a source node $s \in V$, a target node $t \in V$, a set of interfaces $I = \{1, 2, \ldots, k\}$, a sharing function $X: V \times V \rightarrow 2^I$, and an interface bandwidth function $b: I \rightarrow Z_0^+$
Solution	A flow function $f: V \times V \times I \rightarrow Z_0^+$ such that: 1. $f(u, v, i) = -f(v, u, i)\ \forall\ u,v \in V, i \in I$; 2. $f(u, v, i) = 0$ if $X(u, v) = \emptyset\ \forall\ u,v \in V, i \in I$; 3. $\sum_{v \in V: f(u,v,i)>0} f(u,v,i) \leq b(i) \forall u \in V,\ i \in I$; $\sum_{v \in V: f(v,u,i)>0} f(v,u,i) \leq b(i) \forall u \in V,\ i \in I$; 4. $\sum_{v \in V, i \in I} f(u,v,i) = 0 \forall u \in V \setminus \{s,t\}$
Goal	Maximize the total flow from s to t, $F = \sum_{v \in V, i \in I} f(s,v,i) = \sum_{v \in V, i \in I} f(v,t,i)$

Again, considering Figure 5.1 as a sampling network in which the $k = 4$ ordered interfaces (IrDA, Bluetooth, GSM, and Wi-Fi) are associated with the cost function $c(1) = 2$; $c(2) = 6$; $c(3) = 10$; $c(4) = 12$, and bandwidth function $b(1) = 20$; $b(2) = 10$; $b(3) = 3$; $b(4) = 100$, let the PDA on the top left of the figure be the source s and the tablet on the bottom right be the target node t. Then, the resolution of MFMI would provide a maximum bandwidth between s and t of 110. As shown in Figure 5.11, this can be realized by activating two different paths from s to t.

One that exploits a connection via IrDA and then moved to Bluetooth, whereas the other that requires two hops via Wi-Fi. As shown in the definition of the problem, when dealing with MFMI, costs are negligible because the problem only asks for the maximum bandwidth that can be guaranteed between source s and target t, without taking care of the costs. In practical

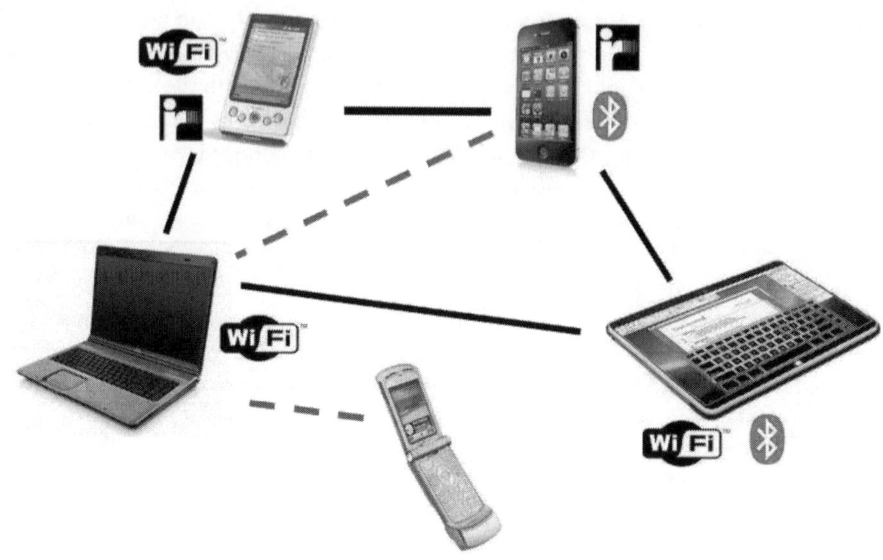

Figure 5.11 A possible solution for MFMI from the instance provided in Figure 5.1 when the PDA on the top left is the source *s*, and the tablet on bottom right is the target *t*.

applications, this is usually not the case. However, studying MFMI may provide an important measure of the network studied.

In practical cases, it is common to ask for a minimum bandwidth guarantee rather than to require the maximum available. This is the case faced by defining MCFMI.

MCFMI: Minimum Cost Flow in Multi-Interface Networks

Input	A set of nodes V, a source node $s \in V$, a target node $t \in V$, a set of interfaces $I = \{1, 2, \ldots, k\}$, a sharing function $X: V \times V \to 2^I$, and an interface bandwidth function $c: I \to Z_0^+$, an interface bandwidth function $b: I \to Z_0^+$ and a bound $B \in Z_0^+$
Solution	An allocation of active interfaces $W_A: V \to 2^I$, $W_A(v) \subseteq \bigcup_{u \in V} X(u,v)\ \forall\, v \in V$ and a flow function $f: V \times V \times I \to Z_0^+$ such that: 1. $f(u, v, i) = -f(v, u, i)\ \forall\, u,v \in V, i \in I$; 2. $f(u, v, i) = 0$ if $W_A(u) \cap W_A(v) \cap X(u, v) = \varnothing\ \forall\, u,v \in V, i \in I$; 3. $\sum_{v \in V: f(u,v,i)>0} f(u,v,i) \le b(i)\ \forall u \in V, i \in I$; $\sum_{v \in V: f(v,u,i)>0} f(v,u,i) \le b(i)\ \forall u \in V, i \in I$; 4. $\sum_{v \in V, i \in I} f(u,v,i) = 0\ \forall u \in V \setminus \{s,t\}$ 5. $\sum_{v \in V, i \in I} f(s,v,i) = \sum_{v \in V, i \in I} f(v,t,i) \ge B$
Goal	Minimize the total cost of the active interfaces, $c(W_A) = \sum_{v \in V} \sum_{i \in W_A(v)} c(i)$

The problem, in fact, introduces an input parameter B that defines such a required minimum bandwidth and hence the aim is to find the cheapest way to guarantee B units of flow between s and t. Considering again the example of Figure 5.11, if $B > 110$, then MCFMI cannot be solved; if $100 < B \leq 110$ then the solution provided in the figure represents the optimum; if $30 \leq B \leq 100$ then the only path established with three Wi-Fi interfaces is required. For smaller vales of B, feasible solutions can be made of different combinations.

For both MFMI and MCFMI problems, we denote by $G = (V,E)$ the graph induced by the sharing function X, also referred to as the *input graph*. Graph G can be easily computed from X in time $O(|V| + |E|)$.

Again, two variants of the MCFMI problem have been considered: the parameter k can be considered as part of the input (unbounded case), or k may be a fixed constant (bounded case).

In terms of computational complexity and approximation bounds, the two problems turn out to be very different between them and with their classic counterpart. However, they share some common characteristics. In particular, some of the algorithms provided rely on a specific graph transformation that turns MFMI and MCFMI into the classic maximum flow and minimum cost flow problems, respectively. The computational complexity and the approximation bounds obtained by applying such a transformation will be studied in the following subsections. The previously mentioned transformation is described next.

5.6.1 Graph Transformation

For each interface of each node, there is an arc that has the same cost and bandwidth of the interface being considered. The head of each such arc is connected to the tail of another arc of the same kind if they share an interface or if they represent different interfaces of the same node. As each of these arcs is associated with the cost and the bandwidth of the interface it represents, the activation of an interface is modeled with the usage of one of these arcs, preserving the bandwidth constraints and the activation costs. Moreover, although the arcs are directed, the possibility to communicate toward and from every node of the original graph is preserved. Given a graph $G = (V, E)$, we denote as $G' = (V', E')$ the directed graph obtained and by b' and c' the associated bandwidth bounds and costs. For each node $v \in V$ and each interface $i \in W(v)$, there are two nodes in V' denoted as $(\bar{v},i)$ and $(\underline{v},i)$: $V' = \left\{(\bar{v},i),(\underline{v},i) \mid v \in V, i \in W(v)\right\} \cup \{\tilde{s},\tilde{t}\}$. The arcs are the following: $A = \{[(\bar{v}, i), (\underline{v}, i)] | v \in V, i \in W(v)\} \cup \{[(\underline{v}, i), \bar{v}, j)] | v \in V, i, j \in W(v) \text{ s.t. } i \neq j\} \cup \{[(\underline{u}, i), (\bar{v}, i)] | i \in X(u, v)\} \cup \{[\tilde{s}, (\bar{s}, i)], [(\underline{t}, j), \tilde{t}] | i \in W(s), j \in W(t)\}$.

An example is given in Figure 5.12, in which node v, for example, is transformed into six nodes and nine arcs.

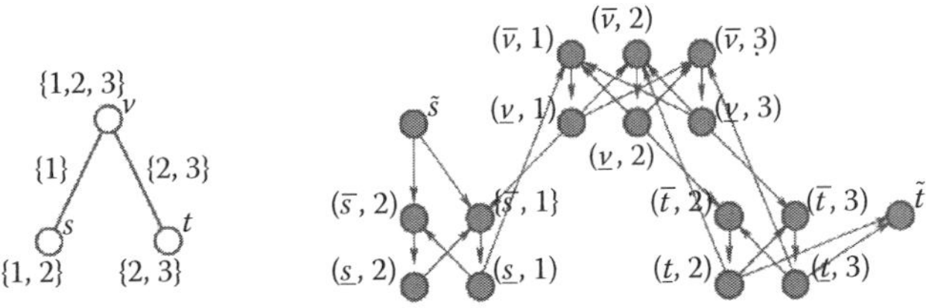

Figure 5.12 The graph G and its transformation in the direct graph G'. $W(s) = \{1,2\}$, $W(v) = \{1,2,3\}$, $W(t) = X(v,t) = \{2,3\}$, and $X(s,v) = \{1\}$.

The bandwidth capacity of each arc $\left[(\bar{v},i),(\underline{v},i)\right]$ is set to $b'\left[(\bar{v},i),(\underline{v},i)\right]=b(i)$ whereas the capacity of each other arc in A is unlimited and it is 0 for each pair in $V \times V/A$. The cost $c'(a)$ of each arc $a=\left[(\bar{v},i),(\underline{v},i)\right]$ is set to $c(i)$ and it is 0 for the remaining arcs. Given a flow function f' from $\tilde{s}$ to $\tilde{t}$ for G', we define a flow function f from s to t in G as follows:

$$f(u,v,i)=\begin{cases} f'\left[(\underline{u},i),(\bar{v},i)\right]-f'\left[(\underline{v},i),(\bar{u},i)\right] & \text{if } i \in X(u,v) \\ 0, & \text{otherwise.} \end{cases}$$

The allocation of active interfaces at node u for MCFMI is defined as $W_A(u)=\{i \in W(u) \mid \exists\, v \in V \text{ s.t. } f(u,v,i) \neq 0\}$. To address MCFMI and MFMI concurrently, we define W_A for MFMI as equivalent to the assignment function W induced by the sharing function X, that is, $W_A(u) \equiv W(u)$. Note that both functions f and W_A can be computed in polynomial time once function f' is known.

By applying the above transformation to an instance I of MFMI or MCFMI, and computing a flow function f and an assignment of interfaces W_A for I by using the above definition on some flow function f' of G', it is possible to obtain a function f, which satisfies properties 1 to 4 needed by both definitions of MFMI and MCFMI.

It follows that graph G' instead of G can be used to cope with problems MFMI and MCFMI. In the next two sections, we investigate such problems in more detail.

5.6.2 Maximum Flow

In this section, it is shown that MFMI can be solved in polynomial time. In particular, an algorithm that makes use of the above graph transformation is described to solve MFMI in polynomial time. Given an instance I_1 of MFMI, the algorithm first transforms the graph G and the function b of I_1 into a graph G' and a function b' as described above, obtaining an instance I_2 of the classic maximum flow problem. Then, in polynomial time, it finds a classic maximal flow function f' for I_2 by using a maximum flow algorithm such as the famous Edmonds–Karp algorithm [34,35]. Finally, the algorithm obtains a maximal flow function f for I_1 from f' by using the transformation given above.

The computational time required by such an algorithm is polynomial because it is given by the cost of transforming I_1 into I_2 and that of solving I_2. Moreover, f is an optimal solution for I_1. In fact, f satisfies properties 1 to 4 of the definition of MFMI and if, by contradiction, we assume that there exists a flow function ensuring a flow greater than that given by f in G, then there would exist a flow function ensuring a flow greater than that of f' in G' (see the article by Ahuja et al. [32] for more details).

5.6.3 Minimum Cost Flow

In this section, we survey the results of the MCFMI problem. In contrast with MFMI, it turns out that, in general, MCFMI is NP-hard and hence we will not be able to find a polynomial-time algorithm. However, in some special cases, we are able to solve the problem polynomially. In more detail, we prove that the problem is NP-hard for any fixed $k \geq 2$ and $\Delta \geq 3$, whereas it is polynomially solvable when $k = 1$, or $\Delta \leq 2$ and $k = O(1)$. Moreover, we show that the problem is not approximable within $\Omega(\log B)$ or $\Omega(\log\log |V|)$ for any fixed $k \geq 3$, $\Delta \geq 3$, unless $P = NP$. We then provide an approximation algorithm with ratio guarantee of $\frac{b_{max}}{M}$, where b_{max} is the maximum communication bandwidth allowed among all the available interfaces and M is the greatest common divisor among the bandwidths allowed by the interfaces and B. Hence, when the bandwidth is constant for all the interfaces, the optimal solution is provided. We also focus on particular cases by providing complexity results and polynomial algorithms for $\Delta \leq 2$. Surprisingly, when k is unbounded and the network is reduced to a single edge, the problem remains NP-hard. Table 5.8 summarizes the results. Finally, we experimentally analyzed the $\frac{b_{max}}{M}$-approximation algorithm showing that, in practical cases, it guarantees a low approximation ratio that is smaller than the theoretical bound.

In an article by D'Angelo et al. [31], it was shown that MCFMI was strongly NP-hard even when restricted to the unit cost case and for any fixed $\Delta \geq 3$ and $k \geq 2$. Moreover, MCFMI cannot be approximated within a factor of $\Omega(\log B)$, or within a factor of $\Omega(\log\log |V|)$, for any fixed $\Delta \geq 3$ and $k \geq 3$, unless $P = NP$. The proofs are based on trasformation from X3C and set cover to MCFMI and, as they are quite technical, we do no report them here and refer the interest reader to the article by D'Angelo et al. [31]. In other words, even if we fix k and Δ, the problem remains NP-hard, except for the cases where $k = 1$ or $\Delta \leq 2$, and inapproximable, except for the cases where $k \leq 2$ or $\Delta \leq 2$. For $k = 1$ and any Δ, the problem is trivial because it admits an obvious solution given by a shortest path connecting s to t of maximum bandwidth $b(1)$. Therefore, we focus on the cases where $\Delta \leq 2$ and k is either bounded or unbounded and we leave the approximability of the cases where $k = 2$ as open problems.

For $\Delta \leq 1$, the input graph can be composed of either one single node or two nodes connected by one edge. In the first case, there are no interfaces to be activated because the source and the destination coincide. In the second case, MCFMI can be solved by an exhaustive search among

Table 5.8 Complexity Results Achieved for MCFMI by Varying on the Maximum Degree Δ and the Number of Available Interfaces k

Δ	k	*Complexity*
$\Delta = 1$	Fixed	Optimally solvable in $O(1)$ time
	Unbounded	NP-hard (equiv. MinKnapsack), $(1 + \varepsilon)$-approx in $O\left(\frac{k^2}{\varepsilon}\right)$
$\Delta = 2$	Fixed	Optimally solvable in $O(\|V\|)$
	Unbounded	NP-hard; $(2 + \varepsilon)$-approx in $O\left(\|V\|\frac{k^2}{\varepsilon}\right)$ for paths
Fixed $\Delta \geq 3$	Fixed $k \geq 2$	NP-hard
	Fixed $k \geq 3$	Not approximable within Ω (log B), or within Ω (loglog $\|V\|$)
Any	$k = 1$	Optimally solvable in $O(\|V\| + \|E\|)$ (equiv. *Shortest Path*)
	Any	$\frac{b_{max}}{M}$-approx (optimal for constant bandwidth)

all the possible combinations of interfaces shared by s and t. The number of such combinations is $O(2^k)$. Among them, a resolution algorithm has to choose the cheapest one that guarantees at least B bandwidth. It follows that if k is bounded, the problem is solvable in $O(1)$ time.

Surprisingly, in the unbounded case, that is, when k is not a given constant, the problem turns out to be already NP-hard using a simple polynomial transformation from the well-known minimization knapsack problem [13,36].

MinKP: Minimization Knapsack

Input	An integer $d \in Z_0^+$ and a set of n items, each one having weight $w_i \in Z_0^+$ and profit $p_i \in Z_0^+$, $i = 1, 2,\ldots, n$
Solution	An allocation of variables $y_i \in \{0, 1\}$, for $i = 1, 2,\ldots,n$, such that $\sum_{i=1}^{n} w_i y_i \geq d$
Goal	Minimize $\sum_{i=1}^{n} p_i y_i$

The MinKP problem is the corresponding minimization version of the knapsack problem. In other words, the goal is to minimize the profits of the items that remain out of the knapsack. If x_i, $i = 1,2,\ldots,n$, are the variables selecting the items for the classic knapsack problem and $c \in Z_0^+$ is its capacity, then the problem can be solved with MinKP, by setting $d = \sum_{i=1}^{n} w_i - c$ and $y_i = 1 - x_i, i = 1,2,\ldots,n$. When $\Delta = 1$, that is when the input graph G consists of a single edge from s to t, the required solution must select a subset of interfaces among the ones shared by s and t in such a way that a bandwidth of B is guaranteed, and the cost for activating such interfaces is minimized. Intuitively, this particular case of MCFMI is equivalent to the MinKP problem.

MCFMI is polynomially equivalent to MinKP in the unbounded case with $\Delta = 1$. It follows that in the unbounded case with $\Delta = 1$, MCFMI is NP-hard. However, as there exists a fully polynomial time approximation scheme (FPTAS) for MinKP, these results imply that such an algorithm can be used to also solve MCFMI. Moreover, the existence of an FPTAS directly implies that it exists as a pseudopolynomial time algorithm for MCFMI. In particular, the FPTAS requires $O\left(\frac{k^2}{\varepsilon}\right)$ time and finds a solution that is a factor $(1 + \varepsilon)$ far from the optimum [31].

For $\Delta = 2$, the input graph of MCFMI is either a path or a cycle. Clearly, because a single edge is a special case of a path, MCFMI remains NP-hard in the unbounded case. For the case of a path, dynamic programming has been exploited to find an algorithm that optimally solves the bounded case and an FPTAS for the unbounded case. Again, the FPTAS implies the existence of a pseudopolynomial time algorithm. In particular, for the bounded case, the dynamic programming algorithm requires $O(|V|)$ time, whereas the FPTAS for the unbounded case requires $O\left(|V|\frac{k^2}{\varepsilon}\right)$ time and finds a solution that is a factor $(2 + \varepsilon)$ far from the optimum.

When the input graph is a cycle because there are two paths from s to t, it is not always clear how the bandwidth B must be split between the two possible ways. However, the problem can be solved in $O(|V|)$ time in the bounded case. Let P_1 and P_2 be the two edge-disjoint paths from s to t composing the input cycle. Because by definition, $b(i) \in Z_0^+$, $1 \leq i \leq k$, the required flow B is provided by summing two integers β_1 and β_2 that are the contributions to the total flow passing via P_1 and P_2, respectively. The values β_1 and β_2 vary among all the integers obtainable by summing the bandwidths provided by each possible subset of interfaces, that is, β_1 and β_2 can assume at most 2^k values. For each subset of interfaces of s and for each subset of interfaces of t, the algorithm for the bounded case on a path is applied to solve the MCFMI instance arising for P_1 with bound β_1, and the one arising for P_2 with bound $\beta_2 = B - \beta_1$. The overall attempts are at most 2^{2k}, each of them requires 2^k tests, one for each possible value of β_1. As $k = O(1)$, then $2^{3k} = O(1)$. Among the obtained solutions, we choose the cheapest one, which guarantees a flow of at least B from s to t. Such an algorithm requires that the algorithm for the bounded case on a path be run $O(1)$ times and hence it requires $O(|V|)$ overall computational time.

In the general case, the graph transformation used to solve the MFMI can be used to find an approximation algorithm. Given an instance I_1 of MCFMI, such an algorithm works in four phases. First, it transforms the graph G and functions b and c of I_1 into a graph G' and functions b' and c' as described previously. Hence, we obtain an instance I_2 of an equivalent problem defined on a directed graph $G' = (V', A)$ without using multiple interfaces but associating costs and bandwidths only with arcs in A. The aim of such a problem is finding the flow function that satisfies flow constraints such that the flow going from the source $\tilde{s}$ to the sink $\tilde{t}$ is greater than or equal to B. Then, the algorithm transforms I_2 into an instance I_3 of the integral minimum cost flow problem, which is polynomially solvable. In the third phase, the algorithm solves I_3 by using a known algorithm and, finally, it transforms the obtained solution for I_3 into a solution for I_2 made of a flow function f'. Function f' can be transformed into a solution for I_1, as described previously, obtaining a flow function f and an assignment of interfaces W_A. In an article by D'Angelo et al. [31], it has been shown that such an algorithm guarantees a $\frac{b_{\max}}{M}$-approximation.

In the article by Ahuja et al. [32], the above algorithm was implemented and tested over several random instances. In the following, we show some of the experimental results presented. In particular, we show some of the results obtained by using the balls-into-bins random instances generator.

The balls-into-bins model is used to simulate devices thrown at random in a two-dimensional space [37,38]. In this model, each instance of MCFMI is made of a graph $G_{\text{BIB}} = (V_{\text{BIB}}, E_{\text{BIB}})$, a set of interfaces $I_{\text{BIB}} = \{1,2,\ldots,k\}$ along with cost and bandwidth functions c_{BIB}, and b_{BIB}, and two allocation functions $W_{\text{BIB}} : V_{\text{BIB}} \rightarrow 2^{I_{\text{BIB}}}$ and $X_{\text{BIB}} : V_{\text{BIB}} \times V_{\text{BIB}} \rightarrow 2^{I_{\text{BIB}}}$. First, nodes in V_{BIB} are generated and a uniformly random position in a unit size square is associated with each of them. The edges and interfaces were processed as follows. For each interface $i \in I_{\text{BIB}}$, the radius $r_i > 0$ of the circle covered by interface i is generated uniformly at random in $\left[\frac{1}{|V_{\text{BIB}}|}, \sqrt{\gamma\frac{\log(|V_{\text{BIB}}|)}{|V_{\text{BIB}}|}} - \frac{1}{|V_{\text{BIB}}|}\right]$. In this way, interfaces cover a circle having an average diameter of $\sqrt{\frac{\gamma \log |V_{\text{BIB}}|}{|V_{\text{BIB}}|}}$. Then, function W_{BIB} is defined by independently assigning the generated interfaces to nodes with a probability of 0.5. Given two nodes $u, v \in V_{\text{BIB}}$, let (x_u, y_u) and (x_v, y_v) be their associated coordinates in the unit square. If $\sqrt{(x_u - x_v)^2 + (y_u - y_v)^2} \le r_i$, for some $i \in W_{\text{BIB}}(u) \cap W_{\text{BIB}}(v)$, an edge $\{u, v\}$ is added to E_{BIB} and interface i is added to $X_{\text{BIB}}(u, v)$, that is, $X_{\text{BIB}}(u, v) = W_{\text{BIB}}(u) \cap W_{\text{BIB}}(v)$. In this way, for large values of $|V_{\text{BIB}}|$ and $\gamma > 4$, there is a high probability of obtaining a connected network. Finally, functions c_{BIB} and b_{BIB} are defined as $c_{\text{BIB}}(i) = r_i^{\alpha}$ and $b_{\text{BIB}}(i) = r_i^{\beta}$, for each $i \in I_{\text{BIB}}$ and for suitable tuning parameters α and β, which are fixed to 1.5 and 2, respectively, in the experiments. Source and target nodes are chosen as the nodes with the biggest Euclidean distance. For each of the defined instances, we considered three values of required flow distributed between the minimal bandwidth assigned to an interface b_{min} and the maximum flow possible F_{max}, computed by the algorithm for maximum flow. That is, we required a flow of $b_{min} + i\frac{F_{max-b_{min}}}{3}$, for $i = 1,2,3$. To measure the approximation ratio in the above settings, we need to know the optimal value of each MCFMI instance. Because it is NP-hard to compute such a value, we measured the ratio between the objective function value computed by our algorithm and a lower bound to the optimal value, obtaining an upper bound to the actual approximation ratio.

Figure 5.13a shows the average values and the standard deviations of the computed upper bounds on the approximation ratio as a function of the number of nodes in the network $|V_{BIB}|$, ranging from 50 to 1000, when the number of k interfaces is 9. The maximum value obtained is 3.12, achieved by an instance of 350 nodes and 5229 edges when the required flow is F_{max}. However, there are very few instances with an approximation ratio in (3,4). In particular, for 3 instances it is in (3,4), for 71 instances it is in (2,3), for 507 instances it is in (1,2) and for all the other 19 instances, it ensures the optimal value. On average, the ratio is always smaller than 2.04. Moreover, these are only the upper bounds to the real ratio as we computed the ratio between the objective function value of our algorithm and a lower bound to the optimum. The curves do not show a strict dependency from the number of nodes $|V_{BIB}|$. Conversely, there exists a small dependency from the required flow, that is, the approximation ratio slightly increases with the required flow.

Figure 5.13b shows the three curves when $k = 3$ and the other parameters are in the same setting as Figure 5.13a. As expected, the approximation ratio is improved here. This is because reducing the number of interfaces implies that the possible overhead at each node is also reduced. In particular, the maximum approximation ratio obtained is 2.71, achieved by an instance of 400 nodes and 5311 edges, when the required flow is F_{max}. The upper bound to the approximation ratio is in (2,3) for 16 instances, in (1,2) for 382 instances and the algorithm ensures the optimal value for the remaining 202 instances. We can conclude that, in graphs G_{BIB}, the approximation ratio is always very small and it depends neither on the number of nodes nor on the number of interfaces, whereas there is a small dependency from the required flow.

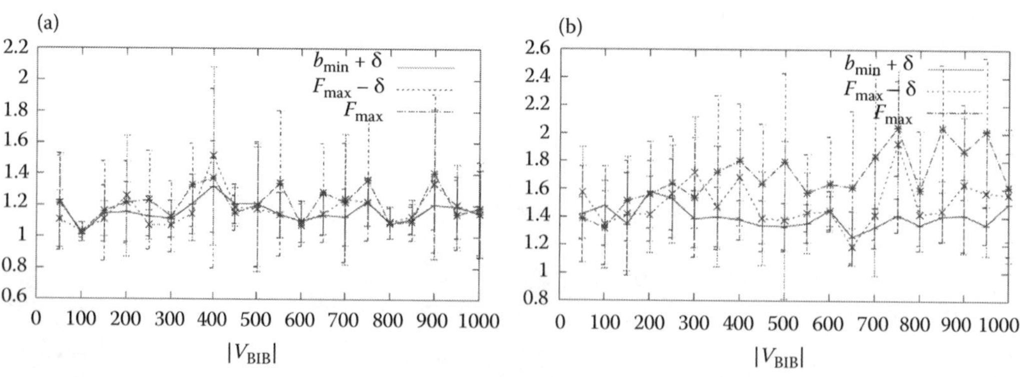

Figure 5.13 Graph G_{BIB}: average upper bounds on the approximation ratio for $|V_{BIB}| \in \{50, 100,..., 1000\}$, $k = 9$ (left) or $k = 3$ (right) and three values of required flow.

5.7 Conclusion

In this chapter, we have shown how wireless networks (including sensor networks) can take advantage of the heterogeneity of the communication interfaces installed on the devices composing the network to minimize energy consumption. In these systems, referred to as multi-interface networks, we have shown how to cope with several problems with the aim of minimizing the energy consumption when the wireless interfaces must be switched-on to accomplish some required connections.

In particular, the chapter focuses on the computational complexity as well as on the algorithmic challenges arising in some fundamental problems of this field. Namely, the chapter deals with the coverage, connectivity, cheapest paths, maximum matching, maximum flow, and minimum cost flow problems. Such problems can be seen as revisited versions of standard algorithmic problems in the context of multi-interface networks. For instance, the connectivity problem corresponds to the minimum spanning tree task, which is well-known in the context of graph theory as well as having a direct correspondence between the cheapest path and the shortest path tree problems.

For each of these problems, the cases where they are polynomially solvable (by giving suitable algorithms), and the cases where they are NP-hard or hard to approximate have been characterized. This has been realized by providing tight lower bounds and polynomial time transformations to well-known basic problems like set cover, vertex cover, minimum spanning tree, shortest path, Hamiltonian path, and the minimum hitting set. In doing so, the study of multi-interface networks also reveals a deep impact from the educational perspective because the reader is required to handle many facets of different and well-known combinatorial problems.

This chapter is intended to provide a wide overview of the work done in recent years on multi-interface networks. For the sake of simplicity, many of the results referred to are explained without providing deep details but emphasizing the practical aspects and giving running examples. In particular, the very simple network provided in the introduction by Figure 5.1 has been used within the presented problems to show how solutions drastically change with respect to the different tasks required on the same network. The interested reader may refer to the wide-ranging bibliography provided to focus on the details of specific results. This also points out the high interest of the scientific community in these kinds of problems, as different research groups have been investigating such topics and the results obtained have been published in highly reputable international journals.

Concerning possible directions for further investigations on multi-interface networks, the design of better approximation algorithms or tighter bounds, where possible, is certainly of interest as well as the experimental study of new heuristics. Moreover, the models as well as the defined problems can be modified or adapted to more practical contexts and different cost functions. A direction that deserves particular attention is certainly the study of the proposed problems from a distributed perspective. In fact, when each node of the input graph is aware only of its own neighborhood, approaches drastically change and suitable solutions must be designed. Collaborative against selfish environments studies also deserve main consideration, depending of what kind of practical network is modeled.

From the above discussion, this survey includes only part of the possible studies on multi-interface networks. In fact, many other directions can be explored to better catch more practical insights of such kinds of networks.

References

1. Friedman R., Kogan A., and Krivolapov Y. On power and throughput tradeoffs of WiFi and Bluetooth in smartphones. In *Proceedings of the 30th IEEE International Conference on Computer Communications (INFOCOM)*, 900–908, 2011.

2. Cormen T. H., Leiserson C. E., Rivest R. L., and Stein C. *Introduction to Algorithms*. McGraw-Hill, New York, 2001.
3. Caporuscio M., Charlet D., Issarny V., and Navarra A. Energetic performance of service-oriented multi-radio networks: issues and perspectives. In *Proceedings of the 6th International Workshop on Software and Performance (WOSP)*, 42–45, 2007.
4. Klasing R., Kosowski A., and Navarra A. Cost minimisation in multi-interface networks. In *Proceedings of the 1st EuroFGI International Conference on Network Control and Optimization (NetCooP)*. LNCS 4465, Springer-Verlag, Berlin Heidelberg, 276–285, 2007.
5. Klasing R., Kosowski A., and Navarra A. Cost minimisation in wireless networks with bounded and unbounded number of interfaces. *Networks* 54(1), 12–19, 2009.
6. Kosowski A., and Navarra A. Cost minimisation in unbounded multi-interface networks, *Proceedings of the 2nd PPAM Workshop on Scheduling for Parallel Computing (SPC)*. LNCS 4967, Springer-Verlag, Berlin Heidelberg, 1039–1047, 2007.
7. Papadimitriou C., and Yannakakis M. Optimization, approximation, and complexity classes. *Journal of Computer and System Sciences* 43(3), 425–440, 1991.
8. Brooks R. L. On coloring the nodes of a network. In *Proceedings of the Cambridge Philosophical Society* 37, 194–197, 1941.
9. Raz R., and Safra S. A sub-constant error-probability low-degree test, and a sub-constant error-probability PCP characterization of NP. In *Proceedings of the 29th Annual ACM Symposium on Theory of Computing (STOC)*, 475–484, 1997.
10. Ausiello G., D'Atri A., and Protasi M. Structure preserving reductions among convex optimization problems. *Journal of Computer and System Sciences*, 21, 136–153, 1980.
11. Kosowski A., Navarra A., and Pinotti M. C. Connectivity in multi-interface networks. In *Proceedings of the 4th Symposium on Trustworthy Global Computing (TGC)*. LNCS 5474, Springer-Verlag, Berlin Heidelberg, 157–170, 2008.
12. Kosowski A., Navarra A., and Pinotti, M. C. Exploiting multi-interface networks: connectivity and cheapest paths. *Wireless Networks* 16(4), 1063–1073, 2010.
13. Garey M. R., and Johnson D. S. *Computers and Intractability, A Guide to the Theory of NP-Completeness*. W.H. Freeman and Company, New York, 1979.
14. Athanassopoulos S., Caragiannis I., Kaklamanis C., and Papaioannou E. Energy-efficient communication in multi-interface wireless networks. In *Proceedings of the 34th International Symposium on Mathematical Foundations of Computer Science (MFCS)*, LNCS 5743, Springer-Verlag, Berlin Heidelberg, 102–111, 2009.
15. D'Angelo G., Di Stefano G., and Navarra A. Minimizing the maximum duty for connectivity in multi-interface networks. In *Proceedings of the 4th International Conference on Combinatorial Optimization and Applications (COCOA)*. LNCS 6509, Springer-Verlag, Berlin Heidelberg, 254–267, 2010.
16. D'Angelo G., Di Stefano G., and Navarra A. Min-max coverage in multi-interface networks. In *Proceedings of the 37th International Conference on Current Trends in Theory and Practice of Computer Science (SOFSEM)*. LNCS 6543, Springer-Verlag, Berlin Heidelberg, 190–201, 2011.
17. D'Angelo G., Di Stefano G., and Navarra A. Minimize the maximum duty in multi-interface networks. *Algorithmica* 63(1–2), 274–295, 2012.
18. Barsi F., Navarra A., and Pinotti M. C. Cheapest paths in multi-interface networks. In *Proceedings of the 10th International Conference on Distributed Computing and Networking (ICDCN)*. LNCS 5408, Springer-Verlag, Berlin Heidelberg, 37–42, 2009.
19. Edmonds J. Paths, trees and flowers. *Journal of Mathematics* 17, 449–467, 1965.
20. Brandstädt A. and Mosca R. On distance-3 matchings and induced matchings. *Discrete Applied Mathematics* 159, 509–520, 2011.
21. Cameron K. Induced matching. *Discrete Applied Mathematics* 24, 97–102, 1989.
22. Kobler D., and Rotics U. Finding maximum induced matchings in subclasses of clawfree and p5-free graphs, and in graphs with matching and induced matching of equal maximum size. *Algorithmica* 37, 327–346, 2003.
23. Golumbic M. C., and Lewenstein M. New results on induced matchings. *Discrete Applied Mathematics* 101, 157–165, 2000.

24. Duckworth W., Manlove D., and Zito M. On the approximability of the maximum induced matching problem. *Journal of Discrete Algorithms* 3, 79–91, 2005.
25. Kanj I., Pelsmajer M. J., Schaefer M., and Xia G. On the induced matching problem. *Journal on Computer System Sciences* 77, 1058–1070, 2011.
26. Kang R. J., Mnich M., and Müller T. Induced matchings in subcubic planar graphs. *Proceedings of the 18th Annual European Conference on Algorithms (ESA).* LNCS 6347, Springer, 112–122, 2010.
27. Zito M. Induced matchings in regular graphs and trees. In *Proceedings of the 25th International Workshop on Graph-Theoretic in Computer Science (WG).* LNCS 1665, Springer-Verlag, Berlin Heidelberg, 89–100, 1999.
28. Kosowski A., Navarra A., Pajak D., and Pinotti M. C. Maximum matching in multi-interface networks. In *Proceedings of the 6th International Conference on Combinatorial Optimization and Applications (COCOA),* LNCS 7402, Springer-Verlag, Berlin Heidelberg, 13–24, 2012.
29. Bertossi A., Navarra A., and Pinotti M. C. Maximum bandwidth broadcast in single and multi-interface networks. In *Proceedings of the 5th International Conference on Ubiquitous Information Management and Communication (ICUIMC)*, Seoul, Korea, 2011.
30. D'Angelo G., Di Stefano G., and Navarra A. Bandwidth constrained multi-interface networks. In *Proceedings of the 37th International Conference on Current Trends in Theory and Practice of Computer Science (SOFSEM).* LNCS 6543, Springer-Verlag, Berlin Heidelberg, 202–213, 2011.
31. D'Angelo G., Di Stefano G., and Navarra A. Maximum flow and minimum-cost flow in multi-interface networks. In *Proceedings of the 5th International Conference on Ubiquitous Information Management and Communication (ICUIMC)*, Seoul, Korea, 2011.
32. Ahuja, R. K., Magnanti, T. L., and Orlin, J. B. *Network Flows: Theory, Algorithms, and Applications.* s.l. Prentice Hall, Upper Saddle River, New Jersey 1993.
33. D'Angelo G., Di Stefano G., and Navarra A. Flow problems in multi-interface networks. *IEEE Transaction on Computers* (in press).
34. Dinic. E. A. Algorithm for solution of a problem of maximum flow in a network with power estimation. *Soviet Mathematics Doklady* 11, 1277–1280, 1970.
35. Edmonds J., and Karp R. M. Theoretical improvements in algorithmic efficiency for network flow problems. *Journal of the ACM* 19(2), 248–264, 1972.
36. Görtz S., and Klose. A. Analysis of some greedy algorithms for the single-sink fixed-charge transportation problem. *Journal of Heuristics* 15(4), 331–349, 2009.
37. Güntzer M. M., and Jungnickel D. Approximate minimization algorithms for the 0/1 knapsack and subset-sum problem. *Operations Research Letters* 26(2), 55–66, 2000.
38. Papadopoulos A., McCann J., and Navarra A. Connectionless probabilistic (CoP) routing: an efficient protocol for mobile wireless ad-hoc sensor networks. In *Proceedings of the 24th IEEE International Performance, Computing, and Communications Conference (IPCCC)*, 73–77, 2005.

Chapter 6

Sensor Bus Architecture for Real-Time Wireless Sensor Networks

Abraham Lamesgin Addisie
Addis Ababa University
Kumudha Raimond
Karunya University

Contents

6.1 Introduction

Nowadays, sensors are becoming smaller, cheaper, more reliable, more power-efficient, and more intelligent. Sensors delivering observations with geographically referenced locations are increasingly used in various applications ranging from environmental monitoring and precision agriculture to early warning systems. The sensors utilized in these applications may be stationary or mobile, either on land, water, or in the air, and could gather data in an *in situ* or remote manner. There is a need for a standard that facilitates the integration of sensors in a platform-independent way so that sensors of different types and from different vendors can be accessed using different applications in an interoperable way.

Sensor web enablement (SWE) is one such standard developed by the Open Geospatial Consortium (OGC) for sharing, finding, and accessing sensors and their data across different applications. It is an infrastructure allowing users to easily share their sensor resources. It hides the underlying network communication details and heterogeneous sensor hardware from the applications built on top of it. Members of the SWE working group have specified interfaces, protocols, and data types that enable the integration of sensors and sensor webs into spatial information infrastructures [1].

In particular, SWE incorporates data models for describing sensors (sensor model language) as well as the sensor data gathered (observations and measurements). The main web service interfaces are the Sensor Observation Service (SOS), the Sensor Alert Service (SAS), and the Sensor Planning Service (SPS). The SOS is designed to access real-time as well as historic sensor data and sensor metadata descriptions. The SAS is used to receive alerts for events that are happening. An example of an event that may cause an alert is the temperature measured by a sensor being greater than 50°C. Although the SOS follows the pull-based communication paradigm, the SAS is capable of pushing sensor data to subscribers. To control and task sensors, the SPS can be used. A common application of SPS is to define sensor parameters such as the sampling rate, mission planning of satellite systems, etc. [1].

6.1.1 Background of the Problem

Nowadays, sensor networks are being used in various areas. Having been deployed in various geographical areas to address the same problem, there is a need to connect this distributed sensor network. The SWE is a solution toward this problem. Even though the SWE, through its services and data representation, provides a standard for an infrastructure of interoperability between the various application users and data sources, it does not fully address the link between the sensor web and the sensor network. It is designed from the application programmer's point of view. The gap between the SWE and the data from the sensor node should be filled for easy deployment of a sensor network.

Because of the missing interoperability between the two layers (sensor network and sensor web), it is currently not possible to dynamically install sensors on-the-fly with a minimum of human configuration effort. Generally, the SWE standards focus on interacting with the upper application level. They are designed from an application-oriented perspective. As a result, the interaction between the sensor web and the underlying sensor network layer has not yet been sufficiently described. The sensor web is based on the World Wide Web (WWW) and its related protocols. On the other hand, sensor network technologies are based on lower-level protocols such as ZigBee [2], Bluetooth [3], the IEEE 1451 standards family, or proprietary protocols [4]. From an application perspective, the SWE services encapsulate the sensor network and hide these lower-level protocols.

Currently, the sensor web and sensor network layers are integrated by manually building proprietary bridges for each pair of web service implementation and sensor types. This approach is cumbersome and leads to extensive adaption efforts. Because the prices of sensor devices are decreasing rapidly, manual integration has become the key cost factor in developing large-scale sensor network systems.

The sensor bus is a relatively recent work toward a better solution to bridging this gap. Only an adapter for the sensors and the SWE services needs to be developed for each unique sensor and service in the system. Although the sensor adaptor will change the communication logic between the sensor network and the sensor bus, the service adaptor will change the communication logic between the sensor bus and the SWE services.

However, this approach has not yet been applied in a real-world scenario to demonstrate its applicability [5]. It has to be applied in such a scenario so that the adoption of this approach, together with the SWE standard services and data encoding, is facilitated.

Also, the performance and scalability of this architecture has not been well evaluated. An evaluation of the sensor bus architecture, including scalability and performance metrics, was planned as a future work by Broring et al. [5]. Because it is one of the main components in the sensor web system, it should be of high performance, scalable, fault tolerant, secure, and easily adaptable. It should also consider unattended sensor nodes, the security of the sensor data requirements, real-time data requirements, and the ease of deploying the sensor network from various vendors.

Therefore, the main focus of this chapter is to show the applicability of the sensor bus architecture in a real-world scenario and in a heterogeneous environment with different sensor networks, and to thereby evaluate the sensor bus architecture and suggest design improvements based on the results.

6.2 Existing Technologies

This has been of particular research interest in the area of middleware for sensor networks. Until now, few researches have been done on middleware that could integrate sensor networks with sensor webs. A survey on middleware for wireless sensor networks (WSN) is provided in an article by Wang et al. [6]. However, most middleware solutions usually focus on lower-level functionality such as cost-efficient message routing rather than reducing the efforts of integrating sensor networks and the sensor web. The global sensor network (GSN) middleware is one of the works which focuses on fast and flexible integration of sensor networks and sensor web [7].

A work based on the SWE standard is the collaborative research project titled the "Sensor-Based Landslide Early Warning System" (SLEWS), which is aimed at the systematic development of a prototype alarm and early warning system. It is an effort to solve the interoperability gap between the sensor web and the sensor network. Here, a lightweight "SWE connector" application providing a generic toolbox that takes raw format sensor data and converts it into an SWE-based data model has been developed. This work primarily depends on the architecture given in Figure 6.1 [8]. However, it can only be implemented for SOS and there is a need for bidirectional communication for services such as the SPS. The work does not take into consideration the SPS, which is one of the SWE services. Although the approach reduced the work to be done toward the integration of the SWE and sensor networks, it still requires additional work for integration from the vendors.

Open Sensor Web Architecture (OSWA), which was proposed at NICTA/Melbourne University, is the OGC's SWE standard–compliant software infrastructure for providing Service-Oriented Architecture (SOA)–based access to and management of sensors. The OSWA is a complete standards-compliant platform for integration of sensor networks and emerging distributed computing platform such as SOA and grid computing. Individual sensor networks can be linked together as services, which can be registered, discovered, and accessed by different clients using a uniform protocol [9].

The various components defined in OSWA are shown in Figure 6.2. The following layers have been defined: SensorWeb Fabric, SensorWeb Core Middleware, SensorWeb User-Level Middleware, and SensorWeb Applications. Fundamental services are provided by low-level components whereas components at a higher level provide tools for creating applications and management of the life cycle of data captured through sensor networks. The project mainly focuses on providing an interactive development environment, an open and standards-compliant SensorWeb services middleware and a coordination language to support the development of various sensor applications.

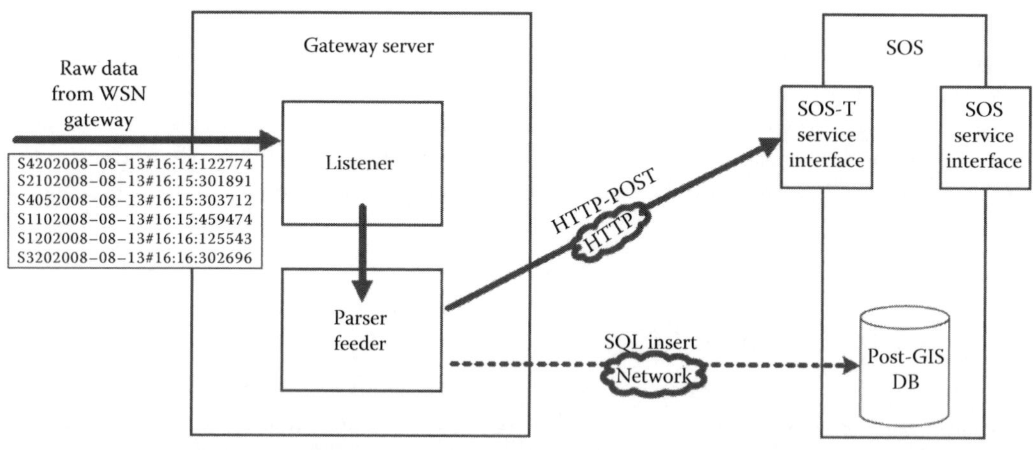

Figure 6.1 SLEWS architecture of SOS. (From C. Arnhardt et al., A sensor based landslide monitoring system integrated in an early warning structure. In *70. Jahrestag der Deutschen Geophysikalischen Gesellschaft (DGG)*, 15–18 March 2010, Bochum, conference transcript, p. 43, 2010.)

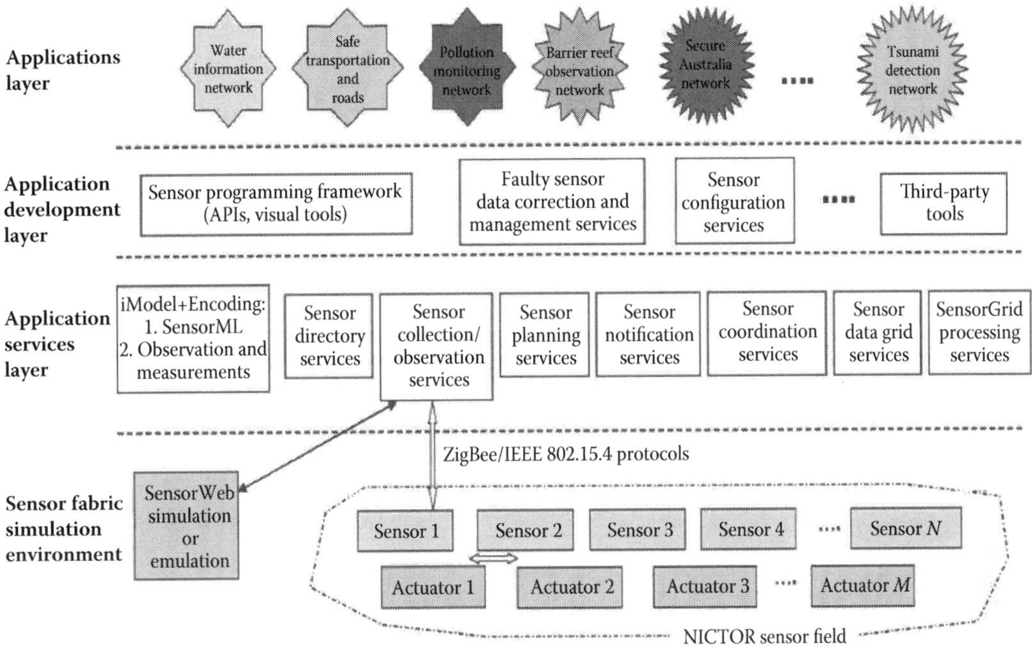

Figure 6.2 Open Sensor Web Architecture overview. (From X. Chu et al., Open Sensor Web Architecture: core services. In *International Conference on Intelligent Sensing and Information Processing (ICISIP 2006)*, IEEE Press, Piscataway, Bangalore, India, pp. 98–103, 2006.)

However, there is still an interoperability gap between the SensorWeb Fabric and the SensorWeb Core Middleware. Although the SensorWeb Fabric layer contains sensor networks from various protocols, the SensorWeb Core Middleware contains various SWE services. The integration of the sensor network to the sensor web needs additional effort.

The sensor bus architecture is by far a better solution toward reducing the integration effort between the sensor network and the sensor web. It establishes an intermediary layer between sensor networks and the sensor web to close the conceptual gap between these two distinct layers resulting from different protocol stacks and data models. This will allow sensor vendors to easily plug-in sensors so that they are subsequently available on the sensor web. Service providers benefit because they can support a wide variety of sensors by adapting their services to the sensor bus [5]. The sensor bus architecture is shown in Figure 6.3.

The sensor bus, as an intermediary layer, acts as a broker by transferring messages between sensors and services and establishes a publish/subscribe mechanism. A client invokes a SWE service, for example, to retrieve observations or to give tasks for a sensor node. The intermediary layer maintains associations with these services as well as with sensor gateways that supply access to connected sensors. The connections are established by adapters plugged into the intermediary layer.

The sensor bus incorporates a common communication infrastructure, a shared set of adapter interfaces, and a well-defined message protocol. The common communication infrastructure is established through a publish/subscribe mechanism based on the underlying messaging

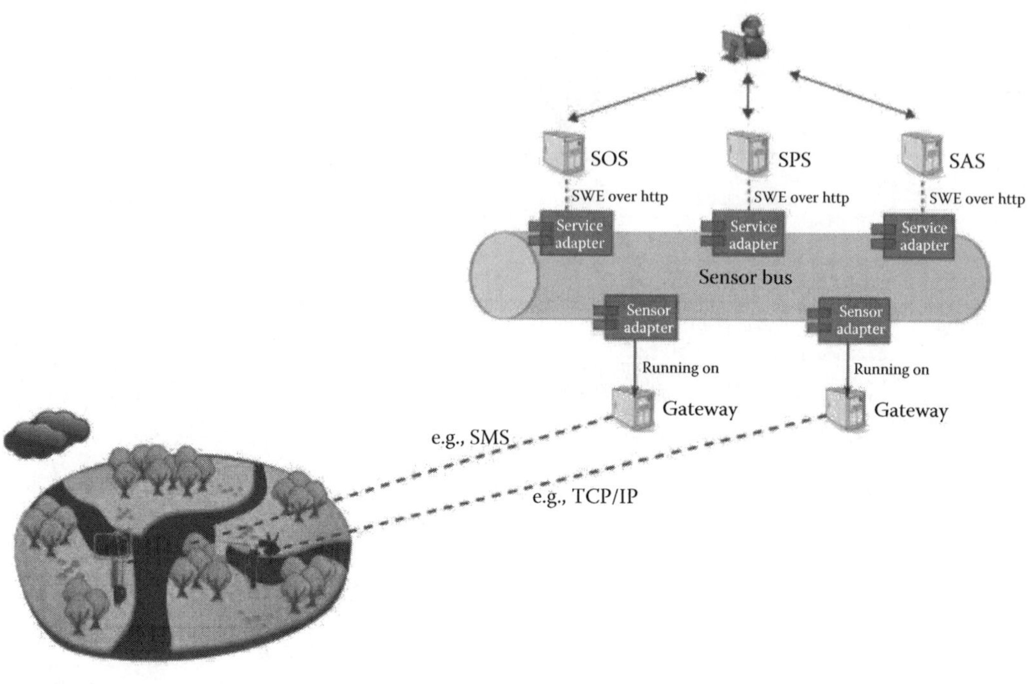

Figure 6.3 Sensor bus architecture. (From 52°North, Sensor bus. http://52north.org/communities/sensorweb/incubation/sensorBus/index.html, 2011.)

technology (e.g., Twitter [23], JMS [21], IRC [22], or XMPP [20]). Services as well as sensors can publish messages to the bus and are also able to subscribe to the bus for receiving messages in a push-based communication style. The underlying messaging technology takes care of forwarding the posted messages to the specific subscribed components. The different components (i.e., sensors and SWE services) can subscribe and publish through interfaces. For these interfaces, pluggable adapters can be developed by sensor vendors or service providers. The adapters convert the service or sensor-specific communication protocol to the internal bus protocol. The adapters of the sensors and SWE services, together with the underlying messaging technology, form the sensor bus [5].

After closely reviewing the reported works related to bridging the interoperability gap between the sensor web and the sensor network, the sensor bus architecture is by far the best solution. However, it has not yet applied in a real-world scenario. Also, evaluation of the architecture including the performance and scalability has not yet been done; although Broring et al. [5] have recommended it for future work. Therefore, the main focus is to evaluate and suggest design improvements on the sensor bus architecture so that optimized architecture will be developed in the future.

To show applicability in a real-world scenario, a temperature sensor network, eZ430-RF2500 from Texas Instruments (Dallas, Texas), was integrated to the sensor bus system. Also, other sensor networks (Sun SPOT and simulated) were integrated into the system to show the heterogeneity of the system.

6.3 Theoretical Concepts of Sensor Bus Architecture

6.3.1 Sensor Network

6.3.1.1 Introduction

A sensor network is a group of specialized transducers with a communication infrastructure intended to monitor and record conditions at diverse locations. Commonly monitored parameters are temperature, humidity, pressure, wind direction and speed, illumination intensity, vibration intensity, sound intensity, power-line voltage, chemical concentrations, pollutant levels, and vital body functions.

A sensor network consists of multiple detection stations called sensor nodes, each of which is small, lightweight, and portable. Every sensor node is equipped with a transducer, microcomputer, transceiver, and power source. The transducer generates electrical signals based on sensed physical effects and phenomena. The microcomputer processes and stores the sensor output. The transceiver, which can be hardwired or wireless, receives commands from a central computer and transmits data to that computer. The power for each sensor node is derived from the electric utility or from a battery.

Potential applications of sensor networks include industrial automation, automated and "smart" homes, video surveillance, traffic monitoring, medical device monitoring, monitoring of weather conditions, air traffic control, robot control, etc.

6.3.1.2 Sensor Network Standard

Several standards are currently either ratified or under development for sensor networks. There are a number of standardization bodies in this field. The IEEE focuses on the physical and MAC layers; the Internet Engineering Task Force works on features above the MAC layer. In addition to these, bodies such as the International Society of Automation provide vertical solutions, covering all protocol layers. There are also several nonstandard, proprietary mechanisms and specifications. Some of the predominant standards include IEEE 1451, ZigBee, and Bluetooth.

However, none of them address the integration of heterogeneous sensor networks to the sensor web such as what hypertext markup language (HTML) and hypertext transfer protocol (HTTP) does for the WWW. There is a need for an open standard for service and data encoding that allows the fast integration of heterogeneous sensor networks to the sensor web. SWE is such a standard, and is explained in Section 6.3.2.

6.3.1.3 Sensor Web

The sensor web is an infrastructure for sharing, finding, and accessing sensors and their data across different applications. The sensor web is to sensors what the WWW is to general information sources, an infrastructure allowing users to easily share their sensor resources. It hides the underlying layers with the network communication details and heterogeneous sensor hardware from the applications built on top of it. The SWE initiative of the OGC standardizes web service interfaces and data encodings for the sensor web. The sensor web concept, according to the SWE initiative of the OGC, is shown in Figure 6.4 [1].

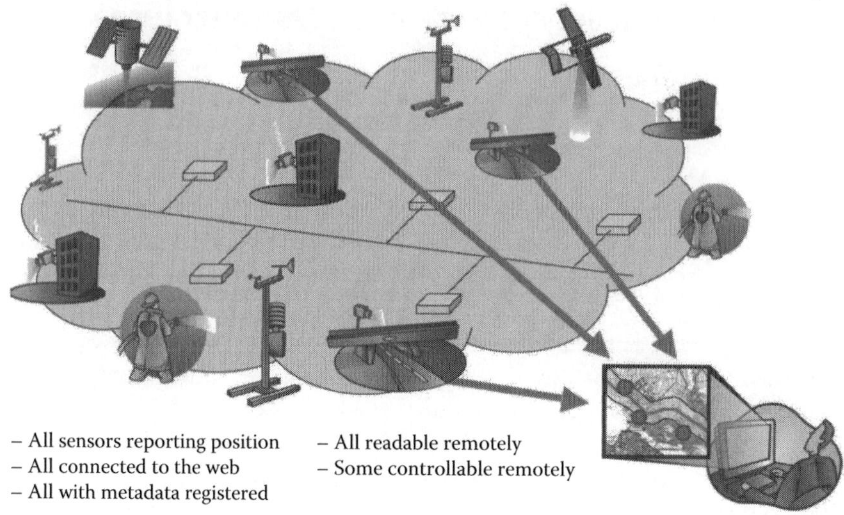

– All sensors reporting position
– All connected to the web
– All with metadata registered
– All readable remotely
– Some controllable remotely

Figure 6.4 The sensor web concept as per SWE initiative. (From M. Botts et al., OGC® Sensor Web Enablement: Overview and High Level Architecture, Open Geospatial Consortium Inc., Wayland, MA, USA: OGC, 2007.)

6.3.2 Sensor Web Enablement

Within the SWE initiative, the enablement of such sensor webs and networks is being achieved through the establishment of several encodings for describing sensors and sensor observations, and through several standard interface definitions for web services. SWE standards that have been built and prototyped by members of the OGC include the following:

- Observations and Measurements Schema (O&M)—Standard models and XML schema for encoding observations and measurements from a sensor, both archived and real-time. [11,12]
- Sensor Model Language (SensorML)—Standard models and XML schema for describing sensors systems and processes; provides information needed for the discovery of sensors, location of sensor observations, processing of low-level sensor observations, and listing of taskable properties [13,14].
- Transducer Markup Language (TransducerML or TML)—The conceptual model and XML schema for describing transducers and supporting real-time streaming of data to and from sensor systems [19].
- Sensor Observations Service (SOS)—Standard web service interface for requesting, filtering, and retrieving observations and sensor system information. This is the intermediary between a client and an observation repository or near real-time sensor channel [15].
- Sensor Planning Service (SPS)—Standard web service interface for requesting user-driven acquisitions and observations. This is the intermediary between a client and a sensor collection management environment [16].
- Sensor Alert Service (SAS)—Standard web service interface for publishing and subscribing to alerts from sensors [17].
- Web Notification Services (WNS)—Standard web service interface for asynchronous delivery of messages or alerts from SAS and SPS web services and other elements of service workflows [18].

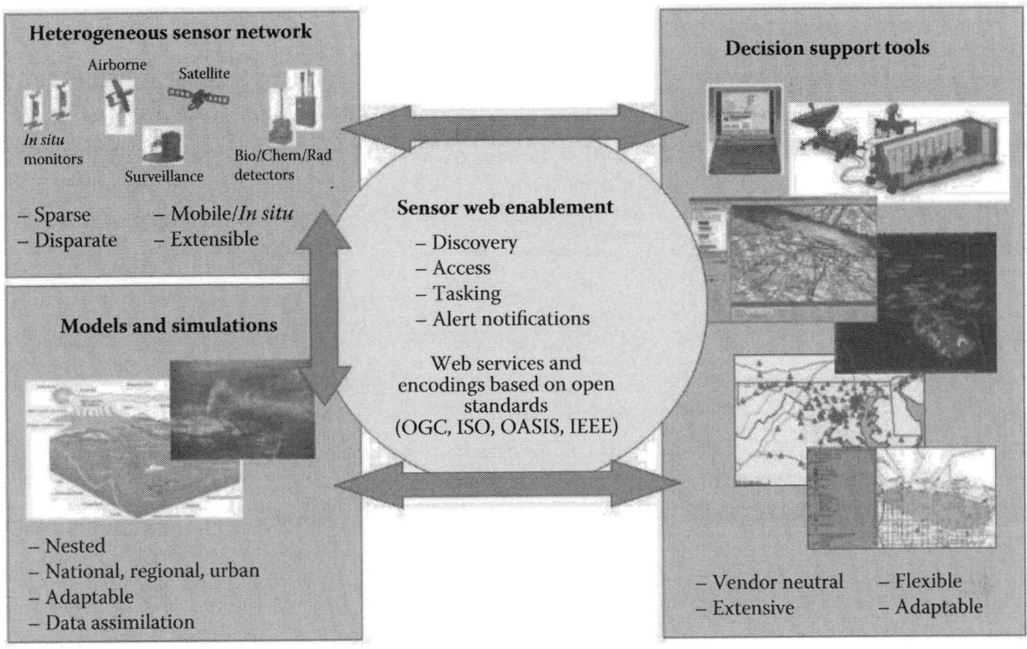

Figure 6.5 The role of the SWE framework. (From M. Botts et al., OGC® Sensor Web Enablement: Overview and High Level Architecture, Open Geospatial Consortium Inc., Wayland, MA, USA: OGC, 2007.)

The models, encodings, and services of the SWE architecture enable the implementation of interoperable and scalable service-oriented networks of heterogeneous sensor systems and client applications, in much the same way that HTML and HTTP standards enable the exchange of any type of information on the Web. The OGC's SWE initiative is focused on developing standards to enable the discovery, exchange, and processing of sensor observations, as well as the tasking of sensor systems. In recent years, these SWE standards have demonstrated their practicability and suitability in various projects and applications. The role that the OGC has targeted within a sensor web is shown in Figure 6.5 [1].

The most important part of the SWE standard service, that is, SOS, SPS, and SAS and the sensor and measurement data encoding (i.e., O&M and sensorML) are discussed in more detail in the following sections.

6.4 System Design, Implementation, and Performance Evaluation

6.4.1 System Design

In this chapter, the sensor bus architecture is adapted for the integration of heterogeneous sensor networks to the sensor web. The total system is composed of the sensor bus, SWE services, and sensor networks. In addition, there are also service adapters (which connect the services to the sensor bus) and sensor adapters (which connect the sensor networks to the sensor bus). The sensor bus is the glue of the system that connects the upper layer (i.e., the services) and the lower layer (i.e., the sensor networks). The SOS runs on a server that is connected to the Internet. The sensor bus (the communication service) can run on the same server as the SOS or on a separate server. The SOS adapter, like the sensor bus, can be run in the same server as the SOS or can be on another server. The SOS adapter will allow the SOS to communicate with various sensors through the sensor bus. It will change the communication logic of the service to the sensor bus. The sensor adapter will allow a sensor to communicate with the SOS through the sensor bus. It will change the communication logic of the sensor network into the communication logic of the sensor bus. A sensor adapter should be developed for each sensor network to be used on this system. The system block diagram representation, which is adapted from Figure 6.2, is shown in Figure 6.6. In this book, only the SOS is considered because it is the most heavily loaded and important service of the SWE services.

A particular scenario of using this system will be as follows: a client will discover a particular service or sensor in the WWW by using the sensor discovery service. Here, the client can use the service's capabilities, location and sensor offering, and other related information provided by the SOS to search for a particular sensor of interest. Then, the client can send an observation request to the SOS. The SOS will search the request from its database and return an observation response. The client application will parse the observation response in XML and present it to the client.

6.4.1.1 Sensor Bus

The interfaces that are used to realize the sensor bus are depicted in Figure 6.7, which is adopted from a previous work implementation [5]. The SensorAdapter (e.g., an adapter for the eZ340-RF2500 sensor platform) and the ServiceAdapter (e.g., adapter for SOS) are used to connect sensors and services, respectively, to the sensor bus. Both interfaces are BusListeners so that they can be

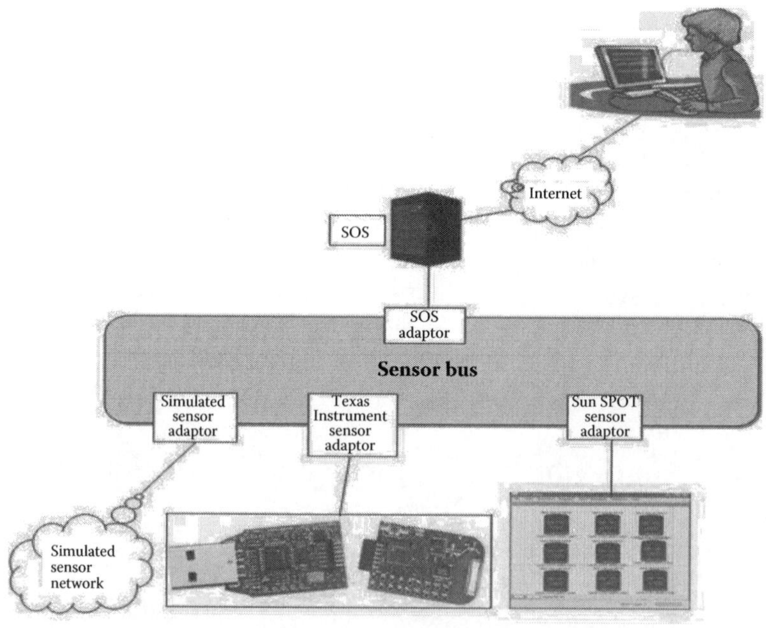

Figure 6.6 System block diagram.

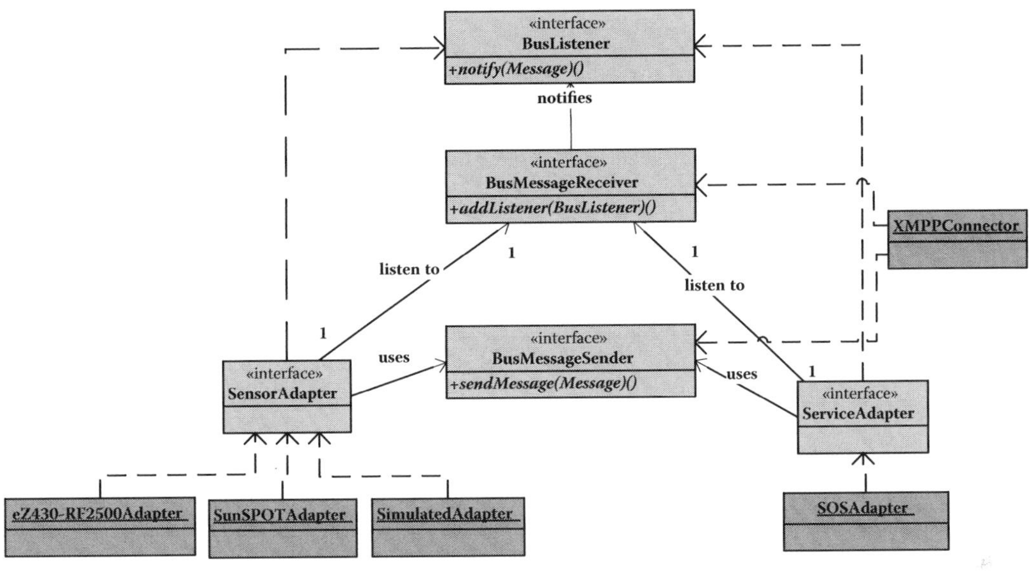

Figure 6.7 Sensor bus core components as UML class diagram.

notified by the BusMessageReceiver for retrieving messages sent over the bus. The SensorAdapter and ServiceAdapter transmit their messages to the bus through the BusMessageSender. The BusMessageReceiver and BusMessageSender are hidden from the underlying communication infrastructure of the bus.

The XMPPConnector in Figure 6.7 is an example implementation of the two interfaces to realize the sensor bus based on XMPP. Because the two interfaces abstract from the communication infrastructure, it is easy to exchange the implementing class and realize the sensor bus based on other messaging technologies (e.g., instant messaging systems). The implementation of BusMessageReceiver calls in case of an incoming message onMessage() is to notify the listeners. The concrete sensor and service adapters, acting as listeners, analyze the incoming message and react to it according to their specifications.

The XMPP was chosen here because it has been fully designed with the sensor bus system and has advantages over other fully integrated Twitter communication services. The Twitter implementation, which is explained by Broring et al. [5], has some functionality limitations. The maximum length of a single tweet is limited to 140 characters. This will limit the size of a single message that a sensor adapter can send per measured data. Also, the maximum number of tweets that can be done per day is restricted to 1000. This will limit the maximum sampling rate of a sensor node to 40 samples/h. This sampling rate is not acceptable in most sensor network applications. However, there are no such limitations on the XMPP implementation. These are some of the reasons why XMPP has been selected for implementation.

The interactions between the sensor network layer, the sensor web, and the intermediary layer are realized through messages [24]. The message format is compact to preserve bandwidth and system resources. Single message fields are divided by a separator sign (the "*" character). Variables are included inside "<" and ">" brackets to show that they act as placeholders for actual values [5]. The necessary messages are listed in Table 6.1.

Table 6.1 Sensor Bus Message Protocol

Type	*Message Protocol*
Service registration	SubscribeService*<service type>*<service URL>*<Service Requirement>
Sensor registration	ConnectSensor*<sensorML document name>
Data publication	PublishData*<sensor id>*<time tag>*<location tag>*<observed property>*<data>
Service disconnection	DisconnectService**<service type>*<service URL>
Sensor disconnection	DisconnectSensor*<sensorML document name>

The message protocol in Table 6.1 is explained in detail as follows:

(a) Service Registration Message

The service adapter, upon initialization, sends the service registration message to the sensor bus communication service, for example, to XMPP. The service registration message starts with the keyword SubscribeService. Then, the service type (which could be SOS, SPS, or SAS) and the service URL (the location where the service is running) should be specified.

For example, the SubscribeService *SOS*http://localhost:8080/52nSOSv3/sos* service requirement is a service registration message. The service type here is SOS and the service location is http://localhost:8080/52nSOSv3/sos. The service requirement is included at the end. An example of a service requirement is: "ssn:hasSurvivalRange":"<swe:QuantityRangedefinition="SurvivalRange_01"><swe:uomxlink="unit:metre"code="m"/><swe:value>100 </swe:value></swe:QuantityRange>". Here, the service is interested, for example, in only the sensor nodes with the capability of working at a depth of 100 m in a river. The sensor bus will register the service adapter after getting this message. Then, it will notify the service adapter about messages published in the communication services.

(b) Sensor Registration Message

The sensor adapter, upon initialization, sends the sensor registration message to the sensor bus communication service (e.g., to XMPP). The sensor registration should start with the keyword ConnectSensor. Then, the name of the sensorML document of the sensor is specified. The sensorML document is used by the service adapter for constructing the sensor registration and sensor data insertion request to the service.

For example, ConnectSensor* eZ430-RF2500_Node1.xml is a sensor registration message. The sensorML document name here is eZ430-RF2500_Node1.xml and the sensor ID is eZ430-RF2500_Node1.

(c) Sensor Data Publication Message

For publishing new data, the sensor adapter transmits the data publication message to the sensor bus communication service (e.g., to XMPP). The sensor data publication message should start with the keyword PublishData. Then, it contains the time when the data was observed, the location where the data is taken, the observed property, and the data itself, respectively, delimited by *.

For example, PublishData *eZ430-RF2500_Node1*2011-09-02T15:52:43.656+03.00*(52.50, 7.06)*http://sweet.jpl.nasa.gov/1.1/property.owl#Temperature</*25.5 is a data publication message. The sensor ID here is eZ430-RF2500_Node1. Date and time are 2011-09-02

and 15:52:43, respectively, when the data is observed. The location where the data is sensed is *(52.50, 7.06) and the observed property is temperature. The measured temperature is 25.5°C.

(d) Disconnect Service and Sensor Messages

The disconnect service and sensor messages will disconnect the service and sensor, respectively, from the sensor bus communication services.

6.4.1.2 Sensor Observation Service

The SOS provides an interface to make sensors and sensor data archives accessible via an interoperable web-based interface. The 52°North SOS version 3.2.0 is used in this work. The 52°North SOS version 3.2.0 implements the SOS implementation specification 1.0.0. The implementation is written in Java and uses the following technical frameworks:

- PostgreSQL database management system with PostGIS extension
- XMLBeans, for parsing and encoding XML requests and responses
- Java Topology Suite, an API of 2D spatial predicates and functions
- Log4J, for enabling the logging at runtime without modifying the application binary
- Apache Maven, for managing the project's build, reporting, and documentation from a central piece of information

This SOS version implements the mandatory operations:

- GetCapabilities, for requesting a self-description of the service
- GetObservation, for requesting the pure sensor data encoded in O&M
- DescribeSensor, for requesting metadata information about the sensor itself, encoded in a sensor model language (sensorML) instance document

And the following transactional operations:

- RegisterSensor, for signing up new sensors and
- InsertObservation, for inserting new observations for registered sensors

6.4.1.3 SOS Adapter

The SOS adapter is used to change the communication logic between the SOS and the sensor bus. A block diagram of the interaction of the adapter with the other system is given in Figure 6.8.

The SOS adapter, upon initialization, will send a service registration message to the sensor bus about a sensor in the sensor bus. On the other side, a sensor adapter, upon initialization, will send a connect sensor message to the sensor bus. The sensor bus will be notified with this message to the SOS adapter. The SOS adapter will use the sensorML document of the sensor and construct the Register Sensor XML request and send it to the SOS. The SOS will register the sensor on its database if it has not been registered, and if it is already registered, then it will return an XML error exception saying that the sensor is already registered.

Once the sensor is registered into the SOS, the next step is to publish its measured data to the sensor bus. The sensor bus will notify the SOS adapter with this message. The SOS adapter will use the sensorML document, construct the insert observation XML request, and send it to the SOS. The SOS will then insert this observation into its database which will be available to the client.

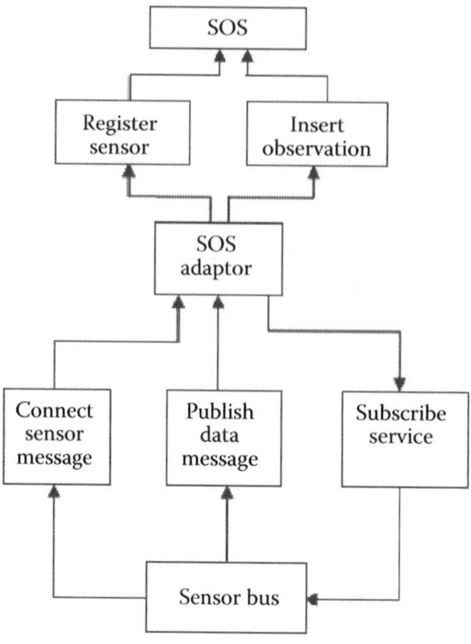

Figure 6.8 SOS adapter interactions with other systems.

6.4.1.4 Sensor Network

Three types of sensor networks are used in this work. These are eZ430-RF2500, Sun SPOT, and simulated sensor networks. Having three different sensor networks in the system shows the heterogeneity of the design, which is the primary goal of the SWE. Each of the sensor network designs is explained as follows.

(a). eZ430-RF2500 Sensor Network

The eZ430-RF2500 sensor network, from Texas Instruments, consists of one sensor node and one base station sensor node. Each sensor node consists of a combination of MSP430 microcontroller and a CC2500 low-power wireless radio. Both sensor nodes have a temperature sensor. The sensor network is shown in Figure 6.9 [25].

Figure 6.9 eZ430-RF2500 sensor network.

This sensor network has a proprietary sensor network protocol called SimpliciTI. SimpliciTI is a proprietary, low-power radiofrequency (RF) protocol targeting simple, small RF networks. The integration of the eZ340-RF2500 to the sensor bus system showed the applicability of the design in a real sensor network application.

(b) Sun SPOT Sensor Network

Sun SPOT is a Java-programmable embedded device. The basic unit includes an accelerometer, temperature and light sensors, radio transmitter, eight multicolored LEDs, two push-button control switches, five digital I/O pins, six analog inputs, four digital outputs, and a rechargeable battery [26].

In this work, the Solarium emulator is used to generate a virtual Sun SPOT instead of a physical sensor board. Also, each virtual Sun SPOT is programmed to run as a temperature and light sensor.

Virtual Sun SPOTs are equipped with a sensor panel where one can set any of the potential sensor inputs (including light level and temperature values). A sensor panel, together with the associated virtual Sun SPOT, is shown in Figure 6.10.

For example, the temperature value of the virtual Sun SPOT shown in Figure 6.11 is set to 25°C and the light value to 450 cd, as displayed in the sensor panel in the left of the virtual Sun SPOT. Using this sensor panel, the values of the light and temperature sensor can be changed as required—emulating a physical sensor board.

(c) Simulated Sensor Network

A simulated sensor node generates a random sensor measured value programmatically using the Java random number generator. An unlimited number of sensor nodes can be added to the sensor network. It uses a timer and schedules the sampling time according to the programmed sampling rate. The sampling rate can be altered programmatically. The random function used to generate the random measured value in the simulated sensor node is Random().nextInt(MAX_VALUE). The ID of the sensor node is also assigned using this random number generator function.

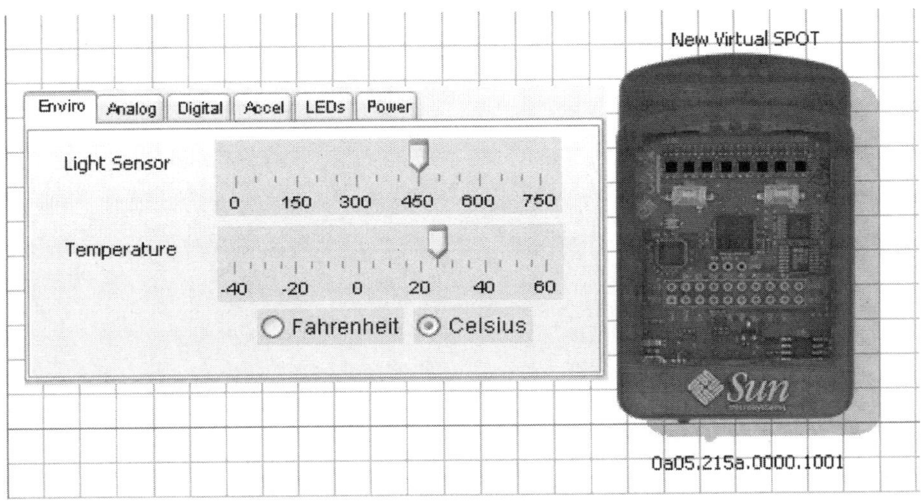

Figure 6.10 Sensor panel for virtual Sun SPOTs.

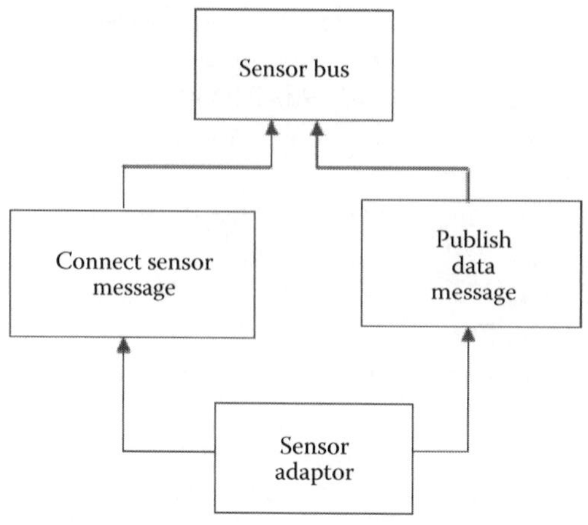

Figure 6.11 Interaction of the sensor adapter with other systems.

6.4.1.5 Sensor Adapter

The sensor adapter is used to change the communication logic between the sensor network and the sensor bus. A block diagram of the interaction of the sensor adapter with the other system is shown in Figure 6.11.

The sensor adapter, upon its initialization, will send a "connect sensor message" to the sensor bus. Then, it will continue to send a "publish sensor data" message to the sensor bus.

In this work, a sensor adapter for the Texas Instruments eZ430-RF2500 has been developed to show the applicability of the design on a real sensor network. The eZ430-RF2500 adapter will send an initialization command to the eZ430-RF2500 sensor network. Then, it will generate the sensor node ID, construct the sensorML document, send it to the FTP server, construct the connect sensor message, and send it to the sensor bus. Then, it will read the measured value of each of the sensor nodes, that is, both the base station sensor node and the non–base station node and construct the publish data message and send to the sensor bus. The implementation of the whole system is explained in detail in the following section.

6.4.2 Implementation

The implementation of the design explained in Section 6.4.1 was done. SOS and client implementation and sensor bus architecture source code from 52°North were used in the implementation. The sensor bus source code was imported to NeatBeans IDE version 7.0. For the implementation of the SOS, an Apache web server was set up. Also, an FTP server was set up to save the sensorML document for each sensor based on their ID. The implementation includes eZ430-RF2500 real sensor network and additional Sun SPOT and simulated sensor networks to test in a real and heterogeneous sensor network environment. Sensor adapters for the eZ430-RF2500, Sun SPOT, and simulated sensor networks and service adapters for the SOS have been implemented.

The SOS, FTP server, sensor adaptors, and SOS adapter are all set up to run in the same local computer. The specifications of the computer used for the implementation and evaluation henceforth of the system are the following:

- Operating system: Windows XP Service Pack 3
- RAM: 1 GB
- Processor: Pentium 4 CPU 3.4 GHZ

6.4.2.1 eZ430-RF2500 Sensor Network

An example of how the Texas Instruments eZ430-RF2500 sensor network system working on this system is explained as follows. An XMPP server is used as a communication service.

When the SOS adapter is started, it will send a service subscribe message to the sensor bus, which actually publishes this message to the XMPP communication service. This message is published on XMPP and taken from the Pidgin XMPP client as shown in Table 6.2.

The table shows that the message type is service subscribe, the service is SOS, and the location of this service. The requirement for this particular service is a sensor with an accuracy of 30% and a survival range up to a temperature of 70°C. These values are available in the sensorML document of each sensor. Then, the SOS adapter waits for a notification of connect sensor and publish data messages from the sensor bus.

Once the eZ430-RF2500 adapter is started, it will build the sensorML document of the sensor and upload it to the FTP server configured for this purpose. Then, it will send the two connect sensor messages (base sensor and non–base station sensor) as shown in Table 6.3 to the sensor bus, which will actually be posted in the XMPP communication service.

These messages will be sent to the SOS adapter, one after the other. The SOS adapter, once notified of the first message, will send the Register Sensor XML request to the SOS. The adapter will use

Table 6.2 SOS Subscribe Message on XMPP


```
(6:14:48 PM) webServiceAccount: Subscribe
Service*SOS*http://localhost:8080/52nSOSv3/sos*{
"namespaces": {
"ssn": "http://purl.oclc.org/NET/ssnx/ssn.owl",
"swt": "http://sweet.jpl.nasa.gov/2.1/propMass.owl",
"ucum": "http://idi.fundacionctic.org/muo/ucum-instances.owl",
"unit": "http://www.w3.org/2005/Incubator/ssn/ssnx/qu/qu-rec20.owl",
"dul": "http://www.loa-cnr.it/ontologies/DUL.owl"
},
"RequiredSensor": {"ssn:observes": "swt:StrengthProperty",
"ssn:hasMeasurementCapability": "<swe:QuantityRange definition =
"Accuracy_01"><swe:uom xlink = "unit:percent" code = "%"/><swe:value>-
30</swe:value></swe:QuantityRange>"
"ssn:hasSurvivalRange": "<swe:QuantityRange definition =
"SurvivalRange_01"><swe:uom xlink = "unit:metre" code =
"m"/><swe:value>- 100 </swe:value></swe:QuantityRange>"
},
"RequiredObservationValue": {"dul:isParameterizedBy": "unit:Celsius "}
}
```

Table 6.3 Connect Sensor Message

First message	ConnectSensor* eZ430-RF2500 _BaseStation.xml
Second message	ConnectSensor* eZ430-RF2500 _Node1.xml

the sensorML document of the sensor to build the register sensor request. It will download the document from the FTP server from which the sensorML document was uploaded. Then, the adapter will send this request to the SOS, and the SOS will register the sensor and return a registration ID.

An example of such a request is shown in Table 6.4. The request is included under the "RegisterSensor" keyword. In this table, only parts of the request are shown. As can be seen from the table, the ID of the sensor node being registered is eZ430-RF2500_Nodel.

Once the sensor node is registered at the SOS, its sensor adapter will continue to post the measured data. For example, the eZ430-RF2500 adapter will send its messages to the XMPP communication service (Table 6.5).

The sensor bus will notify the SOS adapter about these messages, one after the other, based on their sequence. On these messages, the temperature value measured by the eZ430-RF2500_ Nodel sensor node is 25.5 and 26.5, respectively. These messages will be transformed by the SOS adapter to an insert observation request as shown in Table 6.6. The request is included inside the "InsertObservation" keyword.

In Table 6.6, the insert observation request for the first published data message in Table 6.5 is shown. The temperature value measured is 25.5°C, as specified in the results section. This value will be inserted for the sensor eZ430-RF2500_Nodel as it is specified in the AssignedSensorId section. The time at which this data was taken, which is 2011-10-09T18:59:37+03, and the location

Table 6.4 Register Sensor Request to the SOS

```
<sos:RegisterSensor xmlns:sos = "http://www.opengis.net/
sos/1.0"xmlns:sml = "http://www.opengis.net/sensorML/1.0.1"xmlns:om =
"http://www.opengis.net/om/1.0" version = "1.0.0»service =
"SOS"><sos:SensorDescription> <sml:SensorML>
...
<gml:description>A sensor in the Sensor Bus</gml:description>
<sml:keywords>
       <sml:KeywordList> <sml:keyword>sensor bus</sml:keyword>
</sml:KeywordList>/                        </sml:keywords>
<!– sml:identification element must contain the ID of the sensor– >
<sml:identification>
<sml:IdentifierList>
<sml:identifier>
<sml:Term definition = "urn:ogc:def:identifier:OGC:uniqueI>sml:value>
eZ430-RF2500_Node1</sml:value>             </sml:Term>
       </sml:identifier>           ...
</sos:RegisterSensor>
```

Table 6.5 eZ430-RF2500_Node1 Publish Data Messages

First message	PublishData*eZ430-RF2500_Node1*2011-09 02T15:52:43.656+ 03.00*(52.50,7.06)* http://sweet.jpl.nasa.gov/1.1/property.owl#Temperature</*25.5
Second message	PublishData*eZ430-RF2500_BaseStation*2011-09 02T15:53: 43.656+03.00*(52.50,7.06)* http://sweet.jpl.nasa.gov/1.1/property.owl#Temperature</*26.5

Table 6.6 Example of an SOS Insert Observation Request

```
<sos:InsertObservation xmlns:om = "http://www.opengis.net/
om/1.0"xmlns:sa = "http://www.opengis.net/sampling/1.0" xmlns:swe =
"http://www.opengis.net/swe/1.0.1"xmlns:sos = "http://www.opengis.net/
sos/1.0" xmlns:gml = "http://www.opengis.net/gml"xmlns:sml = "http://
www.opengis.net/sensorML/1.0.1"xmlns:xsi = "http://www.w3.org/2001/
XMLSchema-instance"xmlns:xlin = "http://www.w3.org/1999/xlink" version
= "1.0.0"service = "SOS"><sos:AssignedSensorId> eZ430-RF2500_Node1</sos
:AssignedSensorId><om:Measurementxsi:type = "om:MeasurementType">
<om:samplingTime>    <gml:TimeInstantxsi:type = "gml:TimeInstantType">
<gml:timePosition>2011-10-09T18:59:37+03</gml:timePosition> </
gml:TimeInstant> </om:samplingTime> <om:procedure xlin:href =
"eZ430-RF2500_Node1"/> <om:observedPropertyxlin:href = "http://sweet.
jpl.nasa.gov/1.1/property.owl#Temperature"/> <om:featureOfInterest>

<sa:SamplingPointgml:id = "Station_of_sensor_eZ430-RF2500_Node1"
xsi:type = "sa:SamplingPointType">      <gml:description>Station of
sensor eZ430-RF2500_Node1</gml:description>    <gml:name>Station_of_
sensor_eZ430-RF2500_Node1</gml:name>     <sa:sampledFeature/>
<sa:position>         <gml:PointsrsName = "urn:ogc:def:crs:EPSG:4326">
            <gml:pos>7.660052803251871 52.35987390201696</gml:pos>
      </gml:Point> </sa:position> </sa:SamplingPoint>
</om:featureOfInterest> <om:result uom = " Celsius">25.5</om:result>
</om:Measurement>

</sos:InsertObservation>
```

of the sensor when the data was taken, which is approximately (7.7, 52.5), are also shown on the request in the sampling time and position sections, respectively.

Information that is not contained in the publish data message itself but is necessary to construct the InsertObservation request are included in the sensorML document. To access this information, the SOS adapter downloads the sensorML document from the FTP server based on the sensor ID during the construction of every insert observation request. For example, the unit of measure for the temperature (Celsius) is not available in the publish data message in Table 6.5, which has to be taken from the sensorML document to include it in the insert observation request in Table 6.6.

The data are then inserted and stored in the SOS and are henceforth available to clients via the standardized SOS interface. A client can access and retrieve the data in a pull-based manner. An example of getting the last or recent observation request by the 52°North SOS client implementation is shown in Table 6.7. The property used to filter the observation is the sampling time and the

Table 6.7 Get Last Observation Request on the 52°North SOS Client to the SOS

```
<?xml version="1.0" encoding="UTF-8"?>
<GetObservation xmlns="http://www.opengis.net/sos/1.0"
  xmlns:ows="http://www.opengis.net/ows/1.1"
  xmlns:gml="http://www.opengis.net/gml"
  xmlns:ogc="http://www.opengis.net/ogc"
  xmlns:om="http://www.opengis.net/om/1.0"
  xmlns:xsi="http://www.w3.org/2001/XMLSchema-instance"
  xsi:schemaLocation="http://www.opengis.net/sos/1.0
  http://schemas.opengis.net/sos/1.0.0/sosGetObservation.xsd"
  service="SOS" version="1.0.0" srsName="urn:ogc:def:crs:EPSG:4326">
    <offering>House_Temperature</offering>
  <eventTime>
    <ogc:TM_Equals>
      <ogc:PropertyName>om:samplingTime</ogc:PropertyName>
      <gml:TimeInstant>
        <gml:timePosition>latest</gml:timePosition>
      </gml:TimeInstant>
    </ogc:TM_Equals>
  </eventTime>
  <observedProperty>urn:ogc:def:phenomenon:OGC:1.0.30:temperature
  </observationProperty>
  <responseFormat>text/xml;subtype=&quot;om/1.0.0&quot;</
  responseFormat>
<GetObservation>
```

value of this property is the latest shown. This will return the last observation of the temperature of a house as included in the offering section.

To provide data in a push-based way, a SAS can be registered at the Sensor Bus. The SAS receives the incoming data, filters it using certain predefined criteria, and directly forwards it to interested clients. Also, it is possible to change the sampling rate of the sensor by sending the task using SPS. The SAS and SPS are beyond the scope of this work and will be considered in future studies.

6.4.2.2 Heterogeneous Sensor Network

The system in this work is composed of three sensor networks. These are eZ430-RF2500, Sun SPOT, and simulated sensor networks. Together, these sensor networks make a heterogeneous (with different types of sensors) sensor network. An implementation that contains only the eZ430-RF2500 homogeneous (with the same type of sensors) sensor network is given in Section 6.4.2.1.

The connect sensor messages of all the sensor networks in the system, that is, eZ430-RF2500, Sun SPOT, and simulated sensor networks posted on the XMPP communication service are shown in Figure 6.12. The snapshot is taken from a Pidgin XMPP client.

On the right side of Figure 6.12, a list of the sensor networks and services connected to the XMPP is given. Each of the sensor and service adapters must log in and get authenticated by the XMPP before starting to send messages. To do this, an account for the adapters should first be created, and then the adapters will use this account. The security management of the system is done in this way by the underlying communication services.

The SOS adapter will be notified about the connect sensor, one after the other, based on their posting sequence. It will construct the register sensor request and send it to the SOS, which will register the sensor node. An example of part of such a request is given for eZ430-RF2500 (Table 6.4).

An example of publish sensor messages from all the sensor network types posted on the XMPP is shown in Table 6.8. The messages were taken directly from the Pidgin XMPP client and are shown in the table based on their published order. The published time is also included.

The SOS adapter will be notified about the published sensor data messages, one after the other, based on their posting sequence. It will construct the insert observation request and send it to the SOS, which will insert the observation. An example of such a request is given for eZ430-RF2500 (Table 6.6).

The implementation of the design has shown the applicability of the sensor bus in a heterogeneous environment. The design and implementation of the system with SPS and SAS is left for future work. A complete evaluation of the sensor bus architecture has been done. The detailed results and discussion of this evaluation is included in the next section.

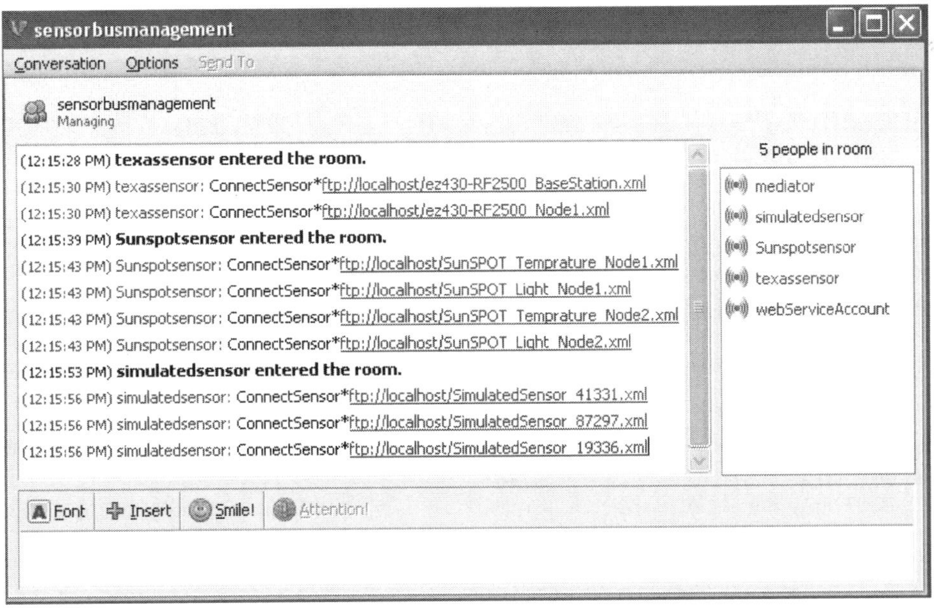

Figure 6.12 Connect sensor messages from heterogeneous network on Pidgin.

Table 6.8 Publish Sensor Messages in the XMPP

Sensor Types	*Published Messages*
eZ430-RF2500	(10:56:51 AM) texassensor: PublishData*ftp://localhost/eZ430-RF2500_Node1.xml*2011-10-16T10:56:51.654+03:00*(52.91171387363812, 7.776307362248074)*http://sweet.jpl.nasa.gov/1.1/property.owl#Temperature*20.3
Simulated	(10:56:52 AM) simulatedsensor: PublishData*ftp://localhost/SimulatedSensor_42159.xml*2011-10-16T10:56:52.232+03:00*(52.375278648425244, 7.475983724573315)*http://sweet.jpl.nasa.gov/1.1/property.owl#Temperature*1
eZ430-RF2500	(10:56:52 AM) texassensor: PublishData*ftp://localhost/eZ430-RF2500_BaseStation.xml*2011-10-16T10:56:52.904+03:00*(52.91171387363812, 7.776307362248074)*http://sweet.jpl.nasa.gov/1.1/property.owl#Temperature*25.9
Sun SPOT	(10:56:53 AM) Sunspotsensor: PublishData*ftp://localhost/SunSpot_Light_Sensor_1.xml*2011-10-16T10:56:53.529+03:00*(52.82201913048282, 7.874168670931996)*Light*450.0
Sun SPOT	(10:56:53 AM) Sunspotsensor: PublishData*ftp://localhost/SunSpot_Temprature_Sensor_1.xml*2011-10-16T10:56:53.529+03:00*(52.82201913048282, 7.874168670931996)*Temperature*21.11

6.4.3 Sensor Bus Architecture: Evaluation and Discussion

6.4.3.1 Introduction

The SWE has shown practical applicability in a number of projects and applications during the realization of the sensor web. However, the integration of the sensor web with the sensor network layer has not been clearly identified by the SWE. There is an interoperability gap between the two layers. The sensor bus is concerned with solutions for easing the integration of heterogeneous sensor networks from different vendors, networking standards and protocols, data outputs, and others. The sensor bus is the main component in the realization of the future sensor web. Therefore, it is necessary to have an efficient sensor bus architecture. Otherwise, it will become a bottleneck in terms of performance, integration time, scalability, reliability, etc., of the total system in the sensor web.

An implementation of the sensor bus on Twitter as a communication service was done by the sensor bus architecture researchers in one of their recent works [5]. The Twitter version of the sensor bus is not applicable for many sensor network applications. The reason being that there are functional limitations associated with Twitter. One limitation is the number of tweets that a Twitter user can make per day, which is limited to only 1000, which will in turn limit the sampling rate of the sensors to 40 samples/h.

In this article, the XMPP version has been implemented. The XMPP does not have the limitations that Twitter has. This will place the XMPP in a better position. However, the XMPP version also has limitations, which are discussed in this chapter. Also, a recommendation is made based on the evaluation results.

6.4.3.2 *Performance Evaluation of the Sensor Bus*

6.4.3.2.1 Average Processing Time

The average processing time is one of the important metrics associated with the performance of the sensor bus. The average processing time for each of the sensor network types given in Table 6.9 is determined after running the sensor bus system for 100 sensor data items with a sampling rate of 1 sample/s. Then, the average processing time for a single sensor data is determined from the above table using the following equation.

$$\text{Average processing time} = \frac{\sum N_i * Tprocavg_i}{\sum N_i} \tag{6.1}$$

This value is determined to be 94.46 ms. An optimization of the SOS adapter to reduce the processing time of the sensor data is recommended. The construction of the insert observation and register sensor requests by the SOS adapter should be optimized to reduce this average processing time.

6.4.3.2.2 Average Response Time

The response time is by far the most important metric in the performance evaluation of the sensor bus architecture. This time should be reasonably small for getting the measured data to the SOS, and then to a client in real-time. Normally, the published sensor data will wait in the XMPP until the SOS adapter is ready to start processing them. This waiting time depends on the average sampling rate of the sensors. The response time is in direct proportion to the wait time. The relationship among response, wait, and processing times is given in the following equation:

$$\text{Response time} = \text{wait time} + \text{processing time} \tag{6.2}$$

Two experiments have been conducted to determine the performance of the sensor bus architecture in terms of the response time of sensor data insertion to the SOS. These experiments are discussed as follows.

In the first experiment, the effect of the sampling rate on average response time for a single simulated sensor was done. The sampling rate of the sensor has been varied to study the effect on the response time of the sensor data. The experiment was run until the number of items published in the sensor data reached 100, and the average response time of the 100-item sensor data has been calculated for the sampling rates selected for the experiment. The result is plotted in Figure 6.13. The full experimental parameters used are given in Table 6.10.

Table 6.9 Average Processing Times for Different Sensor Networks

Sensor Network Types	*Number of Sensors (N_i)*	*Sampling Rate*	*Sensor Property*	*Average Processing Time ($Tprocavg_i$) (ms)*
eZ430-RF2500	2	1	Temperature	99.68
Sun SPOT	2	1	Light and temperature	98.24
Simulated	2	1	Temperature	84.20

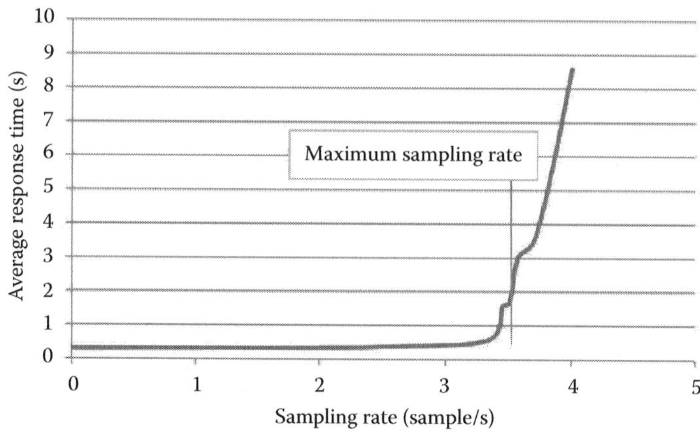

Figure 6.13 Average response time versus sampling rate.

Table 6.10 Experimental Parameters for Average Response Time versus Sampling Rate

Sensor Network Types	*Simulated*
No. of sensors	1
Sampling rate considered (samples/second)	1, 2, 3, 4, 5, 6, 7, 8, 9, 10, 20
Running time	Until 100 sensor data inserted into the SOS

As shown in Figure 6.13, the average response time of the sensor data is constant up to a certain sampling rate, and then it exponentially increases. The exponential increment is caused by the increase in the number of samples per second. There will be more sensor data in the XMPP queue as the SOS adapter and the SOS needs an average processing time to insert these sensor data.

As the response time keeps increasing for the sensor data, the data will wait for a longer time in the queue (that is, in the XMPP) and it will not be available for the client immediately, which will not be acceptable for real-time applications. The average processing time should be optimized for using the sensor bus at high sampling rates. Otherwise, the sum of the sampling rates attached to the SOS adapter should be up to the value where the above graph is constant.

The sampling rate of the sensors affects many aspects of the sensor bus architecture. In this work, the maximum sum for the sampling rate of the sensors that can be connected to a single sensor adapter and SOS adapter is determined. This value is actually the knee point in the above graph in which the exponential increment has started. As shown in the figure, this value is approximately 3.51 samples/s. The same type of experiment has been done using Sun SPOT and heterogenous sensor networks and the results showed the same value. This value is better than the Twitter-based implementation of 40 samples/h, which is approximately 0.01 samples/s [5]. However, considering the possiblity of having a large number of sensors within a sensor network, 3.51 samples/s will still not be applicable for most sensor network applications. Therefore, either a parallel handling of sensors should be possible or the maximum sampling rate needs to be improved.

A multithreaded SOS adapter in which a single SOS adapter instance has to be created for at least each of the sensors in the sensor network is recommended. This will allow sensor registration and sensor data insertion in a parallel way rather than in a sequential way. For a single sensor, the proccessing time of the sensor bus system should be optimized to increase the maximum sampling rate.

In the second experiment, the same experimental parameters shown in Table 6.10 were used. Here, the response time was calculated for each of the inserted sensor data items, whereas in the first experiment, the response time was calculated after inserting 100 sensor data items. The result of the second experiment is plotted in Figure 6.14.

As shown in the above figure, for a sampling rate of 3.57 and 4 samples/s, the response time is increasing very fast. This is due to the increment in the wait time of the sensor data as the sampling rates are very high. A large number of sensor data were published to the XMPP and need to wait in the XMPP queue until the SOS adapter inserts those that are ahead. Whereas for sampling rates of 2, 3.33, and 3.51 samples/s, the response time is decreasing and then almost becomes constant at the same value. This constant value is approximately the average processing time. This time is approximately the time that is taken by a single sensor data to become inserted into the SOS without waiting in a queue. Here, the time taken by the sensor adapter to get it published and the propagation time from the sensor adaptor to the XMPP and then to the SOS adapter is neglected. There is no data in the XMPP queue because the sampling rate is less than or equal to the maximum sampling rate; every sensor data item can get processed and inserted into the SOS before the next sensor data item gets published. The sensor bus is not applicable for sampling rates of sensors such as 3.57 and 4 samples/s, although it is applicable for sampling rates of 2, 3.33, and 3.51 samples/s.

The maximum allowable sampling rate that was determined in the first experiment was 3.51 samples/s. The second experiment can be used to support the value of this sampling rate. It is also shown in the second experiment that for a sensor network with a sampling rate of up to the maximum allowable value, which 3.51 samples/s, then the response time is very small; being approximately the same as the processing time. Whereas for a sampling rate that is more than 3.51 samples/s, the response time keeps increasing very quickly, which is not acceptable for most network applications. The same recommendation, as with the first experiment, is suggested here to cope with the maximum allowable sampling rate.

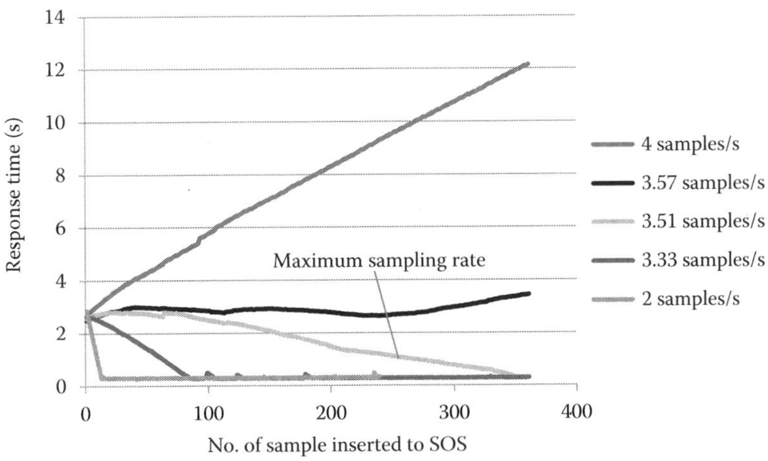

Figure 6.14 Response time versus number of sensor data inserted for different sampling rates.

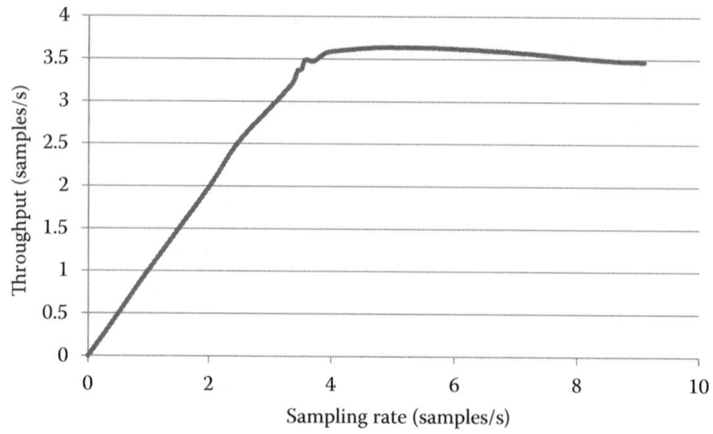

Figure 6.15 Throughput versus sampling rate.

6.4.3.2.3 Throughput

The throughput is an important metric to measure how much useful work, that is, insertion of sensor data is done per unit time. The amount of the throughput of the sensor bus system depends on the sampling rate. An experiment based on the parameters in Table 6.10 has been done to show the relationship between throughput and sampling rate. The results are plotted in Figure 6.15.

The above figure shows that the throughput of the sensor bus system is directly proportional to the sum of the sampling rates of the sensors in the sensor network. This holds true up to the maximum sampling rate, which is 3.51 samples/s. Beyond the maximum sampling rate, the throughput remains constant. The sensor data simply gets published and waits on the XMPP queue.

6.4.3.2.4 Real-Time Sensor Data

Many sensor network applications are based on a real-time sensor data. The sensor data is expected to be inserted into the SOS and subsequently reach the client approximately after the average processing time. This is the minimum amount of time that the sensor bus needs to make the sensor data available for a client after receiving it from a sensor adapter. This is possible if the sum of the sampling rates of the sensor networks connected to a single SOS adapter is less than or equal to the maximum sampling rate. For a sampling rate greater than the maximum, the response time is increasing as discussed in Section 6.4.3.2.2. The sensor data can reach a client long after it has already been measured and published in the XMPP. The sensor data is not up-to-date for a client. This will not be acceptable for real-time critical applications. The recommendation given in Section 6.4.3.2.2 can also be used to alleviate this problem.

6.4.3.2.5 Effect of Initialization Time

The initialization time is the time needed for the sensor adapter to register all of its sensors in the sensor network. The sensor should be registered before its data can be inserted into the SOS database. It takes a considerable amount of time for the SOS adapter to register a sensor. The sensor registration procedure is described in Section 6.4.1.3, which corresponds to the initialization time. An experiment has been done to determine the average initialization time of registering a single sensor. The values determined for different numbers and types of sensor networks are given in Table 6.11.

The average initialization time for a single sensor is determined from Table 6.11 using the following formula:

$$\text{Average initialization time/sensor} = \frac{\sum N_i * \text{Tintavg}_i}{\sum N_i} \quad (6.3)$$

The value is determined to be 2.476 s.

On average, the first sensor data published just after the sensor adapter sent the connect sensor message will wait in the XMPP for the sum of the initialization time of each sensor, and thus the response time for the first sensor data will be high. This time is for a single sensor and it will be even larger if the number of sensors in the system is large. To determine the relationship between the initialization time and the number of sensors, an experiment was conducted (parameters are provided in Table 6.12). The results of the experiment are plotted in Figure 6.16.

The result shows that the total initialization time taken by the sensor adapter is almost equal to the sum of the initialization time for the individual sensors. The effect of the initialization time will propagate to the next published sensor data depending on the sampling rate of the sensor network. To determine this effect, an experiment was conducted (parameters are provided in Table 6.13). The results of the experiment are given in Figure 6.17. This shows the effect of the initialization time for the eZ430-RF2500 temperature sensor on the different sensor data published. First, only for the base station sensor, and then including the other sensors in the sensor network. In the above figure, two plots are included. One is for the eZ430-RF2500 base station only and the other is the eZ430-RF2500 base station with the second node.

Table 6.11 Average Initialization Time for Different Sensor Networks

Sensor Network Types	*Number of Sensors (N_i)*	*Sampling Rate*	*Sensor Property*	*Average Initialization Time, Tintavg$_i$ (second)*
eZ430-RF2500	2	1	Temperature	5.172
Sun SPOT	2	1	Light and temperature	5.420
Simulated	2	1	Temperature	4.265

Table 6.12 Experimental Parameter for Average Initialization Time versus Number of Sensors

Sensor Network Types	*Simulated*
No. of sensors	Variable
Sampling rate considered (samples/s)	1
Running time	Until the first sensor data inserted to the SOS

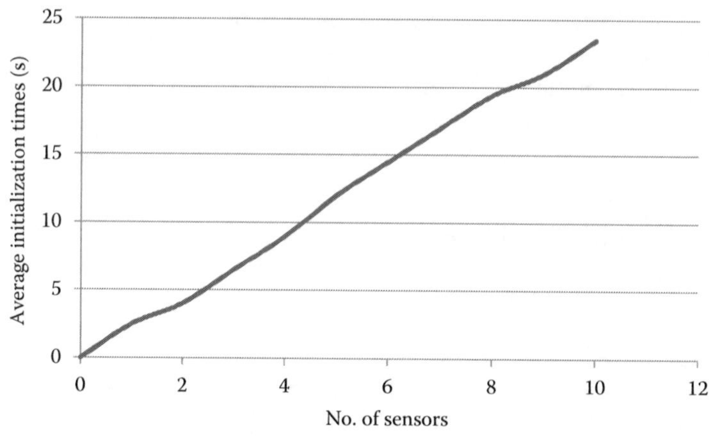

Figure 6.16 Average initialization times versus number of sensors.

Table 6.13 Experimental Parameter for the Effect of Initialization Time on Different Sensor Data

Sensor Network Types	*ez430-RF2500*
No. of sensors	1,2
Sampling rate considered (samples/s)	1
Running time	Until 100 sensor data inserted to the SOS

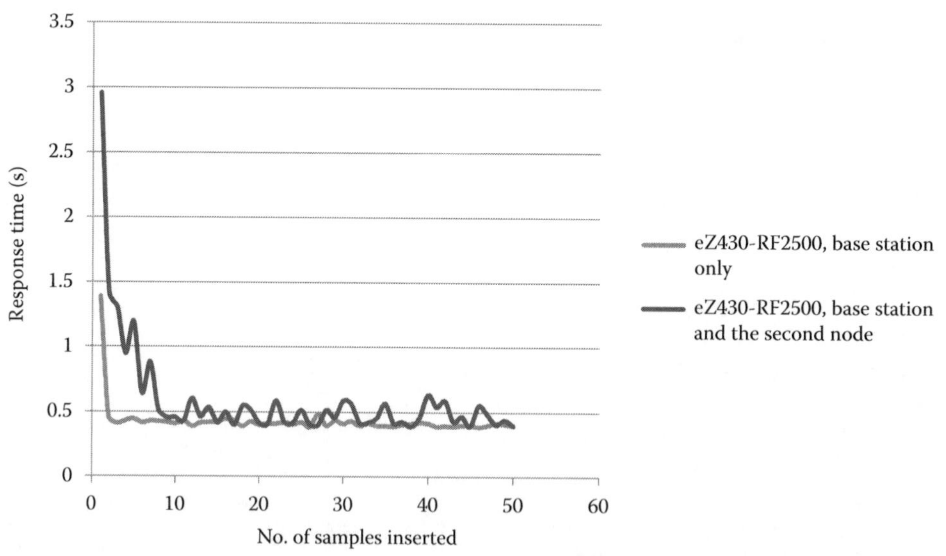

Figure 6.17 Effect of initialization time for eZ430-RF2500.

As shown in the above figure, the response time is very high for the first sensor data and then decreases and reaches an almost constant value. This is because the SOS adapter should register the sensor at the SOS. To do this, it must send a connect sensor message to the sensor bus. Then, the SOS will start the registration of the sensor. In the meantime, the sensor adapter will send a publish sensor message to the sensor bus. However, those messages will wait in the sensor bus queue until the sensor gets registered.

To show the effect of initialization time with respect to the number of sensors, in the above figure, two plots are compared. As can be seen, the response time for the first sensor data is higher in the case of the plot for eZ430-RF2500 with both sensor nodes because the sensor adapter and the SOS needs to perform more tasks, that is, registration of the two sensors before starting to insert the sensor data. The sensor data will wait in the XMPP and will be processed by the SOS adapter when free.

The sensor adapter might be restarted due to power failure, recovery from crashing, administrator intervention, etc. Every time the sensor adapter gets started, then it will send the connect sensor message. This is unnecessary once the sensors are registered because the SOS will reply with an exception stating that it is already registered. This will increase the initialization time that in turn will increase the response time.

To overcome this problem, the following is recommended. The sensor registration status of the sensor adapter should be stored in its own cache, and once it is registered, it would not need to attempt registration of the sensor nodes.

6.4.3.3 Scalability

As already determined, the maximum sampling rate for the sum of individual sampling rates of the sensors in a sensor network attached to a single SOS adapter in the sensor bus is approximately 3.51 samples/s. For a sampling rate of less than 3.51 samples/s, there is no sensor data waiting in the XMPP queue. As already discussed in the previous section, the normal function of the sensor bus is when the sampling rate is less than or equal to the maximum sampling rate.

An experiment based on the experimental parameters given in Table 6.14 has been conducted to determine the scalability of the sensor bus architecture. The experiment has been done so that the response time is equal to the processing time, that is, without a wait time in the XMPP. The results are plotted in Figure 6.18.

The results show that the number of sensors that can be supported by the sensor bus depends on the sampling rate of the individual sensors. As the sampling rate of the individual sensors is increased, the number of sensors that can be supported by a SOS adapter instance decreases. Therefore, to support more nodes, the sampling rate of the individual sensor should be set at a lower rate. This will affect the accuracy of the result with relation to data sampling time. This can be compromised based on the nature of the property being measured and the data requirement at the client's side. For example, the temperature of a location might not change within a couple of

Table 6.14 Experimental Parameters for Scalability

Sensor Network Types	*Simulated*
No. of sensors	Variable, greater than 0
Sampling rate	Variable, less than 7 samples/s
Running time	100 sensor data

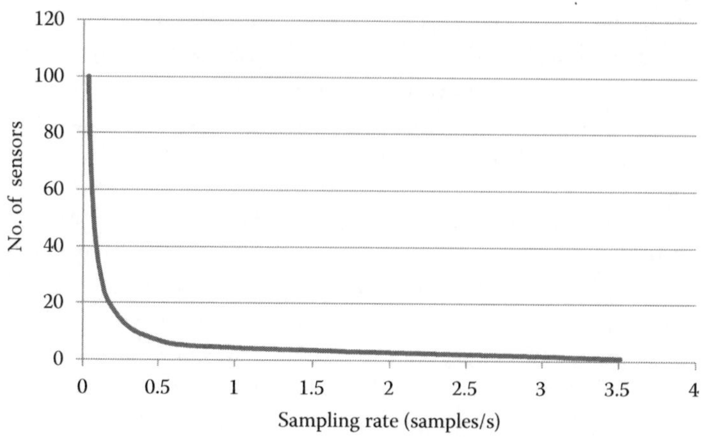

Figure 6.18 The number of sensors versus the sampling rate.

seconds so that the sampling rate can be decreased from 1 to 0.1 sample/s. Here, an assumption is made that the temperature will not be changed within 10 s.

The design recommendation made in the previous section can also be taken here as a possible solution. Having a multithreaded SOS adapter that will allow one SOS adapter instance for each of the sensors in the sensor networks will allow users to overcome the limitation on the sum of the sampling rate of the sensors in the sensor network. However, there will still be limitations on the sampling rate of individual sensors rather than on the sum of the sampling rates. This limitation can be solved by improving the processing time.

6.4.3.4 Reliability and Fault Tolerance

(a) Reliability

Four subsystems can be identified in the sensor bus architecture, which can run independently at different servers. These systems are the following:

- Sensor Adapter
- Instant Messaging Service server (here, XMPP has been used)
- SOS Adapter
- FTP Server—where the sensorML document is stored

While all systems in the sensor bus are running, the sensor bus will continue to insert sensor data. However, the question is, what happens when one of the subsystems is not working fine? There are different cases that could occur during the operation of the sensor bus. One of the subsystems might not work well at a certain time. For some of those cases, the sensor bus will continue to insert sensor data whereas for some of the other cases, it will totally stop and the data might get lost without insertion into the SOS. The reliability of the sensor bus architecture depends on the reliability of those subsystems. These different cases are mentioned in Table 6.15 and are followed by further explanation.

(b) Fault Tolerance

If the active sensor adapter crashes, that is, stops working for any reason, the SOS adapter will continue to insert those measured values published on the sensor bus, assuming that the

Table 6.15 Possible Combination for the Sensor Bus System

Cases	*Sensor Adapter Running*	*Instant Messaging (XMPP) Server Running*	*SOS Adapter Running*	*FTP Server*	*Reliability*
1	True	True	False	True	Not reliable
2	True	False	True	True	Not reliable
3	True	True	True	False	Not reliable
4	False	True	True	True	Reliable
5	True	True	True	True	Reliable

speed of publishing by the sensor adapter is not equal to the speed at which the SOS adapter inserts into the SOS server. The same situation might happen while the sensor is being initialized. This is actually a good fault tolerance feature of the sensor bus.

6.4.3.5 Adaptability and Heterogeneity

(a) Adaptability

As part of this work, Texas Instruments eZ430-RF2500 temperature sensor network was integrated into the sensor bus. A host application that would read the sensor measured data and provide for the sensor adapter was developed. Its sensor adapter was also developed. Compared with the amount of time it would take to adapt to the SOS directly, it can be estimated that this time is reasonably smaller.

(b) Heterogeneity

The main goal of the sensor bus is to ease the task of integrating heterogeneous sensor networks into the sensor bus. An experiment was done to demonstrate its capability toward this. The heterogeneous sensor network was made from eZ430-RF2500, Sun SPOT, and simulated sensor networks. The experimental parameters mentioned in Table 6.16 have been used. The parameters were chosen so that the sum of the sampling rates of the heterogeneous does not exceed the maximum allowable sampling rate determined. The sum of the sampling rates of the heterogonous network mentioned in Table 6.16 is 3.00 samples/s, which is lower than the maximum sampling rate determined (3.51 samples/s).

Table 6.16 Specification of the Sensor Networks Used to Form a Heterogeneous Environment

Sensor Network Type	*Texas Instrument Sensor Network*	*Sun SPOT Sensor Network*	*Simulated Sensor Network*
Sensor property	Temperature	Light and temperature	Temperature
Number of sensors	2	2	2
Sampling rate per sensor (samples/s)	1	0.25	0.25
Sum of sampling rate per sensor network	2	0.5	0.5

The experimental results are plotted as shown in Figure 6.19. In the figure, the plots of the eZ430-RF2500 sensor network with a sampling rate of 1 sample/s (2 sensors = 2 samples/s) per sensor and the Sun SPOT sensor network with a sampling rate of 2.21 samples/s per sensor (2 sensors = 4.42 samples/s) were included to compare the response times of the heterogeneous networks.

As shown in the above figure, the response times for some of the first sensor data were high for the heterogeneous sensor network compared with that of the Sun SPOT and eZ430-RF2500 networks. This is because a total of six sensors needs to be registered so that the initialization time is higher. This will have an effect on the response time of the sensor data.

Otherwise, the response time for the heterogeneous sensor network is decreasing quickly and keeps constant as the eZ430-RF2500 sensor network, and then keeps increasing. Whereas for the Sun SPOT sensor network, the response time is increasing very quickly because of the sampling rate of the sensor network of 4.42 samples/s, which is greater than the maximum allowable sampling rate.

The sensor bus architecture can be used for the real-world eZ430-RF2500 sensor network and for the heterogeneous sensor network configuration given in Table 6.16. In fact, it can be concluded that the sensor bus architecture can work in a heterogeneous sensor network environment for real-time applications.

However, the architecture cannot be used, for example, for the Sun SPOT sensor network. Therefore, the sensor network, whether it is homogeneous (sensor network with the same sensors types) or heterogeneous, should have a sum of the sampling rate that is smaller than the allowable value to be used within a single SOS adapter considering the real-time applications.

Generally, the maximum sampling rate of the sensor bus for a single SOS adapter instance has been determined to be 3.51 samples/s. The response time of the sensor data insertion into the SOS depends on the sum of the sampling rates of the sensors in the sensor network. For a sensor network with a large number of sensors, the sum of the sampling rate will be very large and will make the response time larger. As most sensor network applications work in real-time, a large response time is not tolerable for most applications. The scalability of the system also depends on the sum of the sampling rates. An improvement should be done either to increase this maximum rate or to handle the operation requests of the sensors in parallel. Also, the initiation time of the system will have a considerable effect especially in a sensor network with a large number of sensors. Because there is no need to send a sensor registration message every time the sensor adapter is running, it

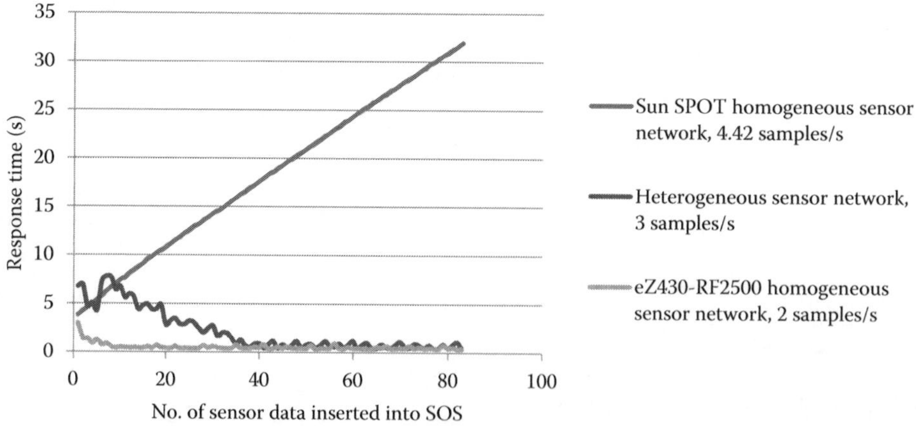

Figure 6.19 The response time versus the number of sensor data for a heterogeneous sensor network.

will be better to have a caching functionality within the sensor adapter so that it will know which sensors are registered and which are not. The interaction and function of the subsystems in the sensor bus should be optimized in such a way that would increase the reliability of the system.

The sensor adapter is designed to send a sensor connect message to the sensor bus. This causes some of the previously published sensor data to wait in the communication service for a long time. This effect is shown in the initialization time for the eZ430-RF2500 sensor network. As it has been suggested that some kind of caching mechanism should be added to the sensor adapter so that it will not send this message once its sensor nodes are already registered.

6.5 Conclusion and Recommendation

The practical implementation of the sensor web system, the SWE standard (which makes use of the sensor bus), is discussed in this chapter. The SOS from the 52°North implementation is used as part of this work. A real temperature sensor network from Texas Instruments, eZ430-RF2500, is integrated into the sensor bus by developing the corresponding sensor adapter. The integration cost and time is estimated to be much less than that of directly integrating into the SOS. A complete evaluation of the sensor bus architecture has been done.

An implementation of a heterogeneous sensor network environment is also done. The eZ430-RF2500, a temperature sensor network, virtual Sun SPOT, each of the sensor nodes with a light and temperature sensor, and a simulated sensor network is used to demonstrate the application of the sensor bus in such a heterogeneous sensor network environment. The performance of the system is evaluated and compared with the homogeneous sensor network. The system behaved in the same way in both cases. This shows the equal applicability of the sensor bus for heterogeneous sensor networks. In fact, the limitations associated with the homogeneous sensor network also hold for the heterogeneous network.

Although the design of the sensor bus toward minimizing the integration effort of new sensors is promising, there are some performance problems that need to be tackled. There is a limitation on the cumulative sampling rate of the entire sensor network. This sampling rate is determined to be 3.51 samples/s. This value indicates that a single sensor node with a sampling rate of more than 3.51 samples/s cannot be integrated at all with the sensor bus architecture considering the real-time requirement of the sensor data. Also, a sensor network with more than one sensor node should have a sampling rate so that its sum is not greater than the maximum sampling rate. Most commonly, a large number of sensor nodes per sensor network are deployed. This will result in a very small maximum sampling rate that the individual sensor node can have. This is not applicable for most sensor network applications. A multithreaded implementation of the SOS adapter is suggested so that a new SOS adapter instance is created for each sensor node in the sensor bus. Therefore, each sensor node can have the maximum allowable sampling rate, that is, 3.51 samples/s. However, to further increase the current maximum allowable sampling rate, an improvement should be done on minimizing the processing time taken by the SOS adapter.

References

1. M. Botts, G. Percivall, C. Reed, and J. Davidson, OGC® Sensor Web Enablement: Overview and High Level Architecture, Open Geospatial Consortium Inc., Wayland, MA, USA: OGC, 2007)
2. ZigBee Specification, ZigBee Alliance Std. 053 474r17, 2008.

3. B. SIG. Bluetooth specification version 3.0, Bluetooth Special Interest Group Std., 2009.
4. K. Lee. IEEE 1451: a standard in support of smart transducer networking. In *Proceedings of the 17th IEEE Instrumentation and Measurement Technology Conference, 2000.* IMTC 2000. vol. 2, pp. 525–528, 2000.
5. A. Broring, T. Foerster, S. Jirka, and C. Priess. Sensor bus: an intermediary layer for linking geosensor networks and the sensor web. In *Proceedings of the 1st International Conference on Computing for Geospatial Research and Application*, COM.Geo'10, 21–23 June 2010, Bethesda, MD, USA; ACM: New York. pp. 1–8, 2010.
6. M.-M. Wang, J.-N. Cao, and J. Li. Middleware for wireless sensor networks: a survey. *Journal of Computer Science and Technology*, pp. 305–326, 2008.
7. K. Aberer, M. Hauswirth, and A. Salehi. A middleware for fast and flexible sensor network deployment. In *32nd International Conference on Very Large Data Bases*, Seoul, Korea, pp. 1199–1202, 2006.
8. C. Arnhardt, T.M. Fernandez-Steeger, and R. Azzam. A sensor based landslide monitoring system integrated in an early warning structure. In *70. Jahrestag der Deutschen Geophysikalischen Gesellschaft (DGG)*, 15–18 March 2010, Bochum, conference transcript, p. 43, 2010.
9. X. Chu, T. Kobialka, B. Durnota, and R. Buyya. Open Sensor Web Architecture: core services. In *International Conference on Intelligent Sensing and Information Processing (ICISIP 2006)*, IEEE Press, Piscataway, Bangalore, India, pp. 98–103, 2006.
10. 52°North, Sensor bus. http://52north.org/communities/sensorweb/incubation/sensorBus/index.html, 2011.
11. M. Fowler. Analysis patterns: reusable object models. Addison-Wesley. Boston, pp. 35–55, 1997. ISBN 0-201-89542-0.
12. Open Geospatial Consortium. Observations and measurements, 2007.
13. Open Geospatial Consortium. OpenGIS Sensor Model Language (SensorML) implementation specification. 2007.
14. L. McKee and M. Botts. A sensor model language: moving sensor data onto the internet Open Geospatial Consortium Inc., Wayland, MA, USA. (http://www.sensorsmag.com/networking-communications/a-sensor-model-language-moving-sensordata-internet-967), 2003.
15. A. Na and M. Priest (Editors). Sensor observation service: OpenGIS® implementation standard, Open Geospatial Consortium Inc., 2007.
16. I. Simonis (Editor). Sensor planning service: OpenGIS® implementation specification. Open Geospatial Consortium Inc., 2007.
17. I. Simonis (Editor). OGC® Sensor alert service implementation specification: candidate OpenGIS® interface standard, Open Geospatial Consortium Inc., 2007.
18. I. Simonis and A.Wytzisk (Editors). Web Notification Service, OpenGIS® Consortium Inc., 2003.
19. S. Havens (Editor). OpenGIS® transducer markup language implementation specification: OpenGIS® implementation specification, Open Geospatial Consortium Inc., 2007.
20. The Extensible Messaging and Presence Protocol (XMPP) Standards Foundation, http://www.xmpp.org, 2011.
21. WebLogic Server Components: Java Messaging Service (JMS), Oracle Corporation, http://docs.oracle.com/cd/E19509-01/820-5892/index.html, 2010.
22. Internet Relay Chat (IRC), http://www.irc.org/, 2011.
23. Twitter, http://www.twitter.com, 2011.
24. A. Broring, T. Foerster, and S. Jirka. Interaction patterns for bridging the gap between sensor networks and the Sensor Web. In *8th IEEE International Conference on Pervasive Computing and Communications Workshops (PERCOM Workshops)*, 29 March 2010–2 April 2010, pp. 732–737, 2010.
25. Texas Instruments development kit, eZ430-RF2500 development kit, http://www.ti.com, 2011.
26. Sun SPOT Java development kit, http://www.sunspotworld.com, 2011.

ROUTING AND TRANSPORT PROTOCOLS

Chapter 7

Quality of Service and Routing Performance Evaluation for IEEE 802.15.6 Body Sensor Networks Using an Accurate Physical Layer

J.-M. Dricot, S. Van Roy, Ph. De Doncker, and A. Nonclercq

Université Libre de Bruxelles

M. Rinaudo

Université Libre de Bruxelles and University of Parma

Gianluigi Ferrari

University of Parma

Contents

7.1 Introduction

7.1.1 Motivation

The current increase in health care spending is one of the many challenges that health care systems have to face. It has led to concerns about long-term sustainability [1]. For instance, inflation-adjusted per capita health expenditures in the United States—which devotes the largest share of its gross domestic product to health care (17.6% in 2009)—have increased from approximately $809 in 1960 to $7375 in 2009, with an average annual growth rate of approximately 4.7% [2]. Another challenge is the aging of the population, which is also accounting for the increase in health care costs. It is expected that the number of persons with ages 60 years or older will increase from 688 million in 2006 to almost 2 billion by 2050, at which point (for the first time in human history) the population of older persons will be larger than the population of children [3]. Those two challenges highlight the need for better health system efficiency.

Recent advances in ultra-low power wireless sensors have fostered research in the field of body-centric wireless networks, also referred to as *body area networks* (BANs) [4–7]. In these networks, a set of nodes (called sensors) is deployed on the human body. They aim at monitoring and reporting several physiological values, such as blood pressure, breath rate, skin temperature, or heart beating rate.

This chapter addresses the development of a specific framework for the accurate computation of the network topology (i.e., routing) suitable for medical applications. First, a comprehensive and detailed analytical framework for BAN performance evaluation is developed, obtaining closed-form expressions for the link probabilities of outage in the context of multiuser communications. This framework encompasses the effects of environment, topology, and traffic intensity.

In the remainder of this section, state-of-the-art medical sensor network applications are presented. The medical best practices and standards in use are presented and the effect of wireless sensor networking on modern medicine is extensively detailed. In Section 7.2, an accurate channel model for on-body communications, based on an extensive measurement campaign using high-accuracy and large bandwidth spectrum analyzers, is presented. In Section 7.3, a link–layer performance is evaluated with a computation of the link probability of success when using multisensor communications. The derivations take into account sensor activity (through its probability of transmission) and the effect of the environment as well. In Section 7.4, the link-level throughput and the delay are defined. The distinction between random access networks (i.e., slotted ALOHA) and time division multiple access networks (i.e., TDMA) is described. Next, in Section 7.5, the routing performance is analyzed and a novel framework for the computation of the optimal BAN topology is introduced. Section 7.6 concludes this chapter and presents open research questions.

7.1.2 Sensor Networks and Their Application in Modern Medicine

BANs allow continuous monitoring of patients at home—compared with short-time monitoring at the hospital and, besides a traditional follow-up of a given pathology for a patient, they may also help in prevention medicine, early detection of pathologies, population screening, monitoring of good adhesion to treatment guidelines, etc. The increase in interest for wearable monitors and Holter monitors may be seen as a first step toward BANs, as they also allow those features. However, current systems are usually cumbersome and wires are impractical. Furthermore, data processing and analysis are usually performed offline, making such devices not well-suited for continuous monitoring [8]. BANs provide a way to overcome those limitations. A BAN may be used in different types of frameworks, including monitoring patients with chronic disease, patients in hospitals, elderly patients at home, or even continuous monitoring of patients in any kind of environment. In particular, it can be used for numerous applications such as arrhythmias, hypertension, diabetes, elderly patient monitoring, and drug delivery through integrated feedback systems (see Aziz et al. [9] for review). Although most applications are focused on surface sensors, some also allow an implanted distributed sensor network. BION, for instance, proposes neural prosthetic interfaces that allow multichannel systems to be assembled from single-channel micromodules [10]. Besides clinical application, BANs may also be instrumental in various other applications, including human–computer interfaces and gaming [11].

As many applications and types of BANs appear, it is important to define some key terminologies and to propose a classificatory framework providing a general understanding of the technical aspects related to these systems [12]. Here, we will define only a few. A BAN may be defined as "a network of communicating devices worn on, around, or within the body, which is used to acquire health-related data and provide mobile health services to the user" [12]. The BAN may be composed of noninvasive (worn on the body) or invasive (worn in the body) sensors and actuators, other network nodes placed around the patient (such as an external sensor informing of the environment of the patient, a video camera informing the posture of the patient, or a screen providing feedback to the patient), as well as a base unit, which are communicating together (intra-BAN communication). Furthermore, extra-BAN communication usually also exists, allowing the system to communicate with a remote user, for instance, a health professional, through a central server system called BackEnd. Numerous BAN systems have been proposed (see, for instance, Aziz et al. [9] and Yuce [13] for reviews), including various physiological monitoring systems for individual or group use in medical centers or in home care [13], computer-assisted physical rehabilitation [14], implantable neural recording [15], and heart activity monitoring [16]. Many systems aiming to monitor patient physiological data share common features and therefore could also be used for a wide range of applications with few modifications.

Building a successful BAN seems to be a challenging task. Most telemedicine applications remain in the pilot phase and never reach proper use in daily practice [17–19]. To be successful, the major determinants may be divided into five classes [19]: (1) technology, (2) acceptance, (3) financing, (4) organization and, (5) policy and legislation. We will address some of the major related challenges.

Biomedical signal acquisition is a difficult task and has been greatly studied [20–23]. The difficulty usually comes from the low amplitude that physiological potentials such as electrocardiographic, electromyographic, or electroencephalographic potentials may show, which makes them sensitive to the noise generated by the device, electrical interference from the surroundings, and artifacts. It is estimated that in most bioelectric measurements, the overall disturbance level should be lower than 1 to 10 μV peak-to-peak [20].

Interferences are due to various unwanted couplings [20]. A capacitive coupling between the patient, the mains, and the earth causes interference currents flowing through the body. Similarly, both the amplifier and the wires connecting the electrodes to the amplifier are capacitively coupled to the mains and to the earth, causing parasitic currents. Magnetically induced interferences are also present due to the loop formed by the measurement wires. By nature, a BAN shows short wires connecting the electrodes and the amplifier, reducing the parasitic coupling; in this regard, it is less prone to interferences.

The noise generated by the device is mainly due to the first stages of the front-end, composed of the electrodes and the amplifier stage. The noise of typical Ag–AgCl electrodes is estimated to range from 1 to 15 μVRMS for electrodes placed on the body surface in a 0.5 to 500 Hz band-pass [24]. Much effort has been made by manufacturers to decrease the input noise of instrumentation amplifiers, even for the low frequencies needed, so that levels lower than 1 μV peak-to-peak may be reached for the typically required bandwidths. Finally, artifacts may affect the quality of the recording. Artifacts may be divided into two categories. First, physiological artifacts could be due to the biopotentials generated by the patients themselves, interfering with the acquisition of the physiological signal of interest. Second, nonphysiological artifacts are due to transitory perturbation such as motion artifacts, impurities, and deterioration of the electrodes and electrostatic artifacts. Among these, motion artifacts are probably one of the most problematic because they are frequent and are not confined to a small spectral band [25,26].

Patient safety, of course, is an issue of major importance for a BAN, as for any other medical device. IEC 60601 International Standards, published by the International Electrotechnical Commission, lists the requirements regarding the safety and effectiveness of medical electrical equipment. Compliance with the IEC 60601 International Standards is a recognized step toward medical device approval in nearly all markets across the world, including the United States, Europe, Canada, Japan, Australia, and other countries [27]. Besides all conventional issues related to safety, the radiation issue that a BAN implies through its communication and the long-term consequences of this radiation on the body should be assessed, particularly in the case of implantable sensors [9].

Safety is of course mandatory under normal operational conditions, but this is also true under first fault conditions. In this regard, even if a medical device may fail, it should do so safely. Besides, it is important to be able to detect a fault in the system when it occurs, to be able to interrupt its use, and to avoid the occurrence of another fault that could lead to a risk for the patient. Fault detection in BAN has been studied for many years [28], and robust fault detection should be designed for any BAN. The concept of essential performance, that is, the performance necessary to achieve freedom from unacceptable risk, has received much attention in the last few years and its application lead to a change in the title of the IEC60601-1 publication [29]. from "Medical electrical equipment, Part 1: General requirements for safety" in the second edition, to "Medical electrical equipment, Part 1: General requirements for basic safety and essential performance" in the third edition (IEC60601-1) [30]. Essential performance is directly related to patient safety as one should understand it by considering whether its absence or degradation would result in an unacceptable risk.

This concept is clearly illustrated in applications such as life-supporting devices for which accuracy is vital, medication administered in a closed-loop system for which overdosing is unacceptable, or preimpact fall detection for which malfunctioning may imply major injury [9,31]. Besides those obvious examples, another example of essential performance is the "correct output of diagnostic information from medical equipment that is likely to be relied upon to determine treatment, where incorrect information could lead to an inappropriate treatment that would present an unacceptable risk to the patient" (IEC60601-1) [30]. Because the purpose of the large majority of medical equipment is to help the medical professional in realizing a diagnostic, such an example nearly always applies. Therefore, the data acquired from the BAN should always be accurate or, at least, it should be made clear when inaccurate information occurs to make sure that it would never lead to an inappropriate treatment. There are many types of inaccurate information, including data that is unavailable, incorrect, or with incorrect timing.

Focusing on BANs, wireless communication is prone to different types of problems leading to inaccurate data, including interference caused by other wireless devices that share identical channels, harsh network environments, low batteries, and unusual motions [28,32]. Various methods have been proposed to achieve robust data transmission, including packet retransmission [13,33], use of a multipath network to avoid disrupted links [9,34], and multicast or broadcast-based routing schemes [35,36]. Local data storage at the node level [26] may also be an option to backup potential data loss during transmission, at the cost of a more complex data transmission (or offline data download), and larger data storage components.

Data synchronization has great importance, as it may affect the synchronization of two physiological signals, or even the integrity of the whole data set. In general, the synchronization timing error between two signals is small in BANs when compared with the length of most physiological patterns. However, one should also take into account that many medical definitions, which affect the corresponding diagnostics, are based on the latency between two physiological events and use a fixed threshold. For instance, let us consider that the occurrence of two physiological events, concurrent within a 5-s window, is considered pathological. Then let us consider that, for a given patient, those two physiological events occur exactly with a 5-s interval and therefore should be considered pathological. In this case, even the smallest synchronization timing error that would increase the time interval between the two events would imply the wrong assumption that no pathological event occurred. This said, one should be able to stand back when analyzing such a recording, and it is assumed that a medical professional would have the insight needed to do so, but automatic signal processing may not. Many algorithms have been proposed to provide time synchronization in BAN, usually achieving average synchronization timing errors of less than 30 μs.

An artifact is also likely to corrupt data [28,32]. Besides the traditional signal processing often proposed to detect and remove artifacts, BAN context awareness may help account for artifacts [9]. Motion artifacts, one of the most problematic types of artifact [25,26], may be detected using motion analysis systems based on kinematic sensors [26,37,38].

As for any medical device, electromagnetic compatibility is an important issue that needs to be assessed, both for the unintentional generation and reception of electromagnetic energy. The aim is to ensure that the BAN does not interfere with other devices or is not disturbed by other devices through spurious emission and absorption of electromagnetic interference. Various standards have been published, depending on the country of use, detailing the technical and performance requirements to which the devices will be tested (e.g., IEC60601-1-2 2007). Compared with traditional wired monitoring devices, the use of wireless communication makes BAN more prone to electromagnetic interference emission and less immune to it, so this issue should be assessed carefully.

Privacy and security issues also need to be considered. The wireless nature of BAN makes them more prone to various security threats. This issue is common to all wireless networks, but the clinical use usually made by the BAN makes it most critical, and has therefore been heavily investigated for those applications [35,39,40]. Threats and attacks include data modification, impersonation attack, eavesdropping, and replaying (the attacker resends a piece of valid information), and security mechanisms include data encryption, data integrity and data origin authentication, authentication, and freshness protection [39]. Various possible solutions for security methods in BAN have been proposed (e.g., see articles by Ng et al. [40] and Dağtas et al. [41]).

Comfort is one of the key features that may be greatly improved by using a BAN. Traditionally, the patient is connected to the (usually large) monitoring device through wires, which reduces the patient's comfort and mobility [11]. Using a BAN is therefore beneficial, as those wires are removed. However, to keep this advantage, the BAN nodes should be small, even forgettable to the user, and allow autonomic donning and doffing [18]. Besides BAN, other architectures also allow the removal of those undesirable wires. For instance, another option is the use of a wearable textile interface that is implemented by integrating sensors, electrodes, and connections in fabric form [42].

Finally, power consumption, due both to data acquisition and transmission, is a key issue in BAN. It has a direct effect on comfort as it characterizes the trade-off between the size of the battery embedded in each node and its battery life. Both low power data acquisition [43–45] and low power data transmission [11,15,46] have been proposed. An interesting new trend is the use of power scavenging to enhance battery life, using surrounding energy sources such as thermal body heat [26] and body movement [47]. Another option, aiming to reduce data transmission power consumption, is to perform data processing (such as feature extraction) within the BAN node, so that only relevant information is transmitted [8]. However, two drawbacks of this method are the increase in power consumption due to data processing, and the potential loss of information occurring during the processing.

At least 10% of the population suffers from a sleep disorder that is clinically significant and of public health importance. To detect and assess sleep disorders, it is common to use a polysomnogram, a medical tool monitoring a wide variety of physiological parameters related to sleep.

Although polysomnography is traditionally performed at the hospital, there is a trend in performing sleep monitoring at home. The benefits over traditional polysomnography include a more comfortable patient's sleep, which is more indicative of his/her normal sleep, a cost that is substantially less expensive (25%–30% of the cost of traditional polysomnography), shorter waiting lists, etc. [48,49]. Even if there is still a debate regarding the validity of available home sleep testing devices, various authors report that the outputs from polysomnography and home sleep tests are not much different [49]. Another issue associated with traditional polysomnography is the long wires connecting the sensors, which can create discomfort and result in cable movement artifacts [11,49].

7.1.3 Medical Scenario of Interest

The portability and wireless nature of the BAN makes it highly suitable for home sleep monitoring. In this sense, various wireless devices and BAN used in sleep monitoring have been previously proposed [11,50,51]. Following those, we propose a BAN used for sleep monitoring. The choice of recorded physiological parameters and positioning of sensors is given in Table 7.1. This is believed to be typical because it is based on technical recommendations such as those made by Patel et al. [49] Chokroverty [52], Quan et al. [53], Patil [54], and Iber et al. [55], and according to the specifications of major manufacturers. The distribution of sensors is represented in Figure 7.1 and will serve as the basis for the performance analysis detailed in Sections 7.4 and 7.5.

Table 7.1 Specifications of a Typical Neurophysiologic Monitoring Device

Acronym	*Name*	*Bandwidth (bits/s)*	*Location*
PHONO	Sound probe	16,000	Neck
EEG	Electroencephalograph	25,600	Head
EOG	Electro-oculogram	6,400	Head
EMG1	Electromyograph	3,200	Chin
EMG2	Electromyograph	3,200	Left leg
EMG3	Electromyograph	3,200	Right leg
ECG	Electrocardiogram	3,200	Heart
NFL	Nasal/oral airflow	600	Nose/mouth
PAP	Positive airway pressure monitoring	600	Nose/mouth
POS	Position	600	Thorax
VAB	Plethysmograph	600	Abdomen
VTH	Plethysmograph	600	Thorax
SPO2	Arterial hemoglobin saturation	600	Finger
PR	Pulse rate	600	Finger
PTL	Plethysmograph	600	Finger
PTT	Pulse transit time	600	Finger
LIGHT	Light	600	Thorax

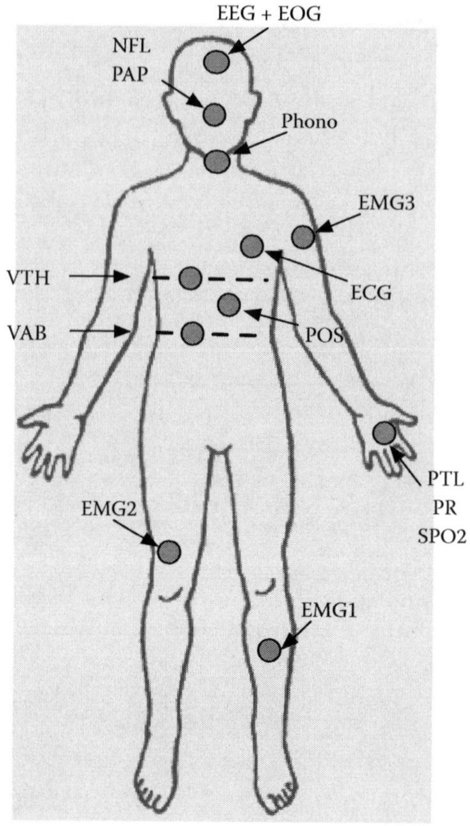

Figure 7.1 Physical model of the medical scenario.

7.2 Physical Channel Modeling for On-Body Communications

To build an accurate model for the on-body propagation, a Rohde & Schwartz ZVA-24 vector network analyzer was used to capture the complex-valued frequency-domain transfer function between 3 and 7 GHz, with a frequency step of 50 MHz. Omnidirectional Skycross SMT-3T010M ultrawideband antennas were used during the entire measurement campaign. Their small size (13.6 mm × 16 mm × 3 mm) and low profile characteristics precisely match the body sensor requirements. These antennas were separated from the body skin by about 5 mm to ensure a return loss value of $S_{11} \leq -9$ dB. Finally, low-loss and phase-stable cables interconnect all components and the IF bandwidth was set to 100 Hz to enlarge the dynamic range to about 120 dB.

The experimental scenario is presented in Figure 7.2 and can be described as follows. The measurements were carried out at approximately 94 cm of the waist of a man (1.87 m, 83 kg) whose body was in a standing position, arms hanging along the side. The transmitter antenna was placed around the body at a distance *d* from the receiver antenna, which is located at the middle axis of the torso.

Figure 7.3 (obtained from our previous work [56,57]) presents the power delay profile and channel gain between two nodes. This information is then used to derive a very accurate propagation model that accounts for the average received power, the statistical variations of the signal

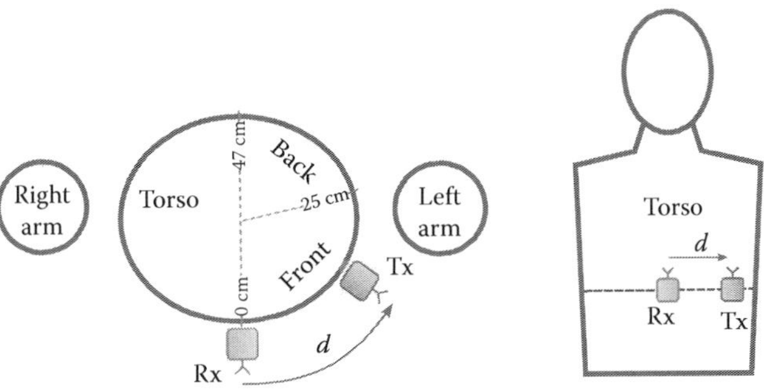

Figure 7.2 Possible positions of a transmitter–receiver pair in a BAN.

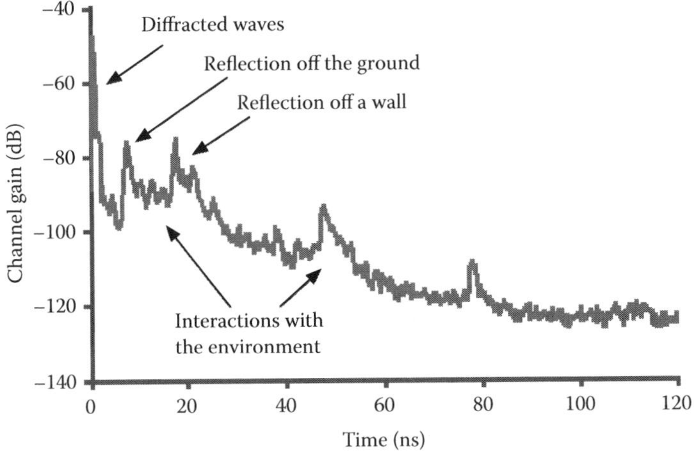

Figure 7.3 Power delay profile as a function of the time in an indoor environment and for $d \leq 25$ cm (body front).

(i.e., shadowing, movements of the limbs, etc.), and the effect of the environment (indoor or outdoor) [57].

The conclusions of this extensive measurement campaign, also highlighted by the model studied by Van Roy et al. [58], can be summarized according to three points. First, there is propagation through the body. However, when high transmission frequencies are considered, the attenuation undergone by these waves is relevant and the corresponding contribution can be neglected.

A second mechanism corresponds to guided diffraction around the body. This mechanism is consistent with surface wave propagation and its properties depend on the body specific characteristics.

Finally, the last propagation contribution comes from the surrounding environment. More precisely, the third propagation mechanism originates from reflections off the body limbs (arms and legs) and the surrounding objects (walls, floor, and ceiling). Obviously, this mechanism is observed only in an indoor environment.

Based on an extensive measurement campaign, we now present accurate statistical models corresponding to the propagation mechanisms described above.

7.2.1 On-Body Propagation (Guided Diffraction)

As previously emphasized by Ryckaert et al. [56] and Van Roy et al. [57], the average received power (on the decibel scale) is the following linearly decreasing function of distance:

$$\mathbb{E}[\mathbf{P}(d)] = P + L_{\text{ref}} + 10\gamma(d - d_{\text{ref}}) \quad d \geq d_{\text{ref}} \tag{7.1}$$

where $\mathbf{P}(d)$ is the instantaneous received power (dimension, W) at distance d (dimension, m), P the transmit power (dimension, W), d_{ref} is a reference distance (dimension, m), L_{ref} is the gain at the reference distance (adimensional, in decibels), and γ is a suitable constant (dimension, m^{-1}). For instance, typical experimental values for these parameters are $d_{\text{ref}} = 8$ cm, $L_{\text{ref}} = -57.42$ dB, and $\gamma = -124$ dB/m [57]. The average received power, in linear scale, can then be expressed as follows:

$$\mathbb{E}[\mathbf{P}(d)] = P \cdot L(d) \quad d \geq d_{\text{ref}} \tag{7.2}$$

with

$$L(d) = \underbrace{10^{(L_{\text{ref}} - 10\gamma d_{\text{ref}})/10}}_{\triangleq L_0} \cdot 10^{\gamma d}$$

$$= L_0 10^{\gamma d} \; d \geq d_{\text{ref}} \tag{7.3}$$

where L_0 is a function of L_{ref}, d_{ref}, and γ.* In Figure 7.4a, the loss L is shown as a function of the distance, considering narrowband transmissions at 5 GHz. More precisely, in Figure 7.4a, experimental measurements (circles) and their linear interpolation (solid line) are shown. Finally, using Equation 7.3 in Equation 7.2, one obtains:

$$\mathbb{E}[\mathbf{P}(d)] = PL_0 10^{\gamma d}. \tag{7.4}$$

Although Expression 7.4 characterizes the average value, it does not provide insights into the instantaneous distribution of the received power. In the study by Van Roy et al. [57], it has been experimentally observed that the on-body propagation channel is characterized by slow large-scale fading (i.e., shadowing). More precisely, the instantaneous received power at distance d can be expressed as follows:

$$\mathbf{P}(d) = PL_0 10^{\gamma d} \mathbf{X}$$

where $\mathbf{X}$ is a random variable (RV) that depends on the channel characteristics. As shown in the study by Takada et al. [59], and as confirmed in our measurements, $\mathbf{X}$ has a log-normal distribution† with parameters μ and σ, where σ_{dB} typically ranges from 4 to 10 dB, μ_{dB} is the average path loss on the link (dimension, dB). Because the loss is accounted for by the term $L(d)$, it follows that $\mu_{\text{dB}} = 0$ dB and the cumulative distribution function (cdf) of $\mathbf{X}$ reduces to the following:

$$F_{\mathbf{X}}(x; 0, \sigma) = \frac{1}{2} - \frac{1}{2}\text{erf}\left(\frac{-10\log_{10} x}{\sigma\sqrt{2}}\right)$$

* Note that, even though Equation 7.3 holds for $d \geq d_{\text{ref}}$, L_0 can be intuitively interpreted as the (extrapolated) gain (adimensional, linear scale) at distance $d = 0$. In other words, L_0 takes into account the loss due to antenna emission.

† Note that we use the $\log_{10}$ variant of the log-normal because the widely used shadowing model uses an additive Gaussian variation expressed in decibels.

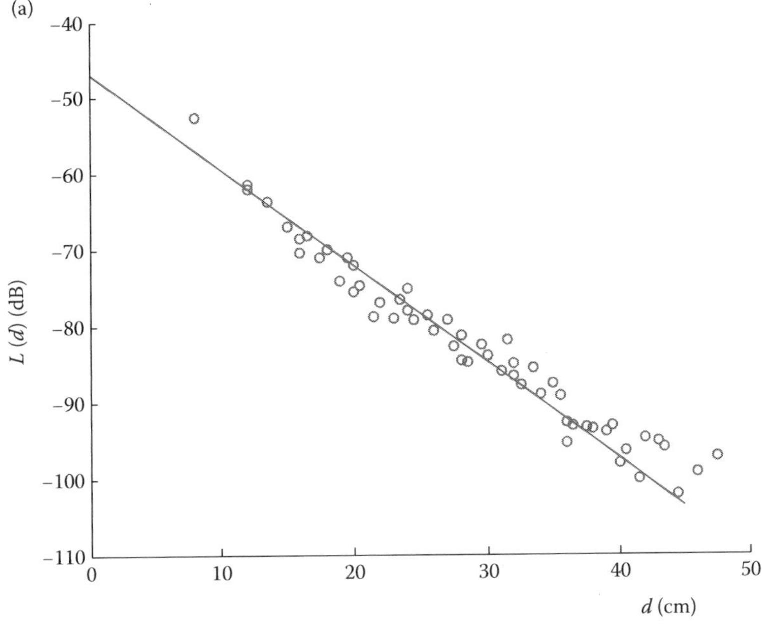

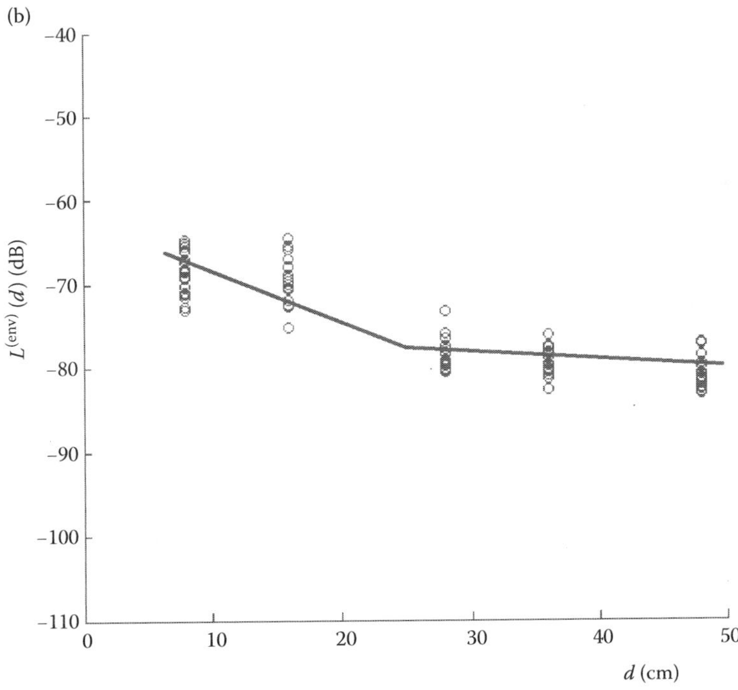

Figure 7.4 Propagation loss as a function of the distance: (a) on-body propagation and (b) propagation through reflections off the environment. In both cases, experimental results (circles) and their linear (or piecewise linear, in b) interpolations are shown (solid line).

with the following corresponding probability density function (pdf):

$$f_X(x;0,\sigma) = \frac{10}{(\ln 10)\, x\, \sqrt{2\pi}\sigma} \exp\left\{-\frac{(10\log_{10} x)^2}{2\sigma^2}\right\}. \tag{7.5}$$

7.2.2 Reflections Off the Environment

The second significant propagation mechanism originates from multiple reflections off the environment. A substantial measurement campaign has shown that the contribution of the environment can be considered, on average, as an additive, constant power when the transmission distance is significant (i.e., when $d > 25$ cm). The obtained results are shown in Figure 7.4b, the power received with reflections from the surrounding environment is shown as a function of the distance. It can be observed that when $d > 25$ cm, the value of the loss is, on average, approximately 78 dB. More precisely, for $d > 25$ cm, the average value of the received power can be expressed, in logarithmic scale, as follows:

$$\mathbb{E}[\mathbf{P}_{\text{env}}] = P_{\text{env}} \triangleq P + L_{\text{dB}}^{(\text{env})} \tag{7.6}$$

where P is the transmit power and $L_{\text{dB}}^{(\text{env})} \simeq -78$ dB. Alternatively, the average received power can be expressed in linear scale as

$$\mathbb{E}[\mathbf{P}_{\text{env}}] = P_{\text{env}} \triangleq P \cdot L^{(\text{env})} \tag{7.7}$$

where $L^{(\text{env})} = 10^{L_{\text{dB}}^{(\text{env})}/10}$. Our measurement campaign has shown that the propagation channel can be accurately characterized as a narrowband Rayleigh block fading. Therefore, the instantaneous received power $\mathbf{P}_{\text{env}}$ has the following exponential distribution [60]:

$$f_{\mathbf{P}_{\text{env}}}(x) = \frac{1}{P_{\text{env}}} \exp\left\{-\frac{x}{P_{\text{env}}}\right\}. \tag{7.8}$$

7.3 Link–Layer Performance

Transmission on a single link of interest strongly depends on the characteristics of the propagation channel and more precisely on its statistical variability, but also on the interference generated by the other surrounding nodes. More specifically, the traffic generated by a single sensing node will be analyzed under the well accepted slotted ALOHA system model. In this model, the nodes are supposed to transmit at discrete moments in time (called time slots) and the probability that a specific sensor has data to send is constant and will be noted as q. The node probability of transmission is a generic value in our analysis but it obviously depends on the medical application considered and can be precisely evaluated for any scenario, as it will be shown later.

We now provide the reader with the closed-form expressions of the link probability of success for a given node in the context of multiuser BANs.

7.3.1 Link Probability of Transmission in Multisensor Communications

The combination of the two propagation mechanisms presented in Subsections 7.2.1 and 7.2.2 allows us to derive a unified propagation model for a generic BAN. It can be observed that the degree of importance of each mechanism depends on the distance between transmitter and receiver. More precisely, in proximity, the dominant propagation mechanism is the on-body propagation described in Subsection 7.2.1. Above the crossover distance $d_{\text{cross}} \approx 25$ cm, the contribution of the environment becomes dominant and the second propagation mechanism, presented in Subsection 7.2.2, is the only relevant one. Therefore, a unified propagation model can be characterized as follows:

- In an outdoor environment, the average received power can be computed using Equation 7.4 (i.e., $\mathbb{E}[\mathbf{P}(d)] \propto P10^{\gamma d}$) and the instantaneous received power is determined by the log-normal fading channel model given by Equation 7.5.
- In an indoor environment:
  - If $d \leq d_{\text{cross}}$, the average received power can be computed using Equation 7.4 (i.e., $\mathbb{E}[\mathbf{P}(d)] \propto P10^{\gamma d}$) and the log-normal fading in Equation 7.5 is used.
  - If d > dcross, the average received power is approximately constant (i.e., $\mathbb{E}[\mathbf{P}(d)] = PL^{(\text{env})}$) and the instantaneous received power, owing to a Rayleigh faded channel model, has the distribution given by Equation 7.8.

In a BAN, all sensors need to transmit to a central controller and, in this sense, the scenario at hand can be interpreted as a multiuser scenario. Transmission over a link of interest is denoted with the subscript "0." Besides the intended transmitter, other nodes may be interfering. Depending on their distance to the receiver, the interfering nodes will be denoted differently. More precisely:

- In an indoor scenario, the interferers located at distances shorter than d_{cross} are referred to as "close-range interferers," their number is indicated as N_{close}, and the generic node will be denoted with a subscript $i \in \mathcal{N}_{\text{close}} \triangleq \{1,2,\ldots,N_{\text{close}}\}$.
- In an indoor scenario, the interferers located at distances longer than d_{cross} are referred to as "far-range interferers," their number is indicated as N_{far}, and the generic node will be denoted with a subscript $j \in \mathcal{N}_{\text{far}} \triangleq \{1,2,\ldots,N_{\text{far}}\}$.
- In an outdoor scenario, the number of interferers is indicated as N_{out}, and the generic node will be denoted with a subscript $k \in \mathcal{N}_{\text{out}} \triangleq \{1,2,\ldots,N_{\text{out}}\}$.

The transmission state of a node at time t is characterized by the following indicator variable:

$$\Lambda(t) = \begin{cases} 1 & \text{if the node is transmitting at time } t \\ 0 & \text{if the node is silent at time } t. \end{cases}$$

Assuming slotted transmissions (i.e., t can assume multiples of the slot time), a simple random access scheme is such that, at each time slot, a node transmits with probability q [61, p. 278]. Therefore, $\{\Lambda_i(t)\}_{t=1}^{\infty}, i \in \mathcal{N}_{\text{close}}$, $\{\Lambda_j(t)\}_{t=1}^{\infty}, j \in \mathcal{N}_{\text{far}}$, and $\{\Lambda_k(t)\}_{t=1}^{\infty}, k \in \mathcal{N}_{\text{out}}$ are sequences of Bernoulli RVs with $\mathbb{P}\{\Lambda_i(t) = 1\} = \mathbb{P}\{\Lambda_j(t) = 1\} = q, \forall t, i, j, k$.

A transmission in a given link is successful if and only if the signal-to-noise and interference ratio (SINR) at the receiver is above a certain threshold θ. This threshold value depends on the receiver characteristics, the modulation format, and the coding scheme, among other aspects. The SINR at the receiving node of the link is given by

$$\text{SINR} \triangleq \frac{P_0(d_0)}{\text{N}_0\text{B}+P_{\text{int}}} \tag{7.9}$$

where $P_0(d_0)$ is the received power from the link source located at distance d_0, N_0 is the power noise spectral density, B the channel bandwidth, and P_{int} is the total interference power at the link receiver, that is, the sum of the instantaneous received powers from all the undesired transmitters. More precisely, in an indoor environment, one has:

$$P_{\text{int}}^{(\text{indoor})} \triangleq \sum_{i=1}^{N_{\text{close}}} \Lambda_i P_i(d_i) + \sum_{j=1}^{N_{\text{far}}} \Lambda_j P_{\text{env}} \tag{7.10}$$

and, in an outdoor environment, one has:

$$P_{\text{int}}^{(\text{outdoor})} \triangleq \sum_{k=1}^{N_{\text{out}}} \Lambda_k P_k(d_k). \tag{7.11}$$

Finally, as typical in the context of BANs, we assume that all nodes use the same transmit power, that is, $P_i(0) = P_j(0) = P_k(0) = P_0(0), \forall i, j, k$.

7.3.1.1 Link Probability of Success with Short-Range Transmission in Indoor Scenarios

The link probability of success for a required threshold SINR value θ in the context of a short, indoor, log-normal faded link is equal to

$$\begin{aligned}
\mathcal{P}_{\text{close}}^{(\text{indoor})} &= \mathbb{P}\{\text{SINR} > \theta\} \\
&= \mathbb{E}_{\mathbf{P}_{\text{int}}}\left[\mathbb{P}\left\{\frac{P_0 L(d_0) X_0}{N_0 B + P_{\text{int}}} > \theta \middle| P_{\text{int}}^{(\text{indoor})}\right\}\right] \\
&= \mathbb{E}_{X,\Lambda,P_{\text{env}}}\left[1 - \mathbb{P}\left\{X_0 \le \theta \frac{N_0 B + P_{\text{int}}^{(\text{indoor})}}{P_0 L(d_0)}\right\}\right] \\
&= \mathbb{E}_{X,\Lambda,P_{\text{env}}}\left[\frac{1}{2} + \frac{1}{2}\text{erf}\left(\frac{-10}{\sigma\sqrt{2}}\log_{10}\left(\theta \frac{N_0 B + P_{\text{int}}^{(\text{indoor})}}{P_0 L(d_0)}\right)\right)\right].
\end{aligned} \tag{7.12}$$

In Appendix A at the end of this chapter, it is shown that

$$\zeta(z;\sigma) \triangleq \frac{1}{2} + \frac{1}{2}\text{erf}\left(\frac{-10\log_{10} z}{\sigma\sqrt{2}}\right) \approx \sum_{m}^{n} c_m \exp(-a_m z)$$

where $\{c_m\}_{m=1}^{n}$ and $\{a_m\}_{m=1}^{n}$, where n is an integer determined by the expansion accuracy and is a suitable coefficient. By using the function $\zeta(\cdot;\cdot)$ and recalling Expression 7.10 for the interference power, the link probability of success (Equation 7.12) can be written as follows:

$$
\begin{aligned}
\mathcal{P}_{\text{close}}^{(\text{indoor})} &= \mathbb{E}\left[\zeta\left(\theta\frac{\mathbf{P}_{\text{int}}^{(\text{indoor})}}{P_0 L(d_0)};\sigma\right)\right] \\
&= \sum_{m=1}^{n} c_m \exp\left(\frac{-a_m\theta N_0 B}{P_0 L(d_0)}\right)\times\mathbb{E}\left[\exp\left(-a_m\theta\sum_{i=1}^{N_{\text{close}}}\frac{L(d_i)}{L(d_0)}\mathbf{X}_i\Lambda_i\right)\right]\times\mathbb{E}\left[\exp\left(-a_m\theta\sum_{j=1}^{N_{\text{far}}}\frac{\mathbf{P}_{\text{env}}}{P_0 L(d_0)}\Lambda_j\right)\right]
\end{aligned}
\tag{7.13}
$$

where, in the last passage, we have used the fact that the RVs $\{\boldsymbol{\Lambda}_i, \boldsymbol{\Lambda}_j, P_{\text{env}}, \text{and } X_i\}$ are independent. The middle term in the second line of the right-hand side of Equation 7.13 can be further expressed as

$$
\begin{aligned}
\mathbb{E}\left[\exp\left(-a_m\theta\sum_{i=1}^{N_{\text{close}}}\frac{L(d_i)}{L(d_0)}\mathbf{X}_i\Lambda_i\right)\right] &= \prod_{i=1}^{N_{\text{close}}}\mathbb{E}\left[\exp\left(-a_m\theta\frac{L(d_i)}{L(d_0)}\mathbf{X}_i\Lambda_i\right)\right] \\
&= \prod_{i=1}^{N_{\text{close}}}\left\{\mathbb{P}\{\Lambda_i=0\}\times 1+\mathbb{P}\{\Lambda_i=1\}\times\mathbb{E}\left[\exp\left(-a_m\theta\frac{L(d_i)}{L(d_0)}\mathbf{X}_i\right)\right]\right\} \\
&= \prod_{i=1}^{N_{\text{close}}} q\int_0^{\infty}\exp(-a_m\theta 10^{\gamma(d_i-d_0)}x)f_{\mathbf{X}}(x)\mathrm{d}x+(1-q).
\end{aligned}
\tag{7.14}
$$

The final integral expression in Equation 7.14 can be numerically computed. The term in the third line of the expression in Equation 7.13 can be expressed as follows:

$$
\begin{aligned}
\mathbb{E}\left[\exp\left(-a_m\theta\sum_{j=1}^{N_{\text{far}}}\frac{\mathbf{P}_{\text{env}}}{P_0 L(d_0)}\Lambda_j\right)\right] &= \prod_{j=1}^{N_{\text{far}}}\mathbb{E}\left[\exp\left(-a_m\theta\frac{\mathbf{P}_{\text{env}}}{P_0 L(d_0)}\Lambda_j\right)\right] \\
&= \prod_{j=1}^{N_{\text{far}}}\left\{\mathbb{P}\{\Lambda_j=0\}\times 1+\mathbb{P}\{\Lambda_j=1\}\times\mathbb{E}\left[\exp\left(-a_m\theta\frac{\mathbf{P}_{\text{env}}}{P_0 L(d_0)}\right)\right]\right\} \\
&= \left[(1-q)+q\int_0^{\infty}\exp\left(-a_m\theta\frac{x}{P_0 L(d_0)}\right)\frac{1}{P_{\text{env}}}e^{-x/P_{\text{env}}}\,\mathrm{d}x\right]^{N_{\text{far}}} \\
&= \left[1-\frac{\theta q}{\dfrac{L_0 10^{\gamma d_0}}{L^{(\text{env})}}+\theta}\right]^{N_{\text{far}}}.
\end{aligned}
\tag{7.15}
$$

Finally, by using Equations 7.14 and 7.15 into Equation 7.13, the link probability of success can be given by the expression in Equation 7.16.

$$\mathcal{P}_{\text{close}}^{(\text{indoor})} = \sum_{m=1}^{n} c_m \underbrace{\exp\left(\frac{-a_m \theta N_0 B}{P_0 L_0 10^{\gamma d_0}}\right)}_{\text{Background noise}} \times \underbrace{\prod_{i=1}^{N_{\text{close}}} \left[q \int_0^{\infty} \exp(-a_m \theta 10^{\gamma(d_i - d_0)} x) f_{\mathbf{X}}(x) \mathrm{d}x + (1-q) \right]}_{\text{Close-range interferers}}$$
$$\times \underbrace{\left[1 - \frac{\theta q}{\frac{L_0 10^{\gamma d_0}}{L^{(\text{env})}} + \theta} \right]^{N_{\text{far}}}}_{\text{Far-range interferers}} \tag{7.16}$$

7.3.1.2 Link Probability of Success with Long-Range Transmission in Indoor Scenarios

The Rayleigh-faded channel model applies to indoor links with length $d > d_{\text{cross}}$. In this scenario, $\mathbb{E}[\mathbf{P}(d)] \approx P_{\text{env}}$ (for both the intended transmitter and interferers) and the link probability of success can be expressed as follows:

$$\begin{aligned}
\mathcal{P}_{\text{far}}^{(\text{indoor})} &= \mathbb{P}\{SINR > \theta\} \\
&= \mathbb{E}_{\mathbf{P}_{\text{int}}} \left[\mathbb{P}\{SINR > \theta\} \mid \mathbf{P}_{\text{int}}^{(\text{indoor})} \right] \\
&= \mathbb{E}\left[\exp\left\{ -\frac{\theta(N_0 B + \mathbf{P}_{\text{int}}^{(\text{indoor})})}{P_{\text{env}}} \right\} \right] \\
&= \exp\left(\frac{-\theta N_0 B}{P_{\text{env}}}\right) \times \mathbb{E}\left[\exp\left(-\theta \sum_{i=1}^{N_{\text{close}}} \frac{P_i L(d_i)}{P_{\text{env}}} \mathbf{X}_i \Lambda_i \right) \right] \times \mathbb{E}\left[\exp\left(-\theta \sum_{j=1}^{N_{\text{far}}} \frac{\mathbf{P}_{\text{env}}}{P_{\text{env}}} \Lambda_i \right) \right].
\end{aligned} \tag{7.17}$$

It can be observed that the terms in the second and third lines at the right-hand side of Equation 7.17 are similar to Equations 7.14 and 7.15. Therefore, by using the same derivation of Subsection 7.3.1.1, with $P_0 L(d_0)$ replaced by $P_0 L^{(\text{env})}$, one has

$$\mathbb{E}\left[\exp\left(-\theta \sum_{i=1}^{N_{\text{close}}} \frac{P_i L(d_i)}{P_{\text{env}}} \mathbf{X}_i \Lambda_i \right) \right] = \prod_{i=1}^{N_{\text{close}}} \left[q \int_0^{\infty} \exp\left(-\theta \frac{L_0 10^{\gamma d_i}}{L^{(\text{env})}} x \right) f_{\mathbf{X}}(x) \mathrm{d}x + (1-q) \right] \tag{7.18}$$

and

$$\mathbb{E}\left[\exp\left(-\theta \sum_{j=1}^{N_{\text{far}}} \frac{\mathbf{P}_{\text{env}}}{P_{\text{env}}} \Lambda_i \right) \right] = \left[1 - \frac{\theta q}{1+\theta} \right]^{N_{\text{far}}}. \tag{7.19}$$

By inserting Equations 7.18 and 7.19 into Equation 7.17, one obtains the final expression (Equation 7.20) for the probability of successful transmission on the link.

$$\mathcal{P}_{\text{far}}^{(\text{indoor})} = \underbrace{\exp\left(\frac{-\theta N_0 B}{P_{\text{env}}}\right)}_{\text{Background noise}} \times \underbrace{\prod_{i=1}^{N_{\text{close}}}\left[q\int_0^{\infty}\exp\left(-\theta\frac{L_0 10^{\gamma d_i}}{L^{(\text{env})}}x\right)f_{\mathrm{X}}(x)\mathrm{d}x + (1-q)\right]}_{\text{Close-range interferers}} \times \underbrace{\left[1-\frac{\theta q}{1+\theta}\right]^{N_{\text{far}}}}_{\text{Far-range interferers}} \tag{7.20}$$

7.3.1.3 Link Probability of Success in Outdoor Scenarios

In these scenarios, the links are subject to log-normal fading and exponential power decreases. The link probability of success can simply be derived by using the derivation in Subsection 7.3.1.1, setting $N_{\text{out}} = N_{\text{close}}$ and $N_{\text{far}} = 0$ (this does not mean that there are not far interferers, but that their propagation model is simply the same of close interferers). Therefore, the computation of the link probability of success $\mathcal{P}^{(\text{outdoor})}$ is straightforward from Equation 7.16, and the final expression is given in Equation 7.21.

$$\mathcal{P}^{(\text{outdoor})} = \sum_{m=1}^{n} c_m \underbrace{\exp\left(\frac{-a_m \theta N_0 B}{P_0 L_0 10^{\gamma d_0}}\right)}_{\text{Background noise}} \times \prod_{i=1}^{N_{\text{out}}}\left[q\int_0^{\infty}\exp(-a_m\theta 10^{\gamma(d_i-d_0)}x)f_{\mathrm{X}}(x)\mathrm{d}x + (1-q)\right] \tag{7.21}$$

7.3.2 Energy Consumption and Minimum Transmit Power

From the above relations, the minimum transmission power on a link can be easily derived. It is defined as the power required to achieve a threshold link probability of success $\mathcal{P}_{\text{th}}$ in a noise-limited regime, that is, when all other nodes are silent. Therefore, by recalling Equations 7.16, 7.20, and 7.21 and setting $N_{\text{out,close,far}} = 0$, one has:

$$P_0^{(\text{indoor})} \geq \begin{cases} \dfrac{\theta k_{\text{b}} TB}{L_0 10^{\gamma d_0}\zeta^{-1}(\mathcal{P}_{\text{th}})} & \text{if } d < d_{\text{cross}} \\ -\dfrac{\theta k_{\text{b}} TB}{\ln \mathcal{P}_{\text{th}}} & \text{if } d \geq d_{\text{cross}} \end{cases} \tag{7.22}$$

where N_0 has been expressed as Tk_{b}, with T being the room temperature (dimension, K) and $k_{\text{b}} = 1.38 \times 10^{-23}$ J/K being Boltzmann's constant, and B is the transmission bandwidth. For the outdoor scenario, one obtains:

$$P_0^{(\text{outdoor})} \geq \frac{\theta k_{\text{b}} TB}{L_0 10^{\gamma d_0}\zeta^{-1}(\mathcal{P}_{\text{th}})}. \tag{7.23}$$

where $\mathcal{P}_{th}$ is the threshold link probability of success. In Figure 7.5, the minimum transmission power is presented as a function of the transmission distance d_0 and the threshold link probability of success $\mathcal{P}_{th}$. It can be observed from Figure 7.5a that, in an indoor scenario, there are reflections off the walls.

Finally, note that in the following, we will consider only interference-limited networks, that is, scenarios in which the conditions in Equations 7.22 or 7.23 are satisfied. Formally, this situation is equivalent to letting the thermal noise $N_0B = 0$ in Equation 7.9.

7.4 Link-Level Throughput and Delay

In this article, we consider multiuser communications in which a transmission will be defined as successful if and only if the associated transmission link is not in an outage. This corresponds to requiring that the (instantaneous) SINR of the link is above the threshold θ, that is,

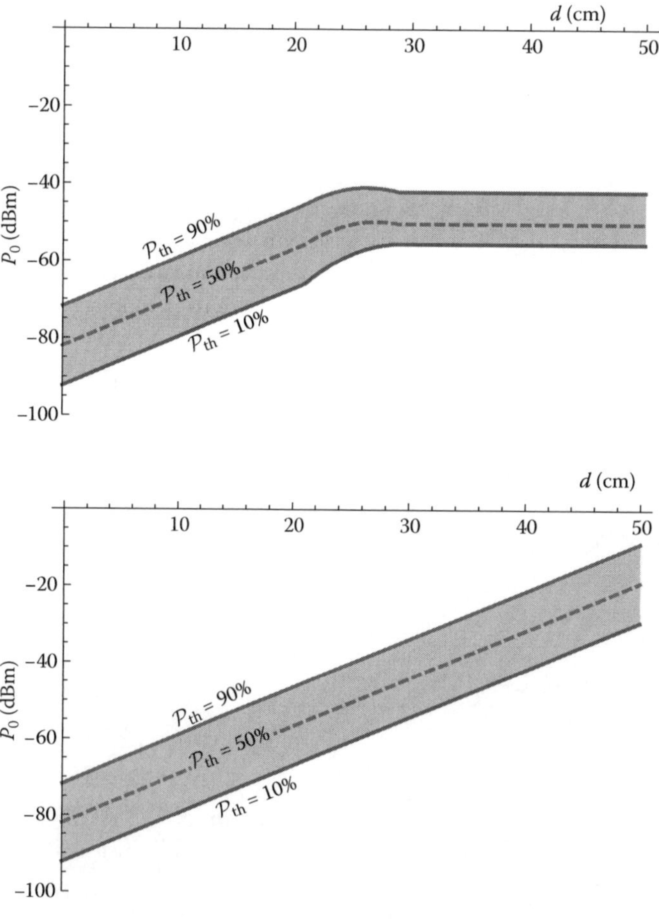

Figure 7.5 Minimum transmission power for a given link probability of success and as a function of the distance.

$$\mathcal{P}^{(\text{link})} = \mathbb{P}\{\text{SINR} > \theta\}. \tag{7.24}$$

Obviously, this probability depends on the physical channel characteristics, the shadowing on the links, the mobility of the nodes (e.g., if the node is located on an arm), and the modulation in use. For instance, in a ZigBee system or, in general, with low-throughput BANs, a value of $\theta = 5$ dB is sufficient. In the article, we refer to the IEEE 802.15.6 standard, which allows the use of two access control schemes: (i) a random, slotted ALOHA access and (ii) a deterministic TDMA access.

In a slotted ALOHA system, the time is divided in equal, discrete elements called *time slots*. During each time slot, a node has a fixed probability of transmitting a packet. In the remainder of this article, this value is noted q (nondimensional). During this time slot, any number of nodes can attempt a transmission and collisions may occur. As a consequence, the delay in a slotted ALOHA system is directly linked to the number of (re)transmissions needed to send (or forward) a packet, which in turn is directly proportional to the link probability of success and the overall activity of the network (through the value of q).

On the other hand, in a TDMA system, the time is exactly divided in a fixed number of slots. Each slot is then allocated to a given a node and it is allowed to transmit only during its time slot. At any other time, the node must remain silent. Therefore, the delay depends on the (deterministic) amount of time to wait before a dedicated slot takes place. In a generic approach, it can be supposed that a relay node will allocate exactly one slot per sensor to receive its data and a slot per sensor for the forwarding uplink. This in turn depends on the routing tree and the quality of service (QoS) requirements.

Finally, we define a probabilistic throughput value as the average probability of successful transmission $\mathcal{P}^{(\text{link})}$ multiplied by the probability that the transmitter actually has a packet to transmit q, that is,

$$\tau \triangleq q\mathcal{P}^{(\text{link})}. \tag{7.25}$$

Obviously, in the slotted ALOHA scheme, the value of $\mathcal{P}^{(\text{link})}$ can be $\mathcal{P}_{\text{close}}^{(\text{indoor})}$, $\mathcal{P}_{\text{far}}^{(\text{indoor})}$, or even $\mathcal{P}^{(\text{outdoor})}$, depending on the scenario and the link of interest. On the other hand, in the TDMA scheme, it is the average activity period, that is, the amount of time that a node is actually allowed to transmit in the time division scheme. We now provide the reader with the exact computation of the delay and the corresponding throughput.

7.4.1 Random Access BANs

The slotted ALOHA multiple access scheme [19] was recently proposed by the IEEE 802.15.6 working group as one of the reference medium access control (MAC) schemes for wireless body networks in the context of narrow-band communications [20]. An example of slotted ALOHA is given in Figure 7.6. In particular, in each time slot, the nodes are assumed to transmit independently with a certain fixed probability [21]. This approach is supported by the observations in studies by Huhta and Webster [22, p. 278], Wood and Ewins [21], and Webster [23], in which it is shown that the traffic generated by nodes using a slotted random access MAC protocol can be modeled with a Bernoulli distribution. In fact, in more sophisticated MAC schemes, the probability of transmission at a node can be modeled as a function of general parameters, such as queuing statistics, the queue-dropping rate, the channel outage probability incurred by fading [24], the adaptation of the sampling rate to the patient's condition [25], the MAC strategy used [26], and

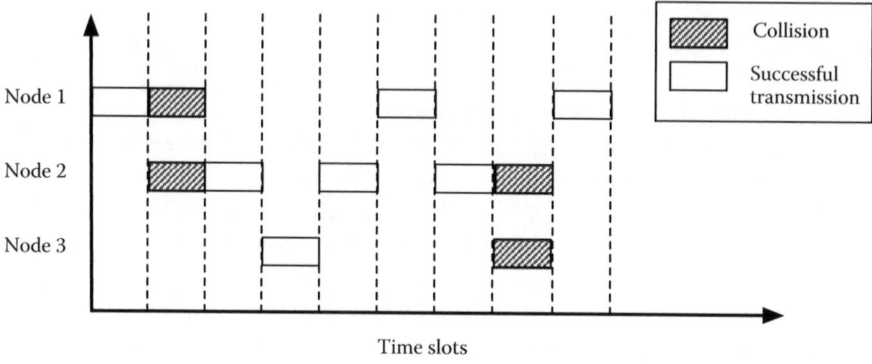

Figure 7.6 Slotted ALOHA scheme.

others. Because the effects of these parameters are not the focus of this study, the interested reader is referred to the existing literature [27–29] for further details.

In a slotted ALOHA system, the delay is directly connected to the number of (re)transmissions needed to send or forward a packet, which depends on the number of nodes and especially their probability of transmission. More precisely, the average number of transmissions can be written as:

$$E[r] = \sum_{r=1}^{\infty} rQ_r$$

where Q_r is the probability if a successful communication takes place exactly at the rth attempt (i.e., $r - 1$ unsuccessful transmissions and a successful transmission). It is expressed as:

$$Q_r = (1 - P_s)^{r-1} P_s$$

where P_s denotes the link probability of success. It can be observed that the variable r is a geometric random variable with parameter P_s. Therefore, the average delay in slotted ALOHA scheme, in terms of time slots, is exactly:

$$D_{\text{ALOHA}} = E[r] = \frac{1}{P_s}. \tag{7.26}$$

By recalling the definition of the link throughput given in Equation 7.25 and using the relation in Equation 7.26, one finally obtains the link throughput

$$\tau_{\text{ALOHA}} = qP_s = \frac{q}{D_{\text{ALOHA}}} \tag{7.27}$$

which shows that the network throughput will directly depend on the value of q (through the value of P_s, which models the number of collisions on a given link).

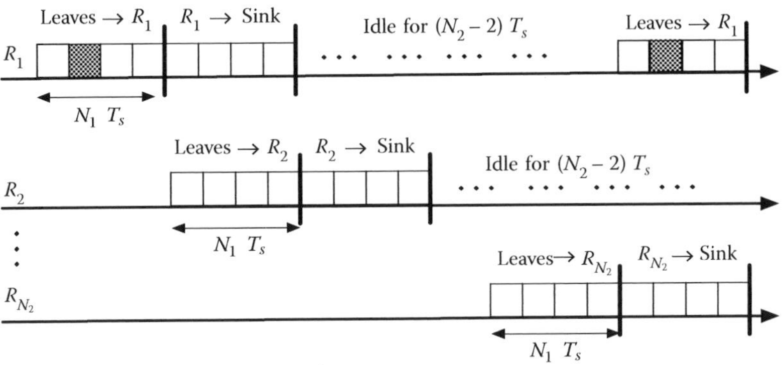

Figure 7.7 TDMA chronogram.

7.4.2 TDMA BANs

In a TDMA system, each node transmits during a dedicated time slot and, as consequence, in an interference-limited regime, one has $P_s = 1$. The delay D_{TDMA} only depends on the amount of time to wait before a dedicated slot takes place. In a generic approach, it can be supposed that a relay node will allocate exactly one slot per sensor to receive its data and a slot per sensor for the forwarding uplink. Note that this is an anticipation on the computation of the route delay. Indeed, on the opposite of the studied random access scheme (i.e., slotted ALOHA), the exact computation of the link delay in TDMA requires that the effect of relaying be taken into account, which in turn depends on the topology of the routes.

In Figure 7.7, the slot allocations (i.e., chronogram) are presented for each relay (noted as $R_1, R_2, \ldots, R_{N_2}$). The time slots have a fixed duration of T_s (dimension, s). As can be seen from Figure 7.7, each relay needs a frame of N_1 time slots to collect the packets (possibly generated) from its N_1 leaves. It then waits another frame (N_1 time slots) to forward them to the sink. At this point, it needs to remain idle for (N_2 – 2) frames, as the sink is busy collecting the packets from the other relays. This corresponds to assuming the same transmission rates at leaves and relays, and the same TDMA-based approach at first and second layers. Therefore, the distance between two consecutive slots assigned to a given leaf is equal to $N_1 \cdot N_2$ slots: when a leaf generates a packet, it needs to wait a number of slots between 0 (its slot is the current one) and $N_1 \cdot N_2 - 1$ (its slot just passed).* As each number of slots has the same probability, the average delay (expressed in time slots) experienced by a given leaf node is

$$\begin{aligned} D_{\text{TDMA}} &= \frac{1}{N_2 \cdot N_1 - 1} \sum_{i=0}^{N_2 \cdot N_1 - 1} i \\ &= \frac{N_2 \cdot N_1}{2}. \end{aligned} \tag{7.28}$$

The above derivation for D_{TDMA} represents an "average" scenario in which a node generates at most a packet in an interval equal to $N_1 \cdot N_2$ time slots. If, on the other hand, more than one

* We assume that packet generation is at the beginning of a slot.

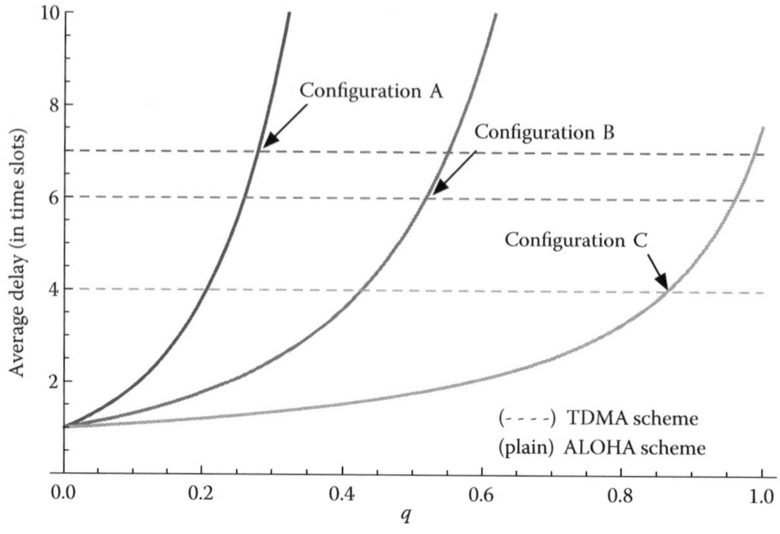

Figure 7.8 Qualitative description of the average delay at the relay nodes for TDMA and slotted ALOHA schemes as functions of sensor probability of transmission *q* and for various network topologies.

packet is generated during a time slot, it means that as soon as a leaf has transmitted a packet to its relay, it is likely to soon generate a new packet, which will wait a period longer than that of Equation 7.28. This waiting period is at most $N_1 \cdot N_2$ time slots. Therefore, in general, it can be stated that

$$\frac{N_2 \cdot N_1}{2} \leq D_{\text{TDMA}} \leq N_2 \cdot N_1.$$

In Figure 7.8, the average delay (expressed in contention time slots) incurred by a leaf to reach its relay is shown in a qualitative manner and as a function of the transmission probability q. Note that in the TDMA case, the expression does not depend on q, because in TDMA systems, a leaf needs to wait for its assigned time slot. Finally, one can anticipate that, as it will be formally demonstrated later, when the node probability of transmission is low, the slotted ALOHA significantly outperforms the TDMA scheme. However, for increasing probability of transmission, that is, for increasing traffic load, there exists a critical threshold above which the TDMA scheme is to be preferred.

7.5 Reliable Routing and Optimal Topology in Multihop BAN

Assuming that all links of a route are independent and that a packet erroneously transmitted on any link of a route is not retransmitted, the route probability of success can be expressed as

$$\mathcal{P}^{(\text{route})} = \prod_{i}^{n_{\text{hops}}} \mathcal{P}_i^{(\text{link})} \tag{7.29}$$

where $\mathcal{P}_i^{(\text{link})}$ is the link probability of success on the ith link of the route and n_{hops} is the route length. In this sense, the derived expression (Equation 7.29) will be an upper bound for successful transmission probability, which corresponds to a real-time streaming of data based on a datagram-oriented transport. More precisely, the connectionless transmission modes [e.g., user datagram protocol (UDP) and real-time streaming protocol (RTSP)] are stateless by nature and allow the nonblocking sending of data stream, which is a requirement in medical applications but, as a consequence, this transport mode does not allow for successful transmission monitoring and resending of nonreceived packets.

QoS in BANs is a nontrivial and relevant question in this very specific context. In a general manner, QoS comprises the requirements on all aspects of a data stream connection, such as the maximum acceptable loss, delay, variability of the service, etc. In the absence of QoS, the delivery network may suffer from dropped packets, excessive latency, jitter, and out-to-order delivery. In the field of medical applications, guaranteed packet delivery is the most important requirement and other issues (such as delay or jitter) can be considered of less importance. First, as discussed previously, medical sensing is more likely to use connectionless transmission modes, such as real-time protocol (RTP; RFC 3350), to ensure a fluid delivery of data packets. The RTP protocols ensure that out-of-order packets are correctly delivered by tagging the packet sequence using an increasing 16-bit sequence number. Second, due to the limited size of the network, jitter (the variability of the data delivery delay) can be fairly neglected because no end-to-end retransmission can take place. Also, once established, the routing tree will be stable for the entire monitoring session, which further guarantees limited jitter. Finally, the existence of a constant delay is not important in a pure monitoring system, that is, a system that performs the collection of vitals without feedback actions. However, as it will be presented later, in a random access network, delay can grow exponentially and even become unbounded. This is not the case in a deterministic, TDMA network. Therefore, specific attention will be paid to the computation of the delay and the comparison of both access methods.

To summarize, the QoS in a medical application can be formulated as the requirement that (i) the number of losses is small and (ii) the delay is noninfinite. To translate these two requirements into equations, we propose the following framework for the computation of the optimal routing tree. On the basis of the expression in Equation 7.29, we define that a route has an acceptable QoS if and only if a minimum end-to-end transmission rate can be achieved, that is,

$$\begin{cases} D^{(\text{route})} < \infty \\ \mathcal{P}^{(\text{route})} \geq \mathcal{P}_{\min}^{(\text{medical})} \end{cases} \tag{7.30}$$

where $D^{(\text{route})}$ is the end-to-end route delay and the value $\mathcal{P}_{\min}^{(\text{medical})}$ depends on the medical application of interest and is defined by the practicians. It is usually high, that is, $\mathcal{P}_{\min}^{(\text{medical})} > 90\%$ is a reasonable expectation.

7.5.1 Routing in BAN Networks

Routing is the process of selecting the optimal path (with respect to an objective function to be maximized or minimized) in a network along which data traffic flows. In the context of a medical monitoring scenario, routing is intended to find the optimal way to send information through the network, from sensors to the sink, under the minimum QoS requirements expressed in Equation 7.30.

In a study by Dricot et al. [62], a preliminary performance analysis of BANs with star topologies was carried out. Indeed, these topologies are simple to implement but not well-suited for medical applications because they exhibit a significant power consumption [63] and cannot perform in-network data aggregation [64,65].

In Figure 7.9, an illustrative medical application is presented. The corresponding routing scheme (plain bold lines) is derived from the connectivity graph (dashed lines). The specific routes are chosen so that the QoS meets the medical requirements. As shown in Figure 7.9, wireless transmissions allow multiple possible paths for the routing of packets from the sensing nodes to the collecting sink. Furthermore, due to the short distance nature of BAN communications, direct transmissions are feasible almost all of the time but they are subject to more interference. Therefore, the objective of the routing algorithm is to determine the most reliable routes, in terms

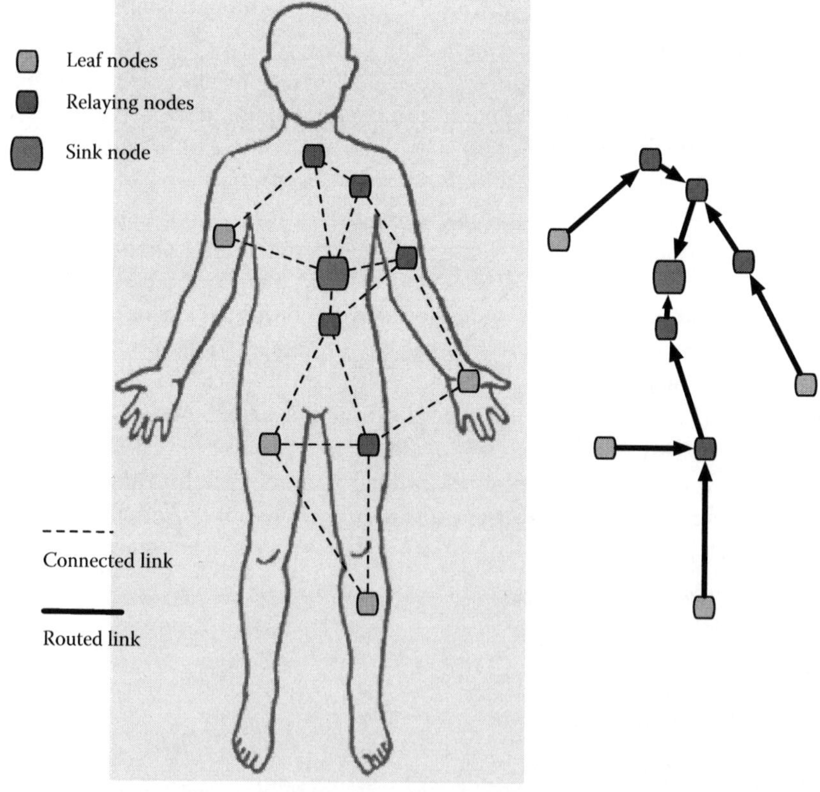

Figure 7.9 Medical application of body sensor networks. Possible links are presented in dashed lines and the routing tree wrt. to QoS requirements is drawn in bold lines.

of the number of hops and relaying nodes, to transport an information stream from a leaf node to the sink. Also, due to the dynamic nature of the network configuration (e.g., arms moving, switching from indoors to outdoors, and others), a dynamic routing algorithm is mandatory, that is, routes cannot be determined a priori for any medical application and forced into the nodes.

Dynamic routing protocols are based on well-known routing tree exploration techniques. These protocols can be divided into two major techniques: link-state and distance-vector. The latter class of protocols was designed for very large architectures, such as the Internet, and is based on a decentralized, step-by-step convergence approach. This class of protocols has less computational complexity but will not be considered here because it exhibits certain limitations that must be avoided in medical applications. These include the well-known count-to-infinity problem (i.e., routing loops) and slow convergence times (due to the iterative nature of the algorithm).

On the other hand, link-state algorithms are performed by each node in the network, that is, every node constructs a map of the connectivity to the network and then independently calculates the best logical path to every possible destination in the network. Practically, the routing task can also be implemented in a centralized node (e.g., the sink node) and the routing tables are subsequently redistributed to the sensing nodes. The incurred overhead of control messages (signaling traffic) is negligible with respect to the data streams' load. Modern link-state algorithms are usually based on variants of the Dijkstra algorithm.

7.5.2 Link-State Routing: Dijkstra's Algorithm

Dijkstra's algorithm [66] is a graph search algorithm that solves the single-source shortest path problem for a graph with nonnegative edge path costs producing the shortest path tree. More precisely, for a given source vertex (node) in the graph, the algorithm computes the path with the lowest cost (i.e., the shortest path with respect to a cumulative cost metric) between that vertex and every other vertex. It can also be used for finding the cost of the shortest path from a given vertex to a specific destination vertex by stopping the algorithm once the shortest path to the destination vertex has been determined. For instance, if the vertices of the graph represent the network connectivity and edge path costs represent the link probability of errors between pairs of nodes connected, Dijkstra's algorithm can be used to find the route with the highest probability of success. This route is denoted as the "optimal route," with respect to a cost function or the "shortest path." Because Dijkstra's algorithm is simple to implement, has a reduced complexity, and is proven to be extremely stable, it is widely used in network routing protocols, most notably in intermediate system–to–intermediate system (IS-IS; RFC 1142) and open shortest path first (OPSF; RFC 2328).

The algorithm works by using the following iterative steps:

1. Assign a tentative distance value to every node: set it to zero for the initial node and to infinity for all other nodes.
2. Mark all nodes unvisited. Set the initial node as current. Create a set of unvisited nodes called the unvisited set consisting of all the nodes except the initial node.
3. For the current node, consider all of its unvisited neighbors and calculate their tentative distances. For example, if the current node A is marked with a tentative distance of 6, and the edge connecting it with a neighbor B has length of 2, then the distance to B (through A) will be 6 + 2 = 8. If this distance is less than the previously recorded tentative distance of B, then overwrite that distance. Even though a neighbor has been examined, it is not marked as visited at this time, and it remains in the unvisited set.

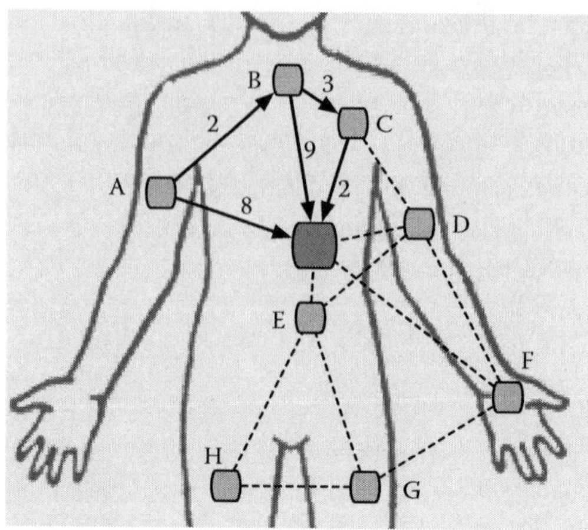

Figure 7.10 Scenario with multiple communication links between the nodes and the sink. On each link, the cost metric is presented.

4. When we are done considering all of the neighbors of the current node, mark the current node as visited and remove it from the unvisited set. A visited node will never be checked again; its distance recorded now is final and minimal.
5. If the destination node has been marked visited (when planning a route between two specific nodes) or if the smallest tentative distance among the nodes in the unvisited set is infinity (when planning a complete traversal), then stop. The algorithm has finished.
6. Set the unvisited node marked with the smallest tentative distance as the next "current node" and go back to step 3.

A pictorial example is presented in Figure 7.10, in which three possible paths connect the sensor node A to the sink node. As can be observed, several paths are possible and, mode specifically, the A sink has a cost of 8, the A-B sink has a cumulative cost of 2 + 9 = 11, and the A-B-C sink has a cost of 2 + 3 + 2 = 7. Therefore, the A-B-C sink is picked up because it is the lowest costing candidate.

7.5.3 Medical Nodes' Maximum Probability of Transmission

In this case, we consider a situation with a limited amount of sensors located on specific parts of the human body. Because the number of sensors in a real-life situation will range between 10 and 18, the density of the corresponding sensor network is low. Moreover, in most medical applications, not all sensors are used to monitor the patient's state, resulting in a sparse deployment that is specific to body sensor networks. For the purpose of this analysis, a particular distribution of nodes has been chosen and is represented in Figure 7.11. In the remainder of this performance analysis, the nodes are numbered as follows: 1, EEG; 2, EMG1; 3, EMG2; 4, EMG3; 5, ECG; 6, NFL; 7, VAB; 8, VTH; 9, PTL; and 10, SINK. The medical purposes of each sensor and its networking characteristics are introduced in Table 7.1.

In medical applications, the node probability of transmission is bound by two constraints: (1) the minimum sampling rate of the medical sensor (reported in Table 7.1) and (2) the minimum

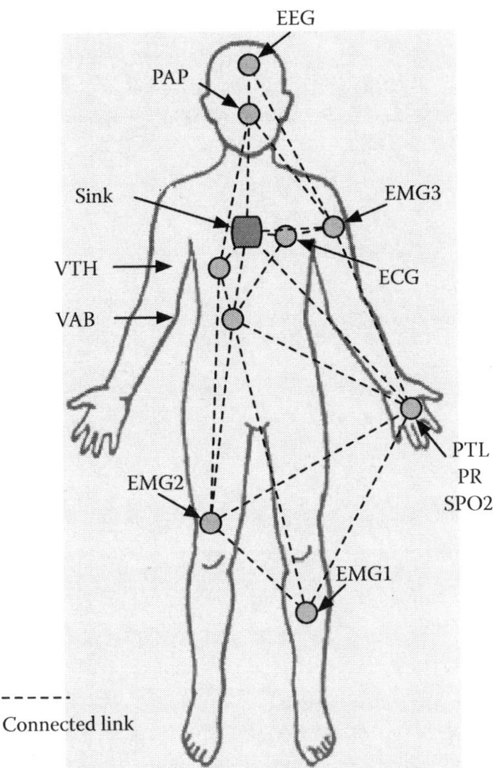

Figure 7.11 Medical scenario of interest for routing performance evaluation and its connectivity topology.

route probability of success, which translates into a maximum node transmission rate. Therefore, a computation of the maximum value for the transmission probability (noted as q_{max}) is the first step before any analysis. The determination of this value is an iterative process, that is, for a given value of the route probability of success $\mathcal{P}_{min}^{(route)}$, the transmission rate of the nodes (through the quantity q) is increased at each iteration until $\mathcal{P}_{min}^{(route)}$ is reached. The corresponding value of the node probability of transmission is then reported as q_{max}.

Note that, to further ensure the robustness of our analysis, the value of $\mathcal{P}_{min}^{(route)}$ is always calculated over the worst possible route. More precisely, all possible routing trees are computed and the value $\mathcal{P}_{min}^{(route)}$ corresponds to the lowest possible route probability of success observed on all possible network configurations. Figure 7.12 presents the medical nodes' maximum probability of transmission in a 10-node setup and as a function of the route minimum probability of success $\mathcal{P}_{min}^{(route)}$. It can be observed that the medical reporting rate decreases exponentially as the value of $\mathcal{P}_{min}^{(route)}$ increases. As a consequence, reducing the packet loss in the entire network is extremely difficult if the route quality is expected or required to be high. Furthermore, for values of $\mathcal{P}_{min}^{(route)} \geq 0.7$, which is in the functioning region of a real-life medical body sensor network, the value of q_{max} decreases linearly. In that region, the value of q_{max} less than or equal to 0.005, meaning that the random access strategy does work, but only if the medical reporting rate is low. In any other case (emergency increase of the sampling of the vitals, high density of nodes, high traffic sensors, etc.), the use of the TDMA strategy is mandatory.

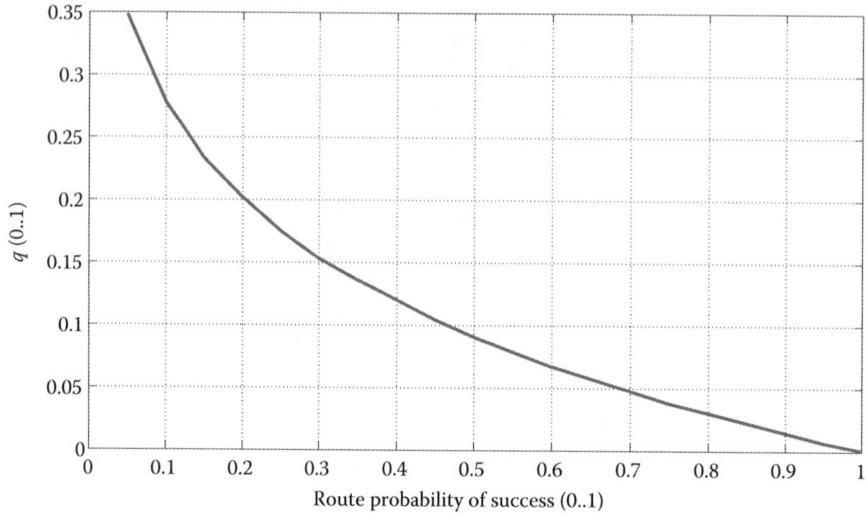

Figure 7.12 Probability of transmission versus minimum route probability of success in medical scenario.

Finally, the simulations were conducted for uplink and downlink traffic. In Figure 7.13, the medical nodes' maximum probability of transmission is reported as a function of the minimum route probability of success for the two traffic modes and in the case of the random access network. It can be observed that the difference is neglectable, and the prior analysis holds both in the uplink and the downlink communications.

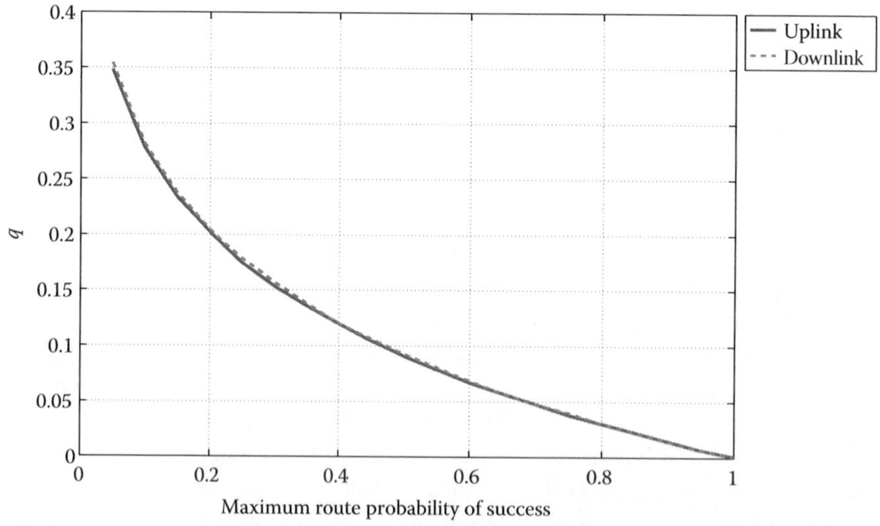

Figure 7.13 Uplink and downlink performance in medical scenario.

7.5.4 Optimum Routing Tree Computation and Corresponding Route Performance

When the computation of the maximum suitable value of $q_{\max}$ is completed, the next step is to derive the optimum routing tree of the random access network. The optimal routing tree is a collection of all the best paths between the central node (i.e., the sink node) and all the other medical sensor nodes, printed in the form of a tree graph in which the sink node is the root and sensors are the leaves.

The computation of the best path to reach every single leaf node is done with the Dijkstra algorithm presented in Section 7.5.2. To properly run the Dijkstra algorithm, one needs to assign a weight value on each link of the network such that a high link probability of success translates into a low link weight. Starting from the link probability of success is not possible because the definition in Equation 7.29 yields

$$\begin{aligned} 1-\mathcal{P}^{(\text{route})} &= 1-\prod_{i}^{n_{\text{hops}}} \mathcal{P}_i^{(\text{link})} \\ &= 1-\left[(1-\mathcal{P}_2^{(\text{link})})(1-\mathcal{P}_1^{(\text{link})})\right] \end{aligned} \tag{7.31}$$

which cannot be used in Dijkstra's algorithm because it is not an additive function of the probability of success of each link. This problem can be solved by considering a logarithmic version of the link costs. More precisely, the objective of Dijkstra's algorithm is to maximize $\mathcal{P}^{(\text{route})}$, which can be seen as an equivalent to minimizing the quantity $1/\mathcal{P}^{(\text{route})}$. Furthermore, by recalling that the logarithm is a continuous, strictly increasing function of its argument, minimizing the quantity $\mathcal{P}^{(\text{route})}$ can be equivalently rewritten as minimizing the function $\log_{10}\mathcal{P}^{(\text{route})}$ and it can be written (in the case of a n-hops scenario) as:

$$\begin{aligned} \max\left\{\log_{10}\mathcal{P}^{(\text{route})}\right\} &\Leftrightarrow \min\left\{\log_{10}\frac{1}{\mathcal{P}^{(\text{route})}}\right\} \\ &\Leftrightarrow \min\left\{\log_{10}\frac{1}{\prod_{i}^{n_{\text{hops}}}\mathcal{P}_i^{(\text{link})}}\right\} \\ &\Leftrightarrow \min\left\{\sum_{i}^{n_{\text{hops}}}\left(-\log_{10}\frac{1}{\mathcal{P}_i^{(\text{link})}}\right)\right\}. \end{aligned} \tag{7.32}$$

The minimization of this quantity can be carried out using Dijkstra's algorithm. Figures 7.14, 7.15, and 7.16 present the optimum routing tree for the uplink and the downlink communications and with $q = 0.1$, $q = 0.2$, $q = 0.33$, respectively.

It can be observed from these figures that, in a random access network, the most performant topology corresponds to a star with direct communications between the medical sensors and the sink node. This can be interpreted as follows. As previously presented in Chapter 2, the wireless propagation channel of BAN is highly specific and exhibits a strong attenuation with respect to

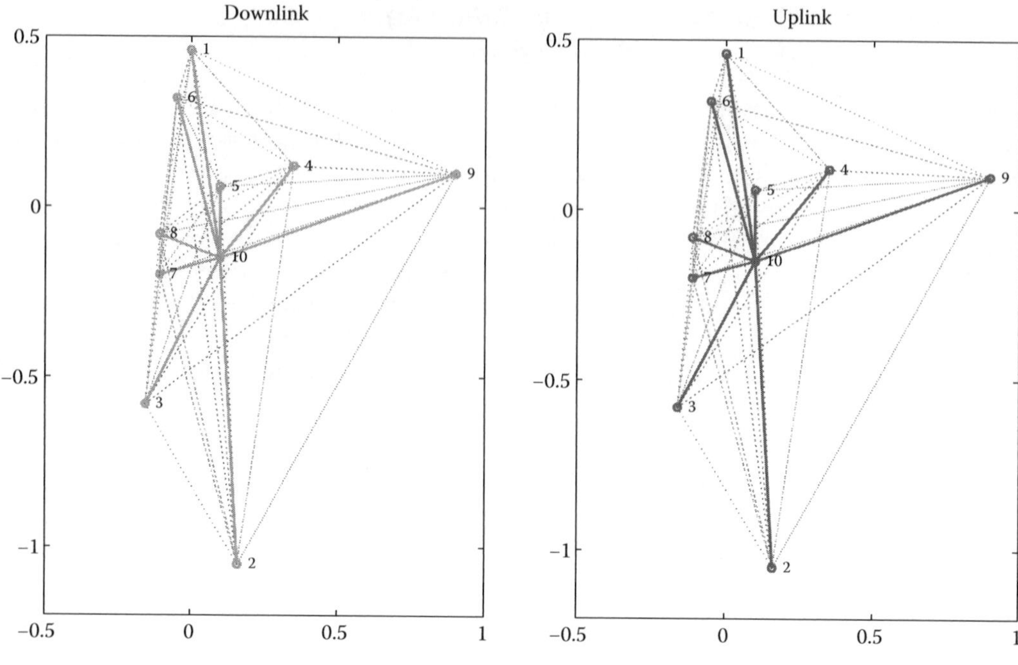

Figure 7.14 Uplink and downlink routing tree of the network in medical scenario with $q = 0.1$.

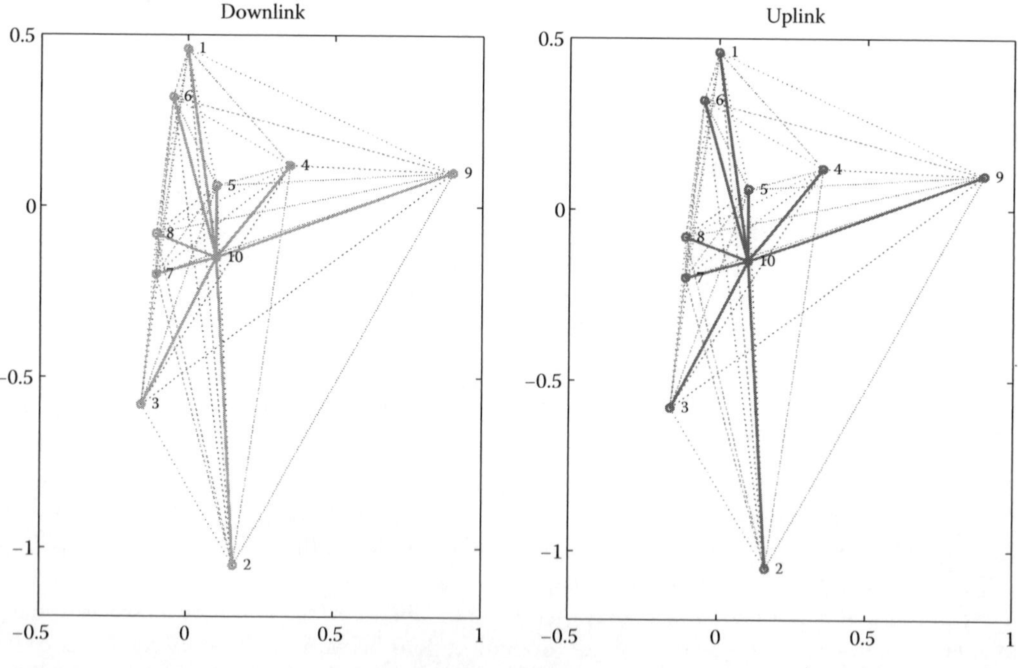

Figure 7.15 Uplink and downlink routing tree of the network in medical scenario with $q = 0.2$.

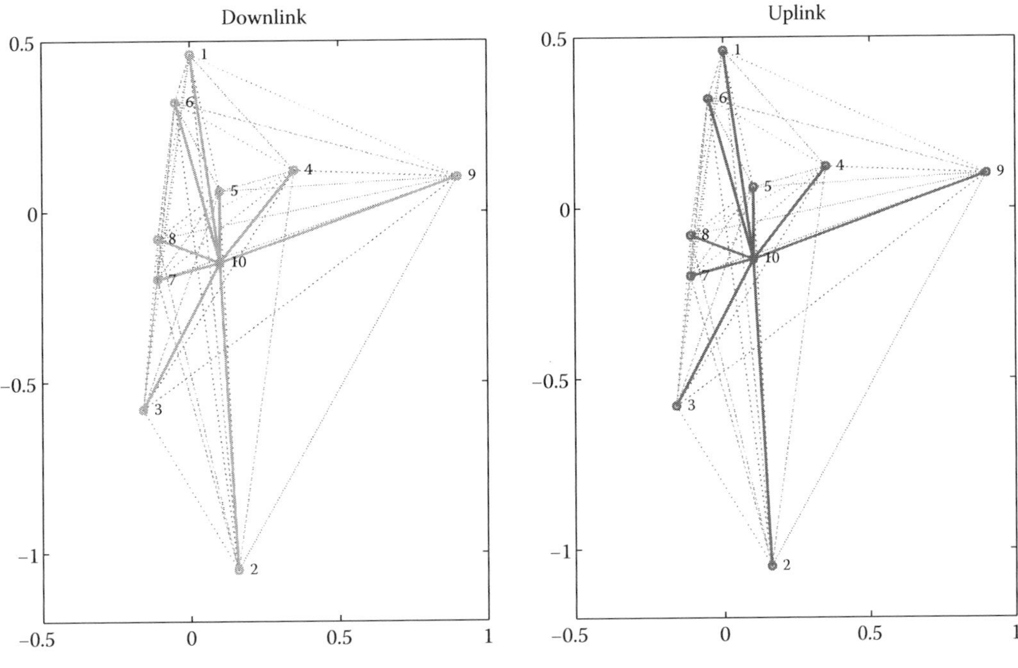

Figure 7.16 Uplink and downlink routing tree of the network in medical scenario with $q = 0.33$.

the distance. Therefore, in a noise-limited regime, far-range interferers (i.e., the nodes at a distance of $d \geq 25$ cm to the receiver) do not interfere with the emitter node of interest. Because the nodes are sparsely deployed over the body, for low values of q (which is the case due to a required high value of $\mathcal{P}_{\text{min}}^{(\text{route})}$) the corresponding total interference on each link is low. More precisely, by recalling the definition in Equation 7.29, one has

$$\mathcal{P}^{(\text{route})} = \prod_{i}^{n_{\text{hops}}} \mathcal{P}_i^{(\text{link})} \approx \left(\mathcal{P}^{(\text{link})}\right)^{n_{\text{hops}}}$$

because $\forall i, j : \mathcal{P}_i^{(\text{link})} \approx \mathcal{P}_j^{(\text{link})} = \mathcal{P}^{(\text{link})}$. Furthermore, because the link probability of transmission is high, $\mathcal{P}_{i,j}^{(\text{link})} \approx 1$ and, finally, one has $\left(\mathcal{P}_i^{(\text{link})}\right)^{n_{\text{hops}}} \approx \mathcal{P}_i^{(\text{link})}$, which demonstrates that a multihopping strategy does not help in improving the route probability of success in the context of low rate BANs.

Finally, Tables 7.2 and 7.3 present a synthesis of the route probability of success as a function of the node medical reporting rate q and for the downlink and uplink scenarios, respectively.

7.5.5 Route Delay and Optimization of the Link–Layer Access Scheme

In the connectionless transport network, in which no traffic is induced by retransmissions, the end-to-end delay is equal to the sum of the delay on each link of the route. In an IEEE 802.15.6 BAN, and as presented in Section 7.4, two strategies can be used: random access transmissions (i.e., slotted ALOHA) and time division transmission (i.e., TDMA). Therefore, the delay of the random access network is

Table 7.2 Downlink Route Performance in a Medical Scenario

			P_{rs}		
Source Node	*Destination Node*	*Distance (m)*	*q = 0.1*	*q = 0.2*	*q = 0.33*
10	1	0.6181	0.5109	0.2453	0.0843
10	2	0.9019	0.5109	0.2453	0.0843
10	3	0.5024	0.5109	0.2453	0.0843
10	4	0.3679	0.5109	0.2453	0.0843
10	5	0.2100	0.6270	0.3866	0.1971
10	6	0.4933	0.5109	0.2453	0.0843
10	7	0.2158	0.6059	0.3595	0.1729
10	8	0.2213	0.5873	0.3362	0.2213
10	9	0.8381	0.5109	0.2453	0.0843

Table 7.3 Uplink Route Performance in Medical Scenario

			$\mathcal{P}^{(route)}$		
Source Node	*Destination Node*	*Distance (m)*	*q = 0.1*	*q = 0.2*	*q = 0.33*
1	10	0.6181	0.5186	0.2534	0.0896
2	10	0.9019	0.5314	0.2675	0.0993
3	10	0.5024	0.5314	0.2675	0.0993
4	10	0.3679	0.5314	0.2675	0.0993
5	10	0.2100	0.6398	0.4032	0.2119
6	10	0.4933	0.5186	0.2534	0.0896
7	10	0.2158	0.5942	0.3439	0.1583
8	10	0.2213	0.5772	0.3232	0.1415
9	10	0.8381	0.5314	0.2675	0.0993

$$D_{\text{ALOHA}}^{(\text{route})} = \sum_{i}^{n_{\text{hops}}} D_{\text{ALOHA}}^{(i)} = \sum_{i}^{n_{\text{hops}}} \frac{1}{\mathcal{P}_i^{(\text{link})}} \tag{7.33}$$

whereas in a TDMA network with N_1 leaves and N_2 relaying nodes, it has been shown in Equation 7.28 that the average delay is deterministic and given by

$$D_{\text{TDMA}}^{(\text{route})} = \frac{N_2 \cdot N_1}{2}.$$

It can be observed from the relation in Equation 7.33 that the random access network will be highly sensitive to link impairment. For instance, let us consider that node 1 is subject to strong interference, that is, $\forall i, \mathcal{P}_1^{(\text{link})} \gg \mathcal{P}_i^{(\text{link})}$. Therefore, Equation 7.33 can be approximated as

$$D_{\text{ALOHA}} = \sum_{i}^{n_{\text{hops}}} \frac{1}{\mathcal{P}_i^{(\text{link})}} = \frac{1}{\mathcal{P}_1^{(\text{link})}} + \sum_{i \neq 1}^{n_{\text{hops}}} \frac{1}{\mathcal{P}_i^{(\text{link})}} \approx \frac{1}{\mathcal{P}_1^{(\text{link})}}$$

and, therefore, the route delay depends on the link subject to the strongest interference level. To avoid unbalancing in the links, relays can be deployed to reduce the link distance and increase the link SINR.

In Figure 7.17, a comparison between the ALOHA and TDMA route delay in a one-hop medical network is presented. It can be observed that there exists a break point that differentiates the two access strategies. More specifically, when $q \leq 0.3$, the random access scheme yields lower end-to-end delay. This is because, for low values of q, the TDMA sensing nodes spend more time waiting their turn before transmitting whereas the ALOHA nodes can transmit their data directly. On the other hand, when $q > 0.3$, the number of collisions in the ALOHA network increases due to the concurrent access by the nodes and the deterministic strategy outperforms the random strategy.

Next, it is of interest to compare between networks implementing the TDMA access scheme with or without relay nodes, that is, for $N_1 \neq 0$ and $N_2 \neq 0$. The results of the corresponding simulation are reported in Figure 7.18. As expected, the relaying nodes further increase the delay and the waiting time at low transmission rates. More specifically, the ALOHA scheme outperforms the TDMA scheme until the node probability of transmission reaches $q = 0.35$.

To summarize, as a TDMA-based scheme has a throughput $\tau = q$, it becomes very attractive for values of q beyond the maximum of the considered slotted ALOHA system, as the latter

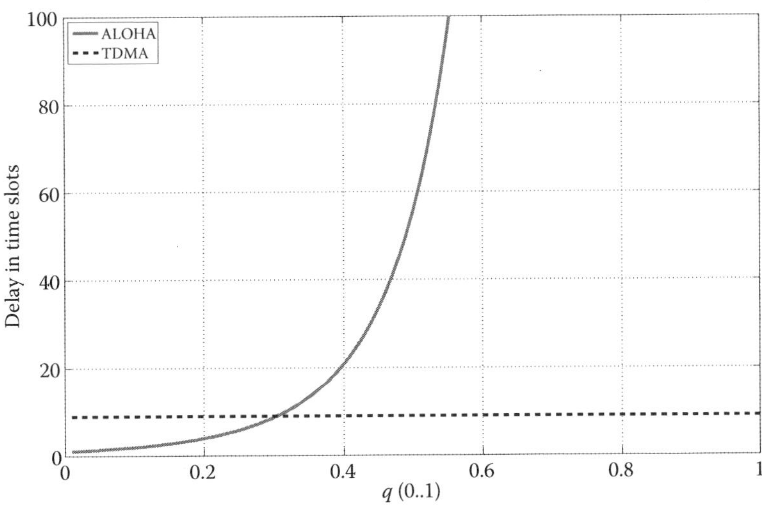

Figure 7.17 Route delay comparison of ALOHA and TDMA in a medical network.

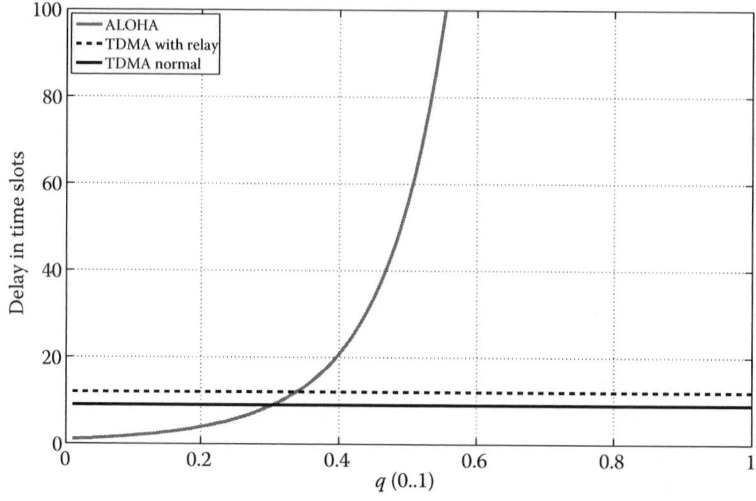

Figure 7.18 TDMA scheme with and without relay comparison in the medical network.

becomes unstable, that is, the value of the delay $D_{\text{ALOHA}} \rightarrow \infty$ because $P_s \rightarrow 0$. In scenarios with low reporting rates, the slotted ALOHA scheme is preferred.

7.6 Conclusions

In this chapter, we have investigated the performance of IEEE 802.15.6 body sensor networks under realistic propagation conditions. More specifically, we have proposed a generic analytical framework for a computation of the link probability of outage in body sensor networks subject to fading interference. The analytical derivation is built on real-life channel measurements, which allows us to accurately model the particularities of the physical propagation mechanisms found in on-body wireless transmissions—strong signal attenuation with respect to the distance and propagation through reflections off the environment.

The link-level performance analysis was conducted under two distinct medium access mechanisms: random access (i.e., slotted ALOHA) and time division multiple access (i.e., TDMA). Both strategies are part of the IEEE 802.15.6 standard and exhibit distinct advantages and shortcomings, depending on the scenario of interest. Regarding the random access approach, it is the simplest to implement and can be used for low throughput body sensor networks. The detailed performance analysis showed that the functioning region (in terms of sensor reporting rate) coincides with the requirements of most medical applications. On the other hand, when the sensor reporting rate is high or if the number of sensors increases, the TDMA strategy is preferred—even if it is more difficult to implement. Indeed, when the load increases in a random access network, the probability of collisions increases proportionally. A strictly defined time ordering of the packet sending times and silence times of each sensor is mandatory. It is important to note that the developments presented in this chapter are generic and can be used to further analyze the performance of future medical sensor networks without loss of generality.

Next, based on the link probability of outage, it has been possible to derive the optimal topology and the performance of the entire medical network. More precisely, a well-known graph technique used in computer networks (i.e., Dijkstra's algorithm) was used to determine the optimal path between any pair of source and destination nodes. Each link was given a weight proportional to its probability of outage and the algorithm was used to determine the routes with the lowest achievable weight, that is, the highest probability of successful end-to-end transmission. A performance analysis of the routing topologies was conducted in the context of a real medical implementation. It has been observed that multihopping (the use of intermediary relay nodes) is not necessary, in general, in a medical sensor network. Moreover, the use of relays may have a detrimental effect on the end-to-end delay and would generate further interference if the random access schemes are in use.

Finally, the topology can be dynamically adapted by the network itself to meet any change in the topology (i.e., addition or removal of nodes) or in the medical reporting rate of the nodes.

Appendix A

The modeling of slow-scale fading as a log-normal distribution (i.e., a zero-mean Gaussian in decibel scale) raises mathematical difficulties, as shown in Equation 7.12. The complementary cdf of a zero-mean log-normal random variable is

$$\zeta(z;\sigma) \triangleq \frac{1}{2} + \frac{1}{2}\,\mathrm{erf}\left(\frac{-10\log_{10} z}{\sigma\sqrt{2}}\right) \tag{7.34}$$

where $\mathrm{erf} \triangleq \frac{2}{\sqrt{\pi}}\int_0^x e^{-t^2}\,dt$ is the error function. The function $\zeta(z;\sigma)$ is shown in Figure 7.19 as a function of z for $\sigma \in \{4,8,12,16\}$ dB. It can be observed that $\zeta(z;\sigma)$ (i) saturates for $z \to \infty$, regardless of the value of ρ, and (ii) has the shape of a decreasing exponential function of z (for a given value of σ).

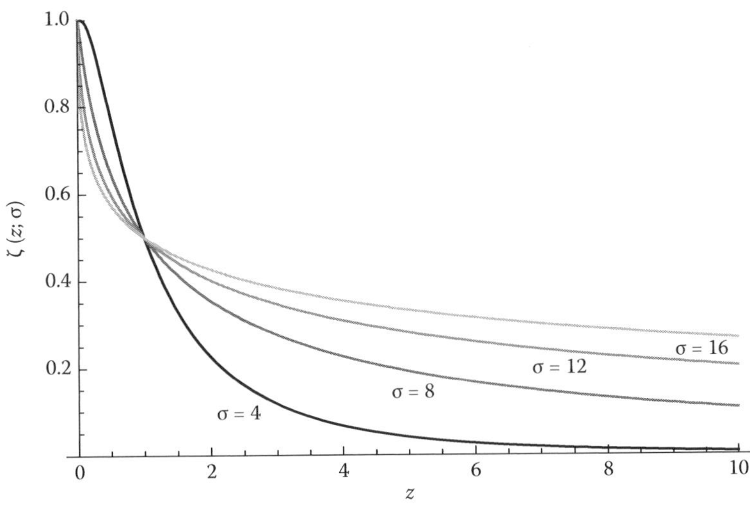

Figure 7.19 The function $\zeta(z;\sigma)$ as a function of z, considering various values of σ (in decibels).

Table 7.4 Coefficients for the Approximation of the ζ Function

	c_1	a_1	c_2	a_2	c_3	a_3	*Residual*
$\sigma = 4$	0.49	0.75	0.49	0.75	0.03	0.16	4.68×10^{-5}
$\sigma = 6$	0.38	0.31	0.56	1.21	0.06	0.07	4.23×10^{-6}
$\sigma = 8$	0.59	1.32	0.34	0.18	0.06	0.02	1.04×10^{-4}
$\sigma = 10$	0.29	0.09	0.65	1.17	0.05	0.01	7.53×10^{-4}
$\sigma = 12$	0.04	0	0.24	0.04	0.70	0.93	3.52×10^{-3}
$\sigma = 14$	0.20	0.01	0.03	0	0.72	0.64	1.03×10^{-2}
$\sigma = 16$	0.18	0.01	0.70	0.49	0.04	0	1.67×10^{-2}

The ζ function can be approximated with a linear combination of negative exponential functions, as in [67,68]:

$$\zeta(z;\sigma) = \sum_{m}^{\infty} c_m \exp(-a_m z) \approx \sum_{m}^{n} c_m \exp(-a_m z)$$

where the coefficients $\{c_m\}_{m=1}^{n}$ and $\{a_m\}_{m=1}^{n}$ depend on σ and can be determined in a least-squares sense with $q \geq 2n$ known points of the ζ function. The Levenberg–Marquardt algorithm [69,70] can be used to determine the coefficients $\{c_m\}$ and $\{a_m\}$ for different values of σ and 10,000 points over the interval $z \in [0,1000]$. The corresponding values are reported in Table 7.4 along with the corresponding residual sum of squares.

References

1. R. Astolfi, L. Lorenzoni, and J. Oderkirk. Informing policy makers about future health spending: a comparative analysis of forecasting methods in OECD countries. *Health Policy (Amsterdam, Netherlands)*, pp. 1–10, June 2012.
2. M. Chernew and J. Newhouse. Health care spending growth. *Handbook of Health Economics*, vol. 2, pp. 1–43, 2011.
3. United Nations Department of Economic and Social Affairs. Population ageing 2006. Wall Chart. 2006.
4. J. Xing and Y. Zhu. A survey on body area network. In *Proceedings of the 5th International Conference on Wireless Communications, Networking and Mobile Computing (WiCOM'09)*, Beijing, China, 2009.
5. E. Monton, J. Hernandez, J. Blasco, T. Herve, J. Micallef, I. Grech, A. Brincat, and V. Traver. Body area network for wireless patient monitoring. *IET Communications*, vol. 2, no. 2, pp. 215–222, 2008.
6. A. Kailas and M. A. Ingram. Wireless communications technology in telehealth systems. In *Proceedings of the IEEE Wireless VITAE Conference*, Aalborg, Denmark, May 2009.
7. D. Bhatia, L. Estevez, and S. Rao. Energy efficient contextual sensing for elderly care. In *Proceedings of the 29th International Conference of the Engineering in Medicine and Biology Society (EMBS 2007)*, Lyon, France, Aug. 2007.
8. A. Milenković and C. Otto. Wireless sensor networks for personal health monitoring: issues and an implementation. *Computer Communications*, vol. 29 no. 13–14, pp. 2521–2533, Aug. 2006.

9. O. Aziz, B. Lo, A. Darzi, G. Z. Yang. Body sensor networks, the ultimate diagnostic tool? *IEE: BSN, Wearable and Implantable Body Sensor Networks*. 2005.
10. M. J. Kane, P. P. Breen, F. Quondamatteo, and G. ÓLaighin. BION microstimulators: A case study in the engineering of an electronic implantable medical device. *Medical Engineering and Physics*, vol. 33, no. 1, pp. 7–16, Jan. 2011.
11. R. F. Yazicioglu, T. Torfs, P. Merken, J. Penders, V. Leonov, R. Puers, B. Gyselinckx, and C. Van Hoof. Ultra-low-power biopotential interfaces and their applications in wearable and implantable systems. *Microelectronics Journal*, vol. 40, no. 9, pp. 1313–1321, Sept. 2009.
12. P. Pawar, V. Jones, B.-J. van Beijnum, and H. Hermens. A framework for the comparison of mobile patient monitoring systems. *Journal of Biomedical Informatics*, vol. 45, no. 3, pp. 544–556, Mar. 2012.
13. M. R. Yuce. Implementation of wireless body area networks for healthcare systems. *Sensors and Actuators A: Physical*, vol. 162, no. 1, pp. 116–129, July 2010.
14. E. Jovanov, A. Milenkovic, C. Otto, and P. C. de Groen. A wireless body area network of intelligent motion sensors for computer assisted physical rehabilitation. *Journal of Neuroengineering and Rehabilitation*, vol. 2, no. 1, p. 6, Mar. 2005.
15. M. Yuce. Wideband communication for implantable and wearable systems. *IEEE Transactions on Microwave Theory and Techniques*, vol. 57, no. 10, pp. 2597–2604, Oct. 2009.
16. Y. Zhang and H. Xiao. Bluetooth-based sensor networks for remotely monitoring the physiological signals of a patient. *IEEE Transactions on Information Technology in Biomedicine*, vol. 13, no. 6, pp. 1040–1048, Nov. 2009.
17. M. Berg. Patient care information systems and health care work: A sociotechnical approach. *International Journal of Medical Informatics*, vol. 55, no. 2, pp. 87–101, Aug. 1999.
18. H. J. Hermens and M. M. R. Vollenbroek-Hutten. Towards remote monitoring and remotely supervised training. *Journal of Electromyography and Kinesiology*, vol. 18, no. 6, pp. 908–919, Dec. 2008.
19. T. H. F. Broens, R. M. H. A. Huis in't Veld, M. M. R. Vollenbroek-Hutten, H. J. Hermens, A. T. van Halteren, and L. J. M. Nieuwenhuis. Determinants of successful telemedicine implementations: a literature study. *Journal of Telemedicine and Telecare*, vol. 13, no. 6, pp. 303–309, Jan. 2007.
20. A. Metting van Rijn, A. Peper, and C. Grimbergen. High-quality recording of bioelectric events: Part 1. Interference reduction, theory and practice. *Medical and Biological Engineering and Computing*, vol. 28, no. 5, 1990.
21. D. Wood and D. Ewins. Comparative analysis of power-line interference between two- or three-electrode biopotential amplifiers. *Medical and Biological Engineering and Computing*, vol. 33, no. 1, pp. 63–68, Jan. 1995.
22. J. C. Huhta and J. G. Webster. 60-HZ interference in electrocardiography. *IEEE Transactions on Biomedical Engineering*, vol. 20, no. 2, pp. 91–101, Mar. 1973.
23. J. G. Webster. *Medical Instrumentation: Application and Design*. Wiley, 2009.
24. M. Fernández and R. Pallás-Areny. Ag-AgCl electrode noise in high-resolution ECG measurements. *Biomedical Instrumentation and Technology/Association for the Advancement of Medical Instrumentation*, vol. 34, no. 2, pp. 125–30.
25. A. Nonclercq and P. Mathys. Quantification of motion artifact rejection due to active electrodes and driven-right-leg circuit in spike detection algorithms. *IEEE Transactions on Biomedical Engineering*, vol. 57, no. 11, pp. 2746–2752, July 2010.
26. B. Gyselinckx, J. Penders, and R. Vullers. Potential and challenges of body area networks for cardiac monitoring. *Journal of Electrocardiology*, vol. 40, no. 6 Suppl, pp. S165–S168, 2007.
27. A. Turnbull. The use of IEC 60601-1 in supporting approvals of medical electrical devices and the role of the new collateral standard IEC 60601-1-9, ENVIRON Technical Report, no. September, 2007.
28. D.-J. Kim and B. Prabhakaran. Motion fault detection and isolation in body sensor networks. *Pervasive and Mobile Computing*, vol. 7, no. 6, pp. 727–745, Dec. 2011.
29. IEC60601-1-2, Medical electrical equipment—Part 1–2: General requirements for basic safety and essential performance—collateral standard: electromagnetic compatibility—requirements and tests. *Technical Report*, International Electrotechnical Commission, Geneva, Switzerland, 2007.

30. IEC60601-1, Medical electrical equipment—Part 1: General requirements for basic safety and essential performance, International Electrotechnical Commission, Geneva, Switzerland, 2005.
31. M. N. Nyan, F. E. H. Tay, and E. Murugasu. A wearable system for pre-impact fall detection. *Journal of Biomechanics*, vol. 41, no. 16, pp. 3475–3481, Dec. 2008.
32. K. Ni, M. Srivastava, N. Ramanathan, M. N. H. Chehade, L. Balzano, S. Nair, S. Zahedi, E. Kohler, G. Pottie, and M. Hansen. Sensor network data fault types. *ACM Transactions on Sensor Networks*, vol. 5, no. 3, pp. 1–29, May 2009.
33. J. Y. Khan, M. R. Yuce, and F. Karami. Performance evaluation of a wireless body area sensor network for remote patient monitoring. In *Conference Proceedings: Annual International Conference of the IEEE Engineering in Medicine and Biology Society*, vol. 2008, pp. 1266–1269, Jan. 2008.
34. D. Ganesan, R. Govindan, S. Shenker, and D. Estrin. Highly-resilient, energy-efficient multipath routing in wireless sensor networks. In *Proceedings of the 2nd ACM International Symposium on Mobile Ad Hoc Networking and Computing—MobiHoc '01*, vol. 1, no. 2, p. 251, 2001.
35. H. Alemdar and C. Ersoy. Wireless sensor networks for healthcare: a survey. *Computer Networks*, vol. 54, no. 15, pp. 2688–2710, Oct. 2010.
36. U. Varshney. A framework for supporting emergency messages in wireless patient monitoring. *Decision Support Systems*, vol. 45, no. 4, pp. 981–996, Nov. 2008.
37. B. Najafi, K. Aminian, A. Paraschiv-Ionescu, F. Loew, C. J. Büla, and P. Robert. Ambulatory system for human motion analysis using a kinematic sensor: Monitoring of daily physical activity in the elderly. *IEEE Transactions on Biomedical Engineering*, vol. 50, no. 6, pp. 711–723, June 2003.
38. T. Klingeberg and M. Schilling. Mobile wearable device for long term monitoring of vital signs. *Computer Methods and Programs in Biomedicine*, vol. 106, no. 2, pp. 89–96, May 2012.
39. M. Al Ameen, J. Liu, and K. Kwak. Security and privacy issues in wireless sensor networks for healthcare applications. *Journal of Medical Systems*, vol. 36, no. 1, pp. 93–101, Feb. 2012.
40. H. S. Ng, M. L. Sim, and C. M. Tan. Security issues of wireless sensor networks in healthcare applications. *BT Technology Journal*, vol. 24, no. 2, pp. 138–144, Apr. 2006.
41. S. Dağtas, G. Pekhteryev, and Z. Sahinoğlu. Real-time and secure wireless health monitoring. *International Journal of Telemedicine and Applications*, doi:10.1155/2008/135808, 2008.
42. R. Paradiso, G. Loriga, and N. Taccini. A wearable health care system based on knitted integrated sensors. *IEEE Transactions on Information Technology in Biomedicine*, vol. 9, no. 3, pp. 337–344, Sept. 2005.
43. B. Fuchs, S. Vogel, and D. Schroeder. Universal application-specific integrated circuit for bioelectric data acquisition. *Medical Engineering and Physics*, vol. 24, no. 10, pp. 695–701, Dec. 2002.
44. N. Van Helleputte, J. Tomasik, W. Galjan, A. Mora-Sanchez, D. Schroeder, W. Krautschneider, and R. Puers. A flexible system-on-chip (SoC) for biomedical signal acquisition and processing. *Sensors and Actuators A: Physical*, vol. 142, no. 1, pp. 361–368, Mar. 2008.
45. J. Van Ham and R. Puers. A power and data front-end IC for biomedical monitoring systems. *Sensors and Actuators A: Physical*, vol. 147, no. 2, pp. 641–648, Oct. 2008.
46. C. K. Ho, T. S. See, and M. R. Yuce. An ultra-wideband wireless body area network: Evaluation in static and dynamic channel conditions. *Sensors and Actuators A: Physical*, vol. 180, pp. 137–147, June 2012.
47. L. Lonys, P. Mathys, and A. Nonclercq. Human energy harvesting used for endoscopic implant power supply. In *Proceedings of Congress of the International Society of Biomechanics*, Brussels, Belgium, 2011.
48. D. Leger, V. Bayon, J. P. Laaban, and P. Philip. Impact of sleep apnea on economics. *Sleep Medicine Reviews*, vol. 16 no. 5. pp. 455–462, Oct. 2012.
49. M. R. Patel, T. H. Alexander, and T. M. Davidson. Home sleep testing. *Operative Techniques in Otolaryngology-Head and Neck Surgery*, vol. 18, no. 1, pp. 33–51, Mar. 2007.
50. M. A. King, M.-O. Jaffre, E. Morrish, J. M. Shneerson, and I. E. Smith. The validation of a new actigraphy system for the measurement of periodic leg movements in sleep. *Sleep Medicine*, vol. 6, no. 6, pp. 507–513, Nov. 2005.
51. A. Astaras, M. Arvanitidou, I. Chouvarda, V. Kilintzis, V. Koutkias, E. M. Sanchez, G. Stalidis, A. Triantafyllidis, and N. Maglaveras. An integrated biomedical telemetry system for sleep monitoring employing a portable body area network of sensors (SENSATION). In *Conference Proceedings: Annual International Conference of the IEEE Engineering in Medicine and Biology Society*, vol. 2008, pp. 5254–5257, Jan. 2008.

52. S. Chokroverty. *Sleep Disorders Medicine: Basic Science, Technical Considerations, and Clinical Aspects, Expert Consult*, 3rd ed. Saunders, Philadelphia, USA, 2009.
53. S. Quan, J. Gillin, and M. Littner. Sleep-related breathing disorders in adults: Recommendations for syndrome definition and measurement techniques in clinical research. Editorials. *Sleep*, vol. 22, no. 5, 1999.
54. S. P. Patil. What every clinician should know about polysomnography. *Respiratory Care*, vol. 55, no. 9, pp. 1179–1195, Sept. 2010.
55. C. Iber, S. Ancoli-Israel, A. Chesson, and S. Quan. *The AASM Manual for the Scoring of Sleep and Associated Events: Rules, Terminology and Technical Specifications*, 1st ed. American Academy of Sleep Medicine, Westchester, 2007.
56. J. Ryckaert, P. D. Doncker, R. Meys, A. de Le Hoye, and S. Donnay. Channel model for wireless communication around human body. *Electronics Letters*, vol. 40, no. 9, pp. 543–544, 2004.
57. S. Van Roy, C. Oestges, F. Horlin, and P. De Doncker. Propagation modeling for UWB body area networks: power decay and multi-sensor correlations. In *Proceedings of the 10th IEEE International Symposium on Spread Spectrum Techniques and Application*, Bologna, Italy, 2008.
58. S. Van Roy, C. Oestges, F. Horlin, and P. De Doncker. A comprehensive channel model for UWB multisensor multiantenna body area networks. *IEEE Transactions on Antennas and Propagation*, vol. 58, no. 1, pp. 163–170, 2010.
59. J. Takada, T. Aoyagi, K. Takizawa, N. Katayama, H. Sawada, T. Kobayashi, K. Y. Yazdandoost, H. B. Li, and R. Kohno. Static propagation and channel models in body area. In *COST 2100 6th Management Committee Meeting*, TD(08)639, Trondheim, Norway, 2008.
60. A. Goldsmith. *Wireless Communications*. New York: Cambridge University Press, 2005.
61. D. Bertsekas and R. Gallager. *Data Networks*, 2nd ed. Prentice-Hall, Upper Saddle River, New Jersey, USA, 1991.
62. J.-M. Dricot, G. Ferrari, S. van Roy, F. Horlin, and P. De Doncker. Outage, local throughput, and achievable transmission rate of body area networks. In *Proceedings of the COST 2100 Meeting*, no. TD(09) 939, Vienna, Austria, Sept. 2009.
63. C. E. Jones, K. M. Sivalingam, P. Agrawal, and J. C. Chen. A survey of energy efficient network protocols for wireless networks. *Wireless Networks*, vol. 7, no. 4, pp. 343–358, 2001.
64. Y.-A. Le Borgne, J.-M. Dricot, and G. Bontempi. Principal component aggregation for energy efficient information extraction in wireless sensor networks. In *Knowledge Discovery from Sensor Data*. Boca Raton, FL: Taylor & Francis/CRC Press. 2007.
65. S. Madden, M. Franklin, J. Hellerstein, and W. Hong. TAG: a Tiny AGgregation service for ad-hoc sensor networks. In *Proceedings of the 5th ACM Symposium on Operating System Design and Implementation (OSDI)*, Boston. 2002.
66. E. W. Dijkstra. A note on two problems in connexion with graphs. *Numerische Mathematik*, vol. 1, pp. 269–271, 1959.
67. B. G. de Prony. Essai expérimental et analytique sur les lois de la dilatabilité des fluides élastique et sur celles de la force expansive de la vapeur de l'eau et de la vapeur de l'alkool, à différentes températures. *Journal de l'École Polytechnique*, vol. 1, no. 2, pp. 24–76, 1795.
68. F. G. Lether. Elementary approximation for erf(x). *Journal of Quantitative Spectroscopy and Radiative Transfer*, vol. 49, no. 5, pp. 573–577, 1993.
69. K. Levenberg. A method for the solution of certain non-linear problems in least squares. *The Quarterly of Applied Mathematics*, vol. 2, pp. 164–168, 1944.
70. D. Marquardt. An algorithm for least-squares estimation of nonlinear parameters. *SIAM Journal on Applied Mathematics*, vol. 11, pp. 431–444, 1963.

Chapter 8

Routing Protocols

Nikhil Marriwala

Kurukshetra University

Contents

8.1 Routing

The process of selecting paths in a network along which to send network traffic is called *routing*. In wireless sensor networks (WSNs), routing is different from conventional routing; there is no infrastructure, the links are unreliable, and the nodes have limited battery power, and thus routing protocols should consume less energy. Broadly speaking, almost all of the routing protocols can be classified according to their network structures as flat-based, location-based, or hierarchical routing protocols. Furthermore, the routing protocols can also be classified according to their operation mode, that is, multipath based, query based, negotiation based, QoS based, and coherent based. The classification of routing protocols is given in the following sections.

8.1.1 Flat Routing Protocols

In flat routing, all the nodes are assigned the same function. In flat routing or data-centric protocols, the aggregate data are sent to the sink node [1] in which intermediate nodes aggregate data that are collected by each sensor node individually in contrast with the address-centric protocol in which each sensor node transmits data to the sink node. Because of the lower amount of transmission to

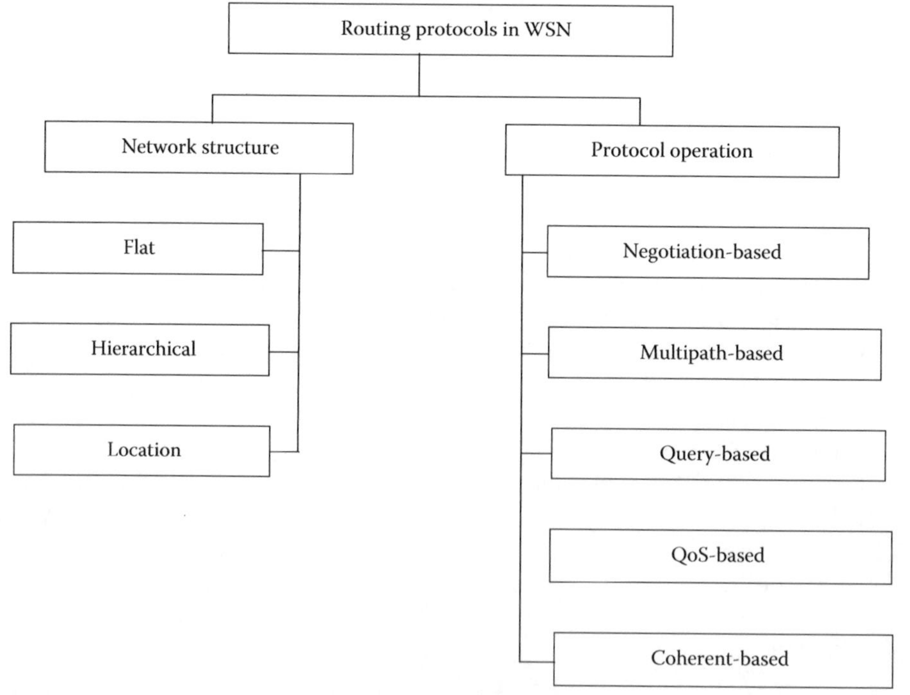

Figure 8.1 Routing protocols in WSN. (From J. N. Al-Karaki and A. E. Kamal, *IEEE Wireless Communications*, vol. 11, no. 6, pp. 6–28, December 2004.)

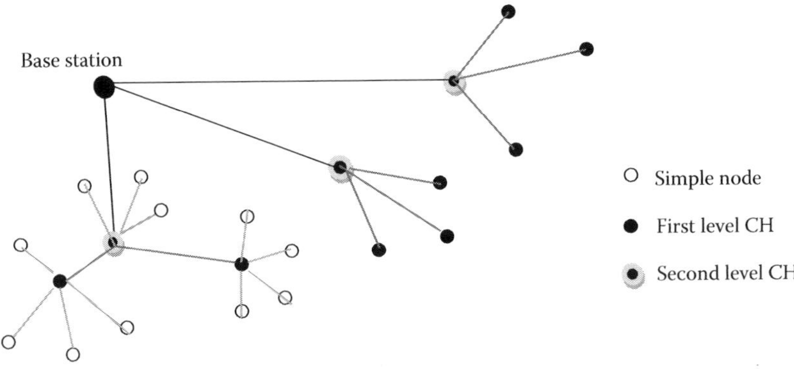

Figure 8.2 Hierarchical clustering in WSN. (From S. K. Singh et al., *International Journal of Computer Science and Engineering Survey (IJCSES)*, vol. 1, no. 2, pp. 63–83, November 2010.)

the sink by sensors in data-centric protocols, it is an energy-saving process. Examples of flat-based routing include sensor protocols for information via negotiation (SPIN), directed diffusion, and rumor routing. In these networks, all nodes play the same role, and there is absolutely no hierarchy.

8.1.2 Location-Based Routing

In location-based routing, the sensor nodes are identified by their geographic location. The position of sensor nodes is required for sensor networks in routing data and to calculate the separation of two nodes to calculate energy consumption. The location of each sensor node is identified by a global positioning system (GPS) receiver, which is built into the sensor node. Examples of location-based routing include geographic and energy aware routing (GEAR), distance routing effect algorithm for mobility (DREAM), and so on.

8.1.3 Hierarchical Protocols

This type of routing method has high energy efficiency in WSNs. It also has special advantages related to scalability and efficient communication. This class of routing protocol attempts to conserve energy by arranging the nodes into clusters [2]. Clustering is an energy-efficient communication protocol that is used to send data to a sink node. Nodes in a cluster transmit to a head node within the proximity, which then aggregates the collected information and forwards this information to a base station. Clustering techniques in WSNs seek to gather data from groups of nodes that elect leaders among themselves. The leader or cluster heads (CHs) have the role of aggregating data and reporting the refined data to the base station. The advantage of this scheme is that it reduces the energy usage and communication costs of each node.

8.2 Design Constraints for Routing in WSNs

Due to the reduced computing, radio, and battery resources for sensors, routing protocols in WSNs (Figure 8.1) are expected to fulfill the following requirements [26]:

- *Autonomy:* The assumption of a dedicated unit that controls the radio and routing resources does not stand in WSNs because it could be an easy point of attack. Because there will not be any centralized entity to make the routing decision, the routing procedures are transferred to the network nodes.
- *Energy efficiency:* Routing protocols should prolong network lifetime while maintaining a good grade of connectivity to allow communication between nodes. It is important to note that battery replacement in the sensors is infeasible because most of the sensors are randomly placed. Under some circumstances, the sensors are not even reachable. For instance, in wireless underground sensor networks, some devices are buried—enabling them to sense even the soil [27].
- *Scalability:* WSNs are composed of hundreds of nodes, thus routing protocols should work with this number of nodes.
- *Resilience:* Sensors may unpredictably stop operating due to environmental reasons or because of battery consumption. Routing protocols should cope with this eventuality so that when a current-in-use node fails, an alternative route could be discovered.
- *Device heterogeneity:* Although most of the civil applications of WSNs rely on homogenous nodes, the introduction of different kinds of sensors could produce significant benefits. The use of nodes with different processors, transceivers, power units, or sensing components may improve the characteristics of the network. Among other advantages, the scalability of the network, energy drainage, and bandwidth are potential candidates that could benefit from the heterogeneity of nodes [28].
- *Mobile adaptability:* The different applications of WSNs could demand nodes to cope with their own mobility, the mobility of the sink, or the mobility of the event to sense. Routing protocols should render appropriate support for these movements.

8.3 Clustered Architecture

Clustering techniques in WSNs are aimed at gathering data among groups of nodes that elect leaders among themselves. The leader or CH has the role of aggregating the data and reporting the refined data to the base station. The advantage of this scheme is that it reduces the energy usage and communication costs of each node.

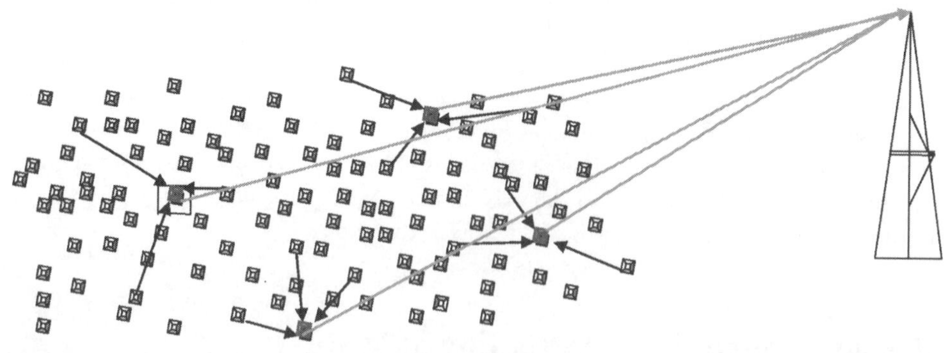

Figure 8.3 Clustering. (From M. Bani Yassein et al., *International Journal of Digital Content Technology and Its Applications*, vol. 3, no. 2, June 2010.)

One of the earliest works using the clustering technique in WSNs is low-energy adaptive clustering hierarchy (LEACH). The other clustering techniques are mostly variants of LEACH with different application scenarios and slight improvements [3]. The hierarchical approach breaks down the network into clustered layers. There is a CH, which is responsible for transmitting data from the cluster to the base station. The data is transmitted from one layer to another, that is, from a lower layer to a higher layer (Figure 8.2).

8.4 Clustering Objective

The main objectives of clustering in WSNs are given in the following sections (Figure 8.3).

8.4.1 Maximizing Network Lifetime

Unlike cellular networks, mobile gadgets (e.g., phones) can easily be recharged constantly after battery drainage, and thus power management in these networks remains a secondary issue. WSNs, however, are heavily constrained in this regard. Aside from being a system with no infrastructure, its battery power is very limited. Most of the sensor nodes are equipped with minimal power sources, thus power efficiency will continue to be of growing concern and will remain one of the main design objectives of WSN. To cope with energy management in WSN, a clustering scheme was pursued to extend network lifetime and help ease the burden of each node transmitting directly to the base station similar to conventional protocols such as direct transmission.

8.4.2 Fault-Tolerance

The failure of a sensor node should have a minimal effect on the overall network system. Because sensor nodes will be deployed in harsh environmental conditions, there is a tendency that some nodes may fail or be physically damaged. Some clustering techniques have been proposed to address the problem of node failure by using proxy CHs in the event of the failure of the originally elected CH, or should there be minimal power for transmission. The adaptive clustering scheme is also used to deal with node failures such as the rotating CH. Tolerating node failure is one of the other design goals of clustering protocols.

8.4.3 Load Balancing

Load balancing technique could be another design goal of clustering schemes. It is always necessary not to overburden the CHs because this may deplete their energies faster. Therefore, it is important to have an even distribution of nodes in each cluster. Especially in cases in which CHs are performing data aggregation or other signal processing tasks, an uneven characterization can extend the latency or communication delay to the base station.

8.5 LEACH

LEACH is the first and most popular energy-efficient protocol in WSN that was proposed for reducing power consumption. The clustering task is rotated in LEACH, and CHs are selected

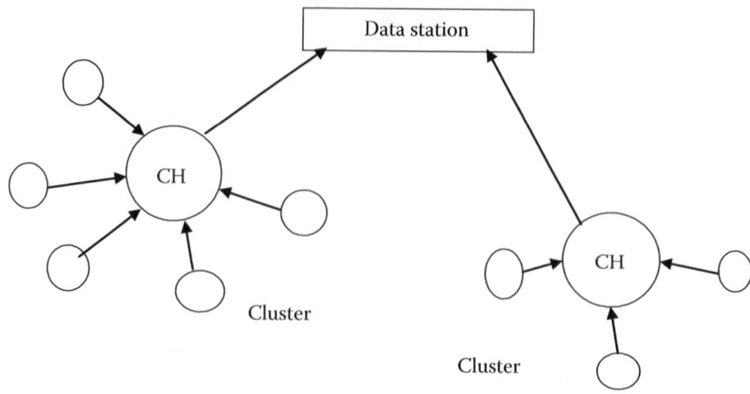

Figure 8.4 LEACH protocol. (From M. Bani Yassein et al., *International Journal of Digital Content Technology and Its Applications*, vol. 3, no. 2, June 2010.)

randomly. LEACH uses clusters to increase the lifetime of the WSN. It is based on an aggregation technique that combines or aggregates the original data into smaller-sized data that carries only meaningful information to all individual sensors [3,4]. LEACH divides the WSN into several clusters. Each cluster has a CH that aggregates data from the cluster nodes and processes that data, which is then transmitted to the base station. LEACH uses a randomized rotation of high-energy CH position rather than selecting in a static manner, to give a chance to all sensors to act as CHs, and avoid the battery depletion of an individual sensor and die quickly due to direct communication, in which the node near the base station depletes energy more quickly. The number of CHs and cluster members generated by LEACH are important parameters for achieving better performance (Figure 8.4).

8.5.1 Design

LEACH organizes nodes into clusters; with one node from each cluster serving as a CH. It randomly selects a predetermined number of nodes as CHs. CHs then advertise themselves and other nodes join one of those CHs whose signal they found to be the strongest (i.e., the CH that is nearest to them). In this way, a cluster is formed. The CH then makes a time division multiple access (TDMA) schedule for the nodes under its cluster. The communication between different clusters is done through CHs in a code division multiple access (CDMA) manner. The CHs collect the data from their clusters and aggregate it before sending it to the other CHs or the base station. After a predetermined time lapse, the cluster formation step is repeated so that different nodes are given a chance to become CHs and energy consumption is thus uniformly distributed (Figure 8.5).

8.5.2 CH Selection Algorithm

LEACH randomly selects a few sensor nodes as CHs, and rotates this role to evenly distribute the energy load among the sensors in the network. In LEACH, the CH nodes compress data arriving from nodes that belong to their respective clusters, and send an aggregated packet to the base station to reduce the amount of information that must be transmitted to the base station (negotiation). WSN is considered to be a dynamic clustering method.

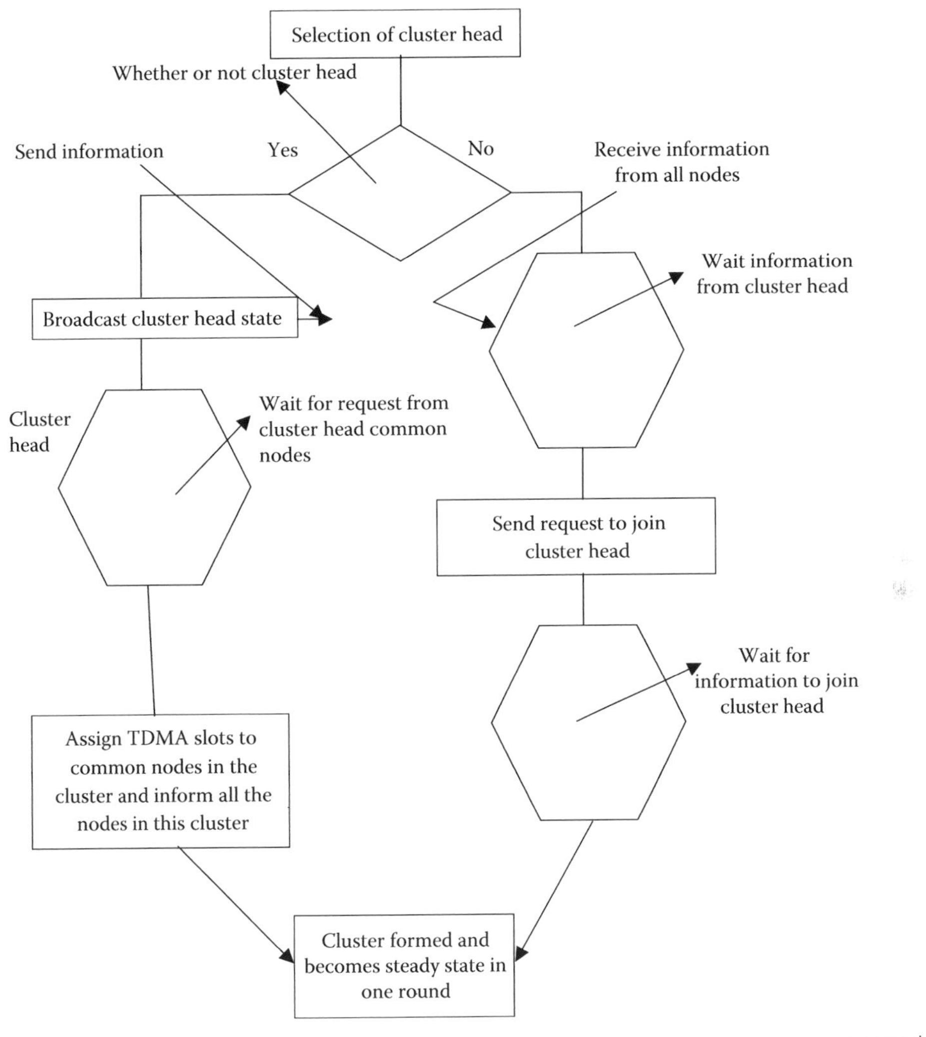

Figure 8.5 LEACH flow chart diagram. (From F. Xiangning and S. Yulin. Improvement on LEACH protocol of wireless sensor network. In *International Conference on Sensor Technologies and Applications in IEEE*, 2007.)

Operation:

It has two phases:

1. Setup state phase
2. Steady state phase
   - In the setup phase, the clusters are organized and CHs are selected.
   - In the steady state phase, the actual data transfer to the base station takes place.
   - The duration of the steady state phase is longer than the duration of the setup phase.
   - During the setup phase, a predetermined fraction of nodes, p, elect themselves as CHs.

– A sensor node chooses a random number, r, between 0 and 1. Let a threshold value be $T(n)$. If this random number is less than a threshold value, $T(n)$, the node becomes a CH for the current round. The threshold value is calculated based on an equation that incorporates the desired percentage to become a CH, the current round, and the set of nodes that have not been selected as a CH in the last $(1/P)$ rounds, denoted by G.

$$T(n) = \frac{p}{i - p\left[r \bmod\left(\frac{1}{p}\right)\right]} \quad \text{if } n \,\varepsilon\, G$$

where G is the set of nodes that are involved in the CH election.

Each elected CH broadcasts an advertisement message to the rest of the nodes in the network that they are the new CHs. All the non-CH nodes, after receiving this advertisement, decide on the cluster to which they want to belong to. This decision is based on the signal strength of the advertisement. The non-CH nodes inform the appropriate CHs that they will be a member of the cluster. After receiving all the messages from the nodes that would like to be included in the cluster, and based on the number of nodes in the cluster, the CH node creates a TDMA schedule and assigns each node a time slot when it can transmit. This schedule is broadcast to all the nodes in the cluster.

During the steady state phase, the sensor nodes can begin sensing and transmitting data to the CHs. The CH node, after receiving all the data, aggregates it before sending it to the base station. After a certain time, the network goes back into the setup phase again and enters another round of selecting new CH.

8.5.3 Cluster Formation Process of LEACH

The setup phase begins with cluster formation. Suppose there are N nodes in the network, to ensure that a certain number of clusters is formed during each round, each sensor elects itself at the beginning of the round with a probability $Pi(t)$, chosen such that the expected number of CH nodes for this round is k_c, the choice of probability is based on the assumption that every node has the same level of energy at the beginning of the network and also that each node has data to send in each round. To complete the setup phase, each node sends a "join request" message after they receive a broadcast from the elected CHs using a nonpersistent carrier sense multiple access (CSMA) media access control (MAC) protocol. The CH creates a TDMA, as shown in the LEACH flow chart, and finally the nodes forming each cluster wait for their schedule before transmission. The steady phase starts immediately after the setup phase. The CH gathers all data from their respective cluster members. The CHs performs data aggregation using signal processing techniques before sending the refined data to the base station in each round. The idea of the TDMA schedule ensures the efficient use of the bandwidth and that the data aggregation process reduces communication cost and energy, thus improving the network's lifetime.

8.6 LEACH Protocol Phases

This protocol is divided into rounds; each round consists of two phases—a setup phase and a steady phase.

8.6.1 Setup Phase

The setup phase is made up of an (1) advertisement phase and a (2) cluster setup phase. Each node decides, independent of the other nodes, if it will become a CH or not. This decision takes into account *when* the node served as a CH for the last round. In the advertisement phase, the CHs inform their neighborhood with an advertisement packet that they have become CHs. Non-CH nodes pick the advertisement packet with the strongest received signal strength [1].

In the cluster setup phase, the member nodes inform the CH that they have become a member of that cluster with a "join packet" that contains their IDs using CSMA. After the cluster-setup subphase, the CH knows the number of member nodes and their IDs. Based on all the messages received within the cluster, the CH creates a TDMA schedule, picks a CSMA code randomly, and broadcasts the TDMA table to cluster members. After that, the steady-state phase begins (Figure 8.6).

8.6.2 Steady Phase

The steady phase is composed of (1) schedule creation and (2) data transmission steps. When data transmission begins, nodes send their data to the CH during the allocated TDMA slot. This transmission uses a minimal amount of energy (chosen based on the received strength of the CH advertisement). The radio of each non-CH node can be turned off until the nodes are allocated a TDMA slot, thus minimizing energy dissipation in these nodes. When the complete data has been received, the CH aggregates this data and sends it to the base station. LEACH is able to perform local aggregation of data in each cluster to reduce the amount of data that is transmitted to the base station. Although the LEACH protocol acts in a good manner, it suffers from many drawbacks such as

- CH selection is random; it does not take into account energy consumption
- It cannot cover a large area

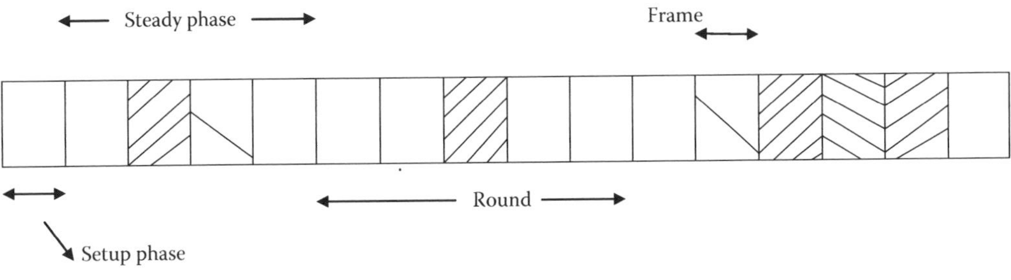

Figure 8.6 LEACH protocol phases. (From M. Bani Yassein et al., *International Journal of Digital Content Technology and Its Applications,* vol. 3, no. 2, June 2010.)

8.7 Power-Efficient Gathering in Sensor Information Systems

A modification in the basic LEACH protocol, power-efficient gathering in sensor information systems (PEGASIS), forms chains from sensor nodes so that each node transmits and receives from a neighbor and only one node is selected from that chain to transmit to the base station. The data is gathered and moves from node to node, becomes aggregated, and is eventually sent to the base station. The chain construction is performed in a greedy way. Unlike LEACH, PEGASIS avoids cluster formation and uses only one node in a chain to transmit to the base station instead of using multiple nodes. A sensor transmits to its local neighbors in the data fusion phase instead of sending directly to its CH as in the case of LEACH [6].

In the PEGASIS routing protocol, the construction phase assumes that all sensors have global knowledge about the network, particularly the positions of the sensors, and use a greedy approach. When a sensor fails or dies due to low battery power, the chain is constructed using the same greedy approach by bypassing the failed sensor. In each round, a randomly chosen sensor node from the chain will transmit the aggregated data to the base station, thus reducing the per round energy expenditure compared with LEACH.

Simulation results showed that PEGASIS is able to increase the lifetime of the network by twice as much compared with the network under the LEACH protocol. Such a performance gain is achieved through the elimination of the overhead caused by dynamic cluster formation in LEACH and through decreasing the number of transmissions and reception by using data aggregation. Although the clustering overhead is avoided, PEGASIS still requires dynamic topology adjustment because a sensor node needs to know about the energy status of its neighbors in order to know where to route its data. Such a topology adjustment can introduce significant overhead especially for highly utilized networks.

8.8 Hybrid, Energy-Efficient Distributed Clustering

Hybrid, energy-efficient distributed (HEED) clustering extends the basic scheme of LEACH by using residual energy and node degree or density as a metric for cluster selection to achieve power balancing. It operates in multihop networks, using an adaptive transmission power in the inter-clustering communication. HEED was proposed with four primary goals, namely, (i) prolonging network lifetime by distributing energy consumption, (ii) terminating the clustering process within a constant number of iterations, (iii) minimizing control overhead, and (iv) producing well distributed CHs and compact clusters.

In HEED, the proposed algorithm periodically selects CHs according to a combination of two clustering parameters. The primary parameter is the residual energy of each sensor node (used in calculating the probability of becoming a CH) and the secondary parameter is the intracluster communication cost as a function of cluster density or node degree (i.e., number of neighbors). The primary parameter is used to probabilistically select an initial set of CHs whereas the secondary parameter is used for breaking ties.

HEED clustering has an improved network lifetime compared with LEACH clustering because LEACH randomly selects CHs (and hence cluster size), which may result in the faster death of some nodes. The final CHs selected in HEED are well distributed across the network and the communication cost is minimized. However, cluster selection deals with only a subset of parameters, which can possibly impose constraints on the system. These methods are suitable for prolonging the network lifetime rather than for the entire needs of the WSN.

8.9 LEACH-C

An enhancement over the LEACH protocol, LEACH-centralized (LEACH-C), is proposed. LEACH-C uses a centralized clustering algorithm, in which an attempt is made to distribute clusters throughout the entire sensor field. As a result of dispersing clusters throughout the network, the LEACH-C protocol records better performance compared with LEACH. The operation of LEACH-C is divided into two phases.

8.9.1 Setup Phase

The base station receives information from each node about their current location and energy level. The nodes may obtain their current location by using a GPS receiver that is activated at the beginning of each round. After that, the base station runs the centralized cluster formation algorithm to determine the clusters for that round. Before running the algorithm that determines and selects the clusters, the base station makes sure that only nodes with enough energy are participating in the CH selection.

8.9.2 Steady Phase

Once the clusters are created, the base station broadcasts the information to all the nodes in the network. Each of the nodes, except the CH, determines the TDMA slot used for data transmission. Then, the node goes to sleep until it is time to transmit data to its CH. LEACH offers no guarantee about the placement or number of CHs (or both). An enhancement over the LEACH protocol was proposed. The protocol, called LEACH-C, uses a centralized clustering algorithm and the same steady-state phase as LEACH. The LEACH-C protocol can produce better performance by dispersing the CHs throughout the network. During the setup phase of LEACH-C, each node sends information about its current location (possibly determined using GPS) and residual energy level to the sink. In addition to determining good clusters, the sink needs to ensure that the energy load is evenly distributed among all the nodes. To do this, the sink computes the average node energy and determines which nodes are below this average energy level.

Once the CHs and associated clusters are found, the sink broadcasts a message that obtains the CH ID for each node. If a CH ID matches its own ID, the node is a CH; otherwise, the node determines its TDMA slot for data transmission and goes to sleep until it is time to transmit its data. The steady-state phase of LEACH-C is identical to that of the LEACH protocol.

8.10 V-LEACH

In this new version of the LEACH protocol, the cluster contains CH (responsible only for sending data that is received from the cluster members to the base station), vice-CH (the node that will become a CH of the cluster in case the CH dies), and cluster nodes (gathering data from the environment and sending it to the CH).

In the original leach, the CH is always "on" and receiving data from cluster members, aggregating these data, and then sending it to the base station, which might be located far away from it. The CH will die earlier than the other nodes in the cluster because of its function of receiving, sending, and overhearing. When the CH dies, the cluster will become useless because the data gathered by the cluster nodes will never reach the base station [3].

In the V-LEACH protocol, aside from having a CH in the cluster, there is a vice-CH that takes the role of the CH when it dies because of the reasons mentioned previously. Cluster node data will always reach the base station; there is no need to elect a new CH each time the CH dies. This will extend the overall network lifetime.

8.11 N-LEACH

The WSN topological models of N-LEACH and LEACH are basically the same, the difference between them is that the N-LEACH algorithm takes into account the difference in the remaining energy after running each round. LEACH assumes that it consumes the same amount of energy in the nodes for each round because the LEACH algorithm operates per round. There are two phases in each round—cluster-established transmission and stable data transmission. The process of cluster-established transmission consists of CH node–formed and spanning tree built parts. The CH nodes of the first round were generated by base stations whereas the CH nodes of the remaining rounds were generated from the node that had the largest residual energy in the last round. The CH node was elected periodically and the cluster node was produced dynamically [5]. Power controls were carried out when the node sent data throughout the whole process. Because the initial energy of all nodes in the initial state was in the same round, the election of the CH node was in accordance with LEACH. The number of CH nodes was determined based on the area of the monitoring location, size, and scale to the WSN. After the CH node in the first round is elected, each CH node sends broadcast messages to all WSNs, the node receives broadcast signals and compares their strengths, then chooses to join the cluster and informs the CH node. The CH node created TDMA schedules for all nodes in the cluster, in accordance with the schedule, the cluster node sends data to the CH node. The CH node integrates all received data and then sends it to the base station. The cluster node sends its residual energy with the last frame of data according to the schedule on its CH node [5]. The CH node compares the residual energy of each node and elects the largest one as the coming CH node in the second round. The CH node broadcasts the ID of the coming CH node in the cluster, the cluster node receives the ID and compares it with its ID; it would be CH nodes of the second round if they were same. The CH node election process in subsequent rounds is the same as in the second round; cluster creation and data transfer processes after CH node election are the same as in the first round.

8.12 Energy Consumption Model for LEACH

We are using the sensor node model shown in Figure 8.3. The sensor node components primarily contributing to the overall energy consumption include the following: the onboard sensors, the analog-to-digital converter unit, the microcontroller, and the communication module (including the sending and receiving modules). Energy consumed during sampling, processing, and logging data by a sensor node is comparable to the energy consumed by the sender and receiver circuits of the sensor node. To run all the components, the microcontroller needs to be awake at all times [21].

Energy consumption by a node includes the following: (i) energy consumption during sensing and logging (including energy consumed by the onboard sensors), (ii) energy consumption during sending and receiving data in our analysis, we ignored the small amounts of energy needed by regular nodes to receive rare and short messages from their CHs, and by CHs to send these messages to regular nodes. Knowing the time spent by each node in each state, as well as the current drawn by the node circuitry in each state [21], we can calculate the average amount of energy consumed by each node.

8.12.1 Average Current Drawn by Regular Nodes

Let ${}_{\text{sen+log}}I^{\text{Reg}}$ and ${}_{\text{snd}}I^{\text{Reg}}$ be the average current used by a regular node during sensing and logging, and the average current used by the node during sending data to its CH, respectively. Then, the average current consumed by a regular node during its awake interval is:

$$I^{\text{Reg}} = \frac{\left(\tau^{\text{Reg}}_{\text{Sen+log}} \times I^{\text{Reg}}_{\text{Sen+log}}\right) + \left(\tau^{\text{Reg}}_{\text{Snd}} \times I^{\text{Reg}}_{\text{Snd}}\right)}{\tau^{\text{Reg}}_{\text{Sen+log}} \times I^{\text{Reg}}_{\text{Snd}}} [\text{A}]$$

8.12.2 Average Current Drawn by CHs

Let ${}_{\text{rcv+log}}I^{\text{CH}}$ be the average current drawn by a CH during receiving messages from regular nodes, ${}_{\text{agg}}I^{\text{CH}}$, the average current drawn during the aggregation of received data, and ${}_{\text{snd}}I^{\text{CH}}$, the average current drawn during sending data to the base station. Then, the average current drawn by a CH during its awake interval is:

$$I^{\text{CH}} = \frac{\left(\tau^{\text{ch}}_{\text{rcv+log}} \times I^{\text{ch}}_{\text{rcv+log}}\right) + \left(\tau^{\text{ch}}_{\text{agg}} \times I^{\text{ch}}_{\text{agg}}\right) + \left(\tau^{\text{ch}}_{\text{Snd}} \times I^{\text{ch}}_{\text{Snd}}\right)}{\tau^{\text{ch}}_{\text{rcv+log}} + \tau^{\text{ch}}_{\text{agg}} + \tau^{\text{ch}}_{\text{snd}}} [\text{A}]$$

References

1. J. N. Al-Karaki and A. E. Kamal. Routing techniques in wireless sensor networks: a survey. *IEEE Wireless Communications*, vol. 11, no. 6, pp. 6–28, December 2004.
2. S. K. Singh, M. P. Singh, and D. K. Singh. Routing protocols in wireless sensor networks—A survey. *International Journal of Computer Science and Engineering Survey (IJCSES)*, vol. 1, no. 2, pp. 63–83, November 2010.
3. M. Bani Yassein, A. Al-zou'bi, Y. Khamayseh, and W. Mardini. Improvement on LEACH protocol of wireless sensor network (VLEACH). *International Journal of Digital Content Technology and Its Applications*, vol. 3, no. 2, June 2010.
4. K. Akkaya and M. Younis. A survey on routing protocols for wireless sensor networks. *Ad Hoc Networks*, vol. 3, pp. 325–349, 2005.
5. F. Xiangning and S. Yulin. Improvement on LEACH protocol of wireless sensor network. In Proceeding SENSORCOMM '07 Proceedings of the 2007 *International Conference on Sensor Technologies and Applications*, IEEE Computer Society Washington, DC, USA, pp. 260–264, 2007.
6. I. Akyildiz, D. Pompili, and T. Melodia. Underwater acoustic sensor networks: research challenges. *Ad Hoc Networks*, vol. 3, pp. 257–279, 2005.
7. J.-H. Chang and L. Tassiulas. Maximum lifetime routing in wireless sensor networks. *IEEE/ACM Transactions on Networking*, vol. 12, no. 4, pp. 609–619, August 2004.
8. O. Younis, M. Krunz, and S. Ramasubramanian. Wireless sensor networks: Recent developments and deployment challenges. *IEEE Network*, vol. 20, no. 3, pp. 20–25, May/June 2006.

9. W. Bo, H. Han-ying, and F. Wen. An improved LEACH protocol for data gathering and aggregation in wireless sensor networks. In *Computer and Electrical Engineering, 2008,* ICCEE 2008. pp. 398–401, 2008.
10. H. Junping, J. Yuhui, and D. Liang. A time-based cluster-head selection algorithm for LEACH. In *Computer and Communication, 2008, ISCC 2008*. pp. 1172–1176, 2008.
11. I. F. Akyildiz, W. Su, Y. Sankarasubramaniam, and E. Cayirci, Georgia Institute of Technology. A survey on sensor networks. *IEEE Communications Magazine*, vol. 40, no. 8, pp. 102–114, August 2002.
12. J. Kim, X. Lin, N. B. Shroff, and P. Sinha. Minimizing delay and maximizing lifetime for wireless sensor networks with any cast. *IEEE/ACM Transactions on Networking*, vol. 18, no. 2, pp. 515–527, April 2010.
13. D. G. Melese, H. Xiong, and Q. Gao. Consumed energy as a factor for cluster head selection in wireless sensor networks. In *2010 6th International Conference on Wireless Communications Networking and Mobile Computing (WiCOM)*, pp. 1–4, 2010.
14. M. J. Hajikhani and B. Abolhassani. Energy efficient algorithm for cluster-head. In *5th International Symposium on Telecommunications 2010*, pp. 397–400, 2010.
15. A. Jangra, Swati, Richa, and Priyanka. Wireless sensor network (WSN): Architectural design issues and challenges. *International Journal on Computer Science and Engineering*, vol. 02, no. 09, pp. 3089–3094, 2010.
16. D. G. Anand, H. G. Chandrakanth, and M. N. Giriprasad. Challenges in maximizing the life of wireless sensor network. *Journal of Advanced Networking and Applications*, vol. 03, no. 01, pp. 999–1005, 2011.
17. H. Al-Refai, A. Al-Awneh, K. Batiha, A. A. Ali, and Y. M. El. Rahman. Efficient routing LEACH (ER-LEACH) enhanced on LEACH protocol in wireless sensor networks. *International Journal of Academic Research*, vol. 3, no. 3, pp. 42–48, May 2011.
18. F. Tashtarian, A. T. Haghighat, M. T. Honary, and H. Shokrzadeh. A new energy-efficient clustering algorithm for wireless sensor networks. 15th International Conference on Software, Telecommunication and Computer Networks 2007 SoftCOM 2007. pp. 1–6, Sept. 2007.
19. S. Deng, J. Li, and L. Shen. Mobility-based clustering protocol for wireless sensor networks with mobile nodes. *IET Wireless Sensor Systems*, vol.1, no.1, pp. 39–47, March 2011.
20. N. Kumar and J. Kaur. Improved LEACH protocol for wireless sensor networks. 7th International Conference on Wireless Communications, Networking and Moblie Computing (WiCOM), 2011 *IEEE*, 978-1-4244-6252-0/11/$26.00 pp. 1–5, 2011.
21. V. Shnayder, M. Hempstead, B. Chen, G. Werner Allen, and M. Welsh. Simulating the power consumption of large-scale sensor network applications. In *SenSys'04 Proceedings of the 2nd International Conference on Embedded Networked Sensor Systems*, Baltimore, Maryland. November 3–5, 2004.
22. B. Li, Q. Wang, Y. Yang, and J. Wang. Optimal distribution of redundant sensor nodes for wireless sensor networks. In *Proceedings of the IEEE International Conference on Industrial Informatics*, pp. 985–989, August 2006.
23. K. Maraiya, K. Kant, and N. Gupta. Application based study on wireless sensor network. *International Journal of Computer Applications*, vol. 21, no.8, pp. 0975–8887, May 2011.
24. B. Baranidharan and B. Shanthi. A survey on energy efficient protocols for wireless sensor networks. *International Journal of Computer Applications*, vol. 11, no. 10, pp. 0975–8887, December 2010.
25. M. Tong and M. Tang. LEACH-B: an improved LEACH protocol for wireless sensor network. 6th International Conference on Wireless Communications Networking and Mobile Computing (WiCOM), 2010.
26. I. F. Akyildiz, W. Su, Y. Sankarasubramaniam, and E. Cayirci. Wireless sensor networks: A survey. *Elsevier Computer Networks*, vol. 38, no. 4, pp. 393–422, 2002.
27. R. Verdone, D. Dardari, G. Mazzini, and A. Conti. *Wireless Sensor and Actuator Networks Technologies, Analysis and Design*. Academic Press, London, 2008.
28. F. Dressler. *Self-Organization in Sensor and Actor Networks*. SensorCOMM 2007 International Conference on Sensor Technologies and Applications 2007. Wiley Series in Communications Networking and Distributed Systems, pp. 260–264, Oct. 2007.

Chapter 9

A Survey of Connected Dominating Set Construction Techniques for Ad Hoc Sensor Networks

Ariyam Das
Yahoo! R&D

Chittaranjan Mandal
IIT Kharagpur

Chris Reade
Kingston University

Contents

9.1 Introduction

Wireless ad hoc and sensor networks are assuming a pivotal role in the next generation of networks in providing flexible deployment and mobile connectivity. They are already being deployed for many applications such as automated battlefield, search and rescue, and disaster relief operations. However, unlike wired networks or cellular networks, no physical backbone infrastructure is installed in wireless ad hoc networks. The wireless networks consist of static or mobile nodes (hosts). Each node, having an omnidirectional antenna, can broadcast messages to all the nodes within its transmission range. Therefore, through broadcasting, a node can reach all of its nearby nodes with one emission. If communicating parties are outside the single hop radio transmission range of each other, then a communicating session is established through multihop links by some intermediate nodes for relaying messages (multihop routing). A simple and intuitive method for multihop routing between nonadjacent nodes in wireless networks is pure flooding, in which a node retransmits a packet only once after receiving it. However, because of the lack of bandwidth available for the wireless channels and the redundant retransmissions generated through pure flooding, the latter is not an efficient communication mechanism in wireless networks. These challenges of low bandwidth availability and memory and battery life limitations of the sensor nodes open up new paradigms for multihop routing in wireless ad hoc and sensor networks. One of the popular means of overcoming these challenges in multihop routing in wireless networks is through the use of a virtual backbone. The existing routing algorithms adopt different methodologies to construct the virtual backbones. We first describe various network models and then, in the following sections, discuss various connected dominating set (CDS) construction techniques and examine their characteristics.

9.2 Network Models

First, we present a network model that is widely accepted and adopted by authors in their respective virtual backbone construction techniques. Given a sensor network containing n nodes, each with an omnidirectional antenna of maximum transmission range, R, the resultant topology of the network can be modeled as a unit disk graph (UDG) in the Euclidean plane. UDG are the intersection graphs of equally sized circles in the plane. They provide a graph theoretic model for broadcast networks. There are mainly three kinds of models in UDGs for representing the ad hoc networks:

1. *Proximity model.* Nodes in the network form the vertices of a graph and the edges between nodes are formed if the Euclidean distance between nodes is below some specified bound d.
2. *Intersection model.* Nodes in the network form the vertices of the graph, and the edges between nodes are formed when circles form around the nodes with a maximum transmission range intersect. It is worth mentioning that tangent circles are also said to be intersecting.
3. *Containment model.* Nodes in the network form the vertices of the graph and each vertex is at the center of a disk formed around the node with a maximum transmission range of R. An edge exists between two vertices, u and v, if and only if $|uv| \leq R$ where $|uv|$ is the Euclidean distance between u and v. In other words, two vertices are connected by an edge if and only

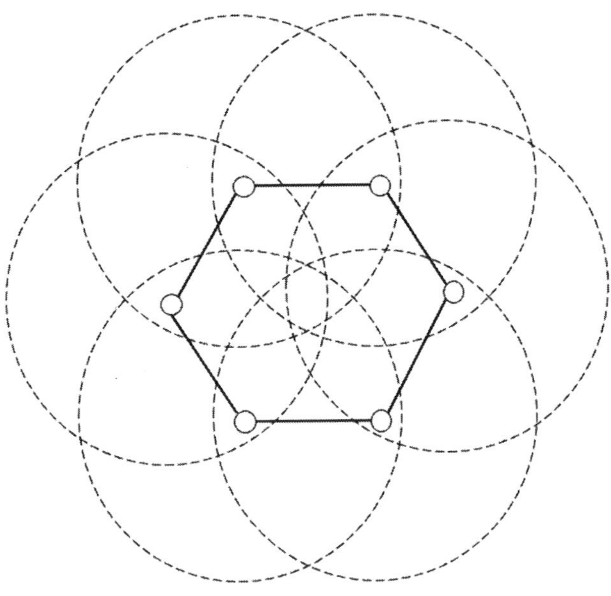

Figure 9.1 UDG containment model.

if one of the corresponding circles contain the other's center and vice versa. For example, Figure 9.1 represents the containment model of UDG. This is the model that the authors have used in their respective approaches toward the virtual backbone construction problem.

We restricted the discussion to UDGs only and skipped other network models such as the radio model, battery model, and others. Interested readers can find the details regarding other network models and their applications in Bulusu et al. [1] and Rakhmatov and Vrudhula [2].

9.2.1 Background

In this section, we provide some preliminary definitions that are relevant to the understanding of the rest of the chapter.

1. *Dominating set.* A dominating set for a graph $G(V,E)$ is a subset $V' \subseteq V$ such that each node in $V - V'$ is adjacent to at least one node in V'. The nodes in the dominating set are called the dominators.
2. *Maximal independent set.* A maximal independent set (MIS) is an independent dominating set in which no two nodes can be adjacent to each other.
3. *Connected dominating set.* A CDS is a dominating set that induces a connected subgraph. The non-CDS nodes in the graph are called dominatees.
4. *Steiner tree.* A Steiner tree in a graph $G(V,E)$ for a given subset of vertices (terminals) $I \subseteq V$, is a tree interconnecting (known as tapping) all the nodes in I using a set of Steiner nodes in $\{V - I\}$.

9.2.2 CDS as Virtual Backbones in Ad Hoc Sensor Networks

A CDS is an excellent candidate for becoming a virtual backbone of a wireless sensor and ad hoc network. The idea of using a CDS as a virtual backbone for routing was first proposed in 1987 by

Ephremides et al. [3]. A CDS offers an optimized way of sending messages in wireless networks. It ensures that, in the absence of any transmission error, it can receive a packet from any node in the network and can retransmit it to any other remote node. If a non-CDS node (non-gateway host) wants to deliver a message to another non-CDS node, it sends the message to a neighboring CDS node (gateway host). Now, the search space for any route is reduced to the CDS. After the message is relayed to the destination gateway host, which is a neighbor of the destination host, the former can broadcast the message to the destination. Only the gateway hosts maintain information regarding the gateway membership of entire CDS and local routing tables. Thus, we see that during routing, broadcasting responsibility lies only on the CDS nodes. We briefly touch on two elementary routing protocols using a CDS [5]: shortest path routing and dynamic source routing.

9.2.2.1 Shortest Path Routing

When a non-gateway host needs to send out a packet, it first sends a request to all the available gateway neighbor hosts. Each neighbor gateway host looks into the gateway membership information of the entire CDS and determines which gateway host(s) are neighboring the destination node. Then, by searching the local routing table, it finds the destination gateway that is a minimum distance away from it and replies with this minimum route length to the source. This distance metric could be the number of hops, estimated round-trip time, and others. The source node will pick up the gateway neighbor with the shortest path to the next hop. The routing tables in the subnetwork domain can be constructed and updated using regular distance vector or link state protocols.

9.2.2.2 Dynamic Source Routing

For a particular source–destination pair, if k_1 is the number of source gateways (dominators adjacent to the source node), and k_2 is the number of destination gateways (dominators adjacent to the destination node), then the source can have a maximum of $k_1 \times k_2$ choices of routes if the latter is a non-gateway host. On the other hand, if the source is a gateway host, it has maximum k_2 choices of routes. The source can pick up any route depending on its transmission criteria. Because a source may have several backup routes, whenever a routing error occurs, instead of initiating a new route discovery procedure immediately, the source gateway will try to use backup routes to transmit the packets. Only when all backup routes fail will the source initiate a route discovery procedure.

As discussed in the previous routing protocols, it is evident that the CDS nodes maintain the necessary routing information. Therefore, reduced CDS size effectively saves storage space. Also, a small-sized CDS makes routing easier because it can reduce transmission interference and the number of control messages. Thus, it is desirable to construct a minimal CDS (MCDS) of the network. However, because computing MCDS is a NP-complete problem [4], the challenge is to approximate optimal-sized CDSs using efficient approximation algorithms in polynomial time. Generally, the quality of the CDS is measured by its approximation factor, which is the ratio of its size to that of the MCDS. The CDS construction cost is measured by the overall message and time complexities of the algorithms. For energy-constrained wireless networks, the challenge is to not only construct thinner CDSs but also construct CDSs with low computation and communication costs. As we will see at the end of the chapter, the computation time of the CDS should also be appreciably small to schedule speedy switches between disjointed CDSs to extend battery lifetime and optimize power consumption. With this formal understanding of the challenges in building a CDS, we shall dissect the different approaches found in the literature on the CDS construction problem in this section.

The algorithms discussed in this section are applicable to both static and mobile wireless networks. In mobile wireless networks, as the connectivity of a mobile node with the rest of the network changes due to its movement, the virtual backbone is frequently disturbed due to node mobility. Therefore, for mobile networks, the CDS construction algorithms are applied together with self-reorganization techniques that allow the constructed virtual backbone to reshuffle itself efficiently with minimal changes, as and when needed. The self-reorganization techniques are beyond the scope of this chapter. Interested readers can find the relevant details in a study by Wu and Li [5].

The existing CDS construction techniques in static wireless networks can be broadly classified into two categories based on the network information they use—*centralized algorithms* and *distributed algorithms*. Some localized approaches also exist for CDS constructions, such as the multipoint relay approach used by Adjih et al. [6], but they lack any approximation analysis of algorithm, making them unsuitable for virtual backbone construction for efficient exploit aggregation. However, for smaller networks, they are used as an alternative to flooding. We shall now discuss the popular algorithms from each of these aforementioned classes with examples.

9.2.3 Centralized Algorithms

In their 1998 study, Guha and Khuller [7] first gave a centralized greedy algorithm for CDS construction in general graphs. They provided a recursive algorithm for determining a CDS by gradually growing a tree. The algorithm is summarized below.

Initially, all the nodes were unmarked (white). A node that had the highest number of white neighbors was selected and marked as black. All its white neighbors were then colored as gray. Among all the gray nodes, one that again had the highest number of unmarked white neighbors was converted to black and its white neighbors were changed to gray, and so on. Instead of selecting just one gray node at any step, the algorithm also applies a look-ahead logic and prefers to select the pair of a gray node and its adjacent white node, if changing both of them to black yields more white nodes than changing their color to gray. At any stage, a tie can be broken according to a predecided rule. The algorithm is terminated when there are no more white unmarked nodes. At any instant, the black nodes form the CDS of the graph induced by the black and gray nodes. Eventually, the nonleaf black nodes form the final CDS. Figure 9.2 illustrates the method behind this algorithm.

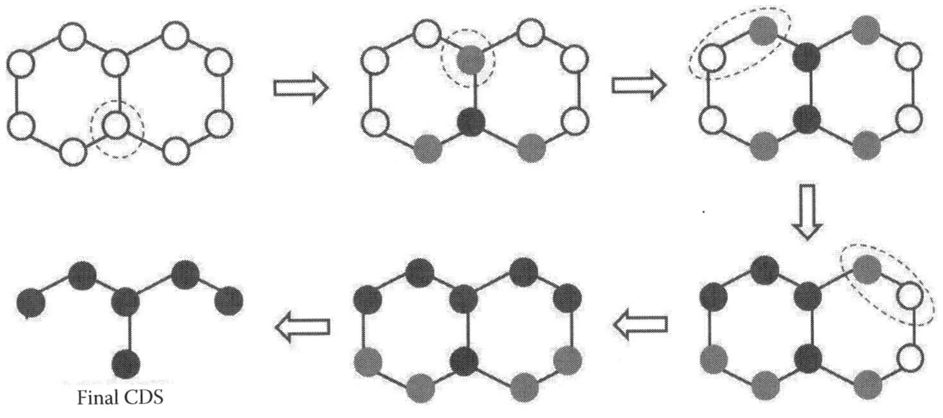

Figure 9.2 Centralized CDS construction.

The algorithm had an approximation ratio of $O(\ln\Delta)$, where Δ is the maximum degree of a node in the graph. It is worth mentioning that because the algorithm is centralized, in general circumstances, it performs quite well in approximating the optimal CDS, at times even outperforming the distributed algorithms. However, because the centralized algorithms require global information from the complete network, they are unsuitable for wireless networks that do not have centralized control. In the latter sections of their article, Guha and Khuller [7] proposed a centralized greedy heuristic to first construct a dominating set, and then interconnecting this dominating set with a Steiner tree, thus forming the CDS of the network. This concept has been extensively used in some of the later works on distributed CDS constructions.

9.2.4 Distributed Algorithms

Before we begin discussing the distributed algorithms for CDS construction, we put forward the following set of premises and assumptions that is common to all the distributed approaches:

1. The nodes do not have any geometric or topological information. They do not even have knowledge of their distances to their neighbors.
2. Each node has a unique ID.
3. Nodes exchange hello messages to identify their neighbors.
4. The resultant topology of the network is modeled as a UDG with the transmission range of the nodes considered to be the same.
5. All the links are bidirectional.
6. The communication overhead due to interference is negligible.
7. Scheduling of transmission is the responsibility of the MAC layer.
8. The computation is partitioned into rounds, where the nodes receive the messages sent in the previous round, execute local computations and send messages to the neighbors in the next round.

All the distributed approaches discussed here for the CDS construction problem follow the above assumptions and require only local information and a constant number of iterative rounds of message exchanges among neighboring hosts.

9.2.5 Unbounded Distributed Construction

The first algorithm for distributed CDS construction was proposed in 1999 by Wu and Li [5]. The algorithm works by first finding a CDS and then pruning certain redundant nodes from the CDS. Initially, each of the nodes exchanges one-hop neighborhood information with all its neighbors. Now, every node can determine if two of its neighbors are mutually adjacent or not. All the nodes, which have at least two nonadjacent neighbors, include themselves in the initial CDS U. In some situations, the initial CDS constructed by this logic becomes trivial and consists of all the nodes in the network. In particular, any vertex-symmetric graph will initially generate a trivial CDS. That is why, in the next step, some redundant nodes are discarded from the CDS, thereby reducing the CDS size. The pruning rule states that any node u in the CDS U is considered as locally redundant and should be pruned if it has either a neighbor in U with a larger ID, which dominates all other neighbors of u, or two adjacent neighbors with larger IDs that, together, dominate all other neighbors of u. The approximation factor of this algorithm was not deduced by Wu and Li [5]. However, in a later report by Wan et al. [8], it was shown that the approximation factor of the above algorithm was $n/2$, where n is the number of nodes in the network. This implies that the

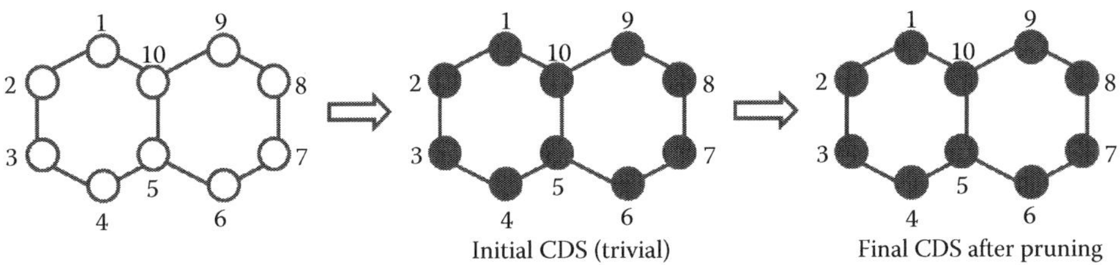

Figure 9.3 Unbounded distributed CDS construction.

above algorithm performs extremely poorly over certain instances. Also, the fact that the pruning rule is also dependent on the IDs of the nodes makes this algorithm undesirable, even for usual situations, where a node can be assigned an ID arbitrarily. Figure 9.3 illustrates the working of this algorithm. Without a constant approximation factor to ensure an upper bound on CDS size, one cannot guarantee to generate a CDS of small size. Also, the message complexity and the time complexity of the approach as reported by Wu and Li [5] are $O(n^2)$ and $O(n^3)$, respectively, which makes the algorithm computationally extremely expensive.

Later on, using the essence of the above algorithm, a modified CDS construction approach was proposed by Stojmenovic et al. [9], in the context of clustering and broadcasting wherein cluster heads were selected to form the dominating set and border nodes were identified to connect the cluster heads. However, the algorithm did not improve the approximation ratio or reduce the CDS construction costs. As shown by Wan et al. [8], this approach has an approximation factor of $n/2$ or n, $O(n^2)$ message complexity, and $\Omega(n)$ time complexity.

The next section discusses a class of algorithms that performs a bounded CDS construction ensuring a constant approximation factor.

9.2.6 *Bounded Distributed Construction*

The first distributed algorithm that ensured a constant approximation factor is the dominating tree construction algorithm in the study by Wan et al. [8]. The important feature of this algorithm is that it guarantees the construction of a MIS, in which the distance between any pair of its complementary subsets is exactly two hops. However, the CDS construction initially requires an arbitrary rooted spanning tree T, which can be constructed by the distributed leader-election algorithm described in the article by Cidon and Mokryn [10] with $O(n)$ time complexity and $O(n \log n)$ message complexity. Given the rooted spanning tree T, the level of a node is the number of hops in T between itself and the root of T, with the root being at level 0. A lexicographic order of ranking is established among all the nodes in which the rank of any node is given by the ordered pair of its level and its ID. Each node identifies its rank in a distributed approach by exchanging a series of messages with its children and parent. Once every node knows its own rank and that of all its neighbors, the root initiates the construction of the MIS by a color-marking process. All nodes are initially marked with a white color. The root first marks itself black and becomes a dominator. It then broadcasts a BLACK message. Upon receiving a BLACK message, a node adds the senders ID to its list of dominators, and if its color is still white, it marks itself gray to become a dominatee and broadcasts a GRAY message that contains its level. Upon receiving a GRAY message, if the rank of the sender is lower than its own, it decrements its number of lower ranked neighbors by 1. If it does not have any lower ranked neighbor, it marks itself black and broadcasts a BLACK message. When a leaf node is marked with either gray or black, it transmits a MARK-COMPLETE message to its *parent*. A parent transmits the MARK-COMPLETE message when it has received the latter from all of its children. This phase ends when eventually the root receives the MARK-COMPLETE from its children. Figure 9.4 shows how the MIS is formed. It can be easily deduced that all the black nodes in this algorithm form a MIS in which every black node will have a two-hop black neighbor and a mutual gray adjacent node. In the next phase, the dominating tree construction begins over the MIS, such that the internal nodes of the dominating tree T would become a CDS. The distributed approach for the dominating tree construction is given subsequently.

Each node maintains a local Boolean variable z, which is initialized to 0 and set to 1 after the node joins the tree T^*. Each node also maintains a local variable parent that stores the ID of its parent in T^* and is initially empty, and a *childrenList* that records the IDs of its children in T^* and is initially empty. The second phase is initiated by a gray neighbor of the root of T, which has the largest number of black neighbors. This can be decided by exchanging one round of messages between the root of T and its gray neighbors and vice versa. The root of T^* first sets $z = 1$ and then broadcasts an INVITE2 message. All other nodes join the tree T^* according to the following principles:

1. Upon receiving an INVITE2 message, a black node with $z = 0$ sets $z = 1$ and sets parent to the ID of the sender, transmits a JOIN message toward the sender, and then broadcasts an INVITE1 message.
2. Upon receiving an INVITE1 message, a gray node with $z = 0$ sets $z = 1$ and sets parent to the ID of the sender, transmits a JOIN message toward the sender, and then broadcasts an INVITE2 message.
3. Upon receiving a JOIN message toward itself, a node adds the ID of the sender to childrenList. Theorem 8 guarantees that whenever there is any black node outside the current T^*, at least one black node would join T^*.

By this logic, eventually all black nodes will join T^*. Consequently, all gray nodes will join T^* eventually. Finally, the internal nodes of T^* form the CDS. Figure 9.5 illustrates how the dominating tree is constructed from the MIS. This particular approach has a $O(n)$ time complexity, $O(n \log n)$ message complexity [9] and an approximation factor of $8|\text{opt}| + 1$, where opt is the size

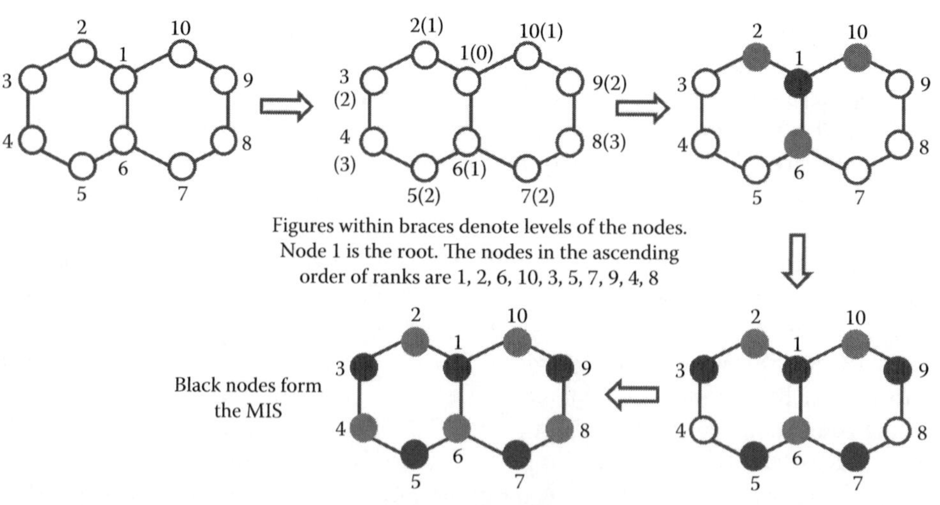
2
10
1
3
9
4
8
6
5
7
2(1)
10(1)
1(0)
3
(2)
9(2)
4
(3)
8(3)
6(1)
5(2)
7(2)
Figures within braces denote levels of the nodes.
Node 1 is the root. The nodes in the ascending
order of ranks are 1, 2, 6, 10, 3, 5, 7, 9, 4, 8
Black nodes form
the MIS

Figure 9.4 Building MIS before dominating tree construction.

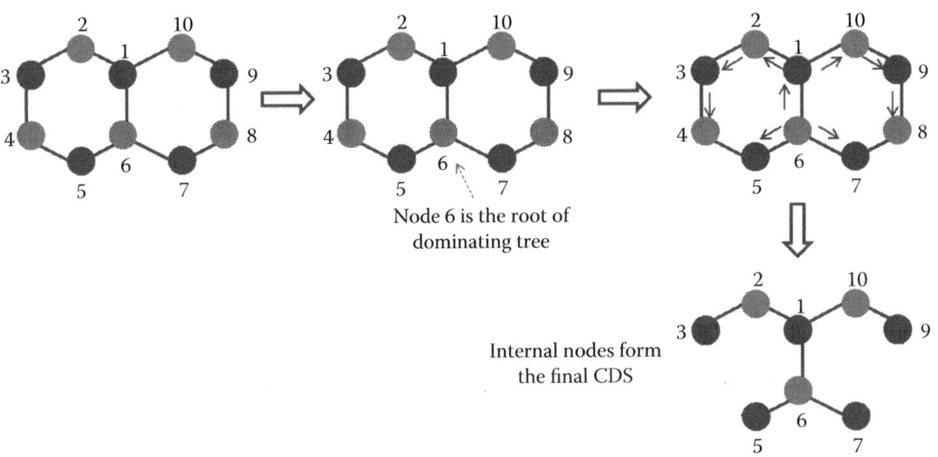
Node 6 is the root of dominating tree
Internal nodes form the final CDS

Figure 9.5 Dominating tree construction from MIS.

of the optimal CDS, as reported by Misra and Mandal [11]. The approximation factor was later improved to 8|opt| by Cardei et al. [12]. Cardei's eight-approximate distributed algorithm grows from a single leader and uses one-hop connectivity information with degree-based heuristics and degree-aware optimization for identifying Steiner nodes as connectors in the CDS construction. The algorithm has $O(\Delta n)$ message complexity, where Δ is the maximum degree of a node, and $O(n)$ time complexity.

The next three distributed algorithms ensure the best approximation factor of $(4.8 + \ln 5)|\text{opt}| + 1.2$ known till date, where opt is the size of the optimal CDS.

The first of this three is the S-MIS algorithm [13]. S-MIS consists of two steps. At the first step, a MIS is constructed with the property that any pair of complementary subsets of the MIS has a distance of exactly two hops. This can be easily assured by using any of the MIS construction algorithms cited by Wan et al. [8], Misra and Mandal [11], and Cardei et al. [12]. Such a MIS indicates that for each pair of nearest dominators in the constructed MIS, the distance between these two nodes is exactly two hops. In the second step, the algorithm employs a greedy heuristic to construct a Steiner tree to tap the nodes in the MIS. Because in a UDG, every node is adjacent to, at most, five independent nodes [14], each Steiner node will be interconnecting a minimum of two and a maximum of five connected components. By the greedy heuristic, the Steiner node connecting the maximum number of different components will be chosen until all the MIS nodes together form a single connected component. The approximation ratio analysis for these stages is discussed later after the other two algorithms. The distributed implementation for the Steiner tree construction [13] is given next.

Let us assume after the MIS construction, that all dominators are marked as black and all dominatees are marked as blue. Each black or blue node carries a z value, which is an identification of the black component that it belongs to, that is, all nodes with the same z value form a black–blue component. Initially, the z value of each black node equals its ID. Gray nodes are ranked based on two values. The first one is the y value, which is the number of the adjacent black nodes in different black–blue components. The second one is its ID. The node with a larger y value is ranked higher. If two gray nodes have the same y value, then the one with a smaller ID is ranked higher. A gray node is adjacent to a black–blue component if it is adjacent to a black node in the black–blue component. A gray node u is a competitor of another gray node v if u and v are adjacent to the same black–blue component. A gray node u is going to change its color to blue if and only if u is ranked higher than every competitor of u. Every gray node keeps two lists, a black list and a competitor list. The black list contains all the adjacent black nodes with their z values, which enables the gray node to compute its y value. The competitor list contains all its competitors and their black lists so that each gray node can also compute the y value of every competitor of it, which enables the gray node to make a decision on whether it should change its color or not. The following events will be triggered under certain conditions:

1. When a gray node u changes its color to blue, all its adjacent black–blue components are merged into one black–blue component and hence their z values should be updated to the same one, say the smallest among them. Meanwhile, all the competitors of u become the competitors of every competitor of u. Therefore, the competitor list of each competitor of u should also be updated. So, after u changes its color, u will send an UPDATE(u) message to all its neighbors. The message contains uID and its two lists.
2. When a black node v receives an UPDATE(u) message, it will update z, send out a COMPLETE(u) message, and pass the UPDATE(u) message to its neighbors other than the nodes that already sent to v the UPDATE(u) or COMPLETE(u) message.
3. When a gray node receives an UPDATE(u) message, it updates both of its black and competitor lists and sends out a COMPLETE(u) message to its neighbors.

The demonstration of a Steiner tree construction is provided in the next algorithm.

The next approach is the collaborative cover heuristic [11], which achieves the same best approximation ratio. In this approach, a MIS is first constructed from a single initiator using effective coverage as a metric instead of degree, ID [12], or other ranking schemes [8]. In the next phase, similar to the S-MIS algorithm, a Steiner tree is also constructed to interconnect the dominators by the same approach as reported by Li et al. [13]. The collaborative cover is particularly useful when the distributions of the nodes are ideal or nearly ideal and uniform. Using effective coverage instead of node degree, collaborative cover can easily identify suboptimal symmetric structures from the node distributions. However, the collaborative cover heuristic has a high message complexity of $O(n\Delta^2)$ and time complexity of $O(n)$. In Figure 9.6, the black nodes represent the dominators and the yellow nodes represent the dominatees, according to the color scheme used by Misra and Mandal [11]. The blue nodes represent the Steiner nodes (connectors). Figure 9.6 shows how the MIS is first constructed and then tapped by the Steiner tree construction.

The last of the distributed algorithms in this context is called pseudo-dominating set construction and Steiner tree spanning (PSCASTS) [15]. PSCASTS provides two fold improvements:

1. Improves the MIS cardinality by a better approximation heuristic
2. Improves the Steiner tree construction by selective pruning of MIS (alternatively called here pseudo-dominating set or PDS) nodes (also known as virtual dominators)

We discuss each of the improvements separately. The last few approaches have constructed a MIS which ensures that for each pair of nearest dominators in the constructed MIS, the distance between them is exactly two hops. However intuitively in a MIS, a node can be separated from its nearest dominator by at most three hops. PSCASTS constructs these MISs with lower cardinalities as PDS to effectively reduce the CDS size further. In the second phase, after Steiner tree construction, some dominators from the MIS can be downgraded to dominatees without any loss in connectivity or coverage of the CDS. This is intuitively true when the neighbors of a dominator

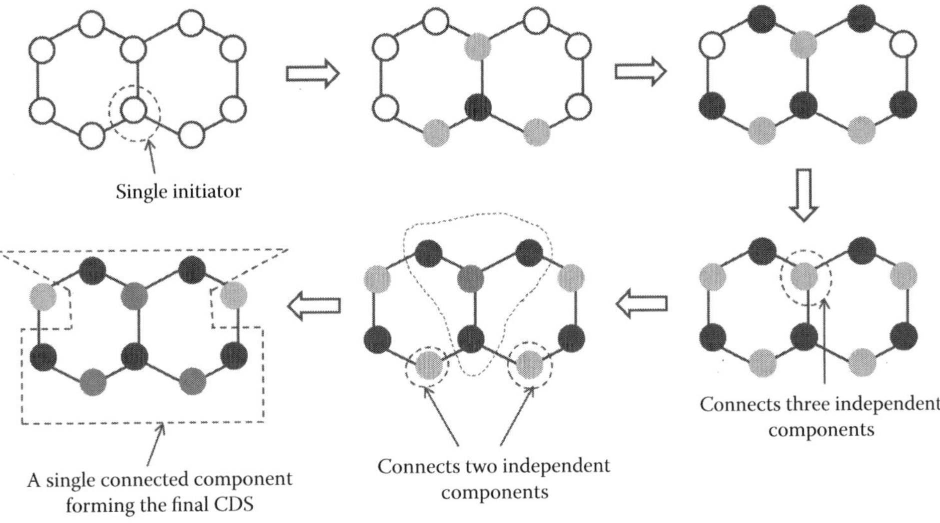

Figure 9.6 Collaborative cover heuristic.

are covered by all the Steiner nodes connecting it. The improved Steiner tree construction selectively discards the redundant dominators or virtual dominators after the initial CDS was formed. The algorithm enjoys the best performance ratio of $(4.8 + \ln 5)|\text{opt}| + 1.2$, with opt being the size of an optimal CDS of the network. It also has the best time complexity of $O(D)$, where D is the network diameter. This stems from the fact that PSCASTS is a multiple-leader initiated process and does not require any leader election, such as that of work done by Wan et al. [8] and Misra and Mandal [11]. The message complexity of the algorithm is $O(n\Delta)$. Figure 9.7 shows the overall workings of the algorithm, following the same color scheme as mentioned by Das et al. [15], in which black nodes represent dominators, yellow nodes represent dominatees, and blue nodes represent connectors.

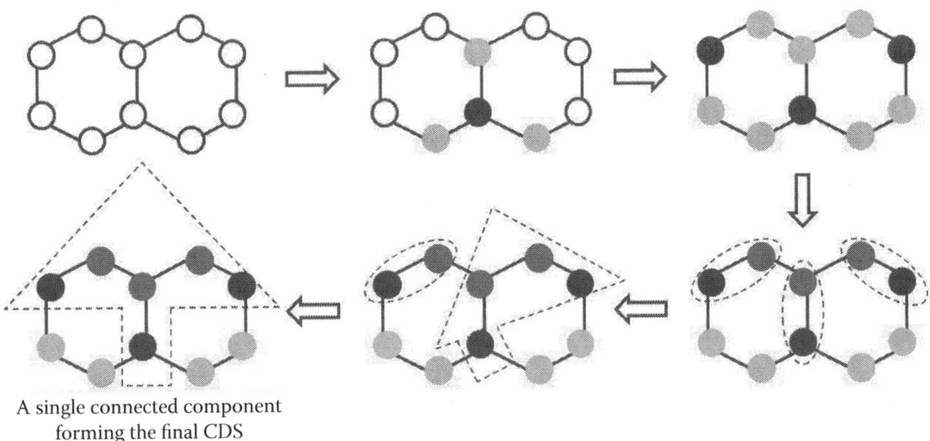

Figure 9.7 PSCASTS.

A simple experiment cited by Das et al. [15] compares the MIS size constructed from collaborative covers and the PDS size constructed from PSCASTS. The results presented in Figure 9.8 substantiate that PDS has smaller cardinality than the MIS selected from collaborative covers, which are some of the best MIS selection algorithms known to date.

The significance of selectively ignoring virtual dominators from PDS is presented in Figure 9.9. The figure shows that nearly all the virtual dominators are discarded after the Steiner tree construction. Occasionally, at most one virtual dominator is retained as a connector for bridging two disjoint components of the CDS. For large network sizes, Figure 9.10 shows that ignoring virtual dominators results in a reduction of around 1/10th of the CDS size.

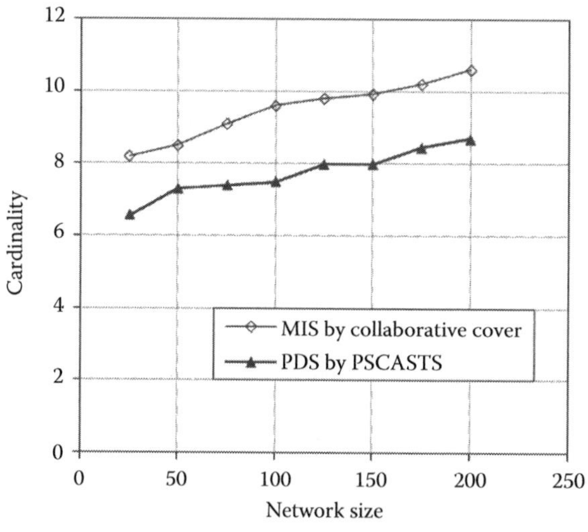

Figure 9.8 Performance comparison of PDS construction phase with MIS selection scheme.

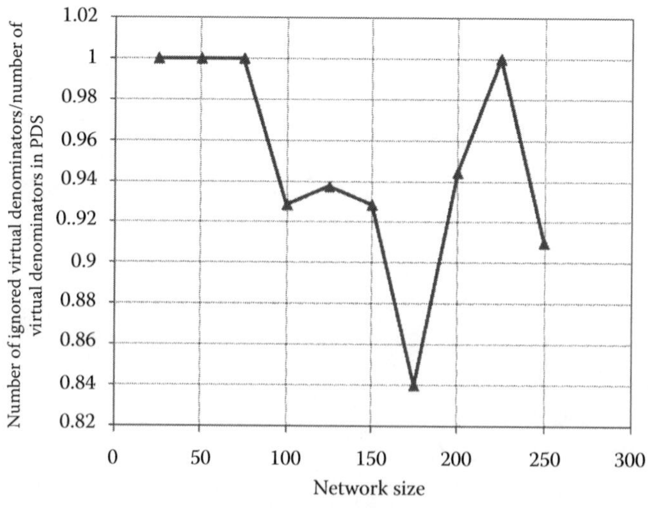

Figure 9.9 Ratio of ignored virtual dominators to total virtual dominators.

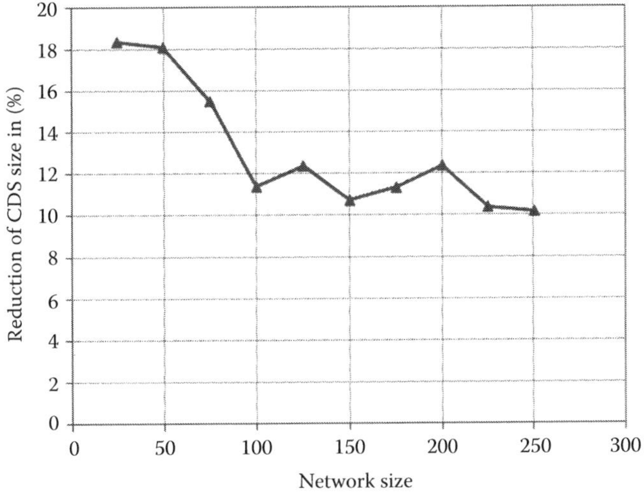

Figure 9.10 Reduction in CDS size by selective removal of virtual dominators.

9.3 Approximation Factor Computation for Bounded Distributed CDS Construction Algorithms

In this section, we provide an idea on how the approximation ratio for the last three distributed approaches was deduced. Before we go further, we put forward a lemma that has been derived using the geometric property of the UDG.

Lemma 9.1

In any UDG, the size of every MIS is upper-bounded by 3.8|opt| + 1.2, where |opt| is the size of the MCDS in this UDG [16]. ■

Another notable observation for the Steiner tree construction is that the size of the Steiner nodes obtained required interconnecting any MIS, in which the two nearest independent nodes are exactly two hops away, is at most (1 + ln 5)|opt|, where |opt| is the size of any optimal CDS [13].

Based on this, we can formulate the following:

$$|CDS| \le |MIS| + |\text{Steiner nodes}|$$

$$|CDS| \le 3.8\,|\,opt\,| + 1.2 + (1 + \ln 5)\,|\,opt\,|$$

$$= (4.8 + \ln 5)\,|\,opt\,| + 1.2$$

Thus, we conclude that the performance ratio of the three algorithms above is (4.8 + ln 5)|opt| + 1.2.

9.4 Performance Comparison for Bounded Distributed CDS Construction Algorithms

In this section, we compare the results of some of the distributed algorithms to assess their performance. The first result analyzes the different distributed approaches with the resultant CDS size. The CDS sizes for the algorithms discussed in several articles [8,11–13,15] are plotted for network sizes of 20, 50, and 100 nodes. The results are presented in Figure 9.11.

The second result that we studied is the number of message exchanges needed for CDS construction. The mean number of messages required for a degree-based approach [12], collaborative cover [11], and PSCASTS [15] is plotted for network sizes varying from 100 to 500 nodes (Figure 9.12). The result is reflective of the discussions regarding message overheads, in the respective works which

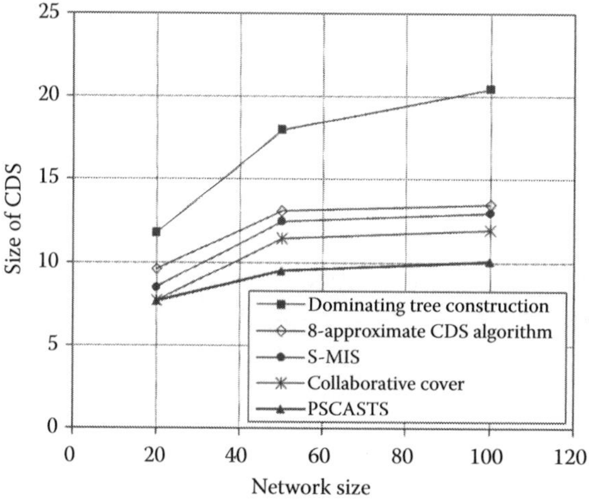

Figure 9.11 Performance comparison of CDS construction algorithms.

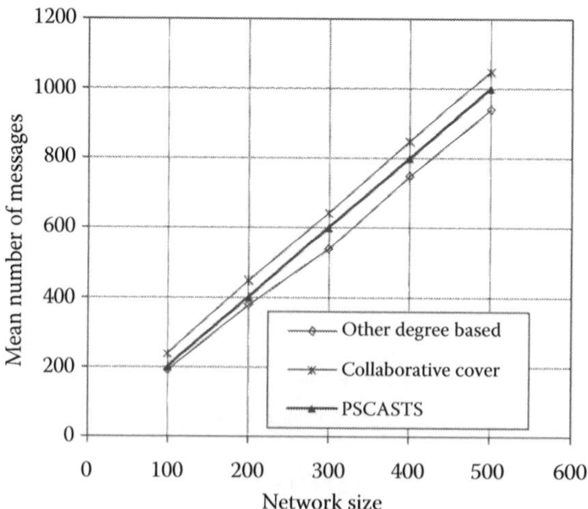

Figure 9.12 Comparison of message exchanges in CDS construction.

state that the message complexity of 8-approximate degree-based CDS scheme and PSCASTS are $O(n\Delta)$ and that of collaborative cover is $O(n\Delta^2)$, where n is the number of nodes in the network and Δ is the maximum degree of a node. Although the theoretical upper bounds on the number of messages for degree-based 8-approximate CDS algorithm [12] and PSCASTS are same, the former actually requires slightly fewer messages than the latter. This is because the 8-approximate CDS technique [12] uses only one-hop connectivity information in contrast with PSCASTS, which uses two-hop neighborhood knowledge to obtain a better CDS.

9.4.1 Aggregation-Based Energy Model

In remote data-gathering applications, a CDS is more particularly used as a data aggregation backbone for in-network data aggregation to optimize network communication, thereby saving communication energy and extending network lifetime. To evaluate the energy profile for data aggregation, we consider the aggregation-based energy model discussed by Misra and Mandal [11]. Let the energy dissipation for aggregation be 5 nJ/bit. This value is drawn from realistic experiments reported in the literature because the energy dissipation required for performing computations to aggregate data is 5 nJ/bit [17]. Table 9.1 summarizes the system parameters used for energy modeling in our simulation.

To evaluate the role of the number of dominators in energy dissipation, we need to compare energy dissipation in the entire network in aggregation CDS with degree CDS. Consider the energy dissipation of nodes in the network represented as E_{dom} for nodes having dominator's role and $E_{non\text{-}dom}$ for the nondominators. The nondominator nodes spend energy $E_{non\text{-}dom}$ to communicate the sensed data to the nearest dominator at distance d within direct transmission radius r_{max} and therefore obey Friss' free space propagation model having an attenuation d^2 with coefficient (α_{friss}). Let E_l be the per bit energy dissipation of transceiver electronics. To transmit a message of m bits at a distance d, the nondominator expends energy:

$$E_{non\text{-}dom} = m \cdot \mathrm{E}_l + m \cdot \alpha_{friss} \cdot d^2 \tag{9.1}$$

Let the dominators dissipate energy E_{dom} in (i) receiving information from dominatees (E_l), (ii) performing aggregation (E_{agg}), and (iii) transmitting aggregate data to the base station ($\alpha_{2\text{-}ray} \cdot d^4$). It may be noted that the average distances d between dominator and base station are much greater than the maximum transmission radius r_{max}. Thus, the network nodes have two modes of communication, that is, a higher range of communication (beyond $d > r_{max}$) and multihop

Table 9.1 Description of Parameters

Parameter	*Value*	*Summary*
E_l	50 nJ/bit	Energy dissipated in transceiver for per bit operation
E_{agg}	5 nJ/bit	Energy dissipated in data aggregation per bit
α_{friss}	10 pJ/bit/m^2	Radio transmitter coefficient for short distances
$\alpha_{2\text{-}ray}$	0.0013 pJ/bit/m^4	Radio transmitter coefficient for longer distances
M	100 m^2	Target area of 100 × 100 m^2
M	1000 bit	Frame size in bit per round of data gathering

communication. Using opportunistic routing if multihop energy dissipation is greater than or higher than the range of direct transmission energy, then higher range transmissions are used which follow the two-ray propagation model with attenuation d^4. Thus, the multihop communication energy is upper-bounded by the energy dissipation of a two-ray propagation model with attenuation d^4. Thus, to transmit m-bit messages after aggregating data from its dominatees in its neighborhood, say |Nbd|, the radio energy E_{dom} expends

$$E_{\text{dom}} = m \cdot E_1 \cdot |\text{Nbd}| + m \cdot E_{\text{agg}} \cdot |\text{Nbd}| + m \cdot \alpha_{2\text{-ray}} \cdot d^4 \tag{9.2}$$

Thus, energy dissipation of a dominator and its dominatee is given by

$$E_{\text{total-dom}} = E_{\text{dom}} + |\text{Nbd}| \cdot E_{\text{non-dom}} \tag{9.3}$$

Therefore, total energy dissipation of a network with |CDS| = k dominators is given by

$$E_{\text{total}} = k \cdot E_{\text{total-dom}} \tag{9.4}$$

Equation 9.4 provides the total energy dissipation of a network in communicating the sensed data to the base station while performing aggregation at the dominators of CDS. Using Equation 9.4, Misra and Mandal [11] conducted an experiment to simulate the CDS construction algorithms for computing network-wide energy dissipation and analyze the effect of the smaller size of CDS on in-network aggregation in energy dissipation of networks. Considering a frame m of size 1000 of sensing data generated from all nodes, which is communicated by our CDS-based aggregation backbone to the base station located centrally inside the target area, the simulation results are captured for a single round of a data-gathering application. The energy dissipation for single-round data communication is compared with the degree-based CDS construction technique of Cardei et al. [12]. The results in Figure 9.13 show the crossover at the early network size of 100 nodes and

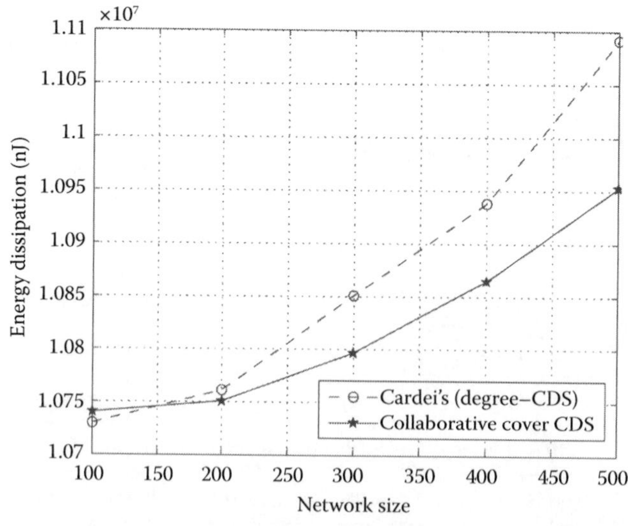

Figure 9.13 Performance comparison of aggregation energy dissipation with degree-CDS algorithm.

beyond the network size of 200 onwards, collaborative cover reduces the dissipation energy substantially of sensed data communication even for a single round. The reduction in the network-wide energy dissipation using collaborative cover heuristics results in an increase of the network lifetime.

Epilogue

CDS construction needs to operate jointly with CDS rotation. Wireless sensor network nodes are operated by batteries, which generally cannot be replenished, thus limiting their lifetime. In addition to this, the backbone nodes have an extra load of computation and communication, thereby depleting their energy resources faster than other nodes of the network. Therefore, CDS construction algorithms should also be equipped to construct disjoint virtual backbones and periodically switch between them, thus effectively performing load balancing and also extending the network lifetime [18].

However, with recent cross-layer protocols in the field of wireless sensor and ad hoc networks, it is assumed that each node performs distributed duty cycle operation, where the duty cycle δ defines the ratio of the time a node is active. Each node is implemented with a sleep frame with length T_S seconds. As a result, a node is active for T_S seconds and sleeps for δT_S seconds. CDS rotation coupled with distributed duty cycles can enhance the network lifetime even more significantly.

References

1. N. Bulusu, D. Estrin, L. Girod, and J. Heidemann. Scalable coordination for wireless sensor networks: Self-configuring localization systems. In *International Symposium on Communication Theory and Applications (ISCTA)*, Lake Ambleside, UK, July 15–20, 2001.
2. D. Rakhmatov and S. Vrudhula. Energy management for battery-powered embedded systems. *IEEE Transactions on Embedded Computing Systems*, vol. 2, no. 3, pp. 277–324, 2003.
3. A. Ephremides, J. Wieselthier, and D. Baker. A design concept for reliable mobile radio networks with frequency hopping signalling. *Proceedings of the IEEE*, vol. 75 pp. 56–73, 1987.
4. B. N. Clark, C. J. Colbourn, and D. S. Johnson. Unit disk graphs. In *Discrete Mathematics*, vol. 86, pp. 165–177, 1990.
5. J. Wu and H. Li. On calculating connected dominating set for efficient routing in ad hoc wireless networks. In *DIALM '99: Proceedings of the 3rd International Workshop on Discrete Algorithms and Methods for Mobile Computing and Communications*, ACM, Seattle, Washington, United States, pp. 7–14, 1999.
6. C. Adjih, P. Jacquet, and L. Viennot. Computing connected dominated sets with multipoint relays. *Ad Hoc and Sensor Wireless Networks*, vol. 1, nos. 1–2, pp. 27–39, 2005.
7. S. Guha and S. Khuller. Approximation algorithms for connected dominating sets. *Algorithmica*, vol. 20 no. 4, pp. 374–387, 1998.
8. P.-J. Wan, K. M. Alzoubi, and O. Frieder. Distributed construction of connected dominating set in wireless ad hoc networks. In *Proceedings of the IEEE INFOCOM '2002*, New York, USA, 2002.
9. I. Stojmenovic, M. Seddigh, and J. Zunic. Dominating sets and neighbor elimination based broadcasting algorithms in wireless networks. In *Proceedings of IEEE Hawaii International Conference on System Sciences*, January 2001.
10. I. Cidon and O. Mokryn. Propagation and leader election in multihop broadcast environment. In *Proceedings of the 12th International Symposium on DIStributed Computing (DISC98)*, Greece, September 1998, pp. 104–119, 1998.
11. R. Misra and C. Mandal. Minimum connected dominating set using a collaborative cover heuristic for ad hoc sensor networks. *IEEE Transactions on Parallel and Distributed Systems*, vol. 21, no. 3, pp. 292–302, March 2010.

12. M. Cardei, M. X. Cheng, X. Cheng, and D.-Z. Du. Connected domination in ad hoc wireless networks. In *Proceedings of the Sixth International Conference on Computer Science and Informatics*, Durham, NC, USA, 2002.
13. Y. Li, M. T. Thai, F. Wang, C. W. Yi, P. J. Wang, and D. Z. Du. On greedy construction of connected dominating sets in wireless networks. *Special Issue of Wireless Communications and Mobile Computing (WCMC)*, vol. 5, pp. 927–932, January 2005.
14. M. V. Marathe et al. Simple heuristics for unit disk graphs. *Networks*, vol. 25, p. 5968, 1995.
15. A. Das, C. Mandal, C. Reade, and M. Aasawat. An improved greedy construction of minimum connected dominating sets in wireless networks. In *Wireless Communications and Networking Conference (WCNC)*, pp. 790–795, 2011.
16. W. Wu, H. Du, X. Jia, Y. Li, and S. C. H. Huang. Minimum connected dominating sets and maximal independent sets in unit disk graphs. *Theoretical Computer Science*, vol. 352, no. 1, pp. 1–7, 2006.
17. W. Heinzelman, A. Chandrakasan, and H. Balakrishnan. An application-specific protocol architecture for wireless microsensor networks. *IEEE Transactions on Wireless Communications*, vol. 1, no. 4, pp. 660–670, 2002.
18. R. Misra and C. Mandal. Rotation of CDS via connected domatic partition in ad hoc sensor networks. *IEEE Transactions of Mobile Computing*, vol. 8, no. 4, pp. 488–499, April 2009.

Chapter 10

Transport Protocols in Wireless Sensor Networks

Prabhudutta Mohanty and Manas Ranjan Kabat
VSS University of Technology

Contents

10.1 Introduction

A wireless sensor network (WSN) is a network of hundreds or thousands of tiny, resource-constrained, inexpensive sensor nodes that can be deployed manually or randomly in a physical space with one or more base stations (sinks). The sensor nodes sense physical information and then process and transmit this sensed information over a wireless medium to the sink. The sink can send queries to the sensor nodes to fetch information. The WSNs have several distinct features [1]. First, it has a unique network topology. The sensor nodes are generally organized in a multihop star-tree topology in which the base station is at the root of the tree. The base station is responsible for gathering information and relaying this information to the external nodes over wired or wireless media. This network topology may vary due to time-varying link conditions and node variations. Second, WSNs are resource constrained. The sensor nodes have limited resources including low processing capability, small memory, low wireless communication bandwidth, and a limited, usually nonrechargeable battery. Third, the WSNs have distinct traffic characteristics. In WSNs, the primary traffic is a many-to-one type of communication, that is, in the upstream direction from the sensor nodes to the base station. Upstream traffic delivery can be classified into continuous, event-driven, query-driven, or hybrid data delivery models. These data delivery models have different quality of service (QoS) and reliability requirements depending on their specific applications. For example, in temperature sensing, event-driven delivery is used, whereas in target tracking, continuous delivery should be used. The sink generates certain downstream traffic only for the purposes of query and control. The query-driven data delivery model is similar to the event-driven model except that the data is pulled by the sink whereas the data is pushed to the sink in the event-driven model. The applications need to receive these desired data reliably and as quickly as possible. Some WSNs apply a hybrid model using a combination of continuous, event-driven and query-driven data delivery models. Another feature of WSN is that the sensor nodes might generate significant redundant data (similar packets from multiple nodes), which can be aggregated using functions such as suppression (eliminating duplicates), min, max, and average [2] so that the number of transmissions would be reduced. This technique has been used to achieve energy efficiency and traffic optimization in a WSN. Finally, WSNs use messages with a very small size to reduce their processing times and conserve energy during data transmission. The need for segmentation is also eradicated in WSNs because of the small message size. The WSNs have attracted many researchers over the years due to these features.

The WSN can be used in different environments, supporting diverse applications from home automation and traffic control to more complex and critical applications like battlefield monitoring, shooter localization, industrial process control, patient observation, habitat supervision, and environmental monitoring, etc. [3]. The practical realization of the current WSN applications depends on the energy-efficient, real-time and reliable communication capabilities of the WSN. Recently, communication protocols [10–18] have been proposed to address the energy-efficient and reliable data communication requirements of WSN. However, in addition to energy efficiency and communication reliability, there exist many proposed WSN applications that have strict delay bounds and hence mandate timely transport of the event features from the field. Many of the potential WSN applications such as real-time target tracking, homeland security, process control, and control of vehicle traffic in highways necessitate reliable event transport to be achieved within a certain application-specific delay bound. For instance, the accuracy and effectiveness of military WSN applications such as border surveillance and intrusion detection are directly related to the timeliness of reliable event detection at the military decision center. The late detection of a certain event leads to the failure of the ultimate objectives of the deployed WSN for such applications. Therefore, the communication protocols that only consider energy efficiency and transport reliability are not capable of addressing the needs of real-time applications. In WSNs, packet loss is caused due to the poor quality of wireless channels, sensor failure, and congestion. Thus, the WSNs must guarantee certain reliability at the packet or application level through loss recovery to guarantee successful and correct delivery of information. Some critical applications need reliable transmission of each packet and thus packet level reliability is needed. Other applications need only a proportionally reliable transmission of total packets and thus application reliability is needed. Most of these applications are deemed useful when these networks collaborate with wired networks or other wireless networks. Commercial deployment of WSN demands that problems such as energy conservation, congestion control, fairness, reliability data dissemination, security [4], and topology management should first be overcome. These problems can be classified as one or several top-down layers—from the application layer to the physical layer—and can be studied separately in each corresponding layer or collaboratively across each layer. For example, congestion control, fairness, and reliability may be involved in only transport layer, but energy conservation may be related to all the layers. Transport control protocols recently gained the attention of some researchers due to the importance of congestion control, reliable data dissemination, and energy conservation for WSNs.

Transport protocols are used to mitigate congestion and reduce packet loss, to provide fairness in bandwidth allocation, and to guarantee end-to-end reliability. There are two transport protocols that are commonly used over the Internet, that is, user datagram protocol (UDP) and transmission control protocol (TCP). Because UDP does not provide reliability or congestion control for data transport, it is not considered as a suitable protocol to transport sensor and control data for WSNs. TCP is the most widely used protocol to provide end-to-end reliable data transmission in the Internet. However, TCP may not be suitable in WSNs to transport data between sensors and the sink. The reasons are summarized as follows:

- Overhead associated with TCP connection establishment may not be supported by event-driven applications.
- Flow and congestion control mechanisms of TCP results in unfair bandwidth allocation and data collections for WSNs.
- TCP is a strictly end-to-end reliability model, which forces all confirmations and retransmissions to follow the complete path between source and destination.
- It is well known that the inability of TCP to distinguish between congestion losses and transmission losses results in poor performance over many wireless networks.
- The end-to-end congestion control in TCP requires a long time to mitigate congestion compared with hop-by-hop congestion control. TCP may lead to high packet loss when congestion occurs.
- TCP guarantees the successful transmission of packets, which is not always required for event-driven applications in WSN.

Although UDP is a connectionless transport control protocol, it is still not suitable for WSNs considering the following reasons:

- UDP does not address any flow control and congestion control mechanisms. If UDP is used for WSNs, it will cause lots of datagram dropping when congestion happens. Thus, UDP is not energy-efficient for WSNs.
- UDP comprises neither ACK mechanism nor any reliability mechanism. The datagram loss can only be recovered by lower medium access control (MAC) algorithms or upper layers including the application layer, which is not energy-efficient for WSNs.

10.2 Guidelines for Designing Transport Protocols in WSNs

Transport protocols for WSNs should provide end-to-end reliability and end-to-end QoS in an energy-efficient manner. The performance of transport protocols for WSNs can be evaluated using metrics such as energy efficiency, reliability, QoS (e.g., packet loss ratio, packet delivery latency), and fairness.

10.2.1 Performance Metrics

- *Energy efficiency:* The sensor nodes are powered by a nonrechargeable battery or capacitors, which have limited energy. Thus, the transport protocol plays a vital role in achieving high levels of performance together with prolonged sensor network lifetime. The packet loss for loss-sensitive applications leads to retransmission and the inevitable consumption of additional battery power. The retransmission can be minimized by controlling the number of packet retransmissions, the distance (e.g., hop) for each retransmission, redundant data transmission and the overhead associated with control messages (CMs).

- *Reliability:* Reliability can be measured with different parameters [5] such as congestion control, which involves congestion detection and avoidance, loss detection and recovery, and packet delivery ratio. In many applications of WSNs, reliability is determined by the quantity of data packets delivered to the base station (BS) rather than the reliability of each data packet delivered to the BS [17]. Reliability can be achieved by retransmitting the data during data loss. The number of retransmissions increases the amount of redundant data and the probability of congestion in the network. The packet loss and poor loss recovery schemes may lead to significant packet delivery delay in the network. Hop-by-hop data recovery maximizes efficiency compared with the end-to-end recovery scheme. Reliability in WSNs can be classified according to the following categories:
  - *Packet reliability:* In loss-sensitive applications, successful transmission of all packets or at a certain success ratio is required.
  - *Event reliability:* In event reporting [32] applications, successful event detection, but not successful transmission of all packets, is required.
  - *Destination reliability:* In some applications, messages might need to be delivered to sensor nodes in a specific subarea, to nodes that cover a specific subarea, or to nodes that are equipped with a particular sensor type such as sensor-to-cluster head and sensor to actor node, etc.
- *QoS metrics:* The QoS metrics include bandwidth, delay, and packet loss ratio. These metrics or their variants could be used for WSNs. For example, depending on the applications, if sensor nodes are used to transmit continuous images for target tracking, then these nodes generate high-speed data streams and require higher bandwidth than most event-based applications. The WSNs may also require timely delivery data for a delay-sensitive application.
- *Fairness:* Sensor nodes are scattered in a geographical area. Due to the many-to-one convergent nature of upstream traffic, it is difficult for sensor nodes that are far away from the sink to transmit data. Therefore, transport protocols need to allocate bandwidth fairly among all sensor nodes so that the sink can obtain a fair amount of data from all the sensor nodes.

10.2.2 Congestion Control

There are two main causes for congestion in WSNs. The first is due to the packet arrival rate exceeding the packet service rate. The second cause is link-level performance aspects such as contention, interference and bit error rate. This type of congestion occurs on the link. Congestion in WSNs must be efficiently controlled, either to avoid or mitigate it. Typically, there are three mechanisms that can deal with this problem: congestion detection, congestion notification, and rate adjustment.

- *Congestion detection:* In TCP, the congestion is observed at the end nodes based on a timeout or redundant acknowledgments. The common mechanisms use queue length [6], packet service time [7], or the ratio of packet service time over packet interarrival time at the intermediate nodes [8]. For WSNs, channel loading can be measured using carrier sense multiple access (CSMA)–like MAC protocols and can be used as an indication for determining congestion.
- *Congestion notification:* After detecting congestion, transport protocols need to propagate congestion information from the congested node to the upstream sensor nodes or the source

nodes that contribute to congestion. The approach to disseminating congestion information can be categorized into explicit congestion notification and implicit congestion notification. The explicit congestion notification uses special CMs to notify the involved sensor nodes of congestion. On the other hand, the implicit congestion notification piggybacks congestion information in normal data packets. By receiving or overhearing such packets, sensor nodes can access the piggybacked information.

- *Rate adjustment:* Upon receiving a congestion indication, a sensor node can adjust its transmission rate. It can be done either by using the additive increase/multiplicative decrease (AIMD) scheme [7,8] or accurate rate adjustment.

10.2.3 Loss Recovery

In wireless environments, both congestion and bit error could cause packet loss. This not only deteriorates end-to-end reliability and QoS but also lowers energy efficiency. Other factors that result in packet loss include node failure, wrong or outdated routing information, and energy depletion. To overcome this problem, one can increase the source sending rate or introduce retransmission-based loss recovery. The first approach works well for guaranteeing event reliability in event-driven applications that require no packet reliability. However, this method is not energy efficient compared with loss recovery. The loss recovery method is more active and energy efficient, and can be implemented at both the link and transport layers. Link layer loss recovery is hop-by-hop whereas the transport layer recovery is usually done end-to-end. Loss recovery consists of loss detection and notification and retransmission recovery.

- *Loss detection and notification:* The packet sequence number attached in each packet header plays a vital role in loss detection. The continuity of the packet sequence number is used to detect packet loss. The loss detection and notification mechanism used is either end-to-end or hop-by-hop. In the end-to-end method, the destination node is responsible for loss detection and notification, whereas in the hop-by-hop method, intermediate nodes detect and notify the packet loss. Hop-by-hop loss detection and notification conserves more energy as compared with the end-to-end method due to the local loss recovery scheme. The sender can be notified of the loss in three ways, that is, acknowledgment (ACK), negative acknowledgment (NACK), and implicit acknowledgment (iACK). ACK and NACK relay special packets, on the other hand, iACK piggybacks ACK in the packet header.
- *Retransmission recovery:* Lost and damaged packets are recovered by retransmitting the packet. Retransmission is either end-to-end or hop-by-hop. In end-to-end retransmission, the source retransmits the lost packet whereas intermediate nodes retransmit the lost packet in the hop-by-hop method.

10.2.4 Design Guidelines

Several factors influence the design aspects of energy efficient transport protocols. Therefore, WSN must consider topology, traffic characteristics, application requirements, resource constants, and fairness. The transport protocol needs to provide high energy efficiency, flexible reliability, and QoS requirements such as throughput, packet loss rate, and end-to-end delay. In this chapter, our major focus is to present various transport protocols designed for WSN.

10.3 Issues and Challenges of Data Transport over WSNs

In this section, we present the major issues and challenges for designing transport protocols for WSN.

10.3.1 Issues

The transport control protocol of WSN should consider the following factors:

- It should provide congestion control mechanism and guarantee reliability. Because data streams are flowed from sensor nodes to the sink in WSNs, congestion might occur around the sink. There are also some high-bandwidth data streams produced by multimedia sensors. Therefore, it is necessary to design effective congestion detection, congestion avoidance, and congestion control mechanisms for WSNs. Although the MAC protocol can recover packets lost from bit errors, it has no way of handling packets lost from buffer overflow. Then, the transport protocol for WSNs should have a mechanism for packet loss recovery such as ACK and selective ACK, which is used in TCP protocols to guarantee reliability. It would be better to use a hop-by-hop mechanism for congestion control and loss recovery because it can reduce packet dropping and conserve energy. The hop-by-hop mechanism can decrease the buffer requirement in intermediate nodes, which is helpful for sensor nodes with limited memory.
- Transport control protocols for WSNs should simplify the initial connecting process or use connectionless protocols to speed up the start and guarantee throughput and lower transmission delay. Most of the applications in WSNs are reactive, passively monitoring and waiting for events to occur before reporting to the sink.
- The transport control protocols for WSNs should avoid dropping packets as much as possible because packet dropping causes energy wastage. To avoid packet dropping, the transport protocol can use active congestion control (ACC). The ACC can trigger congestion avoidance before congestion occurs. An example of ACC is to make the sender (or intermediate nodes) reduce their sending (or forwarding) rate when the buffer size of their downstream neighbors overruns a threshold.
- The transport control protocols should guarantee fairness for different sensor nodes so that each sensor node can achieve fair throughput. Otherwise, the biased sensor nodes cannot report the events in their area and the system may misunderstand that there is no any event in the area.
- It would be better if the transport protocol can enable cross-layer optimization. For example, if the routing algorithm can inform the transport protocol of failure in the route, then the transport protocol will know the packet loss is not from congestion but from route failure and the sender will freeze its status and keep its current sending rate to guarantee high throughput and low delay.

10.3.2 Challenges

In an ideal system, with large bandwidths and very low loss rates, it would not be difficult to provide the required reliability at QoS guarantee. The task of designing transport protocols to provide the above functionalities is challenging because of the following practical factors and limitations:

- *Channel loss:* Due to signal decay and multipath fading effects, the error rates may be quite significant in WSN links. Furthermore, these loss rates are very much a function of the exact location of the nodes as well as the environment, and can therefore fluctuate greatly over space and time.
- *Interference:* The error rates on wireless links are also very much a function of the level of interfering traffic in the vicinity. When traffic rates are high, packet losses can occur, when the signal to inference noise ratio (SINR) drops below the successful reception threshold because of interference.
- *Bandwidth limitation:* Even If individual sensors generate data at a low rate compared with the maximum data rate available, the aggregate traffic of a large-scale network can be large enough to cause congestion in these low-bandwidth networks. This happens particularly when all traffic is headed to the same destination, resulting in bottlenecks near the sink.
- *Traffic peaks:* Although the average data rate in a WSN is low most of the time, there may be a much higher peak traffic rate whenever important events are being detected. Such traffic peaks are likely to be highly correlated in both space and time—resulting in congested hotspots.
- *Node resource constraints:* The intermediate nodes in the network may also suffer from other constraints that can affect transport techniques: (i) low computational processing capabilities that preclude high-complexity approaches, (ii) memory/storage constraints that limit the size of the message buffer (causing packet losses during congestion events), and (iii) energy constraints that limit the amount of possible transmissions.

10.4 Transport Control Protocols for WSNs

There are several transport control protocols [9–40] designed for WSNs. These protocols aim to address congestion control, reliability, fairness, delay, and both congestion control and reliability. These protocols are categorized into three types.

- Protocols for reliability [9–17]
- Protocols for congestion control [18–28]
- Protocols for both congestion control and reliability [32–40]

10.4.1 Protocols for Reliability

Traffic from many WSN applications is considered to be loss tolerant. Loss tolerance in WSNs is due to the dense deployment of sensors and data aggregation properties, giving rise to directional reliability. The design of WSN transport control protocols should exploit directional reliability to minimize the number of transmissions and decrease the computational overhead by lowering the amount of data to be aggregated. Some transport layer protocols [9–17] offer upstream (from sensors to sink) message delivery and some investigate downstream (sink to sensors) reliability.

10.4.1.1 Pump Slowly/Fetch Quickly

The pump slowly/fetch quickly (PSFQ) protocol [9] aims to distribute data hop-by-hop from sink to sensors (downstream) to meet the unique resource challenges of WSNs with a focus on reliability, scalability, and robustness. PSFQ is designed for retasking/reprogramming applications which require reliable delivery of all message segments (i.e., 100% packet reliability in downstream direction). Its focus is on the transport of binary images, such as new sensor control programs used for sensor retasking in the field. PSFQ achieves loose delay bounds while minimizing the loss recovery cost by localized recovery of data among immediate neighbors. It expects low network traffic and does not provide any ACC scheme.

PSFQ contains three protocol functions, that is, message transmission (pump operation), local loss recovery (fetch operation), and selective status reporting (report operation). In the pump operation, sink slowly injects packets into the network until all the data segment has been sent out. The injected packet contains fields like file ID, file length, sequence number, time to live (TTL), and report bit. PSFQ adapts a simple scheduling scheme, which uses two time instants, T_{max} and T_{min}, to control the timely distribution of code segments to all the sensor nodes. This provides basic flow control so that the reprogramming operation does not obstruct the regular operation of the sensor network. A user node sends a packet to its neighbor after every T_{min} time until all the data fragments have been sent out. After receiving this packet, neighbor sensor nodes will check against their local data cache and discard any duplicates. The new message is buffered and TTL is decreased by 1. If the TTL value is not zero and the packets are in sequence, then it schedules to forward the message. To avoid collision, the forwarding of packets to its neighbor is delayed for a random period between T_{min} and T_{max}. A packet propagates outward from the source node up to TTL hops away in this mode.

Reliability in PSFQ is achieved with a NACK based on quick fetch operation (Figure 10.1). A sensor node can go into fetch mode once a loss is detected using the gap sequence. Each message has a sequence number in the message header. If a receiving node determines a gap in the sequence number, it issues NACK in a reverse path to recover the missing fragment. An attempt is made to recover the lost message before the injection interval T_{max} is exceeded and the next packet is sent. PSFQ buffers the messages until the lost data segments have been recovered. In loss recovery technique, the neighbor node simply forwards messages back to the intended destination. Figure 10.1a shows lost packet recovery from the neighbor node. If the lost packet has been removed from the

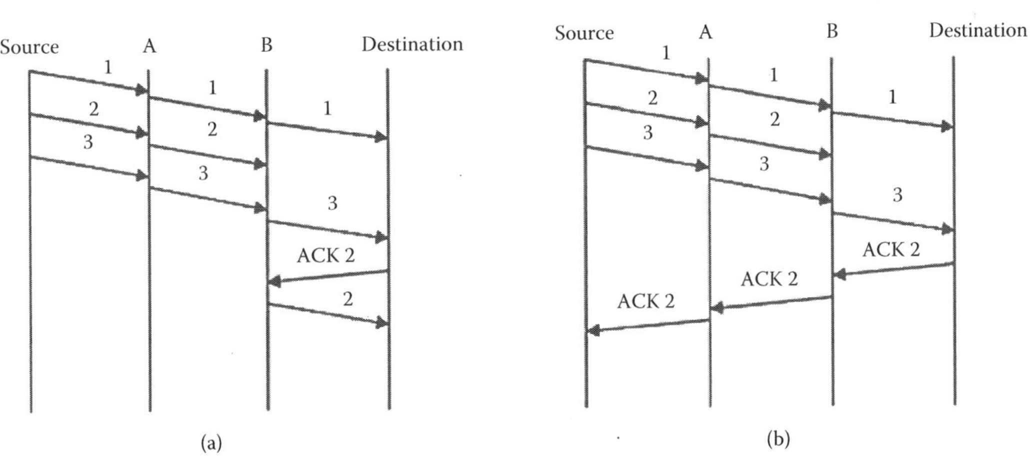

Figure 10.1 (a) A best case scenario. Neighbor B has cached packet 2 and simply forwards it back to the intended destination. (b) A worst case scenario. The source sends a three-part message. Packet 2 is dropped by node B, but by the time ACK 2 reaches B, it has already forwarded packet 3. Thus, the acknowledgment to resend packet 2 has to be sent all the way back to the source. (From J. Jones and M. Atiquzzaman, *Journal of Distributed Sensor Networks*, vol. 3, no. 1, pp. 119–133, 2007.)

cache memory, the NACK must be forwarded all the way back to the source. Figure 10.1b shows lost packet recovery from the source node.

The report operation is designed to feed back data delivery status to users in a simple and scalable hop-by-hop manner. This is an optional operation that minimizes the communication costs of sending a long message for feedback purposes. A node enters the report mode when it receives a data message with a report bit set in the message header. Only the last hop node (i.e., TTL = 1) will respond immediately by initiating a report message that is sent to its parent node (i.e., node from which it receives messages). Each node along the path toward the source node will piggyback their report message by adding its own status information into the report and then propagate the aggregate report toward the user node. To avoid looping, each node will ignore the report if its own ID is found in the report.

PSFQ has several disadvantages such as PSFQ does not address packet losses due to congestion and cannot detect the loss of a single/first packet because it used only NACK and not ACK. Figure 10.2 shows the loss of a single/first packet and proactive fetch after an interval to retrieve the lost last data segment. PSFQ gives good performance in a chain scenario in which a sensor node has only two neighbors, but it floods the network with data packets in the fetch phase using a NACK message in a randomly distributed network. It may not be efficient for reliable transport of multipoint-to-point sensor events (upstream). The PSFQ protocol uses static and slow pumps, which result in larger delay. The hop-by-hop recovery scheme needs more buffers to cache messages.

10.4.1.2 Improved PSFQ

The improved PSFQ [10] is proposed to overcome the shortfalls of PSFQ [9] caused by sequence forwarding and pump and repair operations. The PSFQ strictly follows in-sequence forwarding of data packets, because of which other nodes remain idle until they get in-sequence data packets. This increases the latency of the protocol. PSFQ aggregates the loss of continuous data packets into size 20 windows. PSFQ is unable to recover continuous lost packets when wireless channel error rate is high. This degrades the reliability of PSFQ due to in-sequence forwarding.

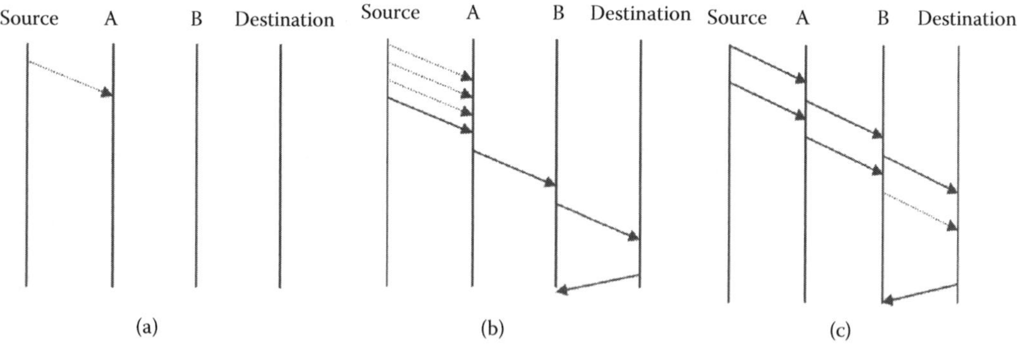

Figure 10.2 (a) A message consisting of a single data segment is sent from the source and is never received at node A. Because no data is ever received at node A, nothing can be recovered. (b) All data segments up to the last data segment are lost. The destination receives the last data segment and is able to NACK for retransmission of all the lost data segments at once. (c) The last data segment is lost. The destination creates a proactive fetch after an interval to retrieve the lost data segment. (From J. Jones and M. Atiquzzaman, *Journal of Distributed Sensor Networks*, vol. 3, no. 1, pp. 119–133, 2007.)

The pump and repair operation does not reschedule transmission if their scheduled time expires. The node will cancel the sending of a data packet that was pending due to the unavailability of radio components. A node treats the unavailability of radio the same as a lost data packet that causes the initiation of loss recovery at the next node. When wireless channel error rate is lower, this loss recovery causes more delay at the target node in getting all the data packets, if the radio component is busy, repairing a particular data packet will fail. PSFQ does not reschedule the data packet for forwarding and thus loss recovery occurs in the next node recovering the packet. This will increase delay at the target node and increase latency significantly at a high error rate in wireless channels.

The improved PSFQ introduces two schemes, that is, out-of-order forwarding and rescheduling of pump and repair packets to overcome the lacunae of PSFQ. The improved PSFQ also solves the problem of NACK implosion, which arises from the proposal itself. Out-of-order forwarding of data packets minimizes the unnecessary delay caused by in-sequence forwarding of data packets. The sensor node performs a duplicity check for the data packets received. All duplicate data packets are discarded whereas new data packets (nonduplicate) are cached and schedule for transmission. The improved PSFQ caches the data in an array called a "buffer" to store and assign sequence numbers for the nonduplicate data (Figure 10.3). This array has two pointers "curForwardIndex," which returns the sequence number of the data packet to be forwarded, and "nextFreeForwardIndex," which shows the location sequence number of the newly arrived data packets. The sequence number of the next data to be forwarded is stored in a variable called "nextToFwd" in PSFQ. The nextToFwd variable is increased by one every time for a successful in-sequence data forwarding or abandoned due to unavailability of radio. When a new data packet arrives, if the buffer is empty and nextToFwd is currently in use, then the node will save the sequence number of the newly arrived data packets at the end of the buffer using nextFreeForwardIndex pointer. Otherwise, it stores the sequence number of the newly arrived packet into nextToFwd and sets a schedule for forwarding packets. After forwarding a packet, the node checks its buffer for the next forwarding until the buffer is empty. The node sorts the array "buffer" when a data packet is forwarded to reduce the gap between the sequence numbers at the next nodes.

In PSFQ, whenever a schedule expires for pump operations, the node checks whether radio components are available or not. If it is unavailable, then the node will reschedule that data packet and nextToFwd will not be updated so that the same packet will be forwarded on the expiry of schedule. The repair operation also reschedules the repair of data packets if radio is unavailable at the time of repairing the data packet. The rescheduling of pump and repair assures 100% packet forwarding and repairing at least once.

In improved PSFQ, out-of-order forwarding results in a number of NACKs from intermediate nodes. Each intermediate node starts loss recovery and sends NACK for gap data packets. This aggravates the NACK implosion problem. A variable called "FullRcvdSeqNum" keeps track of the

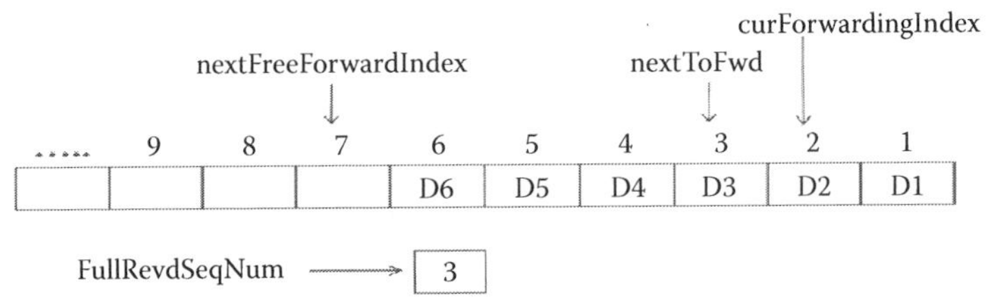

Figure 10.3 Buffer structure.

sequence number for the current data to be sent, which is used to remove the NACK implosion. The sender node piggybacks its FullRcvdSeqNum with every data packet it sends to the receiver node through pump operation. Data packets may be in-sequence or out-of-sequence. The receiver node compares its own FullRcvdSeqNum with the FullRcvdSeqNum of the sender node. If its own FullRcvdSeqNum is less than the FullRcvdSeqNum of the sender node then the receiver node will send NACK only for those data packets that are missing. The receiver node prepares different windows for different sequence number gaps in data packets, which is called *loss aggregation*. It includes all these windows into NACK. The receiver node sends NACK to the sender node. When the sender node receives a NACK, it extracts various windows from NACK. The sender node will process the window only if the sender node has all the data packets specified into the window. The sender node will cancel any window if any of the data packets specified into the window are not available. In this way, the sender node repairs only those data packets that it has. The improved PSFQ protocol improves error tolerance and average latency over PSFQ.

10.4.1.3 Reliable Multiple-Segment Transport

Reliable multiple-segment transport (RMST) [11] is designed to support the reliable transport layer with a hop-by-hop recovery scheme that guarantees transportation of the fragments of data reliably from sensors to the sink (upstream) for applications (Figure 10.4). RMST operates on top of the

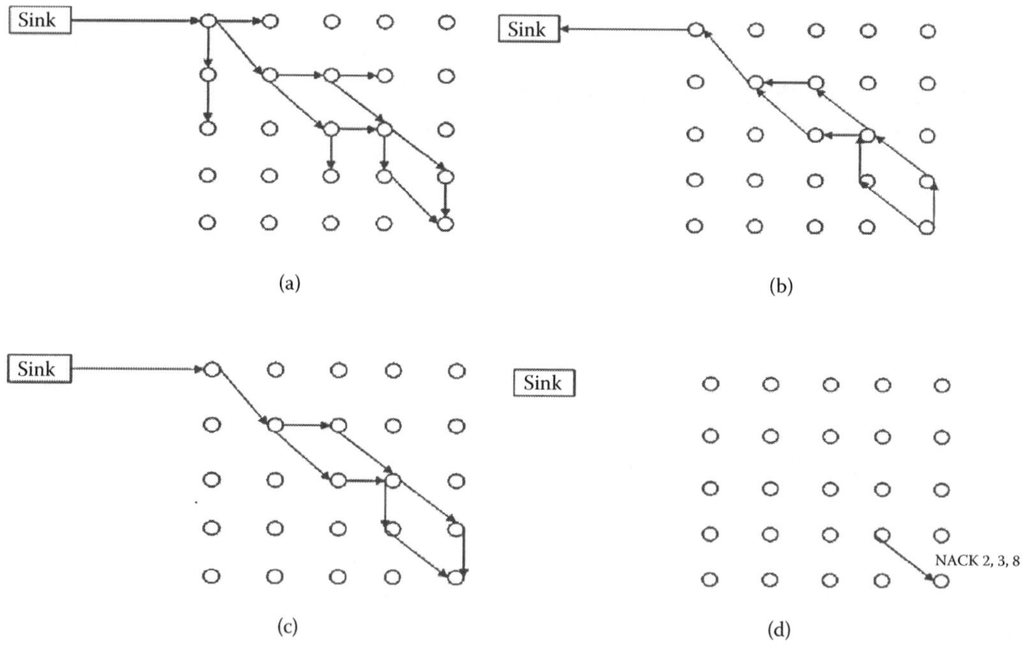

Figure 10.4 An example of the RMST protocol. (a) Interest is disseminated through the network, using directed diffusion. (b) A reinforced gradient path is established between the source and the sink. (c) RMST snoops on the reinforced path at each hop and uses it to establish a backchannel for NACK that needs to be sent in the source's direction. (d) RMST allows nodes to NACK multiple fragments of a message at once. Here, the node is sending NACK 2, 3, and 8 asking for retransmission of packet fragments 2, 3, and 8, respectively. (From J. Jones and M. Atiquzzaman, *Journal of Distributed Sensor Networks*, vol. 3, no. 1, pp. 119–133, 2007.)

gradient mechanism used in directed diffusion [12]. Directed diffusion is used to discover paths from the sensors to the sink. Similar to PSFQ [9], RMST is a NACK-based protocol that primarily uses timer-driven loss detection and repair mechanisms. The basic mechanism for loss detection is a watchdog timer. The main goal of RMST is to minimize the cost of end-to-end retransmission by introducing hop-by-hop retransmission from caches of neighbor nodes. RMST adds two important features to directed diffusion, that is, fragmentation and reassembly of segments, and reliable message delivery. RMST uses the term fragmentation and reassembly, which simply means that the packets originating from a source node (called the RMST entity) are fragmented and then reassembled at the base station. Fragmentation is necessary to adjust the size of the maximum transmission unit (MTU) permissible by the transit nodes. The notion of reliability adopted by RMST is the reliable delivery of fragments originating from any particular RMST entity to the base station.

RMST introduces two modes of operation: caching mode (with hop-by-hop recovery) and noncaching mode (with end-to-end recovery). RMST uses timer-driven mechanisms to detect data loss and send NACK on the way from detecting node to sources (cache or noncache mode). In RMST, loss detection/recovery mechanisms employ a NACK for gap detection to detect and recover lost messages, similar to the scheme used by PSFQ. However, RMST makes no guarantee of in-order message delivery, rendering loss detection is particularly difficult because it is difficult for sensor nodes to determine whether gaps are caused by out-of-order delivery or lost messages. In caching mode, the sink node and all intermediate nodes on an enforced path cache segments and check the cache periodically for missing segments. When a node detects missing segments, it generates a NACK message that travels back to the source along the reinforced path. The noncaching mode reduces the overhead of transport layer. In this mode, the source and the base station maintain the cache and the base station monitors the integrity of an RMST entity in terms of the received fragments. RMST uses selective NACK-based protocol to detect a fragment loss and sends NACK from the detecting RMST node to the source node. Each RMST entity receiving the NACK first looks at its cache to find out the missing segment. In the negative case, it forwards the NACK to the RMST entity down the hierarchy toward the source node.

RMST congestion control mechanisms do not specify any congestion control or detection mechanism. It is concerned solely with reliable data transfer between the sensor nodes and the sink. Any congestion control mechanisms are a by-product of the use of directed diffusion, which offers minimal congestion control. For example, sensor nodes having gradients that show interest in the same information, but have different reporting intervals, may "downconvert" to the lower of the two reporting intervals [12].

RMST is only suitable for the reliable delivery of larger blocks of data consisting of multiple segments such as JPEG images that take advantage of fragmentation at the source and reassembly at the base station. RMST might not be suitable for reliably delivering fragments from multiple RMST entities to the same base station. It cannot ensure the orderly delivery of fragments to the base station. More fragments will cause more contention for the channel, that is, more in-network data flow will happen. Moreover, RMST does not provide any real-time reliability guarantee or application-level reliability and congestion control. RMST does not support effective energy conservation mechanism.

10.4.1.4 GARUDA

GARUDA (a mythological bird that reliably transported gods) [13] is designed to provide downstream (sink to multiple sensor nodes) guaranteed and reliable delivery similar to PSFQ [9], and scalability for control codes and query metadata. GARUDA offers both destination and packet reliability. It is more suitable for query-driven mission critical applications, for example, country border security and remote surveillance applications in which mobile robots are deployed to detect human bodies based on a particular image containing the face queried through a base station, etc.

GARUDA uses three primary phases, that is, wait-for-first-packet (WFP), core election, and loss recovery. The WFP pulse broadcasts from the base station to guarantee the delivery of a single/first packet to all the sensor nodes. The core selection phase of GARUDA involves the construction of a two-tier topology that consists of two layers, that is, core nodes and noncore nodes. The core nodes' selection is based on the hop count. The sensors with hop count multiples of 3 (3^*i, where i is an appositive integer) from the sink are allowed to elect them as the core. Like PSFQ [9], GARUDA uses NACK-based loss detection, notification, and local retransmission for loss recovery. The loss recovery phase of GARUDA utilizes a NACK-based two-stage loss recovery process. The first stage of loss recovery guarantees that all core nodes successfully recover all lost packets from the sink. In the second stage, noncore nodes requests retransmission of lost data from core nodes after the completion of all retransmission of its core node.

The approach used by GARUDA has several disadvantages, for example, it may not be suitable for upstream data reliability and it only supports reliability on the downstream direction from the sink to the sensors. It also does not provide any congestion control mechanism and, in case of a very large WSN, the core construction and loss recovery might be very lengthy. GARUDA only offers reliable transfer of the very first packet without guaranteeing the rest of the packets of a particular message.

10.4.1.5 Distributed TCP Caching

Distributed TCP caching (DTC) [14] is a mechanism based on a TCP adaptation that provides both upstream and downstream reliability with hop-by-hop recovery. DTC improves the transmission efficiency by compressing the headers and balancing the buffer constraints by using probability-based selection for cache points. DTC does not impose any modifications in the end nodes; instead, it resides in the intermediate nodes. It caches the single segment based on the segment selection algorithm that is derived from the highest sequence number observed. The algorithm helps in caching segments that are most probable to be dropped further along the route to the receiver. This significantly improves the efficiency of end-to-end delivery. The caching mechanism may also be easily extended to use more than one segment per node.

The DTC relies mainly on time-outs and duplicate acknowledgments to detect packet losses. Each node estimates the round-trip time (RTT) to the receiver and adapts a retransmission time-out RTT to 1.5× RTT. The RTT is computed during the TCP connection setup phase and the time-out value is set as RTT = 1.5× RTT. The RTT helps to differentiate between an acknowledgment for a lost packet and one for the new segment to avoid extra retransmissions. To avoid retransmissions from the original sender, the response of the DTC for packet loss has to be faster than the regular TCP.

It has also been proposed that the TCP SACK option be applied to optimize the use of the cache. A SACK packet contains information about segments in a cache and an ACK field that contains the sequence number of the last fragment that was received. The SACK field lists the sequence numbers of additional fragments received out-of-order and also works as a multiple NACK by implicitly listing all missing fragments. To understand the SACK mechanism, let us consider that each intermediate node between a source (S) and a destination (D) can store only a single fragment. D periodically sends a SACK packet to S. The intermediate node i along the path to S examines the SACK packet. If the fragment stored by i is acknowledged, then i deletes that fragment from its cache. If the SACK negatively acknowledges a fragment that is stored by i, then i locks the missing fragment and retransmits the missing fragment by inserting its sequence number into the SACK field. Locked data segments should not be overwritten by a TCP segment with higher sequence number. Finally, *i* forwards the SACK packet to the direction of *S*. If an intermediate node can retransmit all missing fragments listed in a SACK, it drops the SACK. Figure 10.5 shows data caching and forwarding in DTC. The data segments D_1 and D_2 are cached in intermediate nodes 5 and 7, and are recovered by destination (sink) through ACK 1 and 2, respectively. The intermediate nodes 5 and 7 clear their buffers by receiving ACK 4.

DTC performance degrades in terms of end-to-end recovery when none of the surrounding nodes have cached the required segment. It always tries to cache more recent segments and old lost segments need to recover from far away nodes. No modification of the congestion control mechanism has been suggested in DTC. The decision for caching (which node will cache which packets) makes the design issue complex. Finally, it does not explicitly address the design challenges of upstream or downstream reliability. If a message transmission from one sensor node fails, this leads to a stop in the transmission of all its previous sensor nodes.

10.4.1.6 Distributed Transport for Sensor Network

Distributed transport for sensor network (DTSN) [15] was developed with the objective of providing an energy-efficient, hop-by-hop reliable transport protocol that focuses on multicast communication in WSN in the fashion of TCP. DTSN provides different grades of reliability in terms of full and differentiated reliability services. The full reliability mode aims to deliver all the packets that need to be delivered to the sink. On the other hand, differential reliability does not grant guaranteed delivery of all packets to the sink. The full reliability level is employed with the help of a selective repeat–automatic repeat request (SR-ARQ) using NACK and positive acknowledgment packet ACK for loss recovery and packet control. Both NACK and ACK are to be sent by the receiver upon request by the sender using an explicit acknowledgment request (EAR), which can be piggybacked on the data packets.

The differential reliability can be achieved by the use of ARQ along with enhancement flow and forward error correction (FEC) strategies. The enhancement flow option consists of buffering only a fraction of data packets at the source (designated as the core) using FEC being transmitted with total reliability. For example, in a still image, the core person of the image with minimum resolution is buffered at the source and transferred with a reliability guarantee. However, the remaining part of the image (image resolution increment) does not provide any reliability

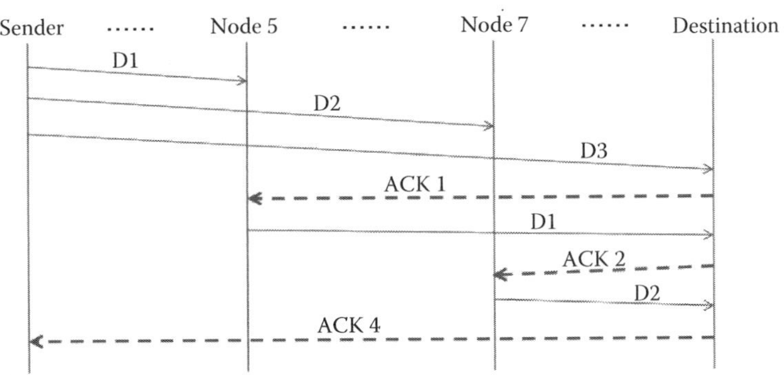

Figure 10.5 Data caching and forwarding in DTC.

guarantee because it is a part of the image with maximum resolution. The enhancement flow, along with the FEC mechanism, achieves high reliability grades by significantly increasing the throughput in comparison with the total reliability service.

In DTSN, a session is a source/destination relationship identified by the tuple (source address, destination address, application identifier, and session number) designed as the session identifier. A randomly selected session number is used to distinguish different sessions between the same source and destination nodes within a session (packets are sequentially numbered). An acknowledgment window (AW) is defined as the number of packets that source sends before generating an EAR. The number of data packets sent in each transmission is equivalent to the size of the AW. The output buffer at the sender works as a sliding window that can span more than the AW. The sender sends out an EAR packet to request feedback from the destination and also starts an EAR timer. The value of both the AW and EAR timer are adjusted according to the individual application. The destination of the session prepares and sends out a feedback packet after receiving the EAR. If no feedback packet is received by the source, then this indicates that there was no packet loss during the last session. The source will free up its output buffer and end the current session. Otherwise, if a NACK is received, then the source will check the bitmap included in the NACK packet, identify the gap(s) in the sequence number, and retransmit the lost packet.

DTSN algorithm does not address congestion detection and control. Caching at intermediate nodes requires a large amount of storage, which is not suitable for resource-constrained WSNs. An efficient cache management mechanism is required to minimize unnecessary cache occupancy. The ACK/NACK mechanism is reserved for high message overhead, which affects the overall energy efficiency. DTSN does not provide the details of how the size of the core data is determined and reliability level is maintained. The protocol does not provide the details that the FEC techniques have used for transferring the core part of the data. Intermediate nodes act as relay nodes whereas source and destination nodes encode and decode data packets. This may not suitable for a sensor network with a large number of hops because this will introduce many entry points for errors and packet loss along the way.

10.4.1.7 Reliable Bursty Convergecast

Reliable bursty convergecast (RBC) [16] was designed for event-driven busty upstream traffic to provide real-time, hop-by-hop packet reliability for continuous data flow. The aim of the RBC protocol is to improve channel utilization to reduce ACK loss, alleviate retransmission-incurred channel contention, and implement the adaptive timer's mechanism for retransmission time-outs (RTOs). The RBC was designed with a windowless block acknowledgment scheme to reduce the packet and acknowledgment loss that guarantees continuous packet forwarding and replicates acknowledgment for packets. The windowless block acknowledgment mechanism detects and notices packet loss more reliably and energy efficiently.

In RBC, every sensor node maintains a priority queue for differentiated contention control to rank the sensor nodes based on the queuing condition. The sensor node assumes that a node is capable of guessing whether or not its neighbor has received and forwarded its supplied packets by listening to the channel and maintaining precise time synchronization for real-time packet transport. It maintains a list of queues as a windowless block to place packets that have been retransmitted a fewer number of times to be transmitted first, if there are some nodes requesting for them. The receiver piggybacks information from all successfully received packets in the header of each data packet to be further forwarded. The sender listens to the information and triggers retransmission if it does not receive any expected information for a specific time interval. Upon the expiry of a timer for each transmitted packet, the sender will start hop-by-hop retransmission. The value of the retransmission timer depends on the queue length at the next hop. RBC proposes intranode and internode packet scheduling to avoid retransmission-based congestion.

In intranode packet scheduling, each node maintains a virtual queue, giving high priority to the packets with fewer retransmissions. In internode packet scheduling, the node with a high rank (node with more packets in buffer) accesses the channel first. The priority scheme may lead to starvation (the low priority to wait forever in the queue) of low priority packets. The lack of an energy model makes it unsuitable for many WSN applications.

RBC supports block acknowledgment, that is, one ACK for every fragment stored by a node. The node will completely empty its cache upon receiving ACK; this leads to a high probability of fragment losses. Another challenge will be to extend the query model for multimedia sensory data communication. In case of a fragment loss in RBC, the receiver would not be able to request the retransmission of a fragment without having a NACK mechanism.

10.4.1.8 Energy-Efficient and Reliable Transport Protocol

Energy-efficient and reliable transport protocol (ERTP) [17] is an upstream (one or more sensor sources transmits data to a sink node), energy-efficient transport protocol designed to provide end-to-end statistical reliability for low data-streaming WSN applications. The reliability of ERTP is determined by the quantity of data packets received at the sink rather than the reliability of each data packet. The main aim of ERTP is to achieve end-to-end reliability by controlling the reliability at each hop and maximize energy efficiency using stop and wait hop-by-hop iACK for loss recovery. ERTP assumes low data rate (negligible collision), overheard packet transmission costs are low for single-hop neighbors and transmission collisions due to transmission from two neighbor nodes at the same time are negligible.

ERTP consists of two components—hop-by-hop reliability and hop-by-hop retransmission time-out. The hop-by-hop recovery algorithm uses iACK for reliability. A node i, after sending a packet to the next node $i + 1$, overhears the next forwarding of a packet by node $i + 1$ and is considered as an iACK to node i. The data packet forwarding continues hop-by-hop up to the sink. The data packet forwarding stops when it reaches the sink. ERTP uses an explicit acknowledgment (eACK) from the sink to the last node that forwards data to sink. The eACK ensures the packet delivery to the sink and retransmission of the packet is denied.

The maximum number of retransmissions of a packet in each node is based on the link loss rate. It is dynamically adjusted in hop-by-hop downstream nodes to forward the packet. ERTP computes the extent of time in which node i is expected to "overhear" the forwarding packet from node $i + 1$, which is called RTO using an algorithm. The RTO dynamically adjusts the node's waiting time for iACK at each hop for the downstream nodes to forward the packet. ERTP significantly decreases energy consumption using hop-by-hop RTO control (Figure 10.6a). A "premature" RTO value for hop-by-hop iACK may increase sensor energy consumption because transmitters will send duplicate packets (Figure 10.6b). A large RTO value tends to increase transmission latency and thus reduces network throughputs (Figure 10.6c). Therefore, to achieve energy efficiency, the hop-by-hop RTO component of ERTP is responsible for adjusting the RTO dynamically.

The simulation results of ERTP shows that it conserves more energy and gives a higher delivery ratio than the protocol using simple eACK. However, hearing all neighbor node traffic increases energy consumption significantly. The protocol ignores energy conservation because the node periodically sleeps, wakes up, and listens.

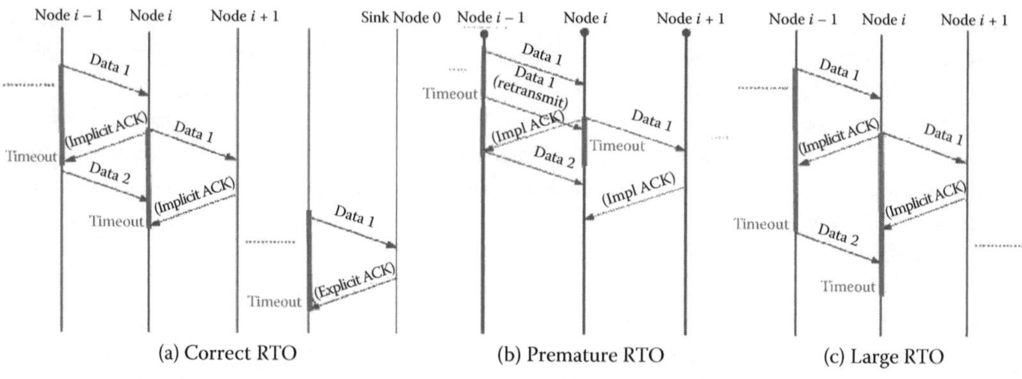

Figure 10.6 Hop-by-hop iACK operation. (From T. Le et al., *Computer Communications*, vol. 32, pp. 1154–1171, 2009.)

10.4.2 Protocols for Congestion Control

Congestion has a significant effect on the performance of reliable transport control because congestion leads to packet losses. The packet loss can also occur due to bad channel errors, collisions, and interference; thus, congestion needs to be explicitly noted. The congestion control mechanism can be classified into upstream or downstream congestion control, end-to-end congestion control, and hop-by-hop congestion control. The hop-by-hop congestion control has faster response and conserves more energy than end-to-end congestion. Many congestion control protocols proposed for WSN mainly address congestion detection, congestion notification, and rate adjustment. Some of these congestion protocols [18–28] are discussed below.

10.4.2.1 Congestion Detection and Avoidance in Sensor Networks

Congestion detection and avoidance (CODA) in sensor networks [18] was designed to solve the upstream (sensors-to-sink) direction congestion control problem. CODA addresses congestion that occurs due to event-driven sensor network applications that cause a sudden burst of traffic from an idle or light load. It comprises three components: receiver-based congestion detection, open-loop hop-by-hop backpressure, and closed-loop end-to-end multisource regulation.

The first step toward congestion control is to accurately and efficiently detect congestion. To do this, CODA senses congestion by a combination of the present and past wireless channel load and buffer occupancy at each receiver. If buffer occupancy or wireless channel load exceeds a predefined threshold value, it infers that congestion happens. Then, the node detecting congestion notifies its upstream neighbor nodes via a backpressure mechanism. If a node receives backpressure signals, then neighbor nodes will be triggered to decrease their output rate by adjusting its sending rates in an AIMD manner or to drop packets based on its local congestion control policy. Finally, CODA uses a closed-loop end-to-end scheme to regulate multisource rates. Before sending a packet, a sensor node senses the channel at a fixed interval and, if it finds the channel busy by more than a predefined number, the source sets the "regulate bit" in the outgoing packet header of the event packets, which is called congestion bit, to inform the base station that it is experiencing congestion. When the base station receives a packet with the congestion bit enabled, it sends back an ACK CM to the source node(s) informing them to decrease their sending rate. If congestion is cleared, then the sink sends an ACK CM to sensors and informs them to increase their rate. CODA uses an AIMD-like mode in TCP protocols to regulate the sensors' rate.

The disadvantages of CODA are that it addresses only unidirectional control from the sensors to the sink (it considers congestion control only). Although it conserves energy, it results in a decrease in reliability with sparse source and high data rate. The congestion detection approach may not provide accurate congestion detection by monitoring queue size. This is because the queue can overflow due to many local conditions and, listening to the communication channel shared among neighbors, can detect collision to avoid congestion. However, listening to channels continuously results in high energy costs. The delay or response time in closed-loop multisource regulation will be increased under heavy congestion because of the high probability of loss of the ACK issued from the sink at this time.

10.4.2.2 Enhanced Congestion Detection and Avoidance

Enhanced congestion detection and avoidance (ECODA) [19] is an efficient congestion detection and avoidance protocol, and provide flexible weighted fairness for multiple classes of traffic. ECODA provides higher energy efficiency and better QoS in terms of throughput, fairness, and delay for multimedia applications that need to transmit large volumes of data concurrently from several sensors. It consists of three mechanisms for congestion mitigation and to achieve fairness, that is, use of dual buffer thresholds and weighted buffer difference for congestion detection, flexible queue scheduler for packet scheduling, and a bottleneck node–based source sending rate control scheme. ECODA maintains the queue with dual buffer thresholds and weighted buffer difference for congestion detection. The two threshold values are Q_{min} and Q_{max} used to differentiate among different buffer states, which maintains three buffer states: accept state, filter state, and reject state (Figure 10.7). In the "accept state," the buffer occupancy is less than Q_{min}, so all incoming packets are accepted. In the "filter state," buffer occupancy is between Q_{min} and Q_{max}. In this state, packets with low dynamic priority in the buffer are dropped or overwritten when the packets with high dynamic priority are received by a node. If the buffer occupancy is greater than Q_{max}, then it is sent to the reject state. Most of the packets with high priority are dropped to increase the buffer length. The packet drop method alleviates congestion. The packets in the buffer can be dropped from the tail or from any position in the queue. If packets at the tail are dropped and the high-priority packet arrives at the tail position, then that high-priority packet will be dropped.

The ECODA protocol achieves fairness through a flexible queue scheduler. Each node maintains two subqueues, that is, one for locally generated traffic and another for route-through traffic. In the route-through traffic, the packets are grouped according to the source and arranged according to their dynamic priority. Packets from both queues are sent alternatively and the round-robin policy is used for sending packets from the queue of route-through traffic.

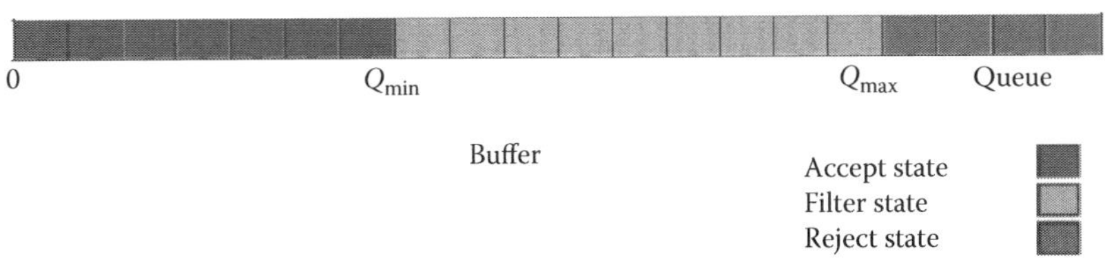

Figure 10.7 Buffer state. (From L. Q. Tao and F. Q. Yu, *IEEE Transactions on Consumer Electronics*, vol. 56, no. 3, pp. 1387–1394, 2010.)

In sensor networks, the data-sending rate of the nodes near the sink is higher than the near-source nodes, which increases the probability of congestion at the sink node. To solve this problem, ECODA uses a method called bottleneck node–based source data sending rate control. In rate control, each node can determine a routing path (or status) from itself to the sink. To indicate the path status from a particular node to the sink node, the maximum data forwarding delay is piggybacked in the data packet's header and this is computed up to source node.

The forwarder can find a better path to forward data and the source data sending rate can be adjusted more accurately and efficiently. When any source node or forwarding node receives a backpressure message, it reduces its data sending rate or adjusts the rate if there are multiple paths. If no backpressure message is received, then the source data sending rate does not increase additively.

The ECODA provides higher throughput than CODA [18]. There are not too many ACKs sent by the sink node, which controls congestion, and the energy of the node is not wasted in transmission of the ACKs. The end-to-end delay in the ECODA is less than CODA. It uses the priority of the data packets and provides weighted fairness. The drawback of ECODA is that if the buffer is in the reject state, there is a possibility of dropping the high-priority data packets.

10.4.2.3 FUSION

FUSION [20] is designed to provide an upstream congestion control mechanism. It consists of three congestion mitigation techniques applied in different layers, that is, hop-by-hop flow control, rate limiting of source traffic in the transit sensor nodes to provide fairness, and prioritized MAC. Using hop-by-hop flow control, a sensor node performs congestion detection and congestion mitigation. Congestion is detected through both queue occupancy and channel sampling techniques. The hop-by-hop flow control scheme in FUSION is similar to the backpressure scheme in CODA. The only difference in FUSION is that each sensor node sets a congestion bit in the header of every ongoing packet instead of using backpressure messages. When a sensor node overhears a packet from its parent node (the node closer to the sink) with the congestion bit is set, it stops forwarding data toward the sink. Rate limiting is a preventive scheme to avoid congestion. A significant amount of energy and bandwidth can be wasted due to congestion and packets drop in a WSN with a large diameter or with many hops. In FUSION, a token bucket scheme is used to regulate the sending rate of each node. A node accumulates one token every time it hears its parent node has forwarded N packets, up to a maximum value. The node is allowed to send only when its token account is above zero, and each packet sending costs one token.

FUSION also incorporates a prioritized MAC scheme to ensure that congested nodes receive prioritized access to the channel. The traditional MAC scheme, such as the CSMA MAC layer, gives all nodes the same chance to compete for channel access. In WSNs, the parent node that is close to the sink may gather traffic from several children nodes for which it overflows if it does not have more chances to transmit its packets. Therefore, it has to drop the packets forwarded from the children nodes. To avoid this problem in FUSION, a random back-off time for each node is introduced, which is related to its local congestion state. The congested node has a better chance to win a contention to drain its buffer faster. This mechanism also allows the children nodes to learn the congestion information indicated by the congestion bit in its packets.

In a WSN with a large diameter or with many hops, if packets are dropped due to congestion, a significant amount of energy and bandwidth can be wasted. In FUSION, each node listens to the traffic when its parent node forwards the estimated total number (N) of unique sources routing through the parent node. The rate-limiting method improves fairness but it is only effective when all nodes have the same traffic load and the routing tree is not significantly skewed.

10.4.2.4 Trickle

Trickle [21] proposes an algorithm for propagating and maintaining code updates in WSNs. It facilitates WSN reprogramming by providing downstream nodes to intelligently infer any new code availability and by subsequently pushing the actual code in a hop-by-hop fashion. Trickle uses the concept of polite gossip to propagate metadata regarding any updated code that needs to be pushed downstream. The detected older metadata of neighbors are updated by the sensor node by broadcasting the appropriate code. On the other hand, if any sensor node receives any newer metadata from its neighbors, they can shorten the broadcast period and therefore broadcast the new code sooner. The metadata are used to describe the code that sensor nodes use, which is usually smaller in size than the code itself.

To understand the trickle update procedure, let us consider two nodes, n_1 and n_2. If node n_1 broadcasts that it has code c_1, but n_2 has code $c + 1$, then n_2 knows that n_1 needs an update. Similarly, if n_2 broadcasts that it has $c + 1$, n_1 knows that it needs an update. If n_2 broadcasts updates, then all of its neighbors can receive the update code without advertising their need. Some of these recipients might not even have heard n_1's transmission. In the example, it does not matter whether it is n_1 or n_2 that transmits first, inconsistency will be detected in either case. In Trickle, it is required that some nodes communicate with other nodes at some nonzero rate, which is called the "communication rate." As long as the network is connected and there is some minimum communication rate for each node, everyone will stay up-to-date. Trickle can scale to thousandfold changes in network density, propagate new code in the order of seconds, and impose a maintenance cost on the order of a few sends an hour.

Trickle guarantees the delivery of metadata about the code; it does not guarantee reliable delivery of the code itself. Trickle also does not provide any mechanism of querying the current code version from any one or from a set of sensor nodes. This makes the base station unaware of the current status of the WSN. Trickle assumes a synchronized environment that is difficult to achieve in WSN. Without synchronization, Trickle can suffer from the short-listen problem (Figure 10.8), in which some subset of motes gossip soon after the beginning of their interval, listening for only a short time, before anyone else has a chance to speak up. This results in redundant transmissions. There is a higher communication overhead to update node status in case of a sparse network. It is also difficult to extract metadata from the code.

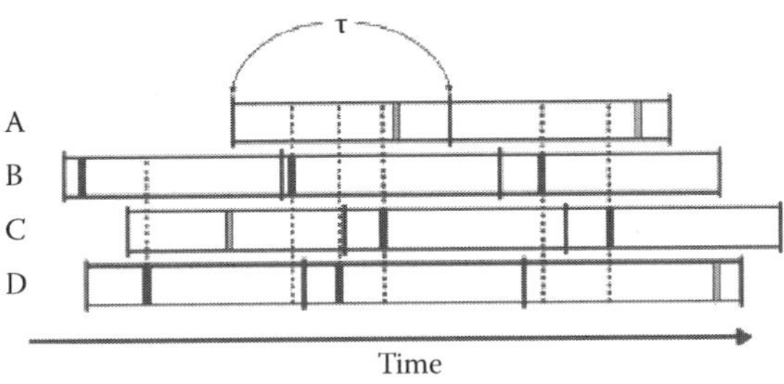

Figure 10.8 The short-listen problem for nodes A, B, C, and D. Dark bars, transmissions; light bars, suppressed transmissions; and dashed lines, receptions. Tick marks indicate interval boundaries. Mote B transmits in all three intervals.

10.4.2.5 Congestion Control and Fairness (CCF)

Congestion control and fairness (CCF) [22] provides hop-by-hop upstream congestion control using a many-to-one distributed and scalable algorithm that not only eliminates congestion but also ensures the fair delivery of packets to the base station. In CCF, fairness is achieved when an equal number of packets are received from each node to the base station or sink. It detects congestion based on packet service time at the MAC layer, and control congestion is based on a hop-by-hop manner. It adjusts traffic rate exactly based on packer service time along with fair packet scheduling algorithms.

In CCF, the congestion control of any particular sensor node is computed by the number of available child nodes (downstream nodes), the average rate at which packets can be sent by it, the per node data packet generation rate by its parent, and the downstream propagation rate. When congestion is experienced, it informs the downstream nodes to reduce their data transmission rate and vice versa (Figure 10.9).

The CCF proposes two concepts to provide fairness for each child, that is, per child packet queues and per child subtree size. In per child packet queues, each sensor node maintains one indexed queue for each of its children along with its own queue. When this node's queue is full or about to become full, nodes allow the queue to empty by reducing the transmission rates of all downstream nodes. The rate updates at a node take the time proportional to the network depth of its subtree (i.e., number of hops from the parent node). Because of this, the rate between neighboring nodes of different depths will be different at any point in time. This means that the nodes will be generating packets at different times. A slight jitter is introduced into the transmission time at the transport layer for neighboring nodes of the same depth. It reduces interference and congestion.

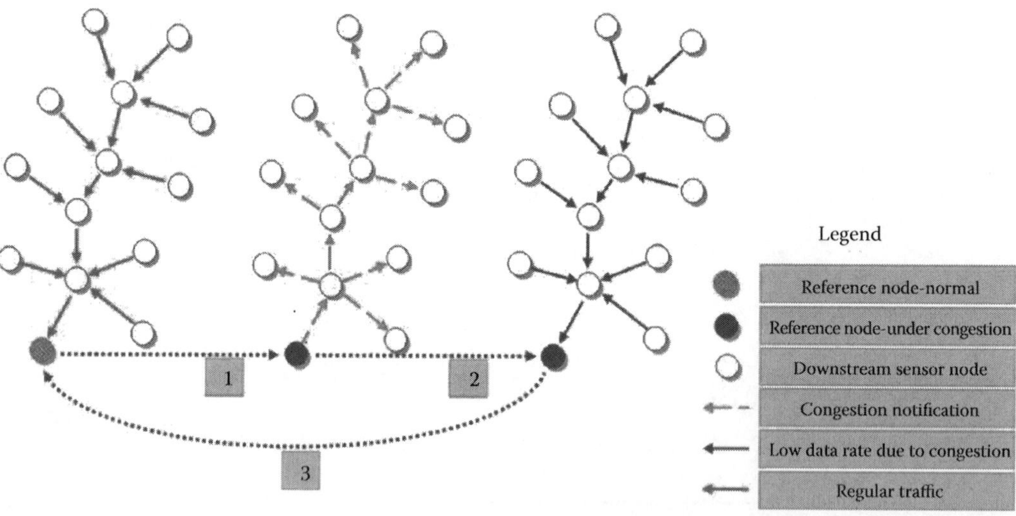

Figure 10.9 1, a node under regular traffic facing congestion informs the downstream nodes to limit their data flow; 2, constrained data flow under congestion; 3, congestion is cleared and all the downstream nodes resume their regular data flow. (From M. A. Rahman et al., Wireless sensor network transport layer: State of the art. In *Sensors*, R. Y. M. Huang (editor), Springer, Berlin Heidelberg, pp. 221–245, 2008.)

The idea of determining the size of a subtree was obtained from the in-network aggregation of data. It defines the parent of a node to be the node to which a packet is destined for transmission. It piggybacks the size of the subtree and data generation rate in the packet header is addressed to their parent node. This piggybacking eliminates the injection of additional packets to the downstream sensor nodes.

The CCF uses only packet service time to detect congestion and therefore it cannot detect either underutilized links or nodes. Another shortcoming of CCF is that it does not provide any reliability mechanism. However, reserving equal resources for each sensor node to provide equal opportunity might be inefficient for many scenarios. For example, some sensor nodes might be capturing events more often than others, such as a video sensor capturing 10 frames per second and thus needs a higher bandwidth and channel access than that required by a static sensor.

10.4.2.6 SenTCP

SenTCP [23] is an open loop hop-by-hop congestion control protocol for upstream traffic flow. It measures the extent to which congestion takes place in every intermediate sensor node, using average local packet servicing time, local packet interarrival time, and buffer occupancy. Packet service time and packet arrival time help in differentiating the factors responsible for the packet loss occurring in wireless environments because arrival time or service time may be small or large in case of congestion.

In case of SenTCP, each sensor node in the intermediate position issues a feedback signal in a backward and hop-wise manner to its neighbors, with local congestion degree and buffer occupancy ratio as its contents. This feedback signal, sent by the intermediate node, is used by the neighboring sensor nodes for adjusting their rate of sending data in the transport layer. The hop-wise control of feedback signal regulates congestion in a quicker way, resulting in a reduction of packet dropping, which helps in preserving energy as well as increasing throughput. However, SenTCP controls congestion without any recovery of loss, and without providing any guarantee on reliability. In comparison, the protocol called FUSION [20] provides fairness while controlling congestion. In this protocol, each sensor node detects congestion by measuring the length of a packet queue. Here, in the header of each outgoing packet, a congestion notification (CN) bit is set, whenever the node detects congestion. The CN bit works as an alarm when overheard by a neighboring node. This prevents the node from forwarding packets to the congested node.

10.4.2.7 Siphon

Siphon [24], is an upstream congestion control protocol that consists of a group of distributed algorithms. It is mainly responsible for discovering virtual sinks (VSs), detecting congestion in sensor networks, and sensitizing the redirection of network traffic. This mechanism is also applicable for controlling the fidelity of applications as well as for avoiding or preventing the network from getting overloaded. Siphon, as its name suggests, detects and prevents the network from being flooded with traffic overload by including sinks that are virtual with a longer range multiradio within the sensor network. VSs can be distributed in a dynamic way to funnelize network traffic events from areas of the sensor field that indicate an overloaded condition. At the congestion point, VSs enable the extra traffic to pass through them so that the desired throughput at the base station can be achieved.

The Siphon design consists of algorithms for VS discovery and visibility scope control, congestion detection, traffic redirection, and finally, congestion avoidance in the secondary network. In VS discovery, a control packet is sent periodically by the physical layer after embedding a signature byte, which consists of the hop count of the sensor nodes using any particular VS. A list of neighbors is maintained by every sensor node, through which the node reaches its parent VS. A list of neighbor VSs is maintained by each VS, wherein each individual VS consists of a dual-radio interface: for communicating with other VSs, a long-range interface is used and a regular low-power radio interface is used for regular sensor nodes. However, for controlling congestion and actuating the infrastructure of the VS, Siphon follows two techniques: node-initiated detection of congestion and physical sink initiated post facto congestion detection.

The level of congestion is measured by a theoretical upper bound of the channel when throughput is exceeded or approached by the load of the local channel. During the growth of buffer occupancy beyond a high water mark, the re-direct algorithm of a sensor node is activated by the node located within the visibility scope of a VS so that the designated traffic is diverted away from the neighborhood utilizing the VS. As a countermeasure to the funneling effect, it is necessary that the overloaded event traffic to be redirected quickly through a propagation funnel. To achieve a balance, it is mandatory for the data at a location to get redirected before the occurrence of congestion in the funnel. In the activation “post facto” of VS infrastructure through inferring congestion at a physical sink, the physical sink is responsible for smart monitoring of the data quality of the event and measured application fidelity and indicating VS signaling after degradation of measured application fidelity below a certain threshold acts as a point for collecting data in the funnel.

The siphoning service takes place only after the measurement of degradation of congestion or fidelity in the primary low-power radio networks. This method is devoid of the requirements of underlying support for detecting congestion at each node. The post facto approach to VS does not deal with transient congestion deep in the network but becomes virtual when congestion occurs closer to the physical sink. Because siphoning of traffic in this approach depends on the perception of performance measurement at the physical sink instead of levels of congestion experienced in the network. Using the signal redirection bit in the header of the network layer enables redirecting traffic in Siphon. However, for setting the redirection bit, the following two approaches may be used: (1) redirection on-demand, in which the redirection bit is set only when congestion is detected; and (2) redirection always on, in which the redirection bit is always set.

The service of traffic siphoning can be executed in parallel with the above mechanisms, either on the primary or secondary networks. Redirecting traffic through VSs becomes less useful when the secondary network is overloaded. This enables a VS to monitor its own level of congestion on both primary and secondary radio channels and hence disable the advertisement related to its existence when either one of its radio network is overloaded. Overloading of both primary and secondary networks leads to an increase of the level of congestion beyond the threshold value. This leads to a triggering of backpressure mechanism. The VSs are less likely to be congested because by using two different radios in channels with different characteristics such as fading, throughput, etc., the VS can send and receive packets at the same time.

10.4.2.8 Congestion Control for Multiclass Traffic

Congestion control for multiclass traffic (COMUT) [25] is a self-organized, distributed, cluster-based mechanism for congestion control in multiple classes of traffic in sensor networks. It proactively monitors congestion within its localized scope and facilitates system-wide rate control.

The cluster plays a vital role in setting the COMUT mechanism. The mechanism is set on a per cluster basis. In COMUT, sensors self-organize into clusters that are governed by an appointed sentinel. Traffic estimates are sent to the sentinel by the sensor nodes through single-hop broadcast. In the sentinel election phase, each sensor node sets a time-out (which is predefined and randomly chosen), if a node receives a sentinel announcement message from a neighboring node before time-out occurs for the node, it selects the sender as its sentinel and joins that sentinel's cluster. Otherwise, upon the expiry of time, the sensor will become a sentinel itself. Each node becomes either a sentinel itself or a member of a cluster headed by a neighbor sentinel. Sentinels proactively monitor network statistics and infer the collective level of congestion. COMUT adapts a routing protocol that combines both proactive and reactive routing approaches.

The level of congestion is determined by the traffic intensity both within and across multiple clusters. The traffic intensity is computed from the number of new incoming and existing flows, the density of nodes in the network, and the computation limitation abilities of sensor nodes. The congestion and rate control is highly dynamic and unpredictable for event-based sensor systems in multiple classes of flows. To support this, COMUT follows a rate adjustment of the sending rate per cluster and a decentralized methodology for intracluster and intercluster, per path estimation of traffic intensity. Based on the congestion state and relative level of importance of the events, the sensor nodes adjust their rates. The timeliness of data delivery for high-importance flows and efficiency is improved by sharing bandwidth between the flows. The rate adaptation policy is based on AIMD and it decreases the data rate to a minimum for less important data flow.

10.4.2.9 Priority-Based Congestion Control Protocol

Priority-based congestion control protocol (PCCP) [26] is a hop-by-hop node priority-based upstream congestion control protocol for WSN. It is designed to work under both single-path routing and multipath routing scenarios. The main goal of the protocol is to allow different priorities to the sensor node depending on its function or location. This protocol attempts to avoid or reduce packet loss while ensuring support for multiple channels. PCCP use an exact rate adjustment. The priority-based fairness of PCCP overcomes the drawbacks caused by the use of non–work conservative scheduling in CCF [22].

The PCCP has three main features, that is, intelligent congestion detection (ICD), implicit congestion notification (ICN), and priority-based rate adjustment (PRA). The ICD mechanism used by this protocol detects congestion based on packet interarrival times along with packet service time. PCCP calculates a congestion degree as the ratio of the average packet service time on the average time of transmission/reception. The congestion value indicates the level of congestion that exists in a particular place and at any given time on the network. The congestion degree is used to achieve exact rate adjustment with priority-based fairness. The PCCP uses ICN by piggybacking the congestion information in the header of the data packets, thus avoiding additional control packets. The ICN mechanism is used in each node of the network to send a notification about the congestion. The notification mechanism is activated if the number of packets received exceeds the maximum, or if the node receives notification from the neighboring node that there is congestion in that area of the network. The node that calculates traffic by adding the priority of upstream nodes will be able to monitor the best level of congestion.

The PRA mechanism used by this protocol adjusts the congestion rate based on a priority system. In PRA, each sensor node is given a priority index. The node with a higher priority index gets more bandwidth and, proportional with the priority index, the nodes with the same priority index get equal bandwidth and a node with sufficient traffic gets more bandwidth than one that generates less traffic. The use of a priority index provides PCCP with high flexibility in realizing weighted proportional fairness. The PCCP imposes hop-by-hop control based on the measured congestion degree as well as the node priority index. The PCCP controls congestion fast and energy-efficiently. Figure 10.10 shows the scheduler proposed for congestion control and the node model in PCCP.

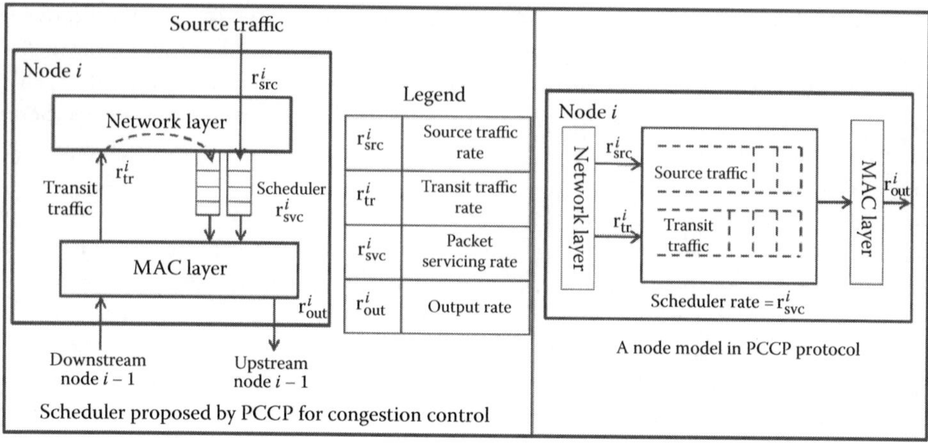

Figure 10.10 Scheduler proposed for congestion control and node model in PCCP.

PCCP does not consider how to choose the priority index for each sensor node, although it is used to enhance congestion control. Congestion detection depends on how quickly packet interarrival and service times can be correctly measured. This technique may suffer in a high loss rate channel.

10.4.2.10 Prioritized Heterogeneous Traffic-Oriented Congestion Control Protocol

Prioritized heterogeneous traffic-oriented congestion control protocol (PHTCCP) [27] is designed to handle both node and link level congestions for diverse data within a single node. PCCP [26] introduces an efficient congestion detection technique addressing both node and link level congestion. However, it doesn't have any mechanism for handling prioritized heterogeneous traffic in the network. The major goals of PHTCCP are to generate and transmit heterogeneous data on a priority basis, adjust the rate while congestion occurs, and ensure efficient link capacity utilization when some nodes in a particular route are inactive or in sleep mode. In PHTCCP, weighted fair queuing (WFQ) is used for the scheduling shown in Figure 10.11.

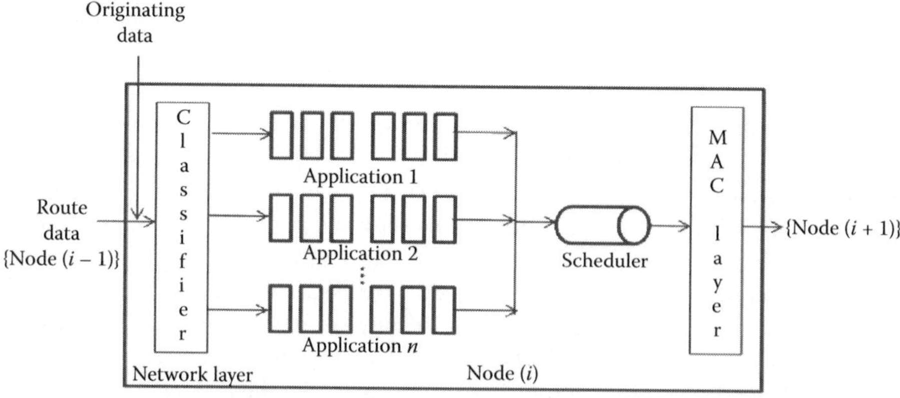

Figure 10.11 Queuing model for a particular node.

In PHTCCP, at each node i, congestion is detected by packet service ratio $\rho(i)$, which is defined as the ratio of average packet service rate (R_s^i) and packet scheduling rate (R_{sch}^i). The packet service rate is calculated as an inverse time interval when a packet arrives at the MAC layer and is successfully transmitted toward the next hop. The packet scheduling rate is computed as the total number of child nodes of the node i. The packet service ratio is an effective measure to detect both node level and link level congestion. If the packet service ratio is equal to 1, then the scheduling rate is equal to the forwarding rate. When this ratio is greater than 1, the scheduling rate is less than the average packet service rate. In both of these cases, the level of congestion decreases when it is less than 1, it indicates link level collisions, which cause the queuing up of packets at the queue.

The PHTCCP uses ICN, which notifies rates using a piggyback technique. Each node i piggybacks its packet scheduling rate of the active child $A(C(i))$ and the weighted average queue length of its active child nodes in its packet header. The rate adjustment procedure is triggered if the value of the packet service ratio goes below a certain threshold (depends on the application requirement).

Rate adjustment in PHTCCP is done hop-by-hop, which ensures the delivery of heterogeneous data to the base station at their desired rates. R_{sch}^i controls the output rate of a node. As the information of packet service ratio for congestion detection is piggybacked, each node i updates its scheduling rate if this ratio goes below the threshold or if there is any change in the scheduling rate of its parent node. The initial scheduling rate is set to r_{sch}^{init}.

10.4.2.11 Fairness-Aware Congestion Control

The fairness-aware congestion control (FACC) [28] scheme allocates a fair share of the available bandwidth to each active flow according to its generating rates. In WSN, the available bandwidth and number of active flows is time varying. Thus, it is very important to share the available bandwidth fairly among the flows. The FACC classifies all intermediate sensor nodes into near-source and near-sink nodes so that the sending rate of each flow is adjusted as early as possible to avoid congestion, and the precious resources at the nodes close to the sink can be saved. The near-source nodes maintain a per flow state and allocate an approximately fair share of available bandwidth to each passing flow. The near-sink nodes use a lightweight probabilistic dropping algorithm based on queue occupancy and hit frequency. The FACC scheme is shown in Figure 10.12. First, the near-sink node sends a warning message (WM) back to the near-source nodes once a packet is dropped at this node. Second, the near-source nodes calculate and allocate approximate fair rate share for each passing flow. Finally, the near-source node sends a CM to notify the designated source node of the updated sending rate. The processes at near-source and near-sink nodes are described as below.

- Near-source node processes

  Due to the time-varying nature of available bandwidth and traffic load in WSN, it is very complicated to implement fair resource allocation. The fairness-aware transmission control mechanism is developed on near-source nodes considering the estimation of available bandwidth, the arrival rate of each flow, and the number of active flows for the particular node.

  - *Estimation of available bandwidth:* The available bandwidth BW_a of each node is estimated after choosing the threshold for optimal channel utilization. The channel utilization and congestion status for IEEE 802.11 MAC is characterized by channel utilization ratio c_b. The channel utilization ratio is defined as the ratio of time intervals when the

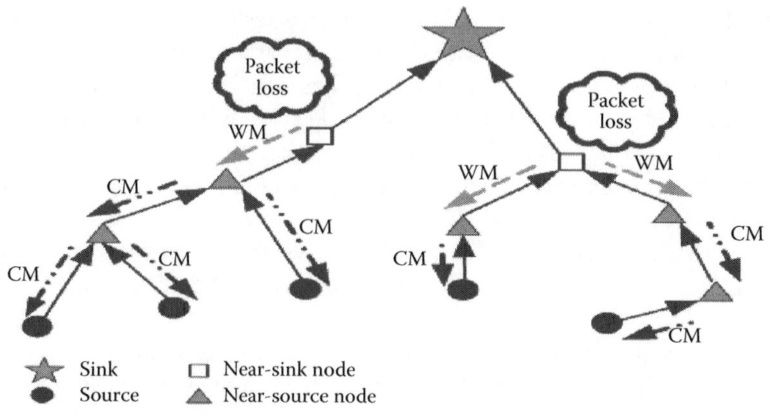

Figure 10.12 Logical framework of FACC. (From X. Yin and X. Zhou, *IEEE Transactions on Vehicular Technology*, vol. 58, pp. 5225–5234, 2009.)

channel is busy due to successful transmission or collision to the total time. The available bandwidth is estimated as shown in the study by Zhai et al. [29] as follows.

$$\mathrm{BW_a} = \begin{cases} 0 & \text{if } c_\mathrm{b} > th_\mathrm{b} \\ \mathrm{BW}(th_\mathrm{b} - c_\mathrm{b})\overline{\mathrm{data}}/\overline{T_\mathrm{s}} & \text{if } c_\mathrm{b} < th_\mathrm{b} \end{cases}$$

where BW is the transmission rate for the data packets, $\overline{\text{data}}$ is the average payload size measured by the channel occupancy time, and T_s is the average time of a successful transmission at the MAC layer. Therefore, as long as c_b does not exceed the threshold th_b, the node will not operate in the overload status and the available bandwidth can be used to accommodate more traffic without causing severe MAC contention.

- *Computation of flow arrival rate:* Let t_i^k be the arrival time of the *k*th packet of flow *i* and let *l* be the packet length. The estimated rate of flow *i*, that is, r_i is updated when the *k*th packet is received as shown in the study by Stoica et al. [30]

$$r_i^k = \left(1 - e^{-\frac{T_i^k}{K}}\right)\frac{l}{T_i^k} + e^{-\frac{T_i^k}{K}} r_i^{k-1}$$

where $T_i^k = t_i^k - t_i^{k-1}$ is the inter packet arrival time and *K* is a constant. The value of *K* should be very carefully chosen so that it can quickly adapt to rate fluctuations, filter noise, avoid system instability, and the estimated rate remains reasonably accurate after the packet traverses multiple links.

- *Estimation of the number of active flows:* The FACC uses one bit in the header of the packet to estimate the number of active flows. The source node sets the number of flows as *J*. The source node sets this bit to 1 if the packets of the current flow are remaining there to be transmitted in the next period; otherwise, this is set to 0. During the fixed control interval T_c, each intermediate node counts the number of flows N according to the source address and excludes the packet with special field 0. Thus, J can be estimated as

$$J = \begin{cases} 2N+1 & \text{a flow is originated} \\ 2N & \text{otherwise} \end{cases}$$

- Transmission control on near-source node
  - *Internode resource allocation:* To determine the available bandwidth for each node, the channel resource ΔS is assigned for each node proportionally to its current traffic load S in T_c. The linear relationship between BW_a and BW is determined from Equation 10.1 as

$$\Delta S = \left(\frac{th_b - c_b}{c_b} \right) S$$

  - *Intranode fair resource allocation:* After calculating the change in total traffic ΔS, all flows passing though that node are assigned a value to achieve both efficiency and fairness. Hence, the total available resource at node is

$$S + S = \left(\frac{th_b - c_b}{c_b} \right) \times S + S = \left(\frac{th_b}{c_b} \right) \times S$$

    Thus, the fair rate share $F(t)$ can be computed as follows:

$$F(t) = \left(\frac{th_b}{c_b} \right) \times S/J$$

  - *Transmission control on near-source node:* To avoid congestion, the rate of flow i should be updated by $\min(r_i^k, F(t))$. When $r_i^k > F(t)$, the near-source node sends a CM to notify the corresponding source node of the updated rate.
  - *Rate update strategy of source nodes:* According to FACC, the near-source node sends a CM to the corresponding source node when the rate of a particular flow exceeds the fair bandwidth share or the near-source node receives a WM from a particular near-sink node. Thus, the intermediate nodes are overwhelmed with many CMs, which degrade the system performance. Let there be m hops from the source to sink for flow f_i. The rate calculated by intermediate nodes for a certain flow is nonincreasing on the path toward the link. If the rate at node at node k is denoted by r_i^k, then r_i^m is the minimum available bandwidth calculated by the mth node along f_i's path. The relaying node relays the CM containing the smallest rate and discards the others. Therefore, the source node updates its sending rate as r_i^m so that the network shall be in good condition.
- Near-sink node processes
  - *Stateless fair queue management mechanism:* In every near-sink node, two thresholds Q_l and Q_h are set for queue occupancy. When a new packet arrives, the near-sink node increments hit frequency $h(t)$ by 1 if the packet is from the same flow as one of the M packets randomly selected from the buffer. Fairness can be achieved by giving more chances to the flows with lower occupancy. The dropping probability p_d of the arriving packet based on the hit frequency $h(t)$ is as follows:

$$p_d = \begin{cases} 0 & \text{if } Q(t) < Q_l \\ h(t)/M & \text{if } Q_l \le Q(t) < Q_h \\ 1 & \text{if } Q(t) \ge Q_l \end{cases}$$

- *Hop-by-hop backpressure:* When a packet loss occurs, the near-sink node generates a WM to send network status information back to the corresponding source node. If the packets are dropped and the queue occupancy is between Q_l and Q_h, then the sending rate of the corresponding source node needs to be reduced. If queue occupancy exceeds Q_h, then the rate of all the passing flows are to be reduced.
- *Fairness of the stateless queue management mechanism:* Let us consider a queue Q with N independent Poisson arrivals, each with rate λ_i. The queuing discipline is first-in, first-out (FIFO) and the mean service time of each packet is assumed to be $1/\mu$. Let the probability that the head of Q is occupied by type i packet is denoted by $p_i.Q$. It is asserted from the well-known PASTA property [31] that $p_i.Q = p_i$ where p_i are the corresponding probabilities for type i at arbitrary time instants. It can be observed that the occupancy probability for each flow is independent of the arrival rate of other incoming flows. Because μp_i is the departure rate of type i packets, the goodput of each flow depends only on its own arrival rate and on the service rate μ. It is concluded that the ratio of dropping probabilities is an indicator for the fairness of the dropping strategy.

In reality, different sensors may be assembled with different onboard apparatuses, and their data may have different priorities. Hence, a weighted rate allocation is required for such difference. Let the weight of sensor x be w_x, then the fair share $F(t)$ should be computed at the near-source node as follows:

$$F(t) = \frac{th_b / c_b \times S}{\sum w_x}$$

The rate allocated to the specific source node should be calculated as follows:

$$F_{new}(t) = F(t) \times w_x$$

10.4.3 Protocols for Both Congestion Control and Reliability

There are several transport protocols that aim to provide both congestion control and reliability for WSN. These protocols can be classified as upstream congestion control with reliability guarantee and downstream congestion control with reliability guarantee. Some of these congestion protocols [32–40] are discussed below.

10.4.3.1 Event-to-Sink Reliable Transport

Event-to-sink reliable transport (ESRT) [31] is a transport protocol that aims at providing both upstream event (sensor-to-sink) reliability and congestion control while maintaining the minimum energy expenditure. It can also reliably deliver multiple concurrent events to the base station.

ESRT guarantees only the end-to-end reliable delivery of individual events, not individual packets from each sensor node. It measures reliability by the number of packets carrying information about a particular event that is delivered to the sink. The main goal of the ESRT is to configure the reporting frequency rate to achieve the desired event detection accuracy with minimum energy consumption.

ESRT uses a centralized rate control mechanism running at the sink. The sink periodically computes the factual reliability r according to successfully received packets in a time interval, then it derives the required sensor report frequency f from r: $f = G(r)$. ESRT informs f to all sensors, using this event reporting frequency sensors can report event and transmit packets with frequency f. The sink periodically controls the event-reporting frequency to achieve reliability. If the required reliability is not achieved, the sink will increase the event-reporting frequency (f) of the nodes and it will reduce (f) when there is congestion. The ESRT node, when it receives the event-reporting frequency, calculates its event-reporting duration and monitors its local buffer level at the end of each reporting interval to guess any possible congestion. In case a sensor node faces congestion, that is, if the buffer overflows, it sets a CN bit in the event report packets. When the sink receives a packet with the CN bit set, it gets an overall view of the congestion level of the network and broadcasts a control signal informing all source nodes to slow down their common reporting frequency. This control signal needs to be broadcast at high energy so that all sources can hear it. ESRT achieves an end-to-end desired reliability through regulating sensor report frequency.

ESRT conserves energy by controlling sensor report frequency and provides reliability for applications. It has certain drawbacks, such as it does not prevent all losses and does not retransmit lost packets. It does not guarantee the delivery of all message segments from all source nodes. ESRT assumes that the sink is one hop away from all the sensor nodes, which might not be applicable to many of the WSN applications. ESRT regulates the reporting frequency of all sensors using the same value, which is unfair because different portions of the network or different individual sensor nodes might face different traffic and therefore contribute different levels of congestion. The sink broadcasts the control signal at high energy, which may disturb any ongoing event transmission. ESRT always regulates all sources regardless of the congestion region.

10.4.3.2 Price-Oriented Reliable Transport Protocol

Price-oriented reliable transport protocol (PORT) [33] provides in-network dynamic rate control and congestion avoidance transport schemes by computing communication cost. PORT conserves energy by adopting low communication price. The price of a node n is computed by the total number of transmission attempts all in-network nodes have made to successfully deliver a packet from the source to sink.

PORT works on a query-driven principle in which the sink floods the task description packet (called interest in) to gather the in-network nodes' neighborhood information. The flooding of the task description packet is called task assignment phase. The nodes that are able to sense the physical phenomenon of interest report to the sink with sensed data packets. The reporting rate of the node can be dynamically adjusted from the feedback about the downstream communication conditions sent by its downstream neighbors. The sink also feeds back new reporting rate requirements to source nodes. The sink uses the communication cost information to slow down the reporting rate if congestion is reported for appropriate sources and increases the reporting rate of other sources that have lower communication cost because reliability must be maintained. The cost of the congested node is increased to slow down its reporting rate. The initial price of a node is set as the hop number between the node and the sink assuming all loss rates are zero. The sensed data reporting rate for a node is described in the interest packet and further rates will be adjusted by the sink. A source node encapsulates its node price in its reporting data packets. In case of node failure, the upstream neighbor shifts the traffic of the failed node to other nodes immediately. Each node employs a timer to detect node failure or quitting of its downstream nodes. If time-out occurs or if any feedback information from a downstream neighbor is received, the downstream neighbor is considered to have failed and the price of the neighbor is set to infinite to avoid routing packets to it. If a new node (which could be a newly awakened node, a newly deployed node, or a node that's recovered from a previous failure) decides to join the routing broadcast to its neighbor nodes about its existence. The neighbor nodes send their prices to the new node. The node selects the lowest prices as its downstream neighbor and sends its own calculated price to those neighboring nodes with higher prices. The neighbors with higher prices will consider the node as a possible downstream neighbor.

10.4.3.3 Sensor Transmission Control Protocol

Sensor transmission control protocol (STCP) [34] is a scalable generic upstream (sensor-to-sink), end-to-end transport protocol that provides both congestion control and reliability based on the application's requirements. STCP considers that a single sensing device is capable of performing multifunctionally by integrating multiple types of sensors (e.g., temperature and humidity measurements). It also assumes that all the sensor nodes within the WSN have clock synchronization. The STCP implements most functionalities at the proxy of the gateway, that is, at the sink because it has unlimited resources compared with sensors. The protocol aims to achieve restricted variable reliability, congestion recognition, and congestion avoidance and maintains multiple applications in the same network. In STCP, sources notify the sink about the congestion by intermediate nodes using a CN bit in packet header that is set based on the queue lengths. The sink then notifies the affected source nodes of the congested paths and finds alternative paths for their packets to be reliably transmitted.

STCP aims to establish an end-to-end connection that is similar to the principle of the TCP three-way handshake protocol. It uses three types of packets: session initiation, data, and ACK. The session initiation packet is transmitted before the data packet to synchronize the sensor node with the sink. In the session initiation packet, the source informs the sink about the number of flows originating from the node, the type of data flow, transmission rate, and requisite reliability. The sink stores all the information, sets the timers and other parameters for each flow after receiving the session initiation packet. The sink acknowledges the session initiation packet to ensure that the association is established. After receiving ACK, the source starts transmitting data packets to the sink. The sink transmits an ACK or NACK in the reverse path depending on the type of data flow. For example, event-driven applications use ACK-based end-to-end retransmissions. On the other hand, continuous data flow applications use NACK-based end-to-end retransmissions. Figure 10.13 shows the packet format of STCP. The first field is the sequence number (16 bits long), which is zero for the session initiation packet and nonzero positive integer for a data packet. The number of flows originating at the node is identified by flows. The clock field indicates the local clock value at the time of transmission. The clock value is helpful to calculate the estimated trip time (ETT) for the node and flow ID. Flow ID is used to differentiate packets from different flows. The flow bit field specifies whether the flow is continuous or event-driven. The packet transmission rate of the source node is specified in the transmission rate field for continuous flows. The reliability field gives the expected reliability required by the flow. The packet header includes a CN bit field for supporting CODA. In the acknowledgment packet, the ACK/NACK bits represent

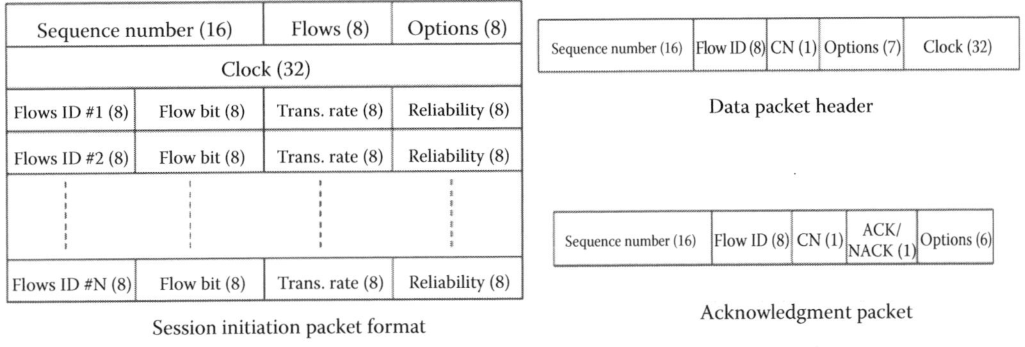

Figure 10.13 Packet form.

a positive or negative acknowledgment. The clock field combines with the sequence number field to avoid issues related to wraparound. The options field is reserved for future purposes. In event-driven flows, the sink is unable to estimate the arrival times of data packets, whereas for continuous flows, the sink calculates the ETT for the next packet when it receives a packet from a sensor node. If a packet from the source is not received within the expected time-out period, then the sink sends a NACK. An expected time for successive packets is computed by any one of the following methods:

- *Time-out* = $T + \alpha \times$ ETT, where T is the time between successive transmissions, α is a positive integer that varies with ETT, and EET indicates current clock value in the packet.
- The second approach is a Jacobson/Karels algorithm, which considers the variance of the trip time.

In STCP, sensor nodes identify the requisite reliability for each flow in the session initiation packet. The sink estimates a running average of the reliability for continuous flows and estimates reliability as a ratio of packets received to the highest sequence numbered for event-driven flows. STCP detects congestion by using two thresholds in its buffer: th_{lower} and th_{higher}. When the buffer reaches th_{lower}, the congestion bit is set with a certain probability. The node assumes high congestion and set the CN bit in every packet it forwards if the buffer reaches th_{higher}. STCP implements the exact CN, the node will set the CN bit in each data packet header. After receiving this packet with a CN field, the sink informs about the congested path by setting the congestion bit in the ACK packet to the source. STCP mitigates congestion either by selecting a different path depending on the routing layer algorithm or by slowing down the transmission rate.

The STCP suffers from long latency in large-scale multihop WSN when the sensing nodes wait for the ACK reply from the sink. Clock synchronization plays an important role in the performance of this protocol but how clock synchronization can be achieved is not clear. A node equipped with multiple types of sensors may have different priorities and can generate sensory data with different features and requirements in terms of loss, bandwidth, and delay requirements so different mechanisms are needed to deal with this diversity. STCP performs well only for single-path routing and multipath routing gives rise to fairness issues as well as network layer issues.

10.4.3.4 Flush

Flush [35] is a reliable high goodput bulk data transport protocol that provides end-to-end reliability. It reduces data transfer time for large-diameter WSNs and adapts to dynamic network conditions. Flush aims to achieve reliable delivery of data using minimal overhead by snooping necessary information instead of using extra control packets and maximizing goodput at a shorter transfer time. To support single data flow and to avoid interpath interference, the sink schedules data transfer for each node in a round-robin fashion. Flush is able to maximize bandwidth through optimal use of pipelining by ignoring interpath interference. The available bandwidth is decided using a local measurement and interference estimation algorithm. After computing the rate, it informs the source and every node between the bottlenecks.

The data transfer is initiated by the sink by sending a request for a data object to a specific source in the network using the underlying delivery protocol. Flush has four phases, that is, topology query, data transfer, acknowledgment, and integrity check. In the topology query phase, the depth of a target node to tune the RTT and a time-out at the receiver are computed. In the data transfer phase, the source sends packets to the sink using the maximum rate that does not cause intrapath interference. The sink keeps track of which packets it receives. The acknowledgment phase begins after the data transfer phase is completed. In the acknowledgment phase, the sink sends the lost packet sequence numbers to the data source for retransmitting the missing packet. This process is repeated until the sink has successfully received all the requested data. After receiving all the requested data, the sink performs an integrity check. The "integrity check" compares the checksum of the data computed at a source node with the checksum computed at the sink. If the integrity check fails, the sink discards the data and sends a fresh request. If the check succeeds, the sink may request the next data object. Integrity is checked at both the packet level and the data objects level.

In the Flush protocol, reliability is achieved with end-to-end selective NACKs and retransmitting the missing data. The sink sends NACK packet for the packets not received from the source. When a source node receives a NACK packet, it retransmits the missing data until it receives all packets successfully. To avoid node inference and improve channel utilization, the rate allocation algorithm follows two basic rules. First, a node allows transmitting only when its successor is free from interference. Second, the sending rate of a node must be less than the sending rate of its successor. Flush uses a queuing technique to buffer packets during transient rate mismatches, which are typically due to changes in link quality. The rate-based flow control algorithms work better than window-based algorithms for multihop wireless flows over unscheduled links. In the Flush protocol, rate plays a vital role; if the rate is too high, then it causes self-interference whereas too low a rate will cause poor capacity utilization.

10.4.3.5 Asymmetric and Reliable Transport

Asymmetric and reliable transport (ART) [36] is designed to provide bidirectional reliability, that is, both upstream end-to-end reliability and downstream query reliability. It also provides upstream congestion control mechanism in a decentralized way and regulates the data flow of intermediate nodes in an energy-efficient way.

A subset of sensor nodes is selected on the basis of their residual energy as essential nodes (E-nodes) to cover the regions (domains) that need to be sensed in an energy-efficient manner. These E-nodes act as cluster heads and are responsible for sensor-to-sink end-to-end reliability and sink-to-sensor query reliability. The remaining nodes are classified as nonessential nodes. A subnetwork comprised of E-nodes transfers data both in the upstream and downstream direction as well as recovering lost fragments. The residual energy of E-nodes is better than the nonessential nodes. ART has four distinct features—similar to ART—nonessential nodes do not face communication overhead, it controls congestion in a distributed manner to regulate the traffic in an efficient way, a smaller number of nodes are involved in loss recovery, and it uses a distributed energy–aware congestion control mechanism.

ART uses both ACK and NACK mechanisms to ensure reliability as it address both downstream and upstream traffic flow. A lightweight ACK mechanism is adopted to guarantee reliability between the E-node and the sink. The E-nodes are primarily responsible for enabling the event notification bit; starting the timer, it sends the first event message to the sink and waits for the ACK. If the ACK is not received by the E-node within a certain period, the event notification bit is re-enabled while retransmitting the message and the timer is reset by the E-node. To ensure query propagation in a reliable manner, two measures are followed by ART. First, the measure is connectionless and reactive wherein the query fragments are sent by the sink ignoring the loss. The receiving E-node takes the responsibility of detecting loss in a query fragment by watching the order of the sequence and thereby, as a measure, the E-node sends back a NACK to the sink. For identifying the last query message of a sequence, ART uses either a poll (P) or final (F) bit. Second, the measure is a connection-oriented communication wherein the detection of loss using the time-out mechanism is handled proactively by the sink. If the E-nodes do not receive ACK from the sink within a particular period, the E-nodes will assume congestion in the network. The E-nodes regulate the flow of data by restraining its neighboring nonessential nodes from sending data until the congestion is cleared.

Reliability and congestion control is performed by ART for E-nodes only and it ignores the reliability of nonessential nodes in the network. It does not guarantee loss recovery from nonessential nodes in the congested network as well as link loss or transmission errors.

10.4.3.6 Loss-Tolerant Reliable Event-Sensing Protocol

Loss-tolerant reliable event-sensing (LTRES) protocol [37] is a distributed data transport protocol applied to continuous surveillance WSN with heterogeneous sensing fidelity requirements over different event data areas. LTRES aims to achieve dynamic event-sensing fidelity requirements with congestion control. It provides loss-tolerant reliable (LTR) data transport services for dynamic event sensing in WSN. The end-to-end reliable data transport requirements can be defined by loss-sensitive reliable (LSR) data transport and LTR data transport. In LSR, each data packet is required to transmit successfully from the source to the destination, whereas in LRT, the receiver defines the reliable data transport for the sender in terms of throughput depending on the application requirement. The sink centrally defines the LTR data transport requirements in terms of required sensing fidelity over an event area, and the source nodes adjust their source rate to meet the required LTR requirement based on the dynamic network condition. The event-sensing fidelity level (ESF_E) is the ratio of observed event-sensing fidelity (OEF_E) at the sink to the desired event-sensing fidelity (DEF_E), that is, $ESF_E = ROEF_E/DEF_E$. OEF is computed from the goodput (packet received from the source at rate) achieved at the sink from the event-sensing source nodes. DEF is determined by the sink based on its computation capability and the event-sensing accuracy requirement. If $ESF_E \geq 1$, reliable event-sensing can be guaranteed by the LTR transport service. A congestion control mechanism should maintain a sustainable ESF_E based on network condition. If ESF_E cannot be supported by the event, then nodes should explore the upper bound of the network capacity to provide best-effort data transport service. Higher event-sensing fidelity means higher event goodput and higher bandwidth requirements. If network capacity is not considered, then it may lead to congestion.

A lightweight, loss rate–based CODA mechanism is used to alleviate the control overhead. The sink detects the congestion and supports the source rates at the aggressive sensor nodes. In LTER, the sink periodically observes the loss rate at each event node to detect congestion. A steady low loss rate is inferred as no congestion or as low congestion rate. If a dynamic increased loss rate is observed, then event nodes are congested. A CN is sent to trigger the congestion avoidance operation. During congestion detection, the event node starts the available bandwidth detection process heuristically. After the bandwidth detection operation, congested event nodes achieve the maximized bandwidth utilization with upstream congestion avoidance. The event node stops the source rate adaptation after the bandwidth detection operation. LTRES enforces fair rate control by adapting a distributed steady state throughput estimated approach. It provides LTR a guarantee with higher bandwidth utilization and fast convergence time.

10.4.3.7 Rate-Controlled Reliable Transport

Rate-controlled reliable transport (RCRT) [38] is an upstream (sensor-to-sink) multipoint-to-point reliable transport protocol, which includes congestion control and explicit rate adaptation functions. RCRT ensures reliability by using explicit end-to-end loss recovery. To ensure end-to-end loss recovery, it implements a NACK-based retransmission mechanism in which each node along the path cache packets to support on-demand loss recovery. The sink performs congestion detection, recovery, and rate adaptation operations centrally. The main design goals of RCRT are to provide end-to-end reliability of all data transmitted by each sensor to a sink, increase network efficiency, support multiple concurrent applications, and allow different applications to choose different policies, minimal sensor functionality, and robustness.

In RCRT, the sink has three distinct logical components, that is, end-to-end retransmission, congestion detection, and rate adaptation. In end-to-end loss recovery, missing packets are requested to retransmit from the source by sending NACK. The sink node keeps a list of lost segments, the missing packet sequence numbers are included in a NACK packet and transmitted to the source along the path. The intermediate node retransmits the requested packets if a packet is available in the cache and forwards the updated NACK toward the source. Cached packets are deleted once they are confirmed by a positive ACK. The RCRT congestion detection scheme is based on time to recover loss. If loss recovery takes more time than that estimated, it is assumed that the network is congested. The sink node maintains a list of out-of-order messages to compute the time to recover loss. The time to recover loss message is set to the multiple of RTTs (i.e., 2× RTT). After detecting congestion, RCRT estimates the total sustainable traffic in the network and uses AIMD to adapt the transmission rate of each source. The data rate for each flow additively increases if congestion is not detected and decreases multiplicatively. RCRT computes the new rate as

$$\text{Increase: } R(t+1) = R(t) + A$$

$$\text{Decrease: } R(t+1) = M(t)R(t)$$

where $R(t)$ denotes the sum of the currently assigned rates $r_i(t)$ for all flows i, A is a constant and $M(t)$ is a function of loss rate. $M(t)$ is computed as $M(t) = p_{i(t)}/2 - p_{i(t)}$ and $p_{i(t)}$ is the loss rate value of the source i at the instant t.

RCRT uses a three-way handshake connection establishment mechanism similar to that of TCP in which the source node initiates flow by first establishing an end-to-end connection with the sink and the third ACK is substituted with the first data packet. The initial RTT is a guess by the RTT during three-way handshake.

In RCRT, reliability depends on the MAC layer retransmission, which is not efficient. A single packet loss may force a rate reduction as congestion detection depends on loss recovery time. RCRT does not address the issue of contention.

10.4.3.8 Tunable Reliability with Congestion Control for Information Transport

Tunable reliability with congestion control for information transport (TRCCIT) [39] provides the desired reliability in varying network conditions and application requirements by using a probabilistic adaptive retransmission hybrid acknowledgment and retransmission timer management mechanism. TRCCIT proactively alleviates network congestion by opportunistically transporting information on multiple paths.

TRCCIT exploits and adapts between the temporal (adaptive retransmissions) and spatial (multiple paths) redundancies inside the network. It uses hybrid acknowledgment (HACK) and local timer management to ensure tunable reliability and provide a comprehensive congestion control mechanism. The TRCCIT includes four basic functions. First, it identifies the problem of iACK loss by addressing HACK for TRCCIT. Second, TRCCIT avoids the problem of retransmission, which is triggered by the source node observing the buffer status of the next hop nodes. It retransmits on the basis of local buffers only. Third, TRCCIT provides statistical bounds for information transport. Finally, TRCCIT provides congestion control on the fly by adapting proactively between single and multiple paths.

TRCCIT achieves the desired application reliability (R_d) by adapting a default single path (SP) for transporting the information at a low information rate and with no congestion. The tunable reliability exploits the temporal redundancies and tunes the number of retransmissions according to R_d and network conditions. The congestion inside the WSN is computed as the ratio of the incoming information rate and the outgoing information rate at the sensor node on a message by message basis. The burst of information gives rise to transient congestion inside the network as the different information flows meet. Congestion is avoided by balancing inflow and outflow traffic at the sensor nodes. Transient congestion is controlled by increasing the outflow traffic at the sensor node to disperse information at the appropriate neighbor nodes. These information transports are carried out on multiple paths. TRCCIT randomly waits before transporting the information to the next hop to avoid transmission failure due to busy channels. It adapts request to send/clear to send (RTS/CTS) and prioritized MAC protocols to provide fair channel access chance for every node.

10.4.3.9 Delay-Sensitive Transport Protocols

The delay-sensitive transport (DST) protocol [40] was designed for several WSN applications that involve event detection and signal estimation/tracking within a certain delay bound. This protocol is very efficient in handling reliable event transport to the sink node within a certain delay bound. The main design components of the DST protocol are reliable event transport and real-time event transport.

In a typical sensor network, the sink is not interested in the individual data sensed by the sensor nodes, rather the interest is in the collective information of sensor nodes within a certain delay bound. Thus, the sensor-to-sink transport paradigm requires a collective event-to-sink reliability rather than individual event-to-sink reliability. The event features are required to be reliably transported to the sink node within a certain delay bound for reliable and timely event detection. This delay bound is called event-to-sink delay, $\Delta e_2 s$. The event transport delay (τ_t) and event processing delay (τ_p) are two components of the event-to-sink delay bound. The event transport delay is the time between the occurrence of an event and its reliable transportation to the sink node, which is the function of buffering delay (τ_b), channel access delay (τ_c), transmission delay (τ_t), and propagation delay (τ_p). The event processing delay is the processing delay experienced at the sink when the desired features of an event are estimated using the data packets received from the sensor node. Then, the sum of τ_t and τ_p should be less than $\Delta e_2 s$ for a reliable and timely event detection.

Based on the event-to-sink reliability and event-to-sink delay bound, the DST protocol computes the delay-constrained reliability indicator (δ_i) at the sink in a decision interval "*i*." The delay-constrained reliability factor (δ_i) is defined as the ratio of the number of received data packets to the minimum number of data packets required for reliable event detection within a certain delay bound at the sink in decision interval "*i*." If $\delta_i > 1$, then it is clear that the event is reliably detected within the delay bound. Otherwise, appropriate action needs to be taken to assure that the desired level of reliability in the event-to-sink communication is reached. For example, the reporting frequency of the sensor nodes can be increased to improve the amount of information transported from the sensor to the sink whereas avoiding congestion in the network.

The DST protocol introduces a time-critical event first (TCEF) scheduling policy for reducing the buffering delay at intermediate sensor nodes. The TCEF scheduling algorithm serves the real-time packets on the principle of earliest deadline first (EDF) service discipline at each sensor node. The elapsed time at each sensor node is measured and the elapsed time to the event packet is piggybacked so that the following sensors can determine the remaining time to the deadline with a globally synchronized clock in the network. In this way, the event packets receive higher priority at the sensor nodes as their remaining time to deadline decreases. However, in severe network congestion, it may become insufficient to provide delay-constrained event reliability. Hence, the DST protocol also considers the event-to-sink delay bounds and congestion conditions in its reporting rate update policies to assure timely and reliable event transport.

Congestion occurs in WSN due to the limited memory of the sensor nodes and the limited capacity of the wireless medium. This leads to the waste of both communication and energy resources. Hence, congestion detection and control plays a major role in achieving reliable event detection and minimizing energy consumption.

The delay-constrained reliability indicator $\delta_i = DR_i/DR^*$ can be determined only at the sink node. Thus, the conventional sender-based congestion detection method cannot be applied here. In addition to that, the sensor node should be aware of the network channel condition around them for efficient congestion control. Because the communication media is shared in WSN, the sensor nodes may experience congestion even if their buffer occupancy is higher. In this context, the DST protocol implements a combined congestion control mechanism in which, if the buffer overflows at any sensor node or if the average node delay is above a certain delay threshold value, then the congestion is notified to the sink node. This notification is done by utilizing the CN bit in the header of the event packet transmitted from the sensor to the sink. The sink can detect the congestion if the CN bit is marked. The sink also determines the network condition from δ_i and adjusts the reporting frequency of the sensors.

The reporting frequency updation police consider five distinct scenarios, that is, early reliability and no congestion condition, early reliability and congestion condition, low reliability and no congestion condition, low reliability and congestion condition, and adequate reliability and no congestion condition.

In early reliability and no congestion condition, $T_i \leq \Delta_{e2s}$ and CN = 0, then no congestion is observed at the sink. T_i is the time needed to provide delay-constrained event reliability for a decision interval i. The $DR_i > DR^*$ because the source nodes transmit event data more frequently than required. To reduce the energy consumption of the sensor, the reporting frequency should be decreased in a controlled way. The updated reporting frequency can be expressed as $f_{i+1} = f_i T_i / \Delta_{e2s}$. In case of early reliability and congestion conditions, $T_i < \Delta_{e2s}$ and CN = 1. When the congestion is observed at the sink and $DR_i > DR^*$, then the reporting frequency is decreased more aggressively to avoid congestion and save energy of the sensor node. The reporting frequency is computed as $f_{i+1} = \min(f_i T_i / \Delta_{e2s}, f_i^{\Delta i / \Delta e2s})$. In case of low reliability and no congestion condition, $T_i > \Delta_{e2s}$ and CN = 0. If no congestion is observed at the sink and $DR_i < DR^*$, then the reporting frequency is increased multiplicatively by computed reporting frequency as $f_{i+1} = f_i DR^*/DR_i$. In case of low reliability and no congestion condition, $T_i > \Delta_{e2s}$ and CN = 0. If no congestion is observed at the sink and $DR_i < DR^*$, then the reporting frequency is increased multiplicatively by computed reporting frequency as $f_{i+1} = f^i DR^*/DR_i$. The low reliability and congestion condition is the worst possible condition where $T_i > \Delta_{e2s}$ and CN = 1 but $DR_i < DR^*$. Because the desired delay constrained event reliability has not been achieved and the network is congested, the reporting frequency is exponentially decreased and updated as $f_{i+1} = f_i^{(DR^*/DR_i/k)}$ where $k > 1$ and k denotes the number of successive decision intervals for which the network remains in the same situation. It includes the current decision interval also. In case of adequate reliability and no congestion condition the frequency of the source node remain unchanged for the next level and computed as $f_{i+1} = f_i$.

DST allows the applications to perform the right action in a timely manner by exploiting both the correlation and collaborative nature of the sensor network. It achieves high performance in terms of reliable event detection, communication latency, and energy consumption. In DST, the whole functionality is totally dependent on the sink to regulate the flow of data, which results in an increase in traffic and more energy usage. It is considered a single event and the reporting rates do not address multiple events happening in parallel, which may be a more practical approach.

10.5 Conclusion

The rapid evolution of WSN has resulted in a new dimension of emerging applications that require the transport layer to be adaptive and one that can take new challenges. In this chapter, we present a survey on transport protocols for WSNs. The unsuitability of the existing traditional protocols such as TCP and UDP for WSNs is analyzed. The ideal transport protocol for WSNs should have the following characteristics: high energy efficiency, flexible reliability, and guaranteed application-dependent QoS. We briefly review the design goals and requirements of transport protocols for WSN. The existing transport protocols for WSNs are briefly reviewed and several problems in the existing protocols are also listed, for example, these transport protocols only consider single path routing. However, multipath routing is also used in the network layer. Therefore, fairness and reliability issues are required to be addressed for this routing. These transport protocols rarely consider cross-layer interactions in a WSN. The link level performance, such as bit error rate, can significantly affect the performance of the transport layer protocol and routing can affect hop-by-hop retransmissions. Therefore, cross-layer optimization is highly desirable.

References

1. C. Wang, K. Sohraby, B. Li, M. Daneshmand, and Y. Hu. A survey of transport protocols for wireless sensor networks. *IEEE Network*, vol. 20, no. 3, pp. 34–40, 2006.

2. B. Krishnamachari, D. Estrin, and S. Wicker. Modeling data centric routing in wireless sensor networks. In *Proceedings of IEEE INFOCOM*, New York, 2002.
3. I. F. Akyildiz, W. Su, Y. Sankarasubramaniam, and E. Cayirci. Wireless sensor networks: A survey. *Computer Networks Journal*, vol. 38, no. 4, pp. 393–422, 2002.
4. C. Karlof and D. Wagner. *Secure Routing in Wireless Sensor Networks Attacks and Countermeasures*. University of California at Berkeley, Berkeley, CA, 2003.
5. A. Ahamed. Energy-efficient and reliable transport protocols for wireless sensor networks: State-of art. *Wireless Sensor Network, Scientific Research*, pp. 106–113, 2011.
6. B. Hull, K. Jamieson, and H. Balakrishnan. Mitigating congestion in wireless sensor networks. In *Proceedings of ACM SenSys '04*, Baltimore, 2004.
7. C. T. Ee and R. Bajcsy. Congestion control and fairness for many-to-one routing in sensor networks. In *Proceedings of ACM SenSys '04*, Baltimore, 2004.
8. C. Wang et al. Priority-based congestion control in wireless sensor networks. *IEEE International Conference on Sensor Networks, Ubiquitous, and Trustworthy Computing*, Taichung, Taiwan, 2006.
9. C. Wan, A. T. Campbell, and L. Krishnamurthy. PSFQ: A reliable transport protocol for wireless sensor networks. In *1st ACM International Workshop on Wireless Sensor Networks and Applications*, New York, USA, pp. 1–11, 2002.
10. C. K. Rupani and T. C. Aseri. An improved transport layer protocol for wireless sensor networks. *Computer Communications*, vol. 34, no. 6, pp. 758–764, 2010.
11. F. Stann and J. Heidemann. RMST: Reliable data transport in sensor networks. In *First IEEE International Workshop on Sensor Network Protocols and Applications*, Anchorage, AK, pp. 102–112, 2003.
12. C. Intanagonwiwat, R. Govindan, D. Estrin, J. Heidemann, and F. Silva. Directed diffusion for wireless sensor networks. *IEEE/ACM Transactions on Networking*, vol. 11, no. 1, pp. 2–16, 2003.
13. S. J. Park, R. Vedantham, R. Sivakumar, and I. F. Akyildiz. A scalable approach for reliable downstream data delivery in wireless sensor networks. In *Proceedings of International Symposium on Mobile Ad Hoc Networking and Computing (MobiHoc)*, Tokyo, Japan, pp. 78–89, 2004.
14. A. Dunkels, J. Alonso, and T. Voigt. Distributed TCP caching for wireless sensor networks. In *Proceedings of the 3rd Annual Mediterranean Ad-Hoc Networks Workshop*, Turkey, 2004.
15. B. Marchi, A. Grilo, and M. Nunes. DTSN: Distributed transport for sensor networks. In *Proceedings of 12th IEEE Symposium on Computers and Communications*, Aveiro, Portugal, pp. 165–172, 2007.
16. H. Zhang, A. Arora, Y.-R. Choi, and M. G. Gouda. RBC: Reliable bursty convergecast in wireless sensor networks. In *Proceedings of the 6th ACM International Symposium on Mobile Ad Hoc Networking and Computing (MobiHoc)*, Urbana-Champaign, IL, pp. 266–276, 2005.
17. T. Le, W. Hu, P. Corke, and S. Jha. ERTP: Energy-efficient and reliable transport protocol for data streaming in wireless sensor networks. *Computer Communications*, vol. 32, pp. 1154–1171, 2009.
18. C. Y. Wan, S. B. Eisenman, and A. T. Campbell. CODA: Congestion detection and avoidance in sensor networks. In *Proceedings of the First International Conference on Embedded Networked SensorSystems (SenSys'03)*, Los Angeles, pp. 266–279, 2003.
19. L. Q. Tao and F. Q. Yu. ECODA: Enhanced congestion detection and avoidance for multiple class of traffic in sensor networks. *IEEE Transactions on Consumer Electronics*, vol. 56, no. 3, pp. 1387–1394, 2010.
20. B. Hull, K. Jamieson, and H. Balakrishnan. Mitigating congestion in wireless sensor networks. In *Proceedings of ACM SenSys '04*, Baltimore, 2004.
21. P. Levis, N. Patel, D. Culler, and S. Shenker. Trickle: A self-regulating algorithm for code propagation and maintenance in wireless sensor networks. In *Proceedings of First Symposium Networked Systems Design and Implementation (NSDI)*, San Francisco, California, 2004.
22. C. T. Ee and R. Bajcsy. Congestion control and fairness for many-to-one routing in sensor networks. In *Proceedings of ACM SenSys '04*, Baltimore, pp. 148–161, 2004.
23. C. Wang, K. Sohraby, and B. Li. SenTCP: A hop-by-hop congestion control protocol for wireless sensor networks. In *Proceedings of IEEE INFOCOM 2005 (Poster Paper)*, Miami, FL, March 2005.

24. C. Y. Wan, S. B. Eisenman, A. T. Campbell, and J. Crowcroft. Siphon: Over-load traffic management using multi-radio virtual sinks in sensor networks. In *Proceedings of ACM SenSys '05*, San Diego, pp. 116–129, 2005.
25. K. Karenos, V. Kalogeraki, and S. V. Krishnamurthy. Cluster-based congestion control for supporting multiple classes of traffic in sensor networks. In *Proceedings of the 2nd IEEE Workshop on Embedded Networked Sensors*, Sydney, pp. 107–114, 2005.
26. C. Wang, B. Li, K. Sohraby, M. Daneshmand, and Y. Hu. Upstream congestion control in wireless sensor networks through cross-layer optimization. *IEEE Journal on Selected Areas in Communications*, vol. 25, no. 4, pp. 786–795, 2007.
27. M. M. Monowar, M. O. Rahman, A.-S. K. Pathan, and C. S. Hong. Congestion control protocol for wireless sensor networks handling prioritized heterogeneous traffic. In *Proceedings of SMPE'08 with MobiQuitous*, Dublin, Ireland, 2008.
28. X. Yin and X. Zhou. A fairness-aware congestion control scheme in wireless sensor networks. *IEEE Transactions on Vehicular Technology*, vol. 58, pp. 5225–5234, 2009.
29. H. Zhai, X. Chen, and Y. Fang. Improved transport layer performance in multihop and ad hoc networks by exploiting MAC layer information. *IEEE Transactions on Wireless Communication*, vol. 6, no. 5, pp. 1692–1701, 2007.
30. I. Stoica, S. Shenker, and H. Zhang. Core-Stateless Fair Queuing: Achieving Approximately Fair Bandwidth Allocation in High Speed Network. Carnegie Mellon University, Pittsburgh, PA. Technical Report, CMU-CS, pp. 98–136, 1998.
31. G. Bolch, S. Greiner, H. Meer, and K. Trivedi. *Queuing Networks and Markov Chains*. Hoboken, NJ: Wiley, 2006.
32. Y. Sankarasubramaniam, O. B. Akan, and I. F. Akyildiz. ESRT: Event-to-sink reliable transport in wireless sensor networks. In *Proceedings of the 4th ACM International Symposium on Mobile Ad Hoc Networking and Computing*, New York, pp. 177–188, 2003.
33. Y. Zhou, M. Lyu, J. Liu, and H. Wang. PORT: A price-oriented reliable transport protocol for wireless sensor networks. In *Proceedings of 16th IEEE International Symposium on Software Reliability Engineering*, Chicago, USA, pp. 10–126, 2005.
34. Y. G. Iyer, S. Gandham, and S. Venkatesan. STCP: A generic transport layer protocol for wireless sensor networks. In *Proceedings of IEEE ICCCN*, San Diego, 2005.
35. S. Kim, R. Fonseca, P. Dutta, A. Tavakoli, D. Culler, P. Levis, S. Shenker, and I. Stoica. Flush: A reliable bulk transport protocol for multihop wireless networks. In *Proceedings of the 5th International Conference on Embedded Networked Sensor Systems*, New York, pp. 351–365, 2007.
36. N. Tezcan and W. Wang. ART: An asymmetric and reliable transport mechanism for wireless sensor networks. *International Journal of Sensor Networks*, vol. 02, nos. 3–4, pp. 188–200, 2007.
37. Y. Xue, B. Ramamurthy, and Y. Wang. LTRES: A loss-tolerant reliable event sensing protocol for wireless sensor networks. *Computer Communications*, pp. 1666–1676, 2009.
38. J. Paek and R. Govindan. RCRT: Rate-controlled reliable transport for wireless sensor networks. *ACM Transactions on Sensor Networks*, vol. 7, no. 3, Article 20, 2010.
39. F. Shaikh, A. Khelil, A. Ali, and V. Suri. TRCCIT: Tunable reliability with congestion control for information transport in wireless sensor networks. In *5th Annual International ICST Wireless Internet Conference (WICON)*, Singapore, pp. 1–9, 2010.
40. V. C. Gungor and O. B. Akan. DST: Delay sensitive transport in wireless sensor networks. In *Proceedings of the 7th International Symposium on Computer Networks (ISCN)*, Istanbul, Turkey, pp. 116–122, 2006.
41. J. Jones and M. Atiquzzaman. Transport protocols for wireless sensor networks: State-of-the-art and future directions. *International Journal of Distributed Sensor Networks*, vol. 3, no. 1, pp. 119–133, 2007.
42. M. A. Rahman, A. E. Saddik, and W. Gueaieb. Wireless sensor network transport layer: State-of-the-art. In *Sensors*, R. Y. M. Huang (editor), Heidelberg, Berlin, Springer-Verlag: pp. 221–245, 2008.

Chapter 11

Energy-Efficient Medium Access Control Protocols for Wireless Sensor Networks

Mohamed Younis
University of Maryland, Baltimore County

Tamer Nadeem
Old Dominion University

Ali Bicak
Marymount University

Kemal Akkaya
Southern Illinois University

Contents

11.1 Introduction

In the last decade, there have been major advances in the development of low-power microsensors. The emergence of such sensors has led practitioners to envision the networking of a large set of sensors scattered over a wide area of interest.[1–5] A typical architecture of a sensor network consists of many sensing devices that are capable of probing the environment and reporting the collected data, using a radio, to a command center (sink).[6,7] Wireless sensor networks (WSNs) serve many civil and military applications such as disaster management, combat field surveillance, and security.[1] In such applications, the sensors are usually powered using small batteries and deployed in an unattended setup. Therefore, replacing the sensors' battery is not possible or practical. Such energy constraints limit the sensors' lifetime and thus make efficient design and management of WSNs a real challenge.

The limitation of energy supply onboard the sensor nodes has motivated a lot of the research on WSNs at all layers of the protocol stack.[8–14] The network and link layers have received the most notable attention, with the bulk of the work focusing on energy awareness and minimization through clever route setup.[10–14] The main idea of energy-aware routing is to minimize transmission power, which is proportional to the distance squared, through the pursuance of multihop data forwarding so that the cumulative transmission energy is reduced compared with direct sensor-sink communication. Energy-efficient link layer protocols tackle the energy wastage due to collisions among the radio transmission of nodes, keeping the receiver unnecessarily active, and the excessive state changes of the radio circuit.[15–17] In this chapter, we concentrate on the minimization of energy consumption at the link layer.

The rest of this chapter is organized as follows: in Section 11.2, we discuss the energy consumption models for radio circuitries identifying the major affecting parameters. Section 11.2 also includes a detailed analysis of the energy implications of the widely used medium access control (MAC) protocols for contemporary wireless networks. We investigate link layer issues for WSNs in Section 11.3 and enumerate the characteristics of ideal MAC protocols. Section 11.4 reports on the state of the research on energy-efficient MAC protocols for WSNs. Finally, we conclude the chapter in Section 11.5 with a summary and a discussion of possible future research directions.

11.2 Analysis of Energy Consumption at the Link Layer

Recent technological breakthroughs in ultrahigh integration and low-power electronics have enabled the development of tiny battery-operated (0.5–1.5 Ah, 1.2–3.6 V)[1,18–20] sensors. Given the interest in using such sensors in unattended deployment (for usually harsh environments), the replenishment of sensor batteries might be impossible and thus sensors are energy-constrained and their lifetime strongly depends on how long their batteries last. In addition to the sensing ability, which is its main task, a sensor typically performs signal processing and data transmission.

Measurements have shown that out of these three major sensor activities, a sensor expends maximum energy during data communication. This involves transmission, reception, and being idle. For example, Stemm and Katz's[21] measurements have concluded that the ratio of the power consumption of the CPU, memory and display systems on a PDA device to its wireless network interface varies from 1:0.97 to 1:1.88 for different devices and network interfaces.[8] In addition, Dam and Langendoen[20] have shown that the ratios of the processor and the radio power for their EYES sensor nodes varies from 1:12.5 when both are in sleep mode, to 1:4.76 when both are in active mode. Therefore, the sensors' MAC protocol should manage the radio in an optimal way to maintain sufficient sensor energy for the required mission.

In this section, we analyze the energy models of the radio circuitry and the different factors that affect the level of energy consumption in wireless communication. In addition, we study and compare the energy required by contemporary link layer protocols that are widely used in wireless data networks and voice communication infrastructures.

11.2.1 Radio Energy Consumption Models

Typically, a radio can operate in four distinct modes of operation: idle, receive, transmit, and sleep. Although it is expected that the radio consumes the most energy in the "transmit" and "receive" modes, running in the "idle" mode is also costly. In most cases, operating in idle mode results in significantly high energy consumption because the radio electronics have to be turned on and continually decode radio signals, even noise, to detect the presence of incoming packets. Different measurements have shown that the energy consumption ratio of these three modes could be as much as 1:1.05:1.4, 1:1:2.7, and 1:2:2.5, respectively.[21–23] It is thus desirable to completely shut down the radio rather than transitioning into the idle mode. However, switching a radio on and off very frequently can sometimes result in even more energy consumption than leaving the transceiver unit in idle mode because of the start-up power. Moreover, as the transmission packet size gets smaller, the transition energy becomes dominant to the energy consumed during receiving and transmitting of packets.[18,19] Therefore, it is important to take this issue into account when designing energy-efficient MAC protocols.

Radio energy consumption, E_{radio}, could be formulated as simply as:[19]

$$E_{radio} = [(cP_{tx}) + b]T,$$

where c is the transmission power coefficient, b is the constant power offset, P_{tx} is the power used in transmitting the signal, and T is the transmission time. A more complex model of E_{radio}, which extends the formulation of Shih et al.[18] can be expressed as:

$$E_{radio} = [P_{tx}(T_{tx} + N_{tx}T_{st}) + P_{out}T_{tx}] + [P_{rx}(T_{rx} + N_{rx}T_{st})] + P_{idle}T_{idle},$$

where $P_{tx/rx}$ is the power consumed by the transmitter/receiver, P_{out} is the transmitter output power, P_{idle} is the power that the radio uses in the idle mode, $T_{tx/rx}$ is the average time a transmitter/receiver is used each second (actual data transmission/reception time), T_{idle} is the average time per second a node is on and idle, T_{st} is the start-up time of the transceiver and $N_{tx/rx}$ denotes the average number of times per second the transmitter/receiver is turned on. $N_{tx/rx}$ mainly depends on the application's traffic model and medium access arbitration scheme. $T_{tx/rx}$ depends on the packet size, the channel data rate and the average number of packets sent/received per second. P_{out} depends on the distance that the signal travels to the destination and the surrounding terrain. P_{idle} is typically very close to P_{rx}. The power consumption when the radio is in sleep mode is usually 1 to 4 orders of magnitude less than P_{idle}.[22,24,25] A detailed analysis of the power consumption of the radio circuit at the level of the individual function is provided by Wang et al.[19] Considering the above discussion and the energy consumption model, the major sources of energy wastage at the link layer could be enumerated as follows.[26–29] The first source is overhearing, meaning the node receives packets that is destined for other nodes. Second, the overhead of sending and receiving MAC packets. The third source is collisions in which multiple packets are transmitted simultaneously, magnifying the signal interference and thus mandating retransmissions. The fourth is wireless noise in which packets get corrupted and need to be retransmitted or to increase the transmission power to overcome the noise level. The fifth cause of energy wastage is the excessive periods of being in an idle state. Finally, frequent switching between modes, especially switching from sleep mode to an active mode, leads to significant energy consumption.

11.2.2 Contemporary MAC Schemes for Wireless Networks

MAC covers two main issues: resource sharing method and multiple access arbitration. A number of MAC schemes are very popular in wireless networks. The most notable MAC schemes are time division multiple access (TDMA), frequency division multiple access (FDMA), and code division multiple access (CDMA). Apart from these, another popular technique is based on random access called carrier sense multiple access (CSMA). In this subsection, we analyze the trade-offs between the performance and energy consumption of these schemes, setting the stage for further analysis in Section 11.3 in the context of WSNs.

11.2.2.1 TDMA

In the TDMA scheme, time is divided into slots and each node is scheduled, typically through a central controller or a base station, to transmit or receive during specific slot(s). In a node's time slot, the full bandwidth of the channel is dedicated to data communication. As the transmission time is inversely proportional to the signal bandwidth, the transmission time (T_{tx}) in the radio energy consumption model will be minimized. Moreover, the node's knowledge of its allocated communication time slots allows the node to transition to the sleep mode during inactive slots. Therefore, the energy wastage due to overhearing and idle mode can be avoided.

However, the TDMA scheme requires maintaining synchronized clocks throughout the network. Due to drifts, finite error can be introduced among nodes' reference clocks, causing collisions among transmitted data packets. Therefore, the base station usually broadcasts periodic synchronization packets. Each node should not miss these synchronization packets and thus be activated before their transmission. Assuming that T_{guard} is the smallest time gap between two consecutive slots and δ is the highest possible clock drift between the clocks of two sensors, nodes must be resynchronized at least once each T_{guard}/δ to avoid packet collisions. In other words, the nodes must be active at least δ/T_{guard} number of times every second to get resynchronized. Based on the radio energy consumption model, the number of times a node is reactivated to receive a packet should be reduced to keep energy consumption at a minimum and thus T_{guard} must be maximized. However, increasing T_{guard} will decrease the effective bandwidth and increase communication latency. Another issue with TDMA is that once the slots are fixed to the existing nodes, the schedule should be changed when new nodes join or when some of the nodes die, which limits flexibility and adaptability.

11.2.2.2 FDMA

The FDMA scheme eliminates the latency concern of time-based medium arbitration by allowing multiple nodes to communicate simultaneously. The total available bandwidth is divided into multiple channels that nodes are assigned to use. Collisions are minimized because nodes do not have to contend for the same channel. However, in a FDMA-based MAC, a lower bandwidth is available for each node and thus the transmission time T_{tx} gets extended, which translates to an increase in the power consumption. On the other hand, because no synchronization mechanism is required in this scheme, N_{rx} becomes minimal, leading to considerable energy savings. A hybrid TDMA/FDMA mechanism could be used to combine the advantages of both schemes and to overcome some of their shortcomings.[18]

11.2.2.3 CDMA

The CDMA scheme enables simultaneous transmissions with minimal interference and allows a node to receive from multiple senders. In CDMA, each node is assigned a unique code sequence for its transmission. A node spreads out its data over the entire channel bandwidth using its code sequence. The receiver's job is to dispread the bits and extracts only the data from the desired sender. Although CDMA allows transmissions to occupy the entire bandwidth of the channel at the same time, the special coding mechanism narrows the bandwidth for the node's data. Therefore, as in FDMA, T_{tx} is extended. In addition to the complexity of the transceiver's circuit in this schema, each node has to know the code sequences of potential senders. In a multihop network, a node relays data and thus needs large memory to tabulate the codes of most if not all other nodes. Finally, the number of codes can be limited, which hinders the scalability of the approach.

11.2.2.4 CSMA

TDMA, FDMA, and CDMA are contention-free schemes in which nodes are assigned to channels that are partitioned (i.e., in time, frequency, or code space) to avoid collisions. On the contrary, CSMA is a contention-based MAC scheme.[30,31] In CSMA, each node is required to keep sensing the medium searching for a free channel for its transmission. When a node has a packet to send, it transmits at the full channel bandwidth. No a priori coordination among nodes or synchronized clocks is required in this scheme. Using CSMA forces the nodes to be awake for a longer time and consequently increases their energy consumption. In addition, data transmissions experience high collisions in dense networks. Increased collisions among nodes make the transmission delay unpredictable and can lead to a high rate of packet drops. On the other hand, CSMA-based medium arbitration is autonomous and does not require external control. It is also flexible in the sense that new nodes can easily join and leave.

11.3 MAC Layer Issues for WSNs

In this section, we analyze the technical considerations for designing efficient MAC protocols for WSNs. We first outline the design goals and then build on the analysis of energy consumption models for radio circuitries, discussed in Section 11.2, to highlight the energy-related trade-offs and enumerate some of the popular performance metrics.

11.3.1 Design Goals for MAC Protocols in WSNs

In this subsection, we briefly discuss the main design goals for the MAC protocols of WSNs. As will become clear, some of these goals may be conflicting and may force a trade-off.

- *Energy-efficiency:* Energy is a scarce resource for sensor networks. As explained previously, medium access is a major consumer of sensor energy, especially for long-range transmission and when the radio receiver is kept on all the time. The output power of the radio transmitter is directly proportional to distance squared and can be significantly magnified in a noisy environment. Energy-aware routing typically pursues multihop paths to optimize the transmission energy.[10–14]

  On the other hand, energy-conscious MAC can save transmission and reception energy by limiting the potential for collisions, minimizing the use of control messages, utilizing most of the available frequency band to shorten the transmission time, turning the radio to low power sleep mode when it is idle, and finally, avoiding excessive transitions among active and sleep states.
- *Scalability:* Most applications of unattended WSNs will involve a large number of nodes. Therefore, the scalability of the employed protocols is crucial. The resources, that is, time and bandwidth, sharing method, and the arbitration strategy have to allow for fair access to the medium and prevent excessive collisions. In addition, the potential for a large set of communicating nodes would impose a restriction on the use of some MAC schemes such as CDMA. Typically, nodes in WSNs relay other sensors' data and can even perform data aggregation. Pursuing a pure CDMA scheme would require a sensor to store many code sequences, which may be impractical for tiny sensor devices with very limited computational resources.

  It is worth noting that the scalability of the link layer protocols is influenced by the network architecture and routing methodology. Hierarchical network structures can allow the use of multiple resource sharing strategies and shape the network flow into patterns that can be exploited at the link layer. For example, grouping sensors into disjointed clusters allows for the designation of nonoverlapping frequency bands to clusters, similar to FDMA, and applying TDMA or CSMA schemes for intracluster communication among sensors. In addition, the methodology for route setup can rule out some MAC schemes. For example, flooding-style data dissemination makes the time-based medium arbitration strategy impractical.

- *Delay predictability:* A number of applications of WSNs, such as target tracking, require delay-bounded delivery of data. Ensuring the timeliness of data reception is typically handled at multiple layers in the communication stack. For example, special consideration at the network layer would alleviate the burden of long queuing times that affect the overall end-to-end delay.[32–34] However, the link layer would play a major role through careful packet scheduling and predictable strategy for medium arbitration.

  The employed MAC scheme determines the schedule for packet transmission not only for the individual node but for the entire network. At the node level, a suitable packet classification and priority mechanism is the base for a service differentiation that allows delay-centric handling of outgoing packets. On the network level, a well-defined and easily enforced strategy is needed to prevent internode competition for medium access from causing contention that makes the time for packet transmission and reception nondeterministic. For example, collision-based medium arbitration mechanisms such as CSMA would not be appropriate for large and densely populated WSNs because it is not known how many times a node will back off until it successfully transmits. On the other hand, reservation-based approaches such as TDMA would be a good match despite their scalability problems.

- *Adaptability:* In most applications of WSNs, traffic density varies significantly over time and from one part of the network to another. Such observations are valid for both event-triggered and query-based models of network operation.[6,13] For example, in a forest monitoring setup, only periodic status updates are sent in normal conditions whereas many sensor reports are generated in case of detecting a fire. In a typical query-based operation, sensors transmit only in response to requests and little traffic is generated otherwise.

  In addition, generated data can be subject to aggregation en route to the sink. Data aggregation can take the form of averaging the reported data, picking the maximum value, removal of redundant reports, etc.[1,10,11] In some cases, the traffic pattern will not change if the aggregating nodes are fixed at the time of route setup and the transformation of the data is many-to-one, for example, averaging multiple readings. However, data aggregation is mostly performed when applicable and thus can cause variability in the traffic flow. For example, the elimination of redundant sensor readings filters out repetitions and prevents resource wastage. Needless to say that dropping the packets of unneeded data depends on the type of sensors and the detected events. The MAC scheme should adapt to such high fluctuation in traffic and should allow medium access rescheduling to efficiently handle bursts of high-priority traffic.

- *Reliability:* Reliable delivery of data is a classic design goal for all network infrastructures. Guaranteed packet delivery is ensured by the careful selection of error-free links, avoidance of overloaded nodes, and the detection and recovery from packet drops. There is usually a trade-off between the control traffic overhead and the level of reliability. For example, acknowledging each packet minimizes the recovery time and limits its scope, at the price of a high control traffic, which can lower the effective link bandwidth and increase end-to-end delay and energy consumption.

  In wireless networks, packet drops are mainly caused by buffer overflow and signal interference. Avoiding buffer overflow is the responsibility of both the routing and the MAC protocols. Balancing load among available routes would reduce the potential for reaching the maximum capacity of the inbound traffic buffer in relay nodes. Meanwhile, the MAC scheme used determines the buffer management strategy and has to ensure a service rate for the outbound flow that is high enough to stop the number of backlogged packets from exceeding the maximum buffer size. Packet drops due to signal interference can be minimized through the use of sufficiently high transmission power and the prevention of contention for medium access among nodes.

11.3.2 Energy Trade-Offs and Metrics

Based on the described design goals for MAC protocols in WSNs, in this subsection, we highlight sample conflicts among some of these goals with respect to the utilization of sensors' energy and enumerate a list of metrics that can be used to assess the performance of MAC protocols.

11.3.2.1 Energy Trade-Offs

Based on the previous discussion about the quality attributes of MAC protocols for WSNs, one can guess that it is difficult to find a protocol that can be very scalable, extremely energy-efficient, flexible and highly adaptable, robust with reliable packet delivery, and predictable with bounded delay. Therefore, it is expected that these attributes will be valued differently in the various applications of WSNs. However, energy efficiency would probably stay among the top attributes given the constrained sensors' energy supply in unattended deployments. The following is a sample of the trade-offs that the designers would encounter with respect to energy when picking an appropriate MAC protocol:

- Although our analysis of the design goals has indicated that contention-based MAC protocols are disfavored due to scalability concerns, delay unpredictability, high signal interference, and increased potential for energy wastage in collisions, they still fit well for ad hoc network formation. In fact, the lack of centralized coordination and the complexity of resource partitioning in many WSN architectures make contention-based approaches one of the more attractive choices.
- One of the approaches for energy saving is to switch the radio circuitry to sleep mode to avoid energy wastage while staying in idle mode and to limit the number of transitions between sleep and active modes. However, to take advantage of such opportunities for energy reduction, the link layer should pursue a time-based medium sharing, for example, TDMA, with accurate clock synchronization so that state transitions can be appropriately scheduled. An alternative is to use separate channels for data and control messages, which requires two radios. Although TDMA and similar approaches prevent medium contention and make the end-to-end delay deterministic in addition to their energy advantage, they are not suitable for many applications of WSNs. Time-based medium sharing does not scale well because the problem of scheduling time slots subject to flow constraints is NP hard. In addition, time-based approaches are often slow to adapt to changes in the traffic flow and density because control messages have to be prescheduled.
- Despite the fact that CDMA is a good match for most of the design goals of the WSNs in terms of collision avoidance (CA), adaptability, and the support of bounded delay, the resources required for implementing CDMA can overburden the design of sensors. For example, it is not expected that sensors would have large memory that can be designated to

storing the codes of all deployed sensors, which limits the scalability of CDMA. In addition, the bit encoding of CDMA extends the transmission time of a message and thus increases energy consumption. Finally, the complexity and cost of the radio circuitry can also be an issue, especially for largely miniaturized implementations and for disposable use of inexpensive sensors. The designer would have to give up some of the CDMA characteristics such as the uniqueness and the length of the employed codes, to allow feasible implementation on tiny sensor nodes and to maintain a conservative usage of the sensors' energy.

Again, the balanced emphasis of the design goals would have to be based on application requirements.

11.3.2.2 MACs' Performance Metrics

To assess and compare the performance of energy-conscious MAC protocols, the following mix of metrics have been deemed indicative by the research community:

- *Ratio of energy wastage to total communication energy:* This metric measures the efficiency of energy utilization. A protocol with low overhead, few collisions, and little idle time would demonstrate small ratio of energy wastage.
- *Ratio of control packets to data packets:* This metric is to assess the control packet overhead with respect to all the data packets sent. The goal of a MAC protocol is to reduce this overhead for increased node/network lifetime and reduced interference/overhearing in the network.
- *Average consumed energy per packet:* This metric is dedicated to assessing the quality of active mode operations. A protocol that has low average energy per packet typically encounters few packet retransmissions and experiences few contention-related collisions. It is worth noting that the use of this metric to capture the performance of the MAC protocol should exclude the effect of the output power, which is proportional to distance and is usually influenced by route setup.
- *Average and standard deviation of node lifetime:* Because the usefulness of the WSN depends on the availability of a sufficiently large number of sensors to cover the area of interest, extending the lifetime of a sensor node is very important. This metric gives a good measure of the overall network lifetime. A slow depletion rate extends the life of the sensor battery and lengthens the duration that a sensor can function. Minimizing the standard deviation of node life makes the sensor coverage and the network connectivity more predictable. It should be noted that many other factors such as the data routing mechanism and the selection of active sensors contribute to this metric. However, an energy-inefficient MAC protocol can still have a noticeable negative effect.
- *Error rate:* This metric indicates the number of lost packets in the WSN. Packet drops are mainly caused by buffer overflow, collisions due to hidden node problem, and signal interference. A MAC protocol, which observes the buffer size limitations and employs an effective packet scheduling at the node and network level, would minimize the occurrence of packet drops. In addition, signal interference can lead to low signal-to-noise ratios and hinder the recipient's ability in decoding the transmitted packet. MAC protocols that reduce the potential for collisions among nodes would experience fewer packet decoding errors.
- *Packet delivery ratio:* This metric is also related to the error rate metric and indicates the ratio of the number of transmitted packets to the number of correctly received packets. The goal is to increase this ratio so that energy, delay, and interference are minimized in the network.

- *Network throughput:* Defined as the total number of packets received at the sink per time unit. Many factors, for example, the setup of the network topology, affect this metric and not just the choice of the MAC protocol. However, the contribution of the MAC protocol is very significant. A high network throughput indicates a small error rate for packet transmission and a low level of contention for medium access.

11.4 State of the Research

Contemporary MAC layer protocols designed for wireless devices such as MACAW[35] and IEEE 802.11[36] are not suitable for WSNs. These schemes consume considerable amounts of energy because they require the sensors to continuously probe the medium. In addition, these schemes require nodes to transmit control packets to avoid collisions. The control packet sizes will be comparable to the size of data packets, which are small in most sensor applications. On the other hand, Bluetooth[37] uses a TDMA-based scheme assuming that all slave nodes are within transmission range of the master node (i.e., Piconet). The transmission range of the nodes is very limited compared with that of the sensors'. Multihop transmissions can be achieved via Scatternets but it is limited and thus energy savings via multihopping is not a goal.

Power management of the radio has gained significant importance in WSNs because the radio is a major consumer of sensors' energy.[38,39] Several methods have been suggested to reduce the energy consumption of the RF circuitry in light of the energy consumption model described in Section 11.2. Energy-conscious MAC protocols for sensor networks found in the literature can be broadly classified into three categories: contention-based, reservation-based, and hybrid protocols. In this section, we report on some of the techniques used in each category. In addition, we also highlight some emerging MAC protocols whose primary goal is not energy efficiency but still provide energy efficiency as a secondary goal. However, we do not discuss the MAC protocols, which have paid little attention to energy while focusing on other performance metrics such as timeliness.[40,41]

11.4.1 Contention-Based Protocols

Contention-based MAC protocols have been the main choice for distributed sensor architectures in which the network infrastructure and access points are not well-defined. Most of the contention-based MAC protocols proposed in the literature follow the operational model of CSMA, incorporating handshaking signals and a back-off mechanism to reduce the probability of collisions. However, the pursued techniques for energy conservation vary. Some focus on energy wastage due to collisions suggesting clever power control to limit the level of interference, the use of different channels for data and control traffic, etc. Other published work exploit energy saving through shortening the time a radio circuitry spends in idle mode. This is typically done by putting the sensors into sleep mode, reducing the duty cycle time during which the sensors need to be active. In this subsection, we discuss the basic idea of some of these techniques.

11.4.1.1 SmartNode

The SmartNode MAC protocol[42] extends the IEEE 802.11 standards. Nodes in the SmartNode network attempt to derive the minimum transmission power to reach the other nodes from the power strengths of received packets. Each node in the SmartNode system maintains a lookup table to store the unique identifiers of the neighbors it knows about with the minimum transmission powers required to reach those neighbors. Before a source node transmits a data packet, it looks up for the destination node in the table. If the lookup fails, the source node transmits the ready-to-send (RTS) packet at the maximum power level; otherwise, the value stored in the lookup table is used for setting the power level of the RTS packet. An additional field is added to the RTS packet to store the used transmission power. Upon overhearing the RTS packet, each neighbor node estimates the minimum transmission power required and stores this value in its lookup table. To maintain compatibility with the IEEE 802.11 protocol, the destination node transmits unmodified clear-to-send (CTS) packet with the same power level indicated in the RTS packet. Hence, the source node is then able to determine the minimum transmission power required for the data packet to reach the destination node and store such value in its lookup table. Although SmartNode reduces the power transmission level, which saves energy, it has some disadvantages. SmartNode still consumes energy for the RTS/CTS control packet transmissions. In addition, using different transmission power levels increases the collision rate in the network.[43] Fairness is another problem in SmartNode because data packets are transmitted with low power, making it easy for them to get corrupted and dropped. It is worth noting that a variant of this power control scheme were pursued in power control medium access control (PCM),[43] power controlled multiple access (PCMA),[44] and dual busy tone multiple access (DBTMA),[45] among others.

11.4.1.2 Power-Aware Medium Access and Signaling Protocol

The power aware medium access and signaling (PAMAS)[16] protocol is a CSMA-based protocol in which the nodes that are not actively transmitting or receiving should power themselves off. The approach requires the nodes to use two separate channels for control and data. The control channel is used for handshaking and the data channel for regular traffic. Using two channels minimizes the potential for collisions. A node senses the data channel and responds to connection requests only if its neighbors are not transmitting or receiving. Senders that cannot establish a connection switch to sleep mode and retry later. The duration of a node's stay in sleep mode is determined based on the exchange of special probe messages on the control channel among nodes in the proximity. Switching nodes that are not participating in communication to sleep mode has been shown to result in energy savings of up to 70%. However, the protocol requires the nodes to sense the medium to transmit and does not eliminate collisions completely. In addition, the protocol requires the nodes to have two separate channels (control and data), which will require two radios at each node—increasing the cost, size, and complexity of the sensor design. The use of multiple radios and channels has recently become a popular choice touted to increase the throughput by clever channel assignment techniques. We will look at the MAC protocols using multiple channels in more detail in the next section.

11.4.1.3 Sensor-MAC

The sensor-MAC[26] (or S-MAC) is a contention-based protocol proposed for energy-constrained networked devices. The S-MAC protocol essentially trades energy for throughput and latency by utilizing the sleep mode of the radio. Basically, nodes are switched to the sleep mode for scheduled periods of time. When a node becomes active, it transmits its ID, indicating its readiness to receive messages. A sender has to wait to hear the ID of the receiver node before transmission. The listen interval of a node is divided into two parts, the first of which is reserved for SYNC packets and the other for receiving RTS packets. Figure 11.1 shows the timing relationship of three possible scenarios for a sender transmission to a receiver. CSMA stands for carrier sense. Sender 1 sends only a SYNC packet whereas sender 2 wants to send data and sender 3 transmits SYNC and RTS packets.

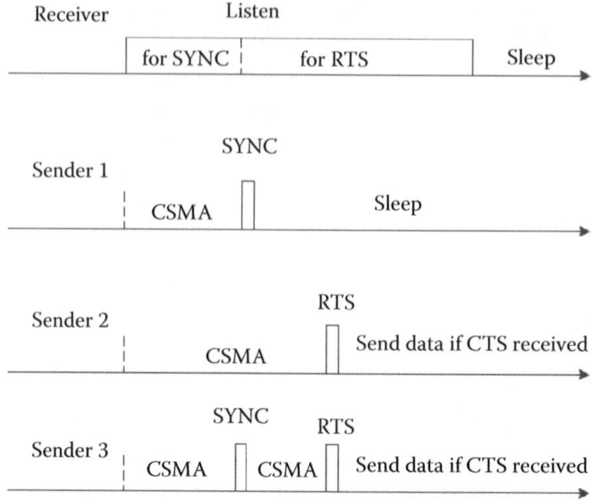

Figure 11.1 S-MAC timing between a receiver and multiple senders.

It is worth noting that a similar mechanism is pursued for Piconet.[46] If multiple senders are waiting, they have to contend for connection with the receiver node. Nodes share their sleep schedules to prevent a sender from spending excessive idle time awaiting a receiver to become active. Relative timestamps are exchanged to mitigate the effect for clock drifts. Throughput is reduced because only the active part of the frame is used for communication. Latency increases because a message-generating event may occur during sleep time and will be queued until the start of the next active part. In addition to latency, this approach sacrifices per hop fairness.

T-MAC[20] extends S-MAC by dynamically adjusting the duration between sleep intervals in which sensors are awake based on the communication of nearby neighbors. To reduce number for switching, T-MAC introduced additional overhead control packets to prevent nodes from early sleep switching. However, the scalability of the S-MAC (and the T-MAC) is questionable. The scheduling of sleep time can be an overburden in large networks and the end-to-end delay for packet delivery can be unpredictable.

11.4.1.4 Berkeley-MAC

Berkeley-MAC (B-MAC)[47] utilizes an adaptive preamble sampling scheme to reduce the duty cycle. It allows each node to have an independent sleep schedule, whereas a sender node precedes the data packets with a preamble that is slightly longer than the sleep period of the receiver. With the extended preamble, the sender is assured that, at some point, the receiver will wake up, detect the preamble, and remain awake to receive the data. B-MAC also supports adaptive control through bidirectional interfaces that allow services to reconfigure the MAC protocol according to their operating conditions.

11.4.1.5 Convergent MAC

Convergent MAC (C-MAC)[48] allows sensors to operate in low-duty cycles by avoiding the periodic synchronization messages used to schedule sleep time. When there are no packets to transmit, C-MAC uses unsynchronized duty cycling. While transmitting, it starts forwarding packets via anycast but then gradually converges to route-optimal unicast with synchronized scheduling.

C-MAC exhibits similar throughput and latency as CSMA/CA using less energy, and outperforms the aforementioned contention-based MAC protocols like S-MAC and B-MAC.

11.4.1.6 X-MAC

X-MAC[49] overcomes the extended preamble and excessive latency issues of duty-cycled MAC protocols by employing a strobe-based preamble approach. It lets the sender transmit a series of short preamble packets in which the receiver gets a chance for early acknowledgment during the pauses in between. Compared with B-MAC, X-MAC provides significant energy savings at both the transmitter and receiver, and also reduces per-hop latency.

11.4.1.7 Informative Preamble Sampling MAC

The informative preamble sampling MAC (IPS-MAC)[50] protocol allows a sender to embed information about its intended receiver while the preamble is transmitted. This results in fewer nodes staying awake for each transmission. A decision-making algorithm is used to help the receiver determine whether that preamble is intended for it. IPS-MAC also optimizes the operating ranges in terms of transmission power with which it can improve the lifetime of nodes by a factor of two over B-MAC.

11.4.1.8 Demand-Wakeup MAC

The demand-wakeup MAC (DW-MAC)[51] protocol uses a scheduling algorithm to have the sensors wake up only on demand during their sleep period, which ensures a collision-free transmission. As traffic load increases, the demand wakeup scheme adaptively increases effective channel capacity, allowing DW-MAC to achieve lower latency under a wide range of traffic loads.

11.4.1.9 Receiver-Initiated MAC

Receiver-initiated MAC (RI-MAC)[52] provides an asynchronous duty cycle scheme via receiver-initiated data transmission. RI-MAC minimizes the time that a sender and receiver occupies the wireless medium for exchanging data. Compared with asynchronous duty cycling approach of X-MAC, RI-MAC achieves higher throughput, packet delivery ratio, and power efficiency. Figure 11.2 provides an overview of the operation of RI-MAC, in which a DATA frame transmission is always initiated by the intended receiver. Each node periodically wakes up and broadcasts a

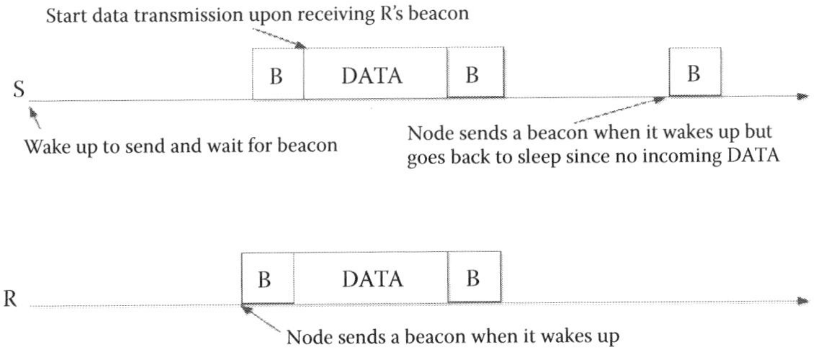

Figure 11.2 Overview of RI-MAC protocols.

beacon. When node "S" wants to send a DATA frame to node "R," it stays active silently and starts DATA transmission upon receiving a beacon from R. Node S later wakes up but goes to sleep after transmitting a beacon as there is no incoming frame for it.

11.4.1.10 Adaptive Transmission Rate

A CSMA-based MAC protocol with an adaptive rate control mechanism was proposed by Woo and Culler[53] for WSNs. In this protocol, a node that has a data packet to transmit senses the medium for a random period. If the medium is idle during this period, a channel reservation process using simple RTS/CTS control packets is started. Otherwise, the node applies a binary exponential back-off scheme and then retries the transmission. During the back-off period, the node does not have to keep sensing the medium and it can turn off its radio to conserve energy. In addition, the protocol eliminates data ACK packets to further reduce energy consumption. To decrease the probability of collisions among nodes that detect the same event, each node, upon detecting the event, applies a random delay before starting the transmission process. A passive rate control mechanism is used to maintain fair bandwidth for both the originating data of a node and the route-thru traffic that passes through it. Basically, each node periodically attempts to inject a packet into the network. When that packet is successfully injected, it signals that the road still has the capacity for more traffic and thus, the node can increase its transmission rate linearly. However, if the injection of that packet was not successful, it signals that the road is jammed and the node decreases its rate of originating data using a multiplicative scheme. Route-thru traffic will adapt to the traffic of the original data using a similar mechanism. If a node injects lots of original traffic into the freeway, the route-thru traffic will be hindered and thus, the rate of transmitting route-thru traffic will decrease. This decrease will propagate deep down into the network which ultimately decreases the amount of aggregate route-thru traffic. Although the approach limits collisions and strives to maintain fair bandwidth splitting among nodes, energy conservation is limited in scope. The issues of overhearing and minimizing the time that the radio is in idle mode are not addressed.

11.4.2 Reservation Based Protocols

Time-based medium access has the potential to capture most of the opportunities for energy optimization in WSNs. As discussed in Section 11.2, energy wastage due to overhearing, collision, idle mode, and transitions between different states can be minimized if medium access is shared on a time basis. In addition, time-based medium arbitration can enhance delay predictability and limit packet drops due to interference and buffer overflow. However, the problem of scheduling access to the medium is NP hard, making the scalability of time-based MAC schemes a major concern. Moreover, distributed time-based medium arbitration typically introduces excessive overhead. In addition, maintaining clock synchronization among nodes is essential to enforce the schedule, which is a nontrivial problem for the resource-constrained sensor nodes.[54] Most of the time-based MAC protocols proposed in the literature have focused on addressing these issues either using reservation requests over preset data routes or pursuing simplified heuristics to tackle the complexity of medium access scheduling. This subsection summarizes some of the published techniques.

11.4.2.1 Collision-Free MAC

Jolly and Younis[55] have proposed an approach for centralized network management setup in which the sink establishes the routes and schedules the sensors' transmission. The basic idea is to take advantage of the flow of routing traffic from the sink and include a reference clock value. Because all the sensors have to switch on to receive route and schedule updates, the overhead that the sensors would have incurred in switching on to listen to synchronization messages will be eliminated. Because the frequency of route updates for some networks may be insufficient to maintain the desired level of clock synchronization, they have exploited the trade-off between the guard time and the frequency of resynchronization. The guard time is a precautionary measure used to tolerate the difference in clock readings of communicating nodes. Increasing the guard time will require a sensor node to be activated earlier than its reception slot to tolerate a clock drift.

11.4.2.2 Single-Sink Setup

The use of reservation requests has been explored for tackling the scalability of time-based medium arbitration.[56] Nodes that have data to transmit make a reservation request to a base station, which responds with a traffic control message indicating medium access schedule. Nodes that are not included in the traffic control message can turn off their radio receivers. The nodes that have been assigned slots transmit in the order scheduled by the base station. The base station trades off latency with energy-efficiency. Although it is better to bundle all transmissions from a node into consecutive time slots, the transmission of other nodes will be delayed. Arisha et al.[57] have pursued a non–reservation based approach that considers the routing paths. A depth and breadth first strategies were investigated for scheduling packets over a multihop paths. Their analysis has concluded that in the absence of buffer size constraints, breadth first can be very energy-efficient. Although the focus on active routes limits the size of the scheduling problem, relying on a single controller and ignoring buffer size constraints raises concerns about applicability of these approaches for large networks.

11.4.2.3 Multicluster Scheduling

Jolly and Younis[55] have proposed a comprehensive approach for energy-efficient, time-based medium arbitration in sensor networks. Scalability is achieved through network clustering. Sensors are grouped around gateway nodes that are less energy constrained than sensors. Each gateway is assigned the responsibility of cluster management, setting up multihop routes and assigning transmission/reception slots to sensors. Time slots are assigned to communicating sensors within the cluster to achieve efficient utilization of the energy resources. A breadth first–based slot assignment heuristics is suggested to conserve the sensors' energy by minimizing the number of transition between active and sleep modes and the duration in which active sensors are idle. Moreover, the slot assignment mechanism observes the buffering limitation at the sensors' node to prevent packet drops. The approach further boosts the effective network bandwidth by allowing the transmissions within the different clusters to overlap. To avoid intercluster interference, the slots assigned to nodes close to the cluster boundaries are further analyzed to detect potential collisions. After each gateway assigns slots to sensors in its cluster, it broadcasts the transmission schedule to peer gateways that manage other clusters. A gateway is elected to perform the multicluster analysis to detect and resolve collision. Gateways can rotate this responsibility to balance the load or use other criteria. Collisions can be simply detected through the comparison of the assignment of each time slot in the different clusters and the transmission range of the designated sensors. Resolving intercluster collisions can lead to modifying the medium access schedule within the individual cluster in a way that diminishes the intracluster MAC level efficiency. Alternative resolution can be through the extension of the TDMA frame size, which may boost the end-to-end latency. A

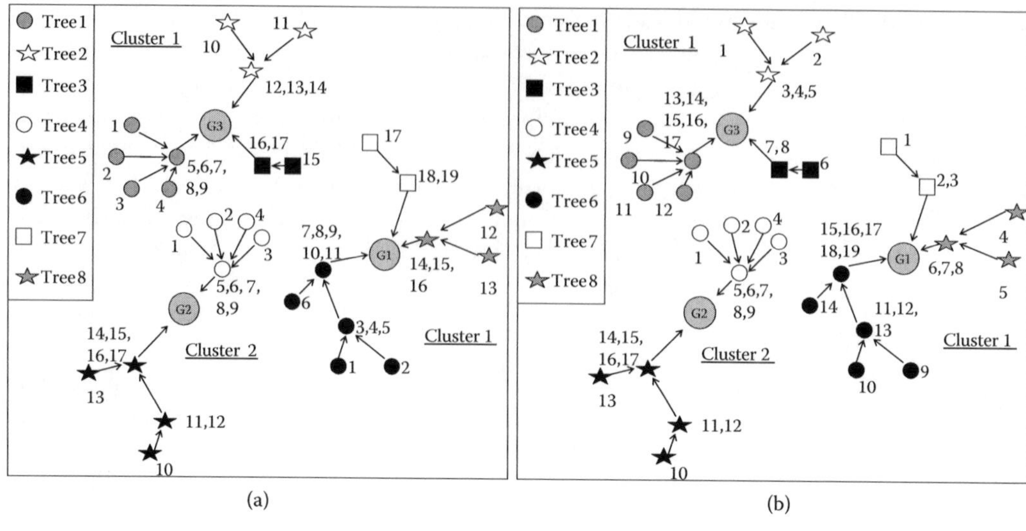

Figure 11.3 Assigned transmission slots are annotated next to each node. Before applying the intercluster collision avoidance heuristics, the transmission of nodes on tree 4 interfere with nodes on trees 1 and 6, and similarly for nodes on trees 3 and 7 (a). Swapping the order of slots allocated to trees in clusters 1 and 3 eliminates the potential of collisions (b).

simple heuristics has been proposed for minimizing the effect of intercluster collision resolution. Basically, slots assigned to routing trees within the same cluster are swapped to prevent the overlap between the transmissions of boundary nodes. Figure 11.3 illustrates the proposed approach through an example.

11.4.2.4 BitMAC

BitMAC[58] is a TDMA-based MAC protocol designed for dense WSNs with low mobility. It allows concurrent transmissions through a "bitwise" transmission scheme, which requires strong synchronization and an on/off keying scheme. BitMAC builds a spanning tree consisting of hierarchically connected star networks in which each star network utilizes an on-demand TDMA schedule. To allow concurrent operation, neighboring star networks are assigned to different channels via a graph coloring algorithm. Drawbacks of BitMAC are its complexity and need for bitwise synchronization.

11.4.2.5 Dynamic, Energy-Efficient MAC

Dynamic, energy-efficient MAC (DEE-MAC)[59] protocol proposes energy savings by forcing the idle listening nodes to sleep according to the synchronization scheme performed at the cluster heads. DEE-MAC operations are comprised of rounds within dynamic clusters. Each round includes a cluster formation phase and a transmission phase. The cluster head builds a TDMA schedule that is broadcast to all nodes, then each node is awakened according to the schedule if it has any data to receive or send. With the help of its clustering and TDMA-based scheme, DEE-MAC reduces the cost of idle listening in large WSNs.

11.4.2.6 Lightweight Medium Access Protocol

The lightweight medium access (LMAC)[60] protocol is a TDMA-based MAC protocol designed for stationary WSNs. It divides time into frames in which each node gets a slot that it can transmit. Every slot is further divided into phases in which one is used to send control messages and the other for data transmission. LMAC reduces the energy wasted in preamble transmissions, and compared with SMAC, it significantly extends the lifetime of nodes.

ML-MAC[61] is an adaptation of LMAC for mobile WSNs. ML-MAC is also a TDMA-based protocol, but in ML-MAC nodes can establish TDMA schedules on demand or join/leave existing schedules while they are moving. Thus, LMAC achieves high channel throughput in mobile WSNs even with heavy load traffic. A control message in ML-MAC contains a byte for the ID of the sender, with five bits for *Slot Number* and three bits for its *Status*. Then, there is the *Occupied Slots* field that identifies the used slots of all neighbors. In the case of node 4 in Figure 11.4, the slots 3, 4, 5, 6, and 8 are marked as used, so the *Occupied Slots* in binary is 00111101. Figure 11.4 shows how slots are chosen for the eight neighbor nodes, in which node 2 is not yet synchronized. Receiving control messages 10000100 (from node 1), 00111000 (from node 3), 00111101 (from node 4), the combined message for node 2 will be 10111101. It means that node 2 can choose between slots 2 and 7. Then, the *Occupied Slots* for the node will be either 11110000 if it chooses slot 2 or 10110010 if it chooses slot 7.

11.4.2.7 Traffic Adaptive MAC

Traffic-adaptive MAC (TRAMA)[62] is another TDMA-based MAC protocol designed to utilize TDMA in an energy-efficient manner. For each time slot, a transmitter is selected via a distributed election algorithm. TRAMA consists of three components: the neighbor protocol (NP), the schedule exchange protocol (SEP), and the adaptive election algorithm (AEA). Nodes start in random access mode, and during this initial period, NP propagates small packets using signaling slots

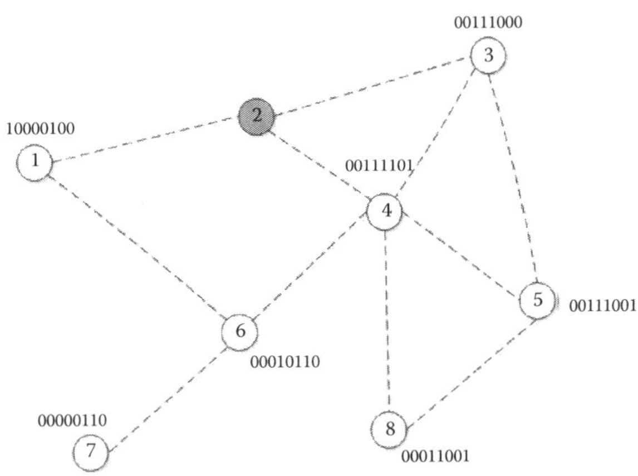

Figure 11.4 Illustrating the adaptive TDMA operation in ML-MAC. (Redrawn from S. Mank et al., An adaptive TDMA based MAC protocol for mobile wireless sensor networks. In *Proceedings of the 2007 International Conference on Sensor Technologies and Applications (SENSORCOMM '07)*. Washington, DC: IEEE Computer Society, 62–69, 2007.)

to collect any updates from neighbors. With the second component, SEP protocol, traffic-based schedule information is maintained among neighbors. SEP packets are exchanged during the scheduled-access periods. The third component, AEA, selects transmitters and receivers according to information obtained from NP and SEP. TRAMA has high delays compared with S-MAC and is therefore only suitable for applications that are not delay sensitive. However, the delivery ratio is much better than that of S-MAC.[63]

11.4.3 Hybrid Approaches

As discussed in Section 11.2, each of the CSMA-, CDMA-, FDMA-, and TDMA-based schemes offer some advantages and experiences some shortcomings with respect to the requirements of MAC protocols for sensor networks. Not surprisingly some researchers have tried combining several of these schemes to better address these requirements. In this subsection, we briefly discuss a sample of such protocols.

11.4.3.1 Power-Aware Reservation-Based MAC

The power-aware reservation-based MAC (PARMAC)[64] is an energy-aware protocol primarily designed for ad hoc networks and is applicable to sensor networks as well. The approach is actually a combination of contention and reservation-based medium arbitration schemes. The network is divided into grids and each node is assumed to reach all the other nodes within its grid. Time is divided into fixed frames. Grids are assigned distinct frames. Each frame is composed of reservation period (RP) and contention-free period (CFP). In each RP, nodes within a grid cell exchange three messages to reserve the slots for data transmission and reception and the exchange of acknowledgements. Data is then sent in the CFP. The clocks of all nodes are assumed to be synchronized. The protocol saves energy by minimizing the idle time of the nodes and allowing the nodes to sleep during a CFP. Moreover, intragrid control packets overhead and packet retransmissions are minimal, achieving significant energy savings. However, intergrid contention is still possible and the efficiency of this approach can significantly diminish if the application requires data exchange among nodes in different grids.

11.4.3.2 Distributed Energy-Aware Node Activation

The distributed energy-aware node activation (DEANA)[65] approach exploits the node's awareness of its neighbors in scheduling transmissions in a noncontending way. A single signaling channel is used and medium is arbitrated in a time-sharing manner. The protocol follows two cycles—a scheduled-access cycle used for data and a random-access cycle. The latter is used for neighbor discovery in case nodes are relocated or if a node is added to or removed from the network. The scheduled-access cycle is partitioned into time slots for the node's transmission. Each time slot is further split into control and data partitions. Nodes contend for time slots. The node that is allocated the slot will use the control portion to broadcast the receiver ID and then follow on with the data. The identified receiver will stay on and the other neighbor nodes will transition to a low-power sleep mode. Again, the clocks of all nodes are assumed to be synchronized. However, the DEANA approach does not consider the state transition energy in the decision of whether to switch to the sleep mode or not. In addition, in a dense node deployment, the protocol can be unpredictable because the set of neighbors will be large, making it hard to expect the outcome of the contention for slots.

11.4.3.3 Self-Organizing Sensor Networks

A self-organization mechanism using a contention-free TDMA medium access protocol for sensor networks has been proposed by Sohrabi and Pottie.[66] At the beginning, nodes operate in random access mode for booting up the network. In this mode, each node listens for other nodes on a fixed channel. When the first node is found, a two-node subnet network is formed and a new TDMA schedule is formed. The subnets continue to grow as they find new unattached nodes or merge with other subnets. Existing TDMA schedules are modified to accommodate the newly formed subnets. After the boot-up period, nodes switch to a TDMA mode in which they follow their own local schedule and go through a sequence of transmission and reception bursts. To make the system scalable, no central entity and no global TDMA schedule are required. Therefore, a node can experience interferences from other nodes that selected the same transmission slots locally. To alleviate such problems, multiple separate channels, that is, different frequencies (FDMA) or distinct spreading codes (CDMA), are used and each node randomly chooses one of those channels for its transmissions.

11.4.3.4 Z-MAC

Z-MAC is a hybrid MAC[67] combining the strengths of TDMA and CSMA. It uses CSMA as the baseline MAC scheme and a TDMA schedule to enhance contention resolution. This design results in high initial overhead, which is amortized over a long period of network operation and eventually improves the throughput and energy efficiency. The protocol uses an efficient and scalable channel scheduling algorithm for channel reuse and slot assignment. Unlike TDMA, a node may transmit during any time slot by performing carrier sensing and transmitting a packet when the channel is clear. By combining CSMA and TDMA, Z-MAC becomes more robust to timing failures, time-varying channel conditions, slot-assignment failures, and topology changes.[68]

11.4.3.5 Emergency Response MAC

Emergency response MAC (ER-MAC)[69] is a hybrid design of TDMA and CSMA, giving it flexibility to adapt to traffic and topology changes. Nodes wake up only at time slots determined by a TDMA-based schedule, and sleep otherwise to conserve energy. In case of an emergency, nodes that participate change their MAC behavior by allowing contention in TDMA slots. This way, ER-MAC outperforms Z-MAC with higher delivery ratio, lower latency, and lower energy consumption.

11.4.3.6 Funneling-MAC

Funneling-MAC[70] is a localized, sink-oriented MAC for boosting fidelity. WSNs exhibit a unique funneling effect as a result of the distinctive many-to-one, hop-by-hop traffic patterns. Therefore, the traffic intensity, collisions, congestion, packet loss, and energy consumption all increase as events move closer toward the sink. Funneling-MAC runs on a network-wide CSMA/CA scheme, with a localized TDMA algorithm overlaid in the funneling region (i.e., within a small area around the sink). In this way, it avoids the scalability issues associated with the network-wide deployment of TDMA. As a result, funneling-MAC improves the throughput and energy efficiency, and significantly outperforms B-MAC and Z-MAC. Figure 11.5 illustrates the concept.

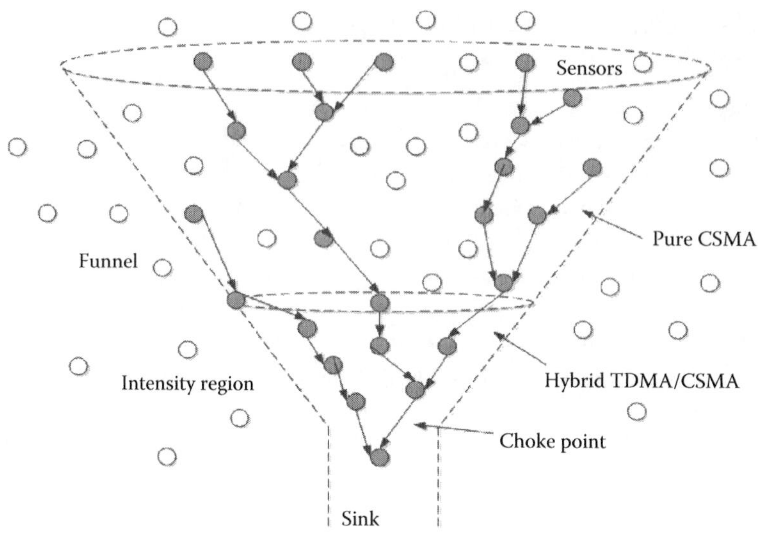

Figure 11.5 Illustrating the funneling effect in WSNs (Redrawn from G.-S. Ahn et al., Funneling-MAC: a localized, sink-oriented MAC for boosting fidelity in sensor networks. In *Proceedings of the 4th International Conference on Embedded Networked Sensor Systems (SenSys '06)*. New York, NY: ACM. 293–306, 2006.)

11.4.3.7 IEEE 802.15.4 Standard

Most of the current sensor products (e.g., Micro Motes) in the market employ an IEEE standard for medium access. IEEE 802.15.4[62] is designed for low-rate wireless personal area networks (WPAN). It has been widely used as the de facto standard along with the ZigBee[71] standard at the network layer in WSNs. The main idea of IEEE 802.15.4 is based on CSMA/CA, which is a contention-based approach. However, there is also an optional CFP for guaranteed access to provide certain delay guarantees. In the CFP, a TDMA-based approach is followed. The IEEE 802.15.4 standard works on different node topologies, namely, star and peer-to-peer. In these topologies, there is a PAN coordinator that can communicate with all the sensors in one hop. Two versions of CSMA/CA based on slotted or unslotted access can be used. In case of slotted CSMA/CA, PAN coordinator arranges everything from synchronization to beacon announcements. The standard mainly saves energy by reducing the transmission range of sensors and the amount of data transmitted. Thus, it cannot handle applications with higher data rates.

11.4.4 Emerging MAC Protocols

Emerging MAC protocols for WSNs propose more innovative designs in terms of combining features or resources in multiple layers or channels. We review these recently emerging MAC protocols under three groups, namely, multichannel, cross-layer, and quality of service (QoS)–based approaches.

11.4.4.1 Multichannel MAC Protocols

Multichannel MAC protocols recently become more popular with the availability of multichanneling capability in the sensor radios. The idea is to use different channels simultaneously to boost the network throughput. Although the use of multiple channels can provide opportunities for parallel transmissions in the same neighborhood, it raises the issue of careful channel assignment. In this subsection, we summarize some sample MAC protocols that utilize multichannels. One key observation is that although energy efficiency was not the primary goals of these protocols, multichannel protocols are more energy-efficient than single-channel MAC protocols under heavy traffic conditions.

11.4.4.1.1 Y-MAC

Y-MAC[72] is a TDMA-based multichannel MAC protocol that utilizes a lightweight channel-hopping scheme enabling multiple nodes to transmit simultaneously over multiple channels. Y-MAC avoids redundant channel assignments by not allocating fixed channels to the nodes. Initially, messages are exchanged on the base channel. When a traffic burst occurs, a receiver and potential senders hop to one of the other available channels, according to the hopping sequence. Because these messages are carried over additional channels, each node is guaranteed to receive at least one message on the base channel. This helps Y-MAC achieve both high performance and energy efficiency under heavy traffic conditions.

11.4.4.1.2 Efficient Multichannel MAC

The efficient multichannel MAC (EM-MAC)[73] protocol proposes an adaptive receiver-initiated multichannel rendezvous scheme. Combining this scheme with a random channel hopping algorithm, it optimizes the selection of channels. This better channel utilization helps EM-MAC improve the transmission efficiency while maintaining low energy consumption.

11.4.4.1.3 Packets in Pipe

Packets in pipe (PIP)[74] is a multichannel, TDMA-based MAC protocol for efficient and scalable bulk transfer in WSNs. With its centralized but multichannel and multihop connection scheme, PIP achieves better throughput under variable error rates even without any flow control. PIP uses a technology similar to 802.15.4, which has 16 different channels in the 2.4 GHz ISM band. Channel-hopping introduces a small but noticeable overhead, yet the benefits of multichannel operation far outweigh it. PIP can also be integrated with duty cycling, and support streaming data with little overhead.

11.4.4.2 Cross-Layer Approaches

The idea in cross-layer approaches is to utilize the available information at different layers of the protocol stack so that the performance of the MAC protocols can be further improved. Information from both the physical and network layers can be used in these approaches. Recent cross-layer MAC protocols demonstrate that cross-layer approaches can achieve far better performance than protocol layers working in isolation. Some of the sample protocols are discussed in this subsection.

11.4.4.2.1 BulkMAC

BulkMAC[75] is a duty cycling based cross-layer MAC protocol for WSNs, and supports the transmission of multihop multiple packet flows during a single sleep period. By improving channel allocation using the upper routing layer information, BulkMAC achieves far better throughput compared with routing-enhanced MAC (RMAC) and pipelined RMAC (PRMAC) protocols.

11.4.4.2.2 Low-Latency Sensor MAC

Low-latency sensor MAC (LLS-MAC)[76] is an adaptive low-latency MAC protocol based on cross-layer architecture. With its adaptive cross-layer design, LLS-MAC minimizes end-to-end delay even under heavy traffic. LLS-MAC adopts the periodic sleep schedule and synchronization scheme specified in S-MAC, but for continual data transmission, it accesses the interface queue and adjusts the transmission order to allow the sender to transmit packets continually when the interface queue is too long.

11.4.4.2.3 Cross-Layer MAC

Cross-layer MAC (CL-MAC)[77] protocol uses two adjacent layers (MAC and network) to conserve energy. The basic idea behind CL-MAC is to wake up only nodes on a routing path from the source to the base station (sink) by exploiting routing information while the remaining nodes stay in sleep mode.

11.4.4.3 QoS-Based Approaches

As WSNs evolve toward higher transmission rates, recent designs consider sensing and transmission of real-time traffic, such as audio and video. Such multimedia traffic requires QoS guarantees at different layers including the MAC layer. QoS-based MAC protocols should not only meet the performance requirements but also consider energy efficiency. In this subsection, we describe some of the recent MAC protocols that can provide certain QoS.

11.4.4.3.1 QUATTRO

QoS-capable cluster-based, time-shared, routing-assisted (QUATTRO) MAC protocol[78] proposes an architecture in which the MAC and routing protocols collaborate to discover and reserve routes, to organize nodes into clusters, and schedule medium access in a time-shared fashion. As a consequence, not only is QoS achieved but also great energy savings are achieved by eliminating collisions and considerably reducing idle listening.

11.4.4.3.2 Diff-MAC

Diff-MAC[79] is a QoS-aware MAC protocol with differentiated services and hybrid prioritization for wireless multimedia sensor networks. Diff-MAC increases channel utilization with effective service differentiation mechanisms while providing fair and fast delivery of the QoS-constrained data. It adapts its duty cycle according to the dominating traffic, and also adjusts its contention window size to minimize the latencies. In addition to these adjustments, Diff-MAC can also prioritize data packets among nodes or queues on each node. With the help of a built-in fragmentation feature, Diff-MAC divides multimedia data into smaller chunks and reserves the medium to send these packets as a burst. Compared with S-MAC, Diff-MAC shows significant improvements, in terms of latency, data delivery, and energy efficiency.

11.5 Conclusion and Open Research Issues

WSNs have been drawing increased attention in recent years. Sensors in such systems are typically disposable and expected to last until their energy is drained. Therefore, energy is a very scarce resource for such sensor systems and has to be managed wisely to extend the life of the sensors for the duration of a particular mission. Energy conservation is generally targeted at all layers of the communication stack. This chapter has been dedicated to the link layer analyzing the requirements, identifying energy-related design parameters, exploring trade-offs, and reporting on advances made in the research community.

Generally, an efficient MAC layer protocol for WSNs should have the following characteristics:

- The protocol should be scalable because most applications of WSNs involve a large set of sensor nodes.
- Collision among the transmissions of various nodes should be avoided. Collisions lead to packet drop and thus reduce throughput and cause energy wastage.
- Energy consumed by the radio circuit in idle mode is almost equal to that consumed in active state. Consequently, idle mode of operation and transmission overhearing among sensors should be minimized.
- To limit energy consumption during idle time, the sensors are typically switched to a sleep mode when not in use. However, active to sleep transitions and vice versa consume considerable amounts of energy. Therefore, an efficient protocol should minimize such transitions.
- The protocol should minimize overhead. For example, control packets overhead and active sensing of the medium, typically performed by contention-based protocols, are inefficient in terms of energy consumption.
- Packet drop due to limited buffer capacity should be prevented.
- The protocol should adapt to changes in the network topology and all sensors should have a fair chance of transmitting.
- Transmission of messages should be reliably handled while striving to maintain bounded latency.

Although most published works have tried to address some of these design goals, there is still room for future research. For example, the analyses and discussions in this chapter have made it clear that there is no ideal strategy for resource sharing and medium access arbitration. We envision the pursuance of hybrid approaches to be the most promising design methodology. Although there have been few attempts such as those summarized in Section 11.4.3, extensive research is required to address the above efficiency requirements in a comprehensive way. Scalability issues are most notable among possible future research directions given the growing ambition for very large deployments of tiny sensors and the widening scope of use in many application domains.

Other avenues for research include the combined handling for QoS and energy constraints and the development of innovative radio designs that balance the desire for flexibility and robustness while maintaining acceptable circuit complexity. For example, little effort has been dedicated to exploring the effect of the directional antenna on the performance of sensor networks. Directional antennas can increase the spatial reusability of the medium and limit the scope of the overhearing problem.

Some of the possible future research issues can be listed as follows:

- The focus of future research should be on finding a balance between minimizing delay and control overhead, guaranteeing some kind of QoS, and also optimizing power usage; a combination of requirements for WSNs. The recent hybrid MAC protocols are promising in achieving these goals with better energy efficiency. The cross-layer designs have the most potential in achieving the same goals. By combining the information gathered at routing and link layers, WSNs can optimize path determination without giving in on the energy required for control overhead.
- Energy efficiency is the main design objective of most MAC protocols, but the reliable delivery of data in real-time is essential for certain time-critical applications. Cross-layer architectures can improve reliability by combining the error correction mechanisms utilizing various layers and encoding of the radio signals.
- A recent phenomenon in the MAC layer design is the possible use of network coding. The idea is to reduce the number of transmissions by applying smart network coding techniques at the nodes. With some additional processing overhead, the number of transmissions can be reduced, which leads to minimal collisions in the network.
- WSN security at MAC layer to protect against eavesdropping and malicious behavior has to be studied further. There have been several recent proposals for integrating security but more schemes incorporating lightweight encryption and authentication are needed. Moreover, new designs should consider the new privacy and trust models against malicious nodes that could take on Sybil, traffic analysis, or denial of service attacks.
- Recently, there has been a growing interest in the inter-networking of multiple network segments and forming the internet of things (IoT).[80] Handling the protocol diversity and combining multiple segment-centric optimizations to serve the overall networking goal is a very challenging problem that warrants extensive research efforts.
- Last but not least, future research should focus on experimenting on real sensor platforms. Most of the published WSN protocols have been evaluated through simulations with a number of specific assumptions. However, the performance of these architectures needs to be evaluated and comparatively analyzed within real WSN settings.

References

1. I. F. Akyildiz, W. Su, Y. Sankarasubramaniam, and E. Cayirci. Wireless sensor networks: A survey, *Computer Networks*, 38(4), 393–422 (2002).
2. S. Tilak, N. B. Abu-Ghazaleh, and W. Heinzelman. A taxonomy of wireless microsensor network models. *ACM Mobile Computing and Communication Review (MC2R)*, 6(2), 1–8 (2002).
3. G. J. Pottie and W. J. Kaiser. Wireless integrated network sensors. *Communications of the ACM*, 43(5), 51–58 (2000).
4. J. Rabaey, M. Ammer, J. Silva Jr., D. Patel, and S. Roundy. PicoRadio supports ad hoc ultra low power wireless networking. *IEEE Computer*, 33(7), 42–48 (2000).
5. R. Min, M. Bhardwaj, S. Cho, E. Shih, A. Sinha, A. Wang, and A. Chandrakasan. Low power wireless sensor networks. In *Proceedings of International Conference on VLSI Design*, Bangalore, India (January 2001).
6. D. Estrin, R. Govindan, J. Heidemann, and S. Kumar. Next century challenges: Scalable coordination in sensor networks. In *Proceedings of the 5th IEEE/ACM Annual Conference on Mobile Computing and Networks (MobiCOM'99)*, Seattle, WA (August 1999).

7. K. Sohrabi, J. Gao, V. Ailawadhi, and G. J. Pottie. Protocols for self-organization of a wireless sensor network. *IEEE Personal Communications*, 7(5), 16–27 (2000).
8. A. Buczak and V. Jamalabad. Self-organization of a heterogeneous sensor network by genetic algorithms. In *Intelligent Engineering Systems* through *Artificial Neural Networks*, edited by C. H. Dagli et al., 8 ASME Press, (1998).
9. R. Burne et al. A self-organizing, cooperative UGS network for target tracking. In *Proceedings of SPIE Conference on Unattended Ground Sensor Technologies and Applications II*, Orlando, FL (April 2000).
10. K. Akkaya and M. Younis. A survey on routing protocols for wireless sensor networks. *The Journal of Ad Hoc Networks*, 3(3), pp. 325–349 (2005).
11. C. Intanagonwiwat, R. Govindan, and D. Estrin. Directed diffusion: A scalable and robust communication paradigm for sensor networks. In *Proceedings of the 6th IEEE/ACM Annual Conference on Mobile Computing and Networks (MOBICOM'00)*, Boston (August 2000).
12. R. Shah and J. Rabaey. Energy aware routing for low energy ad hoc sensor networks. In *Proceedings of IEEE Wireless Communications and Networking Conference (WCNC)*, Orlando, FL (March 2002).
13. W. Heinzelman, A. Chandrakasan, and H. Balakrishnan. Energy-efficient communication protocols for wireless microsensor networks. In *Proceedings of the 33rd Hawaii International Conference on System Sciences (HICSS)* (January 2000).
14. M. Younis, M. Youssef, and K. Arisha. Energy-aware routing in cluster-based sensor networks. In *Proceedings of the 10th IEEE/ACM Symposium on Modeling, Analysis and Simulation of Computer and Telecommunication Systems (MASCOTS'02)*, Fort Worth, TX (October 2002).
15. G. Jolly and M. Younis. Energy efficient arbitration of medium access in sensor networks. In *Proceedings of the IASTED Conference on Wireless and Optical Communications (WOC 2003)*, Banff, Canada (July 2003).
16. S. Singh and C. S. Raghavendra. PAMAS: power aware multi-access protocol with signaling for ad hoc networks. *ACM Computer Communications Review*, 28(3), 5–26 (1998).
17. S. Xu and T. Saadawi. Does the IEEE 802.11 MAC protocol work well in multihop wireless ad hoc networks? *IEEE Communications Magazine*, 39(6), 130–137 (2001).
18. E. Shih, S. Cho, N. Ickes, R. Min, A. Sinha, A. Wang, and A. Chandrakasan. Physical layer driven protocol and algorithm design for energy-efficient wireless sensor networks. In *Proceedings of the 7th ACM/IEEE Conference on Mobile Computing and Networks (MOBICOM'01)*, Rome, Italy (July 2001).
19. A. Wang et al. Energy-efficient modulation and MAC for asymmetric microsensor systems. In *Proceedings of ISLPED 2001*, Huntington Beach, CA (August 2001).
20. T. Dam and K. Langendoen. An adaptive energy-efficient MAC protocol for wireless sensors networks. In *Proceedings of SenSys'03*, Los Angles (November 2003).
21. M. Stemm and R. Katz. Measuring and reducing energy consumption of network interfaces in handheld devices. *IEICE Transactions on Communications*, E80-B(8), 1125–1131 (1997).
22. M. Miller and N. Vaidya. Minimizing energy consumption in sensor networks using a wakeup radio. In *Proceedings of the IEEE International Conference on Wireless Communications and Networks (WCNC'04)*, Atlanta, GA (March 2004).
23. O. Kasten. Energy consumption. http://www.inf.ethz.ch/~kasten/research/bathtub/energy_consumption.html, Eldgenossische Technische Hochschule Zurich.
24. MICA2 Mote datasheet. www.xbow.com.
25. National Semiconductor Corporation. LMX3162 single chip radio transceiver, *Evaluation Notes and Datasheet* (March 2000).
26. W. Ye, J. Heidemann, and D. Estrin. An energy-efficient MAC protocol for wireless sensor networks. In *Proceedings of the IEEE INFOCOM*, New York City, (2002).
27. C. Jones, K. Sivalingam, P. Agrawal, and J. C. Chen. A survey of energy efficient network protocol. *Wireless Networks*, 7, 343–358 (2001).
28. P. Lettieri and B. Srivastava. Advances in wireless terminals (I), *IEEE Personal Communications Magazine*, 6 (February 1999), 6–19 (1999).
29. P. Havinga, G. Smit, and M. Bos. Energy efficient wireless ATM design. In *Proceedings of wmATM'99*, San Jose, CA (June 1999).
30. J. Walrand. Communication networks, 2nd edition, McGraw-Hill, Boston, MA (1998).

31. C. Sakr and T. Todd. Carrier-sense protocols for packet-switched smart antenna base-station. In *Proceedings of International Conference on Network Protocols (ICNP'97)*, Atlanta, GA (1997).
32. K. Akkaya and M. Younis. An energy-aware QoS routing protocol for wireless sensor networks. In *Proceedings of the IEEE Workshop on Mobile and Wireless Networks (MWN 2003)*, Providence, RI (May 2003).
33. T. He, J. A. Stankovic, C. Lu, and T. Abdelzaher. SPEED: a stateless protocol for real-time communication in sensor networks. In *Proceedings of the International Conference on Distributed Computing Systems (ICDCS)*, Providence, RI (May 2003).
34. K. Akkaya and M. Younis. Energy-aware routing of delay-constrained data in wireless sensor networks. *Journal of Communication Systems, Special Issue on QoS Support and Service Differentiation in Wireless Networks*, 17(6), 663–687, 2004.
35. V. Bhagwan et al. MACAW: a media access protocol for wireless LANs. In *The Proceedings of ACM SIGCOMM Conference*, London, UK, 212–225 (1994).
36. http://grouper.ieee.org/groups/802/11/main.html.
37. Bluetooth. http://www.bluetooth.com.
38. V. Raghunathan. Energy-aware wireless microsensor networks. *IEEE Signal Processing Magazine*, 19(2), 40–50, (March 2002).
39. E. Shih et al. Energy-efficient link layer for wireless microsensor networks. In *The Proceedings of the Workshop on Very Large Scale Integration (WVLSI'01)*, Orlando, FL (April 2001).
40. M. Caccamo, L. Y. Zhang, L. Sha, and G. Buttazzo. An implicit prioritized access protocol for wireless sensor network. In *Proceedings of the IEEE Real-Time Systems Symposium (RTSS'02)*, Austin, TX (December 2002).
41. C. Lu, B. Blum, T. Abdelzaher, J. Stankovic, and T. He. RAP: a real-time communication architecture for large-scale wireless sensor networks. In *Proceedings of the IEEE Real-Time and Embedded Technology and Applications Symposium (RTAS2002)*, San Jose, CA (September 2002).
42. E. Poon and B. Li. SmartNode: Achieving 802.11 MAC interoperability in power-efficient sensor networks with dynamic range adjustments. In *Proceedings of the IEEE International Conference on Distributed Computing Systems (ICDCS)*, Providence, RI (May 2003).
43. E. Jung and N. Vaidya. A power control MAC protocol for ad hoc networks. In *Proceedings of the 8th ACM/IEEE Conference on Mobile Computing and Networks (MOBICOM'02)*, Atlanta, GA (September 2002).
44. J. Monks, V. Bharghavan, and W. Hwu. A power controlled multiple access protocol for wireless packet networks. In *Proceedings of IEEE Conference on Computer Communications (INFOCOM)*, Anchorage, AK (April 2001).
45. S.-L. Wu, Y.-C. Tseng, and J.-P. Sheu. Intelligent medium access for mobile ad hoc networks with busy tones and power control. *IEEE Journal on Selected Areas in Communications*, 18(9), 1647–1657 (2000).
46. F. Bennett et al. Piconet: Embedded mobile networking, *IEEE Personal Communications Magazine*, 4(5), 8–15 (1997).
47. J. Polastre, J. Hill, and D. Culler. Versatile lower power media access for wireless sensor networks. In *Proceedings of the ACM Conference on Embedded Network Sensor Systems (SenSys)*, Baltimore, MD, 95–107 (2004).
48. S. Liu, K-W. Fan, and P. Sinha. CMAC: an energy-efficient MAC layer protocol using convergent packet forwarding for wireless sensor networks. *ACM Transactions on Sensor Networks (TOSN)*, 5(4), 1–34 (November 2009).
49. M. Buettner, G. V. Yee, E. Anderson, and R. Han. X-MAC: a short preamble MAC protocol for duty-cycled wireless sensor networks. In *Proceedings of the 4th International Conference on Embedded Networked Sensor Systems (SenSys'06)*, ACM, New York, 307–320 (2006).
50. F. Ahdi, W. Wang, V. Srinivasan, and K.-C. Chua. IPS-MAC: an informative preamble sampling MAC protocol for wireless sensor networks. *Journal of Wireless Networks*, 16, 5, 1373–1387 (July 2010).
51. Y. Sun, S. Du, O. Gurewitz, and D. B. Johnson. DW-MAC: a low latency, energy efficient demand-wakeup MAC protocol for wireless sensor networks. In *Proceedings of the 9th ACM International Symposium on Mobile Ad Hoc Networking and Computing (MobiHoc'08)*, ACM, New York, 53–62 (2008).

52. Y. Sun, O. Gurewitz, and D. B. Johnson. RI-MAC: a receiver-initiated asynchronous duty cycle MAC protocol for dynamic traffic loads in wireless sensor networks. In *Proceedings of the 6th ACM Conference on Embedded Network Sensor Systems (SenSys'08)*, ACM, New York, 1–14 (2008).
53. A. Woo and D. Culler. A transmission control scheme for media access in sensor networks. In *Proceedings of 7th ACM/IEEE Annual Conference on Mobile Computing and Networks (MobiCOM01)*, Rome, Italy (July 2001).
54. J. Elson and K. Romer. Wireless sensor networks a new regime of time synchronization. In *Proceedings of the First Workshop on Hot Topics in Networks (HotNets-I)*, Princeton, NJ (October 2002).
55. G. Jolly and M. Younis. An energy efficient, scalable and collisionless MAC layer protocol for wireless sensor networks. *Journal of Wireless Networks and Mobile Computing*, 5(3), 285–304 (May 2005).
56. P. Havinga and G. Smit. Energy-efficient TDMA medium access control protocol scheduling. In *Proceedings of the Asian International Mobile Computing Conference (AMOC 2000)*, Penang, Malaysia (2000).
57. K. Arisha, M. Youssef, and M. Younis. Energy-aware TDMA-based MAC for sensor networks. In *Proceedings of the IEEE Workshop on Integrated Management of Power Aware Communications, Computing and Networking (IMPACCT 2002)*, New York, (May 2002).
58. M. Ringwald and K. Romer. BitMAC: a deterministic, collision-free, and robust MAC protocol for sensor networks. In *Proceedings of the 2nd European Workshop on Wireless Sensor Networks (EWSN 2005)*, Istanbul, Turkey, 57–69 (January 2005).
59. S. Cho, K. Kanuri, J.-W. Cho, J.-Y. Lee, and S.-D. June. Dynamic energy efficient TDMA-based MAC protocol for wireless sensor networks. In *Autonomic and Autonomous Systems and International Conference on Networking and Services (ICAS-ICNS 2005)*, 48 (October 2005).
60. L. F. W. Van Hoesel and P. J. M. Havinga. A lightweight medium access protocol (LMAC) for wireless sensor networks: reducing preamble transmission and transceiver state switches. In *Proceedings of the 1st International Workshop on Networked Sensing Systems (INSS)*, 205–208 (2008).
61. S. Mank, R. Karnapke, and J. Nolte. An adaptive TDMA based MAC protocol for mobile wireless sensor networks. In *Proceedings of the 2007 International Conference on Sensor Technologies and Applications (SENSORCOMM'07)*. IEEE Computer Society, Washington, DC, 62–69 (2007).
62. V. Rajendran, K. Obraczka, and J. J. Garcia-Luna-Aceves. Energy-efficient collision-free medium access control for wireless sensor networks. In *Proceedings of the 1st International Conference on Embedded Networked Sensor Systems (SenSys'03)*, ACM, New York, 181–192. (2003).
63. M. Anwander. Comparison of TDMA and contention based MAC protocols on embedded sensor boards (ESB), Masters's Thesis, University Bern, Berne, Switzerland (2006).
64. M. Adamou, I. Lee, and I. Shin. An energy efficient real-time medium access control protocol for wireless ad-hoc networks. In *WIP Session of IEEE Real-Time Systems Symposium (RTSS'01)*, London, UK (December 2001).
65. V. Rajendran, J. J. Garcia-Luna-Aceves, and K. Obraczka. An energy-efficient channel access scheduling for sensor networks. In *Proceedings of the 5th International Symposium on Wireless Personal Multimedia Communication (WPMC)*, Honolulu, HI (2002).
66. K. Sohrabi and G. J. Pottie. Performance of a novel self-organization for wireless ad-hoc sensor networks. In *Proceedings of the IEEE Vehicular Technology Conference*, Amsterdam, the Netherlands, 1222–1126 (1999).
67. I. Rhee, A. Warrier, M. Aia, and J. Min. Z-MAC: a hybrid MAC for wireless sensor networks. In *Proceedings of the 3rd International Conference on Embedded Networked Sensor Systems* (*SenSys'05*), ACM, New York, 90–101 (2005).
68. J. Yick, B. Mukherjee, and D. Ghosal. Wireless sensor network survey. In *Computer Networks*, 52(12), 2292–2330 (August 2008).
69. L. Sitanayah, C. J. Sreenan, and K. N. Brown. Emergency response MAC protocol (ER-MAC) for wireless sensor networks. In *Proceedings of the 9th ACM/IEEE International Conference on Information Processing in Sensor Networks (IPSN'10)*, ACM, New York, 364–365 (2010).
70. G-S. Ahn, S. G. Hong, E. Miluzzo, A. T. Campbell, and F. Cuomo. Funneling-MAC: a localized, sink-oriented MAC for boosting fidelity in sensor networks. In *Proceedings of the 4th International Conference on Embedded Networked Sensor Systems (SenSys'06)*, ACM, New York, 293–306 (2006).
71. Z. B. Alliance. Draft standard: 02130r4ZB-NWK-network layer specification (March 2003).

72. Y. Kim, H. Shin, and H. Cha. Y-MAC: an energy-efficient multi-channel MAC protocol for dense wireless sensor networks. In *Proceedings of the 7th International Conference on Information Processing in Sensor Networks (IPSN'08)*, IEEE Computer Society, Washington, DC, 53–63 (2008).
73. L. Tang, Y. Sun, O. Gurewitz, and D. B. Johnson. EM-MAC: a dynamic multichannel energy-efficient MAC protocol for wireless sensor networks. In *Proceedings of the Twelfth ACM International Symposium on Mobile Ad Hoc Networking and Computing (MobiHoc'11)*, ACM, New York, (2011).
74. V. Gabale, K. Chebrolu, B. Raman, and S. Bijwe. PIP: a multichannel, TDMA-based MAC for efficient and scalable bulk transfer in sensor networks. *ACM Transactions of Sensor Networks,* 8, 4, Article 28 (September 2012).
75. T. Canli, M. Hefeida, and A. Khokhar. BulkMAC: a cross-layer based MAC protocol for wireless sensor networks. In *Proceedings of the 6th International Wireless Communications and Mobile Computing Conference (IWCMC'10)*, ACM, New York, 442–446 (2010).
76 Z. Tang and Q. Hu. An adaptive low latency cross-layer MAC protocol for wireless sensor networks. In *Proceedings of the 2009 Eighth IEEE International Conference on Dependable, Autonomic and Secure Computing (DASC'09)*, IEEE Computer Society, Washington, DC, 389–393 (2009).
77. B. Kechar, A. Louazani, L. Sekhri, and M. F. Khelfi. Energy efficient cross-layer MAC protocol for wireless sensor networks. In *Proceedings of the Second International Conference on Verification and Evaluation of Computer and Communication Systems (VECoS'08)*, British Computer Society, Swinton, UK (2008).
78. J. Ruiz, J. R. Gallardo, L. Villasenor-Gonzalez, D. Makrakis, and H. T. Mouftah. QUATTRO: QoS-capable cross-layer MAC protocol for wireless sensor networks. In *Proceedings of the 28th IEEE Conference on Global Telecommunications (GLOBECOM'09)*, Mehmet Ulema (Ed.). IEEE Press, Piscataway, NJ (2009).
79. M. A. Yigitel, O. D. Incel, and C. Ersoy. Diff-MAC: a QoS-aware MAC protocol with differentiated services and hybrid prioritization for wireless multimedia sensor networks. In *Proceedings of the 6th ACM Workshop on QoS and Security for Wireless and Mobile Networks (Q2SWinet'10)*, ACM, New York, (2010).
80. O. Hersent, D. Boswarthick, and O. Elloumi. *The Internet of Things: Key Applications and Protocols,* 2nd ed., John Wiley and Sons, West Sussex, UK (2012).

ENERGY SAVING APPROACHES

Chapter 12

Pulse Switching: A Packetless Networking Paradigm for Energy-Constrained Monitoring Applications

Qiong Huo, Bo Dong, and Subir Biswas
Michigan State University

Contents

12.1 Introduction

In wireless sensor networks, the main tasks of the sensors are detection, monitoring, and tracking of events. Events could consist of the measurement of physical phenomena, such as temperature, humidity, and acidity. Events can also be triggered through the detection of changes in a subject's behavior, such as a target approaching or leaving an area.

This chapter will introduce a new energy-efficient alternative called pulse switching to replace traditional packet switching. In the proposed pulse switching paradigm, an event can be coded as a single pulse for event monitoring and target tracking applications in sensor networks. Such an event pulse is then transported multihop to a sink while preserving the event's localization information. The resulting operational lightness, leveraged via zero collision, zero buffering, no addressing, no packet processing, and ultralow communication and energy budgets makes the protocol applicable for severely energy-constrained monitoring applications.

12.1.1 Event Monitoring in Sensor Networks

For specific applications, we only need the binary information of the event, for example, the presence or the absence of the target, or the increase or decrease of physical parameters. An example application is structural failure detection in which, while monitoring a bridge, it is often sufficient for a sensor to generate an event to indicate a crack in its vicinity. Sending an event, indicating the presence of the crack, to a sink would require single-bit information transport. Another application with binary information usage is moving target tracking. The presence and timing of a target is informed (by individual sensors) to a sink node as individual pulses. Then, the sink can postprocess such binary information to construct the intruder's trajectory with a resolution that is bounded by the event localization resolution of the proposed paradigm.

12.1.2 Energy as the Primary Design Constraint

It is well-known that energy efficiency is a significant concern in sensor networks. Under the specific applications above, packets might be energy-inefficient to transport binary data in sensor networks. Sending an event to a sink ideally requires a single-bit information transport for which the traditional mode of packet communication can be highly energy-inefficient. Such inefficiency stems from communication, processing, and buffering overheads of a large number of bits within the payload, header, and the synchronization preambles [1] in each packet.

Very few efforts exist in the literature on packetless networking using pulse communication. An article by Jain et al. [2] reduces the preamble and header overheads of packet communication by aggregating payloads from multiple short packets into a single large packet that is routed to a sink. Although the energy costs are reduced, aggregation still requires the inherent packet overheads. For target tracking applications, Wang et al. [3] propose a binary sensing model in which each sensor returns only one-bit information regarding a target's presence or absence within its sensing range. Although this binary sensing saves energy to some degree, the approach by Wang et al. [3] also uses a packet abstraction. The objective of our work is to fully replace packets by routable pulses.

Fragouli and Orlitsky [4] developed models for energy and delay bounds for bit (i.e., packet based) and pulse communications in single hop networks. The main results in their study [4] are that the worst case energy performance of pulse communication can be substantially better than that of packet-based communication, although with a possibly worse delay performance. A notable limitation is that their study does not provide mechanisms for scaling these results for multihop networks. Also, no MAC and routing protocol details are provided. This limitation is addressed in our work through the design of a MAC-routing framework that can be used for practical implementation of multihop pulse switching with sensor cell–based event localization.

12.1.3 Pulse Switching as an Energy-Efficient Alternative to Packets

12.1.3.1 Objective

The objective of this chapter is to develop an ultralight pulse switching protocol framework for resource-constrained sensors in event monitoring and target tracking applications. The key idea is to introduce a new abstraction of pulse switching to replace the traditional packet switching. In the proposed pulse switching paradigm, such an event can be coded as a single pulse, which is then transported multihop while preserving the event's localization information. The resulting operational lightness, leveraged via zero collision, zero buffering, no addressing, no packet processing, and ultralow communication and energy budgets makes the protocol applicable for severely resource-constrained sensor devices such as radiofrequency (RF) identifiers (IDs) operating with tight energy budgets, often from harvested energy [5].

12.1.3.2 Challenges

The primary challenges for pulse networking are (1) how to transport localization information using a single pulse, (2) how to route a pulse multihop without being able to explicitly code any information within the pulse, and (3) how to cope with pulse loss and false-positive detection errors.

12.1.3.3 Contributions

Ultrawideband (UWB) impulse radio (IR) technology is used for implementing the abstraction of single pulse transport. A key architectural novelty in this work is to integrate a pulse's (i.e., an event) location of origin within the MAC-routing protocol syntaxes. More specifically, by observing the time of arrival of a pulse with respect to the MAC-routing frame, a sink can resolve the corresponding event location with a preset resolution. The problem for multihop pulse routing is addressed by introducing a novel wave front routing protocol. Synchronized pulse waves are created in the network so that a pulse can simply "ride" synchronized phase waves across different hop-distance nodes from a sink to get delivered to the sink. This chapter explores the architectural solutions to address these three fundamental protocol challenges for pulse switching.

The contributions of this chapter include (1) a new pulse-switching protocol paradigm and its associated MAC and routing syntaxes for multihop operations, (2) a hop-angular framework for event localization, (3) an implementation approach using UWB-IR, and finally (4) an analytical and simulation framework for performance characterization and comparison with packet-based event monitoring in the presence of various types of pulse loss and pulse detection errors.

Note that the proposed pulse switching architecture is targeted mainly to small sensor networks with few tens of sensors distributed within a restricted geographical area. As a result, a number of proposed protocols may not scale well for large networks. However, the protocols can enable intrusion detection and event monitoring for targeted applications such as structural health monitoring (SHM) [1] for aircraft wings, bridges, and other small structures.

12.2 Pulse Switching Protocol

12.2.1 Pulse Switching Using UWB-IR

UWB-IR is used for implementing the abstraction of single pulse transport. Ultra-narrow pulses (i.e., fine time resolution), carrierless transmissions, low transmit power, and low hardware complexity make UWB-IR an ideal tool for pulse switching. This section presents the usage details of UWB-IR for pulse switching.

12.2.1.1 UWB Slotting for Pulse Switching

The ability to transmit and receive a single pulse without per pulse synchronization overhead is a key physical layer requirement for supporting the frame structure described above. UWB-IR [6,7] technology can be used for implementing such framing because (1) it can support single-pulse transmission and (2) the technology is mature enough for practical system [8] implementation. The top graph in Figure 12.1 depicts a UWB implementation of the required slot structure in Figures 12.6 and 12.8. A typical UWB pulse width is 1 ns, and the pulse repetition period T_b is 1000 ns [6], which determines the slot size in this case. This large difference between the pulse width and the slot size minimizes the overlapping probability between pulses in adjacent slots in the presence of multipath delay [9].

Multiband orthogonal frequency-division multiplexing (MB-OFDM) [10] is an alternative to IR for UWB implementation. Unlike IR, which uses very short pulses with relatively low energy, in MB-OFDM the UWB frequency spectrum is divided into multiple nonoverlapping bands. Within each band, OFDM-based transmission is used. Although the MB-OFDM–based approach can also be used for the proposed system, we chose IR because of the simplicity of its implementation (i.e., because of the lack of mixers in the IR RF hardware).

12.2.1.2 Modulation and Synchronization

Unlike packet-based UWB, the pulses in different slots are not correlated among themselves. Instead, each individual pulse carries information about an event by itself. This is why the usage of impulses in this architecture does not require any modulation scheme such as the UWB pulse position modulation (PPM), which is commonly used for packet transport over UWB-IR.

For packet transport, nodes use pseudorandom time-hopping sequence (THS) [7] for implementing PPM. This requires a large preamble bit sequence for synchronizing the THS phase sequence between the transmitter–receiver pairs on a per packet basis. Such synchronization overhead can be up to 512 to 600 pulse repletion periods or bit durations as reported by the IEEE P802.15 Working Group for Wireless Personal Area Networks [11]. For pulse communication in

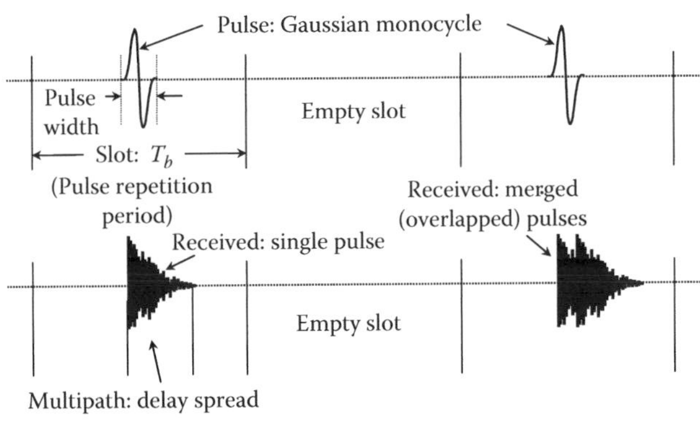

Figure 12.1 Pulse switching with unmodulated UWB impulses.

this architecture, because synchronization is achieved through the sink (Section 12.2.2), no per pulse synchronization preambles are needed.

12.2.1.3 Energy Budget

The simple all-digital baseband operation of the UWB-IR enables it to be implemented in low-cost CMOS logic [8]. For example, with 0.18 μm CMO-based UWB, power consumptions of 4 nJ for each pulse transmission (average 4 mW with 1000 ns pulse repetition period), 8 nJ for each pulse reception (average 8 mW), and idling consumption of 8 mW can be typical [8]. It should be noted that the transmission of a pulse using the baseband UWB would have the exact same energy expenditure for a bit transmission using THS-based PPM modulation during packet transmissions.

12.2.1.4 Sources of Error

The primary sources of errors for UWB-IR are large multipath delay spread (see the bottom graph in Figure 12.1), noise, and interference. Large multipath delay spreads can cause overlapping pulses, leading to errors in pulse detection. As shown in Figure 12.1, very large slot size (i.e., 1000 ns) can be chosen in comparison to the pulse width (i.e., 1 ns) to minimize overlapping by constraining the multipath delay spread within a slot. Pulse losses due to the misdetection of overlapped pulses need to be dealt with in the proposed pulse switching architecture.

The other primary contributors to error are interferences including multiuser interference (MUI) and narrowband interference (NBI). MUI can occur when the pulse switching system coexists with other spread-spectrum users in a UWB system, and NBI can occur when pulse switching would coexist with signals from conventional communication systems. The effects of NBI can be reduced significantly by using a rejection filter [12]. Although the effect of MUI in UWB-IR networks is generally less harmful than in narrowband networks, MUI can still degrade performance by creating the near–far effect. An article by Flury and Le Boudec [13] proposes a receiver to mitigate MUI through a combination of statistical interference modeling and thresholding.

The above factors can result in one of two types of detection errors at a UWB receiver [14]. These are false-positive pulse errors or pulse loss errors. A series of measures to minimize the effects of these errors are presented in Section 12.7.

12.2.2 Event Localization Methods

As shown in Figure 12.2, a network may contain arbitrarily distributed sensors that detect events and send corresponding pulses to a sink. Besides, the network is logically divided into a fixed number of overlaid event areas, as shown in Figures 12.3 and 12.5. Each sensor is preprogrammed with its specific event area ID. The sink is assumed to be capable of making high-power transmissions with full network coverage for frame-synchronizing the sensors as described in Section 12.2.2. The location of the sink is not necessarily at the center of the network.

An event is localized at the spatial resolution of an event area. If sensors from multiple event areas detect such an event, the sink can resolve the event spanning multiple such cells. The sensors are not individually addressed and, therefore, no per sensor addressing is necessary at the MAC or routing layers. An event is always identified by its original event area ID, not the sensor of origin. Event localization resolution depends on the size and shape of the event areas. More importantly, the joint MAC-routing layer design is also closely related to the event localization method. This

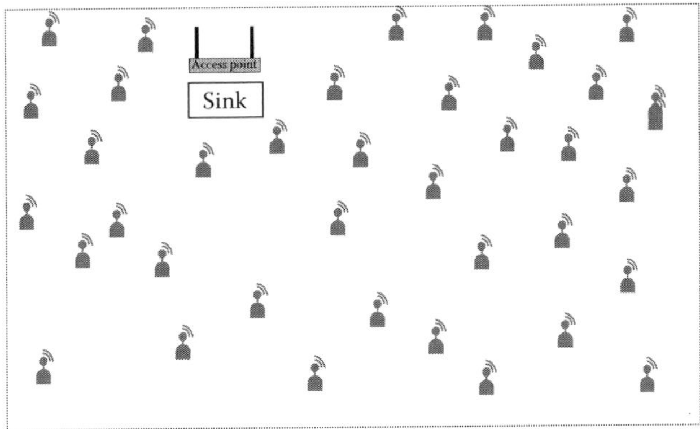

Figure 12.2 Network model.

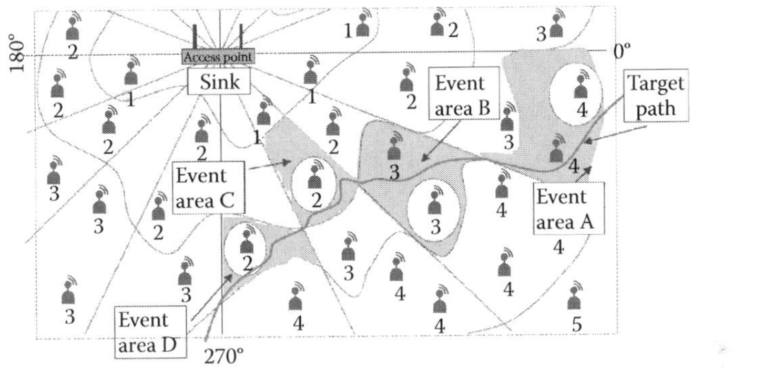

Figure 12.3 Network model with hop-angular event localization.

chapter will describe two event localization methods, (1) hop-angular event localization [15] and (2) cellular event localization [16], and their corresponding pulse-switching protocols.

12.2.2.1 Hop-Angular Event Localization

Depending on the node locations and the transmission range (assumed to be nonuniform), each node resides at a certain hop-distance from the sink. In Figure 12.3, the hop-distance for each node is marked under the node. The concept of a hop-angular event area is introduced for event localization. Event areas are angular sectors in the hop-angular event localization. For example, there are sixteen 22.5°-wide sectors in Figure 12.3. With a predefined sector width (α°), the location of a sensor can be represented by the tuple {*sector-id*, *hop-distance*}. For example, the location of the encircled sensor in event area B in Figure 12.3 can be represented as {15, 3}, meaning that the node is located in the 15th sector, with a hop-distance of 3 from the sink.

The concept of an event area does not assume any specific shape (i.e., circular or otherwise) of a node's transmission coverage area. It could be of any arbitrary shape, as shown in Figure 12.3. Although the angle for a node is preprogrammed at the deployment time, its hop-distance can be

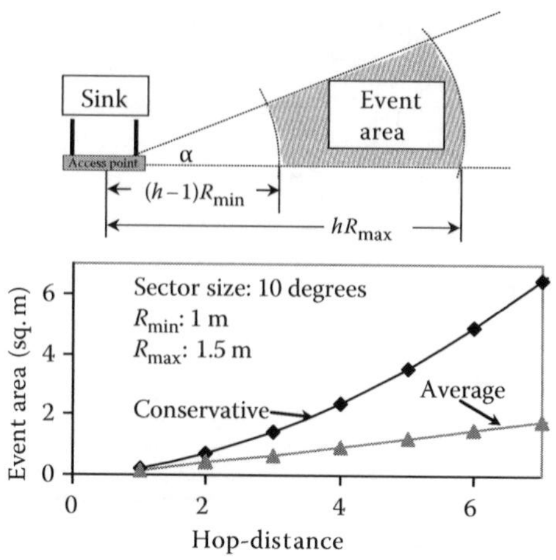

Figure 12.4 Hop-distance event localization.

dynamically discovered using the process outlined in Section 12.3.2. The {*sector-id, hop-distance*} tuple indicates an event area whose size determines the event localization resolution. This tuple for an event's origin is carried to the sink by the corresponding pulse. Additionally, the sink might successively receive multiple tuples that represent a moving target's locations in time. For example, the target path in Figure 12.3 can be represented as [{16, 4}, {15, 3}, {14, 2}, {13, 3}], meaning that such a target goes through event areas A–B–C–D sequentially.

Consider the example event area identified by the tuple {α, *h*} in the top portion of Figure 12.4. With a sector width of α, and R_{min}, R_{max} representing the known minimum and maximum wireless transmission range, the most conservative (coarse) localization resolution can be expressed as the largest possible event area: $A_{conservative} = \{h^2 R_{max}^2 - (h-1)^2 R_{min}^2\}\alpha\pi/360$. The average resolution is $A_{average} = \{h^2R^2 - (h - 1)^2R^2\}\alpha\pi/360$, where $R = (R_{min} + R_{max})/2$. For example, with transmission range spanning between 1 and 1.5 m, in a network with a sector width (i.e., α) of 10°, the size of an event area that is 5 hop-distances away is approximately 3.5 m^2. For the SHM application on a bridge, this means that a structural crack can be localized within an approximately 3.5 m^2 area. For a given α and transmission range, because this resolution reduces with higher hop-distances, the maximum network size will be determined based on the desired resolution.

12.2.2.2 Cellular Event Localization

As shown in Figure 12.5, geographically neighboring sensor cells (i.e., arbitrarily shaped and placed) are predefined and individually represent event areas. Each sensor cell has a unique Cell-ID, and each sensor is preprogrammed with its own Cell-ID and with a list of Cell-IDs of all the geographically neighboring sensor cells.

An event is localized at the spatial resolution of a sensor cell. Note that the proposed architecture does not assume geometrically regular cells. As shown in the figure, irregular cells of different shapes and sizes can coexist.

Event localization resolution depends on the size and shape of the event area cells. For example, with regular hexagonal cells with a side length of *l*, the localization resolution is approximately 2.6l^2 square units. By shrinking the side length *l*, it is possible to improve spatial resolution, which is independent of the wireless transmission range of the sensors.

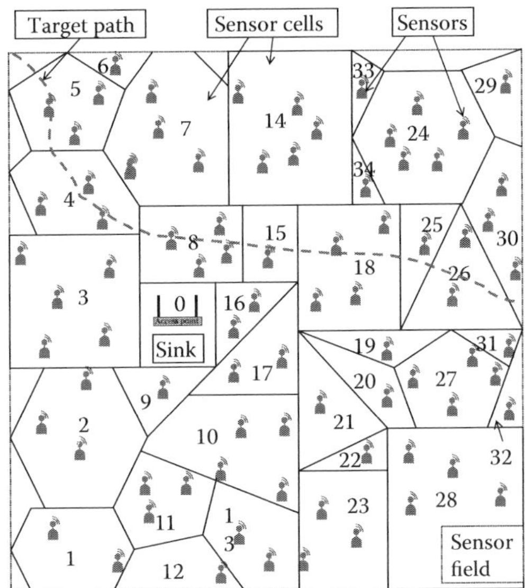

Figure 12.5 Network model with arbitrarily defined sensor cells.

Unlike the transmission hop-distance–dependent event localization described in an article by Huo et al. [15], localization resolution in this sensor cell–based architecture can be uniform across the entire network as long as the cells are of equal size and shape. In other words, no matter where an event occurs with respect to the sink, it can be localized with the same spatial resolution as long as the sensor cells are designed to be of uniform size and shape.

12.2.3 Pulse as Protocol Data Unit

Upon detection of an event, a sensor node generates a single pulse that needs to be transported multihop to a sink. A pulse is able to represent (a) the very occurrence of the event, and (b) its location of origin. The key idea behind pulse switching is to allow a receiver to localize an event by observing the temporal position of a received pulse with respect to a synchronized frame structure. Each time slot in such a synchronized frame is preassigned to a specific location or a special functionality. The details of frame structures for both event localization methods are shown in Sections 12.3 and 12.4.

An event can result in multiple pulses generated by all sensors detecting the event, and all such pulses need to be transported to the sink. With the localization information for each such pulse, several application level conclusions can be derived at the sink by correlating multiple event pulses. For example, while monitoring a bridge, by correlating the time and approximate location of a fatigue event, it is possible to study the dynamics of structural failure.

12.3 Hop-Angular Pulse Switching Protocol

12.3.1 Joint MAC-Routing Frame Structure

Nodes are frame-by-frame time-synchronized by the sink, and they maintain joint MAC-routing frames (Figure 12.6) in which each slot is used for sending a single pulse. The slot includes a guard

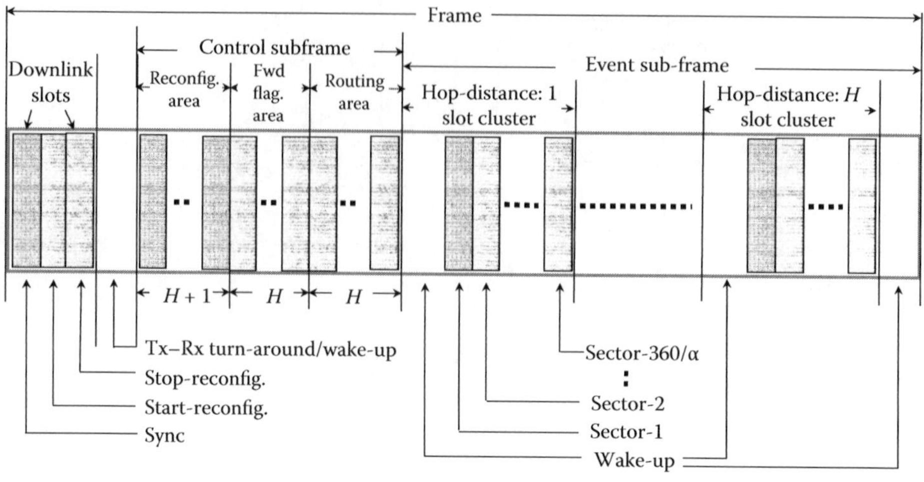

Figure 12.6 MAC-routing frame for hop-angular localization.

time to accommodate the cumulative clock-drift during a frame, which can be very small for RF technology such as UWB-IR, as the frame size itself can be ultrashort (μs) for UWB. As shown in Figure 12.6, the frame contains an uplink part and a downlink part. The uplink part contains a control subframe and an event subframe. The downlink part of the frame contains a synchronization slot in which the sink transmits a full-power pulse to make all nodes frame-synchronized. The two following downlink slots and the reconfiguration part of the uplink control subframe are used for hop-distance discovery. The reconfiguration area in the control subframe has $(H + 1)$ slots, where H is the maximum hop-distance. The forwarding flag area is designed for pulse compression. The H slot wide-routing area of the control subframe is used for energy management. These functions will be explained in detail in the next section.

The event subframe contains H slot clusters, each cluster containing $360/\alpha$ slots, where α corresponds to the sector width. Each slot within a cluster corresponds to a specific {*sector-id*, *hop-distance*} tuple. Meaning that, for each event area, represented by {*sector-id*, *hop-distance*}, there is a dedicated slot in the event subframe. An event-originating node transmits a pulse during the dedicated event subframe slot that corresponds to the {*sector-id*, *hop-distance*} of the node's event area. While routing the pulse toward the sink, at all intermediate nodes it is transmitted at the same event subframe slot that corresponds to the {*sector-id*, *hop-distance*} of its event area of origin. In other words, while being forwarded, the transmission slot for the pulse at all intermediate nodes does not change with respect to the frame. This is how information about the location of origin of an event is preserved during routing. Upon reception, the sink can infer the event area of origin from the {*sector-id*, *hop-distance*} value corresponding to the slot at which the pulse is received. The role of the control subframe in Figure 12.6 will be described in the next section.

12.3.2 Hop-Distance Self-Discovery

The sink initiates a reconfiguration phase by sending a full-power start-reconfigure pulse. It then transmits a regular-power (i.e., as used by other nodes) pulse at the first slot in the reconfiguration area in the control subframe. Nodes that receive this pulse conclude that they are 1 hop-distance away from the sink. All hop-distance 1 nodes send a pulse in the second slot of the reconfiguration area during the next few frames. Nodes receiving these pulses conclude that they are in hop-distance 2. This process continues for Hc frames ($c > 1$) after which all nodes are done discovering their individual hop-distances. After Hc frames, the sink ends the reconfiguration process by sending a full-power pulse in the stop-reconfigure slot. Although H frames can be sufficient for the hop-discovery, the factor c is introduced for redundancy to cope with pulse losses. To accommodate any network changes, the above hop-distance discovery process needs to be periodically executed.

12.3.3 Pulse Forwarding

When a pulse is transmitted by a node at hop-distance h, only its neighboring nodes at the hop area $(h - 1)$ forward it toward the sink. Meaning that the nodes at hop areas h and $h + 1$ should ignore the pulse. This logic ensures that a pulse is eventually delivered to the sink. While transmitting a pulse in the event subframe (Section 12.3.1), its transmitter also sends a pulse in the corresponding slot of the forwarding flag area of the control subframe. That is, while forwarding a pulse by a hop area h node, it sends a pulse in the hth time slot of the forwarding flag area. By looking at the received pulse in the forwarding area, all the receivers of the pulse can decide if it should be discarded or forwarded toward the sink. This can ensure that a pulse from hop area h should be forwarded only by nodes in hop area $h - 1$.

12.3.4 Synchronized State Transitions

Nodes synchronously transition in a frame-by-frame interface-sleep (IS)–listen-only (LO)–transmit-listen (TL) state cycle, which ensures that nodes that are not forwarding pulses can sleep. Nodes at the same hop area cycle in-phase, but those with different hop areas cycle out-of-phase. When the hop-distance h nodes transmit (i.e., in state TL), the hop-distance h and $h - 1$ nodes (i.e., in state LO) listen, and the hop-distance $h + 1$ nodes sleep (i.e., in state IS). This ensures that the nonparticipating nodes can save energy by turning their radio interfaces off.

Immediately after the hop-discovery reconfiguration process (Section 12.3.2) is terminated, a node at hop-distance h decides its state phase by computing h modulo 3. The outcomes 0, 1, or 2 cause the node's state to be initialized as LO, TL, or IS, respectively. During the subsequent frames, the state machine indefinitely cycles in the sequence IS-LO-TL. Irrespective of its current state phase, a node is always awake at the end of a frame for receiving the frame synchronization pulse from the sink. Consider a pulse being generated at node C {*sector-id*, 2; *hop-distance*, 3} in Figure 12.7. The maximum hop-distance is 4, and with an α of 90°, the maximum number of sectors is 360/90 = 4. The forwarding flag area of the control subframe and the event subframe are shown in Figure 12.7b and c.

When node C is in TL state and has an event to send, it sends a pulse in the corresponding slot {*sector-id*, 2; *hop-distance*, 3} in the event subframe. As shown in Figure 12.7d, transfer of the pulse from C to B occurs during frame 1. In frame 2, B forwards it to A, and finally during frame 3, the pulse is delivered to the sink during the same {*sector-id*, 2; *hop-distance*, 3} slot. All three frames used by nodes C, B, and A look the same except for the forwarding flag area.

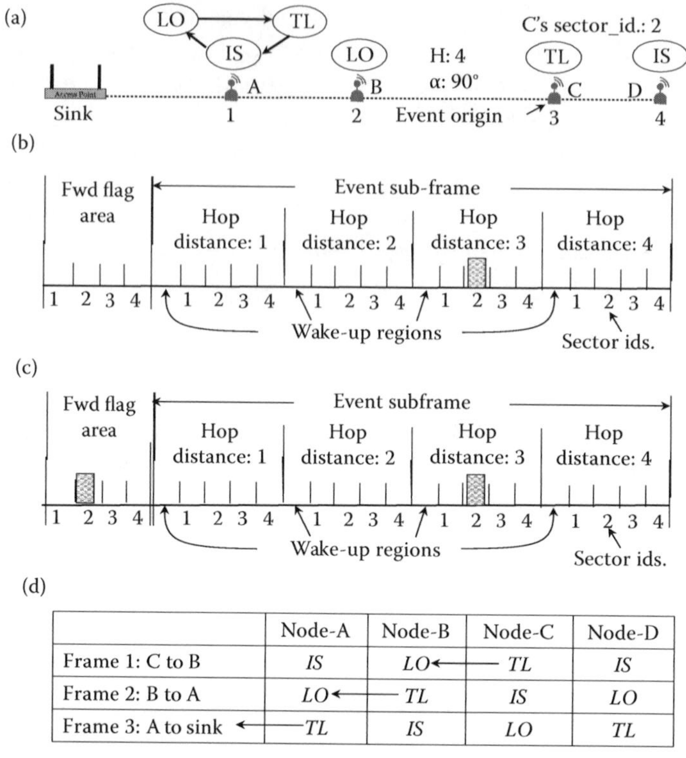

	Node-A	Node-B	Node-C	Node-D
Frame 1: C to B	IS	LO←	— TL	IS
Frame 2: B to A	LO←	— TL	IS	LO
Frame 3: A to sink ←	—TL	IS	LO	TL

Figure 12.7 Synchronized state transitions. (a) Sensor hop-distance and alternating state phases. (b) Frame level activities at hop-distance 3. (c) Frame level activities at hop-distance 2. (d) State transitions sequence.

12.4 Cellular Pulse Switching Protocol

12.4.1 Joint MAC-Routing Frame Structures

Nodes in the proposed cellular system are frame-by-frame time-synchronized by the sink as the proposed hop-angular system, and they maintain MAC-routing frames (Figure 12.8), in which each slot is used for sending a single pulse. The slot includes a guard time to accommodate the cumulative clock-drift during a frame, as shown in Figure 12.1. Because of UWB-IR's very short frame size (μs), clock-drift can be very small.

As shown in Figure 12.8, each frame contains a downlink part and an uplink part in which pulses can be sent either a downlink or an uplink. The downlink part of the frame includes a synchronization slot where the sink periodically transmits a full-power pulse to frame-synchronize all nodes in the network. The following discovery area in the frame is designed for adaptive route discovery, as described in Section 12.4.2. In the uplink part, there are two components: (a) a control area for energy management and (b) a localization area representing an event's location of origin and the routing table information. The operational details for each of these areas are presented in the next section.

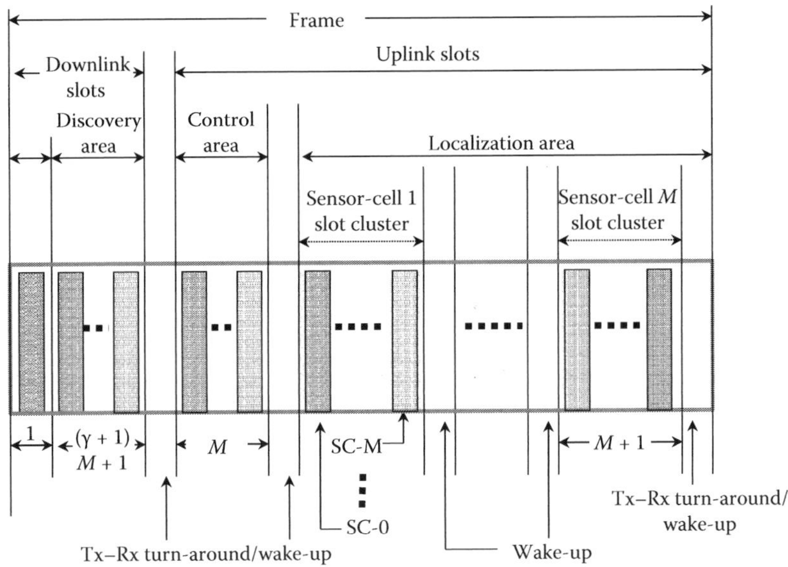

Figure 12.8 MAC-routing frame for cellular localization.

12.4.2 Route Discovery

This is a continuous background process that creates and maintains the routing table in each sensor in terms of the next-hop Cell-IDs. For routing to work, each sensor needs to be aware of its own Cell-ID and those of all the geographically neighboring cells. A sensor should also have wireless reachability to at least another sensor in one of its neighboring sensor cells. The structure of the discovery area in the MAC-routing frame is as follows.

This area of the frame contains $\{(\gamma + 1)M + 1\}$ slots, where M is the total number of sensor cells and γ is the transceiver turnaround time (between transmission and reception modes) as a multiple of slot duration. As shown in Figure 12.9, the first slot is allocated to the sink, and the rest of the slots are allocated to all M sensor cells in contiguous slot clusters. Within a cluster, the first γ slots are designated as dummy slots during which transceiver turnaround happens, and the $(\gamma + 1)$th slot is designated as a normal slot, during which discovery-related transmission happens as follows.

Once in every discovery cycle (termed as F_{discv} frames), the sink initiates a routing discovery phase by sending a regular power (as opposed to a full-power synchronization pulse) discovery pulse in the first slot of the discovery area. Upon receiving that pulse, all sensors within the sink's neighboring sensor cells register the reception time and the Cell-ID (retrieved from the temporal location of the pulse) of the sensor cell (which is the sink in this case) from which the discovery pulse was received. The sensors then forward the pulse in the normal slots of their corresponding clusters in the discovery area of the next frame. Formally stated, when a sensor in Cell-ID m receives a discovery pulse in the normal slot of the nth slot cluster, the sensor checks if the sensor cell of Cell-ID n is one of its neighboring cells. If not, it simply ignores the pulse. If it is, it registers the time of reception and the sender Cell-ID n, and forwards the pulse in the normal slot of the mth slot cluster in the discovery area of the next frame. This time stamp and forwarding process continues until the discovery pulse is flooded throughout the entire network.

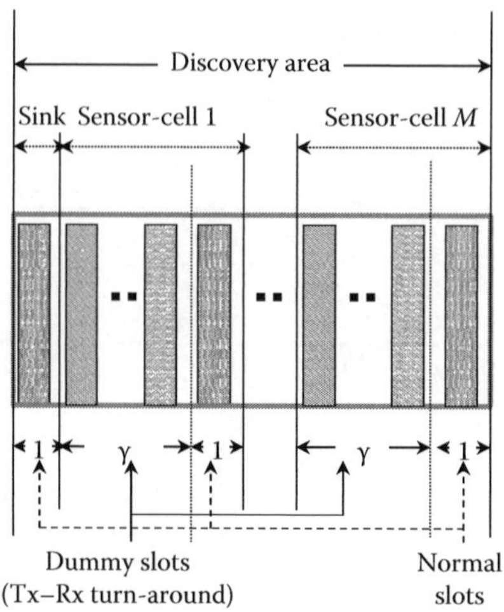

Figure 12.9 Structure of discovery area.

At the end of each discovery phase, each node develops a routing table, which is a list of time stamps and the corresponding Cell-IDs of it neighboring cells, indicating which sensor cells forwarded the discovery pulse at what time. In the absence of queuing delay, the entry with the earliest time stamp would indicate the shortest hop next-hop sensor cell to reach the sink. If multiple entries contain the same time stamp then all the corresponding Cell-IDs would indicate shortest hop next-hops to the sink. A node may have one or multiple shortest hop next-hop sensor cells depending on the shape, size, and relative locations of the cells with respect to the sink's location. The fact that each sensor has wireless reachability to at least another sensor in one of its neighboring sensor cells ensures at least one routing table entry to reach the sink.

Note that the above discovery process and the proposed architecture do not assume any specific shape of a sensor cell and node's transmission coverage area. To accommodate network connectivity and the resulting topology changes, the above discovery protocol needs to be periodically executed as a background process.

12.4.3 Pulse Forwarding

The localization area of the uplink part of the frame is key to pulse forwarding. In a network with M sensor cells, this area contains M slot clusters in which each such cluster contains (M + 1) individual slots. Each slot cluster corresponds to a specific Cell-ID, which represents an event's cell of origin. For example, when a node in a sensor cell with Cell-ID i senses an event, it transmits a pulse in the slot cluster corresponding to the Cell-ID i. In a given slot cluster, each slot corresponds to a specific Cell-ID, which represents the next-hop cell for a transmitted pulse within the slot cluster. For example, when a pulse is transmitted at the (j + 1)th slot of the slot cluster corresponding to Cell-ID i, this means that the pulse originated within the sensor cell of Cell-ID i and is forwarded to the next-hop sensor cell with Cell-ID j. Sensor cells i and j in this case are assumed to be geographically adjacent, and cell j is expected to be geographically closer to the sink compared with cell i if cell j is the best next-hop cell in the case of no fault-occurrence. Note that the first slot in each cluster of localization area represents a sink with ID 0.

Pulse forwarding decisions are made based on a sensor's routing table. The routing table maintains a sorted list of next-hop cells based on the hop-counts of the corresponding resulting routes. For routing with no diversity, a node chooses the best next-hop sensor cell from the routing table and forwards the pulse. With nonzero route diversity (parameterized as δ), a pulse is forwarded to the top δ, next-hop sensor cells from the routing table. Consider an example in which a sensor node in cell r receives a pulse that originated from cell i. With route diversity δ set to 2, the sensor finds the top two next-hop cells in its routing table to be cell j and cell k. Upon receiving the pulse, the sensor forwards it by sending two pulses—one in the $(j + 1)$th slot and the other in the $(k + 1)$ th slot of the slot cluster corresponding to the Cell-ID i.

It is important to note that the slot cluster for a pulse remains unchanged during the entire forwarding of the pulse from the sensor of origin all the way to the sink node. For example, a pulse that originates from the sensor cell of Cell-ID i is always transmitted within the slot cluster corresponding to the Cell-ID i of successive frames during all the hops until the sink. This way, the localization information of the pulse (i.e., of the corresponding event) is preserved until the pulse arrives at the sink. What changes in a hop-by-hop basis is the specific slot within the slot cluster depending on the specific next-hop sensor cell. For example, if the pulse from cell of Cell-ID i is forwarded first to a sensor cell with Cell-ID m (based on its routing table entry) then the sensor of origin transmits the pulse in the $(m + 1)$th slot of the slot cluster corresponding to the Cell-ID i. Now, if a sensor in the sensor cell with Cell-ID m finds the next-hop cell to be of Cell-ID n, then it forwards the pulse in the $(n + 1)$th slot of the slot cluster corresponding to the Cell-ID i in one of the future frames. This way, when the pulse arrives at the sink, it is still a part of the slot cluster i, which indicates the pulse's cell of origin.

Because the above pulse-forwarding process is executed on a per pulse manner, a large number of pulses can be simultaneously and independently routed in different parts of a frame depending on their cells of origin and the respective next-hop cells.

12.4.4 Protocol State Machine

Each sensor node maintains two separate state machines for route discovery and pulse forwarding processes, respectively. Each state machine has three states: transmission (T) state, listening (L) state, and sleeping (S) state. During both processes, the initial state of each node is in L state. State transition is triggered by pulse reception on a node in the last frame. Additionally, a node can switch between T and L states within the discovery area of the same frame, but should keep either T state or L state within the localization area of the same frame. The usage of state S will be described in Section 12.5.

In route discovery process, a sensor node switches to T state in its corresponding specified normal slot of the discovery area of a frame, when it needs to forward the discovery pulse. After forwarding this discovery pulse, the node switches back to L state in the upcoming normal slot of the same frame. Suppose a node in the sensor cell of Cell-ID m transits from L state to T state in the normal slot of the mth slot cluster of the discovery area of a frame, when it sends out a discovery pulse. This node goes back to L state in the rest of the normal slots of the same frame. In other words, except for the normal slot of the mth slot cluster of the discovery area of this frame, the node always keeps L state in the rest of normal slots of the same frame. It helps the node to receive route discovery information as much as possible. The dummy slots in each slot cluster of the discovery area of a frame are reserved for state transition time between T state and L state.

For the pulse forwarding process, when a sensor node is transmitting a pulse generated from its own sensor cell or forwarding a pulse originally from another sensor cell, the node keeps T state in the localization area of the current frame. Once pulse forwarding is done, the node state will switch back to L state in the localization area of the next frame.

12.5 Energy-Saving Measures

To further improve the energy efficiency of pulse switching protocols, this section will discuss intraframe interface shutdown, constrained routing, and pulse compression measures to save more energy.

12.5.1 Intraframe Interface Shutdown

Both hop-angular pulse switching (HPS) and cellular pulse switching (CPS) protocols are able to use intraframe interface shutdown measures to save idling energy. This subsection will introduce such measures for HPS and CPS, respectively.

12.5.1.1 Intraframe Interface Shutdown in HPS

In an IS frame, a node sleeps for the entire uplink duration. However, in the LO and TL frames, a node needs to remain awake during the uplink control subframe slots. During the event subframe, however, the node can sleep except during the slots it transmits or expects to receive pulses. During a TL frame, a node can both transmit and receive. From the transmission standpoint, because a node knows the {*sector-id*, *hop-distance*} information about a pulse that needs to be transmitted, it can simply wake up before the appropriate slot cluster and sleep after the transmission. From the reception standpoint, however, without the knowledge of the expected pulses, sleeping is not possible. The same applies for the LO frames in which all a node does is listen for expected receptions. We incorporate the following mechanism for enabling intraframe sleep for both TL and LO frames.

Whenever a pulse is transmitted in the event subframe, an associated pulse is also transmitted in the corresponding slot within the routing area of the control subframe. As shown in Figure 12.10b, because the main pulse is transmitted in the event subframe slot {*sector-id*, 2; *hop-distance*, 3}, the corresponding routing area pulse is transmitted in the third hop-distance slot within the routing area of the control subframe. This ensures that a TL or LO state node is informed about an impending pulse arrival in the event area of a frame. This way, the node can now wake up before the appropriate slot clusters and sleep after the corresponding event area pulse is received. Such intraframe sleep can further reduce the idling consumption beyond what is accomplished by synchronized state transitions.

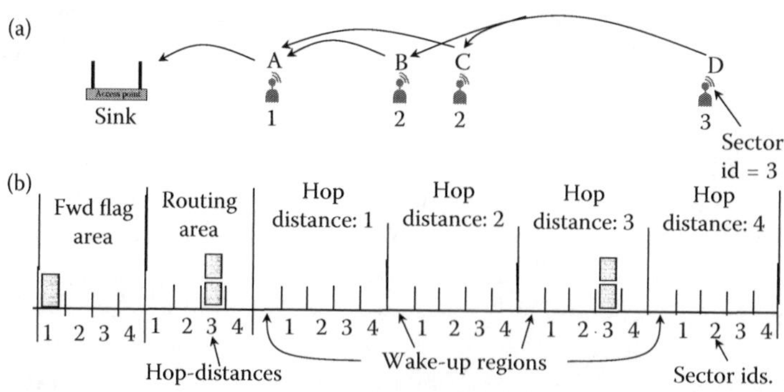

Figure 12.10 Route diversity and pulse merging/aggregation for HPS. (a) Fault tolerance due to route diversity. (b) Intra-event pulse merging at node A.

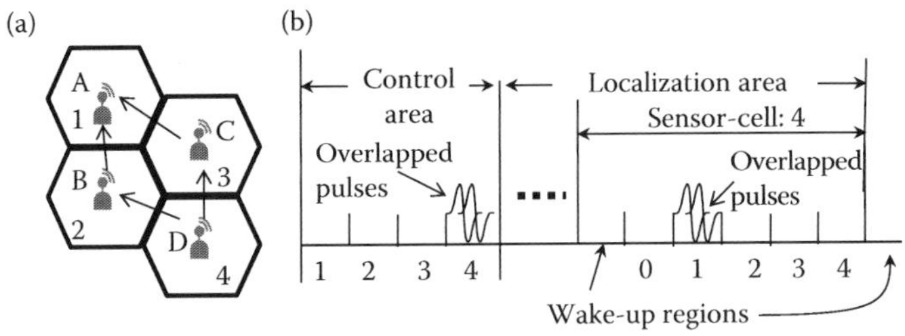

Figure 12.11 Pulse merging due to overlapped pulse reception for CPS. (a) Fault tolerance due to route diversity. (b) Intra-event pulse merging at node A.

12.5.1.2 Intraframe Interface Shutdown in CPS

Protocol syntaxes are added for nodes to be able to selectively turn off their RF interfaces during appropriate parts of the frame. As for transmissions in the pulse forwarding process, a node needs to be awake only during the slot clusters (of the localization area) at which it needs to transmit. During the other slot clusters, the node can simply keep the interface in sleep mode to save energy.

However, considering the asynchronous nature of pulse receptions, a node cannot sleep during all the nontransmission slot clusters in the localization area. To address this, a control area (Figure 12.8) is added at the beginning of the uplink part of the frame. The slots in the control area of a frame are used for notifications about the impending receptions that are expected during the slot clusters of the localization area of the frame. When a node plans to send a pulse originally from the sensor cell of Cell-ID m during the mth slot cluster of the localization area of a frame, it also sends a pulse in the mth slot in the control area of the same frame. All nodes remain awake in the control area, thus ensuring the reception of notification about impending transmissions in the mth slot cluster of the localization area of the frame. Based on this information, the node can remain awake during the mth slot cluster of the localization area. To summarize, a node can turn its RF interface off during all slot clusters of the localization area of a frame except the ones in which it either intends to transmit or expects to receive based on the notification received in the control area. Such intraframe interface sleep can reduce the idling energy consumption. Additionally, the node's state in the control area of a frame is the same as that in the mth slot cluster of the localization area of the same frame.

As shown in Figure 12.11, node B transmits the main pulse in the fourth slot cluster of the localization area, meaning that the original Cell-ID was 4, and another corresponding control area pulse in the fourth slot within the control area. Such a control area pulse from node B informs node A (in the L state) of an impending pulse arrival in the localization area of a frame. Then, node A wakes up before the appropriate slot clusters and sleeps after the corresponding event area pulse is received.

12.5.2 Constrained Routing

12.5.2.1 Constrained Routing in HPS

The extent of sector constraints during pulse forwarding in HPS can be parameterized using δ, which represents the ratio of the angular resolution α and an angle γ. The angle γ is the sector width beyond which a pulse may not be flooded while forwarding. For a given α, the minimum and the maximum values of γ are α and 180°, respectively. The corresponding δ values are 1 and 180°/α. When δ is 1, routing is maximally constrained, indicating minimum communication energy consumption, and the maximum susceptibility to errors due to the minimum pulse redundancy.

12.5.2.2 Constrained Routing in CPS

In CPS, route diversity, termed as δ, represents the number of next-hop sensor cells for a node in pulse forwarding. The maximum value of δ is equal to the maximum number of neighbor cells for a sensor cell. It depends on the shape and the size of sensor cells. When δ is 1, routing is maximally constrained as well as in HPS.

12.5.3 Pulse Aggregation and Compression

12.5.3.1 Aggregation via Pulse Merging

For pulse switching, collisions between pulses may not necessarily lead to information loss. Instead, the pulses within the same time slot of a frame can be recognized as an overlapped pulse, no matter if it is in HPS or CPS. As long as the RF hardware can detect the presence of this overlapped pulse, the routing continues. Pulse merging mainly arises from route diversity in the pulse forwarding process. In fact, this pulse merging and route diversity provides inherent in-network aggregation for events from the same event area.

For example, in HPS, because the pulse originating from D is transmitted by B and C on the same slot in the event subframe, the receiver A detects RF signals for a merged pulse in that slot (Figure 12.10). Similarly, in CPS, in which the Cell-ID of each node is marked at the bottom of the node, a pulse originates at sensor D in cell 4 and gets forwarded to two neighboring cells 2 and 3 (Figure 12.11). Sensors B and C in those cells independently forward the pulse to cell 1. Meaning that both the sensors send a pulse in the second slot of the fourth slot cluster in the localization area of the same frame. As a result, node A in cell 1 receives two overlapping pulses in the same slot. Note that pulse stacking, as shown in Figures 12.10 and 12.11, is just a representation of the fact that multiple pulses are transmitted during the same slot.

12.5.3.2 Pulse Compression

Due to pulse transmission redundancies, we propose the pulse compression algorithm to further improve the energy efficiency of pulse switching. Pulse compression, including spatial compression and temporal compression, can be applied to both HPS and CPS. In this section, we take HPS as an example. We call the baseline HPS protocol without compression as PSP-B, and the HPS protocol with compression as PSP-C.

12.5.3.3 Spatial Compression

12.5.3.3.1 Compression during Pulse Generation

Although multiple sensors can generate pulses for the same event within an event area, ideally, only one such pulse from the event area should be forwarded to the sink to inform about the event. Spatial pulse compression accomplishes this as follows. Upon detecting an event, a sensor node k defers its pulse generation for a back off period that is randomly distributed between 0 to R frames. If another node within the same event area generates a pulse before its deferring period is over, then after hearing that pulse, node k cancels its own transmission. Note that all nodes within an event area are able to listen to other nodes' transmissions. Thus, it would lead to very few pulses (often a single pulse) generation per event per event area.

12.5.3.3.2 Compression during Pulse Forwarding

Figure 12.11a demonstrates the forwarding envelope (with different δ) for a pulse that is generated in hop area 5. Observe that whereas increasing δ reduces the envelope size laterally by preventing pulses from being forwarded beyond limited sectors, there can still be a considerable amount of pulse redundancy. Spatial pulse compression using the random back off strategy can also used for eliminating this remaining transmission redundancy during pulse forwarding. When a node from the origin event area transmits a pulse, it can be received by multiple nodes in the next lowest hop area. Upon receiving the pulse, each such receiver follows the back off process as used during the pulse generation portion to ensure that only a minimal number of them (ideally 1) end up forwarding the pulse toward the sink. As shown in Figure 12.11b, due to spatial compression the number of forwarded pulse transmissions is significantly reduced compared with the case without compression in Figure 12.11a. Similar to HPS, spatial compression applied into CPS also reduces the amount of pulse redundancies in each sensor cell of the pulse routes.

12.5.3.4 Temporal Compression

When a node detects multiple events in quick succession (e.g., detecting a moving target multiple times while it is within the detection range), it generates multiple pulses, each of which are independently forwarded to the sink. Temporal pulse compression can be used to remove this redundancy by ensuring that a node generates, at most, one pulse within a certain time duration. This duration is referred to as a deactivation period. The length of deactivation needs to be dimensioned based on the expected speed of movement of the moving targets. Temporal compression is also applicable while pulse forwarding so that a node forwards, at most, one pulse originating from the same event area within a deactivation period.

12.6 Physical Layer Cooperation Mitigation Measures

Pulse merging, as explained in Section 12.5.3.1, can give rise to an undesired effect of node cooperation [17] when multiple nodes simultaneously transmit pulses in a slot. Such simultaneous transmissions can increase the effective transmission power, thus extending RF transmission range. Node cooperation can affect both hop-distance discovery and pulse forwarding in HPS. However, CPS is immune to such a node cooperation phenomenon. This section will introduce the measures for mitigating the effects of node cooperation in HPS and also explain the reason for immunity of CPS to node cooperation.

12.6.1 Mitigating Cooperation in HPS

12.6.1.1 Mitigating Cooperation during Hop-Distance Discovery

During hop-distance discovery in HPS (Section 12.3.2), after the nodes at hop-distance 1 have discovered their own hop-distance, they send simultaneous discovery pulses in the same slot in the reconfiguration area. Unintended node cooperation due to the energy aggregation of all such pulses can cause them to reach nodes that are beyond a hop-distance of 2. This can give rise to faulty hop-distance recovery, leading to possible pulse forwarding failures. Such problems in hop-distance discovery due to node cooperation can happen at all hop-distances except at a hop-distance of 1. The following mechanisms are proposed for minimizing the effects of node cooperation by reducing the chances of overlapping pulses during hop-distance discovery.

The number of slots in the reconfiguration area (Figure 12.6) is increased from $H + 1$ to $HM + 1$. The first slot is still allocated to the sink, and the nodes in each hop-distance are allocated a cluster of M slots instead of a single slot. Additionally, a slot is functionally divided into N_p pulse positions or mini-slots. The hop-discovery process in Section 12.3.2 is augmented with the following new rule. A node in the hop-distance h generates a pulse after (1) randomly selecting one of the M slots allocated for the hth hop-distance, and (2) randomly selecting one of N_p pulse positions within the slot selected in step 1.

With this rule, the probability of nonoverlapping pulses generated by N_h nodes in hop-distance h can be estimated as $P_h^{\text{dif}} = 1 - \sum_{k=2}^{N_h} \binom{N_h}{k} (\hat{p})^k (1-\hat{p})^{N_h-k}$, in which the second term represents the probability of having at least two overlapping pulses among N_h nodes, and $\hat{p}$ represents the probability of having exactly one pulse in the slot cluster (i.e., M slots) allocated for the nodes in hop-distance h. $\hat{p} = MN_p p_c (1-p_c)^{M-1} p_s (1-p_s)^{N_p-1}$, where $p_c = 1/M$ and $p_s = 1/N_p$. For a typical UWB pulse width of 1 ns and pulse repetition period of 1000 ns [6], the typical value of N_p is 1000, with which the quantity p_h^{dif} approaches 1. This means that with the augmented hop-distance discovery rule, the probability of nonoverlapping pulses generated by the nodes in a hop-distance is near 100%. Therefore, the effects of physical layer node cooperation on hop-discovery can be mostly mitigated.

In very rare occasions, when a node may still receive overlapping pulses (a receiver node cannot detect overlapping pulses), the following additional rule can further reduce the possibility of faulty hop-distance discovery. If a node receives multiple pulses in different hop-distance slots (i.e., hop-distances $1,\ldots, h-1$) in the reconfiguration area, the node decides its own hop-distance as $\max\{1,\ldots, h-1\} + 1 = h$. For example, consider the rare occasion in which two nodes at hop-distance 2 send out overlapping pulses, which can reach a node with hop-distance 4. This happens due to node cooperation caused by energy aggregation of the overlapping pulses. Without such cooperation, the receiver node in question should have received discovery pulses only from nodes in hop-distance 3 and not hop-distance 2. With cooperation, however, the receiver node ends up receiving pulses from nodes in hop area 2 as well as correct discovery pulses from hop area 3. According to the max rule, the node interprets its own hop-distance as 4, which is one more than the maximum of 2 and 3. This way, the effects of node cooperation are fully mitigated during hop-distance discovery for HPS.

12.6.1.2 Immunity to Node Cooperation during Pulse Forwarding

Wave front pulse forwarding, as proposed in Section 12.3.3, is inherently immune to node cooperation due to its hop-distance synchronized state transitions. Consider a cooperation situation in which multiple nodes in hop-distance $h + 1$ (at state T) simultaneously forward a pulse for an even that was generated at a higher hop-distance event area. During this transmission, nodes in h, $h - 1$, and $h - 2$ hop-distances are in states L, S, and T, respectively. This means, due to node cooperation even if the pulse reaches areas with hop-distances $h - 1$ and $h - 2$, it will be ignored simply because the nodes in those areas are not in a listen state. Only nodes in hop-distance h will receive it, which is what was intended.

12.6.2 Immunity to Node Cooperation in CPS

The proposed sensor cell–based event localization in CPS is able to avoid the side effect of node cooperation. Because of this, sensor cell–based event localization is unrelated to varied wireless transmission ranges. Additionally, the pulse forwarding rule described in Section 12.4.3 is constrained by the neighborhood limitation of sensor cells, which is also irrelative to wireless transmission range.

12.7 Performance Evaluation

We developed an event-driven C^{++} simulator that implements MAC framing and pulse routing using the UWB-IR model as presented in Sections 12.1 through 12.6. This section evaluates the performances of our proposed HPS and CPS protocols under two different scenarios—a static event scenario and a moving target scenario.

12.7.1 HPS

The network terrain size is set to 10×10 m^2 with 441 sensor nodes, and a sink node, which is placed at the center of the terrain.

12.7.1.1 Static Event Scenario

This subsection presents the results of when a single pulse is generated for a single static event from within an event area.

12.7.1.1.1 Pulse Transmission Count

For PSP-B and PSP-C, Figure 12.13a reports the number of forwarding transmissions across different hop areas when an event is created in hop area 5 in the 441 node network. The numbers are reported for two different sector constraints ($\delta = 0.2$ and 1). With PSP-B, observe how the number of pulse transmissions maximizes at an intermediate hop-distance for both δ values, confirming the lateral fan-out and convergence seen as the routing envelopes in Figure 12.12.

With PSP-C, the number of pulse transmissions in all hop areas are reduced to a very small value due to the back off process described in Section 12.5.3.2. With such spatial pulse compression, the minimum number of forwarding transmissions per event area is 1, and it stays around 1 for a sufficiently large back off counter R. Because of this small spread in the forwarding transmission count, unlike PSP-B, the PSP-C lines in Figure 12.13a do not show an obvious maximum.

12.7.1.1.2 Route Diversity

The route diversity factor β ($\beta \geq 1$) represents the number of forwarding transmissions for a pulse from hop-distance h, normalized by h, which is the minimum number of required transmissions. Figure 12.13b demonstrates β with increasing sector constraint δ for two different angular resolutions of α = 15° and 30°. A larger α means more nodes are involved in forwarding (Figure 12.12a), leading to higher

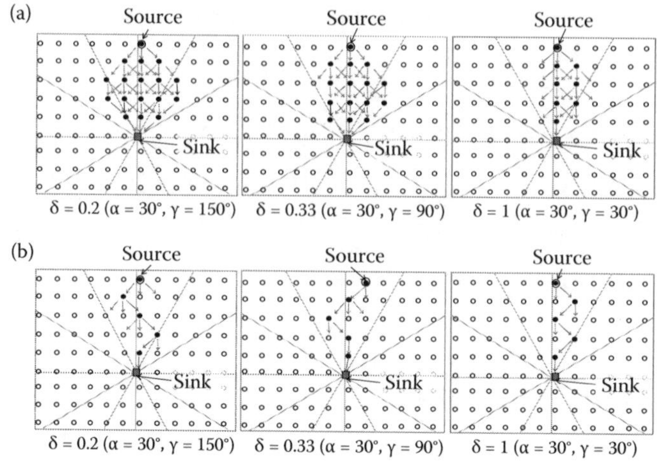

Figure 12.12 Effects of pulse compression on pulse forwarding in HPS. (a) Without pulse aggregation (PSP-B). (b) With pulse aggregation (PSP-C). Pulse routing envelopes with different sector constraints for a given angular resolution in the cases of compression and no compression.

route diversity. Observe that β reduces for PSP-B as the sector constraint is increased. This is the result of reduced forwarding transmissions. It follows that when α is lower, the overall route diversity is also lower.

Because of compression, PSP-C demonstrates much lower and δ-insensitive route diversity. As shown in Figure 12.13b, the low diversity is due to the transmission reduction caused by the back off–based compression, and the δ-insensitivity is because of the number of forwarding nodes in this case, which is already very low (i.e., nearly one node per hop area).

12.7.1.1.3 Error Analysis

Primary sources of errors for UWB-IR are multipath delay spread (Figure 12.1), noise, and interference. Multipath delay spreads can cause overlapping pulses, leading to errors in pulse detection. As shown in Figure 12.1, very large slot size (i.e., 1000 ns) should be chosen in comparison to the pulse width (i.e.,

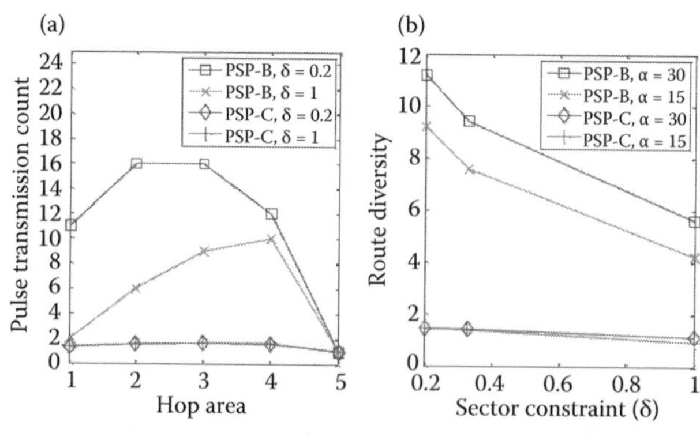

Figure 12.13 (a) Pulse transmission count in each hop area along the route with different sector constraints in the cases of compression and no compression. (b) Effects of sector constraint on route diversity with different angular resolutions in the cases of compression and no compression.

1 ns) for minimizing the overlapping errors by constraining the multipath delay spread within a slot. The other primary contributors to error are interferences including MUI and NBI. The factors described above can result in one of two types of detection errors [14], namely, pulse loss and false-positive errors.

12.7.1.1.4 Effects of Pulse Loss Error

Pulse losses can manifest in the form of unreported events. We define pulse loss rate (PLR) as the probability that a pulse is lost in a given time slot due to multipath, channel noise, or interferences. Event loss rate (ELR) is the probability that a pulse is not reported to the sink. Figure 12.14a depicts the simulation results of PLR versus ELR for a single event generated in hop area 5 of the 441-node network. Observe that for practical range of PLR [18], the ELR for PSP-B remains vanishingly small and it is generally insensitive to the value of PLR. This is mainly because of the huge redundancy in pulse transmissions (i.e., route diversity) for PSP-B as demonstrated in Figure 12.13. The physical implication of this result is that at the cost of transmission redundancy, the baseline version of PSP offers very high immunity from pulse losses.

With spatial compression enabled (i.e., PSP-C), the ELR is higher and it shows clear sensitivity to PLR. The higher ELR is caused by almost zero transmission redundancy (i.e., near-one β in Figure 12.13b), which leads to a high probability of event losses when pulses are lost. It should be noted that although PSP-C is less immune to pulse losses compared with PSP-B, the resulting ELR is still low for practical PLR values [18].

12.7.1.1.5 Effects of False-Positive Errors

If pulses are erroneously detected [14] by a node in state LO or TL such that a false-positive pulse in the control subframe corresponds to another false-positive pulse in the event subframe with a corresponding forwarding flag (Figure 12.6), then an event is falsely detected at that node. Once such a false-positive event is generated, it is forwarded all the way to the sink, leading to a false-positive event reporting. Let FPPR (false-positive pulse rate) be the probability that a false-positive pulse is generated due to faulty UWB detection in a given time slot. We intend to determine the quantity FPEGR (false-positive event generation rate), which corresponds to the probability of at least one false-positive event generation per frame per node at a given hop area.

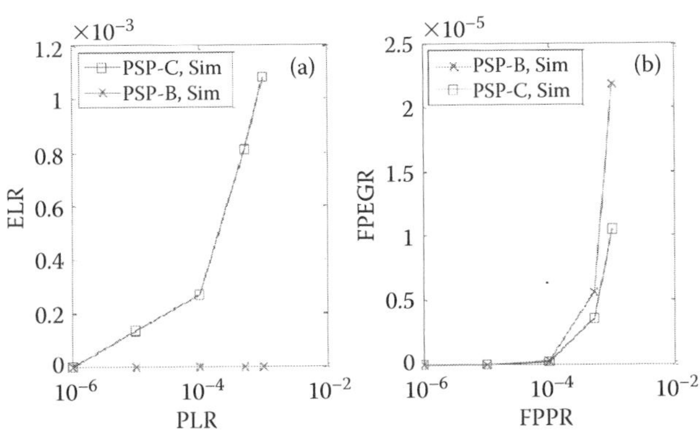

Figure 12.14 (a) Effects of pulse loss rate (PLR) on event loss rate (ELR) in the cases of compression and no compression. (b) Effects of false positive pulse rate (FPPR) on false positive event generation rate (FPEGR) in the cases of compression and no compression.

The effects of FPPR on FPEGR in hop area 3 both in PSP-B and PSP-C are shown in Figure 12.14b. Hop area 3 was chosen because it represents the middle of the experimental network. Observe that in both PSP-B and PSP-C, FPEGR is extremely small with a practical range of FPPR [18], which is less than 10^{-4}. This indicates that both with and without compression, the proposed pulse switching protocol is fairly immune to false-positive errors. Also observe that for larger FPPR (i.e., larger than 10^{-4}), when the compression is turned on, PSP-C shows less sensitivity to false-positive errors compared with PSP-B.

12.7.1.1.6 Energy Consumption

A distributed TDMA [19] with sink-rooted minimum spanning tree for packet routing has been used as the representative protocol to compare its energy consumption with that of the proposed pulse switching. TDMA was chosen because of its high energy efficiency compared with random access mechanisms. An event detected by a sensor is reported to the sink using min-hop routing along the minimum spanning tree. A packet contains the minimum amount of information to represent a {*sector-id*, *hop-distance*} event area and also a per packet preamble [20].

Based on the UWB specification [8], the transmission and reception consumptions are set to 4 and 8 nJ per pulse, respectively. Because a pulse transmission using the baseband UWB has the same energy expenditure for PPM bits in a packet, the same 4 and 8 nJ values are used for both pulse and bit (in packets) transmission and reception.

Figure 12.15 reports on the communication power consumption for both pulse and packet transport with varying event rates. Observe that the consumption is linearly dependent on the event rate λ for both pulse and packet scenarios. The slope of the TDMA graph in Figure 12.15 is noticeably higher than that for both PSP-B and PSP-C, and it is mainly due to the per packet preamble and payload overheads. Also observe that the slope of the PSP-C is lower than that of PSP-B, indicating the reduction of energy expenditure as a result of spatial pulse compression. This result further validates the low route diversity of PSP-C in comparison to PSP-B. Overall, Figure 12.15 validates the primary premise of pulse switching that it can transport multiple point-to-point binary events at a lower energy budget compared with traditional packet switching.

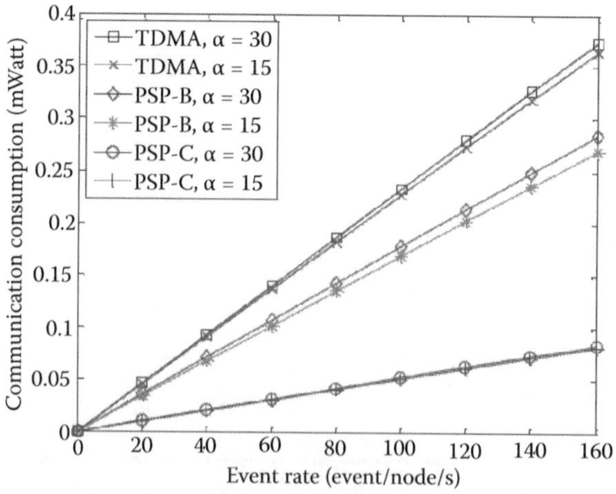

Figure 12.15 Communication power consumption per node.

12.7.1.2 Moving Event Scenario

This subsection presents results for scenarios in which a moving target causes the sensor on its path to generate a series of pulses that are transported to the sink to estimate the target's trajectory. Results are obtained with the following parameterization: (1) target speed in meters per frame duration, (2) sensor detection range that represents the radius of detection, (3) sensor detection rate in sensing per frame duration, and (4) target trajectory path. As shown in Figure 12.16, three diagonal (D1–D3), three horizontal (H1–H3), and three circular (C1–C3) trajectories were used for the presented results. Once a target enters the network and follows one of the above trajectories, the affected nodes (i.e., within the sensing radius) generate pulses at the specified detection rate. The pulses are then transported to the sink using PSP-B or PSP-C.

Figure 12.17 shows the pulse transmission counts for various target trajectories (for PSP-B and PSP-C) when the detection range and detection rates were chosen as 0.75 m and 5 times/frame,

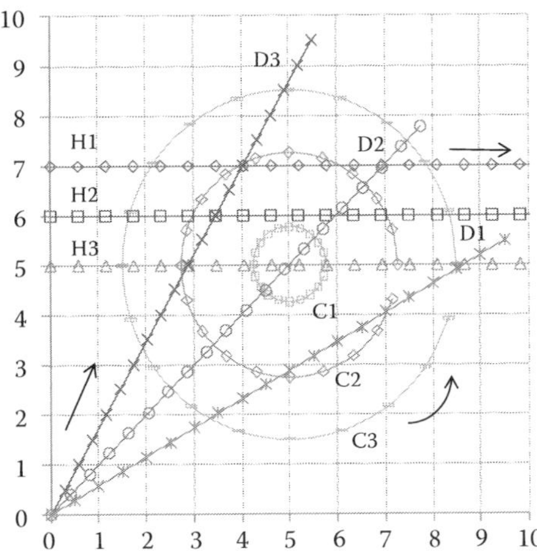

Figure 12.16 Experimental trajectories.

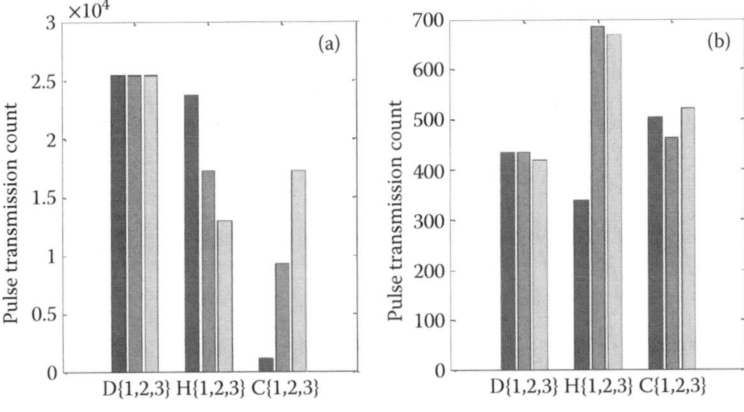

Figure 12.17 Pulse transmission count for (a) no compression (PSP-B) and (b) compression (PSP-C).

respectively. The target speed was set to 0.0116 m/frame for the diagonal and horizontal trajectories, and 0.314 m/frame for the circular trajectories.

As expected, the PSP-C protocol significantly reduced the forwarded pulse count compared with PSP-B. This huge reduction in this mobile target case is contributed not only by spatial compression as observed in the static single event case but also by the temporal compression as described in Section 12.5.3.2.

Because pulses generated in higher hop areas result in more forwarding transmissions, with PSP-B, the horizontal trajectory H1 generates more pulses than H2, and H2 generates more pulses than H3 (Figure 12.17a). Similar observations can be also made for trajectories C1 to C3, among which the trajectories with higher hop radii generate more pulse transmissions. For the diagonal trajectories D1 to D3, however, there is no significant difference in forward transmission count because all those trajectories have the exact same hop-distance properties. As shown in Figure 12.17b, PSP-C, the protocol with compression, can report the moving target's locations with approximately 38.59 times (on an average) fewer forwarding transmissions compared with PSP-B. Because of the randomness of the back off process used in spatial compression, the transmission count numbers in PSP-C do not follow a clear monotonic trend as seen for PSP-B in Figure 12.17a.

Figure 12.18a reports the effects of target speed on the forwarding pulse count. For the results shown, which are for trajectory D3, observe that the pulse count reduces with increasing target speed for both PSP-B and PSP-C. This is because for a given event detection range and event detection rate, the number of pulses generated by each source sensor (on and around the trajectory) is less for a target moving at higher speed. Also, as expected, irrespective of the speed, the pulse-count for PSP-C is lower than that of PSP-B, further validating the effectiveness of temporal compression for moving targets.

The effects of deactivation period, a key parameter used for temporal pulse compression (Section 12.5.3.2), are shown in Figure 12.18b. These results are for the trajectory C3. Observe that as the deactivation time increases, the number of forwarding transmissions decreases because of the higher level of compression. The flip side of the higher deactivation period is the possibility of unreported events for high-speed targets. In other words, for a given detection range and rate,

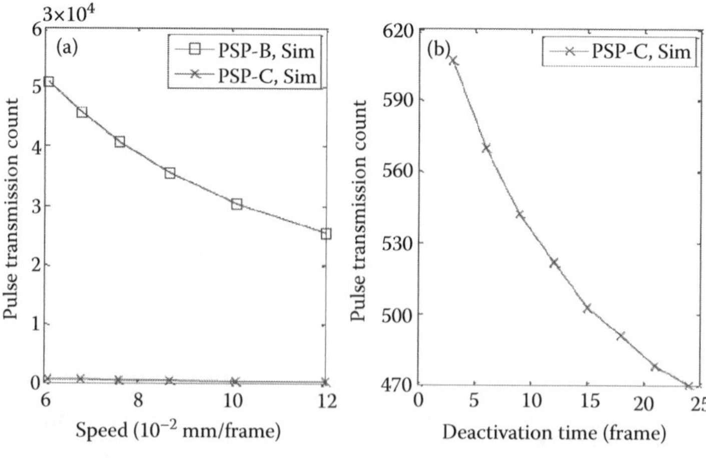

Figure 12.18 (a) Effects of target speed on pulse transmission count in the cases of compression and no compression. (b) Effects of deactivation time on pulse transmission count.

a sensor may not report successive events when they are generated in quick succession due to high target speed. Therefore, there exists a trade-off between compression performance and the maximum supported target speed. The length of deactivation needs to be dimensioned based on the expected speed of the moving targets in a sensor field.

Compression in PSP-C is accomplished during pulse generation and forwarding, both spatially and temporally. To track the breakdown of the spatial and temporal compressions, we measured the compression factor by turning the deactivation (i.e., spatial compression) on and off for the circular trajectory C3. With no temporal compression, the spatial compression factors in this case were observed to be 3.91 and 23.99 during generation and forwarding, respectively, leading to a resulting compression factor of 32.04. With temporal compression activated, the factors were 4.11 and 25.1 during generation and forwarding, leading to an overall compression factor of 34.2.

12.7.2 CPS

The network terrain size is set to 10×10 m^2 with evenly distributed sensor nodes and a sink node placed at the lower left corner of the terrain. Sensor nodes are grouped into 115 regular hexagonal sensor cells, each of which contains five nodes on average.

12.7.2.1 Pulse Transmission Count

Figure 12.19a reports the number of pulse transmissions in sensor cells that are at specific cell counts away from the sink node. An event is generated at a cell which is 15-cell count away from the sink node. The event is then routed (with route diversity set to 1) to the sink using the presented CPS protocol. The resulting number of transmitted pulses within the sensor cells along the route is plotted in Figure 12.19a. The *x* axis corresponds to the distance of the cells along the route expressed as the cell count from the sink. Cell count 1 corresponds to the sensor cell nearest to the sink and the highest cell count corresponds to the source cell.

For comparison purpose, we also present pulse transmission counts for the HPS reported by Huo et al. [15], and applied to the same network. The *x* axis for HPS corresponds to the distance of the cells along the route expressed as hop-count from the sink. Similar to CPS, hop-count 1 corresponds to the hop-distance nearest to the sink and the highest hop-count corresponds to the source area for the HPS scenario. Note that although the event is generated at 10 hop-distance

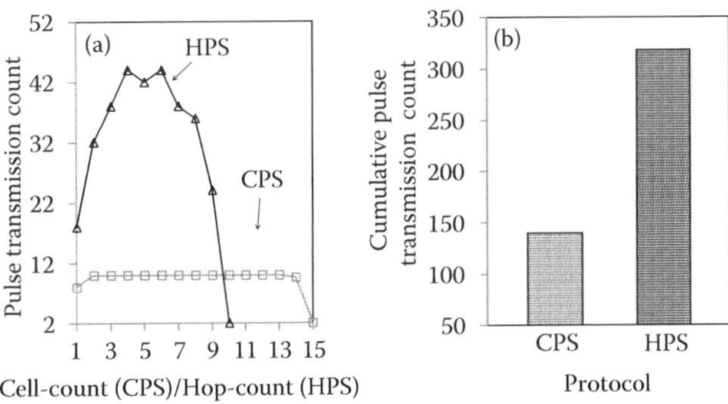

Figure 12.19 (a) Pulse transmission count in each cell-count along the route of a single event for HPS and CPS. (b) Cumulative pulse transmission count of a single event for HPS and CPS.

away, the shortest route for CPS is 15 sensor cells–long in this case. This is because the size of a hop area in HPS is larger than that of a sensor cell in CPS, and thus, the hop-count in HPS is smaller than the cell count in CPS for the same distance. For HPS, the resolution and the sector constraint are set to 30° and 1, respectively.

With CPS, the number of pulse transmissions in a sensor cell is equal to the number of forwarding nodes in the corresponding sensor cell multiplied by the number of pulses in a frame. As shown in Figure 12.19a, the line corresponding to the CPS is flat across different cell counts except for the cell counts 1 and 15. All five nodes in the sensor cell corresponding to the cell counts from 2 to 14 participate in forwarding pulses. Therefore, multiplying the node count 5 by 2, which is the number of pulses in a frame (i.e., one in the control area and the other in the localization area), results in 10 pulse transmissions. Pulse transmission count in the cell count 15 is 2 because only one node in the source cell sends a pair of pulses in a frame. The sensor cell in cell count 1 has four nodes. This is the reason why the pulse transmission count in the cell count is 1 to 8, which is slightly smaller than 10.

Observe that the number of pulse transmissions in HPS maximizes at certain intermediate hop-counts. This is caused by the following: in HPS, pulse routing is implemented in the form of constrained flooding with very few nodes participating in forwarding near the source and the sink nodes, and large number of nodes participating in the middle. This, coupled with the fact that the number of pulse transmissions in each hop-count is linearly dependent on the number of forwarding nodes in the corresponding hop-distance, explains the pulse count maximization near the hop-count range of 4 to 6.

More importantly, the pulse transmission counts in HPS are much larger than that in CPS for up to 10 hop-counts. It is because the hop area in HPS includes more nodes than a sensor cell in CPS.

Figure 12.19b reports the cumulative pulse transmission count along the entire route in both CPS and HPS scenarios. The cumulative pulse transmission count of a single event originating from the source node toward the sink in HPS is 2.23 times greater than that in CPS.

These indicate that CPS is able to achieve better energy efficiency than HPS by being able to transport events to the sink with a lower number of pulse transmissions. Because HPS was proven to be more energy-efficient than packet switching [15], these results demonstrate that CPS can provide a more energy-efficient solution by replacing packet switching in applications involving binary event sensing. Additionally, because the number of sensors within a sensor cell in CPS is generally smaller than the number of sensors within HPS hop areas, CPS offers better localization resolution compared with HPS.

12.7.2.2 Different Source Cells

Figure 12.20a shows the cumulative pulse transmission count as a function of the distance between the sink node and the sensor cell of event generation. Such distance, represented as a cell count, varied from 1 to 15. Pulse transmission count results are collected for two different route diversity (i.e., δ) values of 1 and 2. As expected, Figure 12.20a demonstrates how the pulse transmission count increases with increasing distance (i.e., cell count) between the source sensor cell and the sink node. The line with $\delta = 1$ shows a linear relationship between the pulse transmission count and the cell count, whereas the line with $\delta = 2$ demonstrates a more steep linearity. Additionally, pulse transmission count with $\delta = 2$ is significantly higher than that with $\delta = 1$ for a given distance.

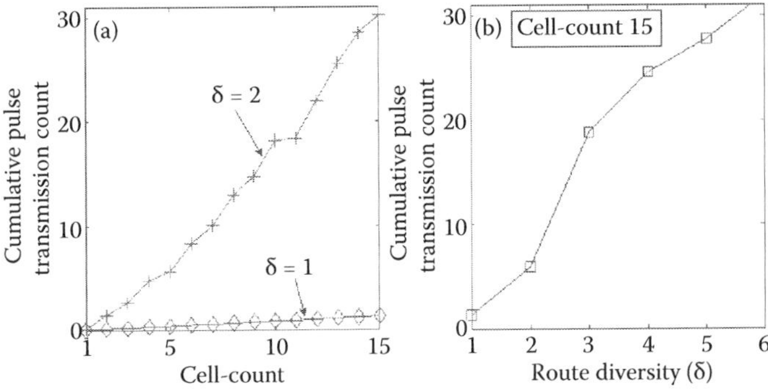

Figure 12.20 (a) Effects of distance between the event origin and the sink on cumulative pulse transmission count with different route diversities. (b) Effects of route diversity on cumulative pulse transmission count for a single event.

12.7.2.3 Effects of Route Diversity

Figure 12.20b reports cumulative pulse transmission count with different route diversities applied to the switching/routing process for a pulse generated at a distance corresponding to cell count of 15. Route diversity δ varied from 1 to 6. The figure shows that with increasing route diversity, the pulse transmission overhead increases. Eventually, the CPS process approaches effective pulse flooding for very high δ values. From Figure 12.20b, it can be observed that the increment in pulse transmission count is generally most prominent when δ is changed from 2 to 3 for a cell count of 15. Furthermore, Figure 12.20a shows that the cumulative pulse transmission count is higher with a greater δ for any given cell count. Hence, the observations from Figure 12.20 suggest that route diversity should be kept below 2 to keep the energy costs under control.

12.7.2.4 Error Analysis

This section reports the effects of pulse detection errors, both in terms of false-positive pulses and missed pulses, on pulse reporting using the proposed CPS architecture. When a node in the listen state L receives a false-positive pulse in the control area, which corresponds to another false-positive pulse in the localization area (Figure 12.8a), an event is falsely detected at that node. Once such a false-positive event is generated, it is forwarded all the way to the sink, leading to a false-positive event reporting. Let FPPR be the probability that a false-positive pulse is generated due to faulty UWB detection in a given time slot. We intend to determine the FPEGR, which corresponds to the probability of at least one false-positive event generation per frame per node at a given sensor cell. The effect of FPPR on FPEGR in any sensor cell is shown in Figure 12.21a. Observe that FPEGR remains extremely small for the practical range of FPPR, which is usually lower than 10^{-4} [18]. This indicates that the proposed CPS protocol is fairly immune to false-positive errors.

Pulse losses can be manifested in the form of un-reported events. We define PLR as the probability that a pulse is lost in a given time slot due to multipath, channel noise, or interferences. ELR is the probability that a pulse is not reported to the sink. Figure 12.21b depicts the results of PLR versus ELR for a single event generated in a specific sensor cell. The route diversity δ is set to 1.

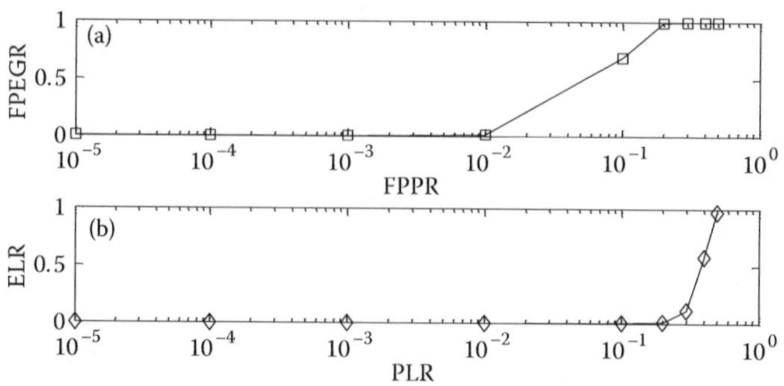

Figure 12.21 (a) Effects of pulse loss rate (PLR) on event loss rate (ELR). (b) Effects of false positive pulse rate (FPPR) on false positive event generation rate (FPEGR).

Observe that for the practical range of PLR (less than 10^{-4}) [18], the ELR remains vanishingly small and it is generally insensitive to the value of PLR. When the route diversity δ is larger than 1, the system's reliability against pulse loss errors would be better than the case when the route diversity δ was 1 due to more pulse transmission redundancies.

12.8 Conclusion

A novel pulse switching protocol for ultra-lightweight networking applications has been developed in this chapter. The joint MAC-routing architectures for pulse switching with hop-angular and cellular event localization were presented. The key contribution of the presented architecture is to combine hop-angular and cellular event localization with a pulse switching protocol in a manner that allows a receiver to localize an event by observing the temporal position of a received pulse with respect to a synchronized frame structure. Through simulation-based experiments, it is shown that the proposed pulse switching architecture can be an effective means for transporting information that is binary in nature with high energy efficiency and strong reliability against errors.

References

1. C.R. Farrar, G. Park, D.W. Allen, and M.D. Todd. Sensor network paradigms for structural health monitoring. *Journal of Structural Control and Health Monitoring*, vol. 13, no. 1, pp. 210–225, 2006.
2. A. Jain, M. Gruteser, and D. Grunwald. Benefits of packet aggregation in ad-hoc networks, Technical Report CU-CS-960-03. Department of Computer Science, University of Colorado, Boulder, CO, August 2003.
3. Z. Wang, E. Bulut, and B. Szymanski. A distributed target tracking with binary sensor networks. In *IEEE International Conference on Communication Workshops*, pp. 306–310, 2008.
4. C. Fragouli and A. Orlitsky. Silence is golden and time is money: Power-aware communication for sensor networks. In *Proceedings of the Allerton Conference on Communication Control and Computing*, Monticello, Illinois, USA, 2005.

5. E. Minazara, D. Vasic, and G. Poulin. Piezoelectric diaphragm for vibration energy harvesting. *Ultrasonics*, vol. 44, no. 1, pp. e699–e703, 2006.
6. S. Haykin and M. Moher. *Modern Wireless Communications*, Prentice-Hall, Inc., Upper Saddle River, NJ, 2004.
7. M.Z. Win and R.A. Scholtz. Impulse radio: How it works. *IEEE Communication Letters*, vol. 2, no. 2, pp. 36–38, 1998.
8. B. Poucke and B. Gyselinckx. Ultra-wideband communication for low-power wireless body area networks. *Industrial Embedded Systems Resources Guide*, Industrial Embedded Systems, 2005.
9. M. Ghavami, L.B. Michael, and R. Kohno. *Ultra Wideband Signals and Systems in Communication Engineering*. John Wiley & Sons. Ltd., England, 2004.
10. H. Nikookar and R. Prasad. *Introduction to Ultra Wideband for Wireless Communications*. Springer, 2008.
11. IEEE P802.15 working group for Wireless Personal Area Networks (WPANs) DS-UWB physical layer submission to 802.15 Task Group 3a, 2005.
12. T. Ikegami and K. Ohno. Interference mitigation study for UWB radio. In *Proceedings of the 14th IEEE International Symposium on Personal, Indoor and Mobile Radio (PIMRC)*, Beijing, China, pp. 583–587, 2003.
13. M. Flury and J.-Y. Le Boudec. Interference mitigation by statistical interference modeling in an impulse radio UWB receiver. In *IEEE International Conference on Ultra-Wideband (ICUWB)*, Waltham, MA, USA, pp. 393–398, 2006
14. H. L. Van Trees. *Detection, Estimation, and Modulation Theory. Part I: Detection, Estimation, and Linear Modulation Theory*. John Wiley & Sons, Ltd., England, 2001.
15. Q. Huo, S. Biswas, and A. Plummer. Ultra wide band impulse switching protocols for event and target tracking applications. *2011 8th Annual IEEE Communications Society Conference on Sensor, Mesh and Ad Hoc Communications and Networks (SECON)*, Salt Lake City, Utah, USA, pp. 197–205, 27–30 June 2011.
16. Q. Huo and S. Biswas. A novel concept of UWB pulse switching in sensor networks. *The Eighth International Conference on Wireless and Mobile Communications, ICWMC 2012*, June 2012, Venice, Italy.
17. A. Nosratinia, T.E. Hunter, and A. Hedayat. Cooperative communication in wireless networks. *IEEE Communications Magazine*, vol. 42, pp. 74–80, 2004.
18. Yu. Andreyev, A. Dmitriev, E. Efremova, A. Khilinsky and L. Kuzmin. Qualitative theory of dynamical systems, chaos and contemporary wireless communications. *International Journal of Bifurcation and Chaos*, vol. 15, no.11, pp. 3639–3651, 2005.
19. Z. Chen and A. Khokhar. Self-organization and energy-efficient TDMA MAC protocol by wake up for wireless sensor networks. In *Proceedings of 1st IEEE Conference Sensor and Ad Hoc Communication and Networks (SECON '04)*, pp. 335–341, 2004.
20. Z. Yuanjin, C. Rui, and L. Yong. A new synchronization algorithm for UWB impulse radio communication systems. In *Proceedings of the International Conference on Communication Systems (ICCS)*, Singapore, 2004.

Chapter 13

Handling the Energy in Wireless Sensor Networks

Vivek S. Deshpande
MIT College of Engineering

Contents

13.1 Introduction

Nowadays, wireless sensor networks (WSN) are very popular for its specialty. WSN applications have a wide variety of domains, and could range from military applications to farming applications. From surveillance systems for enemies or threats to precision agriculture in which a farmer can control the temperature, humidity, etc., are just a few examples of WSN applications. The field of health and medicine has plenty of challenges in which WSNs could play an important role in monitoring and disseminating data to a base station. Every application of WSN is composed of a set of sensor nodes and a base station called the sink. It is a sort of distributed system in which all the nodes can work together to convey the data to the sink (Figure 13.1). The entire node senses the data, depending on the application, and sends it to the sink. The data may reach the destination in a single hop or through multiple hops. While considering the data dissemination, one must consider the data reliability, congestion status, required delay, and many more parameters [1].

All the parameters mentioned are generally called quality of service (QoS) parameters. While maintaining the QoS, one must consider the energy consumed in the process [2,3]. All the QoS parameters must be tested for all types of data such as periodic data, nonperiodic data, bursty data, transient data, etc. All the nodes in the WSN send their data with a particular frequency called a reporting rate. This rate can be controlled to maintain the congestion. At the same time, by adjusting the reporting rate, one can also control the data's reliability [4,5]. The QoS parameters must be taken care of, when one is designing WSNs. The existing protocols provide either congestion control or reliability [6]. A few protocols such as STCP provide both congestion control along with reliability of data.

The WSN is a collection of sensors that communicate with each other by sending data. These sensors are very small devices distributed in a large geographical area called the sensing field. It has sensing ability (to sense environmental conditions), processing ability, and some memory to store the data. It can send and receive data within a particular communication range. The job of the sensors is to sense the environmental conditions, process that data, and send this information to the base station. There are different types of sensors such as seismic, thermal, visual, and infrared sensors. These sensors monitor a variety of conditions such as temperature, humidity, and the characteristics of objects and their motion. These different types of sensors are used in a number of applications such as fire or flood detection, smoke detectors, measuring temperature, humidity in the air, vehicular movement, pressure, soil makeup, noise levels, or mechanical stress levels. It can also be used in military, health, chemical processing, and disaster relief scenarios. Many home appliances such as refrigerators or vacuum cleaners also include sensors.

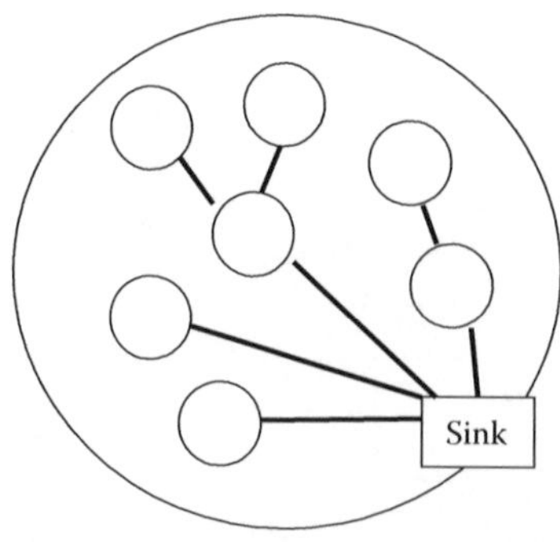

Figure 13.1 Structure of WSN.

There are different issues related to the sensor network. For example, orderly transmission, flow and congestion control, loss recovery, QoS, fairness, reliability, and energy efficiency. Congestion is a major issue upon which the rate of energy consumption depends. If congestion in the network increases, the node buffer fills up. This results in packet dropping, and thus, the sender has to retransmit those dropped packets. This means that all of the nodes that are in the path from the source to the destination consume some energy in the retransmission of those packets. The energy requirements are increased in such applications in which the reliability requirement is higher [7].

Different protocols are available to address these different issues [8]. Much work has been done thus far. Handling these issues is important, but there is no single technique or protocol that will address all these issues. Hence, which protocol should be used is totally dependent on the application. Sensors are powered by batteries. Small sensors have small batteries, which are limited. The sensors use energy in sensing the environment, processing that data and transmitting it to the sink node, as well as forwarding the packets that are received from other nodes. Actually, the energy requirements and consumption of different nodes in the network are different. The near-sink node forwards more traffic toward the base station and thus is depleted of energy sooner than the others, whereas nodes that are far away from the sink have to forward only their own data and thus require less energy. Sensors with the highest rate of energy usage per data cycle are denoted as critical nodes [9]. Such critical nodes limit the overall network lifetime. The lifetime of the network depends on the sensor lifetime, which can be defined in different ways, for example, the time when the first sensor runs out of energy, when the first region gets disconnected, when a certain percentage of sensors die, or when all the sensors run out of energy [10].

The battery limits the lifetime of the entire network. If any sensor node lost its energy, other nodes near it will start to be depleted of energy. This could result in loss of network connectivity, coverage and reliability, and availability of data. The energy in the nodes is also consumed due to other reasons, for example, retransmission due to any failure such as link failure, congestion, or full buffer as well as improper transmission policies, lack of or absence of cooperation between the nodes, and improper node deployment strategies. If the nodes are far away from each other, then it also requires more energy for data transmission. This improper energy utilization and imbalanced energy consumption can directly affect network lifetime. It also causes the wastage of residual energy. Often, the amount of residual energy is much higher. In an article by Wenguo and Tiande [11], the authors studied the nonuniform property of each nodes' energy consumption and the residual energy consumption of these sensor networks. Therefore, to improve sensor network lifetimes, this precious resource should be saved and consumed properly.

Sometimes, the sensors can have rechargeable batteries that can be charged through natural means, such as solar energy, or that can be replaced. This battery recharging technique is called energy scavenging. It is a good way to overcome the problem of limited battery lifetime. However, these sensors are often used in applications in which it is difficult to change or recharge the battery. And that is why the sensor network lifetime is still a critical issue. Much more work is available in the literature related to reducing energy consumption and increasing the lifetime of the network. Different techniques are available to reduce the energy consumption of the nodes. This chapter studies such strategies that reduce power consumption and ultimately improve network lifetime.

There are some easy methods to decrease the energy used for communication such as reducing the amount of data that is transmitted, reducing the number of reporting sensors, and shortening the communication range [12]. Because the energy requirements of nodes are different, if we assign the energy according to requirement, then it is possible to avoid the wastage of residual energy. Nonuniform energy assignment achieves both energy efficiency and energy balance simultaneously [10]. However, it is very difficult to calculate the energy requirement of nodes and assign them accordingly. The sensors are cooperative in nature, which can be taken advantage of in sensor networks. In an article by Fang et al. [13], the authors proposed a data transmission policy called energy-efficient cooperative communication (EECC). This policy takes advantage of the cooperative nature of sensors for energy-efficient transmission.

13.1.1 Topology

The topology of a network depends on different factors such as the nodes' mobility, failure and duty cycles, directional antennas, and transmission power [7]. There are different topologies and different routing strategies available that can achieve energy efficiency and balanced energy consumption. In a study by Ramesh et al. [14], the authors compared balanced and progressive topologies for sensor networks. Both of the balanced and progressive topologies provide energy efficiency, but which one should be used depends on the network size.

13.1.2 Routing

The topology of the network affects routing. In the literature, many protocols have been proposed on energy efficient routing. In an article by Ren et al. [15], the authors proposed an energy-balanced routing policy. Sometimes, due to uneven deployment, the network may have holes. Hwang et al. [16] proposed a protocol that provides energy-efficient routing for sensor networks with holes.

13.1.3 Mobility

By keeping some nodes mobile, the amount of energy that is consumed during data transmission can be reduced. When the network is static, that is, all the nodes are placed on a fixed position, all the data is forwarded to the sink node using the same path. Therefore, only those nodes that are on the route consume more energy than other nodes. Also, the sensor network has a convergent nature. All nodes forward traffic toward the sink node. Therefore, nodes that are near the sink have higher energy consumption than others. If the nodes are mobile, the route for the data will change. If the sink node is mobile, it can move and collect the data from the sensors themselves. This reduces the number of hops for data toward the sink and balances the energy consumption of all of the nodes. In a study by Kumar and Sivalingam [17], the authors proposed a communication strategy that could reduce this delay. Instead of mobile sinks or mobile nodes, the authors used a collector node, which is mobile. This node moves and collects data from the sensors. When this mobile data collector comes within range of the sink node, it forwards the collected data to that sink.

13.1.4 Duty Cycling

If the sensor nodes have a periodic job, that is, if it is not continuous in nature, then there is no use turning the node on. Turning off the node when it is no longer in use is also an option that would reduce the wastage of energy. The node should be on only when there is data to be sent or received. When the job is over, the node is placed in low power mode or turned off. This technique of energy saving is called duty cycling. However, in this technique, coordination among the sensors' sleep and wake-up times is necessary.

Abusaimeh and Yang [18] proposed a scheme to balance the energy consumption of the network. Herein, when the node has an energy consumption speed that is faster than the threshold value, the node is put off. The sensor with data to send first checks the energy consumption speed of the next hop node. If it is higher than the threshold value, the next hop node is put off and the sensor looks for a new route for data transfer. By putting the node off for a certain period, this scheme achieves balanced energy consumption.

13.1.5 Adaptive Sampling

As we know, the sensors consume energy in the sensing and transmission of data; thus, the more data there is to send, the more energy that is consumed. Sensors sense and transmit data periodically. The network lifetime can be increased by reducing the traffic generation rate. Often, the data to be sent is similar in periodic transmissions. Here, energy is wasted when it is consumed during the transmission of repeated packets. To avoid this, sampling can be done. An efficient adaptive sensing and adaptive sampling strategy reduces the number of samples which in turn reduces the amount of data to be processed and transmitted [19].

13.1.6 Transmission Policy

Like the amount of data to be transmitted, the transmission range also affects the energy consumption of the node. More energy is consumed in transmitting data for long distances. Accordingly, energy consumption is reduced by reducing the transmission range. Varying transmission range over time attains more uniform traffic and energy usage distribution among sensors [9].

There are different transmission policies—single hop, multihop, and hybrid transmission policy. In an article by Azad and Kamruzzaman [9], the authors proposed three different transmission policies, namely, fixed hop size (FHS), synchronous variable hop size (SVHS), and asynchronous variable hop size (AVHS). These policies can achieve uniform energy distribution in the network.

13.1.7 Deployment Strategy

One factor affecting energy is the deployment strategy, that is, how the nodes are deployed in the network. The deployment strategy changes according to some parameters such as the number of sink nodes in the network, node mobility, power assignment of the sensor, placement or location of nodes, mobility options (i.e., whether nodes and sinks are static or mobile), and traffic patterns. By varying these parameters, the deployment strategy can be changed. Cheng et al. [10] proposed and evaluated the performance of different deployment strategies. Each of these strategies has its own advantages and disadvantages but which strategy to use is totally dependent on the application.

13.1.8 Clustering

The sensors are grouped into clusters. Each cluster has assigned a node called the cluster head. Each individual node within a cluster transmits to the cluster head instead of transmitting to the sink node directly. This cluster head aggregates the data, reduces redundancies in the data, and then transfers this data to the base station. Clustering has several benefits such as reduced redundancy and reduced data to transfer. This saves the battery life of the cluster head. Data is transmitted to the cluster head and thus the transmission range of the sensors is reduced, thereby reducing the energy needs of the sensor. Generally, nodes that are rich in resources are selected as the cluster head. The role of the cluster head can be rotated among all the nodes in network. This can help achieve balanced energy consumption. In an article by Park et al. [20], the authors proposed an energy-efficient fair clustering scheme. In their study, the cluster head node is at the center of a cluster, whereas in a study by Wei et al. [21], the authors proposed an energy-efficient clustering scheme in which for each data collection round, a probability scale (i.e., the ratio of initial energy level to the average initial energy of the network) is calculated. Using these values, the cluster head candidates are selected. For each data collection round, the node with more resources is chosen for data transmission.

13.1.9 Multiple Base Stations

Increasing the number of sink nodes in the network will help in data transmission. This can reduce transmission distance between sensors, that is, the number of hops, and reduce the burden on other nodes up to some extent. Also, the chances of retransmission can be reduced.

WSNs have some problems to overcome, for example, energy conservation congestion control, reliability data dissemination, and security [1]. These problems are often involved in one or several layers top-down from the application layer to the physical layer, and can be studied separately in each corresponding layer, or collaboratively across each layer. One of the more important problems, congestion control, might be involved in the transport layer only, but energy conservation may be related to the physical layer, data link layer, network layer, and higher layers. Many researchers have recently turned their attentions to transport layer protocols, which are important for reliable data dissemination and energy conservation for WSNs. Congestion causes many problems when sensors receive more packets than that its buffer space allows, the excess packets have to be dropped and the energy consumed by the sensor nodes on the packet is wasted, and if additional packets have been sent, the more waste there is, which in turn diminishes the network throughput and reliable data transmissions. Congestion control studies how to recover from congestion. Congestion avoidance studies how to prevent congestion from happening. For this, we have to monitor the parameters that can help us avoid congestion in WSN, which is the subject of this chapter.

WSNs consists of a small microcontroller fitted with sensors and some means of communication radios. They are distributed over a wide area and transmit the gathered data to one or many central nodes called a sink, also known as the base station. WSNs are one of the emerging research areas that provide designated services such as disaster prevention, environmental monitoring, medical monitoring, habitat monitoring, military surveillance, inventory tracking, intelligent logistics, and health monitoring. These networks deliver numerous types of traffic; from simple periodic reports to unpredictable bursts of messages. Because of this, congestion control in WSNs is a very rare event. Congestion occurs in WSN due to radio channel interference, addition and removal of sensor nodes, and lastly, sensed events that cause bursts of messages. But always, it consumes energy.

In their article, Qiang and Yu [6] discussed congestion detection and congestion avoidance techniques. For congestion detection, ECODA uses dual-buffer threshold values and weighted buffer difference, that is, queue length difference in particular time intervals. The queue scheduler maintained in ECODA is based on priority of traffic; therefore, only low-priority packets are dropped. To avoid sink side congestion, which can happen most probably due to the nature of most WSNs, a bottleneck node–based source data sending rate control is used. It uses overheard delay information for detecting bottlenecks and, based on that, rate adjustment is done. The main advantage of this algorithm is that upon the occurrence of congestion, packets are dropped based on priority as it uses priority-based queue scheduling. Thus, the high-priority packets are not dropped. Bottleneck node–based source data sending rate control uses backpressure-based overhearing; therefore, it saves energy. However, after Q_{max}, that is, when queue utilization is higher than the upper threshold value, then even the highest priority packets get dropped to control congestion which wastes energy.

Alam and Hong [22] suggested a rate control method to control congestion in the network. They used a reservation-based MAC retransmission efficiency, which increases the delivery ratio and transmission efficiency. A centralized rate control mechanism ensures a fair reporting rate from each node. The authors even proposed a multisink environment by simulating all the mechanisms simultaneously and showing good performance results. The authors also achieved hop-by-hop reliability by acknowledgement at the MAC layer. A weighted proportional back off is used instead of binary exponential back off for MAC to reduce the number of collisions rather than avoiding them.

Performance metrics of energy and transmission efficiency were achieved using the proposed algorithm. This algorithm runs at the sink and hence it is a somewhat slower process to recover from congestion but the sink node's energy will be consumed to a large extent.

In EECC, Enigo and Ramchandran [23] designed the algorithm to adjust to the reporting rates of the sources. It also adapts to wireless communication conditions, minimizes energy consumption of data transport, and is a congestion control mechanism at the source, which reacts based on the sum of the node weights (NWs) at each node. EECC achieves congestion control along with energy saving. NW is a product of buffer size and channel busyness ratio. The channel busyness ratio is defined as the ratio of time intervals when the channel is busy due to successful transmission or collision to the total time. It controls congestion at the source and it reacts based on the sum of the nodes' weights at each node. The new transmission rate is calculated and rate adjustment takes place upon receiving the new rate. Here, the clustering algorithm runs at the sink and sensor nodes within the same cluster work alternatively to save energy. Hence, the average energy consumed by the nodes is reduced. This is achieved by keeping the packet delivery ratio untouched. It requires an efficient MAC protocol to work in the power-saving mode and at the same time the sink has to perform reclustering in case of changes in the time series reading of nodes.

Before we discuss the issues of processing data in the network, first we will explore how data is processed within the node [24]. The nodes' responsibility is to process the data packet, which is queued in the input buffer. Based on the logic of queuing, the packet leaves the queue for processing and is fetched into the node processing zone (Figure 13.2).

Furthermore, the packet is unpacked and data is kept ready for actual processing. Depending on the policy, aggregation will be done. This may include functions such as averaging, summing, among others. The routing strategy will be decided according to the networks' or clusters' policy, and the packet is then cached. This is required for hop-by-hop reliable communication strategy. The lost packets may be called by the next node to the previous node. Nowadays, cached packets

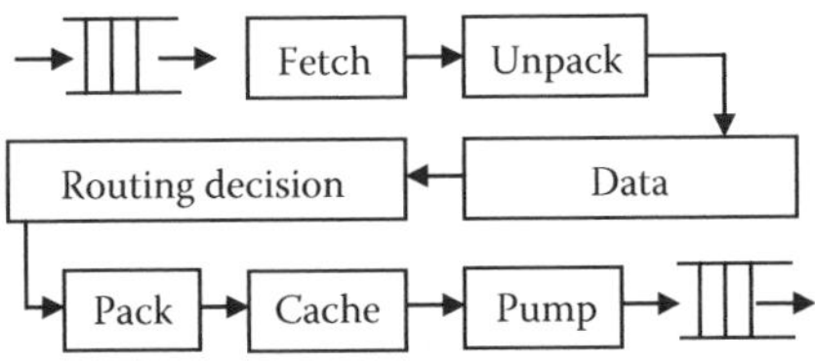

Figure 13.2 Logical data processing.

will be served to the next node. The packet is packed and finally, it is pumped into the output buffer. The abovementioned process offered to each packet takes some time. Every block in the chain will take time to process the current packet. The main challenge in this case is to reduce the service time to process the packet. The service time depends on the frequency of operation for the processor used in the node. WSN has two types of data flow, continuous data flow and event-driven data flow. Continuous data flows happen slowly, whereas in event-driven data flow, data packets must have the highest priority. These packets must reach their destination or base station (sink) within minimum amount of time. Here, processing delays play an important role.

If the frequency of operation is increased, then the amount of heat generated by the processor will be higher. On the other hand, the power taken from the battery will increase because power consumed by the processor and other allied peripherals will be also be higher, which reduces the network lifetime of that particular node. If the node processes data with less service time, the throughput of the node will increase and the overall performance of the network will increase [16].

Ouyang et al. [25] concentrated on congestion control along with fairness. The authors used an active bandwidth proportional (ABPS) technique to mitigate the congestion. Changes in the reporting rate are made as data travels toward the sink. Local source control provides good control over congestion. The authors show satisfactory results using bursty traffic. Energy utilization will be the main advantage of this control. The energy is fairly utilized throughout the network. The congestion scenario becomes more difficult when nodes that are near the sink increase their reporting rates. These nodes are heavily loaded with the traffic. This may cause congestion near the sink.

Ee and Bajcsy [26] discussed various types of congestion, which are related with the different losses occurring in the network (these may be lost packets). The authors insisted on dealing with congestion caused by simultaneous transmission. They also considered full queues for bursty traffic. The authors suggested a phase-shifting method instead of simultaneous transmission. A backpressure method was used to suggest the congestion happening by queue occupancy method. Epoch based solution was recommended to provide a fair chance for all packets originating from each node. Implementation results also show that there was good performance of algorithm for both congestion control and fairness. The authors did not consider radio signal strength, which is not uniform. Self-interference due to multipath effects must also be focused on.

There are many energy-efficient protocols for WSNs with mobile nodes. The sensor network has a convergent nature. Many nodes sense and forward data to another node that then forwards this data to a base station. Thus, this forwarder node consumes more energy and is depleted earlier. Often, the sensor network topology changes from time to time. Therefore, it is difficult to predict which node will consume more energy and which will consume less. Thus, the idea of our proposed work is to use a mobile node to alleviate the load of the node whose energy consumption speed is faster.

We are going to consider two-ray ground model as shown in Figure 13.3. In the figure, the source node is transmitting data to a destination node. Let the circled node n have an energy consumption speed that is faster than the other nodes due to the higher work load. This increases imbalance energy consumption of the network. The node with the faster energy consumption speed, that is, n will die earlier than the others.

Therefore, to balance the energy consumption among all nodes, the consumption speed of node n should be reduced. We proposed an idea to use mobile node nearby node n. This mobile node transfers data and reduces the burden on node n.

Initially, the data sender checks the energy consumption speed of the next hop node. If this speed exceeds a certain threshold, then a cooperative node is called within that region. The sender

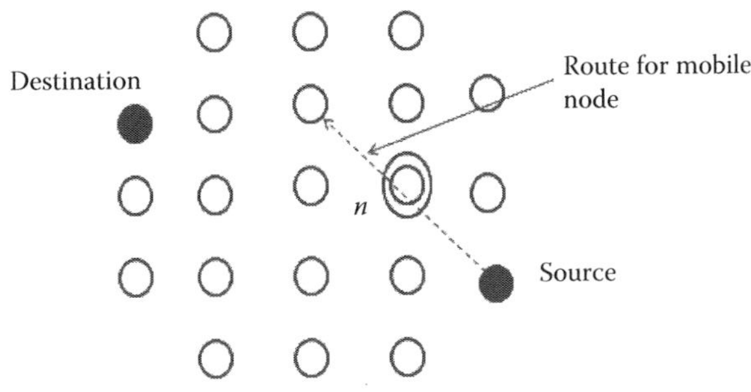

Figure 13.3 Two-ray grid network model.

transmits the traffic through this cooperative node. If the energy consumption speed is not exceeded, the data sender transmits to the next hop node. We are going to evaluate the performance of the proposed scheme. In future work, we will implement and find out the results of the abovementioned protocol.

13.2 Experimental Results

By reviewing the literature, one has to be very keen on energy while developing one's own protocols. The protocols must consume less energy itself. To throw some light on the energy considerations, we tested the hypothesis by using the NS2 and the test was setup as follows: number of nodes, 30; initial energy, 1 J; routing protocol, AODV; MAC protocol, 802.11; packet size, 50 bytes; reporting rate, 50 packets/s; localization, random initially. The parameters considered here are QoS parameters such as node density. These parameters may affect congestion as one of its root causes [24].

13.2.1 Node Density

Node density depends on the number of nodes per unit area in the physical sensor network. If we consider the two cases for density, then one of them is *sparse* and the other is *dense*. In the sparse network, the nodes are deployed in such a fashion that no two nodes are close to each other. It also depicts that the probability of finding their neighbor in the WSN sparse networks is low. Although, on other hand, in dense networks the nodes are very closely located to one another. Hence, the probability of finding their neighbor is very high. As node density increases, the probability of finding a neighbor increases. Hence, the performance of the routing protocol will improve drastically. This may reduce the overall energy consumption in the network, which is otherwise wasted in the routing decisions of network layer protocols (Figure 13.4). With increased density, neighbors can be found at a shorter distance from the node. This will require a smaller amount of transmission power and hence transmission energy is saved. On the other hand, when density increases, the number of packets generated by the different nodes will increase. This will lead to network congestion. Hence, an increase in congestion will unnecessarily consume more energy, and network lifetime will be reduced. An increase in the node density will also raise the problem of implosion. Duplicate packet generation due to implosion will inevitably consume more energy for the transmission of more copies of the same packets into the network.

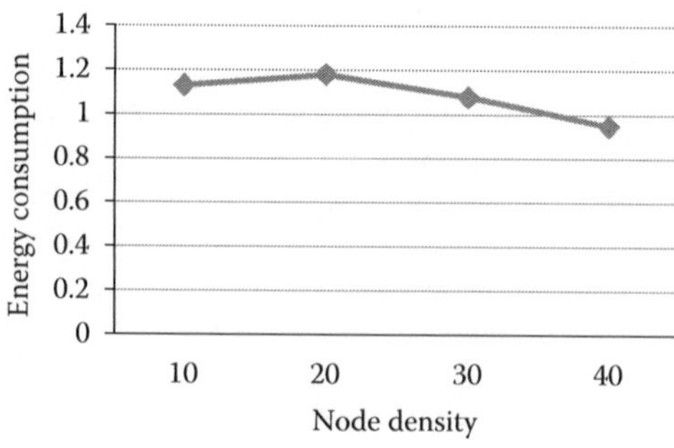

Figure 13.4 Graph of average energy consumption against density.

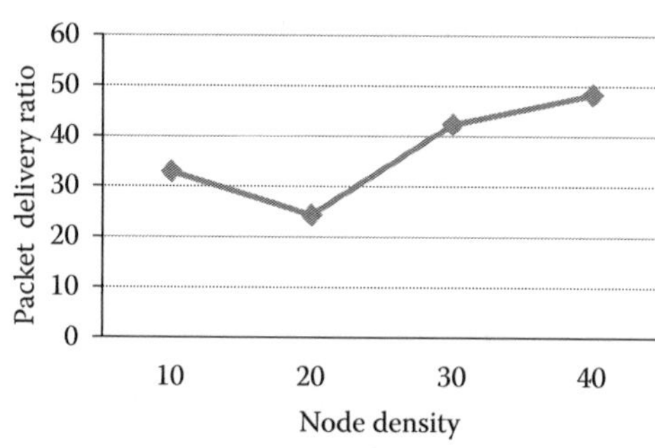

Figure 13.5 Graph of packet delivery ratio against density.

Controversially, when node density is reduced, the node may be less loaded. The throughput of the node will be reduced (Figure 13.5). Hence, in the sparse network, the throughput is much lower as compared with dense networks. The chances of congestion are very low and the lifetime of the network and node will automatically increase. In the sparse network, nodes are placed at longer distances from one another. It requires a large amount of power to transmit to neighbors and hence it will consume large amounts of energy. In some cases, redundant nodes are introduced into the sensor network. These nodes will only relay the data rather that generate the data. Because the sensing circuit is absent, these types of nodes will consume less energy. By increasing the number of redundant nodes, the density of the network will increase (this will increase the routing performance). By varying the duty cycle of the status of the liveliness of the nodes, we can achieve the required node in active or passive states. When a redundant node is not required in the network, one may keep it in sleep or passive mode.

13.2.2 Reporting Rate

As the rate of sending packets increases, the number of packets generated increases. This increases the energy consumption of the node. The node will die earlier and because of this, the lifetime of the node will be reduced. There is a chance of congestion happening in the network as the rate of packets increases [12]. As congestion occurs, the packets get lost in the network and the energy, which is utilized for the transmission of those packets, is wasted.

With the increase in the rate of packet transmission, the input to the node will be increased. If the node is not able to service these input flows, then there may be chances of accumulating the packets in the input buffer only. If the number of incoming packets still increases, it may overflow the buffer. The packets may get dropped out of the buffer. This scenario again leads to wastage of the energy that is required to transmit these dropped packets [17].

The MAC protocol used in WSNs is called carrier sense multiple access collision avoidance (CSMA/CA). This is the integral part of the 802.11 protocol. Every time the node wants to send a packet, the node must check the wireless media. If the media is busy, then the node must wait for some arbitrary time. After a lapse of this arbitrary time, the node will again check the media for its busyness. Every time data must travel from the application layer to the physical layer, it consumes energy. As the rate of transmission intentionally increases, the energy consumed in the layered structure becomes larger and will consume even more energy. With the help of the cross-layer approach, one can reduce the energy consumption in the layered architecture.

13.2.3 Energy Consumption as a Function of Reporting Rate

Energy consumption is very much dependent on the reporting rate. The number of packets generated in the network is totally dependent on the reporting rate of each node. The nodes much process these large numbers of packets. Large amounts of energy are required to process and retransmit these packets (Figure 13.6). As the graph shows, as the number of packets generated per second increases, the amount of energy consumed will also increase. It enjoys an exponential relationship.

13.2.4 Dropped Packets as a Function of Time

As time lapses, more packets get dumped in the network. A node does not process this large number of packets as fast (Figure 13.7). The result of this is that the nodes' buffer gets full and overflows afterward. If we continue with the same scenario, then the number of dropped packets will increase exponentially.

13.2.5 Energy Consumption as a Function of Dropped Packets

The dropped packets are a waste of the network's resources (shown in Figure 13.8). A lot of energy is wasted in the transmission of the packets that get dropped due to different reasons.

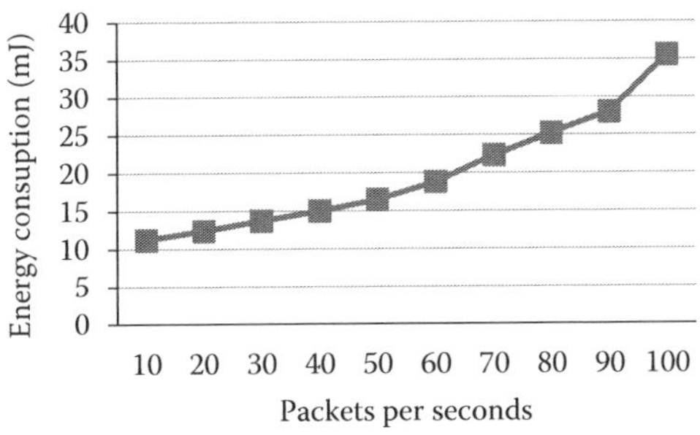

Figure 13.6 Energy consumption as a function of reporting rate.

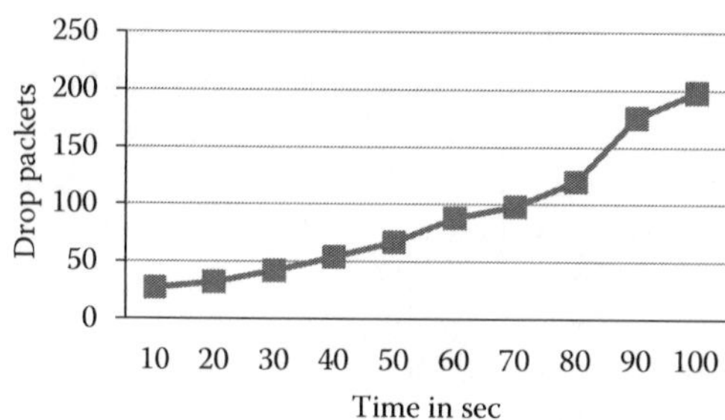

Figure 13.7 Dropped packets as a function of time.

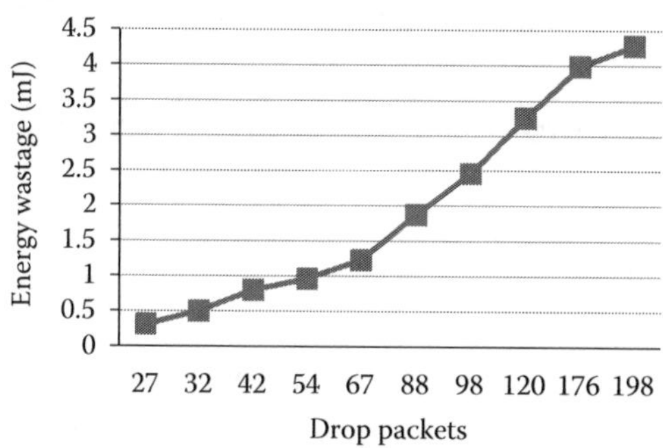

Figure 13.8 Energy consumption as a function of dropped packets.

Congestion is one of the vital reasons for the wastage of the energy. As the number of dropped packets increases, the energy wastage also increases exponentially. Due to congestion, the node buffer is full. More packets are dumped in the network, but these packets do not get served by the node. The buffer gets overflowed and dropped as RED is adopted by the buffer policy.

13.2.6 Packet Size

Different sources of data form different traffic patterns. They may contribute variable size data. Some of the data can have small packet size whereas some of them are of a large size. In both cases, the energy is dependent on fitting the size to the links between the nodes. If the link has a sufficient enough bandwidth to carry the large packets, data is easily carried up to the destination. But if it fails to do so, then it will result in congestion. Packet size is mostly dependent on the type of the data that is disseminated to the destination [25].

For the text or normal data type, the packet size will be smaller. Data types such as multimedia require very large size packets. As the packet size is large, the probability to congest the network will be high. The congestion in turn will have different nodes to consume more energy. This may waste the energy that will be required to generate the packet and send it to its destination. For smaller packet sizes, the energy required to generate will be small. The chances of the networks to congest will be very low. If we want to have more reliable transmission, then the required packet must be retransmitted again if it did not reach its destination. This process of retransmission will consume more energy. For small packets to retransmit requires a smaller amount of energy than the large packet retransmission. Hence, it is recommended that one should design the sensor networks to have smaller data sizes for their packets.

The error control in large packets is very difficult and requires large amounts of complexity. The processor will consume more energy because complexity is very high. Controversially, as far as packet size is concerned, for small packet sizes, the overheads on the node as well as on the network will be larger. Every time we have added the header to every small packet. This packet may be unpacked and processed at every intermediate node. This will consume even more energy. It will also add some delays in the delivery of the packets up to the destination. However, if we consider the large packet sizes, these supposed overheads are somewhat less and will consume a smaller amount of energy at the intermediate node in the networks.

13.2.7 Energy Consumption as a Function of Packet Size

Figure 13.9 shows the behavior of energy consumed with respect to the packet size. As packet size increases, the energy required to transmit these packets increases. However, this increase in energy consumption is not as sharp as in the changes for reporting rate, and only varies from 15 to 25 mJ. This means that even if packet size changes the number of overheads to transmit, it does not change drastically. The reporting rate is adjusted to 50 packets/s for the said scenario.

The reporting rate is the one of the main root causes of congestion [24,25]. For applications with slow packet rates, the node may give good performance; however, as the reporting rate increases, the number of packets arriving at the node also increases. If the node cannot serve these packets quickly, then chances are that the buffer will be filled. Generally, Drop tail or RED technology is used in buffer management. The latest packet arriving at the node may get dropped. In the simulation, the reporting rate will be controlled by adjusting the interarrival time. As interarrival time increases, the reporting rate decreases. As the reporting rate decreases, the degree of congestion also decreases exponentially (Figure 13.10).

The degree of congestion depends on two basic factors. One is interarrival time and the other is service time of the node. The graph shows the behavior of the congestion for different service times. As service time decreases, there are fewer chances for congestion.

There are two basic types of reliabilities. One is normalized reliability and other is desired reliability. Reliability is generally defined as the ratio of received packets to the total number of packet sent. This is treated as normalized reliability. Desired reliability is dependent on the application. It is the ratio of the desired number of packets received to the total number of packets sent. As the

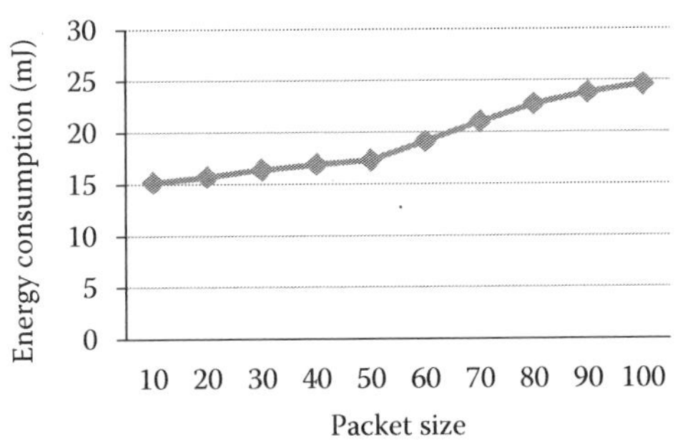

Figure 13.9 Energy consumption as a function of packet size.

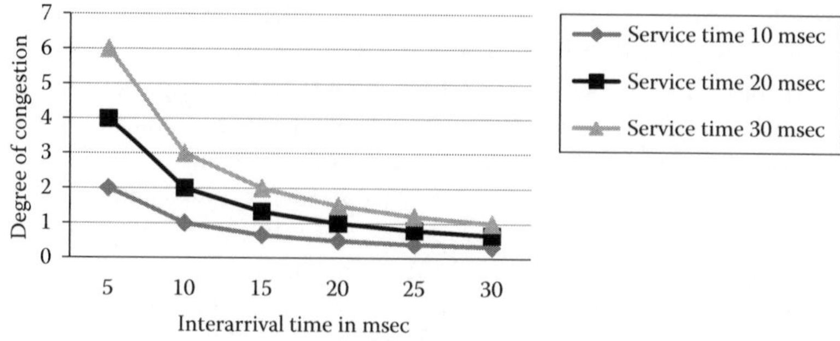

Figure 13.10 Congestion as a function of reporting rate or interarrival time.

number of received packets increases, the reliability increases linearly (Figure 13.11). It can also been seen that as the number of desired packets increases, the reliability decreases.

Fairness is defined as how the resources are evenly distributed throughout the network [26]. With the additive increase and multiplicative decrease (AIMD) techniques, one can achieve traffic fairness. For controlling the congestion in the network, flow rate control will mostly be used. This may increase or decrease the reporting rate of the previous node, which may arise in the unfair traffic patterns in the network (Figure 13.12). The graph shows that as we increase the flow rate, the fairness index may tend to unity, which is desirable. However, this also increases congestion in the network. This will be an indication for the wastage of energy.

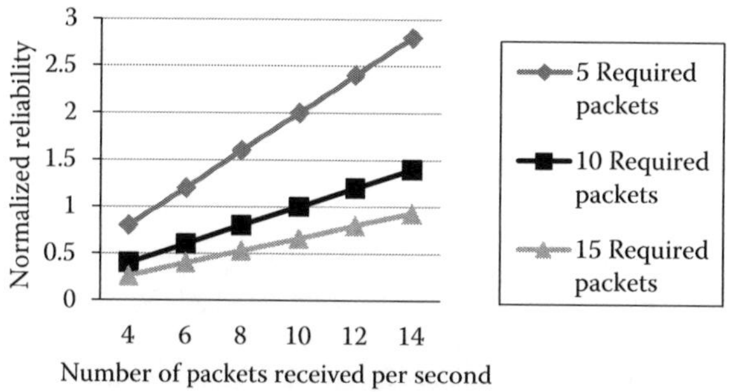

Figure 13.11 Reliability as a function of received packets.

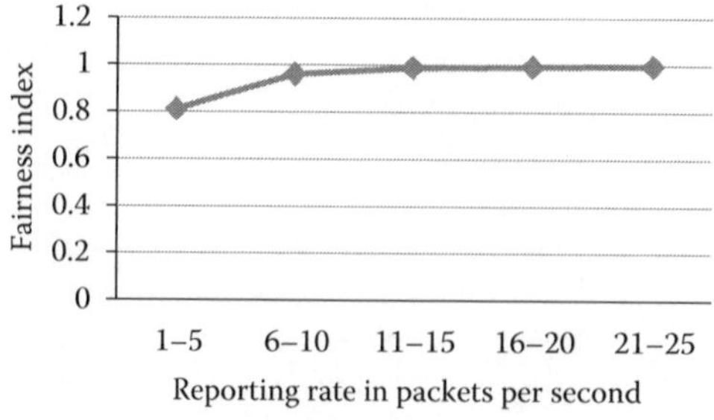

Figure 13.12 Fairness as a function of reporting rate.

13.2.8 Network Topology

Topology is a very important parameter to watch out for. The WSN's topology coincides with the energy in the WSN. The way by which the nodes are deployed in the network is called topology, and there are three basic types, namely, chain, grid, and random.

In the chain topology, the nodes are deployed in the form of a chain. The node that is situated near the sink may exert high traffic (this is one node in the chain). This node must relay the heavy traffic or a large number of packets to the sink and consume more energy (Figure 13.13).

For the grid topology, the nodes are deployed in a uniform manner. All the nodes are placed in a rows and column fashion. The distance between each node is the same as with any other. If the sink is placed at the center, then this scenario will give rise to a somewhat equally distributed traffic pattern. This may have fair distribution of traffic and energy throughout the network. If the sink is situated at any of the corners, then it will give rise to uneven traffic flow in the network. The farther nodes will send data packets which in turn will travel through the nodes near the sink. Nodes that are one hop away will route the maximum amount of traffic and will consume more energy than the other nodes in the network. If nodes are randomly distributed then, depending on the location of the sink, the energy distribution varies (Figure 13.14). The sink node's placement at any position in the network will give rise to different flow patterns. The nodes that are closer to the sink will exert a maximum amount of traffic compared with any other nodes in the network. In either case of topology, the node that is nearer to the sink will be heavily loaded and consume more energy than the others, and hence die earlier, creating a big gap in the network.

The discrimination between flat and hierarchical topology also leads to different consumptions in energy. The flat topology behaves similar to a chain, grid, or random deployment of nodes in the network. However, in the case of hierarchical topology, the logical structure of the tiered architecture gives rise to a special case for consideration. A group of nodes will form a cluster. Within the cluster, there is a cluster head, which takes care of gathering the data from the nodes that are within its own cluster. Therefore, the nodes will always have a cluster head very close to them. It will result in less energy being spent on transmission. The cluster head then transmits the collected data to the sink. The overall amount of energy consumed in the network clustering is much less compared with its consumption in the flat-type structure of sensor networks.

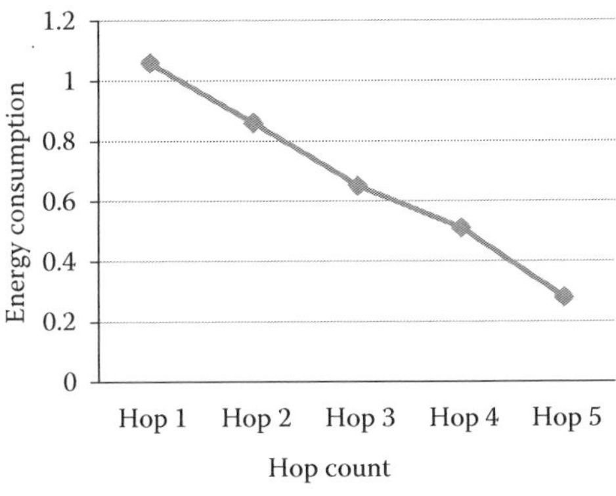

Figure 13.13 Average energy consumption as a function of hop count for chain topology.

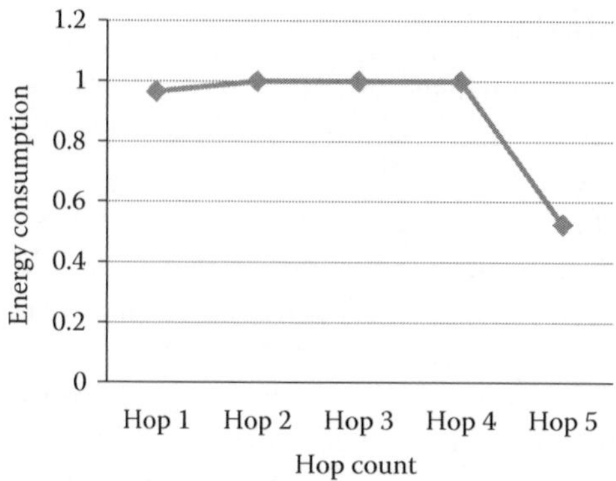

Figure 13.14 Average energy consumption as a function of hop count for random topology.

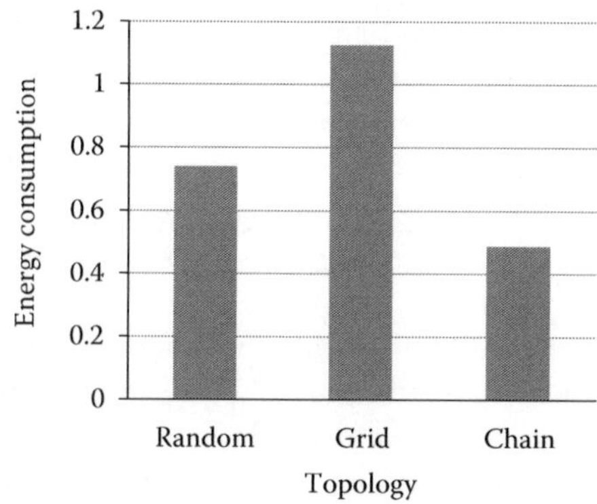

Figure 13.15 Average energy consumption as a function of topology.

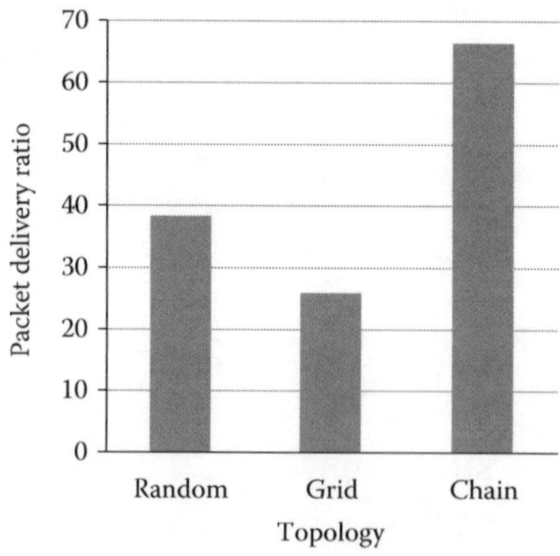

Figure 13.16 Packet delivery ratio as a function of topology.

The graph of energy consumption against topology (Figure 13.15) clearly indicates that grid topology consumes more energy, whereas chain topology consumes less energy than the other two. It can also be seen that the packet delivery ratio for chain topology will be larger than chain or random topologies (Figure 13.16). The random topology will be having moderate energy consumption and packet delivery ratio. This is mainly because of the AODV routing protocols dynamically decides the route of the data flow very easily in the random topology. The chain will be having fixed route and the grid will always follow the same shortest path to the sink.

13.3 Conclusion

The simulation results clearly show that the random topology is best suited for WSNs. The optimum energy is consumed and at the same time the packet delivery ratio will be fairly moderate in random topology. A node density of 20 will be best suited for the network to be congestion-free. It also gives less packet drop and reliable data delivery. These results will give guidelines to potential researchers to work their research ideas for minimizing the energy.

WSNs are composed of the deployment of huge amounts of sensor nodes with many resource constraints, which may cause congestion in the network. The parameters described in this chapter are some of the indicators of energy consumption. By monitoring various aspects of WSN, such as localization, traffic or load distribution, node processing time, etc., we will be able to achieve a more reliable network.

The QoS parameters are of prime importance to be considered while designing the WSNs. In this chapter, congestion, reliability, and fairness are considered. By analyzing the results, one can see how these three QoS members behave with the reporting rate. By studying these graphs, other researchers may design their own sensor networks with high reliability at the cost of very low congestion. The throughput of the network will also increase and we can achieve the near-unity fairness factor.

The energy of sensor nodes in the sensor network is a very precious resource. It is difficult to recharge the battery of the sensors. The faster energy consumption speed of a single node can imbalance energy consumption. This can directly affect the entire lifetime of the entire network. Therefore, energy should be consumed in the proper manner. Balanced energy consumption in the network definitely increases the network lifetime. The congestion in the network causes packet loss and retransmission, which results in wastage of the energy of the sensors. The use of mobile nodes that can be sink nodes or simple forwarder nodes in the network also helps data transfer. This alleviates the load of the sensors and lessens the chance of congestion. Thus, in this way, the use of mobile nodes, cooperative nodes, proper deployment strategies, and the scheme for balanced power consumption speed, transmission policies, and also clustering policies can reduce the energy consumption and increase the network lifetime.

Researchers are designing lightweight, energy-aware congestion control and avoidance algorithms based on (and analyzing) the localization, traffic or load distribution, node processing times, and other parameters.

References

1. M.A. Rahman, A. El Saddik, and W. Gueaieb. *Wireless Sensor Network Transport Layer: State of the Art*, Springer-Verlag, vol. 21, pp. 221–245, April 2008.

2. A. Ayadi. Energy-efficient and reliable transport protocols for wireless sensor networks: State-of-art. *Wireless Sensor Network*, vol. 3, no. 3, pp. 106–113, 2011.
3. F. Yunus, N.N. Ismail, S.H.S. Ariffin, A.A. Shahidan, N. Fisal, and S.K. Syed-Yusof. Proposed transport protocol for reliable data transfer in wireless sensor network (WSN). In *IEEE 4th International Conference on Modeling, Simulation and Applied Optimization (ICMSAO)*, pp. 1–7, 2011.
4. J. Paek and R. Govindan. RCRT: Rate-controlled reliable transport protocol for wireless sensor networks. *ACM Transactions on Sensor Networks (TOSN)*, vol. 7, no. 3, 2010.
5. R. Wang, T. Taleb, A. Jamalipour, and B. Sun. Protocols for reliable data transport in space Internet. *IEEE Communications Surveys and Tutorials*, vol. 11, no. 2, pp 21–32, 2009.
6. L. Qiang and F.Q. Yu. ECODA: enhanced congestion detection and avoidance for multiple class of traffic in sensor networks. *IEEE Transaction on Consumer Electronics*, vol. 56, no. 3, pp. 1387–1394, 2010.
7. D.Q. Bai, I.-Y. Kong, and W.-L. Hwang. Trade-off between reliability and energy-efficiency in transport protocol for wireless sensor networks. In *SICE-ICASE International Joint Conference*, Busan, pp. 429–434, October 18–21, 2006
8. I.F. Akyildiz, W. Su, Y. Sankarasubramaniam, and E. Cayirci. Wireless sensor networks: a survey. *Computer Networks*, vol. 38, no. 4. pp. 393–422, 2002.
9. A.K.M. Azad and J. Kamruzzaman. Energy-balanced transmission policies for wireless sensor networks. *IEEE Transactions on Mobile Computing*, vol. 10, no. 7, pp. 927–940, July 2011.
10. Z. Cheng, M. Perillo, and W.B. Heinzelman. General network lifetime and cost models for evaluating sensor network deployment strategies. *IEEE Transactions on Mobile Computing*, vol. 7, no. 4, pp. 484–497 April 2008.
11. Y. Wenguo and G. Tiande. The non-uniform property of energy consumption and its solution to the wireless sensor network. In *Second International IEEE Workshop on Education Technology and Computer Science*, vol. 2, pp. 186–192, 2010.
12. Z. Tan, Y. Liu, and Z. Zhang. Performance requirement on energy efficiency in WSNs. In *3rd International IEEE Conference on Computer Research and Development (ICCRD)*, vol. 3, pp. 159–162, March 2011.
13. W. Fang, F. Liu, F. Yang, L. Shu, and S. Nishio. Energy-efficient cooperative communication for data transmission in wireless sensor networks. *IEEE Transactions on Consumer Electronics*, vol. 56, no. 4, pp. 2185–2192, November 2010.
14. M.V. Ramesh, A.G. Sreedevi, K. Kamalanathan, and P.V. Rangan. Energy comparison of balanced and progressive sensor networks. In *Annual IEEE Conference on Wireless and Optical Communications*, Taiwan, pp. 93–98, April 19–21, 2012.
15. F. Ren, J. Zhang, T. He, and S.K. Das. EBRP: energy-balanced routing protocol for data gathering in wireless sensor network. *IEEE Transactions on Parallel and Distributed Systems*, vol. 22, no. 12, December 2011.
16. S.-F. Hwang, H.-H. Lin, and C.-R. Dow. An energy-efficient routing protocol in wireless sensor networks with holes. In *Fourth International Conference on Ubiquitous and Future Networks*, Phuket, Thailand, pp. 17–22, 2012.
17. A.K. Kumar and K.M. Sivalingam. Energy-efficient mobile data collection in wireless sensor networks with delay reduction using wireless communication. In *Second International IEEE Conference on Communication Systems and Networks*, pp. 1–10, January 2010.
18. H. Abusaimeh and S.-H. Yang. Balancing the power consumption speed in flat and hierarchical WSN. *International Journal of Automation and Computing*, vol. 5, no. 4, pp. 366–375, October 2008.
19. C. Alippi, G. Anastasi, M. Di Francesco, and M. Roveri. Energy management in wireless sensor networks with energy-hungry sensors. *IEEE Instrumentation and Measurement Magazine*, vol. 2, no. 2, pp. 16–23, April 2009.
20. S. Park, W. Lee, and D.-H. Cho. Fair clustering for energy efficiency in a cooperative wireless sensor network. In *75th IEEE Conference on Vehicular Technology*, Yokohama, pp. 1–5, 2012.
21. D. Wei, Y. Jin, S. Vural, K. Moessner, and R. Tafazolli. An energy-efficient clustering solution for wireless sensor networks. *IEEE Transactions on Wireless Communications*, vol. 10, no. 11, pp. 1–11, November 2011.

22. M. Alam and C.S. Hong. Congestion aware rate controlled reliable transport in wireless sensor networks. *IEICE Transaction Communication*, vol. E92B, no. 1, pp. 184–199, January 2009.
23. V.S.F. Enigo and V. Ramchandran. EECC: Energy efficient congestion control protocol for wireless sensor networks. In *IEEE International Advance Computing Conference*, 2009, IACC 2009, Patiala, India, pp. 1573–1578, 2009.
24. V. Deshpande and P. Sarode. Root cause analysis of congestion in wireless sensor networks. *International Journal of Computing Application*, vol. 18, no. 18, article 6, pp. 27–30, 2010.
25. Y. Ouyang, F. Ren, C. Lin, T. He, C. Li, Y. Hu, and H. Wen. A simple active congestion control in wireless sensor networks. *IEEE International Conference on Mobile Ad Hoc and Sensor Systems*, MASS 2007, pp. 1–7, 2007.
26. C.T. Ee and R. Bajcsy. Congestion control and fairness for many-to-one routing in sensor networks. In *Sensys 04*, Baltimore, USA, ACM Press, pp. 148–161, 2004.

Chapter 14

Cooperative Systems in Wireless Sensor Networks

Rawya Yehia Rizk
Port Said University

Contents

14.1 Introduction

Wireless sensor networks (WSNs) are generally composed of battery-powered modules. They have to operate without battery replacement or recharging for a long time. Consequently, minimizing the energy consumption of a WSN is a critical issue in system design. A module's energy efficiency directly affects its useful lifetime. When a module dies, not only does it stop collecting data but also the network loses the module's ability to relay data. Therefore, energy efficiency affects how long both individual sensors and the entire network will continue to function. Therefore, it is important to take energy efficiency into account in all aspects of module and network design.

Several studies on the energy efficiency of multiple-input/multiple-output (MIMO) systems have been done. Energy-efficient communication techniques typically focus on minimizing the transmission energy only, which is reasonable in long-range applications where the transmission energy is dominant in the total energy consumption. However, in short-range applications such as sensor networks in which the circuit energy consumption is comparable to or even dominates the transmission energy, different approaches need to be taken to minimize the total energy consumption.

The performance of virtual MIMO in WSNs depends on the structure of the network layer and data link layer. There are several approaches for implementing virtual antenna arrays in WSNs. Although the core implementation of a virtual antenna array or cooperative transmission lies in the physical layer, there is a deep dependency on the higher layers (network and data link) to implement this issue. In a cognitive network framework, the network components can modify the operational parameters to respond to the needs of a particular environment.

This chapter presents the concept of cooperative MIMO, the energy consumption techniques, the energy model, and a complete study of the parameters that affect the system in all its situations.

14.2 Motivation of Cooperative MIMO in WSNs

The cooperation of sensor nodes in WSN allows resource saving. Virtual MIMO concepts are applied in WSNs for energy-efficient communication to save energy and increase reliability. Multiantenna systems have been studied intensively in recent years due to their potential to dramatically increase the channel capacity in fading channels. The MIMO system has found great importance in improving the performance of wireless communication. It has been shown that MIMO systems can support higher data rates under the same transmit power budget and bit error rate (BER) performance requirements as a single-input/single-output (SISO) system. An alternative view is that for the same throughput requirement, MIMO systems require less transmission energy than SISO systems.

Hence, it is tempting to believe that MIMO systems are more energy-efficient than SISO systems. However, the physical implementation of multiple antennas at a small-sized sensor node may not be feasible. The direct application of multiantenna techniques to sensor networks is impractical due to the limited physical size of a sensor node, which typically can only support a single antenna. In addition, the circuit energy consumption of a MIMO system is higher for a SISO system because it has multiple radio frequency (RF) chains and requires more signal processing.

As a result, a cooperative MIMO in which multiple inputs and outputs are formed via cooperation is a solution for energy-efficient communication.

14.3 The Concepts of Cooperative MIMO

Various MIMO systems provide a great improvement of energy efficiency in low-range communications for WSNs. In cooperative MIMO systems, a group of sensors cooperate to transmit and receive data. Within a group, sensor nodes can communicate with relatively low power as compared with intergroup communication.

Although the participation of multiple transmitters and receivers in a transmission saves significant energy in long-range communications, the increase in the number of transmitters and receivers also increases the circuitry power consumption. As a result, the energy optimization techniques have to be adapted with the environment [1–6].

14.3.1 Forms of MIMO

14.3.1.1 Multiantenna Types

Multiantenna MIMO (or single-user MIMO) technology is based on the number of input and output signals of a system. Figure 14.1 shows the different antenna configurations [7]:

- Single-input/single-output (SISO):
  SISO techniques have a single input signal and a single output signal.
- Single-input/multiple-output (SIMO):
  SIMO techniques have a single input signal (the transmitter has a single antenna) and more than one output signal.

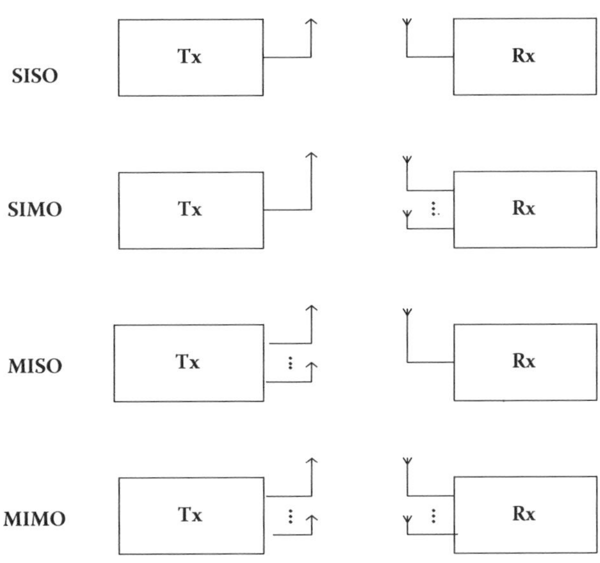

Figure 14.1 Different forms of MIMO and their configurations. (From J.M. Chung et al., *IEEE Transactions on Vehicular Technology*, vol. 61, no. 9, pp. 4069–4078, 2012.)

- Multiple-input/single-output (MISO):
  MISO techniques have multiple input signals and a single output signal (the receiver has a single antenna).
- Multiple-input/multiple-output (MIMO):
  MIMO techniques have multiple input and output signals, and form the most general of the four classes. They have the dual capability of combining the SIMO and MISO technologies.

14.3.1.2 Multiuser Types

- Multiuser MIMO (MU-MIMO):
  In multiuser communications, multiple users can simultaneously share the same time and frequency intervals. This is known as space division multiple access (SDMA), and can be implemented in a wireless local area. Multiuser MIMO algorithms are developed to enhance MIMO systems when the number of users, or connections, are greater than one. Multiuser communication is of significant value as it provides huge capacity gains over single-user communication.
- Cooperative MIMO (CO-MIMO):
  Cooperative MIMO utilizes distributed antennas that belong to other users.
  Conventional MIMO systems require both the transmitter and receiver of a communication link to be equipped with multiple antennas. Due to the size, cost, and hardware limitations of WSNs, cooperative MIMO aims to utilize distributed antennas on multiple radio devices to achieve considerable performance gains similar to those provided by conventional MIMO systems.

14.3.2 Basic Building Blocks

Figure 14.2 shows the basic building blocks that comprise a MIMO communication system. In the figure, x and y represent transmitted and received signal vectors, respectively. First, the

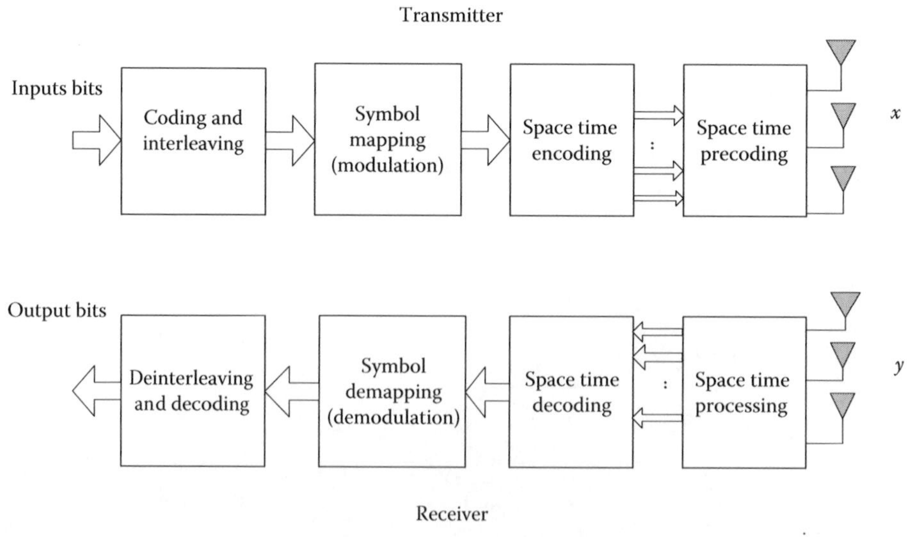

Figure 14.2 Building blocks of MIMO. (Reprinted from *Wireless Communications and Networking*, V.K. Garg, pp. 245–246. Copyright 2007, with permission from Elsevier.)

information to be transmitted is encoded and interleaved. The symbol mapper maps the encoded information into data symbols. These data symbols are then fed into a space–time encoder, which creates some spatial data streams. The data streams are then transmitted by different antennas. The transmitted signals propagate through channels and are received by receiving arrays.

The receiver then collects the signals at the output of each receive antenna element and reverses the transmitter operations to decode the data: receive space–time processing, followed by space–time decoding, symbol remapping, deinterleaving, and decoding. Deinterleaving in the receiver is the reverse operation of interleaving. The major drawback of deinterleaving and interleaving is that they introduce delay [8].

14.3.3 Advantages of Multiple Antennas

MIMO systems constructively explore multipath propagation using different transmission paths to the receiver. These paths can be exploited to provide a redundancy of transmitted data, thus improving the reliability of transmission or increasing the number of simultaneously transmitted data streams and increasing the data rate of the system. Thus, MIMO channels are well-known to provide a number of advantages over conventional single-antenna channels, which have been traditionally described by the diversity, multiplexing, and array gain and interference reduction [9].

14.3.3.1 Array Gain

Array gain indicates an improvement of the signal to noise ratio (SNR) at the receiver compared with traditional systems with one transmit and one receive antenna [10]. The said improvement can be achieved with a correct processing of the signals at the transmitter or at the receiver side, so the transmitted signals are coherently combined at the receiver. To achieve array gain at the transmitter antenna array, the channel state information (CSI) has to be known at the transmit side, whereas for the exploitation of antenna array gain at the receiver, the channel has to be known at the receiver side.

14.3.3.2 Diversity Gain

Diversity is used in wireless systems to combat fading. Diversity gain is the improvement in link reliability obtained by receiving replicas of information signals through fading link. With an increasing number of independent copies (the number of copies is often referred to as the diversity order), the probability that at least one of the signals is not experiencing a deep fade increases, thereby improving the quality and reliability of reception. A MIMO channel with M_t transmit antennas and M_r receive antennas potentially offers $M_t \times M_r$ independently fading links, and hence a spatial diversity order of M_tM_r [11]. A comparison between the different spatial diversity schemes is shown in Table 14.1.

14.3.3.3 Multiplexing Gain

Multiple antenna systems are capable of establishing parallel data streams through different antennas. This is done to increase the data transfer rate. This process is called multiplexing gain. Under suitable channel conditions, such as rich scattering in the environment, the receiver can separate the data streams [12]. In general, the number of data streams that can be reliably supported by a MIMO channel is equivalent to the minimum number of transmit antennas and the number of receive antennas. The spatial multiplexing gains increase the capacity of the wireless network.

Table 14.1 Comparison of Main Spatial Diversity Schemes [11]

Schemes	M_t	M_r	*Function*	*Diversity Gain*
SISO	1	1	No transmit or receive diversity	No diversity gain
SIMO	1	>1	Receive diversity	Diversity proportional to M_r
MISO	>1	1	Transmit diversity	Diversity proportional to M_t
MIMO	>1	>1	Multiple antennas at both the transmitter and receiver	Diversity proportional to $M_t \times M_r$

14.3.3.4 Interference Reduction

Interference in the wireless network results from multiple users sharing time and frequency resources. Interference may be mitigated in MIMO systems by exploiting the spatial dimension to increase the separation between users. Interference reduction and avoidance improves the coverage and the range of the wireless network [13].

14.4 Energy Consumption Enhancement Techniques

A number of techniques are proposed for energy consumption improvement. It has been proven by many lecturers that cooperative MIMO systems achieve great improvement in the performance of WSNs [3–6]. In these systems, multiple inputs or outputs (or both) are formed via cooperation to combine the physical implementation of small-sized sensor nodes and high energy efficiency. In a study by Jayaweera [14], they used a large collection of low-end data collection sensors connected over a wireless link with a high-end data-gathering node (DGN), which may act as a lead sensor or a fusion center.

Clustering is the process of grouping the sensor nodes in a densely deployed large-scale sensor network. This technique can effectively reduce the energy consumption of sensor nodes and has been widely used in WSNs. A variety of clustering protocols have been proposed to address the energy efficiency problem [15–25]. There are some issues involved with the process of clustering in a WSN. First, the number of clusters should be formed; second, the number of nodes should be taken into a single cluster; and third, an important issue is the selection procedure of cluster heads (CH) in a cluster.

Combining and compressing the data belonging to a single cluster is known as data aggregation in a cluster-based environment. Data aggregation protocols aim at eliminating redundant data transmission and thus improve the lifetime of energy-constrained WSN. Data aggregation at the CH has been presented in studies by Gai et al. [26] and Wang et al. [27].

Several techniques consider a node selection approach to minimize the amount of energy consumed. In a study by Ahmed et al. [28], the nodes were selected on the basis of geometric locations. In articles by Islam and Kim [29,30], the selection was based on channel gain parameters in both MISO and MIMO, respectively. The same authors [31,32] noted that the selection was based on a combination of channel estimate energy, residual energy, intersensor distance, and geographical location of the sensors.

Table 14.2 presents how some important works fit the major aspects of energy consumption. These aspects include cooperative MIMO, data aggregation, clustering, and node selection.

Table 14.2 Classification of Literature Proposals

Proposal	*Cooperative MIMO*	*Clustering*	*Data Aggregation*	*Node Selection*
[3]	√			
[4]	√			
[5]	√			
[6]	√			
[14]	√			
[15]		√		
[16]		√		
[17]		√		
[18]		√		
[19]		√		
[20]		√		
[21]		√		
[22]		√		
[23]		√		
[24]		√		
[25]		√		
[26]	√		√	
[27]	√		√	
[28]	√			√
[29]	√			√
[30]	√			√
[31]	√			√
[32]	√			√

14.5 Energy Efficiency in Cooperative Systems

In this section, the energy model for the MIMO system is presented. The variations of the model in the case of data aggregation or node selection in the transmitter and receiver are included.

14.5.1 System Model

In the model shown in Figure 14.3, it is considered that a sensor network is composed of multiple clusters of nodes. The nodes within the same cluster are closely spaced and cooperate in single

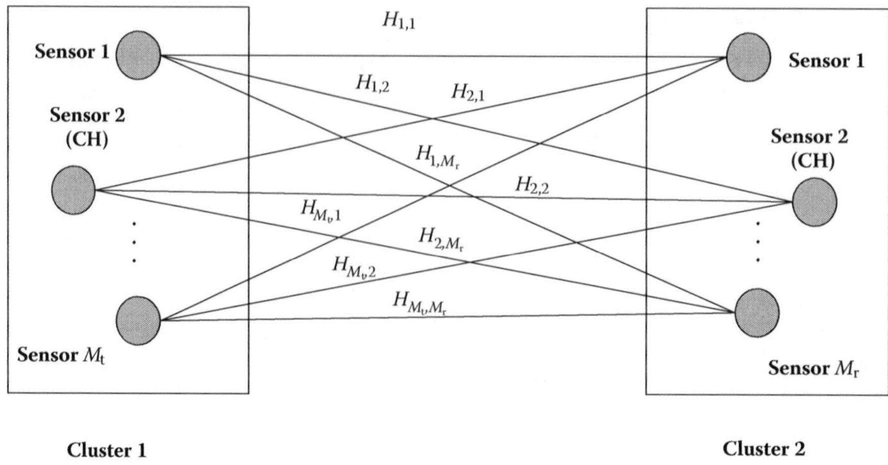

Figure 14.3 System model.

transmission or reception (or both). Each cluster is controlled by a special node called a CH. It acts as the coordinator for cooperative transmission in the cluster-based WSN.

It is assumed that a system with narrowband, perfect synchronization exists between transmitters and receivers for wireless sensor nodes, and frequency-flat Rayleigh fading channels. This means that the channel gain between each transmitter antenna and each receiver antenna is scalar. Therefore, the fading factors of the MIMO channel can be represented as a scalar matrix.

It is assumed that the number of transmitting sensors M_t, the number of receiving sensors M_r, and each sensor contains one antenna. The received discrete-time signal is attenuated by a channel matrix H of scalar fading coefficients. The channel matrix of a MIMO system can be expressed as:

$$H = \begin{bmatrix} H_{1,1} & H_{1,2} & \cdots & H_{1,M_{Br}} \\ H_{2,1} & H_{2,2} & \cdots & H_{2,M_{Br}} \\ \vdots & \vdots & \ddots & \vdots \\ H_{M_{Bt},1} & H_{M_{Bt},2} & \cdots & H_{M_{Bt},M_{Br}} \end{bmatrix} \tag{14.1}$$

where each element in H is a zero-mean circulates symmetric complex Gaussian random variable with unit variance.

For cooperative MIMO approach, each of the M_t sensors on the transmitting side cooperate by first broadcasting its information to all the other local nodes using different time slots. After each node receives all the information bits from other nodes, they encode the transmission sequence according to the Alamouti diversity codes. On the receiving side, the M_r nodes join the cooperative reception to do the joint detection. It has been shown that for Rayleigh fading channels, MIMO systems based on Alamouti schemes can achieve lower probability of error than SISO systems due to the diversity gain and the array gain [26].

The number of sensors that are selected to transmit data from all active sensors is defined as M_{Bt} out of M_t available sensors and the number of sensors that are selected to receive data in the case of receiving cluster is M_{Br} out of M_r available sensors.

A low-energy adaptive clustering hierarchy (LEACH) protocol is considered. It is a cluster-based structure that uses a time division multiple access (TDMA)–based media access control (MAC) protocol. The operation of LEACH is divided into periods, and each period consists of a setup phase and a steady-state phase. During the setup phase, nodes communicate with short messages and are organized into clusters with some nodes selected as CHs. After the setup phase, each CH sets up TDMA schedules for the nodes in its cluster. Sensor nodes send any data they generate to the CHs according to the TDMA schedule [33].

In this model, a sensor with high residual energy is deployed as a CH and it remains the CH until the network dies. The CH broadcasts its status to other sensors in the network. Each sensor node determines which cluster it wants to belong to by choosing the CH that requires the minimum communication energy.

14.5.2 Energy Model

In the energy consumption model, all signal processing blocks at the transmitter and the receiver are taken into consideration. However, to keep the model from being overcomplicated at this stage, baseband signal processing blocks (e.g., source coding, pulse-shaping, and digital modulation) are intentionally omitted. Figures 14.4 and 14.5 show the analog transmitter and receiver circuit blocks, respectively. The same transmitter and receiver blocks shown in the study by Cui et al. [3] are used.

The total power consumption for a single node consists of two main parts, the power consumption of all power amplifiers P_{PA} and the power consumption of all other circuit blocks P_{C} [3,34]:

$$P_{\text{T}} = P_{\text{PA}} + P_{\text{C}} \tag{14.2}$$

The power consumption of all power amplifiers is [34]:

$$P_{\text{PA}} = (1 + \alpha)P_{\text{out}} \tag{14.3}$$

where $\alpha = \frac{\xi}{\eta} - 1$ with η is the drain efficiency and ζ is the peak to average power ratio, which is dependent on the modulation scheme and the associated constellation size. Throughout this chapter, M-array quadrature amplitude modulation (MQAM) scheme is considered [26]. Thus, $\zeta = 3\left(\frac{\sqrt{M}+1}{\sqrt{M}-1}\right)$ with $M = 2^b$ where b is constellation size. P_{out} is the transmitted power. When

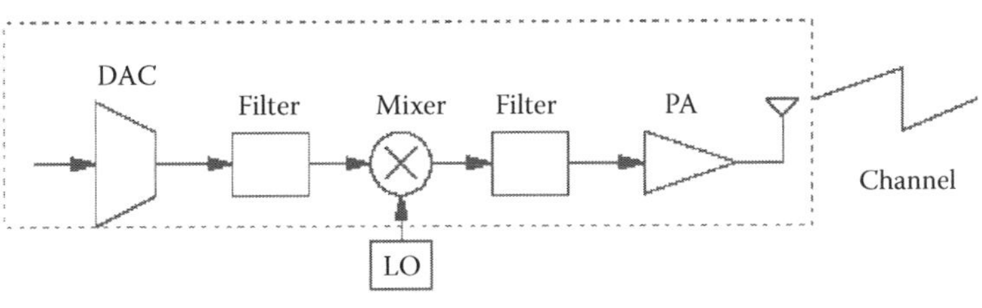

Figure 14.4 Transmitter circuit blocks. (From S. Cui et al., *IEEE Journal on Selected Areas in Communications*, vol. 22, no. 6, pp. 1089–1098, August 2004.)

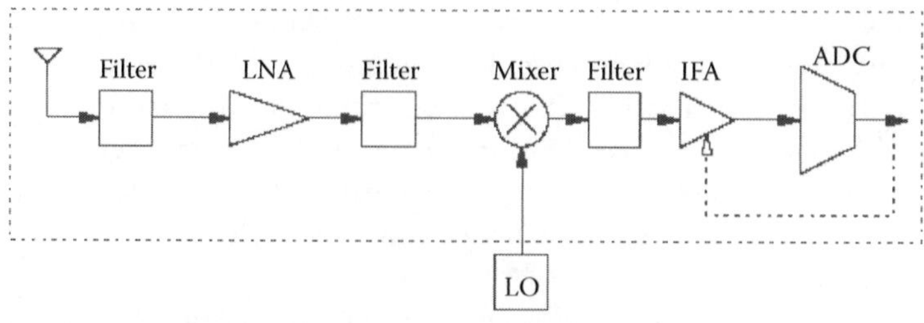

Figure 14.5 Receiver circuit blocks. (From S. Cui et al., *IEEE Journal on Selected Areas in Communications*, vol. 22, no. 6, pp. 1089–1098, August 2004.)

the channel only experiences a kth power channel loss, it can be calculated using the link budget relationship [35] as follows:

$$P_{\text{out}} = \overline{E_{\text{b}}} R_{\text{b}} \times \frac{(4\pi)^2 d^k}{G_{\text{t}} G_{\text{r}} \lambda^2} M_{\text{l}} N_{\text{f}} \tag{14.4}$$

where $\overline{E_{\text{b}}}$ is the average energy per bit required at the receiver for a given average BER specification, R_{b} is the transmission bit rate, d is the long-haul transmission distance, G_{t} and G_{r} are the transmitter and receiver antenna gains, respectively, λ is the carrier wavelength, M_{l} is the link margin compensating for the hardware process variations and other background noise, and N_{f} is the receiver noise figure. It is defined as $N_{\text{f}} = \frac{N_{\text{r}}}{N_0}$ where N_{r} is the power spectral density (PSD) of the total effective noise at the receiver input, and N_0 is the single-sided thermal noise PSD at room temperature.

The average energy per bit $\overline{E_{\text{b}}}$ for a given BER requirement is obtained as:

$$\overline{E_{\text{b}}} = \frac{2}{3}\left(\frac{\overline{P_{\text{b}}}}{\frac{4}{b}\left(1-\frac{1}{2^{\frac{b}{2}}}\right)}\right)^{-\frac{1}{M_{\text{t}}}} \frac{2^b-1}{b} M_{\text{t}} N_0 \tag{14.5}$$

where $\overline{P_{\text{b}}}$ is the average BER.

The power consumption of the circuit block can be expressed as [3]:

$$P_{\text{C}} = M_{\text{t}}(P_{\text{DAC}} + P_{\text{mix}} + P_{\text{filt}}) + 2P_{\text{synth}} + M_{\text{r}}(P_{\text{LNA}} + P_{\text{mix}} + P_{\text{IFA}} + P_{\text{filtr}} + P_{\text{ADC}}) \tag{14.6}$$

Equation 14.6 includes the power consumption values of the mixer P_{mix}, the frequency synthesizer P_{synth}, the active filters at the transmitter and the receiver side P_{filt}, and P_{filr}, respectively, the low-noise amplifier P_{LNA}, the intermediate frequency amplifier P_{IFA}, and the D/A and A/D converters P_{DAC} and P_{ADC}, respectively.

The total energy consumption per bit is given as [3]:

$$E^{\text{t}} = \frac{P_{\text{PA}} + P_{\text{C}}}{R_{\text{b}}} \tag{14.7}$$

R_b can be replaced by the effective bit rate of the system $R_e^{eff} = \frac{F - pM_t}{F} R_b$ when pM_t training symbols are inserted in each block to estimate the channel. The block size is equal to F symbols and can be obtained by setting $F = [T_c R_s]$, where R_s is the symbol rate and T_c is the fading coherence time, which can be estimated as $T_c = \frac{3}{4 f_m \sqrt{\pi}}$ [36], where the maximum Doppler shift f_m is given by $f_m = \frac{\nu}{\lambda}$, and ν is the velocity. The total energy consumption is estimated by multiplying E_{bt} by the number of bits L to be transmitted.

Therefore, by using Equations 14.3 and 14.7, the total energy per bit becomes:

$$E^t = \frac{1}{(R_b / R_{eff})} \left[\frac{3}{\eta} \frac{(4\pi)^2 d^k M_l N_f}{G_t G_r \lambda^2} \left(\frac{\sqrt{M} + 1}{\sqrt{M} - 1} \right) \overline{E_b} + \frac{P_C}{bR_s} \right] \quad (14.8)$$

The total energy consumption is estimated by multiplying E^t by the number of bits to be transmitted.

14.5.2.1 Data Aggregation

In data aggregation, the CH node collects the data from multiple local sensors for aggregation. Therefore, the aggregated data are transmitted to a remote destination node. The total energy consumption is then given as [26]:

$$E_{DA+MIMO} = \sum_{i=1}^{M_t - 1} L_i E_i^t + E_{da} \sum_{i=1}^{M_t} L_i + (M_t - 1) E_i^{t0} \sum_{i=1}^{M_t} L_i y_i + E_b^r \sum_{i=1}^{M_r} L_i y_i + \sum_{j=1}^{M_r - 1} E_j^r n_r N_s \quad (14.9)$$

$$E_i^t = \frac{1}{(R_b / R_{eff})} \left[\frac{3}{\eta} \frac{(4\pi)^2 d_m^k M_l N_f}{G_t G_r \lambda^2} \left(\frac{\sqrt{M_i} + 1}{\sqrt{M_i} - 1} \right) \overline{E_{i,b}} + \frac{P_{i,C}}{b_i R_s} \right] \quad (14.10)$$

$$E_j^r = \frac{1}{(R_b / R_{eff})} \left[\frac{3}{\eta} \frac{(4\pi)^2 d_m^k M_l N_f}{G_t G_r \lambda^2} \left(\frac{\sqrt{M_j} + 1}{\sqrt{M_j} - 1} \right) \overline{E_{j,b}} + \frac{P_{j,C}}{b_j R_s} \right] \quad (14.11)$$

where, E_i^t is the local transmission energy consumption per bit for aggregation on the transmission side ($i = 1,2,\ldots M_t - 1$), E_{da} is the energy dissipation of data aggregation, the same energy per bit E_i^{t0} is the local energy consumption per bit for transferring the aggregated data to the all sensors, E_b^r is the energy consumption per bit for the long-haul transmission, E_j^r is the local energy consumption per bit for joint detection on the receiving side ($j = 1,2,\ldots .M_r - 1$), y_i is the correlation for each node that's defined as the percentage of remaining data after aggregation and it reflects the data among different sensors, and L_i is the number of transmission bits for each node.

$N_s = \frac{\sum_{i=1}^{M_t} \lambda_i L_i}{b_m}$ is the total number of symbols received by each node at the receiving side with b_m as the optimal constellation size (bits per symbol) for MIMO transmission. The $M_r - 1$ assisting nodes first quantize each symbol they receive into n_r bits, and then transmit all the bits using MQAM to the destination node to do the joint detection. d_m is the local transmission distance.

Therefore, total energy consumption for MISO, SIMO, and SISO are given, respectively, as [26]:

$$E_{\mathrm{DA+MISO}} = \sum_{i=1}^{M_t-1} L_i E_i^{t} + E_{\mathrm{da}} \sum_{i=1}^{M_t} L_i + (M_t - 1) E_i^{t0} \sum_{i=1}^{M_t} L_i y_i + E_b^{r} \sum_{i=1}^{M_r} L_i y_i \tag{14.12}$$

$$E_{\mathrm{DA+SIMO}} = \sum_{i=1}^{M_t-1} L_i E_i^{t} + E_{\mathrm{da}} \sum_{i=1}^{M_t} L_i + E_b^{r} \sum_{i=1}^{M_r} L_i y_i + \sum_{i=1}^{M_r-1} E_h^{r} n_r N_s \tag{14.13}$$

$$E_{\mathrm{DA+SISO}} = \sum_{i=1}^{M_t-1} L_i E_i^{t} + E_{\mathrm{da}} \sum_{i=1}^{M_t} L_i + E_b^{r} \sum_{i=1}^{M_r} L_i y_i \tag{14.14}$$

The energy efficiency is calculated with respect to SISO as:

$$\text{Energy efficiency} = \frac{E_{\mathrm{SISO}} - E_{\mathrm{MIMO}}}{E_{\mathrm{SISO}}} \tag{14.15}$$

14.5.2.2 Node Selection

The total energy in the case of node selection is [32]:

$$\begin{aligned} E_{\mathrm{DA+MIMO}} = E_{\mathrm{DA+MISO}} &= \sum_{i=1}^{M_t} \frac{L_i}{F} E_{\mathrm{ch}} + L_{\mathrm{ch}} \sum_{i=1}^{M_t-1} \frac{L_i}{F} E_i^{t} \\ &+ L_{\mathrm{ch}} \frac{L_i}{F} E_s^{t} + \sum_{i=1}^{M_t-1} L_i E_i^{t} + E_{\mathrm{da}} \sum_{i=1}^{M_t} L_i + (M_t - 1) E_i^{t0} \sum_{i=1}^{M_t} L_i y_i \\ &+ L_c P_s \frac{L_i}{F} \sum_{i=1}^{x} E_i^{t0} + E_b^{r} \sum_{i=1}^{M_r} L_i y_i \end{aligned} \tag{14.16}$$

where, E_{ch} is the channel estimation energy, L_{ch} is the number of bits needed to transmit the channel estimation result. Data size L_i is divided by the frame size F to find out the number of channel estimations required for the transmitted data size. The second term is due to the transfer of channel estimation result to their own CH. $L_{\mathrm{ch}} \frac{L_i}{F} E_s^{t}$ is the term required to transmit the channel estimation result to the receiving CH for channel estimation purposes, P_s is the probability that a selected sensor is changed in the next frame and is chosen as $(1/M_t)$, and L_c is the bit length of a command signal.

For a SISO scheme, there is no burden for channel estimation and the CH transmits all the aggregated data directly to the destination node without any cooperation [27]. That is

$$E_{\mathrm{DA+SISO}} = E_{\mathrm{DA+SIMO}} = \sum_{i=1}^{M_t-1} L_i E_i^{t} + E_{\mathrm{da}} \sum_{i=1}^{M_t} L_i + E_b^{r} \sum_{i=1}^{M_r} L_i y_i \tag{14.17}$$

14.6 Effects of WSN Parameters

There are many factors that affect energy consumption in WSNs. These parameters are geographical location of the sensors, intersensor distance in a cluster, channel estimate energy, power circuit, channel loss, mobility factor, and residual energy. Each of the following subsections describes the effect of one of these parameters on energy consumption and determines the degree of its effect on the overall system [37].

To perform this study, some other system variables have to be adjusted, such as optimum constellation size (b), number of training symbols per block (p), and correlation (y). The adjustment of these system variables helps in determining the exact effect WSN parameters have on energy consumption. Table 14.3 summarizes the values of system parameters. Figures 14.6 through 14.9 and Table 14.4 show the adjustment of these variables.

Table 14.3 System Parameters

Parameter	*Value*	*Parameter*	*Value*
G_t G_r	5 dBi	P_{ADC}	15.437 mW
M_l	40 dB	E_{ch}	28 μJ/bit/signals
N_o	−171 dB/Hz	E_{da}	5 nJ/bit/signals
f_r	2.5 GHz	L_{ch}	8
N_f	10 dB	L_c	8
λ	0.12 m	n_r	10
η	0.35	f_{cor}	1 MHz
$\overline{P_b}$	1e-3	I_0	10 μm
L_i	10 Kb	L_{min}	0.5 μm
P_{synth}	50.0 mW	$n_1 = n_2$	10
P_{mix}	30.0 mW	B	10 kHz
P_{LNA}	20 mW	C_p	1 pF
$P_{filt} = P_{filr}$	2.5 mW	β	1
P_{IFA}	3 mW	V_{dd}	3 V
P_{DAC}	6.698 mW		

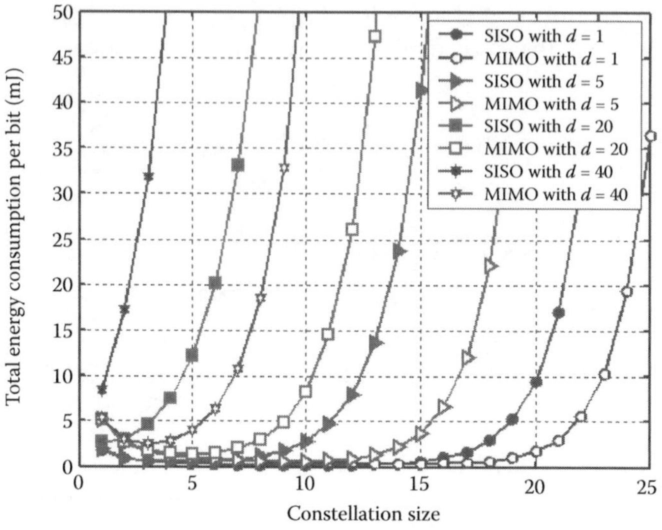

Figure 14.6 Total energy consumption versus constellation sizes.

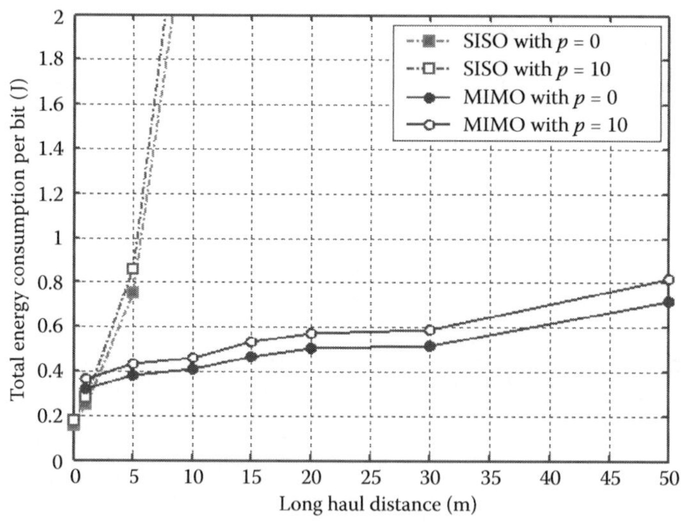

Figure 14.7 Total energy consumption for different training symbols.

The constellation size is shown for different transmission distances in Figure 14.6. In both SISO and MIMO cases, optimized constellation size is used according to the different communication distance. Therefore, communication energy consumption is minimized under its constellation size. For example, a large amount of energy can be saved for MIMO by using a constellation size equal to 20 in the case d = 1. As a result, the optimized constellation sizes for different transmission distances are summarized in Table 14.4.

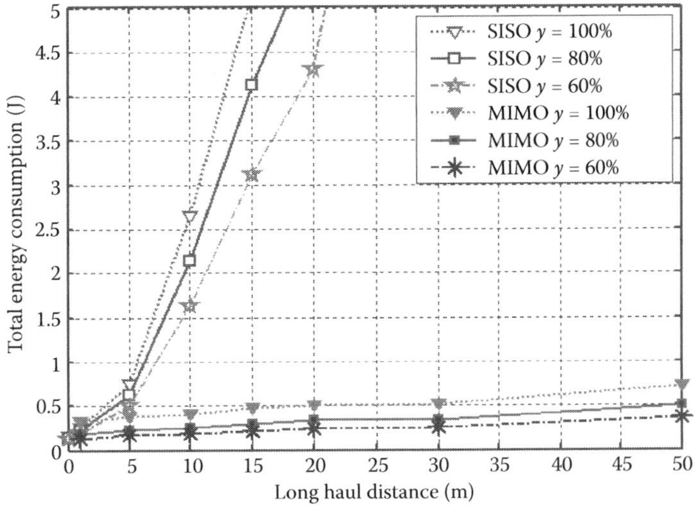

Figure 14.8 Total energy consumption for different *y*.

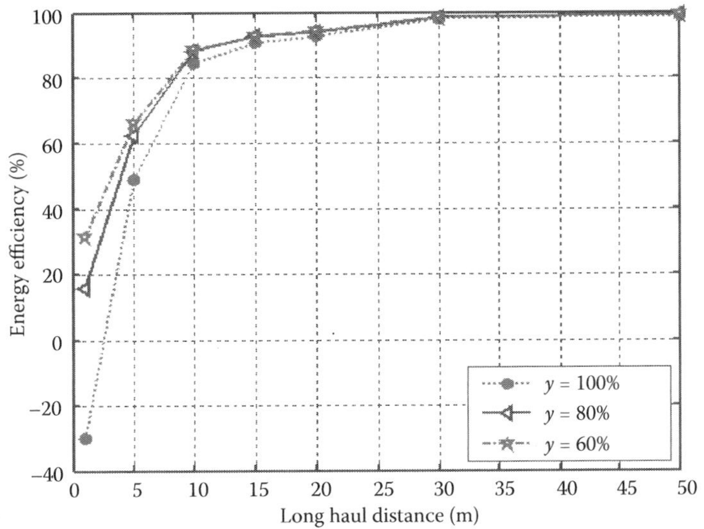

Figure 14.9 Energy efficiency for different *y*.

Figure 14.7 shows the total energy consumption at training symbols per block (p = 0 and 10). It is shown that training symbols per block have unnoticeable effects in the case of SISO. In the case of MIMO, p = 0 gives lower energy consumption. Therefore, it is considered as the optimum value of our study. The effects of correlation on total energy consumption and energy efficiency are shown in Figures 14.8 and 14.9, respectively. Because it is difficult to achieve high correlation, moderate correlation at y = 60% is considered as the beginning value. It is shown that y = 60%

Table 14.4 Optimum Constellation Sizes for Cluster-to-Cluster Communication

d (m)	*1*	*5*	*10*	*15*	*20*	*30*	*50*	*70*	*100*
$b_{5\times5}$	23	17	14	13	12	10	9	8	7
$b_{1\times5}$	13	8	7	7	6	5	4	3	2
$b_{5\times1}$	22	10	5	3	2	2	2	2	2
$b_{1\times1}$	10	4	2	2	2	2	2	2	2

Note: $M_t = 5$, $M_r = 5$.

decreases the energy consumption compared with y = 80% or 100%. This is because the total energy consumption is reduced as the data becomes more correlated. Therefore, the value of 60% is considered in the rest of our study.

14.6.1 Effect of Long-Haul Distance

Figure 14.10 shows the effect of the long-haul distance on total energy consumption. As expected, the total energy increases with the increase of the long-haul distance. It is shown that MIMO outperforms the other techniques.

14.6.2 Effect of Local Transmission Distance

The effect of local transmission distance on total energy consumption is shown in Figure 14.11. At d_m less than 20 m, the differences between the various techniques are unnoticeable. As d_m increases, the diversity increases. In general, it is shown that the effect of the local distance on energy is less than the effect of the long-haul distance shown in Figure 14.10.

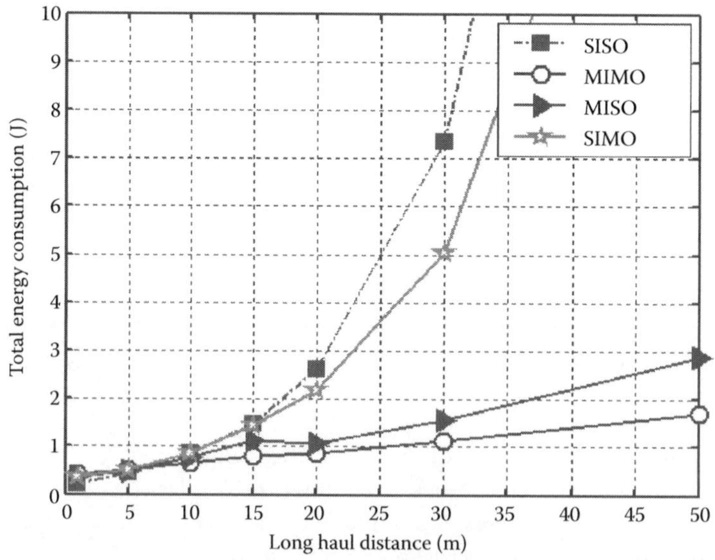

Figure 14.10 Total energy consumption versus long-haul distance.

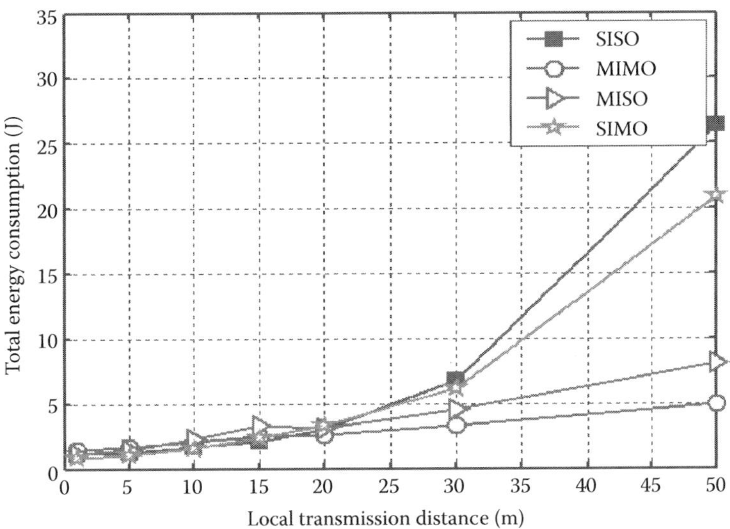

Figure 14.11 Total energy consumption versus local distance.

14.6.3 Effect of Residual Energy

Residual energy is different for different types of nodes. For normal nodes, the residual energy is [32]:

$$r_{e(n)} = r_{e(n_0)} - L_r E^t - M_t L_r E_{rel} - M_t L_r E_b^r \tag{14.18}$$

where $r_{e(n_0)}$ is the residual energy in the previous round for the normal node, E_i^t is the energy needed per bit for local transmission, E_{rel} is the energy needed per bit for local reception for transmitting cluster, E_b^r is the energy needed per bit for long-haul transmission and L_r is the bit size in a single round. These energies can be obtained from:

$$E_{rel} = (P_{LNA} + P_{mix} + P_{IFA} + P_{filtr} + P_{ADC} + P_{synth})/R_b \tag{14.19}$$

$$E^t = \frac{3(4\pi)^2 d_m^k M_l N_f}{G_t G_r \lambda^2} \left(\frac{\sqrt{M}+1}{\sqrt{M}-1} \right) \overline{E_b} + E_{cir} \tag{14.20}$$

$$E_b^r = \frac{3(4\pi)^2 d^k M_l N_f}{G_t G_r \lambda^2} \left(\frac{\sqrt{M}+1}{\sqrt{M}-1} \right) \overline{E_b} + E_{cir} \tag{14.21}$$

where E_{cir} is the circuit energy consumption during transmission. It can be expressed as:

$$E_{cir} = (P_{DAC} + P_{mix} + P_{filt} + P_{synth})/R_b \tag{14.22}$$

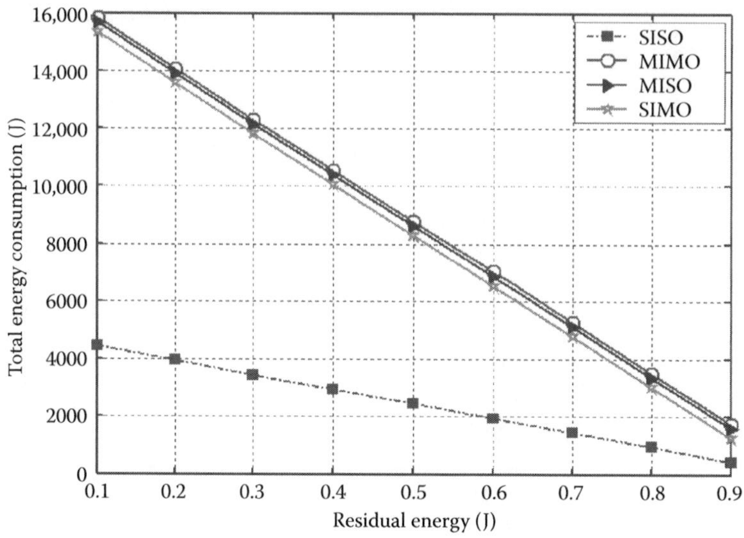

Figure 14.12 Total energy consumption versus residual energy.

The residual energy of the CH is given by:

$$r_{e(h)} = r_{e(h_0)} - M_t L_r [E_{rel} + E_{da} + E^t + E_b^r] \tag{14.23}$$

where $r_{e(h_0)}$ is the residual energy in previous round for the CH.

Figure 14.12 shows the effect of residual energy on total energy consumption. It is assumed that the CH starts with a residual energy of 3 J and the normal nodes start with 1 J. It is clear that by increasing the residual energy, the energy of the network decreases. SISO has less total energy especially at low residual energy.

14.6.4 Effect of Channel Estimate Energy

Figure 14.13 shows the total energy consumption versus channel estimate energy for MIMO, SIMO, and MISO techniques. It is shown that the effect of channel estimate energy is unnoticeable on total energy consumption. Therefore, the channel estimate energy is considered as the parameter with the least effect on energy consumption.

14.6.5 Effect of Mobility

The mobility factor is a measure that reflects the degree of node mobility through the concept of the remoteness of the nodes. Let $M_i(t) = 1, 2, 3, \ldots, M - 1$ where M is the number of nodes, represents the location vector of node i at time t and $d_{ij}(t) = | M_j(t) - M_i(t) |$, the distance from node i to j at time t. Then the remoteness from node i to node j at time t is $R_{ij}(t) = F(d_{ij}(t))$, where F is the function of remoteness. For a simple choice of F as an identity function, remoteness is just the distance between the nodes. As a node moves relative to the other nodes, remoteness remains proportionate to its previous values. But as the node moves in a manner in which its speed and angular deviation from the current state are not predictable, remoteness changes in time. Thus, the

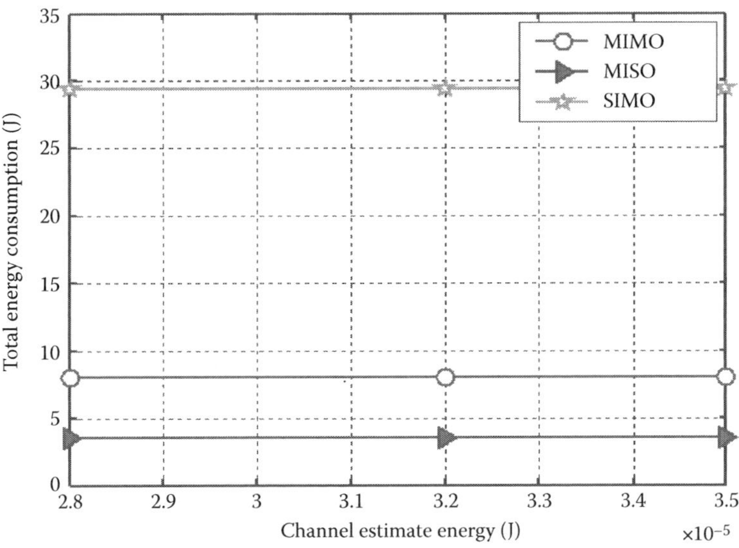

Figure 14.13 Total energy consumption versus channel estimate energy.

definition of relative mobility measure in terms of the remoteness of a node as a function of time with respect to its immediate neighbors is [38]:

$$M_i(t) = \frac{1}{M_t - 1} \sum_{i=0}^{M_t - 1} d_{ij}(t) \tag{14.24}$$

Therefore, the mobility of a wireless network in terms of the time derivatives of the remoteness $M(t)$ is [38]:

$$M(t) = \frac{1}{M_t - 1} \sum_{i=0}^{M_t} M_i(t) \tag{14.25}$$

These mobility measures are normalized by the number of nodes. Thus, the mobility factor reflects the actual link change rate. Figure 14.14 shows the network with and without a mobility factor. It is assumed that $M_t = 5$, $M_r = 5$, the distance between two nodes is equal to 10 m and the percentage of mobile sensor nodes is 20% of the total nodes. This means that, in this case, there is only one mobile node from a total of five nodes. The mobility is considered either positive (+ve), which means that the mobile nodes move toward the CH, or negative (−ve), which means that the mobile nodes move far away from the CH. The mobility factor in this figure is assumed to be 10.5 for +ve mobility and 14 for −ve mobility. It is shown that the +ve mobility decreases the total energy consumption because the distance between the mobile nodes and the CH is decreased. On the contrary, the −ve mobility increases the total energy consumption. It is also clear that a mobility factor of 14 affects the system more than a mobility factor of 10.

Figures 14.15 and 14.16 show the effect of the percentage of mobile nodes to the total number of nodes at different network sizes (n = 10, and 20 nodes). It is considered in these figures that the nodes move with +ve mobility. It is shown that energy consumption is increased with an increase

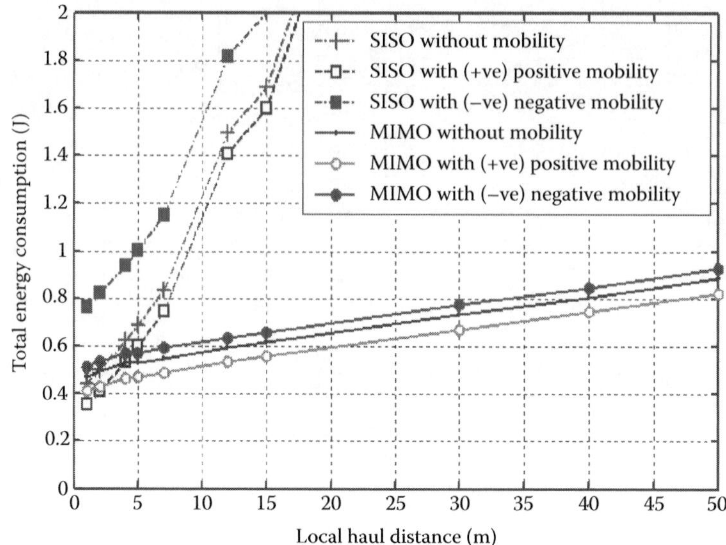

Figure 14.14 Total energy consumption versus distance without and with mobility factor parameter.

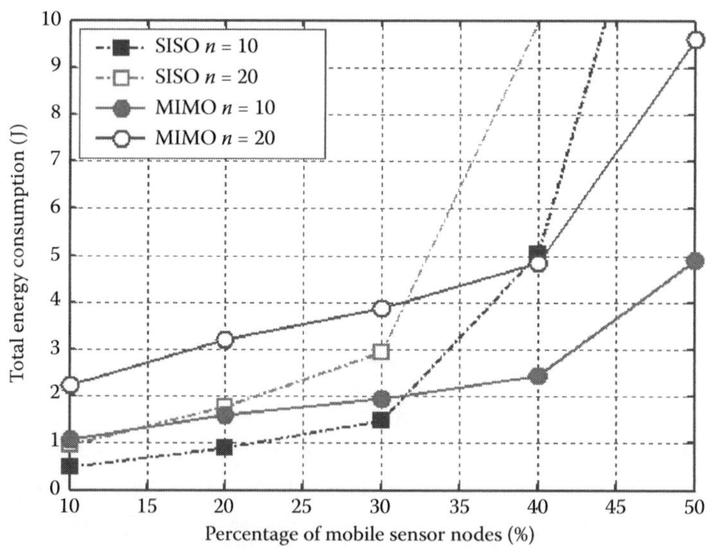

Figure 14.15 Total energy consumption for different network sizes.

of the network size. When the number of mobile sensor nodes is increased, the network becomes more dynamic and thus more energy is consumed.

Figure 14.17 shows the total energy consumption of SISO and various virtual MIMO techniques such as (5 × 5), (4 × 4), and (2 × 2) versus mobility factor. It is assumed that the nodes move with +ve mobility factor of 6. The results show that SISO performs better than MIMO techniques

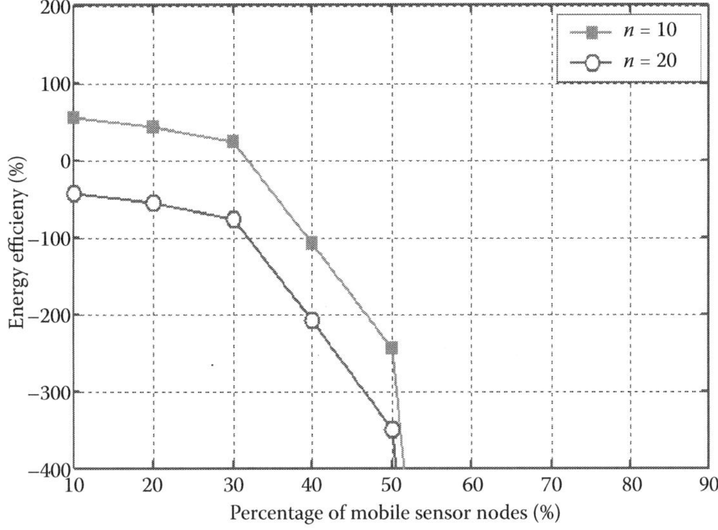

Figure 14.16 Energy efficiency versus percentage of mobile sensor nodes for different network sizes.

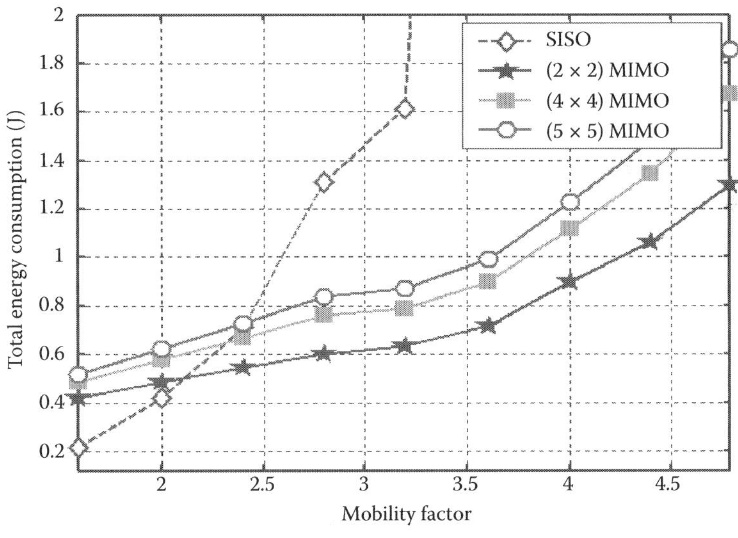

Figure 14.17 Total energy consumption versus for mobility factors.

for smaller mobility factor; however, at mobility factor greater than 2, the MIMO technique consumes less energy. The energy consumption is decreased as the number of transmitters and receivers is decreased.

An energy efficiency comparison is shown in Figure 14.18 for rate-optimized MQAM with a reference SISO system. It is shown that a selective approach such as (4 × 4) MIMO

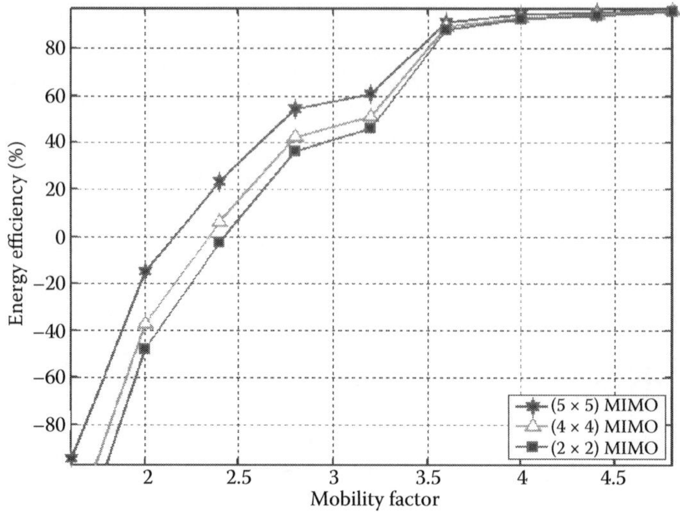

Figure 14.18 Energy efficiency for various mobility factors.

and (2 × 2) MIMO is more energy-efficient than the existing unselected approach [(5 × 5) MIMO]. Because in the unselected approach, all the sensors in a cluster are used for transmission without considering their parameter conditions. This makes this technique inefficient. Figure 14.19 shows the effect of mobility factor on total energy consumption for SISO and all MIMO techniques. As shown previously, SISO achieves lower energy consumption at low mobility factors.

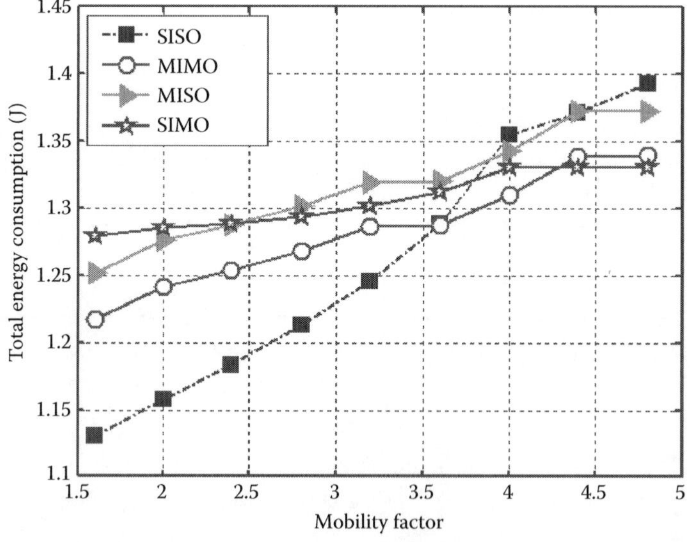

Figure 14.19 Total energy consumption versus mobility factors for different techniques.

14.6.6 Effect of Power Circuit

Power circuits include ADC and DAC. Many different techniques are used to reduce power consumption at the circuit level. Some of the main techniques are ADC resolution and DAC resolution. Therefore, resolution has a valuable effect on the power circuit.

The value of P_{DAC} and P_{ADC} can be calculated as follows [39]:

$$P_{\text{DAC}} \approx \beta\left(\frac{1}{2}V_{\text{dd}}I_0(2^{n_1}-1)+n_1C_p(2B+f_{\text{cor}})V_{\text{dd}}^2\right) \tag{14.26}$$

$$P_{\text{ADC}} \approx \frac{3V_{\text{dd}}^2L_{\text{min}}(2B+f_{\text{cor}})}{10^{-0.1525n_2+4.838}} \tag{14.27}$$

where V_{dd} is the power supply, L_{min} is the minimum channel length for the given CMOS technology, f_{cor} is the corner frequency of the $1/f$ noise, C_{p} is the parasitic capacitance of each switch, n_1 and n_2 are resolution of DAC and ADC, respectively, B is the bandwidth, β is a correcting factor to incorporate some second-order effects, and I_0 is the current source corresponding to the least significant bit (LSB). In this work, we used the estimation model in the study by Lauwers and Gielen [40].

Total energy consumption as a function of distance with different power circuits that have different resolutions is shown in Figure 14.20. The effect of resolution is clear. Less resolution means less power circuits, which then results in less energy consumption. The same result is shown in Figure 14.21, where resolution = 6 achieves more energy saving than resolution = 10. Figure 14.22 shows the effect of resolutions for the total energy consumption at different techniques. The increase of the resolutions of ADC and DAC increases the total energy consumption at various techniques.

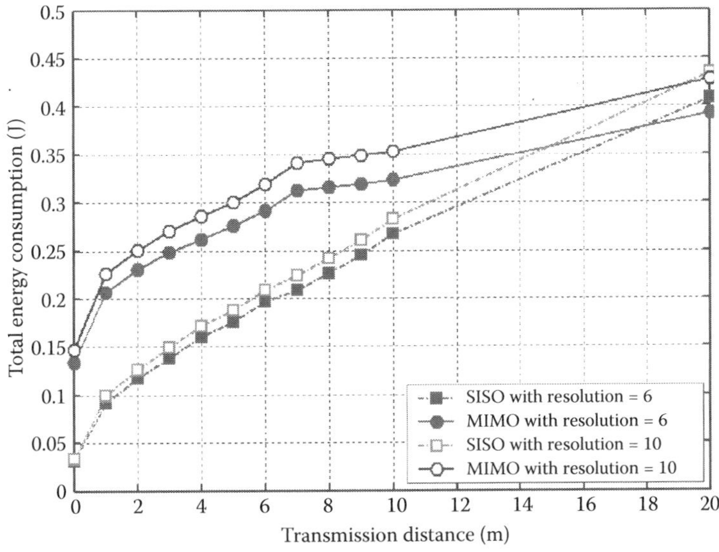

Figure 14.20 Total energy consumption as a function of distance with different power circuits.

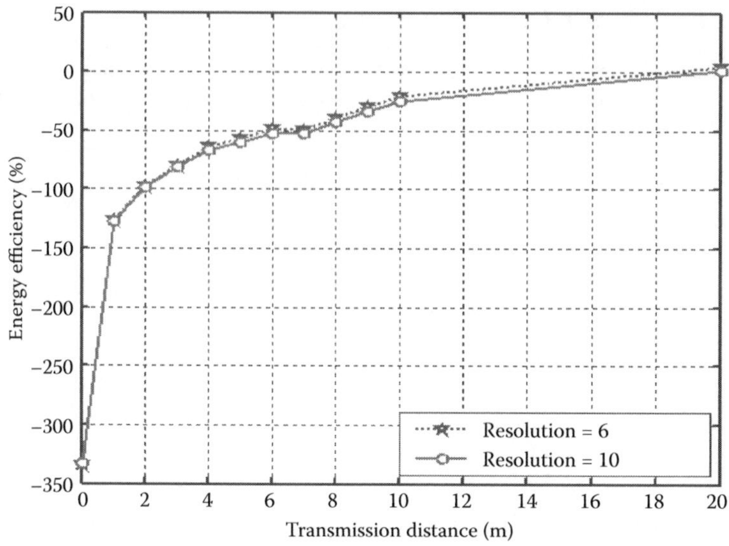

Figure 14.21 Energy efficiency as a function of transmission distance with different power circuits.

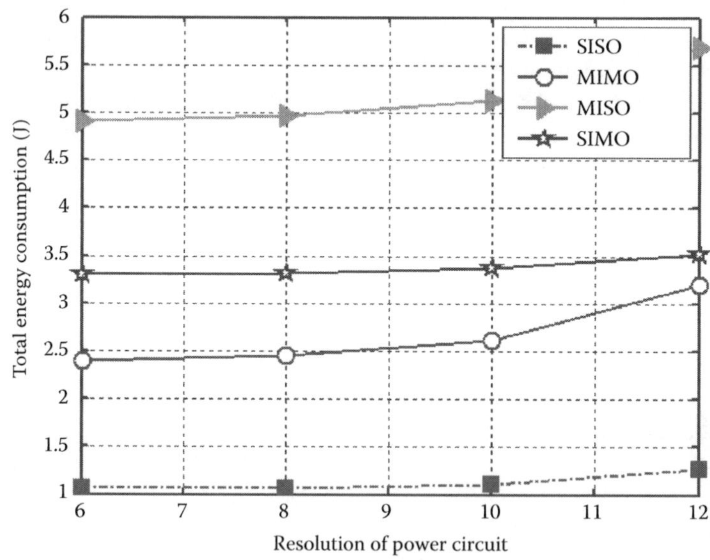

Figure 14.22 Total energy consumption as a function of resolutions for different techniques.

14.6.7 Effect of Channel Loss

The total energy consumption of MIMO and SISO techniques are represented versus the long-haul transmission distance d in Figure 14.23 with $M_t = 5$, $M_r = 5$ and d_m between the nodes is 5 m. It is shown that MIMO outperforms SISO at a different channel loss parameter (k). By the increase of k, SISO consumes more energy and the diversity between MIMO and SISO gets higher

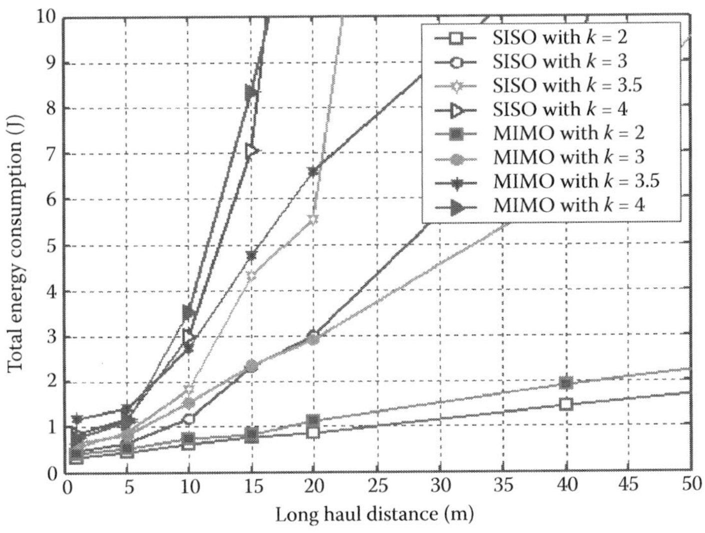

Figure 14.23 Total energy consumption as a function of long-haul distance at various channel loss parameters.

than lower values of k. Figure 14.24 shows the enormous energy savings of a virtual MIMO. For example, when $k = 3$, MIMO offers 30% of energy savings compared with SISO at $d = 50$ m. At the same distance and $k = 4$, 87% of the energy can be saved. So, the maximum achievable energy saving increases as k increases. Figure 14.25 shows the effect of channel loss at total energy consumption for various techniques.

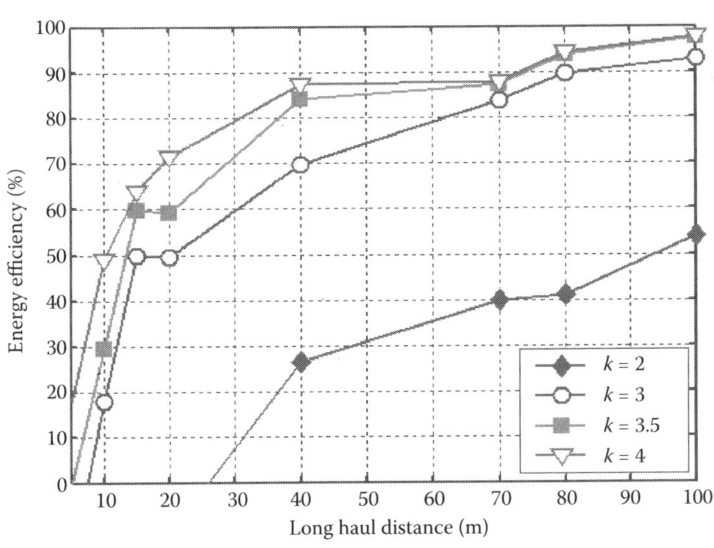

Figure 14.24 Energy efficiency as a function of long-haul distance with different channel loss parameters.

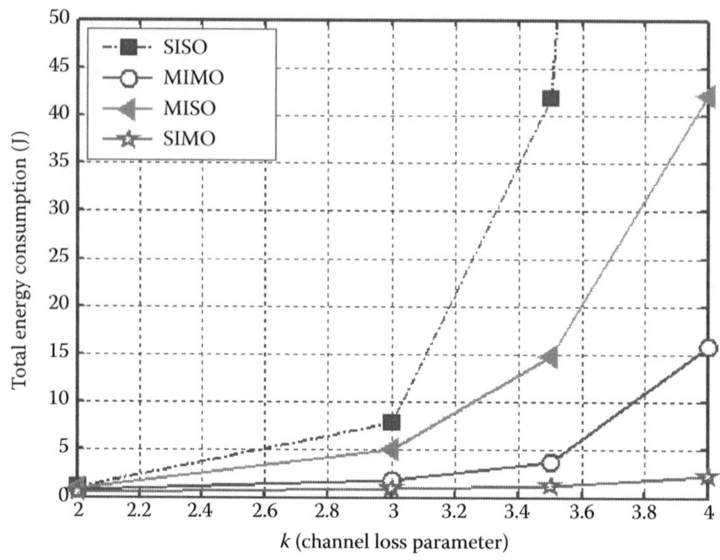

Figure 14.25 Total energy consumption as a function of channel loss parameter at various techniques.

References

1. M. Francesco, G. Anastasi, M. Conti, S. Das, and V. Neri. Reliability and energy-efficiency in IEEE 802.15.4/ZigBee sensor networks: An adaptive and cross-layer approach. *IEEE Journal on Selected Areas in Communications*, vol. 29, no. 8, pp. 1–18, September 2011.
2. M. Mohaisen, H. An, and K. Chang. Detection techniques for MIMO multiplexing: A comparative review. *Transactions on Internet and Information Systems*, vol. 3, no. 6, pp. 647–666, 2009.
3. S. Cui, A.J. Goldsmith, and A. Bahai. Energy-efficiency of MIMO and cooperative MIMO techniques in sensor networks. *IEEE Journal on Selected Areas in Communications*, vol. 22, no. 6, pp. 1089–1098, August 2004.
4. S. Cui and A.J. Goldsmith. Cross-layer design of energy-constrained networks using cooperative MIMO techniques. *Signal Processing*, Elsevier, vol. 86, pp. 1804–1814, August 2006.
5. S.K. Jayaweera. Energy analysis of MIMO techniques in wireless sensor networks. In *Proceedings of the 38th Annual Conference on Information Sciences and Systems (CISS '04)*, Princeton, NJ, March 2004.
6. S.K. Jayaweera. Energy efficient virtual MIMO-based cooperative communications for wireless sensor networks. In *Proceedings of the 2nd International Conference on Intelligent Sensing and Information Processing and Information Processing (ICISIP'05)*, Chennai, India, January 2005.
7. J.M. Chung, J. Kim, and D. Han. Multihop hybrid virtual MIMO scheme for wireless sensor networks. *IEEE Transactions on Vehicular Technology*, vol. 61, no. 9, pp. 4069–4078, 2012.
8. V.K. Garg. *Wireless Communications and Networking*. Elsevier, USA, pp. 245–246, 2007.
9. T.M. Duman and A. Ghrayed. *Coding for MIMO Communication Systems*, John Wiley, England, 2007.
10. L.G. Ordonez, D.P. Palomer, and J.R. Fonollosa. On the diversity, multiplexing, and array gain tradeoff in MIMO channels. In *Proceedings of IEEE International Symposium on Information Theory (ISIT)*, pp. 2183–2187, Austin, Texas, USA, June 2010.
11. M. Ahmad, E. Dutkiewicz, X. Huang and M. Suaidi. Cooperative MIMO systems in wireless sensor networks. In *Radio Communications*, A. Bazzi (Ed.), InTech, 2010. Available from: http://www.intechopen.com/books/radio-communications.

12. S. Sandhu, R. Nabar, D. Gore, and A. Paulraj. Introduction to space–time codes. Smart Antenna Research Group, Stanford University, Technical Report, 2004.
13. H. Bölcskei and A.J. Paulraj. Multiple-input multiple-output (MIMO) wireless systems. In *The Communications Handbook*, 2nd ed., J. Gibson (Ed.), CRC Press, Boca Raton, FL, USA, 2002.
14. S.K. Jayaweera. Virtual MIMO-based cooperative communication for energy-constrained wireless sensor networks. *IEEE Transaction Wireless Communication*, vol. 5, no. 5, pp. 984–989, May 2006.
15. Y. Yuan, M. Chen, and T. Kwon. A novel cluster-based cooperative MIMO scheme for multi-hop wireless sensor networks. *EURASIP Journal on Wireless Communications and Networking*, vol. 2006, no. 2, pp. 38–47, 2006.
16. W. Heinzelman, A. Chandrakasan, and H. Balakrishnan. An application specific protocol architecture for wireless micro sensor networks. *IEEE Transactions on Wireless Communications*, vol. 1, no. 4, pp. 660–670, 2002.
17. Y. Liu, N. Xiong, Y. Zhao, and A. Vasilakos. Multi-layer clustering routing algorithm for wireless vehicular sensor networks. *IET Communications*, vol. 4, no. 7, pp. 810–816, 2010.
18. K. Lin, L. Wang, and K. Li. Multi-attribute data fusion for energy equilibrium routing in wireless sensor networks. *KSII Transactions on Internet and Information Systems*, vol. 4, no. 1, pp. 5–24, 2010.
19. S. Deng, J. Li, and L. Shen. Mobility-based clustering protocol for wireless sensor networks with mobile nodes. *IET Wireless Sensor Systems*, vol. 1, no. 1, pp. 39–47, 2011.
20. G. Ahmed, N.M. Khan, and R. Ramer. Cluster head selection using evolutionary computing in wireless sensor networks. In *Proceedings of Progress in Electromagnetic Research Symposium*, Hang Zhou, China, pp. 883–886, March 2008.
21. B. Ashish and J. Raman. The criteria require for cluster head gateway selection in integrated mobile ad hoc network. *International Journal of Engineering Science and Technology (IJEST)*, vol. 3, no. 7, pp. 5452–5458, July 2011.
22. H. Yang and B. Sikdar. Optimal cluster head selection in the LEACH architecture. In *Proceedings of the 26th IEEE International Performance and Communications Conference (IPCCC2007)*, New Orleans, LA, April 2007.
23. K. Taewook, Y. Jangyu, L. Hoseung, L. Icksoo, K. Hyusook, L. Byunghwa, L. Byeongjik, and H. Kijun. A clustering method for energy efficient routing in wireless sensor networks. In *Proceedings of the 6th WSEAS International Conference on Electronics, Hardware, Wireless and Optical Communications*, Corfu Island, Greece, pp. 133–138, February 2007.
24. A. Khan, M. Imran, and B.A. Abdullah. Energy efficient technique for cluster head selection and data gathering in wireless sensor network. In *Proceedings of International Conference on Information and Communication Technologies*, Pakistan, 2008.
25. T. Thumthawatworn, P. Pakdeepinit, T. Yeophantong, S. Charoenvikrom, and J. Daengdej. Method for cluster heads selection in wireless sensor networks. In *Proceedings of the IEEE Aerospace Conference*, vol. 6, pp. 3615–3623, March 2004.
26. Y. Gai, L. Zhang, and X. Shan. Energy efficiency of cooperative MIMO with data aggregation in wireless sensor networks. In *Proceedings of the IEEE Wireless Communication and Networking Conference (WCNC 2007)*, Hong Kong, pp. 791–796, March 2007.
27. C.L. Wang, Y.W. Huang, and Y.C. Huang. An energy efficient cooperative SIMO transmission scheme for wireless sensor network. In *Proceedings of the IEEE International Conference on Communications (ICC '09)*, pp. 1–5, Dresden, Germany, June 2009.
28. I. Ahmed, M. Peng, and W. Wang. Exploiting geometric advantages of cooperative communications for energy efficient wireless sensor networks. *International Journal of Communications, Network and System Sciences*, vol. 1, pp. 55–61, 2008.
29. M.R. Islam and J. Kim. Cooperative technique based on sensor selection in wireless sensor network. *Advances in Electrical and Computer Engineering*, vol. 9, no. 1, pp. 56–62, 2009.
30. M.R. Islam and J. Kim. Channel estimated cooperative MIMO in wireless sensor network. *IETE Technical Review*, vol. 25, no. 5, pp. 234–243, September–October 2008.
31. M.R. Islam and J. Kim. Development of selection function for the application in cooperative multiple input single output at energy aware wireless sensor network. *IETE Technical Review*, vol. 26, no. 6, pp. 453–460, November–December 2009.

32. M.R. Islam and J. Kim. On the cooperative MIMO communication for energy-efficient cluster-to-cluster transmission at wireless sensor network. *Annals of Telecommunications (Springer)*, vol. 65, no. 5–6, pp. 325–340, 2010.
33. W. Heinzelman, A. Chandrakasan, and H. Balakrishnan. Energy-efficient communication protocols for wireless micro sensor networks. In *Proceedings of the Hawaii International Conference on Systems Sciences*, Island of Maui, January 2000.
34. S. Cui, A.J. Goldsmith, and A. Bahai. Modulation optimization under energy constraints. In *Proceedings of the IEEE International Conference on Communications (ICC '03)*, Alaska, May 2003.
35. J.G. Proakis. *Digital Communications*, 4th ed. McGraw-Hill, New York, 2000.
36. T.S. Rappaport. *Wireless Communications Principles and Practices*, 2nd ed. Prentice Hall, Upper Saddle River, NJ, 2002.
37. R. Rizk, S.M. Magdy, and F.W. Zaki. Energy efficiency of virtual multi-input, multi-output based on sensor selection in wireless sensor networks. *Wireless Communication Mobile Computing*, doi: 10.1002/wcm.2310, 2012.
38. G.S. Kumar, M.V. Vinu Paul, G. Athithan, and K.P. Jacob. Routing protocol enhancement for handling node mobility in wireless sensor networks. In *Proceedings of 2008 IEEE Region 10 Conference (TENCON2008)*, pp. 1–6, Hyderabad, India, November 2008.
39. S. Cui, A.J. Goldsmith, and A. Bahai. Energy-constrained modulation optimization. *IEEE Transaction on Wireless Communication*, vol. 4, no. 5, pp. 2349–2360, September 2005.
40. E. Lauwers and G. Gielen. Power estimation methods for analog circuits for architectural exploration of integrated systems. *IEEE Transactions on VLSI Systems*, vol. 10, pp. 155–162, April 2002.

Chapter 15

Evolution of Virtual Clustering in Wireless Sensor Networks

Asis Kumar Tripathy and Suchismita Chinara
National Institute of Technology Rourkela

Contents

15.1 Introduction

Wireless sensor network (WSN) refers to a group of spatially dispersed and dedicated sensors for monitoring and recording the physical conditions of the environment and organizing the collected data at a central location. WSNs measure environmental conditions such as temperature, sound, pollution levels, humidity, wind speed and direction, pressure, illumination intensity, vibration intensity, power line voltage, chemical concentrations, and vital body functions. WSNs provide reliable monitoring from faraway distances. These networks are basically data-gathering networks in which data are highly correlated and the end user needs a high-level description of the environment by the sensing nodes. It emerged due to advancements in micro-electromechanical systems and digital electronics, which enabled the development of low-cost, low-power, multifunctional sensors that are small in size and communicate in short distances [1]. Initially, it was used in military operations for surveillance activity on borders, although nowadays WSNs span a huge area of implementations.

Figure 15.1 shows the working procedure of a WSN; how sensor nodes are sensing data and transmitting those to a base station (BS) via a wireless link. With the proliferation of automated devices and the development of wireless technologies, WSNs have gained worldwide attention in recent years. WSNs are an exciting emerging domain of deeply networked systems of low-power wireless nodes with a tiny amount of CPU and memory for high-resolution sensing of the environment. The wireless nodes are nothing but a large number of low-cost, multifunctional sensor nodes that are deployed in a region of interest. The sensor nodes not only sense but also process the data to make itself meaningful by using its embedded microprocessors and communicate those

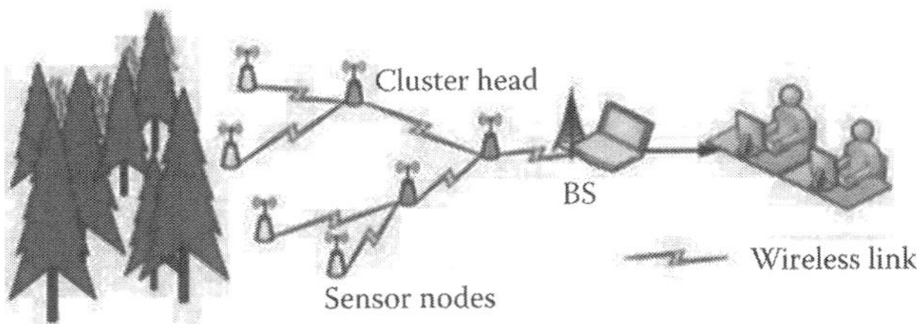

Figure 15.1 Architecture of a typical wireless sensor network.

meaningful data through its transceiver [2]. They communicate over short distances via a wireless medium and collaborate to accomplish a common task, for example, environmental monitoring, battlefield surveillance, industrial process control, etc. WSNs are made up of a large number of inexpensive devices that are networked via low-power wireless communications [3–5]. Due to the networking capability that fundamentally appears in a sensor network, it overcomes the flaws present in a mere collection of sensors by enabling cooperation, coordination, and collaboration among sensor assets [6]. WSN technology is expected to have a significant effect on our lives in the twenty-first century by harvesting advancements in the past decade in microelectronics, sensing, analog and digital signal processing, wireless communications, and networking. WSNs fundamentally differ from general data networks such as the Internet and, as such, they require the adoption of a different design paradigm [7,8].

Often, WSNs are application-specific; they are designed and deployed for special purposes to solve some intended applications. In the context of WSNs, the broadcast nature of the medium must be taken into account. Figure 15.2 gives an extensive idea of collecting data by a hierarchical manner. The WSN is very much useful in this scenario because of the wireless communication between the sensors. Because of the battery-operated sensors, energy conservation is one of the most important design parameters, since replacing batteries may be difficult or impossible in many applications [9]. Thus, sensor network designs must be optimized to extend the network lifetime. In view of energy consumption in a WSN, data transmission is the most important with respect to others. A clustering organization can be classified as intracluster communication (single hop or multihop) or intercluster communication [10]. Researchers have shown that multihop communication between a data source and a BS is usually more energy-efficient than direct transmission

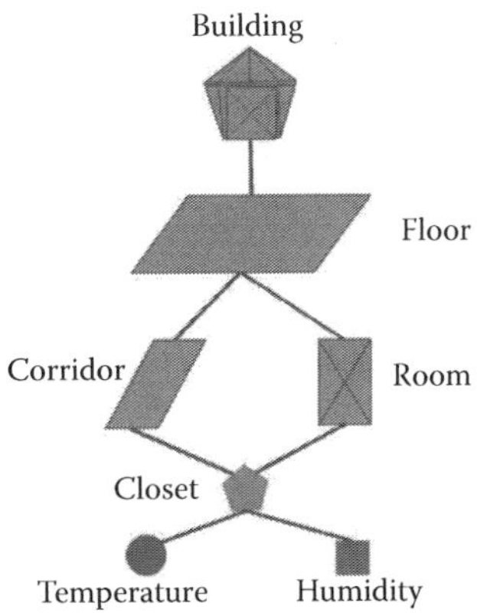

Figure 15.2 Example of sensing hierarchical data.

because of the characteristics of wireless channels [11]. Although many protocols proposed in the literature reduce energy consumption on forwarding paths to increase energy efficiency, they do not necessarily extend network lifetime due to the continuous many-to-one traffic pattern. In a sensor node, energy consumption can be "useful" or "wasteful" [12,13]. Useful energy consumption can be due to the following items: transmitting/receiving data, processing query requests, and forwarding queries/data to neighboring nodes. Whereas wasteful energy consumption can be due to the following items: idle listening to the media, retransmitting due to packet collisions, overhearing, and generating/handling control packets [14,15].

Apart from energy consumption issues, data congestion, data loss, and collision problems are also experienced in the WSNs. Data congestion and collision occurs when each and every sensor starts to communicate and engage in data transmission at the same time stamp in the networks. One of the available solutions to this problem is segregating the sensors into logical groups or clusters. Typically, sensors are grouped into clusters; one node is elected as the cluster head (CH) in each cluster. The CHs are responsible for collecting, processing, and sending information to the BSs [16–18]. Some advantages of a cluster-based network include the following:

- Clustering enhances the lifetime of the network by efficient use of limited energy in sensors.
- Data aggregation is simplified by decreasing the number of redundant packets at the CHs via intracluster communication.
- Reduced size of routing table at each individual node by localizing the route set up within the cluster.
- Due to intercluster communication, clustering can conserve communication bandwidth.
- Clustering schemes reduce the topology maintenance overheads among the sensors [19].

Disadvantages:

- Additional overheads during CH selection, assignment, and cluster formation process
- Computation overhead due to reformation of clusters as and when required

15.2 Clustering in WSN

This section provides elaborative knowledge about the clustering process in WSNs. The requirement of clustering in the sensors and the cluster formation techniques are outlined here. After creating a cluster, the job is not over, it must be maintained (i.e., also described briefly in this section). The important factors that affect the clustering process are network architecture, deployment of nodes, data processing, load balancing, fault tolerance, increasing connectivity, and reduced delay. Minimal cluster count, maximizing network longevity, scalability, and hardware constraints are briefly described in this section.

Modeling of a WSN is highly affected by certain parameters such as network architecture, deployment of nodes, data processing, load balancing, fault tolerance, connectivity, scalability, network longevity, production costs, stability, operating environment, mobility, sensor network topology, hardware constraints, and power consumption. Figure 15.3 shows a normal clustering approach of sensor nodes, in which at least one member from the cluster is designated as the CH.

CHs are responsible for coordination among the sensor nodes within their clusters and aggregation of their data (intracluster coordination), and communication with other CHs or external observers on behalf of their clusters (intercluster communication) [20]. As a CH needs to perform more load than other sensor nodes, it may consume energy at a much faster rate. A dynamically

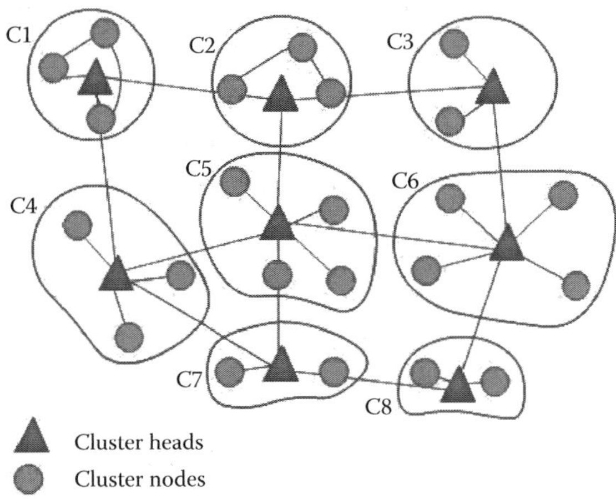

Figure 15.3 Clustering of nodes.

changing CH algorithm for WSNs has been proposed to distribute energy consumption as evenly as possible. In most scenarios, the following assumptions are made:

1. The BS is located far from the sensor nodes and is stationary.
2. All sensor nodes in the network are homogeneous and energy constrained.
3. All sensor nodes are able to reach the BS.
4. Sensor nodes have no location information.
5. The propagation channel is symmetric.
6. CHs perform data compression.

The lifetime of a sensor network is defined by using three metrics: first node dies (FND), half of the nodes alive (HNA), and last node dies (LND).

15.2.1 Clustering Requirement

To achieve scalability and efficiency of WSN, we use a very common approach known as clustering. Clustering is defined as a mechanism that creates logical groups among nodes. One node is elected as the leader of the cluster, which communicates the information from the member nodes to the BS. First of all, one sensor is elected as a CH based on some parameters such as residual energy, distance from the BS, etc., and after the CH is elected, the fellow members are appointed. To maintain the load-balancing among the nodes and to enhance the lifetime of the network, reclustering is done periodically so that the load on the CH is reduced [21]. Dynamic clustering is very useful where the nodes are prone to changing environments. Combined with a dynamic operation, clustering is very useful for achieving good scalability and high energy efficiency. The clustering technique is one of the effective approaches used to save energy in WSNs. In each cluster, sensors are given different roles to play, such as CH, ordinary member node, or gateway node. Most of the algorithms aim to extend the network lifetime by balancing energy consumption among nodes and by distributing the load among different nodes from time to time. During the reformation of clusters, the CH is changed along with the members affiliated to it. Clustering provides resource utilization and minimizes energy consumption in WSNs by reducing the number of sensors that take part in long-distance transmission. In energy-constrained WSN, energy efficiency is given the primary concern to extend.

15.2.2 Cluster Formation

To divide the network into logical clusters is always a complex job. Different parameters such as stability, load balancing, maximum cluster size, maximum number of hops to the nearest CH, and nearest CH are chosen in different ways to form the clusters in the network. Figure 15.4 shows the transmission between different clusters when they are present in different hierarchies.

15.2.3 Cluster Maintenance

The objective of cluster maintenance is to preserve the existing clustering structure as much as possible. When a CH reaches a certain level of energy after that, reclustering is started to enhance the lifetime of the network. The requirement for the re-election of CHs arises when the current heads fail to cover all the nodes in the network. Sometimes, a node may move away from the transmission range of all the current CHs and becomes an orphan node. In some of the algorithms, reclustering is performed from time to time to balance the load among different nodes.

15.2.4 Factors Affecting Clustering

15.2.4.1 Network Architecture

Network architecture is dependent on the implementation of the application of WSNs. Some parameters are highlighted according to their requirements on the network. A network consists of sensors, BSs, and in some processes where sensors may be static or mobile, it depends on the application. It is always very difficult to create clusters on a mobile network as the sensors are moving. In this type of dynamic network, a periodic alert to the BS, which creates huge traffic at the sink, is required. CHs periodically rotate according to the residual energy label of the sensors and continuous events make the cluster-based network stable.

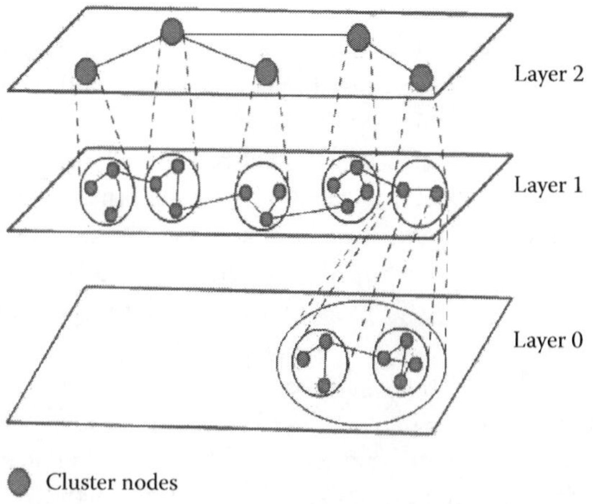

Figure 15.4 Transmission between different layers.

15.2.4.2 Deployment of Nodes

Depending on the need of the network, clustering topological deployment can be done in various different ways. They can be deployed by

- Dropping from a plane
- Delivering in an artillery shell, rocket, or missile
- Throwing by a ship board
- Placing in factory
- Placing one-by-one either by a human or by robot

Many different schemes are used for initial deployment, which will reduce the installation cost, increase the flexibility of the network, and promote self-organization and fault tolerance. In self-organized deployments, sensors are manually placed and data are routed through predetermined paths. Sensor nodes are scattered randomly and create an ad hoc network. Thus processing and energy efficiency is very uncertain on a nonuniform–based network. Thus, intercluster connectivity is a vital aspect. After deployment, topology changes are caused by the mobility of the sensors' position, jamming noise, moving obstacles, available energy, malfunctions, and task details. The redeployment process can be taken when some nodes are faulty or dead. After redeployment, the network is organized accordingly so that there will no failure on the network.

15.2.4.3 Data Processing

Network data processing is required to reduce redundant data in the network, which helps in decreasing the number of transmissions. Hence, in a multihop network, local data processing is crucial in minimizing power consumption. Sensor nodes must have some computational power to communicate with the surroundings. The processor aggregates data from different nodes by eliminating duplicates, minimum, maximum, and average. This process can be used to achieve energy efficiency and traffic optimization of the routing protocols.

15.2.4.4 Load Balancing

Load balancing is one of most important aspect in WSNs. Load balancing is done by CHs, in which CHs are selected among sensors. CHs try to set up symmetric clusters to balance the load, but it is a bit difficult to build such a network in a mobile setup. However, if we try to construct the cluster with similar number of nodes, then this issue may be resolved. Load balancing is a more vital issue in WSNs in which CHs are picked from the available sensors from the same cluster. At the time of data aggregation by the CH, other nodes should prepare the data report, which becomes ready almost at the same time, for further processing at the BS or at the next tier in the network.

15.2.4.5 Fault Tolerance

In WSNs, some sensors may malfunction because of lack of power, physical damage, or environmental interference. In this type of situation, the overall network should not affect and tolerate the failure. Reclustering is a way to overcome this issue but it is difficult to recluster in a shorter time stamp. We can take backup CHs, which will operate on failure time. For example, if sensors are being deployed in a house to keep track of humidity and temperature levels, the fault tolerance requirement may be low because this kind of sensor networks is not easily damaged or interfered with by environmental noise. On the other hand, if sensors are being deployed in a battlefield for surveillance and detection, then the fault tolerance has to be high because the sensed data are critical and sensors can be destroyed by hostile actions. As a result, the fault tolerance level depends on the application of the sensor networks, and the schemes must be developed with this in mind.

15.2.4.6 Increased Connectivity and Reduced Delay

In many applications, inter-CH connectivity is an important aspect—unless CHs have very long-haul communication capabilities. This is particularly true when CHs are picked from the sensors' population. The connectivity goal can be limited to just ensuring the availability of a path from every CH to the BS or to be more restrictive by imposing a bound on the length of the path. When some of the sensors assume the CH role, the connectivity objective makes network clustering one of the many variants of the connected dominating set problem. On the other hand, when data latency is a concern, intracluster connectivity becomes a design objective or constraint.

15.2.4.7 Minimal Cluster Count

Cluster count is directly proportional to delay. The network designer often likes to employ the least number of these nodes because they tend to be more expensive and vulnerable than sensors. For example, if CHs are laptop computers, robots, or a mobile vehicle, there will be some inherent limitation on the number of nodes. The limitation can be due to the complexity of deploying these types of nodes, for example, when the WSN is to operate in a combat zone or a forest. In addition, the size of these nodes tends to be significantly larger than sensors, which makes them easily detectable. Node visibility is highly undesirable in many WSNs applications such as border protection, military reconnaissance, and infrastructure security.

15.2.4.8 Maximizing Network Longevity

Wireless sensors are small-sized, battery-operated sensors, so they have limited energy storage. Because sensors are energy-constrained, the network's lifetime is a major concern. It is not practicable to recharge or replace their batteries after exhaustion. When CHs are richer in resources than sensors, it is imperative to minimize the energy for intracluster communication. The clustering algorithms are more energy-efficient compared with the direct routing algorithms. This can be achieved by balancing the energy consumption in sensors by optimizing cluster formation, periodically re-electing CHs based on their residual energy, and efficient intracluster and intercluster communication. Combined clustering and route setup has also been considered for maximizing the network's lifetime. The energy limitation on nodes results in a limited network lifetime for nodes in a network. Clustering schemes help prolong the network lifetime of WSNs by reducing the energy usage in the communication within and outside clusters.

15.2.4.9 Scalability

The number of sensors deployed is in the order of hundreds or thousands. Depending on the application, the number may reach an extreme value of millions. They must also utilize the high density nature of the sensor networks. The density can range from a few sensors to a few hundred sensors in a region.

15.2.4.10 Hardware Constraints

A sensor node is made up of four basic components—a sensing unit, a processing unit, a transceiver unit, and a power unit. They may also have additional application-dependent components such as a location-finding system, a power generator, and a mobilizer. Sensing units are usually composed of two subunits: sensors and analog to digital converters (ADCs). The analog signals produced by the sensors are based on the observed phenomenon, which are converted to digital signals by the ADC, and then fed into the processing unit.

Thus, it is common that a sensor node has a location-finding system. The nodes must

- Consume extremely low power
- Operate in high volumetric densities
- Have low production cost and be dispensable
- Be autonomous and operate unattended
- Be adaptive to the environment

The processing unit, which is generally associated with a small storage unit, manages the procedures that make the sensor node collaborate with the other nodes to carry out the assigned sensing tasks. A transceiver unit connects the node to the network. One of the most important components of a sensor node is the power unit. Power units may be supported by a power scavenging unit such as solar cells. There are also other subunits, which are application-dependent. Most of the sensor network routing techniques and sensing tasks require the knowledge of location with high accuracy.

15.3 Protocol Overview

This section gives a detailed knowledge about some of the well-known clustering algorithms in the field of WSN. All the algorithms considered here are energy efficient. Most of the clustering algorithms proposed in recent times are ancestor of these protocols. The model used in the low-energy adaptive clustering hierarchy (LEACH) protocol is the most efficient method used for energy consumption calculation.

15.3.1 LEACH

LEACH [1] protocol distributes the energy load evenly among the sensors present in the network by changing the role of cluster heads periodically based on certain probability. In some clustering algorithms, the CHs were fixed for the cluster and thus there may be a chance that a sensor with low battery power will become a CH and will die soon. LEACH provides a solution for this kind of problem by electing a CH from time to time so that the battery power of a particular node will not be wasted. This algorithm uses data fusion at the CH so that it can send the compressed data to the BS to minimize the utilization of battery power. Sensors elect themselves to be local CHs at any given time with a certain probability p. Each sensor node determines to which cluster it wants to belong to by choosing the CH that requires the minimum communication energy to reach. The role of being a CH is rotated periodically among the nodes of the cluster to balance the load. To determine the number of clusters, it depends on several parameters such as network topology and the relative costs of computation versus communication. The LEACH algorithm is partitioned into four rounds such as advertisement phase, cluster setup phase, schedule creation, and data transmission. In each round, there is a certain probability that a sensor can become a CH.

15.3.1.1 Advertising Phase

The decision to become a CH is made by the node n choosing a random number between 0 and 1. If the number is less than a threshold $T(n)$, the node becomes a CH for the current round.

The threshold is set as:

$$T(n) = \begin{cases} \dfrac{p}{1 - p * \left(r * \bmod \dfrac{1}{p} \right)}, & \text{if } n \in G \\ 0, & \text{otherwise} \end{cases}$$

where p = the desired percentage of CHs, r = the current round, and G is the set of nodes that have not been CHs in the last $\frac{1}{p}$ rounds.

15.3.1.2 Cluster Setup Phase

After the CH becomes selected, all other sensors will join the cluster and send a message to the CH regarding their membership in that cluster by using CSMA MAC protocol.

15.3.1.3 Schedule Creation

After the completion of a setup phase, the CH creates a time division multiple access (TDMA) slot for all the cluster members so that they can transmit within that time frame. The schedule is broadcast to all the sensors present in that cluster.

15.3.1.4 Data Transmission

Once the clusters are created and the TDMA schedule is fixed, data transmission can begin. After receiving all the data from the cluster members, the CH node performs signal processing functions to compress the data into a single signal and transmits the composite signal to the BS. In LEACH, each sensor can directly transmit to the CH, which means that it uses one-hop technology for transmission.

At the completion of the CH selection process, every node that was selected to become a CH advertises its new role to the rest of the network. Upon receiving the CH advertisements, each remaining node selects a cluster to join. The selection criteria may be based on the received signal strength, among other factors. The nodes then inform their selected CH of their desire to become a member of the cluster. Upon cluster formation, each CH creates and distributes the TDMA schedule, which specifies the time slots allocated for each member of the cluster. Each CH also selects a code division multiple access (CDMA) code, which is then distributed to all members of its cluster. The code is selected carefully so as to reduce intercluster interference. The completion of the setup phase signals the beginning of the steady-state phase. During this phase, nodes collect information and use their allocated slots to transmit the data collected to the CH. This data collection is performed periodically.

15.3.2 Hybrid, Energy-Efficient, Distributed Clustering

Hybrid, energy-efficient, distributed (HEED) clustering [2] is a distributed clustering scheme in which CHs are selected periodically according to a hybridization of the node residual energy and a secondary parameter, that is, intracluster communication cost. HEED selects the CH that has the highest residual energy and requires the minimum distance for communication. Intracluster communication cost is a function of cluster properties, that is, cluster size and whether or not variable power levels are permissible for intracluster communication. If the power level used for intracluster communication is fixed for all nodes, then the cost can be proportional to either

(i) Node degree, if the requirement is to distribute load among CHs, or

(ii) $\frac{1}{\text{node degree}}$, if the requirement is to create dense clusters

The average of the minimum power levels required by all M nodes within the cluster range to reach the CH is,

$$\text{AMP} = \frac{\sum_{i=1}^{M} \min p_i}{M}$$

where min p_i denotes the minimum power level required by a node v_i, $(1 < i < M)$, and M is the number of nodes within the cluster range.

15.3.2.1 Initialization Phase

In HEED, clustering is triggered in every $T_{CP} + T_{NO}$ second to select new CHs. Where T_{CP} is time required to create a cluster and T_{NO} is the time interval between the end of a T_{CP} and start of a subsequent T_{CP}. In each iteration before the start of execution, each node sets its probability of becoming a CH, CH_{prob} as,

$$\text{CH}_{\text{prob}} = C_{\text{prob}} * \frac{E_{\text{residual}}}{E_{\text{max}}}$$

where

C_{prob} = initial percentage of CHs among all n nodes
E_{residual} = estimated current residual energy in the node
E_{max} = maximum energy

The HEED algorithm gets terminated in $O(1)$ iterations.

15.3.2.2 Repetition Phase

In repetition phase, every sensor goes through several iterations until it finds the CH that will use the least transmission power (cost). If it hears from no other CH, the sensor elects itself as a CH and sends an announcement message to its neighbors informing them about the change of status. Finally, each sensor doubles its CH_{prob} value and goes to the next iteration of this phase. It stops executing this phase when its CH_{prob} reaches 1.

15.3.2.3 Finalization Phase

At last, each sensor makes a final decision on its status. A node can either elect to become a CH according to its CH_{prob}, or join a cluster according to overheard CH messages within its cluster range.

HEED has a worst-case processing time complexity of $O(n)$ per node, where n is the number of nodes in the network. Also, it has a worst-case message exchange complexity of $O(1)$ per node, that is, $O(n)$ in the network. The probability of becoming CH for two nodes within each others' cluster range is very minimal. The HEED protocol, which is terminated after a constant number of iterations, is independent of network diameter.

15.3.3 Distributed Weight–Based Energy-Efficient Hierarchical Clustering

Ding et al. [10] proposed a distributed weight-based energy-efficient hierarchical clustering (DWEHC) method to achieve better cluster size balance and optimizing clusters such that the minimum energy topology would be maintained. DWEHC makes no assumptions on the size and density of the network. This algorithm is implemented by each node individually. The nodes that use DWEHC follow a hierarchical structure for clustering. The number of levels in the hierarchy depends on the cluster range and the minimum energy required to reach the CH. Within a cluster, TDMA is used for transmission, that is, within a particular time frame, one sensor can send the data to the CH.

The weight is calculated by each sensor after locating the neighbours in its sensing area. The weight is a function of the sensors' reserve energy and proximity to its neighbors. The node having highest weight is elected as CH and the remaining nodes become cluster members. Each node in the network is either a CH or a child (first level, second level, etc.). DWEHC follows the steps mentioned below to complete the algorithm

- *Relay*: Here, the authors only concentrated on path loss due to the dependability of two sensors using distance by assuming that all the sensors have similar antenna heights.
- *Relay region*: Let S be the sender node and R be the relay node, the nodes in the relay region can be reached with the least energy by relaying through R.
- *Enclosure region*: The enclosure region is the complement of the relay region.
- *Neighbors*: These are the nodes that do not need relaying when a node s transmits to the others that can receive directly.
- *Cluster range*: The radius of the cluster, that is, the highest distance between a node and the CH inside the cluster.
- *Weight used in CH election*: Weight is calculated based on parameters such as the distance between the node and the receiver, the residual energy of the node, and the initial energy of that node.
- *Levels in a cluster*: Here, each cluster is multilevel, so the number of levels in a cluster depends on the cluster range and the minimum energy path to the CH.

DWEHC is fully distributed over the whole network in which every node is covered by only one CH. The CHs are distributed in such a way that when two nodes are within each others' cluster range, the probability of both of them becoming CHs is very small. The complexity of broadcast message exchange is $O(1)$ for each node.

15.3.4 Energy Dissipation Forecast and Clustering Management

Proposed an energy dissipation forecast and clustering management (EDFCM) protocol for heterogeneous networks to provide a longer lifetime and more reliable transmission service. Different from other energy-efficient protocols that consider residual energy and energy consumption rate in the nodes, the process of CH selection in EDFCM is based on a method of one-step energy consumption forecast. Besides, the management nodes play a cooperative role in the selection of CHs to make sure that the number of CHs per round is optimum. The algorithm tries to balance energy consumption round by round, which will provide the longest stable period for the networks.

In actual heterogeneous application scenes, the node functioning as a CH has more residual energy than the others in a previous round, even though it may die or consume much more energy in the operation of the next round due to computational heterogeneity. For further considerations, because the nodes are deployed uniformly in application scenes and the number of non-CH nodes per round in a cluster is almost the same, the energy dissipation of a CH will only be relative to the locations of nodes in a cluster. We can think that the energy dissipations in those sequent rounds are correlative. EDFCM uses the average energy consumptions of two types of CHs in previous rounds as forecast values for their energy consumptions in the next round. The more residual energy in a node after the operation of the next round, the higher probability the node will be selected as a CH. Contributions of EDFCM provide the longest stability period (when the first node is dead) and improve the scheme of clustering management in LEACH and LEACH-based algorithms. EDFCM yields a longer stability period and much more effective messages transmitted to the BS, compared with other typical clustering protocols, and the number of clusters per round in EDFCM is stable.

15.3.5 Energy-Efficient Unequal Clustering

Li et al. [20] proposed energy-efficient unequal clustering (EEUC), an energy-efficient clustering protocol for periodic data-gathering application in WSNs. Here, the authors tried to remove the hotspot problem that arises in multihop routing. The hotspot problem arises when the CHs closer to the data sink dies due to the burden of heavy relay traffic. The CHs nearer to the BS are heavily loaded with network traffic and loses energy quickly as compared with the CHs farther from the BS. To solve this kind of problem, the authors proposed such a clever algorithm so that the clusters closer to the BS are expected to have smaller cluster sizes, thus they will consume lower energy during the intracluster communication, and can preserve some more energy for the intercluster relay traffic. After the network deployment, the BS broadcasts a "hello" message to all the nodes present in the network with certain power levels. Then, all the nodes calculate the approximate distance from the BS, which then helps the algorithm in making clusters of unequal size.

Figure 15.5 illustrates an overview of the algorithm in which circles of different sizes denote clusters of unequal size with respect to the distance of the nodes from the BS. The responsibility of being a CH is rotated among sensors in each data-gathering round to distribute the energy

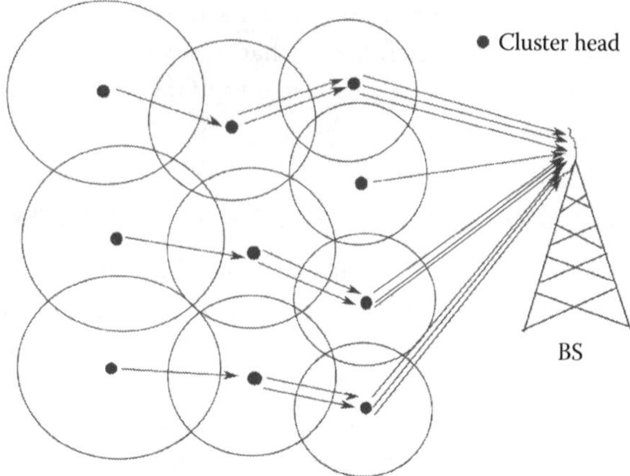

Figure 15.5 Energy-efficient unequal clustering.

consumption across the network. The figure shows that the size of the cluster decreased when the distance between the CH and BS decreases. This algorithm is based on distributed CHs, in which CH selection is primarily based on the residual energy of each node. Throughput shows that unequal clustering improves network lifetime and balances the energy consumption in the network over LEACH and HEED.

15.4 Analysis of the Existing Algorithms

In this section, the advantages and the disadvantages of the existing clustering algorithms are figured out. The scope of improving the performance of the clustering algorithms is also outlined. The attributes used for classification of the algorithms are described. A comparative analysis of different characteristics is also summarized.

15.4.1 Scope and Functionality

The main purpose of clustering for sensor networks is to support the development, maintenance, deployment, and execution of sensing-based applications. This includes mechanisms for formulating complex high-level sensing tasks, communicating this task to the BS, coordination between sensor nodes to split the task and distribute it to the individual sensor nodes, and data fusion for merging the sensor readings of the individual sensor nodes into a high-level result. Moreover, appropriate abstractions and mechanisms for dealing with the heterogeneity of sensor nodes should be provided. All mechanisms provided by a middleware system should respect the design principles sketched above and the special characteristics of WSN, which mostly boils down to energy efficiency, robustness, and scalability. The scope of clustering for WSN is not restricted to the sensor network alone but also covers devices and networks connected to the WSN. Classic mechanisms and infrastructures are typically not well-suited for interaction with WSN. One reason for this is the limited resources of a WSN, which may make it necessary to execute resource-intensive functions or store large amounts of data in external components. This may result in a close interaction of processes executing in the WSN and a traditional network. One example of such "external" functionality is so-called virtual counterparts, outside components that augment real-world objects with information processing capabilities. Thus, clustering for sensor networks should provide a holistic view on both clustered WSN and traditional networks, which is a challenge for architectural design and implementation. Traditional networks are designed to accommodate a wide variety of applications without necessarily needing application knowledge. Clustering for WSN, however, has to provide mechanisms for injecting application knowledge into the infrastructure and the WSN.

Data-centric communication mandates a communication paradigm that more closely resembles content-based messaging systems than traditional remote procedure call (RPC)–style communication. Moreover, event-based communication matches the characteristics of WSN much better than traditional request–reply schemes. In general, communication-specific and application-specific data processing is much more integrated in WSN rather than in traditional systems. The design principle adaptive fidelity algorithms requires the infrastructure to provide appropriate mechanisms for selecting parameters or whole algorithms, which solve a certain problem with the best quality under given resource constraints. In traditional systems, each computing device belongs to someone who is responsible for configuration, maintenance, and error handling. In contrast, WSN nodes must operate unattended, which means that clustering for WSN has to provide new levels of support for automatic configuration and error handling. Because WSN process real-world data, the concepts of physical time and location play a much more important role than in traditional computing systems. Time and location of sensed real-world events are key elements for fusing individual sensor readings to obtain a high-level sensing result. Some application areas might even pose real-time requirements on WSN. Therefore, support for time and location management should by tightly integrated into a clustered infrastructure for WSNs.

Here, it is summarized by the set of attributes that can be used to categorize and differentiate clustering algorithms of WSNs (Table 15.1).

15.4.2 Cluster Properties

Quite often, clustering schemes strive to achieve some characteristics for the generated clusters. Such characteristics can be related to the internal structure of the cluster or how it relates to others. The following are the relevant attributes:

- *Cluster count*: CHs are predetermined in some of the published approaches, thus the number of clusters is preset. CH selection algorithms generally pick CHs randomly from the deployed sensors, hence, yielding a variable number of clusters.
- *Intracluster topology*: Some clustering schemes are based on direct communication between a sensor and its designated CH, but sometimes multihop sensor-to-CH connectivity is required.
- *Intercluster connectivity*: The CHs of different clusters can communicate directly with single-hop communication, but sometimes multihop CH-to-CH connectivity is required.
- *Stability*: CHs send aggregated data to the BS directly or indirectly with help of other CH nodes. This means that there exists a direct link or a multihop link.

Table 15.1 Analysis of the Existing Algorithms

Algorithm	*Pros*	*Cons*
LEACH	• Load balanced clusters • Uses random probability for each sensors • Low message overhead • Stability increased by balancing	• Doesn't check the residual energy of the network • Uses centralized approach
HEED	• Balanced clusters • Low message overhead • Uniform and nonuniform node distribution • Intercluster communication explained • Outperforms generic clustering protocols on various factors	• Repeated iterations complexes algorithm • Decrease of residual energy forces to iterate the algorithm • Nodes with high residual energy, one region of a network
DWEHC	• Hierarchical clusters • Balanced cluster size • Intercluster communication using TDMA	• Calculating weight is difficult • Number of iterations the algorithm uses • Algorithm is implemented by each node
EDFCM	• Provides longer lifetime • CHs per round are optimum • Stability increased by balancing • Energy consumption round by round	• Uses energy consumption statistics of the previous round • Requires more memory to store the previous data
EEUC	• Removes the hotspot problem • CH chooses a relay node from its adjacent nodes • Uses distributed CHs • Increases network lifetime	• Location of the CH is precomputed • Each node calculates their distance from the BS

15.4.3 CH Capabilities

The following attributes of the CH node are differentiating factors among clustering schemes:

- *Mobility*: CH may be stationary or mobile. In most cases, they are stationary, but sometimes CHs can move within a limited region to reposition themselves for better network performance.
- *Node types*: Generally, sensor nodes among the deployed sensors are designated as CHs, but sometimes sensor nodes equipped with significantly more computation and communication resources are selected as CHs.
- *Role*: Some of the main roles of the CHs are simply relaying the traffic, aggregation, or fusion of the sensed data.

15.4.4 Clustering Process

- *Methodology*: After some of the rounds are completed, the CH selection should be based on the energy remaining in the sensors.
- *Objective of node grouping*: This is another important parameter that considers the energy consumption to classify the algorithms. The average energy is used as the reference energy for each node. It is the ideal energy that each node should own in the current round to keep the network alive.
- *CH selection*: The CH may be selected by some random probability or by some specific formula.
- *Algorithmic complexity*: This is the computational time requirement to complete the clustering process.

Note that the transmission distance over which the sensors send their data to their CH is smaller compared with their respective distances to the global BS. Because a network is characterized by its limited wireless channel bandwidth, it would be beneficial if the amount of data transmitted to the BS can be reduced. To achieve this goal, a local collaboration between the sensors in a cluster is required to reduce bandwidth demands. The sensors within a cluster can be scheduled using TDMA or CDMA to avoid collisions in a cluster. At the beginning of a sensing task, a sensor transmits its sensed data when its value is higher than the hard threshold specified by its CH (Table 15.2).

15.5 Proposed Protocol

This section describes an improved clustering protocol proposed by the authors. This protocol gives better performance as compared with some existing algorithms. The prerequisites of the algorithm are also described so that a better understanding of the knowledge can be grasped before reading the actual protocol. Also, the motivation for algorithm and the network model is briefly described in this section.

15.5.1 Staggered Clustering Protocol: An Efficient Clustering Approach for WSN

Clustered protocols have the advantage of minimizing long-distance communication with the BS through the optimized utilization of CH nodes and consequently reducing the energy consumption of the network. Data aggregation is performed at the CHs, which can be simplified by decreasing the number of redundant packets via intracluster communication. Due to the limitation of sensor node energy, emphasis is given to energy-efficient routing protocols to prolong the lifetime of sensor networks. Efficient CH election algorithms are highly desired to balance the distribution of energy load and avoid the degradation of network longevity due to the premature battery drain of any node. Although the energy consumption varies greatly between nodes in different roles, the roles must be rotated periodically. In view of energy consumption in a WSN, data transmission is the most important with respect to others. Within a clustering organization, intracluster communication can be single-hop or multihop, as well as intercluster communication. Researchers have shown that multihop communication between a data source and a BS is usually more energy-efficient than direct transmission because of the characteristics of a wireless channel. Useful energy consumption

Table 15.2 Classification of Algorithms Based on Clustering Attributes

	Cluster Properties				*CH Capabilities*			*Clustering Process*			
Clustering Algorithms	*Cluster Count*	*Intracluster Topology*	*Intercluster Connectivity*	*Stability*	*Mobility*	*Node Type*	*Role*	*Methodology*	*Objective of Node Grouping*	*CH Selection*	*Algorithmic Complexity*
LEACH	Variable	Fixed (one-hop)	Direct link	Provisioned	Stationary	Sensor	Relaying	Distributed	Save energy	Random	Constant
HEED	Variable	Fixed (one-hop)	Direct link/ multihop	Assumed	Stationary	Sensor	Aggregation and relaying	Distributed	Save energy	Random	Constant
DWEHC	Variable	Adaptive (Multilevel)	Direct link	Provisioned	Stationary	Sensor	Aggregation and relaying	Distributed	Save energy	Random	Constant
EDFCM	Variable	Adaptive	Direct link	Provisioned	Stationary	Sensor	Aggregation and relaying	Distributed	Scalability and fault tolerance	Random	Variable
EEUC	Variable	Fixed (one-hop)	Direct link	Provisioned	Relocatable	Sensor	Aggregation and relaying	Distributed	Save energy	Random	Constant

can be due to the following items: transmitting/receiving data, processing query requests, and forwarding queries/data to neighboring nodes Sensor networks are application-specific. A network is usually designed and deployed for a specific application. The design requirements of a network change with its application. The number of clusters present in the network is one of the key parameters that determines the lifetime of the sensor network. If the number of clusters is much smaller, then non-CH nodes are likely to spend too much energy transmitting data to their CHs because most of the clusters will be of a large size. In an article by Heinzelman et al. [1], a LEACH protocol was proposed in which each member in a cluster will take the role of CH in a random, rotating way. LEACH requires knowledge of the location information of each node. In a study by Akyildiz et al. [3], a HEED clustering approach for ad hoc sensor networks has been proposed as a topology control algorithm. HEED assumes that wireless sensor nodes equip radiointerfaces with multiple power levels. Therefore, distance information can be inferred from the power level in HEED and no explicit localization information is needed. The major contribution of that work is that the connectivity of clustered networks can be asymptotically guaranteed with appropriate bounds on node density and intracluster and intercluster transmission ranges.

Therefore, to minimize energy consumption in the network and to maximize the lifetime of the network, this article proposes a staggered clustering protocol (SCP), which solves the problems cited above. In addition, two clustering metrics are defined to select the best set of CHs, namely, the power level of each node and the total communication cost in the network.

15.5.2 Motivations

Multidimensional data usually have intrinsic relationships to the physical domains where they are produced, and the changes in data values are closely related to the changes occurring in its corresponding physical world. For example, the remaining energy level of a sensor is a function of several key system and network parameters such as location, distance to the current CH, and the current remaining energy, as well as to the changes in its surrounding environment such as channel quality, the number of neighboring nodes, topology management, etc. In this sense, the remaining energy data is multidimensional data, which can be thought of as a point on a nonlinear manifold in a large observation space formed by the set of physical parameters (dimensions) of that sensor, in which a manifold is defined as a locally Euclidean topological space but one that is not globally so.

15.5.3 Network and System Models

In this work, we consider a WSN similar to the network model used by Akyildiz et al. [3], with the following properties:

- All the sensor nodes are uniformly distributed in a field and they are stationary.
- There is only one network sink existing in the sensor network. The BS does not have an energy constraint. For example, the sink can be connected to AC power or some sort of energy harvesting system such as solar panels.
- Initially, each sensor node has the same energy capacity, but the energy consumption of each sensor node is different. The batteries cannot be changed after the sensors are deployed.
- There is no coverage hole existing in the sensor network and all the sensors can communicate with the network sink through routing protocols.
- Only sensing activities are considered in the network, meaning that traffic only flows unidirectionally, from the sensors to the sink.

Once the clustering process is completed, each CH will update its records on the cluster members based on the traffic received. In practice, once a sensor platform is chosen, the energy consumption of transmitting, receiving, and sleeping will be defined. In each round, the CH will report the remaining energy of each sensor in its cluster to the BS. Then, the BS will run data-mining algorithm to infer the node location and choose a node to be the CH for the next round.

15.5.4 Clustering Parameters

In a clustered network, the cost is divided into intracluster and intercluster costs. The intracluster communication cost is from the nodes inside a cluster to the CH. SCP attempts to maintain the constraint of well-balanced energy consumption in the network. The nodes that have more residual energy at the beginning of each round have more chance to become a CH. This protocol mainly concentrates on the average energy level of each node, which can be calculated by dividing the energy level of each node with the energy level of the neighbor nodes.

Lemma 1

Network density: Suppose nodes are deployed in an area A, $\rho(x, y)$ is defined as the network density, with the property:

$$\iint_A \rho(x, y)\, dxdy = 1$$

■

Lemma 2

Network energy intensity: Network energy intensity $E(x, y)$ is defined as the energy distribution of the network. Suppose the nodes have the same initial energy E_O, then given a region D,

$$E(x, y) = E_O N \iint_D \rho(x, y)\, dxdy$$

■

This algorithm uses normal power calculation (NPC) to evaluate the power level of each node$_i$.

$$\mathrm{NPC}_i = \frac{\sum_{j \in \mathrm{nbr}_i} E_j^{\mathrm{current}}}{|\mathrm{nbr}_i| \cdot E_i^{\mathrm{current}}}$$

nbr_i is the set of neighbors of node$_i$, which are located in the detective range of node$_i$, and $|\mathrm{nbr}_i|$ is the total number of nodes in the neighbor list. E_i^{current} is the current residual energy of node$_i$.

NPC reflects the power distinction between $node_i$ and its neighbors. If NPC_i is more than zero, $node_i$ is a high-energy node, which means that $node_i$ has more energy than its neighbors; if NPC_i is less than zero, $node_i$ is a low-energy node, and should have less opportunity to be the CH node.

In heterogeneous sensor networks, both the average power distinction and communication cost for CH selection should be considered,

$$T_{\text{cost}} = \text{NPC}_i \cdot \frac{\sum_{j \in \text{nbr}_i} d_{i,j}^2}{|\text{nbr}_i|}$$

$D_{i,j}$ is the distance between $node_i$ and $node_j$, and this value should be computed by receiving sensitivity. T_{cost} provides a unified criterion for all nodes to select CH nodes, which means that all nodes could use T_{cost} to select CH nodes, which are the nodes with high energy and low communication costs.

The radio model utilized in SCP is similar to that of LEACH. The energy consumed by the radio in transmitting L bits of data over a distance d is given by the following:

$$E_{\text{Tx}}(L,d) = \begin{cases} L \times (E_{\text{elec}} + \in_{\text{fs}} \times d^2), & \text{if } d \le d_o, \\ L \times (E_{\text{elec}} + \in_{\text{mp}} \times d^4), & \text{if } d \ge d_o \end{cases}$$

where E_{elec} is the energy dissipated per bit to run the transmitter or the receiver circuit. The parameters $\in_{\text{fs}}$ and $\in_{\text{mp}}$ depend on the transmitter amplifier model we use.

15.5.5 Staggered Clustering Protocol

The optimal probability of a node to become a CH is a function of spatial density when nodes are uniformly distributed over the network. This clustering is optimal in the sense that energy consumption is well distributed among all sensor nodes and the total energy consumption is minimal. The set of all CH nodes is denoted by CH, in which CH ∈ N, and N is the set of all nodes including CH nodes and non-CH nodes. Let us assume E_0 is the initial energy of each normal node, m fraction of advanced nodes among normal nodes are equipped with α times more energy than the normal nodes.

In the clustering process, the distance of each node to every other node is calculated to detect the neighbors. After neighbor detection is done, each node broadcasts its current energy information to all other nodes, that is, to the neighbors. Then, after getting the energy information from the neighbors, each node calculates the distance by comparing the signal strength they have received. When nodes have sufficient information about its neighbors, such as distance and current energy, nodes calculate T_{cost} about itself and broadcast its T_{cost} to its neighbors. According to the T_{cost}, each node selects the candidate node which has the minimal T_{cost}, and sends elect_msg to the candidate node. The nodes that receive the most elect_msg from its neighbors will announce that the CH nodes are elected, and all non-CH nodes chose the one nearest to the CH to join the cluster.

Algorithm

1. for $r = 1$ to $r_{\max}$
2. for $i = 1$ to n

3. identify neighbors of each $node_i$
4. $node_i$ broadcast (energy_msg)
5. calculate NPC and T_{cost}
6. if $node_i.E \leq 0$
7. dead = dead + 1
8. end if
9. $node_i$.broadcast (T_{cost})
10. $node_i$.send (elect_msg) to all neighbors
11. if $node_i$.receive (elect_msg) from $node_j$
12. then $node_j$.ticket =$node_j$.ticket + 1
13. end if
14. if Max($node_j$.ticket_msg)
15. then CH←$node_j$
16. end if
17. if $\exists node_i$ is not CH, associate with the nearest CH
18. end for
19. end for

15.5.6 Simulation and Result Discussion

We simulate a WSN of 100 nodes in a 200 × 200 square area using MATLAB®, and the sink node is located in the center of the area. We assume that m is the percentage of the nodes that are equipped with a times more energy than the normal nodes. The initial energy of normal node is 0.1 J; therefore, the initial energy of the advanced node is 0.5 (a + 1) J (Table 15.3).

We compare the performance of SCP protocol with LEACH and distributed energy efficient clustering algorithm (DEEC) in the same heterogeneous setting, where m = 0.2 and a = 4.

Figure 15.6 shows that SCP extremely extends the stable region by 152.16% compared with LEACH and by 69.17% compared with DEEC. On the other hand, SCP increases the ratio of

Table 15.3 Simulation Parameters

Parameter	*Value*
Area	200 × 200 m^2
Nodes	200
BS	(100, 100)
Initial energy	2 J
E_{elec}	50 nJ/bit
ϵ_{fs}	10 pJ/bit/m^2
ϵ_{mp}	0.0013 pJ/bit/m^4
E_{fusion}	5 nJ/bit/message
Packet size	2000 bits
Broadcast message	50 bytes

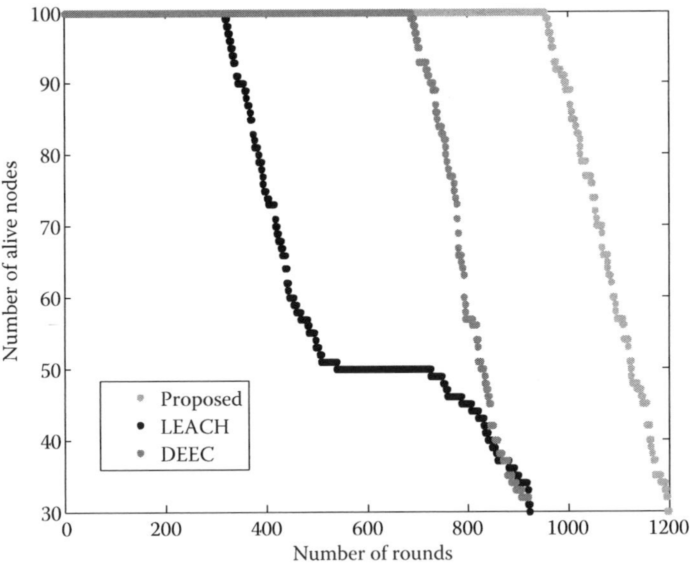

Figure 15.6 Lifetime of the network.

the stable region in network lifetime. SCP selects high-energy nodes to be CHs for load balancing, and low-energy nodes spend less energy than high-energy nodes. Therefore, SCP avoids the premature death of low-energy nodes and prolongs the stable region of the WSNs.

Figure 15.7 shows that the residual energy of SCP is higher than LEACH and DEEC. Residual energy is the ratio of actual energy and the amount of energy left with the node. The node that

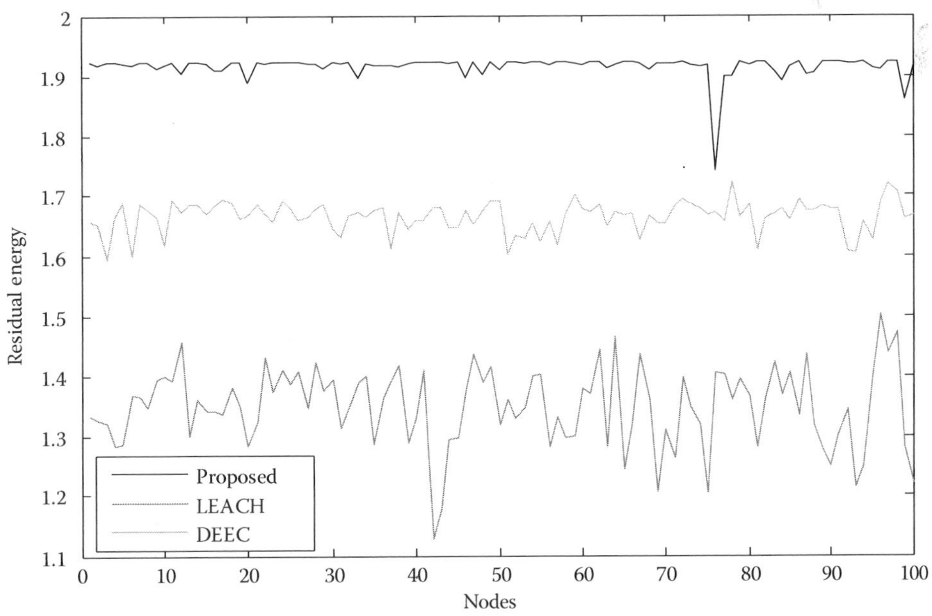

Figure 15.7 Residual energy of the network.

has more residual energy will have a better chance to become the CH. LEACH uses random probability to select CH nodes, so residual energy has no role in the selection of CHs. However, in DEEC, the advanced nodes have more opportunities to be the CH, but the arithmetic still use random mechanisms to select CH, so the residual energy parameter plays a small role in the selection of CHs. SCP has the best performance in residual energy because SCPs do not use a random mechanism for CH selecting, thus SCP could accurately select the high-energy node with low communication cost to be the CH, and implement load balancing. We also analyze the performance of SCP with different m and a, the experimental results show that the stable region of SCP is far greater than that of LEACH and DEEC, even in the homogeneous networks ($a = 0$).

15.6 Application of WSN in Watering

As water supplies become scarce and polluted, there is an urgent need to irrigate more efficiently to optimize water use. The system is fully adaptive not only to environmental conditions but also to the specific water needs that different plants may have.

15.6.1 System Description

The system architecture includes sensor motes, soil humidity sensors, mote-driven electrovalves that control the water flow toward the plants, and a Java application running on a PC, which collects data from the sensor network and stores them in a MySQL database. For the purposes of this study, we have used a garden containing different plants with highly diverse watering needs; a geranium that has very limited watering needs (once a week), a lavender that, under normal weather conditions, has medium watering needs (three times a week), and a mint that requires regular watering (in high temperatures during summertime, even twice a day). The soil humidity of each area is monitored by a mote equipped with a soil humidity sensor [22]. The watering of each area is controlled by corresponding mote-driven electrovalves. When the soil of the garden is too dry, then the corresponding mote, monitoring soil humidity, informs the mote that drives the corresponding electrovalve to start watering the area. When the soil humidity returns to normal levels, the soil monitoring mote signals the electrovalve mote to cut off water supply for that area [23].

Throughout the operation of the system, the levels of soil humidity in each area are forwarded to the BS by the corresponding motes. The sink is a mote connected to a PC on a USB port that acts as a gateway for the rest of the motes. When it receives a soil humidity measurement, it forwards this to the PC where a Java application receives data and stores them in a MySQL database for postprocessing. Apart from soil humidity, measurements concerning temperature are also forwarded to the BS and logged at the database.

15.6.2 Hardware Description

15.6.2.1 Sensor Motes

For this implementation, TelosB motes can be used. The TelosB motes are ZigBee-compliant, small, lightweight, and when using energy-saving protocols, can be powered with two AA batteries for several weeks or even months. These characteristics make them ideal for our smart garden watering system as they can easily be deployed everywhere while being independent of power installations.

Figure 15.8 The EC-5 soil sensor shouldered on a TelosB mote.

15.6.2.2 Soil Sensors

The EC-5 soil humidity sensor was used for soil monitoring (Figure 15.8). It consists of a cable, which on one end has two prongs and on the other end has three wires. The prongs are pushed inside the potting soil and the three wires of the other end are connected to the 10-pin expansion connector of TelosB motes. The bare wire is connected to the ground pin, the red one is connected to the ADC channel pin (programmed as input), and the white wire is connected to the VCC pin.

15.6.2.3 Electrovalves

To control the irrigation process, solenoid valves can be used. A solenoid valve is an electromechanical valve that is controlled by an electric current. The electric current runs through a solenoid, which is a wired coil wrapped around a metallic core. The solenoid creates a controlled magnetic field when electrical current is passed through it. This magnetic field affects the state of the solenoid valve, causing the valve to open or close. The electric valves operate with a 9 to 32 V battery.

In future work, solar panels can be used along with rechargeable batteries to make the system self-sustaining in terms of energy consumption. There are plans to incorporate the ability to be managed remotely to the system. This will be done by representing sensor motes as resources that will allow access and control to the system with the use of web services (e.g., via Android smartphones).

15.7 Conclusion and Future Scope

Clustering in WSNs has attracted significant attention over the past few years. Clustering is not a young field in WSN; it has seen a lot of research activity in the recent past. Although a lot of issues have been solved by the existing algorithms, new challenges continue to appear. By surveying the algorithms, it has been observed that the WSN clustering algorithms achieved a great improvement with energy efficiency, scalability, robustness, fast convergence, and adaptability. Many clustering algorithms tried to solve the problem of load balancing by using the energy consumption attribute as a key term at the time of CH selection. In recent times, researchers have been trying to elect the CHs based on residual energy as a key parameter. The reason for focusing on residual energy is the limited availability of battery power for a sensor. To maintain the balance of residual energy between the CHs and the normal nodes, there is a need for reclustering from time to time.

Load balancing is also an important parameter to increase the lifetime of the network. Previously, authors tried to concentrate on location-based clustering, that is, dividing the network with some hierarchical structure, so that at the time of communication, the nodes at higher hierarchy become responsible for transmitting to the BS. Some authors have also concentrated on the stability of the network.

The effect of the network model on the pursued approaches and stating their strength and limitations have been shown in this article. Currently, so many researchers are working in this regard to increase the lifetime of the network using different parameters such as residual energy, perfect load balancing, fault tolerance of the network, and synchronization in the network. This protocol not only takes care of the efficient load balancing but also minimizes the cost of communication energy to increase the lifetime of the network. This protocol does not need any global energy knowledge at the clustering process. As long as the nodes exchange local information, CH nodes could be selected. SCP is scalable to a large number of nodes because it does not require any prior information. It mainly depends on local information sharing such as energy information and communication cost. Simulation results are discussed to describe the effect of CH selection, cluster density, and frequency of re-election. In addition to energy constraints, quality of service metrics such as delay, data loss tolerance, and network lifetime are carefully handled to expose reliability issues for the clustering algorithm.

References

1. W.B. Heinzelman, A.P. Chandrakasan and H. Balakrishnan. An application-specific protocol architecture for wireless microsensor networks. *IEEE Transactions on Wireless Networking*, 1 (4), 660–670 (2002).
2. O. Younis and S. Fahmy. HEED: A hybrid, energy-efficient, distributed clustering approach for ad Hoc sensor networks. *IEEE Transactions on Mobile Computing*, 3 (4), 366–379 (2004).
3. I.F. Akyildiz, W. Su, Y. Sankarasubramaniam and E. Cayirci. Wireless sensor networks: A survey. *Computer Networks*, 38 (4), 393–422 (2002).
4. X. Wang and G. Zhang. DECP: A distributed election clustering protocol for heterogeneous wireless sensor networks. In *Proceedings of the 7th international Conference on Computational Science, Part III: ICCS 2007*. ICCS'07, Beijing, China, pp. 105–108 (2007).
5. S. Chinara and S. Rath. A survey on one-hop clustering algorithms in mobile ad hoc networks. *Journal of Networks System Management, Springer*, 17, 183–207 (2009).
6. L. Qing, Q. Zhu and M. Wang. Design of a distributed energy-efficient clustering algorithm for heterogeneous wireless sensor networks. *Computer Communications*, 29 (12), 2230–2237 (2006).
7. R. Rajagopalan and P. Varshney. Data-aggregation techniques in sensor networks: A survey. *IEEE Communications Surveys and Tutorials*, 8 (4), 48–63, Fourth Quarter (2006).
8. Y.T. Hou, Y. Shi and H.D. Sherali. On energy provisioning and relay node placement for wireless sensor networks. *IEEE Transactions on Wireless Communications*, 4 (5), 2579–2590 (2005).
9. B. Gong, L. Li, S. Wang and X. Zhou. Multihop routing protocol with unequal clustering for wireless sensor networks. In *ISECS International Colloquium on Computing, Communication, Control, and Management, 2008*. CCCM'08. Guangzhou, China, vol. 2, pp. 552–556, August (2008).
10. P. Ding, J. Holliday and A. Celik. Distributed energy efficient hierarchical clustering for wireless sensor networks. In *Proceedings of the IEEE International Conference on Distributed Computing in Sensor Systems (DCOSS05)*, Marina Del Rey, CA, June (2005).
11. T. Kaur and J. Baek. A strategic deployment and cluster-header selection for wireless sensor networks. *IEEE Transactions on Consumer Electronics*, 55 (4), 1890–1897, November (2009).
12. A.A. Abbasi and M. Younis. A survey on clustering algorithms for wireless sensor networks. *Computer Communications*, 30 (14–15), 2826–2841 (2007).

13. B. Elbhiri, R. Saadane and D. Aboutajdine. Stochastic and equitable distributed energy efficient clustering (SEDEEC) for heterogeneous wireless sensor networks. *International Journal of Ad Hoc and Ubiquitous Computing*, 7 (1), 4–11 (2011).
14. G. Xin, W. HuaYang and B. DeGang. EEHCA: Energy efficient hierarchical clustering algorithm for wireless sensor networks. *Information Technology Journal*, 7 (2), 245–252 (2008).
15. V. Katiyar, N. Chand and S. Soni. A survey on clustering algorithms for heterogeneous wireless sensor networks. *International Journal of Advanced Networking and Applications*, 2 (04), 745–754 (2011).
16. E.P. de Freitas, T. Heimfarth and C.E. Pereira. Evaluation of coordination strategies for heterogeneous sensor networks aiming at surveillance applications. In *Proceedings of IEEE Sensors (SENSORS)*, Christchurch, New Zealand, pp. 591–596 (2009).
17. P. Neamatollahi, H. Taheri, M. Naghibzadeh and M. Yaghmaee. A hybrid clustering approach for prolonging lifetime in wireless sensor networks. In *International Symposium on Computer Networks and Distributed Systems (CNDS)*, February 23–24 (2011).
18. D. Kumar, T.C. Aseri and R.B. Patel. EEHC: energy efficient heterogeneous clustered scheme for wireless sensor networks. *Computer Communications*, 32(4), 662–667 (2009).
19. S. Yi, J. Heo, Y. Cho and J. Hong. PEACH: power-efficient and adaptive clustering hierarchy protocol for wireless sensor networks. *Computer Communications*, 30, pp. 2842–2852 October (2007).
20. C. Li, M. Ye, G. Chen and J. Wu. An energy efficient unequal clustering mechanism for wireless sensor networks. In *Proceedings of 2005 IEEE International Conference on Mobile Ad Hoc and Sensor Systems Conference, MASS'05*, pp. 604–611 (2005).
21. M.C.M. Thein and T. Thein. An energy efficient cluster-head selection for wireless sensor networks. In *Proceedings of International Conference on Intelligent Systems, Modeling and Simulation*, Liverpool, UK, pp. 287–291 (2010).
22. C. M. Angelopoulos, S. Nikoletseas and G. Constantinos. A smart system for garden watering using wireless sensor networks. In *14th ACM International Conference on Modeling, Analysis and Simulation of Wireless and Mobile Systems*, Miami, FL (2011).
23. M. Dursun and S. Ozden. A wireless application of drip irrigation automation supported by soil moisture sensors. *Scientific Research and Essays*, 6 (7), 1573–1582 (2011).

MOBILE AND MULTIMEDIA WSN

Chapter 16

GPS-Free Indoor Localization for Mobile Sensor Networks

Yuanfang Chen and Lu Chen
Dalian University of Technology

Lei Shu
Guangdong University of Petrochemical Technology

Contents

16.1 Introduction

Wireless sensor networks (WSNs) have attracted a great deal of research interest during the last few years. Localization is a key aspect of such networks [1–3] because knowledge of a sensor's location is critical to processing information originating from this sensor, actuating responses to the environment, or making inferences regarding an emerging situation, etc. In recent years, mobility has become an important area of research for the localization community [4–6], and some applications for mobile localization have been proposed, for example, location-based human behavior monitoring (belonging to "human dynamics", which is currently a hot topic) [7,8]. The mobility of WSN's deployment was initially regarded as having several challenges which are needed to be overcome, including connectivity, coverage, and energy consumption [9,10]. However, recent studies have been showing mobility in a more favorable light [11,12], and some complicating issues can be solved effectively by mobile entities [13,14]. In addition, mobility enables sensor nodes to target and track moving phenomena such as vehicles.

Nowadays, the most widely used method for mobility localization is based on Global Positioning System (GPS) [15–17]. The accuracy of GPS is within 10m and, admirably, GPS is free to use anywhere on the planet. However, there are several situations in which GPS is not reliably workable because GPS requires line of sight to multiple satellites. In mobile sensor networks that are deployed indoors, downtown, underground, or in off-planet environments, GPS cannot be used. Furthermore, although the GPS receiver is available for mote-scale devices, it is still relatively expensive (because specialized equipment is still needed) and is energy-inefficient. Therefore, GPS is undesirable for many deployments. Some applications use schemes that are proposed for static sensor networks. However, the accuracy of these schemes is not high, the speed of localization is not enough, the communication overhead between sensor nodes is costly, or these schemes cannot meet the requirement of enlarging the sensing region when they are used in dynamic sensor networks with mobile nodes. Furthermore, in some static schemes, the locations of anchors need to be measured (the workload is heavy for large-scale networks).

Thus, a designed localization scheme for mobile sensor networks is necessary. A Monte Carlo Localization (MCL) scheme specifically designed for mobile sensor networks has been proposed [18]. In the MCL, each unknown node collects the locations of its one-hop and two-hop anchor nodes via message exchange, and constructs a new possible location set in each time slot. The possible location set is a collection of coordinates in which the unknown node may be located. The possible locations are also constrained by the communication range of anchor nodes and the moving region of location set in the previous time slot. Therefore, the localization error with low anchor density in the MCL is high. Mobile and Static sensor network Localization (MSL) [19] is another range-free scheme using the Monte Carlo method. MSL improves localization accuracy by using the location estimation of all neighbors for an unknown node (not just anchor nodes). As another improvement, a distributed localization scheme has been proposed based on the Monte Carlo method [20]. It improves the localization error of previous work. The above schemes are time-consuming because they need to keep sampling and filtering until enough samples are obtained to construct a new possible location set in each time slot. Therefore, the communication cost is high and the speed of localization is not fast enough. The Dynamic Triangular (DTN) location method is another type of localization scheme for mobile sensor networks [21]. DTN requires at least three sensor nodes to estimate the location of a mobile node. The DTN discards the worst Received Signal Strength Indicator (RSSI), which is measured by a sensor node and uses the other sensor nodes to estimate a location. The DTN chooses the sensor node that receives the greatest RSSI to take as the master node, and assumes the mobile node's location within the mapping circle of the master node. The radius of the mapping circle is the estimation distance d between the mobile node and the master node. The DTN finds the angle θ on the mapping circle by using a cost function to pick one that best matches the observed distance. However, the DTN needs three sensor nodes within the transmission radius of a mobile node for localization.

Therefore, in this chapter, we propose a GPS-free, real-time, low-cost (for communication), and anchor-free localization scheme called GPS-Free Indoor Localization (GFIL) to solve indoor and high-precision localization problems (Figure 16.1).

First, a mobile node moves according to a special trajectory in the deployment area, and the mobile node dynamically calculates its real-time locations according to its start coordinate (which can be set by manual operation) and its mobile information (e.g., movement speed, elapsed time, and steering angle). Then, the mobile node broadcasts its current coordinates to unknown nodes within its transmission radius, and finally the unknown nodes use these broadcast coordinates to calculate their locations according to a posterior probability formula.

In summary, the contributions of this chapter are shown as follows:

(1) A GPS-free and anchor-free indoor localization scheme is proposed. Recently, the interest for indoor localization has been increasing [22]. Google Maps launched an indoor mapping functionality to figure out how to find ATMs inside a mall, or your flight gate once you are inside an airport. However, Google Maps is based on floor plans that are provided by those businesses (e.g., IKEA, Macy's, and airports), and some businesses or governments do not want to provide their floor plans for security or other reasons. Google also claims 5m to 10m accuracy inside buildings such as airports and shopping malls. Our scheme is

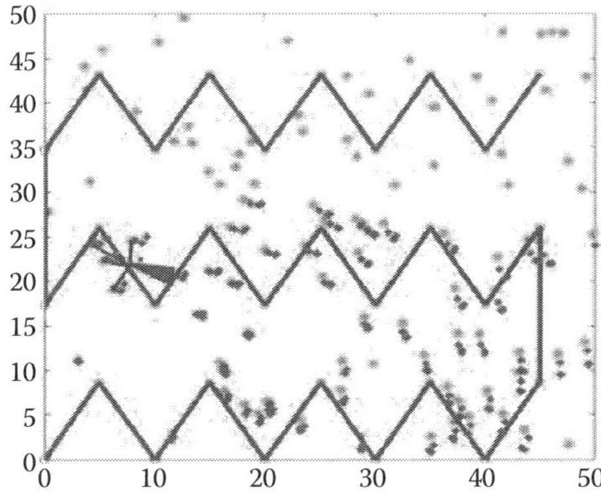

Figure 16.1 The indoor and high-precision localization problem and our scheme. The gray points are unknown nodes, and the black points are the estimated coordinates of these unknown nodes; the black triangle is a mobile node that can connect with the unknown nodes within its transmission radius.

GPS-free, anchor-free, and even floor plan–free. Moreover, our localization scheme is high-precision: from extensive simulations, we can find that centimeter-level localization error can be reached under some parameter configurations.

(2) The GFIL scheme can perform real-time coordinate calculations. This character decreases the communication cost of a network. For example, anchor-based localization schemes (e.g., DV-Hop) need wide-ranging communication to spread the coordinate information of anchors. Compared with these schemes, our scheme only needs currently observed information to dynamically calculate the coordinates of the mobile node and unknown nodes.

(3) A detailed analysis and proof about the effect of mobile node's trajectory on localization are shown in this chapter.

This chapter is organized as follows: Section 16.2 states related work in more detail. Section 16.3 describes our models. The details of our GFIL scheme and correlative discussion are discussed in Section 16.4. Simulation results and analyses are shown in Section 16.5. Finally, Section 16.6 summarizes this chapter and predicts future research work.

16.2 Related Work

Currently, a kind of highly sensitive GPS receiver has been developed based on chip and processing power technologies, which is able to receive satellite signals in most indoor environments and attempts to determine three-dimensional locations indoors [23–25]. However, a proper coverage that is a requirement for satellites to locate a receiver has not been achieved with all current designs of this highly sensitive GPS receiver, and such a GPS receiver is expensive and energy-inefficient.

The localization schemes that can be used in an indoor environment can be divided into two categories: range-based and range-free [26]. The range-based localization schemes depend on calculating the absolute distance or angle between two nodes. Each node can estimate distance by Time Of Arrival (TOA) technology, which obtains distance information between nodes via signal propagation time. GPS is a famous TOA-based public location service system [27]. Time Difference Of Arrival (TDOA) technology for estimating the distance between two communication nodes has been widely used in the localization of WSNs. One example of important work that is based on TDOA technology is AHLos [28,29] (it employs TDOA technology in infrastructure-free sensor networks). Like the TOA technology, the TDOA also relies on extensive and energy-consuming hardware, making it less suitable for low-power sensor nodes. To augment and complement the TDOA and the TOA technologies, an Angle Of Arrival (AOA) technology has been used [30]. It allows nodes to estimate and map relative angles between neighbor nodes. Similar to the TOA and the TDOA, the AOA requires additional hardware, and the sensor node usually equips directional antenna. In an article by Peng and Sichitiu [31], sensor nodes use the RSSI value to infer physical distance. After getting the distance information, the node estimates its location by triangulation or rigidity [32,33]. In RSSI-based localization schemes, either the theoretical or empirical model is used to translate signal strength into distance estimate. However, some problems make the distance estimate inaccurate, for example, for RF-based localization systems [34,35], the problems include multipath fading, background interference, and irregular signal propagation. To mitigate the effect of these problems, two-phase refinement positioning [29,36] and parameter calibration methods [37] have been proposed; these methods take advantage of averaging, smoothing, and alternate hybrid techniques to reduce error to within acceptable limits. RSSI-based schemes have demonstrated effectiveness in simulations and in controlled laboratory environments but "whether the distance can be accurately determined based on signal strength in a real and uncontrollable environment" remains questionable.

On the contrary, range-free schemes estimate the position of a sensor node by network connectivity instead of absolute distance. Each sensor node confirms its connectivity with its neighbors by transmitting packets and estimates its location by gathering information from neighboring nodes. In most range-free localization schemes, each unknown node estimates its position by collecting the locations of anchor nodes, thus these schemes are known as anchor-based localization, and they can be used in an indoor environment. According to processing techniques, anchor-based schemes can be classified into two categories: centroid-based and hop count–based.

First, centroid-based schemes estimate the location of a sensor node by calculating the centroid positions of proximate anchors. This requires a large number of anchors. In an article by He et al. [26], an anchor and centroid-based localization scheme called Approximate Point-In-Triangulation (APIT) is proposed. In this work, each node first tests whether it is inside a particular triangle formed by three anchors (the is called Point-In-Triangulation test (PIT), which are from all audible anchors. Afterward, the APIT repeats the PIT test with different audible anchor combinations until all combinations are exhausted or the required accuracy is achieved. The node's position is estimated to be the center of intersection of all triangles within which the node has been identified. The APIT scheme significantly improves the previous centroid-based range-free schemes. However, it relies on the number of audible anchor nodes for the unknown node and the scheme requires information exchange between the node and its neighbors, thereby increasing the communication overheads of the node.

Grid scan is inspired by the APIT. First, each unknown node obtains the intersection region of its one-hop anchors' transmission area as the estimate region of its location. Second, the region is divided into a grid array (by examining the distance between any of the one-hop and two-hop anchors, the grid can be confirmed [38]) and, if possible, every grid places some sensor nodes. Finally, the estimated location is obtained by averaging the locations of the possible grids.

Second, in hop count–based schemes, the hop count value is increased at every intermediate hop. The classic anchor-based, hop-counting, and range-free scheme is DV-Hop [39]. First, it uses a distance vector routing method to propagate packets and each node maintains a hop counter denoting the minimum number of hops to each anchor; when a node receives a new anchor node information packet, if its hop count is smaller than the stored hop count for that anchor, the recipient updates its hop count value and forwards this packet by increasing the hop count value. Then, the nodes compute "average distance per hop" using the anchors' coordinates and hop count information between anchor nodes. Finally, based on the received locations of anchor nodes and the corresponding average distance per hop, every unknown node calculates the coordinate.

MDS-MAP is a multidimensional scaling (MDS)–based, hop count–based, and range-free localization scheme [40]. It determines the positions of nodes based on which nodes are within the communication range of the others. In addition, if distances between neighboring nodes can be measured, this distance information is also useful in this scheme. The MDS-MAP is able to generate relative node positions' maps when there are no "anchor" nodes that have known absolute coordinates in the network, and when the positions of a sufficient number of anchor nodes are known, for example, three anchors for two-dimensional localization and four anchors for three-dimensional localization, the MDS-MAP can determine the absolute coordinates of all nodes in the network. The MDS-MAP is creditable, when nodes are positioned in relatively uniform space, especially when the number of anchor nodes is low, because the MDS-MAP uses the distance or connectivity information between all nodes at the same time, compared with previous triangulation-based schemes that locate one unknown node at a time and only use the information between the unknown nodes and the anchor nodes. However, like many existing schemes, the MDS-MAP does not work well on irregularly shaped networks in which the distance of shortest path between two nodes does not well correlate with their true Euclidean distance.

A bounding box [28,41] can work well on irregularly shaped networks, and is a kind of computationally simple scheme, when the "unknown nodes' ranges to several anchor nodes" is given. Each node assumes that it lies within the intersection of its anchors' bounding boxes. The bounding box for an anchor node s_a is centered at the anchor position (Xs_a, Ys_a), and has height and width: $2Ds_a$, where the Ds_a is the node's distance measurement to this anchor node. The position of a node is the center of intersection of several anchor bounding boxes.

However, the performance of anchor-based localization schemes depends on the content of received packets; therefore, these presented schemes perform well only when a large percentage of anchors are available, when graph connectivity is high, and when precise range measurements can be determined.

The MCL scheme for mobile sensor networks has been proposed in literature [18]. There are two basic assumptions in the MCL to make the localization problem for mobile sensors simple: (1) the time is divided into several time slots; and (2) the maximum moving distance for every sensor node in each time slot does not exceed v_{max}. In the MCL, each anchor node periodically sends its actual location to two-hop neighbors. If an unknown node receives messages from anchor nodes, its location is restricted by these anchors. After gathering the locations of the anchors, the unknown node starts to estimate its location. The process for estimating the location of an unknown node includes three phases: initialization phase, prediction phase, and filtering phase. During the initialization phase, each unknown node constructs a possible location set including some coordinate points, called samples, used to represent its possible located positions. First, each unknown node in each time slot t establishes a new possible location set L_t. Afterward, in the prediction phase, each unknown node updates its possible location set, restricted by the circle with radius v_{max} and centered on each sample of time slot $t - 1$. Finally, in the filtering phase, the impossible samples located outside of the intersectant region of communication ranging at least approximately two anchor nodes away, will be removed from the new location set L_t. To obtain enough samples, the prediction and filtering phases will be repeated until the number of samples is enough for the set L_t. The estimated location of the unknown node is determined by the average of these samples.

The family of MCL schemes have some common drawbacks. First, in low anchor density, each unknown node gets less anchor information, and the estimated location error becomes large. Second, in high anchor density, each unknown node gets more location restrictions from the anchors, and the possible located region of the unknown node becomes small. However, if the overlapping region in the communication range of the anchors is fairly small, a few samples may be enough to represent the possible located positions of an unknown node, and it is difficult to find many valid samples in a small region. Therefore, too many close samples in a small region cannot increase the localization accuracy, and instead causes unnecessary waste of memory and computation overhead.

MSL [19] is another range-free scheme that uses the Monte Carlo method. In MSL, the authors improve on MCL by using the information from all one-hop and two-hop neighbors of unknown nodes and anchor nodes. This improvement results in faster convergence of the localization error and better estimation of the locations. However, there are two disadvantages for the MSL scheme.

First, MSL has lower location accuracy in a higher mobility environment. Second, MSL spends a lot of communication costs in forwarding location information.

16.3 Our Models

Our network consists of two parts—some static unknown nodes (these unknown nodes can also be mobile nodes, but in this chapter, we fix them for facilitating the study of indoor localization problem) and a mobile node. The unknown nodes are to be located and the mobile node can be worn by a person moving through a building.

16.3.1 Network Model

A number of unknown nodes are installed at fixed and unknown locations randomly. A key role of these static unknown nodes is to receive the data packets coming from the mobile node. The mobile node worn by people has two important roles. First, the number of data packets being transmitted from the mobile node needs to be minimized to save energy. Second, the data packets have to be reliably transferred to the static unknown nodes within a one-hop range of the mobile node. In this way, we guarantee a reliable transmission and a maximum transmission rate between the mobile node and the static unknown nodes with minimal data packet overhead.

Our WSN can be modeled as a directed graph $G = (S, L)$, where the $S = \{s_{\mathrm{m}}, s_1,\ldots, s_n\}$ denotes the set of nodes that include a mobile node s_{m} and n unknown nodes: $s_1, s_2,\ldots, s_n$, and L is the set of transmission links that are directed. A transmission link exists between the mobile node and an unknown node, if and only if the unknown node can be found by the mobile node and can receive data packets from the mobile node. Moreover, the transmission radiuses of all nodes are same and are denoted as r. The mobile node needs to maintain the data that include these fields: start coordinate $(x^s_{\mathrm{m}}, y^s_{\mathrm{m}})$, move speed v_{m}, elapsed time dt_{m}, and steering angle θ_{m}. The transmitted data packet has the following fields: current, real-time coordinate of the mobile node $[x^c_{\mathrm{m}}(t), y^c_{\mathrm{m}}(t)]$, distance between the mobile node and the observed unknown node at time t: $d_{\mathrm{m,u}}(t)$, and the deviation angle between the mobile node and the observed unknown node at time t: $\theta_{\mathrm{m,u}}(t)$.

16.3.2 Observation Model

The observation model of the mobile node is:

$$z_t = \begin{pmatrix} \sqrt{\left[x_{s_i}(t) - x^c_{\mathrm{m}}(t)\right]^2 + \left[y_{s_i}(t) - y^c_{\mathrm{m}}(t)\right]^2} \\ a\tan 2\left[\dfrac{y_{s_i}(t) - y^c_{\mathrm{m}}(t)}{x_{s_i}(t) - x^c_{\mathrm{m}}(t)}\right] - \theta_{\mathrm{m}} \end{pmatrix}. \tag{16.1}$$

Where $\left[x_{s_i}(t), y_{s_i}(t)\right]\left(i = 1,\ldots,N_t\right)$ are the coordinates of observed unknown nodes, which are to be calculated, and the N_t is the number of observed unknown nodes, in the time slot t.

From Formula 16.1, we can find that the observation of the mobile node in time slot t includes two parts: distance and angle ($z_t = \begin{pmatrix} d_{\mathrm{m,u}}(t) \\ \theta_{\mathrm{m,u}}(t) \end{pmatrix}$; this is shown in Figure 16.2).

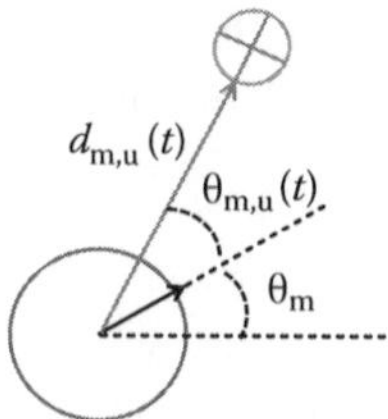

Figure 16.2 Observation model. The gray circle denotes the observed unknown node.

16.3.3 Motion Model

The motion model of the mobile node can be described as follows:

$$u = \begin{pmatrix} v_m dt_m \cos(\theta_m) \\ v_m dt_m \sin(\theta_m) \\ \omega_m dt_m \end{pmatrix}. \tag{16.2}$$

Where the ω_m is the angular speed of the mobile node.

16.4 Our Scheme

In this section, we present our scheme and explain how it increases localization accuracy. Moreover, we provide a detailed analysis and proof about the effect of mobile node's trajectory on localization.

16.4.1 Our Localization Problem Description

Our localization problem can be described in the following way: the mobile node moves from the start point with coordinate (x_m^s, y_m^s) through a sequence of controls, $c_1, c_2, \ldots, c_t$. As it moves, it observes the nearby unknown nodes, for example, in the time slot t, it observes the unknown node s_u, and the observation measurement can be denoted as $z_t = \begin{pmatrix} d_{m,s_u}(t) \\ \theta_{m,s_u}(t) \end{pmatrix}$. Our localization problem is concerned with estimating the coordinates of the unknown nodes from the real-time coordinates of the mobile node and the observation measurements. Between the real-time coordinate estimations of the mobile node is a conditional independence relation.

16.4.2 Our GFIL Scheme

The GFIL scheme consists of three steps, and different from the MCL and MSL, a communication only happens between one-hop nodes, and the communication is real-time, without any communication message forwarded by other intermediate nodes.

Step 1: A mobile node moves along a special trajectory and calculates its current coordinate dynamically. Using the start coordinate (x_m^s, y_m^s), movement speed v_m, elapsed time dt_m, and steering angle θ_m of the mobile node, according to the formulae: $x_m^c = \cos\theta_m \times v_m \times dt_m + x_m^s$ and $y_m^c = \sin\theta_m \times v_m \times dt_m + y_m^s$, we can obtain the current coordinate of the mobile node $[x_m^c(t), y_m^c(t)]$.

Step 2: The mobile node broadcasts its current coordinate to unknown nodes within its transmission radius. The broadcast packet is $([x_m^c(t), y_m^c(t)]; z_t)$.

Step 3: Each unknown node uses the received broadcast packets to calculate or adjust its coordinate using a posterior probability. Because the coordinate estimation of an unknown node is based on the information from the mobile node, the coordinate estimation of an unknown node can be formulized as:

$$p\left(\left[x_{s_k}(t), y_{s_k}(t)\right] \middle| (x_m, y_m)^t, z^t\right) = p\left(z_t \middle| \left[x_{s_k}(t), y_{s_k}(t)\right], \left[x_m^c(t), y_m^c(t)\right]\right)$$
$$p\left(\left[x_{s_k}(t), y_{s_k}(t)\right] \middle| (x_m, y_m)^{t-1}, z^{t-1}\right). \quad (16.3)$$

where the $(x_m, y_m)^t = \left\{[x_m^c(1), y_m^c(1)], [x_m^c(2), y_m^c(2)], [x_m^c(3), y_m^c(3)], \ldots, [x_m^c(t), y_m^c(t)]\right\}$, the $z^t = \{z_1, z_2, z_3, \ldots, z_t\}$, the $k = \{1, \ldots, N\}$, and N is the number of unknown nodes. According to the observation model, motion model, and their distributions, Formula 16.3 can be written in Gaussian form as:

$$p\left(\left[x_{s_k}(t), y_{s_k}(t)\right] \middle| (x_m, y_m)^t, z^t\right) = \exp\left\{-\frac{1}{2}\left[\left(z_t - \hat{z}_t - G_{s_k}\left(\left[(x_{s_k}(t), y_{s_k}(t)\right] - \mu_{t-1}\right)\right]^T\right.$$
$$\left. R_t^{-1}\left[z_t - \hat{z}_t - G_{s_k}\left(\left[x_{s_k}(t), y_{s_k}(t)\right] - \mu_{t-1}\right)\right] - \frac{1}{2}\left(\left[(x_{s_k}(t), y_{s_k}(t)\right] - \mu_{t-1}\right)^T \sum_{t-1}\left(\left[x_{s_k}(t), y_{s_k}(t)\right] - \mu_{t-1}\right)\right\}.$$
$$(16.4)$$

The mean μ_{t-1} and covariance $\sum_{t-1}(.)$ are obtained using the standard Extended Kalman Filter (EKF) measurement update formula [42].

Moreover, an example about the data acquisition process is shown as follows: the mobile node moves from $[x_m^c(1), y_m^c(1)]$, and as it moves, it observes nearby unknown nodes. In the time slot $t = 1$, it observes the unknown node s_1 (the coordinate is $[x_{s1}(1), y_{s1}(1)]$), and the observation measurement is z_1. Afterward, $t = 2$, it observes the other unknown node s_2, and $t = 3$, it observes s_1 again. From the coordinates of mobile node and measurements z^t, we can estimate the coordinates of the observed unknown nodes, and the individual coordinate estimation for the unknown nodes is independent from each other.

16.4.3 Localization Accuracy

Our scheme only uses local connectivity in each time slot. The probability that the mobile node can move a distance d from I to II, without changing connectivity, is equal to the probability that there is at least one unknown node in the black shadow region (shown in Figure 16.3).

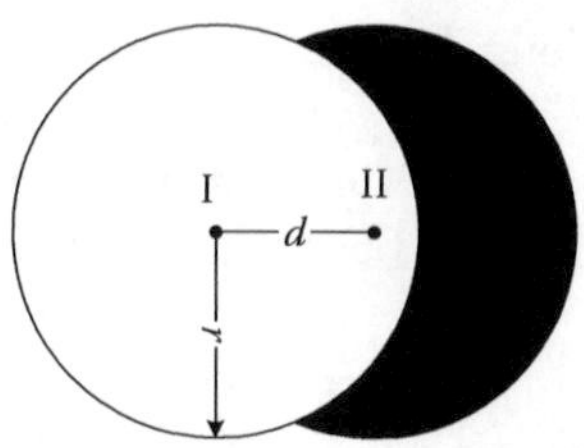

Figure 16.3 Accuracy limit. The black shadow region can affect connectivity between the mobile node and unknown nodes, when the mobile node moves a distance *d* from I to II.

The area of black shadow region is approximately $2rd$. Because our scheme only uses the one-hop mobile node information to achieve localization, the lower bound on the expected location error is $E(d)=\int_0^{\infty} d\cdot d(1-e^{-\rho_n 2rd})=\frac{\pi r}{2n_d}$, where the n_d represents the average number of nodes in a node's communication range. However, the real location error will be much higher than this lower bound.

16.4.4 Move Trajectory Discussion of Our Scheme

We use a mobile sensor node to move according to a special trajectory (we test three common move trajectories [43–45]: grid, equilateral triangle, and randomization, which are shown in Figure 16.4) in the deployment area, to implement the localization of unknown nodes.

Which kind of traveling trajectory of mobile sensor node can optimize the localization? To solve this problem, two essential requirements need to be considered. First, the traveling node can cover an entire WSN during the localization process of unknown nodes. Second, the traveling trajectory length of the mobile node should be as short as possible to guarantee the efficiency of a scheme. Therefore, the geometric problem of the optimal broadcast position about the mobile node is how to use the minimum number of broadcast messages to cover all unknown nodes in the deployment area, and each unknown node receives a broadcast message at least once. The optimal coverage without a hole needs to be guaranteed for the localization of unknown nodes in a deployment area.

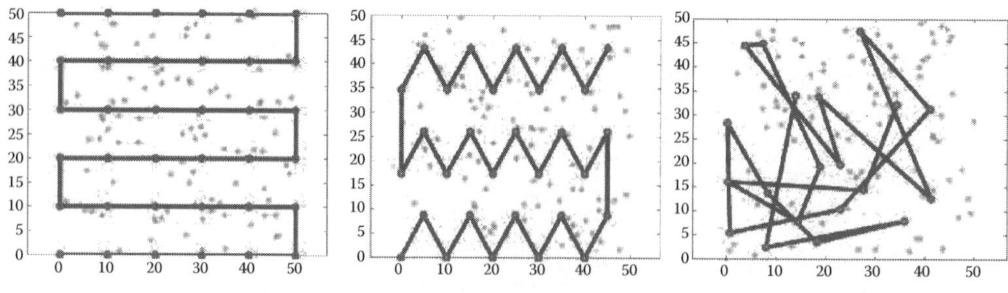

Figure 16.4 Three common move trajectories: grid, equilateral triangle, and randomization.

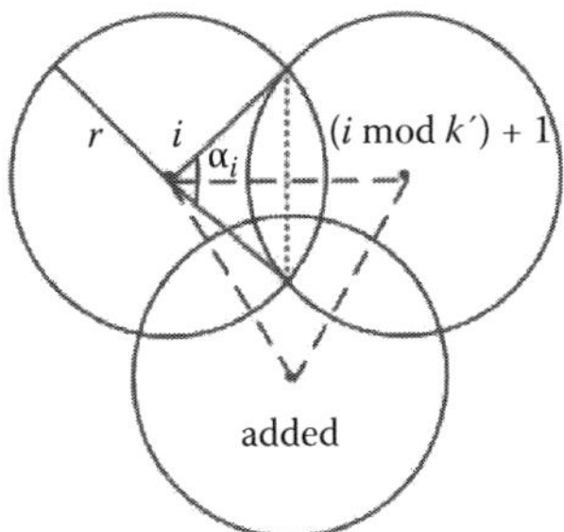

Figure 16.5 An example of how to minimize the overlap of disks *i*, (*i* mod *k*) + 1, and "added" without a hole.

To totally cover the deployment area R, sufficient broadcasts must be sent by the mobile node inside the area R and their coverage areas (we can use a disk to denote the coverage area of a broadcast) intersect with each other. Given that the overlap between the ith disk and $[(i \bmod k') + 1]$th disk (they are adjacent disks and we assume the radius of these disks $r = 1$; Figure 16.5) is $(\alpha_i - \sin\alpha_i)$, $1 \le i \le k'$, where the k' is the number of broadcast messages of the mobile node, received by one unknown node, and we ignore the overlap caused by nonadjacent disks, the total overlap is $L = \sum_{i=1}^{k'} (\alpha_i - \sin\alpha_i)$. Our optimal coverage problem can be formulated as:

$$\min \sum_{i=1}^{k'} (\alpha_i - \sin\alpha_i), \text{ subject to } \sum_{i=1}^{k'} \alpha_i = (k'-2)\pi \,. \tag{16.5}$$

Using the Lagrangian multiplier method, we can get the solution of abovementioned optimization problem: $\alpha_i = \frac{(k'-2)\pi}{k'}$, and $L = (k'-2)\pi - k' \sin\left[\frac{(k'-2)\pi}{k'}\right]$. So the overlap per disk is $\frac{L}{k'} = \pi - \frac{2\pi}{k'} - \sin\left[\frac{(k'-2)\pi}{k'}\right]$, monotonically increase with k', and $k' \ge 3$. Moreover, when $k' = 3$, the optimal solution is $\alpha_i = \pi/3$, and when $k' > 3$, the overlap per disk is always higher than that in the case of $k' = 3$. This implies that the center of three disks should form an equilateral triangle with edge $\sqrt{3}$, and when $r \ne 1$, the edge of the equilateral triangle is $\sqrt{3}r$. According to the above analytic result and in studies by Zhang and Hou [46], Zalyubovskiy et al. [47], and Han et al. [48], we can get the following theorem:

Theorem 1

To minimize the overlap of two disks, only one disk should be used and the center of these three disks should form an equilateral triangle with side length $\sqrt{3}r$, where the r is the radius of disks. ■

From Theorem 1, we can find that the equilateral triangle traveling trajectory can make the mobile node use the least number of broadcast messages to cover all unknown nodes in the deployment area, without a hole.

16.5 Simulation and Results

In this section, we introduce our simulations of the localization accuracy and the communication cost of GFIL scheme with different parameter configurations, and then we also show the comparative results about the localization accuracy of GFIL scheme and nine other up-to-date schemes, under different topologies.

16.5.1 Simulation Setup

Our simulations are performed using MATLAB®. We performed the simulation study in a 50 × 50 m^2 region in which unknown sensor nodes are distributed. Each data point shown below is an average of 100 simulation runs. Moreover, the calculation formula of error is: $\text{error} = \sum_{i=1}^{n} \frac{\Delta r_i}{nR_{\max_i}}$, where the $\Delta r_i = \sqrt{\Delta x_i^2 + \Delta y_i^2}$, the n is the number of unknown nodes, and $R_{\max_i}$ is the maximum transmission radius of unknown nodes (all transmission radiuses of sensor nodes are the same and invariable in our experiments).

16.5.2 Simulation Results and Analyses for the GFIL Scheme

16.5.2.1 Effect of the Transmission Radius

First, to analyze the effect of transmission radius on localization error, we vary the transmission radius from 4 to 12 m. The simulation results are shown in Figures 16.7 through 16.10, with different move-speed-relative control noises and three kinds of traveling trajectories, under four kinds of topologies (the parameters and their values are listed in Table 16.1).

We add move-speed-relative and steering-angle-relative control noise, when the mobile node moves, and the move-speed-relative control noise is denoted as σV and the steering-angle-relative control noise can be calculated using the formula $\sigma G = (3.0 \times \pi/180)$. Combining the move-speed-relative and steering-angle-relative control noise, we can obtain the final control noise and its

Table 16.1 Simulation Parameters

Parameter	*Value*
Transmission radius (m)	4, 5, 6, 7, 8, 9, 10, 11, 12
Move-speed-relative control noise (σV)	0.5, 1, 2, 3, 4
Topology	square random, square regular, C random, C regular
Number of unknown nodes	100

calculation formula as $Q = [\sigma V^2\ 0;\ 0\ \sigma G^2]$ (it is a 2 × 2 matrix). The four kinds of topologies are shown in Figure 16.6.

The experiment's results are shown in Figures 16.7 through 16.10. In these results, first, we can find that along with the control noises' decrease and the transmission radiuses of nodes increase, the trend of localization accuracy is increasing for different traveling trajectories and different topologies, because the control noise and the transmission radius affect the number of received data packets from the mobile node by unknown nodes, for example, if the transmission radius of the mobile node is long enough and can cover the whole deployment area, each unknown node can receive abundant data packets from the mobile node and then, according to our localization scheme, the unknown nodes obtain enough information to adjust their coordinates to achieve the desired accuracy. Second, the localization accuracy of random traveling trajectory is higher, compared with the traveling trajectories: grid and equilateral triangle (this result is useful in practical applications because, in most cases, the traveling trajectory of person with a sensor node is random). We guess that maybe the random traveling trajectory is better at avoiding the boundary effect (boundary effect: a mobile node will cover fewer unknown nodes when it is placed near the boundary of a network than when it is placed at the central zone). Finally, it is worthwhile to note that an appropriate move speed can improve the localization accuracy (we will discuss this problem in the next section).

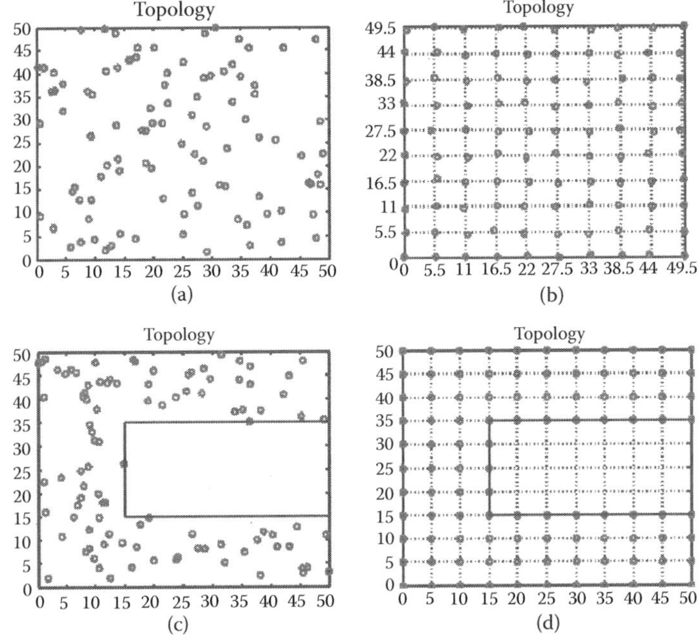

Figure 16.6 The topologies: 100 sensor nodes (dots) are deployed in a 50 × 50 m² area. (a) In square random topology, sensor nodes are deployed randomly and uniformly; (b) in square regular topology, sensor nodes are deployed regularly and on a grid; (c) in C random topology, sensor nodes are deployed in a C-shaped square area, randomly and uniformly; (d) in C regular topology, sensor nodes are deployed in a C-shaped square area, regularly and on a grid.

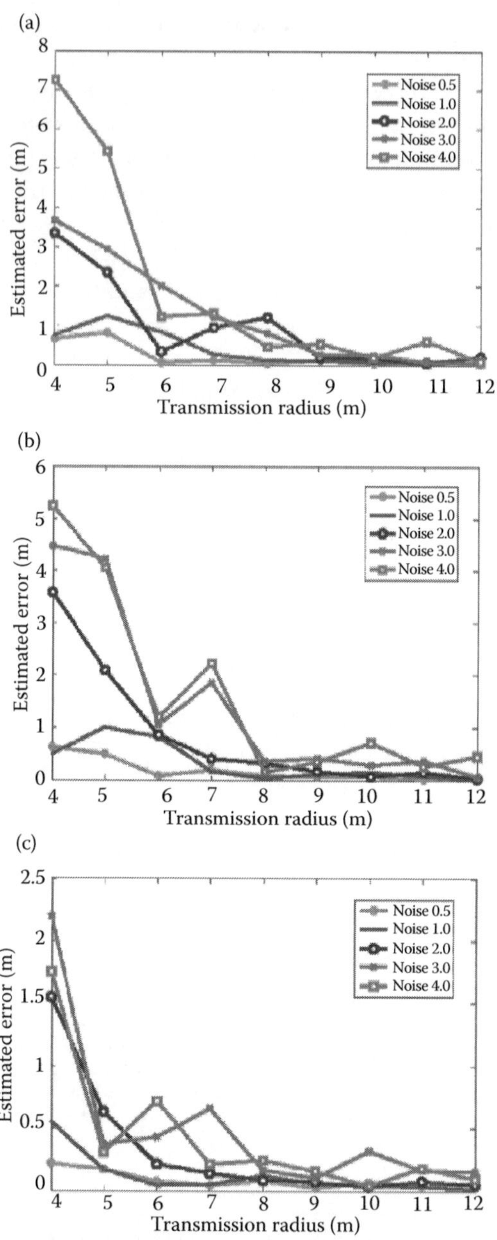

Figure 16.7 Estimated error changes with an increase in transmission radius, under a square random topology, when the move speed is 1, the traveling trajectory is grid (a), equilateral triangle (b), or random (c).

Second, we analyze the effect of the transmission radius on the communication cost for our localization scheme. In our scheme, with regard to communication cost, only the mobile node broadcasts its locations to one-hop neighbors with each observation. Therefore, we can use the formula: com_cost = sum_time/dt_observe, to perform an approximate calculation for the communication cost, where the "sum_time" is the total travel time of the mobile node, and the "dt_observe =

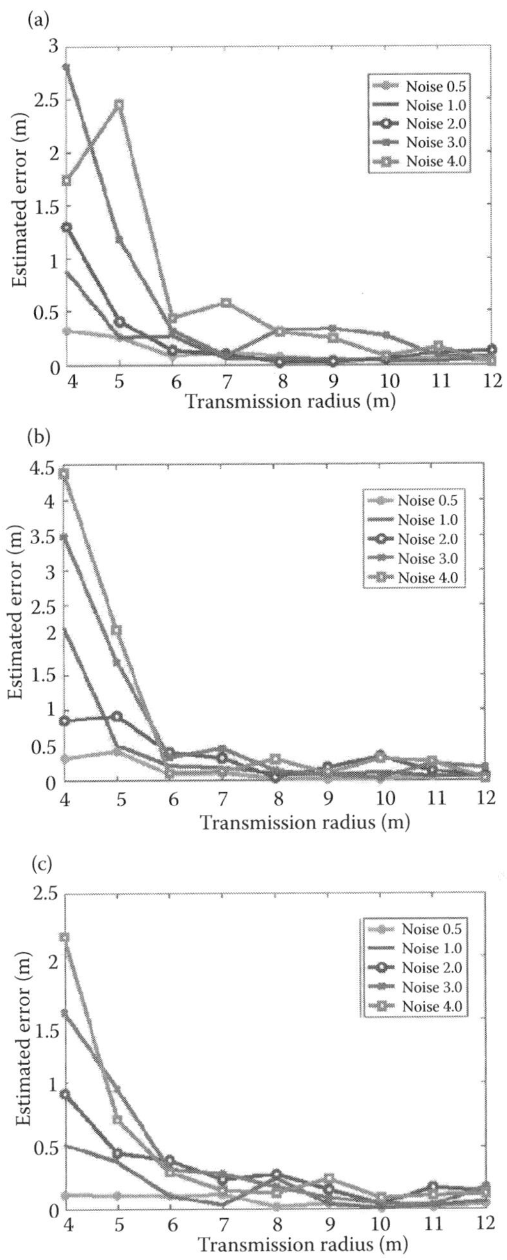

Figure 16.8 Estimated error changes with an increase in transmission radius, under a square regular topology, when the move speed is 1, the traveling trajectory is grid (a), equilateral triangle (b), or random (c).

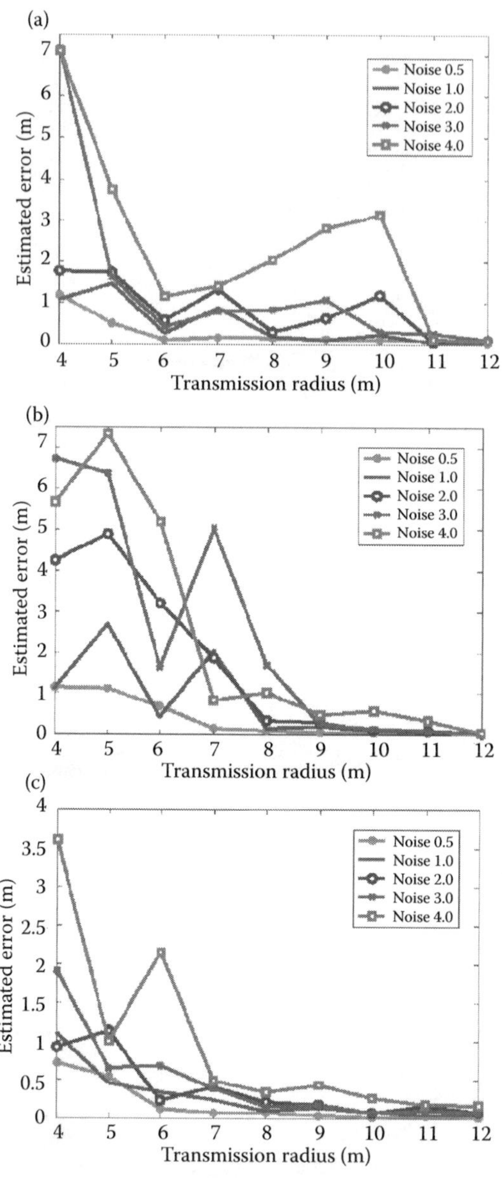

Figure 16.9 Estimated error changes with an increase in transmission radius, under a C random topology, when the move speed is 1, the traveling trajectory is grid (a), equilateral triangle (b), or random (c).

8*dt_controls" denotes the time interval between observations, and the "dt_controls" is the time interval between control signals that are used to control the move of the mobile node. Figures 16.11 through 16.14 show the results under different topologies.

From these figures, we find that the communication costs of equilateral triangle traveling trajectory are less than that of two other traveling trajectories. Therefore, the results of these experiments confirm the theorem that is shown in Section 16.4.4.

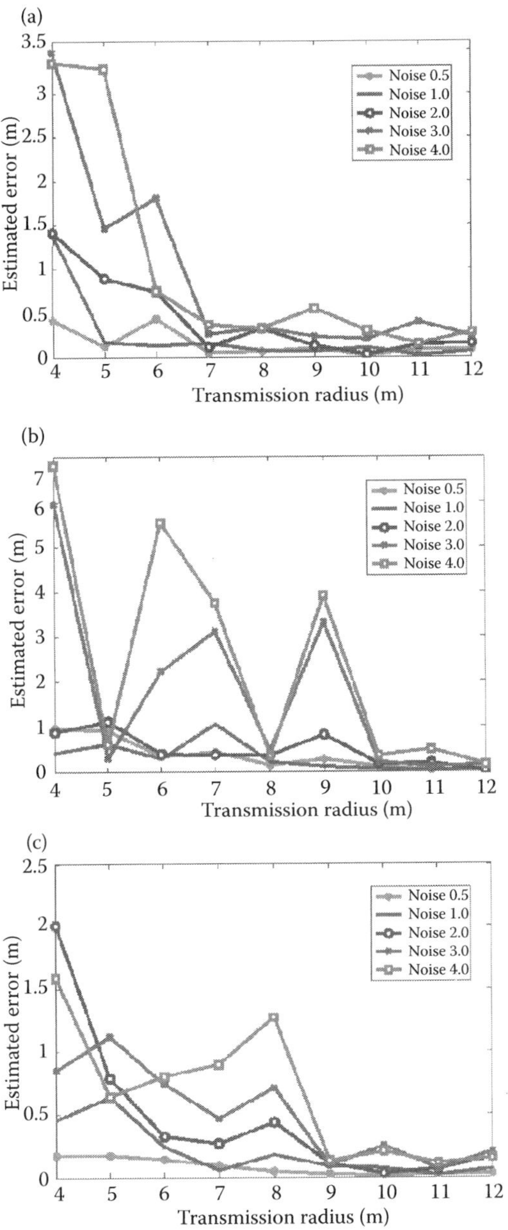

Figure 16.10 Estimated error changes with an increase in transmission radius, under a C regular topology, when the move speed is 1, the traveling trajectory is grid (a), equilateral triangle (b), or random (c).

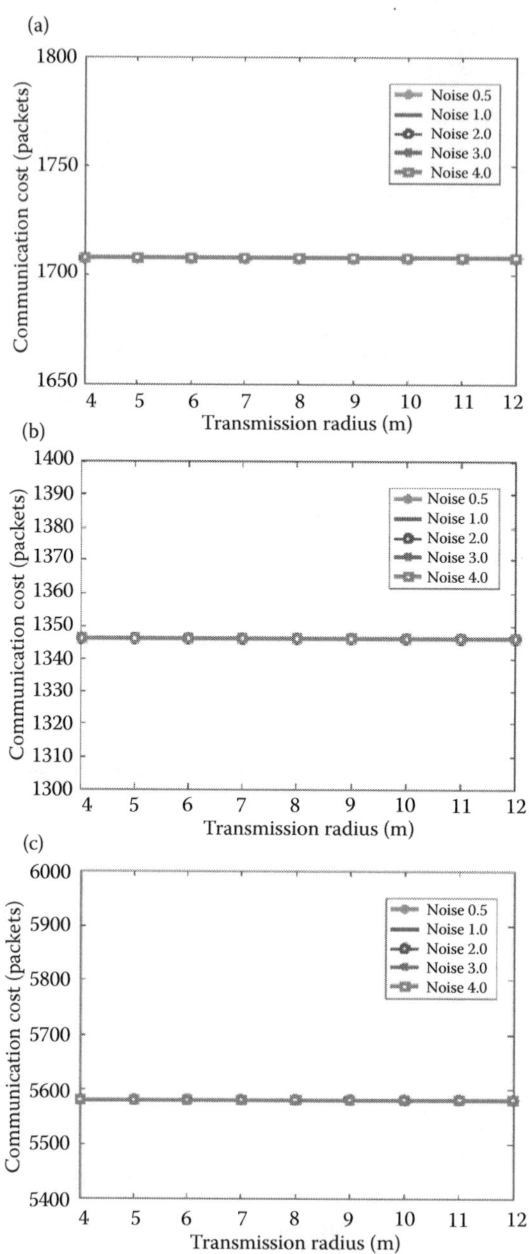

Figure 16.11 Communication costs are not changed along with an increase in transmission radius and for different noises, under a certain traveling trajectory, when the topology is square-random. However, the communication costs are different with different traveling trajectories; the traveling trajectory is grid (a), equilateral triangle (b), or random (c).

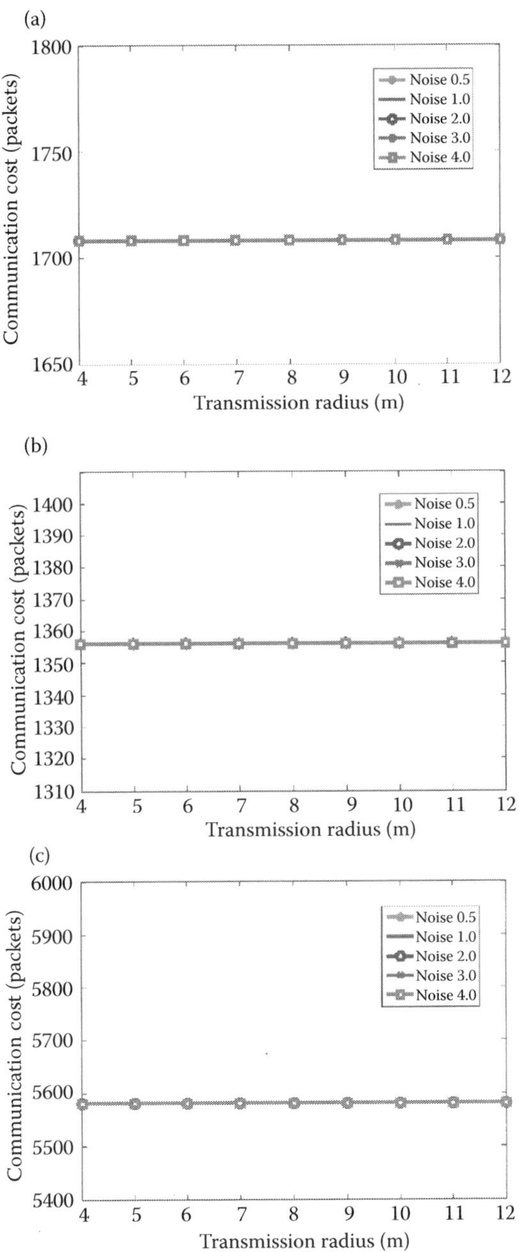

Figure 16.12 Communication costs are not changed along with an increase in transmission radius and for different noises, under a certain traveling trajectory, when the topology is square-regular. However, the communication costs are different with different traveling trajectories; the traveling trajectory is grid (a), equilateral triangle (b), or random (c).

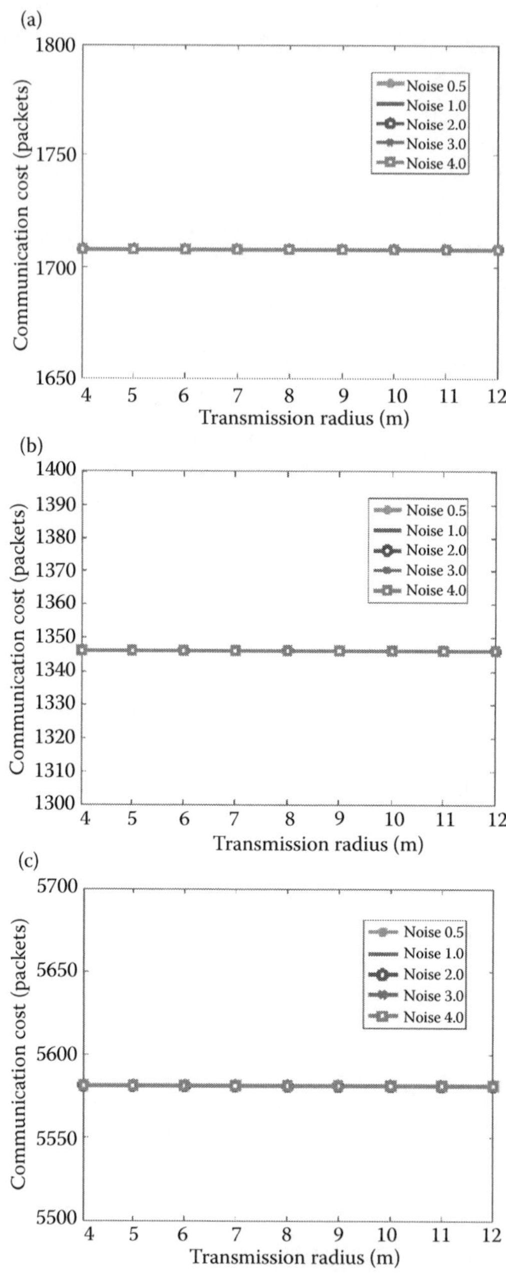

Figure 16.13 Communication costs are not changed along with an increase in transmission radius and for different noises, under a certain traveling trajectory, when the topology is C random. However, the communication costs are different with different traveling trajectories; the traveling trajectory is grid (a), equilateral triangle (b), or random (c).

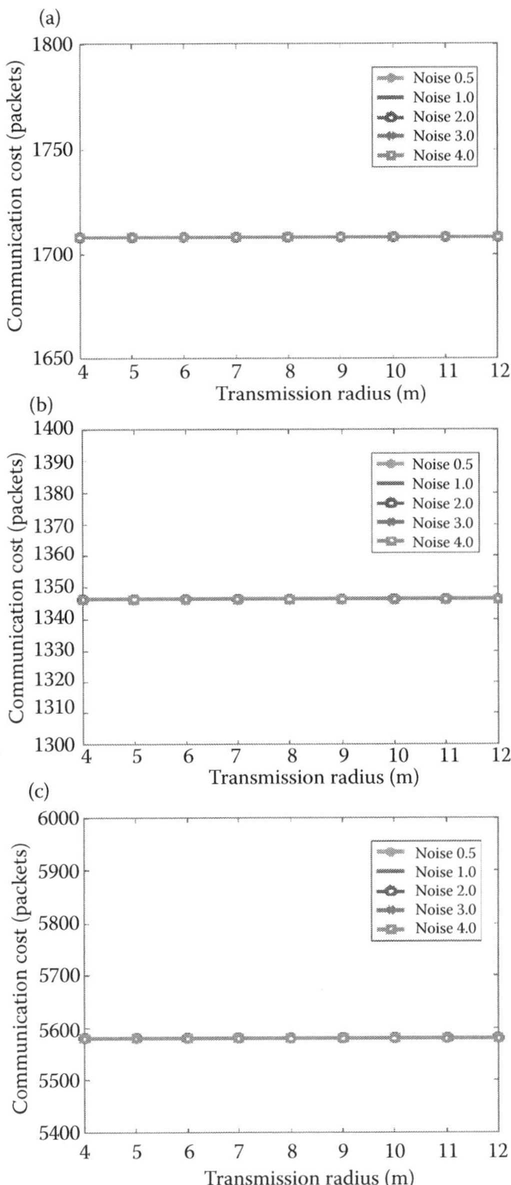

Figure 16.14 Communication costs are not changed along with an increase in transmission radius and for different noises, under a certain traveling trajectory, when the topology is C regular. However, the communication costs are different with different traveling trajectories; the traveling trajectory is grid (a), equilateral triangle (b), or random (c).

16.5.2.2 Effect of the Move Speed

From previous experiments, we find that an appropriate move speed for the mobile node can improve localization accuracy. In this experiment, the values of move speed (v_m) are set as 0.5, 1, 2, 3, and 4 m/s. The results are shown in Figures 16.15 through 16.18.

When the traveling trajectory is random, the estimated error is smaller for each move speed value. Moreover, an appropriate point is existent for best localization accuracy, under different topologies and different traveling trajectories.

Under different topologies, the communication costs of equilateral triangle traveling trajectory are less than that of two other traveling trajectories (Figures 16.19 through 16.22). Therefore, these experimental results for move speed confirm the theorem that is shown in Section 16.4.4.

16.5.3 Comparative Simulation Results and Analyses

In the following experiments, we compare the GFIL scheme with nine other up-to-date localization schemes (e.g., centroid, bounding box, grid scan, APIT, DV-Hop, MDS-MAP, Amorphous, MCL, and DTN) in accuracy, with different topologies. Because some schemes need anchor nodes to attain localization, we set up some sensor nodes in the network as anchor nodes, and the percentage of anchor nodes is 50%. Moreover, the parameter configuration of our scheme is v_m = 1, and $\sigma V = 0.5$.

16.5.3.1 Square Random Topology

Based on the square random topology, we test the accuracy of 10 localization schemes with varying numbers of sensor nodes, and the results are shown as follows.

From Figure 16.23, we can find that the connectivity of network with 50 sensor nodes is not enough for the MDS-MAP scheme. The MDS-MAP is based on a fully connected network to achieve localization. Therefore, the nodes can be located until a fully connected topology is obtained (the transmission radiuses of nodes are equal to 11 m), and the estimated error is 1.7840 m. Moreover, when the transmission radius is less than 5 m, the APIT scheme cannot achieve localization (the estimated error is infinity). However, the localization accuracy is good about our GFIL scheme under different transmission radiuses.

When there are 100 nodes in a network, the connectivity of the network is better than the situation of Figure 16.23. When the transmission radius is greater than 9 m, the MDS-MAP can attain localization, and along with an increase in transmission radius, the estimated error decreases. Moreover, our GFIL can always achieve effective localization.

The connectivity of our network is increased, along with an increase in the number of sensor nodes. In Figure 16.25, when the transmission radius is greater than 7 m, a fully connected topology can be obtained and the MDS-MAP can attain localization. As always, our GFIL can get better and stable localization accuracy under different transmission radiuses.

When there are 200 sensor nodes, the connectivity of our network keeps on getting better: after 6 m about the transmission radius, the MDS-MAP can work, and get better localization accuracy. Moreover, the localization accuracy keeps on improving in our GFIL scheme compared with the accuracy shown in Figures 16.23 through 16.25.

The comparative result of the estimated error for the 10 schemes, at 250 sensor nodes, is shown in Figure 16.27. From Figures 16.23 through 16.27, we can conclude that the connectivity of a network is important for a localization issue. However, our scheme is connectivity-irrespective.

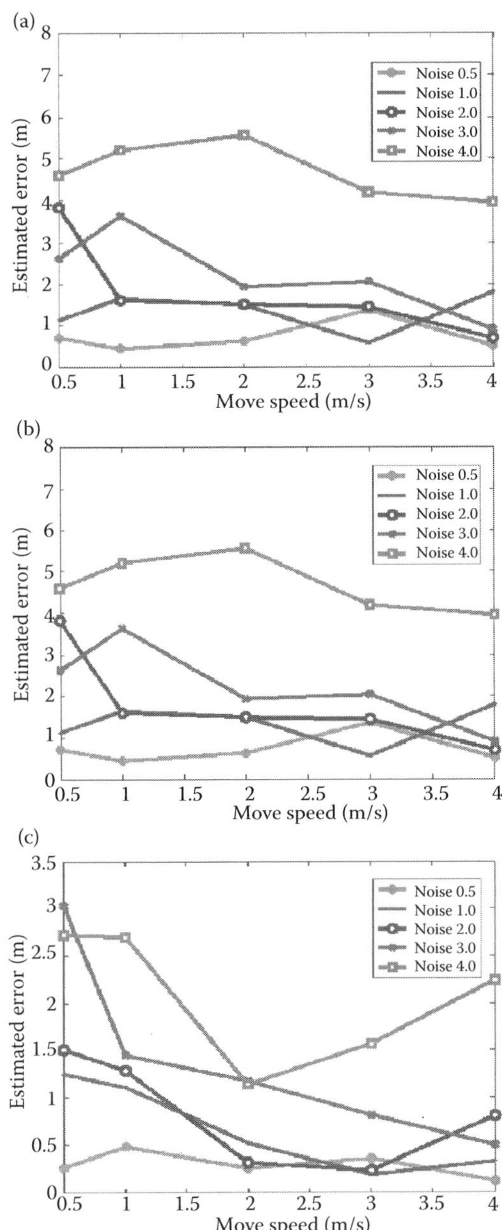

Figure 16.15 Estimated error changes with move speed increase, when the topology is square random and the transmission radius is 4 m; the traveling trajectory is grid (a), equilateral triangle (b), or random (c).

Our scheme can always achieve effective and good localization accuracy, even if the connectivity of a network is not good.

From comparative results (Figures 16.23–16.27), we can distinctly observe that the localization accuracy of our scheme is better than in the nine other schemes, for example, the localization error of our scheme can be reduced to less than 0.2 m when the transmission radius is 4 m for

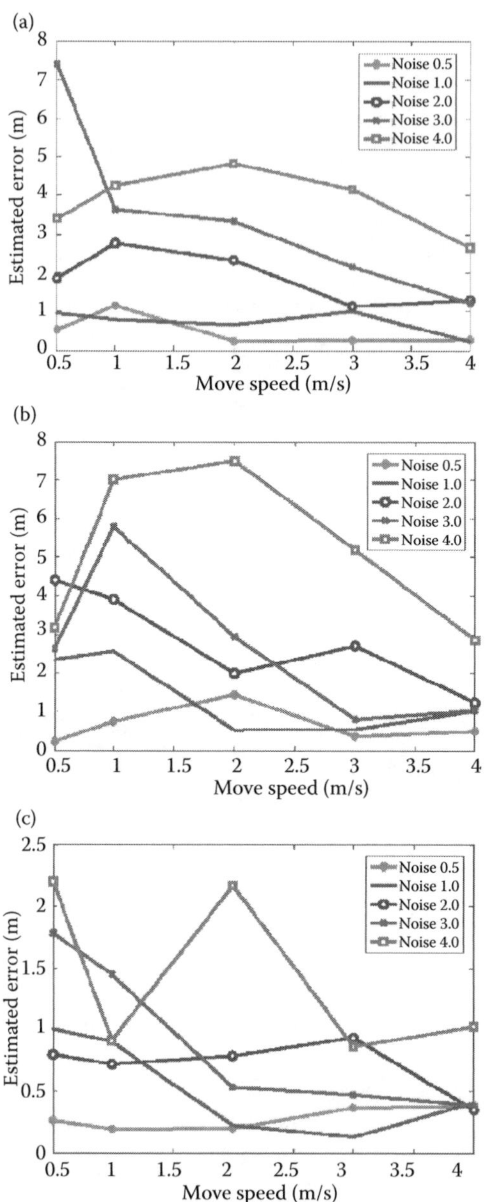

Figure 16.16 Estimated error changes with move speed increase, when the topology is square regular and the transmission radius is 4 m; the traveling trajectory is grid (a), equilateral triangle (b), or random (c).

all sensor nodes. It is worth mentioning that some localization schemes cannot locate unknown nodes in some special situations, for example, when the transmission radius is less than 5 m and there are 50 sensor nodes, the localization error of APIT is infinity because the unknown nodes cannot be covered by the transmission coverage of anchor nodes. However, our scheme is coverage-irrespective. Our scheme can always achieve effective localization, even if the coverage of a network is not good.

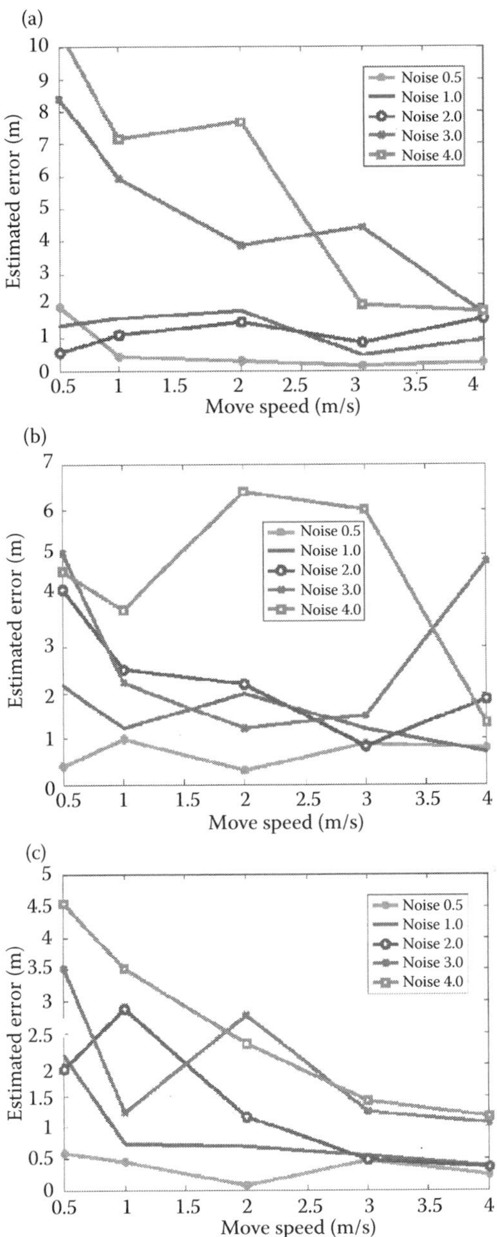

Figure 16.17 Estimated error changes with move speed increase, when the topology is C random and the transmission radius is 4 m; the traveling trajectory is grid (a), equilateral triangle (b), or random (c).

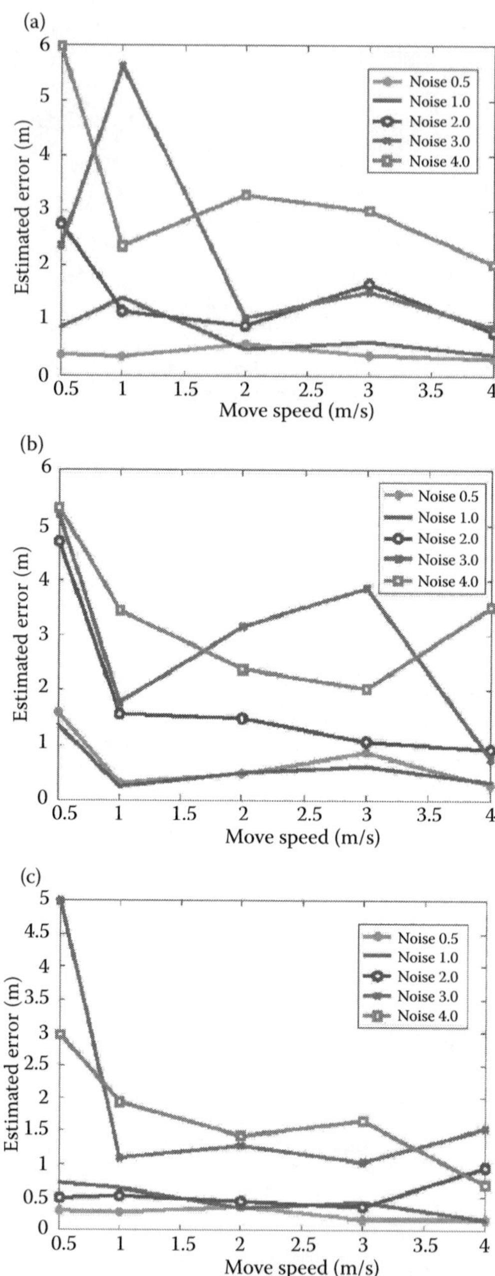

Figure 16.18 Estimated error changes with move speed increase, when the topology is C regular and the transmission radius is 4 m; the traveling trajectory is grid (a), equilateral triangle (b), or random (c).

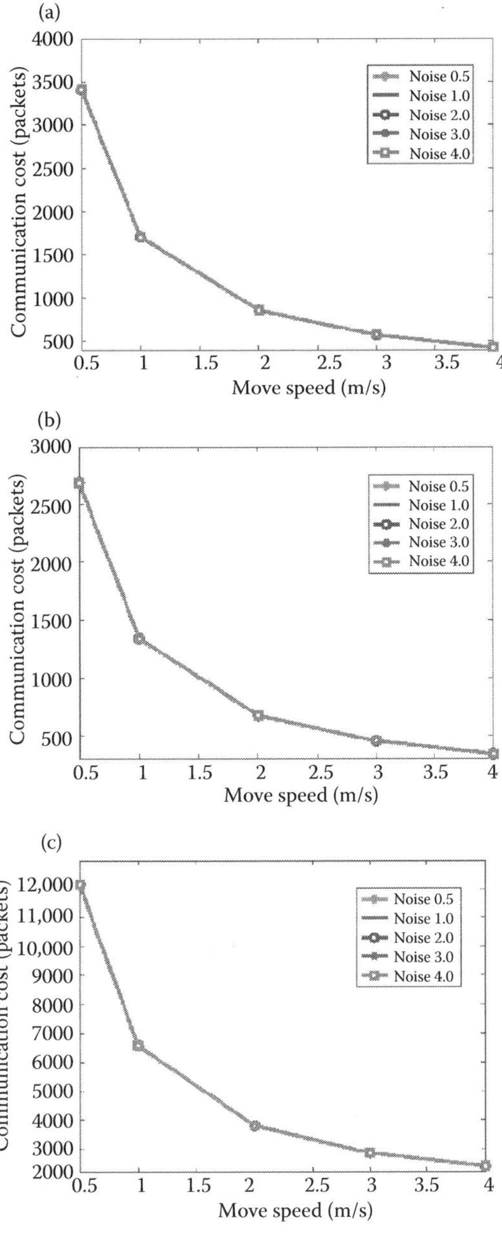

Figure 16.19 Communication costs are decreased along with the move speed increase, and for different noises; the communication costs are same, under a certain traveling trajectory, when the topology is square random. However, the communication costs are different with different traveling trajectories; the traveling trajectory is grid (a), equilateral triangle (b), or random (c).

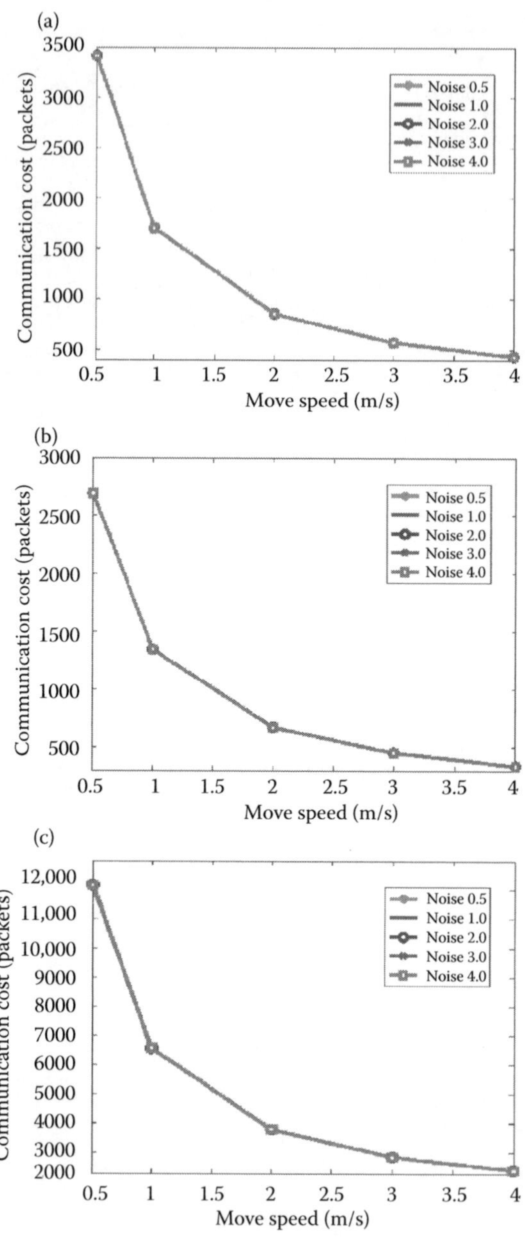

Figure 16.20 Communication costs are decreased along with the move speed increase, and for different noises; the communication costs are same, under a certain traveling trajectory, when the topology is square regular. However, the communication costs are different with different traveling trajectories; the traveling trajectory is grid (a), equilateral triangle (b), or random (c).

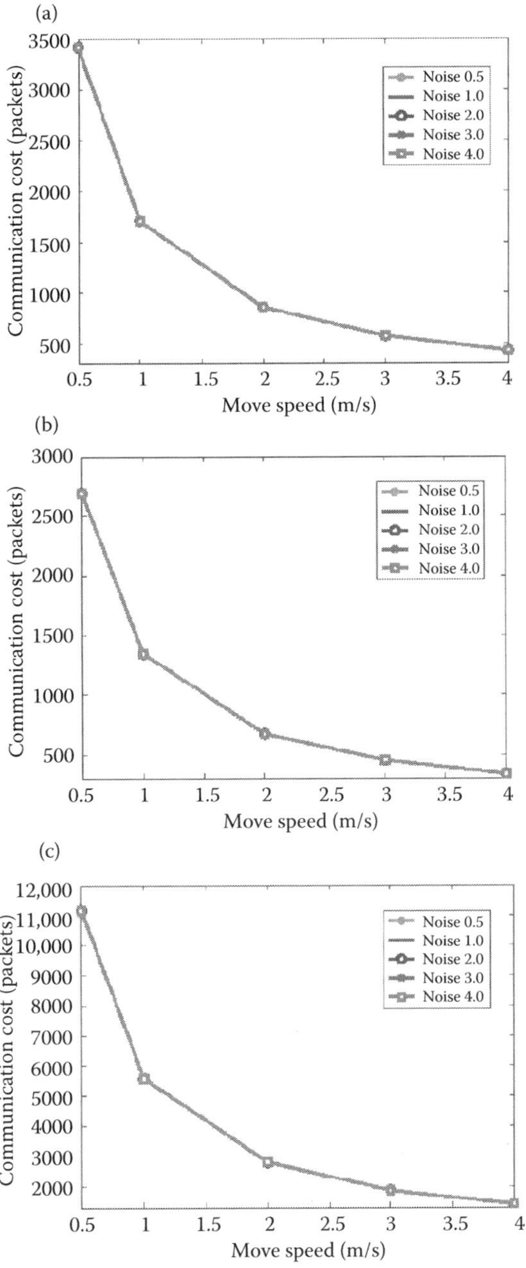

Figure 16.21 Communication costs are decreased along with the move speed increase, and for different noises; the communication costs are same, under a certain traveling trajectory, when the topology is C random. However, the communication costs are different with different traveling trajectories; the traveling trajectory is grid (a), equilateral triangle (b), or random (c).

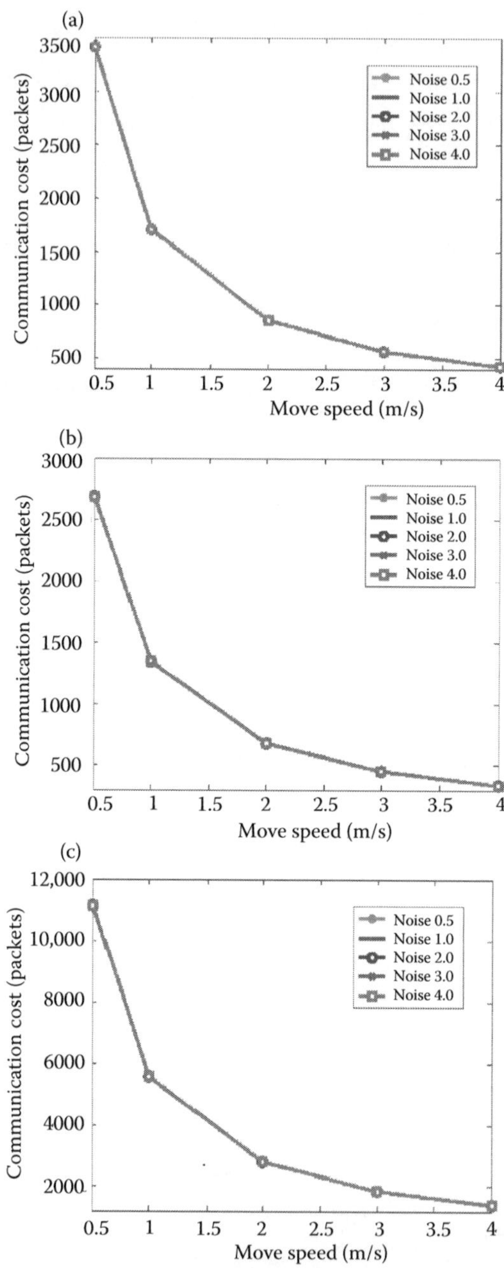

Figure 16.22 Communication costs are decreased along with the move speed increase, and for different noises; the communication costs are same, under a certain traveling trajectory, when the topology is C regular. However, the communication costs are different with different traveling trajectories; the traveling trajectory is grid (a), equilateral triangle (b), or random (c).

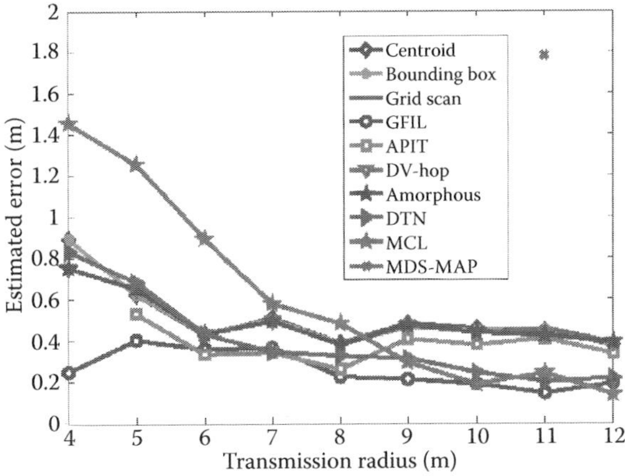

Figure 16.23 Comparative results for the GFIL scheme and nine other localization schemes when there are 50 sensor nodes.

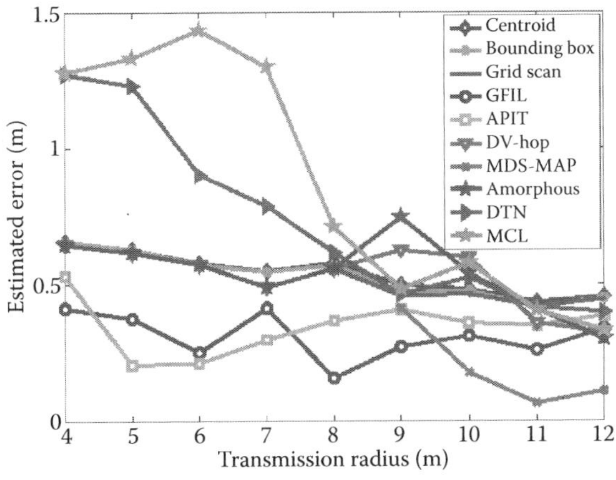

Figure 16.24 Comparative results for the GFIL scheme and nine other localization schemes when there are 100 sensor nodes.

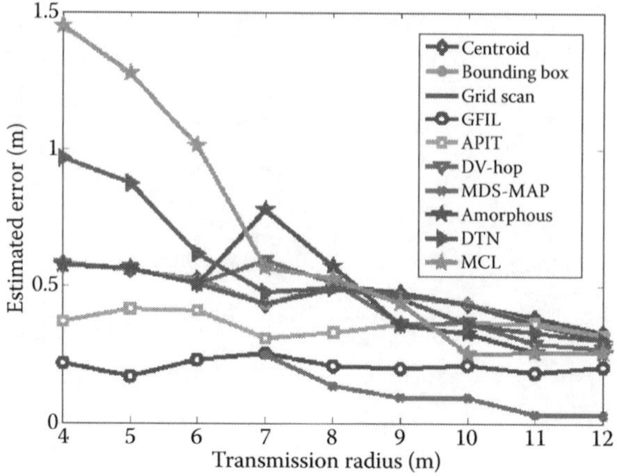

Figure 16.25 Comparative results for the GFIL scheme and nine other localization schemes when there are 150 sensor nodes.

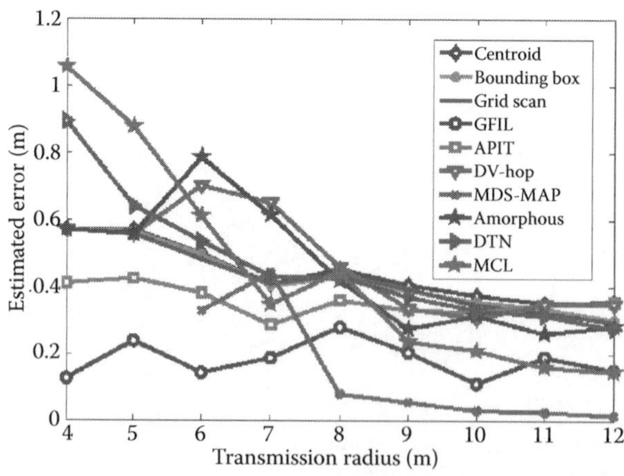

Figure 16.26 Comparative results for the GFIL scheme and nine other localization schemes when there are 200 sensor nodes.

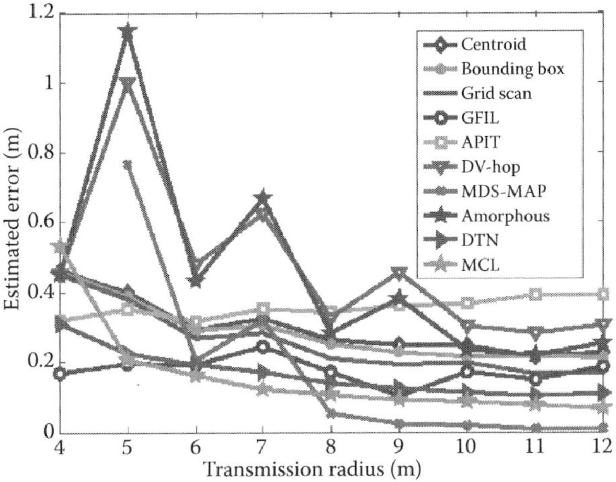

Figure 16.27 Comparative results for the GFIL scheme and nine other localization schemes when there are 250 sensor nodes.

16.5.3.2 Square Regular Topology

In the square regular topology, the distances between sensor nodes are almost equal. Based on this type of network topology, we test the accuracy of 10 localization schemes with varying numbers of sensor nodes, and the results are shown as follows: when there are 50 sensor nodes in the network (Figure 16.28); 100 sensor nodes in the network (Figure 16.29); 150 sensor nodes in the network (Figure 16.30); 200 sensor nodes in the network (Figure 16.31); and 250 sensor nodes in the network (Figure 16.32).

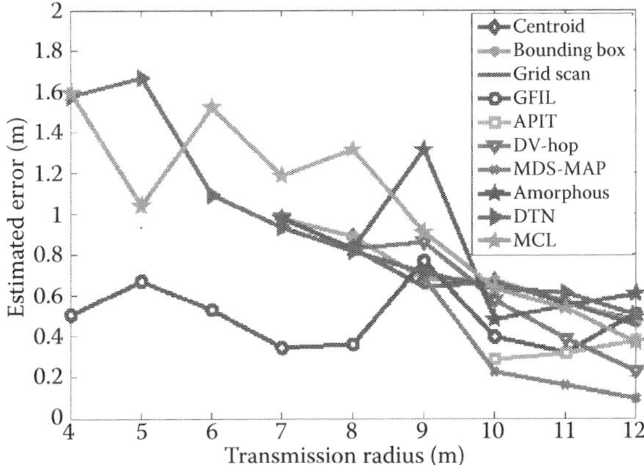

Figure 16.28 Comparative results about the GFIL scheme and nine other localization schemes when there are 50 sensor nodes.

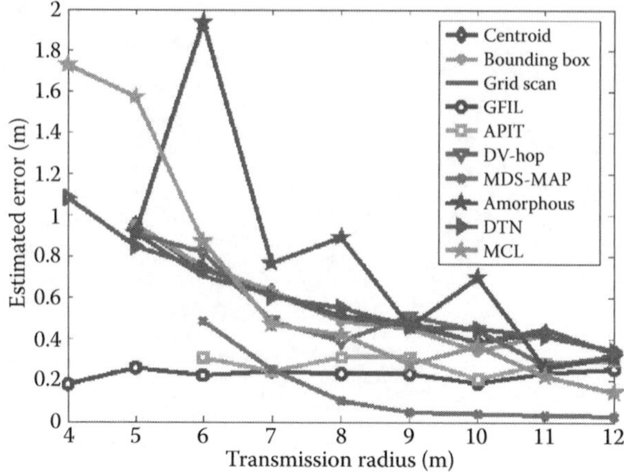

Figure 16.29 Comparative results for the GFIL scheme and nine other localization schemes when there are 100 sensor nodes.

First, along with an increase in the number of sensor nodes, the connectivity of network is improved. Even if the transmission radiuses of sensor nodes are small, the connectivity-based localization schemes can still work well and get better localization accuracy, for example, in the MDS-MAP, when there are 250 sensor nodes and the transmission radius is 5 m, the estimated error is close to 0.05 m. Second, the GFIL scheme can achieve better localization accuracy, steadily, for different transmission radiuses and different "the number of sensor nodes".

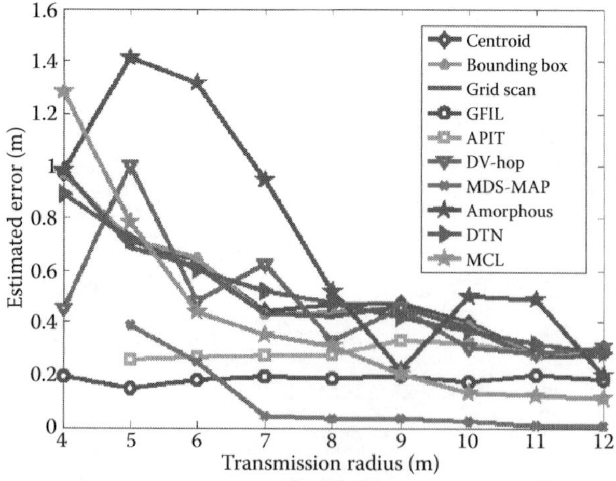

Figure 16.30 Comparative results for the GFIL scheme and nine other localization schemes when there are 150 sensor nodes.

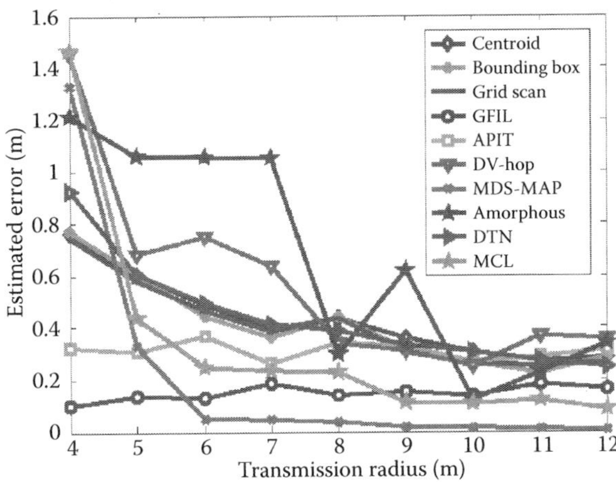

Figure 16.31 Comparative results for the GFIL scheme and nine other localization schemes when there are 200 sensor nodes.

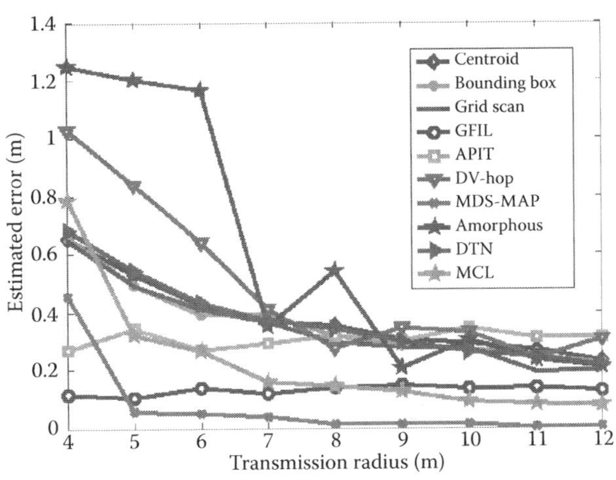

Figure 16.32 Comparative results for the GFIL scheme and nine other localization schemes when there are 250 sensor nodes.

16.5.3.3 C Random Topology

By comparing and analyzing the accuracy of 10 localization schemes, based on C random topology with varying numbers of sensor nodes, we can get some useful results to direct improvement in the accuracy of localization.

The comparative results regarding accuracy for 10 localization schemes are shown here for instances when there are 50 sensor nodes in the network (Figure 16.33), 100 sensor nodes in the

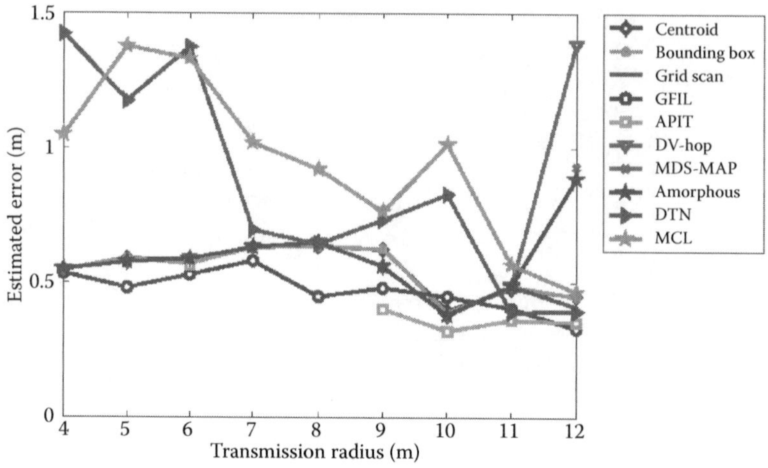

Figure 16.33 Comparative results for the GFIL scheme and nine other localization schemes when there are 50 sensor nodes.

network (Figure 16.34), 150 sensor nodes in the network (Figure 16.35), 200 sensor nodes in the network (Figure 16.36), and 250 sensor nodes in the network (Figure 16.37).

We can distinctly observe that the effect of irregular topology on accuracy is less for our GFIL scheme compared with the nine other localization schemes. Therefore, our scheme is effective for irregular network topology.

16.5.3.4 C Regular Topology

Based on C regular topology, we compare and analyze the accuracy of 10 localization schemes with varying numbers of sensor nodes. Because the C regular topology is more special than three

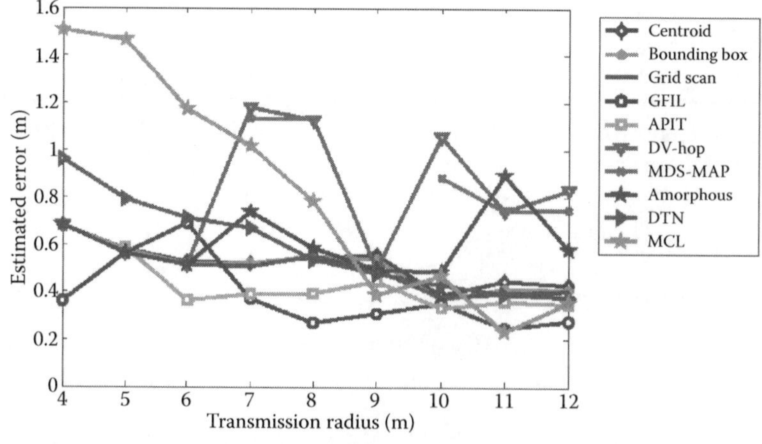

Figure 16.34 Comparative results for the GFIL scheme and nine other localization schemes when there are 100 sensor nodes.

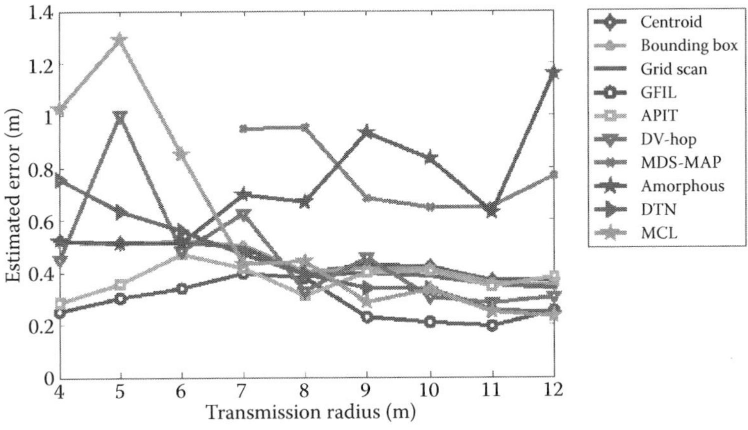

Figure 16.35 Comparative results for the GFIL scheme and nine other localization schemes when there are 150 sensor nodes.

other kinds of topologies—it is irregular and the distances between the sensor nodes are almost equal—it is worth analyzing to get some special and useful results.

The comparative results regarding accuracy for 10 localization schemes are shown for instances when there are 50 sensor nodes in the network (Figure 16.38), 100 sensor nodes in the network (Figure 16.39), 150 sensor nodes in the network (Figure 16.40), 200 sensor nodes in the network (Figure 16.41), and 250 sensor nodes in the network (Figure 16.42).

Under different "the number of sensor nodes", the localization accuracy of the C regular topology is better than the accuracy of random types of topologies. From the comparative results (Figures 16.38–16.42), we can distinctly observe that the effect of irregular topology on accuracy is less for our GFIL scheme compared with nine other localization schemes. Therefore, our scheme

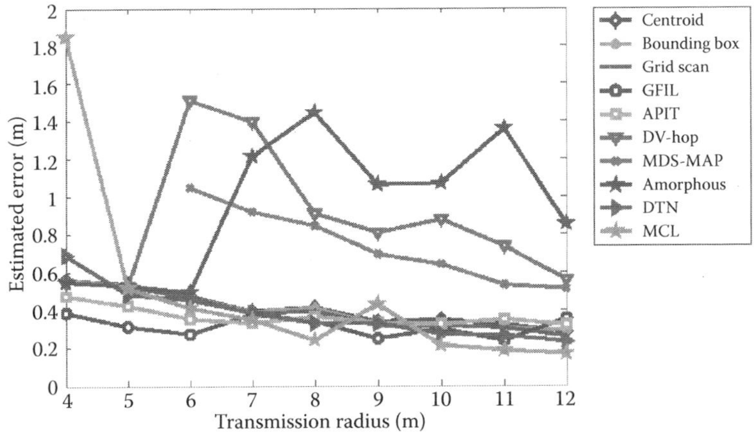

Figure 16.36 Comparative results for the GFIL scheme and nine other localization schemes when there are 200 sensor nodes.

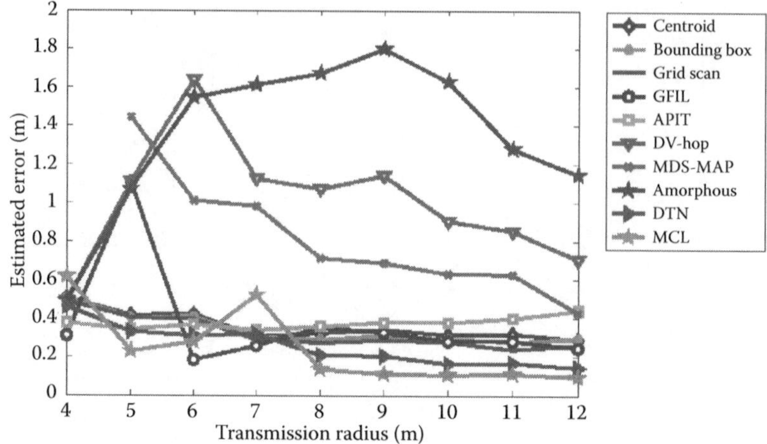

Figure 16.37 Comparative results for the GFIL scheme and nine other localization schemes when there are 250 sensor nodes.

is effective for irregular network topology and the localization effectiveness of our scheme is stable for different transmission radiuses.

16.6 Conclusion and Future Work

In this chapter, we propose a high-accuracy GFIL scheme, which only needs a mobile node. And the real-time coordinate of the mobile node is calculated from the real-time traveling information

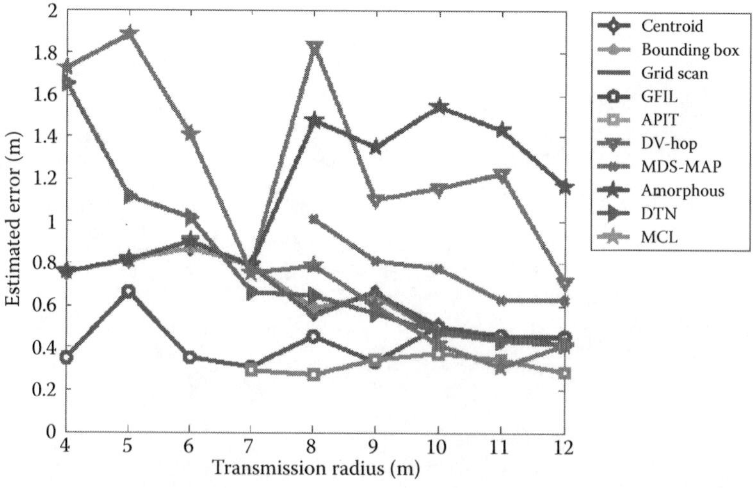

Figure 16.38 Comparative results for the GFIL scheme and nine other localization schemes when there are 50 sensor nodes.

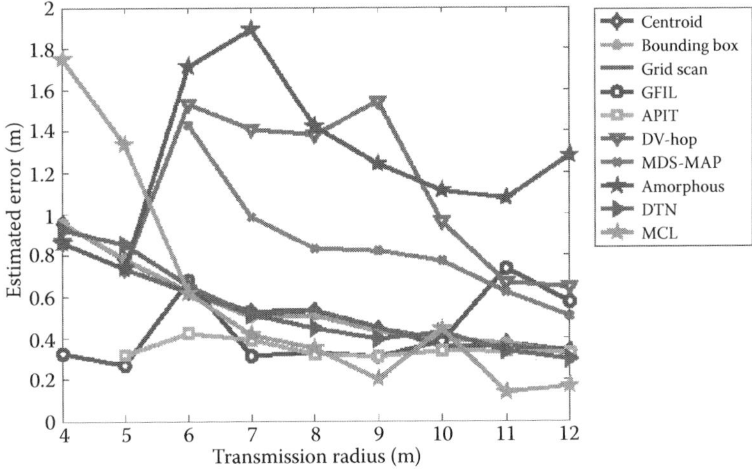

Figure 16.39 Comparative results for the GFIL scheme and nine other localization schemes when there are 100 sensor nodes.

of the mobile node (e.g., movement speed, elapsed time, and steering angle) and the mobile sensor's start point coordinate. Unknown nodes only receive broadcast information from the mobile node to attain localization; therefore, the communication overhead of the whole network is low. Moreover, the localization accuracy of the GFIL scheme is high (up to 60% improvement: the estimated localization error 0.5 m is a wonderful value in the current localization research area; however, a localization error of 0.2 m can be reached for our scheme).

We also prove that the equilateral triangle traveling trajectory can make the mobile node use the least number of broadcast messages to cover the deployment area of unknown nodes, without a hole.

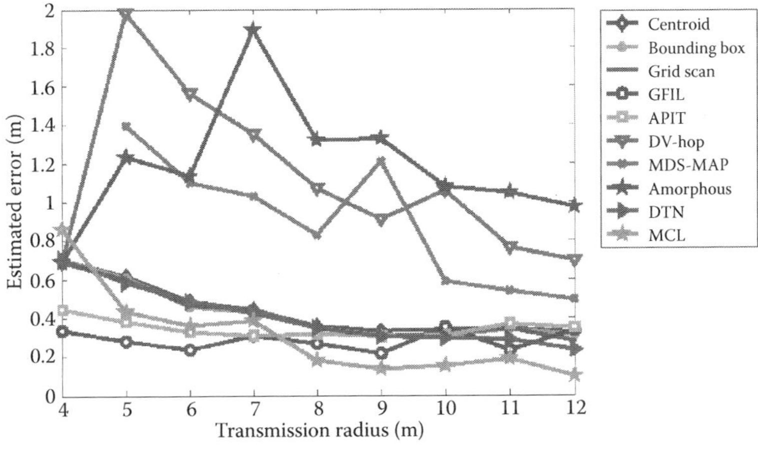

Figure 16.40 Comparative results for the GFIL scheme and nine other localization schemes when there are 150 sensor nodes.

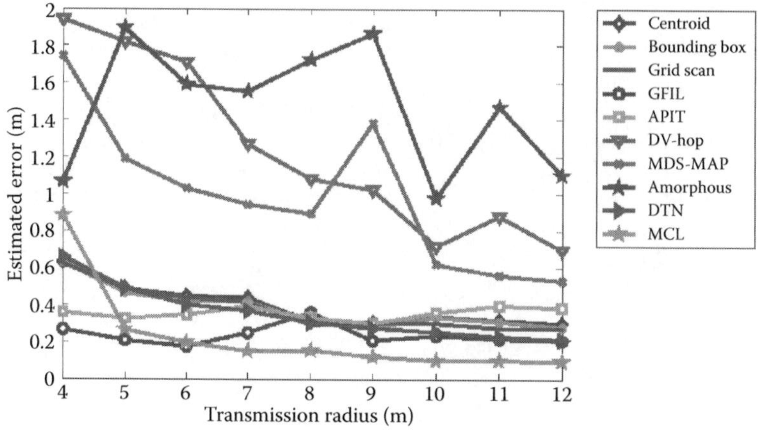

Figure 16.41 Comparative results for the GFIL scheme and nine other localization schemes when there are 200 sensor nodes.

From the simulations, we can find that the connectivity of a network, which belongs to the MAC layer property, can affect the accuracy of localization; thus, we will perform research about the effects of the MAC layer on localization accuracy. Moreover, because energy and network lifetime are important for WSNs, we will do some investigations and evaluations about the energy efficiency of our localization scheme.

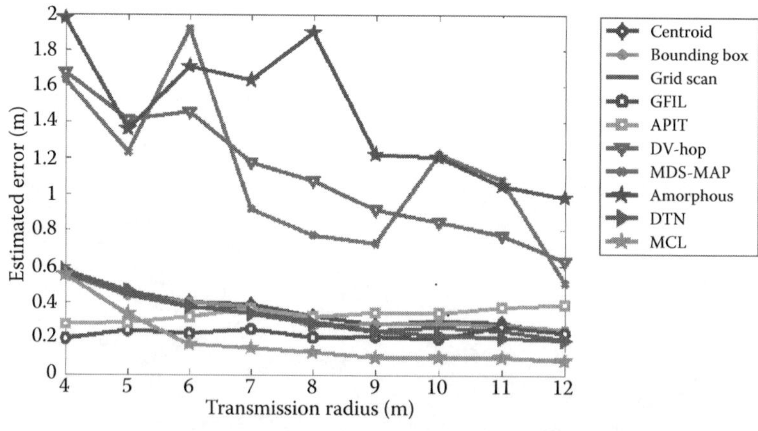

Figure 16.42 Comparative results for the GFIL scheme and nine other localization schemes when there are 250 sensor nodes.

References

1. J. Jiang, G. Han, H. Xu, L. Shu, and M. Guizani. LMAT: Localization with a mobile anchor node based on triangulation in wireless sensor networks. In *The IEEE Global Communication Conference (Globecom 2011)*, Houston, TX, December 5–9, 2011.
2. Y. Chen, L. Shu, M. Li, Z. Fan, L. Wang, and T. Hara. The insights of DV-based localization algorithms in the wireless sensor networks with duty-cycled and radio irregular sensors. In *International Conference on Communications (ICC 2011)*, Kyoto, Japan, June 5–9, 2011.
3. Z. Fan, Y. Chen, L. Shu, L. Wang, T. Hara, and S. Nishio. Impacts of duty-cycle and radio irregularity on HCRL localization in wireless sensor networks. In *Proceedings of the 18th International Conference on Network Protocols (ICNP 2010)*, Kyoto, Japan, October 5–8, 2010.
4. I. Constandache, S. Gaonkar, M. Sayler, R.R. Choudhury, and L. Cox. EnLoc: Energy-efficient localization for mobile phones. *Proceedings of INFOCOM Mini Conference (INFOCOM 2009)*, Rio de Janeiro, Brazil, pp. 2716–2720, April 19–25, 2009.
5. H. Chen, B. Liu, P. Huang, J. Liang, and Y. Gu. Mobility-assisted node localization based on TOA measurements without time synchronization in wireless sensor networks. *Mobile Networks and Applications*, vol. 17, no. 1, pp. 90–99, 2012.
6. X. Li, N. Mitton, I. Simplot-Ryl, and D. Simplot-Ryl. Dynamic beacon mobility scheduling for sensor localization. *IEEE Transactions on Parallel and Distributed Systems*, vol. 23, no. 8, pp. 1439–1452, 2012.
7. L. Isella, J. Stehle, A. Barrat, C. Cattuto, J.F. Pinton, and W. Van den Broeck. What's in a crowd? Analysis of face-to-face behavioral networks. *Journal of Theoretical Biology*, vol. 271, no. 1, pp. 166–180, 2011.
8. C. Song, Z. Qu, N. Blumm, and A.L. Barabasi. Limits of predictability in human mobility. *Science*, vol. 327, no. 5968, pp. 1018–1021, 2010.
9. X. Wu, L. Shu, M. Meng, J. Cho, and S. Lee. Coverage-driven self-deployment for cluster based mobile sensor networks. In *Proceedings of the 2006 IEEE International Conference on Computer and Information Technology (CIT 2006)*, Seoul, Korea, 2006.
10. X. Wu, Y. Niu, Lei Shu, J. Cho, and Y. Lee. Relay shift based self-deployment for mobility limited sensor networks. In *Proceedings of the 3rd International Conference on Ubiquitous Intelligence and Computing (UIC 2006 acceptance rate: 31%)*, Wuhan, China, 2006.
11. E. Ekici, Y. Gu, and D. Bozdag. Mobility-based communication in wireless sensor networks. *IEEE Communications Magazine*, vol. 44, no. 7, pp. 56–62, 2006.
12. P. Xu, X.H. Xu, S.J. Tang, and X.Y. Li. Truthful online spectrum allocation and auction in multi-channel wireless networks. *INFOCOM*, Shanghai, China, pp. 26–30, April 10–15, 2011.
13. S.A. Munir, B. Ren, W. Jiao, B. Wang, D. Xie, and J. Ma. Mobile wireless sensor network: Architecture and enabling technologies for ubiquitous computing. In *Proceedings of the 21st International Conference on Advanced Information Networking and Applications Workshops*, AINAW, pp. 113–120, 2007.
14. L. Shu, Y. Chen, T. Hara, M. Hauswirth, and S. Nishio. The new challenge: Mobile multimedia sensor networks. In *Indersicence, International Journal of Multimedia Intelligence and Security*, vol. 2, no. 2, pp. 107–119, 2011.
15. K.F. Ssu, C.H. Ou, and H.C. Jiau. Localization with mobile anchor points in wireless sensor networks. *IEEE Transactions on Vehicular Technology*, vol. 54, no. 3, pp. 1187–1197, 2005.
16. A. Ledeczi and M. Maroti. Wireless sensor node localization. *Philosophical Transactions of the Royal Society A: Mathematical, Physical and Engineering Sciences*, vol. 370, no. 1958, pp. 85–99, 2012.
17. D. Koutsonikolas, S.M. Das, and Y.C. Hu. Path planning of mobile landmarks for localization in wireless sensor networks. *Computer Communications*, vol. 30, no. 13, pp. 2577–2592, 2007.
18. L. Hu and D. Evans. Localization for mobile sensor networks. In *Proceedings of the 10th Annual International Conference on Mobile Computing and Networking (MobiCom 2004)*, Philadelphia, USA, pp. 45–57, September 26–October 1, 2004.
19. M. Rudafshani and S. Datta. Localization in wireless sensor networks. In *Proceedings of the International Conference on Information Processing in Sensor Networks (IPSN)*, Cambridge, Massachusetts, pp. 51–60, April 25–27, 2007.
20. J.P. Sheu, W.K. Hu, and J.C. Lin. Distributed localization scheme for mobile sensor networks. *IEEE Transactions on Mobile Computing*, vol. 9, no. 4, pp. 516–526, 2010.

21. R.C. Luo, O. Chen, and S.H. Pan. Mobile user localization in wireless sensor network using grey prediction method. In *Industrial Electronics Society, IECON 2005, 31st Annual Conference of IEEE*, Raleigh, NC, pp. 2680–2685, November 6–10, 2005.
22. S.A. Mitilineos, D.M. Kyriazanos, O.E. Segou, J.N. Goufas, and S.C.A. Thomopoulos. Indoor localization with wireless sensor networks. *Progress in Electromagnetics Research*, vol. 109, pp. 441–474, 2010.
23. J. Zhang, B. Li, A.G. Dempster and C. Rizos. Evaluation of high sensitivity GPS receivers. In *Proceedings of the International Symposium on GPS/GNSS*, Taipei, Taiwan, pp. 410–415, October 26–28, 2010.
24. M.N. Shah, W.T. Riley, and D.G. Farmer. High sensitivity satellite positioning system receiver. *World Intellectual Property Organization*, Geneva, Switzerland, WIPO Patent No. 2011143604, November 18, 2011.
25. N.F. Krasner. Fast acquisition, high sensitivity GPS receiver. *European Patent Office*, Munich, Germany, European Patent EP 2293106, March 9, 2011.
26. T. He, C. Huang, B.M. Blum, J.A. Stankovic, and T. Abdelzaher. Range-free localization schemes for large scale sensor networks. In *Proceedings of the 9th Annual International Conference on Mobile Computing and Networking*, San Diego, CA, pp. 81–95, September 14–19, 2003.
27. B. Hofmann Wellenhof, H. Lichtenegger, and J. Collins. Global positioning system: Theory and practice. Springer, Wien (Austria), 5th edition, 406 pages, February 8, 2001.
28. A. Savvides, C. Han, and M. Strivastava. Dynamic fine-grained localization in ad-hoc networks of sensors. In *Proceedings of the 7th Annual International Conference on Mobile Computing and Networking*, Rome, Italy, pp. 166–179, 2001.
29. A. Savvides, H. Park, and M. Srivastava. The bits and flops of the n-hop multilateration primitive for node localization problems. In *First ACM International Workshop on Wireless Sensor Networks and Application*, Atlanta, GA, pp. 112–121, September 28, 2002.
30. D. Niculescu and B. Nath. Ad hoc positioning system (APS) using AOA. In *Twenty-Second Annual Joint Conference of the IEEE Computer and Communications*, San Francisco, CA, vol. 3, pp. 1734–1743, March 30–April 3, 2003.
31. R. Peng and M.L. Sichitiu. Robust, probabilistic, constraint-based localization for wireless sensor networks. In *Proceedings of the Annual IEEE Conference on Sensor and Ad Hoc Communications and Networks (SECON)*, Santa Clara, CA, pp. 541–550, September 26–29, 2005.
32. D.K. Goldenberg, P. Bihler, M. Cao, and J. Fang. Localization in sparse networks using sweeps. In *Proceedings of ACM MobiCom*, Los Angeles, CA, pp. 110–121, September 24–29, 2006.
33. N.B. Priyantha, H. Balakrishnan, E.D. Demaine, and S. Teller. Mobile-assisted localization in wireless sensor networks. In *Proceedings of the IEEE INFOCOM*, Miami, FL, vol. 1, pp. 172–183, March 13–17, 2005.
34. P. Bahl and V. Padmanabhan. RADAR: An in-building RF-based user location and tracking system. In *Proceedings of 2000 Nineteenth Annual Joint Conference of the IEEE Computer and Communications Societies*, Tel Aviv, Israel, vol. 2, pp. 775–784, March 26–30, 2000.
35. J. Hightower, G. Boriello, and R. Want. SpotON: An indoor 3D location sensing technology based on RF signal strength. University of Washington CSE Report #2000-02-02, February 2000.
36. C. Savarese, J. Rabay, and K. Langendoen. Robust positioning algorithms for distributed ad-hoc wireless sensor networks. In *USENIX Technical Annual Conference*, Monterey, CA, June 2002.
37. K. Whitehouse and D. Culler. Calibration as parameter estimation in sensor networks. In *First ACM International Workshop on Wireless Sensor Networks and Application*, Atlanta, GA, September 2002.
38. Z. Yan, Y. Chang, Z. Shen, and Y. Zhang. A grid-scan localization algorithm for wireless sensor network. In *Proceedings of IEEE International Conference on Communications and Mobile Computing*, Yunnan, China, vol. 2, pp. 142–146, January 6–8, 2009.
39. D. Niculescu and B. Nath. DV based positioning in ad hoc networks. *Telecommunication Systems*, vol. 22, issue 1–4, pp. 267–280, January 2003.
40. Y. Shang, W. Ruml, Y. Zhang, and M. Fromherz. Localization from mere connectivity. In *ACM MobiHoc*, Annapolis, MD, pp. 201–212, 2003.
41. S. Simic and S. Sastry. Distributed localization in wireless ad hoc networks. UC Berkeley ERL Report, 2002.

42. S.J. Julier and J.K. Uhlmann. Unscented filtering and nonlinear estimation. *Proceedings of the IEEE*, vol. 92, no. 3, pp. 401–422, 2004.
43. J.S. Kim and B.K. Kim. Minimum-time grid coverage trajectory planning algorithm for mobile robots with battery voltage constraints. In *Proceedings of IEEE International Conference on Control Automation and Systems (ICCAS)*, Gyeonggi-do, Korea, pp. 1712–1717, October 27–30, 2010.
44. G. Han, H. Xu, J. Jiang, L. Shu, T. Hara, and S. Nishio. Path planning using a mobile anchor node based on trilateration in wireless sensor networks. *Wireless Communications and Mobile Computing*, Early View (Online version of record published before inclusion in an issue), October 6, 2011.
45. D. Koutsonikolas, S.M. Das, and Y.C. Hu. Path planning of mobile landmarks for localization in wireless sensor networks. *Computer Communications*, vol. 30, no. 13, pp. 2577–2592, 2007.
46. H. Zhang and J.C. Hou. Maintaining sensing coverage and connectivity in large sensor networks. *Ad Hoc and Sensor Wireless Networks*, vol. 1, nos. 1–2, pp. 89–124, 2005.
47. V. Zalyubovskiy, A. Erzin, S. Astrakov, and H. Choo. Energy-efficient area coverage by sensors with adjustable ranges. *Sensors*, vol. 9, no. 4, pp. 2446–2460, 2009.
48. D. Han, W. Li, and Z. Li. Semantic image classification using statistical local spatial relations model. *Multimedia Tools and Applications*, vol. 39, no. 2, pp. 169–188, 2008.

Chapter 17

Mobile Wireless Sensor Networks: A Cognitive Approach

M. Sujeethnanda, Sumit Kumar, and G. Ramamurthy
IIIT Hyderabad

Contents

17.1 Introduction

Recent advances in wireless communications and digital electronics have enabled the development of low-cost, low-power, sensor nodes that are small in size and can communicate over a radio link in short distances. These tiny sensor nodes, which consist of sensing, data processing, and communicating components, are potentially low-cost solutions to a variety of real-world challenges [1].

Sensor networks are deployed in one of the following two ways [2]:

- Sensors can be deployed far away from the sensing location, that is, something known by sense perception. In this approach, large sensors that use some complex techniques to distinguish the targets from environmental noise are required.
- The sensing element of sensor nodes can be deployed in the sensing location. The positions of the sensors and the network topology are designed carefully. They transmit the time series of sensed variables to the central nodes in which data fusion and aggregation can be performed.

A wireless sensor network (WSN) is composed of a large number of sensor nodes that can communicate over radio link, which are then deployed either in the actual site of measurement or very close to it.

The deployment of sensor nodes does not need to be predetermined. These sensor nodes are randomly deployed in inaccessible terrains. This means that sensor network protocols and algorithms should have self-organizing capabilities. The unique feature of sensor networks is the cooperative effort of the sensor nodes. These nodes have small onboard processing elements. They send only the required and partially processed data to the fusion nodes or higher capability nodes.

The above described features ensure a wide range of applications for sensor networks. Some of the application areas are health, military, and security. For example, physiological data about a patient can be monitored remotely by a doctor. Although this is more convenient for the patient, it also allows the doctor to better understand the patient's current condition. We can say that, soon, WSNs will be an integral part of our lives, more so than the present-day personal computers.

Realization of these sensor network applications requires wireless ad hoc networking techniques. However, the protocols and algorithms that have been proposed for traditional wireless ad hoc networks are not well-suited for the unique application requirements of sensor networks. To illustrate this point, the differences between sensor networks and ad hoc networks [3] are outlined below:

- The number of sensor nodes in a sensor network can be several orders of magnitude higher than the nodes in an ad hoc network.
- Sensor nodes are densely deployed.
- Sensor nodes are prone to failure.
- The topology of a sensor network changes very frequently.
- Sensor nodes mainly use broadcast communication paradigms whereas most ad hoc networks are based on point-to-point communications.
- Sensor nodes are limited in power, computational capacities, and memory.
- Sensor nodes may not have global identification (ID) because of the large amount of overhead and the large number of sensors.

Because large numbers of sensor nodes are densely deployed, neighbor nodes may be very close to each other. Hence, multihop communication in sensor networks is expected to consume less power than the traditional single-hop communication. Multihop communication can also effectively overcome some of the signal propagation effects experienced in long-distance wireless communication.

One of the most important constraints on sensor nodes is the low power consumption requirement. In many applications, sensor nodes carry limited, irreplaceable, power sources. Therefore, whereas traditional networks aim to achieve high quality of service (QoS) provisions, sensor network protocols must focus primarily on power conservation. They must have trade-off mechanisms that give the end user the option to choose network lifetime, lower throughput, or higher transmission delay.

17.2 Types and Applications of WSNs

Sensor networks may consist of many different types of sensors such as seismic, low sampling rate magnetic, thermal, visual, infrared, acoustic, and radar, which are able to monitor a wide variety of ambient conditions [4].

Sensor networks can be used for continuous sensing, event detection, location sensing, and actuating some control action. The microsensing elements and wireless communication capabilities of these nodes have many application areas. The applications can be of military, environment, health, home, security, and other commercial areas.

Sensor networks consist of various different sensors that measure ambient conditions. These sensors are temperature, humidity, vehicular movement, lighting conditions, pressure, soil makeup, noise levels, object detection, strain and stress gauge, tachometers to measure speed, etc. The architecture of the sensor node is illustrated in Figure 17.1.

The general WSN is illustrated in Figure 17.2.

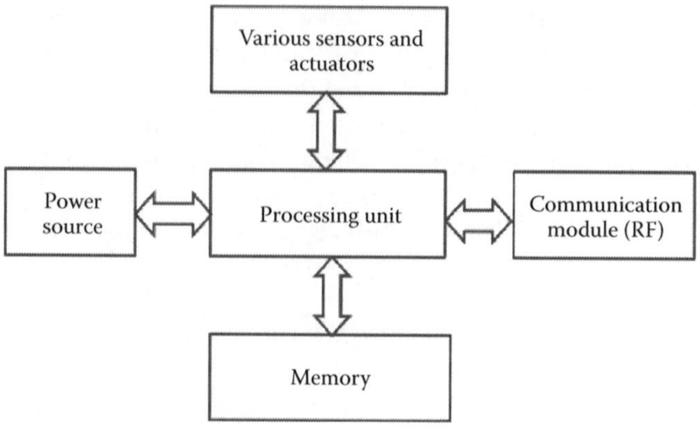

Figure 17.1 Architecture of sensor node.

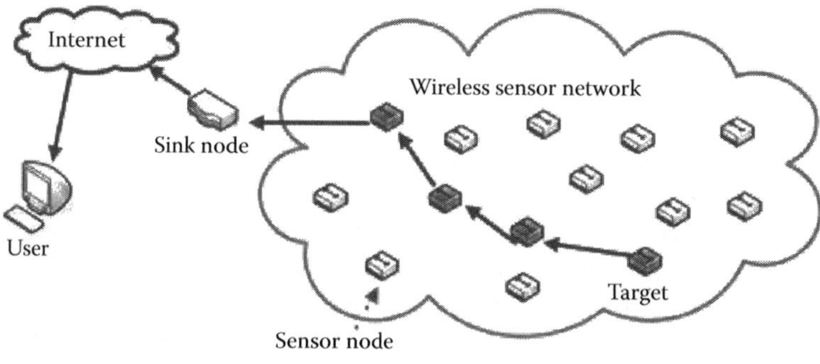

Figure 17.2 General structure of WSN.

Current deployment of wireless sensor nodes can be classified mainly in to following types: (1) terrestrial WSN; (2) underground WSN; (3) underwater WSN; (4) multimedia WSN; and (5) mobile WSN.

Terrestrial WSNs consists of hundreds wireless sensor nodes deployed for sensing a phenomenon, either in ad hoc or preplanned manner. For these kinds of networks, reliable communication in dense environment and power conservation are very important. For terrestrial WSNs, energy can be conserved by optimal routing, short transmission range, reduced data redundancy, and minimizing delays [5–11].

Underground and underwater WSNs consist of a number of sensor nodes deployed underground or underwater. These kinds of networks are expensive compared with terrestrial networks in terms of equipment, deployment, and maintenance. They are expensive due to the extra cost incurred in choosing the appropriate parts for reliable communication through soil, rocks, water, and other minerals encountered between the sensor node and the sink node. Replacement of energy sources is also a difficult task in these kinds of networks. Once deployed, it is almost impossible to replace them.

Multimedia WSN have come into existence to monitor and track events in the multimedia format. The data in these networks include audio, video, and images. These networks require high bandwidth, high energy consumption, QoS provisioning, high data rates, and high data processing and compression techniques. Video transmissions require high bandwidth to transfer data from the source node to the sink node, hence higher data rates lead to higher energy consumption. QoS provisioning is a challenging task in multimedia WSNs. It is important that QoS is required for reliable content delivery. The detailed descriptions of QoS and the factors affecting it are discussed in later sections of this chapter.

Mobile WSNs (MWSNs) consist of sensor nodes that can move on their own and collect the sensed phenomenon, and then communicate this sensed phenomenon. The key difference between static nodes and mobile nodes is that mobile nodes have the capability of repositioning and self-organizing themselves in the network. In static WSN, data can be transferred using fixed routing mechanisms whereas data transmission must be done using dynamic routing in MWSN. MWSNs pose more challenges in terms of deployment, localization, navigation, control, coverage, energy, maintenance, data processing, self-organizing, etc. MWSN applications include real-time monitoring of hazardous minerals, disaster areas, military surveillance, and tracking. Design issues and applications of MWSNs are explained in the latter sections of this chapter. WSNs are classified as shown in Figure 17.3 based on the applications' requirements.

17.2.1 Environmental and Health Monitoring

Environmental monitoring can be further classified into two types:

1. Indoor environment monitoring
2. Outdoor monitoring

The deployment of sensor nodes in an indoor environment can be used optimally to monitor light, temperature, air streams, and indoor air pollution. For example, UC Berkeley deployed 50 "Smart Dust motes" throughout the Departments of Electrical and Computer Science to monitor light and temperature [13]. Moreover, a lot of energy is wasted in buildings due to unnecessary heating or cooling. WSNs can be used for a healthier environment and a greater level of comfort for residents. Other indoor applications may be for fire detection and earthquake damage sites

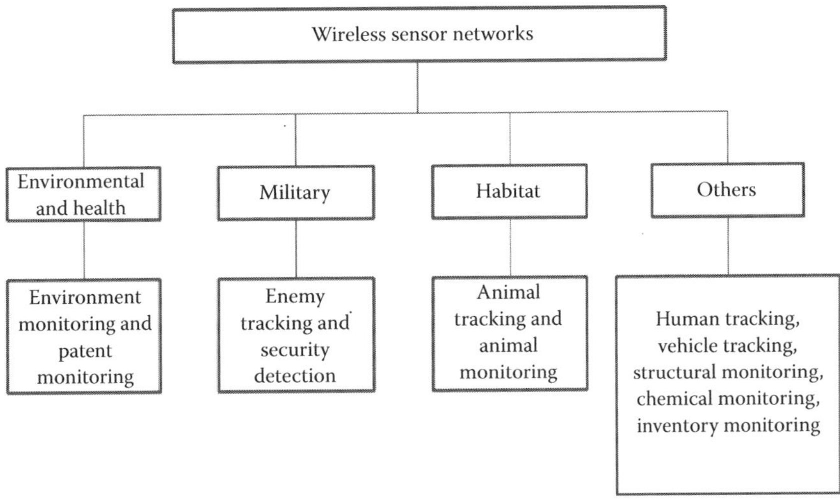

Figure 17.3 Classification of WSN.

[14]. In addition to the systems described above, civil engineering research has shown that it is possible to monitor the health of structures based on vibrations [15].

Outdoor monitoring is a vast area for WSN applications. The most popular example for outdoor monitoring networks is the Great Duck Island project [16]. This network was used for habitat monitoring. Other applications pertaining to outdoor monitoring have been implemented for environmental observations and forecasting weather phenomenon. An early implementation of sensor networks was the Automated Local Evaluation in Real-Time (ALERT) system developed by the National Weather Service for monitoring rainfall and predicting floods in California in the 1970s [17]. Other outdoor applications include agricultural applications such as precision farming and irrigation management. These applications enable an efficient and economic way to use water resources by monitoring soil, air humidity, etc., and also helps in frost detection and as a warning for pesticide application and disease detection. Precision agriculture is mainly used for lowering costs and increasing yield quality [18].

Health care applications are used for patients' health monitoring systems. Systems such as FireLine, Heart@Home, and LISTSENse are some of the examples of applications of WSNs in health care. The main objective of these applications is to provide better service and increase the personal health of the patients.

17.2.2 Military Applications

Military applications are closely related to WSNs. Regarding military applications, the areas of interest are enemy tracking and battlefield surveillance [19,20]. For example, a number of motes are deployed in the battlefield to detect the intrusion of hostile units. Then, the defense system is able to respond with preventive measures [21].

The Ohio State University has also developed a similar application called *A Line in the Sand*, which refers to the 90 nodes that are deployed and capable of detecting metallic objects. The ultimate objective of this application is to detect objects with metallic content such as vehicles, armed soldiers, etc. [22]. The above examples mostly refer to wartime scenarios, their results can also be used in applications such as homeland security and property protection. These activities perhaps may be the future of WSNs.

17.2.3 Habitat Monitoring

Habitat monitoring applications include animal tracking and animal monitoring. The famous ZebraNet [23] application is deployed at the Sweetwater Game Reserve in Kenya to monitor and collect the movement data from zebras. wildCENSE [24] is another application used for monitoring the movement of animals.

17.2.4 Other Applications

Applications such as inventory management, asset tracking, fleet management, etc., come under this category. Delivery and distribution systems are another area of applications for WSNs [25,26].

17.3 MWSNs Challenges and Design Issues

WSNs have gained much attention in the last few years. "One deployment and multiple applications" is the motto of WSNs nowadays. Therefore, special capabilities are required for handling multiple applications. However, it is not possible to run all the programs on these embedded nodes, which have tight memory constraints. The use of mobile agents seems to be an effective idea to overcome this situation. The basic MWSN is shown in the figure below. These mobile agents are special nodes that migrate between nodes and perform the task allotted to them. These mobile agents are very useful in efficient data fusion and dissemination in WSNs [27–29]. The benefits of mobile agents in WSNs are that they reduce bandwidth consumption by moving the data processing elements to the location of the sensed phenomenon, further reducing the energy consumption of the nodes. They also ensure uniform energy consumption among the sensor nodes, that is, they remove the battery drainage problem in sensor nodes near the sink node because they have to transfer more data compared with the nodes that are far away from the sink node.

Thus, with the introduction of mobile agents in WSN, a new paradigm called MWSNs came into existence. The architecture of MWSNs is shown in Figure 17.4. Furthermore, MWSNs can have three paradigms

1. Mobile sensor nodes, static sink node
2. Static sensor nodes, mobile sink node
3. Mobile sensor nodes, mobile sink node

The first paradigm can be explained as where the source nodes, which sense the sensing phenomenon, are moving around in the entire sensor field region and reporting the sensed phenomenon to the sink node in a periodic manner. The second paradigm is where the source nodes, which sense the phenomenon, are static and the sink node moves around the entire region to collect the sensed phenomenon. The third paradigm is where both the source nodes as well as sink nodes will be moving in the sensor field and, when both come into range, where they exchange the sensed phenomenon between them.

The potential applications of MWSNs are mainly where we need to track, monitor, or collect data from real-time environments. Thus, the applications include hospitals, agricultural fields, military, inventory, and asset tracking. MWSNs can be used in hospitals where doctors carry handheld embedded devices or PDAs and do their rounds attending patients. Each patient is equipped with sensor nodes to collect information; data regarding the patient's vital parameters can be transferred to the doctor's handheld device for further examination as well as for recordkeeping. Health status monitoring applications such as MobiHealth [30], was one of the early studies proposing the mobility aspects that enabled personalized health care systems. AlarmNet [31] is a combination of MWSN and IP networks. MWSN and mobile sensors enable the measurement of vital parameters of the patient and IP networks are responsible for passing the information to the health care professional.

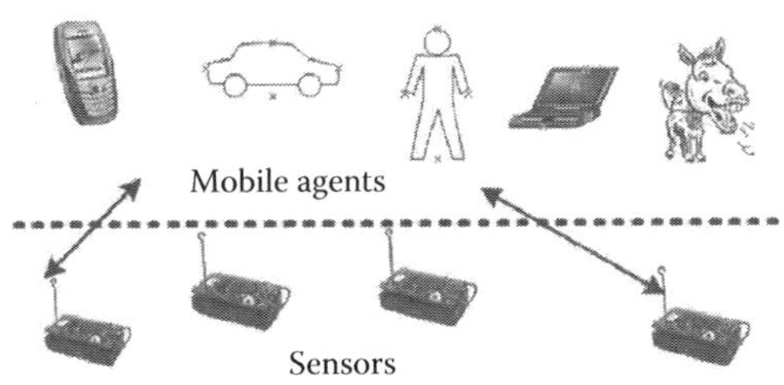

Figure 17.4 Mobile WSN.

MWSNs can be used in precision agriculture. The data collected by the sensors deployed in an agricultural field is very useful for the future. Based on past data, the farmers can opt for different crops for different seasons. Thus, collecting data from farms and storing them on an embedded device is not a feasible option. With the help of mobile sink nodes, which can travel through the entire farm periodically in different seasons, information is collected and furthermore that information can be sent to an expert's location. There, the data can be preserved for longer periods for further use.

Other applications include habitat monitoring, in which sensor nodes are attached to animals to monitor their movements and to track them in different seasons. These kinds of applications are very useful for knowing the migration strategies of animals with respect to various seasons of the year. Inventory and fleet management for the bigger companies is very complex and very tiresome work. These can be automated with the help of MWSNs. The network can keep track of the inventory and fleet easily with the help of these nodes. The widest area in which MWSNs could be feasible is in military applications. There, WSNs are often used for detecting enemy intrusion, tanker movements, and soldier's movements. All these areas need MWSN platforms as a base to perform the tasks in efficient manner. The challenges in MWSNs are as follows.

17.3.1 Resource Constraints

The design and implementation of any WSN application suffers from resource constraints such as energy, memory, and processing. The MWSN will also have the same constraints as far as resources are concerned.

17.3.2 Dynamic Topologies

In MWSN, the constant change in the positions of various sensor nodes, due to mobility, changes the network topology very frequently. Furthermore, sensor nodes may also be subject to environmental changes like dust, vibrations, wind, and other conditions that cause change in the network topology. Also, harsh conditions may cause malfunctions in some sensor nodes.

17.3.3 Quality of Service

The QoS in MWSN refers to how data reaches the mobile agents and how accurate the data is, as well as the latency of the data reaching the mobile sink node. The most difficult challenge in MWSN is QoS provisioning. These problems are due to constant changes in network topology and how fast the network is self-organizing and transferring the data to the mobile agents.

17.3.4 Variable-Link Capacity

Compared with wired and static WSN, the wireless link depends on the interference level when deployed in harsh environments. In addition, they exhibit abnormal characteristics over time and space due to obstructions. Because of the presence of mobile agents, links are varying continuously, making QoS challenging.

17.3.5 Ad Hoc Nature

MWSN contains a large number of sensor nodes deployed randomly over the field. The lack of predetermined infrastructure and mobility makes maintaining network connectivity a major issue. To overcome the challenges present in constructing MWSNs, the following design goals need to be followed.

17.3.6 Design Goals

17.3.6.1 Resource-Efficient Design

In every type of WSN, energy efficiency is important to increase the lifetime of the network. This can be accomplished by implementing energy-aware routing and energy-saving modes of the sensor nodes.

17.3.6.2 Adaptive Network Operation

Adaptability in the MWSN is very crucial. It enables networks to adapt to dynamic or varying wireless channel conditions in mobile environments. It is also important for gaining the new connection requirements needed for the network to maintain connectivity issues.

17.3.6.3 Secure Design

During the design of the security issues for MWSN, issues such as robustness of the communication with varying topologies, secure routing, resilience to node capture, secure group management, and intrusion detection should be addressed. The overhead associated with security issues should be balanced with the energy and QoS requirements of the networks.

17.3.6.4 Self-Configuring and Self-Organizing Networks

The dynamic topology caused by node mobility and failure needs self-organizing and self-configurable architectures. It should be noted that with the help of self-configurable and self-organizing WSN, the addition and deletion of nodes is possible without affecting the objective of the application.

It is very challenging to meet all the design goals simultaneously. Different applications have different requirements on design objectives. Therefore, there should be a balance between the different parameters when designing the MWSN networks.

17.3.7 Design Issues

To design a MWSN that meets the above design goals, we consider some important design issues.

17.3.7.1 Scalable Architectures

The MWSN supports various heterogeneous network applications with various requirements. It is very much necessary to develop flexible and suitable architectures for various applications. The flexible and hierarchical networks enhance robustness and reliability in the network. The newly developed architectures must support or must be interoperable with existing network topologies such as Internet-based networks, cellular networks, etc.

17.3.7.2 Hardware and Software Design

Hardware design is the most challenging issue in the sensor network. The hardware architecture of mobile sensor nodes is composed of four basic components:

1. Sensor
2. Processing unit
3. Transceiver unit
4. Mobile unit

Sensors and actuators are the physical devices that respond to the change in the sensing phenomenon and convert them into the appropriate signals. These signals are supplied to the processing unit for further processing. The processing unit is responsible for processing the data, storing the data, and issuing the control commands to the actuators. As a whole, the processing unit controls the functionality of entire components in the sensor network. The transceiver is the component responsible for connecting the sensor nodes to the network and is also responsible for maintaining network connectivity. The mobile agents in the network are responsible for covering the entire region and collecting the data from the entire sensor field. The operations carried out by all the components in the node or network consume power. Therefore, the selection of low-power and low-cost equipment is necessary. Power can be conserved by enabling sleep modes for the sensor nodes and enabling idle and sleep cycles for the transceiver. The power source for mobile agents can be replaced or recharged as they will be moving and are easily accessible compared with the other static sensor nodes.

The application software of sensor networks must be designed to be very user-friendly as well as being easily assessable. The application programming interface (API) does this job. The API enables rapid development and deployment of the networks. Furthermore, operating systems for sensor networks are available. TinyOS is one of the earliest and most popular operating systems for sensor networks [32]. A recent development in these operating systems is the introduction of mobility patterns as well as support for various network protocols and multitasking. Simulators like NS2 and NS3 will help in the prototyping stage of the network before actual deployment by simulating the entire network [33,34]. These simulators are mainly used ad hoc and sensor networks. Recent developments include WSNs and cognitive networks.

17.3.7.3 Itinerary Planning

Itinerary is defined as the route followed by mobile agents to collect data from the sensor field. Itinerary planning helps mobile agents to select the set of nodes to be visited by the mobile agent and determines the path in an energy-efficient manner. The order in which the mobile agents cover the sensor region will have a considerable effect on the energy consumption of the network. Finding the best possible path is a challenging task. For a better understanding of itinerary planning, we should know how the sensor fields are represented, divided, and how data is collected. The basic tasks are leveling, sectoring, and clustering. These terms are explained briefly to better understand itinerary planning. For detailed descriptions, please refer to the study by Hajela et al. [35].

Leveling. The entire sensor field is divided into a number of levels. Leveling can be done by using the number of hop counts required to transmit data from a source to the base station (BS) or the various power levels to which that BS can transmit.

Sectoring. A BS using a directional antenna will transmit the signals with maximum power and divide the sensor field into equiangular sectors with an angle of θ (let θ = 45°). So that each node knows its level ID and sector ID.

Clustering. Clusters can be formed based on the signal strength and the cluster head will be decided by using a round-robin technique [36]. The entire sensor field is divided into levels and sectors as shown in Figure 17.5.

Itinerary planning can be categorized as static planning, adaptive planning, or hybrid planning. Static planning is when the path followed by the mobile agents is predetermined and the mobile agents will follow that path to cover the entire sensor field to collect data. Adaptive planning is when the path followed by the mobile agents is determined dynamically based on the current network status. Hybrid planning is the combination of both static planning and adaptive planning. The path followed by the mobile agent is predetermined but, based on the network status, the path can be dynamically decided.

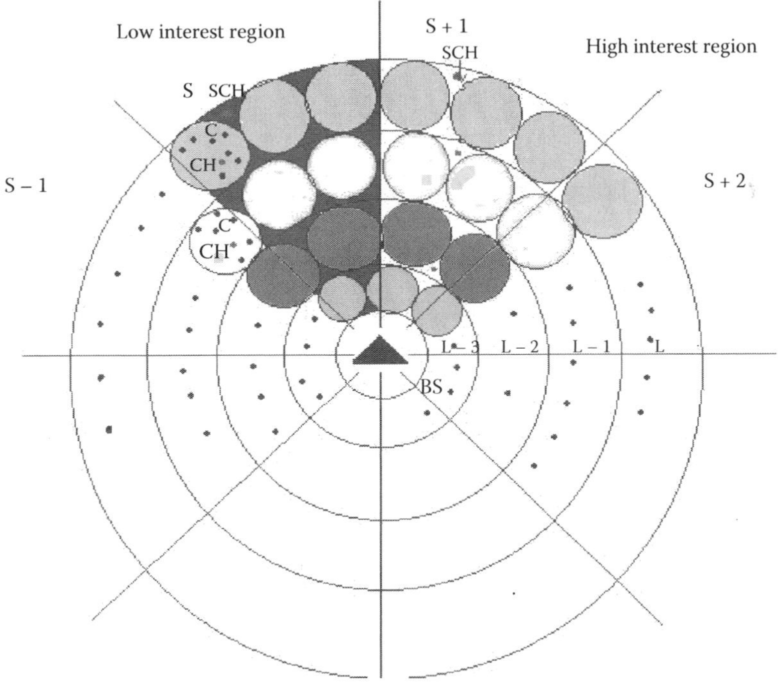

Figure 17.5 Sensor field after leveling sectoring and clustering.

17.4 QoS in MWSN

MWSNs pose more challenges compared with the regular WSNs, wireless local area networks (WLAN), and mobile ad hoc networks (MANET). The following are the challenges posed by MWSN [1,37]:

- A sensor node suffers from very limited power source.
- A sensor network topology faces frequent changes due to external forces such as animals, humans, power, or software failure.
- A sensor node does not have a global ID, which makes most of the current network protocols inapplicable to WSN.
- Sensor networks mainly operate without any human intervention and they should be self-configurable and self-organizing.
- Sensor nodes are densely deployed increasing redundancy and collisions.
- Sensor nodes normally use the broadcast communication model, whereas traditional networks use point-to-point communication.

For all the above reasons, implementing QoS in WSNs differs from regular QoS implementations in other types of networks. The next section is a discussion of QoS in MWSNs in general followed by some challenges in deploying normal QoS mechanisms in MWSNs.

QoS is the ability to provide different performance levels to different applications and users, or to guarantee certain performance measures for latency, jitter, throughput, bit error, error rate, packet dropping, etc. Delay is the time that the network spends to deliver a data packet from the source to the destination. Jitter, in turn, is the delay between two consecutive packets in the same frame. Packet dropping determines the maximum amount of data packets that can be lost in the frame to provide good QoS.

QoS metrics are important if the network limitation is limited resources. QoS is not only dependent on the abovementioned performance measures but also depends on some nonfunctional parameters. The nonfunctional parameters include energy or power, reliability, robustness, security, scalability, mobility, cost-effectiveness, delay, etc.

The main factors affecting QoS parameters can be represented in the form of Figure 17.6.

- *Power*. This considers the most critical limitation. Therefore, almost every protocol proposed considers the energy problem. The main power consumer, as discussed previously, is communications. Therefore, implementing a data aggregation and local data fusion technique for achieving a better service (QoS) is always the price of energy [38].
- *Bandwidth*. Bandwidth is one of the factors affecting QoS parameters; thus, the lack of bandwidth presents more difficulties in achieving QoS in WSN. Techniques for utilizing bandwidth efficiently based on the nature of the data stream are necessary to overcome the scarcity of bandwidth.
- *Memory size*. The limitation of memory size affects most applications to enhance WSN networking capabilities. In some cases, local memory is not enough to load the whole QoS technique to implement QoS measures.
- *Lifetime*. The nature of WSN's lifetime is limited because most nodes operate on unchangeable power sources such as a battery. Another reason is the ease of node damage. Attempts to recharge the battery using solar or wind power have been proposed.

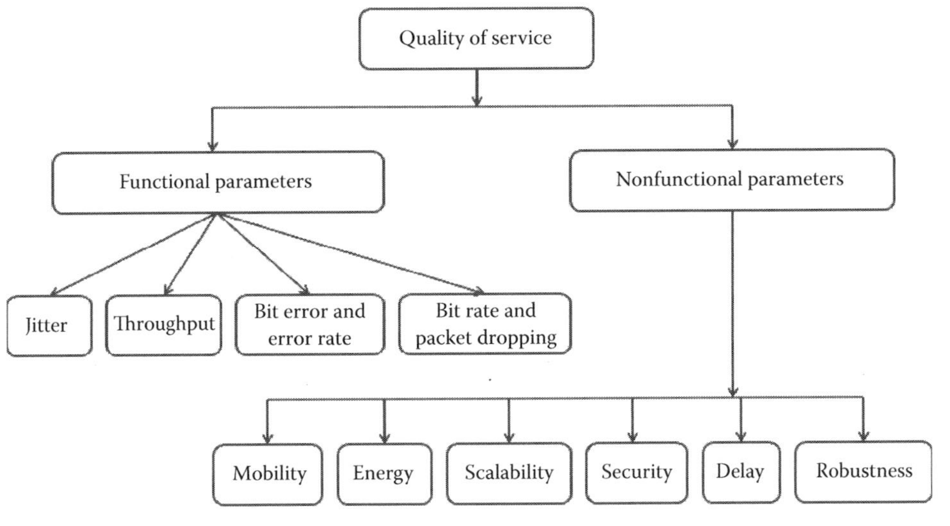

Figure 17.6 QoS and factors affecting QoS.

- *Redundancy.* It may help achieve reliability but it may add overhead and consume power to aggregate traffic to the sink. Also, it may add some sort of latency and complexity to QoS design [39].
- *Application diversity.* WSNs are considered to be application-specific rather than general purpose. They carry only the hardware and software actually needed for an application. The vast number of applications in WSN offers different QoS requirements.

The major challenges in the implementation of QoS in WSNs are resources, that is, energy, bandwidth, and transmission capacity of sensor nodes. These are some extreme resource constraints; among these constraints, efficient energy utilization is a considerable factor because installed batteries' replacement and recharging is highly expensive and difficult. Usually, QoS are affected by the mobility of sensor nodes, that is, a change in network topology due to this link failure and node failures occur frequently. Adaptability and self-organizing properties will help in avoiding the failures that are caused by mobility; however, implementation of the properties in WSNs that influence QoS is another challenge.

In WSNs, air acts as a transmission medium. It is less reliable and has high security risks. For instance, if a sensor network is deployed in a hostile and hazardous environment, and in coexistence with other WSNs, the other networks significantly cause interference in the newly deployed network. Thus, the data received may be faulty or unpredictable. To avoid these instances, in the process of building a sensor network, care should be taken to consider factors such as robustness, reliability, and security.

Another challenging part is QoS providing multiple BSs deployed according to the nature of applications. Currently, most of our sensor networks have only one BS or sink node.

Soon, WSNs will be a part of our everyday life. These networks have to coexist with huge numbers of WSNs, cellular networks, vehicular networks, etc. There is a lot of demand for spectrum. Nowadays, spectrum is becoming very scarce. Soon, we will need to share and utilize the spectrum very efficiently. This spectrum sharing and utilization lead to a new concept called cognitive networks. Cognitive networks are designed to use the spectrum more efficiently and in an opportunistic way. The primary user (PU) and secondary user (SU) share the spectrum for better utilization of the spectrum. The ability to sense the environment and adapting to the environment for achieving optimal performance is the design motto of cognitive networks. A detailed description of cognitive networks can be found in the following section.

17.5 Cognitive Radio

Currently, we are living in an era of conventional fixed spectrum assignment policy, in which regulatory authorities (government bodies) assign the spectrum chunks to service providers on a long-term basis for large geographical regions. Once these spectrum chunks are leased, they can be used only by the licensed users. Even though they are vacant for some time or for some geographical region, they cannot be used by unlicensed users. However, it has been found that those licensed spectrum chunks are underutilized as they remain unused 90% of the time temporally and spatially [40]. Considering this scenario, the federal communications commission (FCC) and the research community is focusing on "dynamic spectrum allocation" [41]. The goal of dynamic spectrum allocation is to remove regulatory barriers and facilitate the development of secondary markets in spectrum usage rights among the wireless radio services. Dynamic spectrum allocation technology senses open channels and allows devices to communicate in unused parts of the spectrum. These unused parts of the spectrum, also called spectrum holes/white space, can be visualized in Figure 17.2. Dynamic spectrum allocation is implicitly required and can be achieved with the use of cognitive radios (CR) to improve spectral efficiency. CRs help in the efficient use of the existing spectrum through opportunistic access to licensed bands without interfering with the PUs. CR is an intelligent wireless communication system that is aware of its surrounding environment. It works on the idea of intelligent signal processing and decision making to enable the radio to not just utilize the spectrum efficiently but also to manage/adapt the other wireless parameters as required. CR is able to dynamically reconfigure its center frequency, waveform design, time diversity, and spatial diversity options to achieve efficient communication without interfering with PUs (Figure 17.7).

A simple model of a cognitive radio in which the cognitive engine interacts with the radio is shown in Figure 17.8.

The cognitive engine is the heart of the cognitive radio and is responsible for all intelligence, adaptation, and controls.

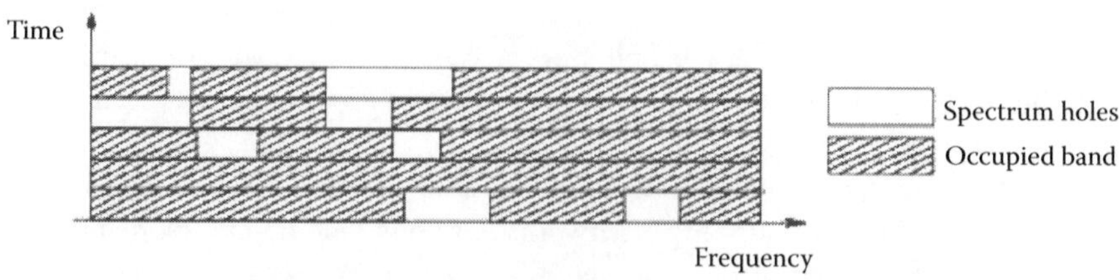

Figure 17.7 Spectrum holes in between the occupied frequency bands.

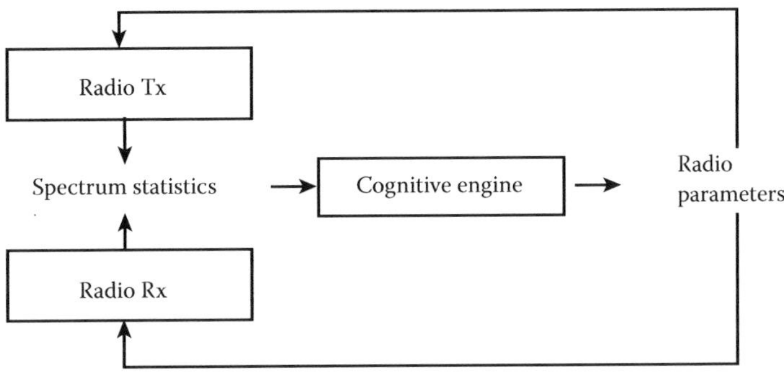

Figure 17.8 Simplified model of cognitive radio.

17.6 CR-Based WSN

In both CR and WSN, sensing tasks are performed to collect information from the operating environment about spectrum occupancy and environmental parameters, respectively, and then appropriate actions are taken accordingly. CR-based WSN (CWSN) speaks about the application of CR in WSN and not only in the PHY and MAC layers but also in all the layers and thus employs a cross-layered approach. It is promising as well as challenging for WSNs to adopt CR technology. It enables the WSN to sense spectrum holes and utilize vacant frequencies to improve spectrum utilization. It is also capable of increasing its own QoS and throughput by adaptively and cognitively changing various transmission and reception parameters such as transmitted power, operating frequency, modulation, pulse shape, symbol rate, coding technique, and constellation size. A wide variety of data rates and QoS can be achieved, improving power consumption and network lifetime in a CWSN.

CR technology in WSNs can provide access not only to new spectrum bands but also to spectrum bands with better propagation characteristics. Generally, the lower frequencies have better propagation characteristics than the higher ones (refer to the free space propagation formula). The operation of WSNs at lower frequency bands allows range extension and higher energy efficiency. This helps in getting simpler topology as well as fewer sensor nodes to cover a given area because, with lower frequencies, the transmission range of the same node with the same transmission power is increased. Higher transmission range improves several important factors in WSNs including network connectivity, lifetime, and end-to-end delay.

Some of the advantages of using low frequency for transmission are:

- Higher transmission range
- Fewer sensor nodes required to cover a specific area
- Lower energy consumption
- Smaller number of hops to the destination
- Lowered end-to-end delay

Another advantage of using CR technology in WSNs is that data from various sensor nodes, which are not spatially correlated or which have low redundancy, can be transmitted to the sink simultaneously in different channels (noncontiguous transmission). This reduces the delay and enables the sink to monitor a large number of nonspatially correlated information in a real-time manner.

17.7 CWSN Architecture

CWSN is also like WSN in the sense that it consists of several tiny sensor nodes with all the constraints that a normal WSN has, especially the "limited battery energy." They differ in their transceiver hardware architecture and states. In a CWSN, the hardware also consists of a cognitive module that is responsible for spectrum sensing and adaptively changing the transmission parameters in a reconfigurable transceiver (the reconfigurable transceiver is another advantage of the CWSN). Apart from these, there is also a cognitive engine in the cognitive module, which is responsible for learning and achieving cognition to make the changes automatically without human intervention. This cognitive engine is also responsible for controlling the changes that should take place in the transmitter and receiver.

Cognitive engine works on the concept of cognition cycle (CC) [42]. States of a CC are shown in Figure 17.4. The cognitive engine is composed of six main states: observe, orient, act, decide, plan, and learn. It enables the nodes to achieve context-awareness and intelligence so that it can be aware of its operating environment to sense for the white spaces, and use them in an intelligent and efficient manner. With regard to WSN, this cognitive engine may also assist in intelligent localization, routing, and scheduling in WSN (Figure 17.9).

In contrast to WSN, the CWSN nodes have an additional state called sensing state where they keep sensing their environmental parameters. Spectrum sensing state consumes a lot of energy

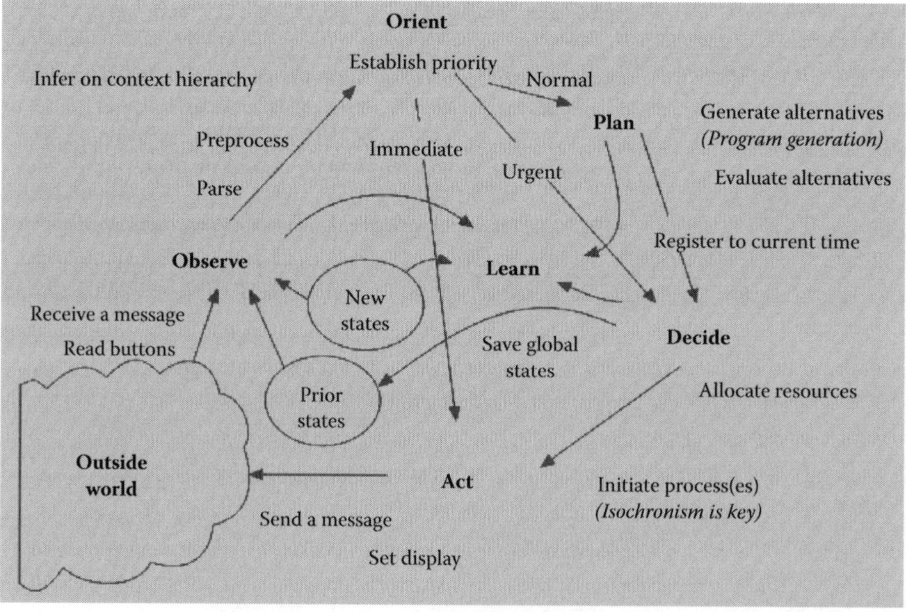

Figure 17.9 Cognitive cycle. (Redrawn from I. Howitt and J. Gutierrez. IEEE 802.15.4 low rate-wireless personal area network coexistence issues. In *Proceedings of IEEE WNCN* 2003.)

because it is directly related to transmission and reception, which is the most energy consuming activity of a CWSN/WSN node [44] (Figure 17.10).

Spectrum sensing can be done in distributed as well as centralized fashion as shown in Figure 17.6. In the distributed scheme, all the nodes keep sensing the spectrum environment on their own and compete with other sensor nodes to grab the unoccupied spectrum. However, the problem in this architecture is that all the nodes essentially need a spectrum-sensing module, which may not be economically feasible. In the centralized scheme, there is one network coordinator that has the responsibility of spectrum sensing as well as spectrum scheduling, that is, allotting the free channels to the needy nodes according to their need in preferential order.

However, in centralized control there has to be a separate control channel, which could become a problem. In this control channel, there is a broadcast channel switch command according to which the sensor nodes change their Tx/Rx frequencies. The control channel could be from a licensed or ISM band. But there is always a fear that the control channel could somehow get faded, and in case this occurs, the entire network will be in chaos.

In a common WSN scenario in which there are a large number of sensor nodes, it may not be feasible to have a spectrum-sensing module on each node. The feasible architecture will be to make only a few specialized nodes capable of performing the spectrum sensing. The network co-coordinator or the cluster head can do this task in realistic deployment. If we take a deterministic deployment scenario like a home or office, then easily a few specialized nodes capable of spectrum sensing can be deployed especially for spectrum sensing [45]. Because spectrum sensing is a repetitive process, which would consume extra energy from battery-powered sensors, implementing spectrum sensing in all nodes in a WSN may not be efficient in terms of energy consumption (Figure 17.11).

Sensing the target frequency/frequencies can be fixed or vary depending on the changing environmental and mobility condition of the nodes. Here, the cognitive engine can play an important role. After collecting spectrum sensing data over a long period, the cognitive engine will be able to know at what time, which channel, or which specific band has to be sensed, instead of sensing all the channels every time. The channel occupancy pattern also varies spatially. Hence, the cognitive engine can assist the spectrum-sensing nodes to decide which channel/channels to sense at a particular time and in a particular geographical region.

Sensing duration depends on the accuracy of sensing and also the required probability of detection. The longer the sensing duration, the more complex the algorithm for it, and thus the

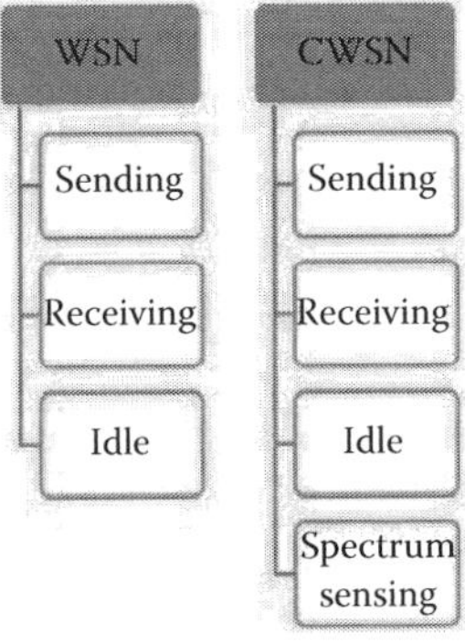

Figure 17.10 Environmental parameter sensing in CWSN. (Redrawn from J. Mitola III and G.Q. Maguire. Cognitive radio: making software radios more personal. *IEEE Personal Communications*, vol. 6, pp. 13–18, 1999.)

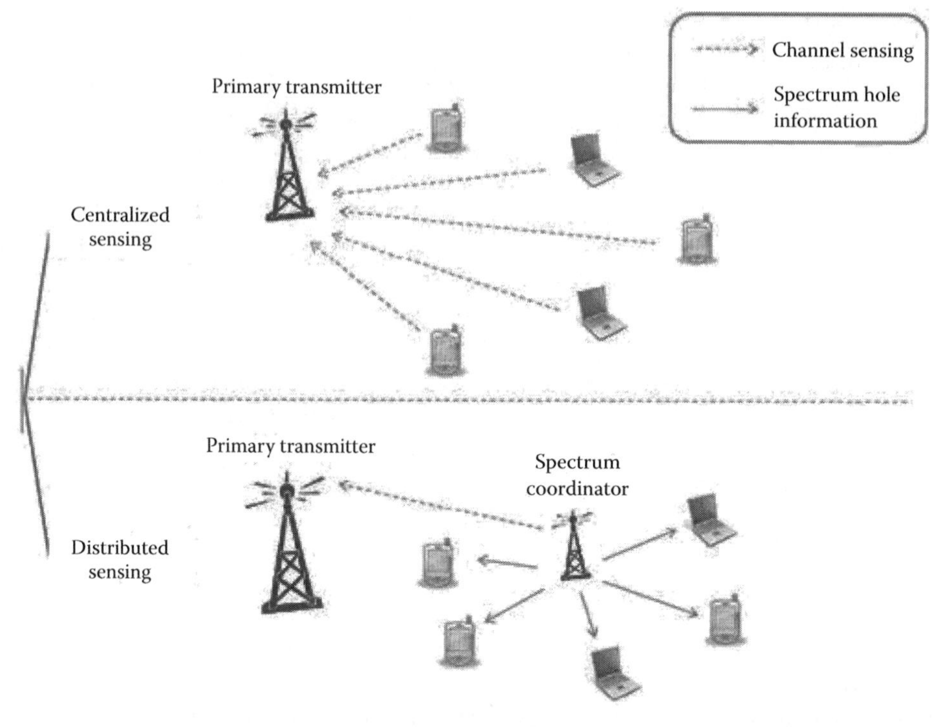

Figure 17.11 Centralized and distributed sensing. (Redrawn from D. Cavalcanti, S. Das, J. Wang et al. Cognitive radio based wireless sensor networks. In *Proceedings of the 17th International Conference on Computer Communications and Network*. IEEE Press. pp. 1–6, 2008.)

more energy it will consume. There has to be a trade-off between sensing duration and sensing accuracy. During sensing, there has to be network-wide quiet period, that is, nodes have to suspend their transmission following some schedule, which would have been prebroadcast well ahead of time by the coordinator to avoid any overlap; otherwise, false alarms may occur, indicating that some channel is occupied because of its own transmission or by some of the member sensor nodes.

Spectrum sensing has to be done only during network-wide quiet periods. This is usually done to detect incumbents at low incumbent detection threshold (IDT) values and avoid false alarms (event when a system detects a nonexistent PU). All nodes are given a fixed schedule of quiet periods that they have to follow essentially. These quiet periods can be scheduled well ahead of time by a broadcast through the network coordinator so that all nodes can adjust their transmission to avoid overlaps with scheduled quiet periods [46]. This will require a very tight and proper coordination among the nodes in WSN quiet period (QPs) [47].

QPs may be a critical issue for high-throughput networks, but in WSN, the traffic load is typically much lower and it won't be that much of an overhead because most of the time, the sensors would be in a "stand by" or "sleep" mode [48].

Now, we should be acquainted with some terms related to spectrum sensing, which may be helpful for further readings. There have been some limits set by the FCC for IDT, probability of detection (PD), probability of false alarm (PFA), maximum probability of false alarm, channel move time (CMT), and channel closing transmission time (CCTT) for CR networks. The same set of limits would be applicable to a CWSN. However, some parameter values can be relaxed for a WSN due to the much lower transmission power used by the WSN transmitters compared with the 802.22 devices for which these protocols and regulations are intended [46].

Modifying the existing 802.15.4 protocol (i.e., ZigBee) to suit the CWSN physical and MAC layer will be a very good approach. The 802.15.4 standard defines 16 channels, each at 2 MHz bandwidth, in the 2.4 GHz band [49], among which, only four are nonoverlapping with the 802.11 channels (at 22 MHz bandwidth) in the same band [46] (Figure 17.12).

Whenever an incumbent signal is detected well above IDT, WSN has to switch to a backup channel within the CMT to avoid interference. This requires the coordinator to broadcast the channel switch command. The channel switch command also consists of the scheduled switching times for the nodes. This command indicates the WSN nodes to switch their channels to the free available channel as specified by the coordinator. In case a free channel is not available, they have to utilize the backup channel. This switching is a responsibility held by the coordinator, which decides how to allocate channels, that is, scheduling of free channels among the needy nodes. The coordinator can use its own spectrum-sensing results (centralized spectrum sensing) or it may use the spectrum-sensing reports (distributed spectrum sensing) from other specialized spectrum-sensing nodes.

In practice, the coordinator has to make a list of backup channels available as there may be a good probability that many backup channels are available at a particular time and geographical region. Generally, in WSN a very fast incumbent recovery mechanism is not required. Once the PUs come into the picture, the nodes have to vacate the channel very fast. However, in some WSNs, in which there is very tight delay requirement, there has to be provision for backup

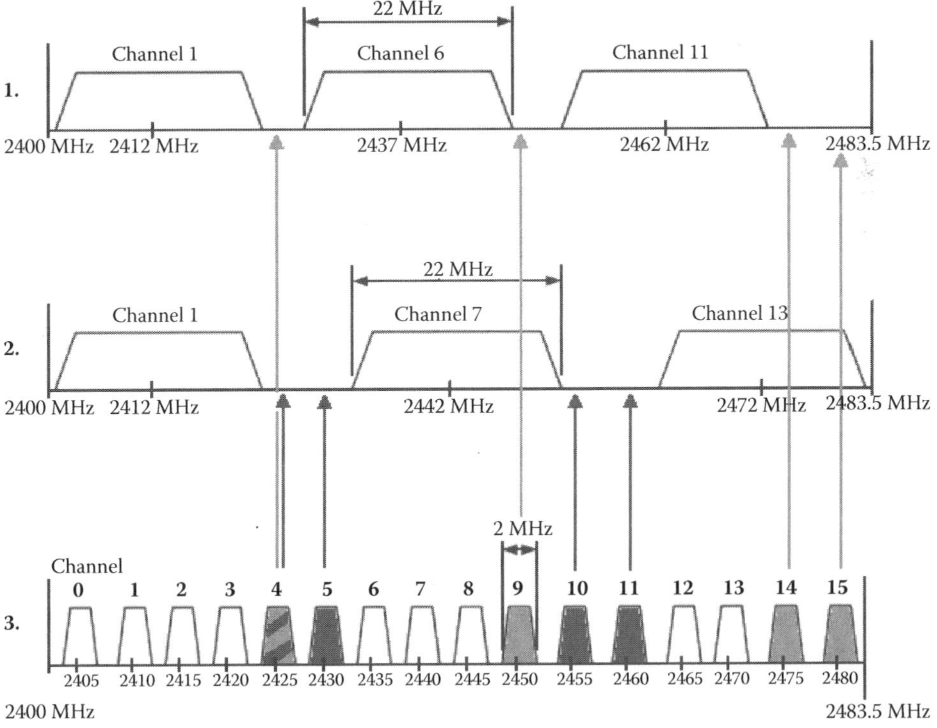

Figure 17.12 Overlap between ZigBee and WLAN channels.

channels. This backup channel can be either from the licensed band shared with the PUs or from the unlicensed band shared with other SUs. Sensing of the backup channel has to be done regularly to make sure that the backup channel is readily available and clean.

17.8 Spectrum-Sensing Schemes for CWSN

Energy-efficient spectrum-sensing techniques are required to meet the power constraints of the CWSN. Especially in the case of mobile cognitive wireless sensor networks (MCWSN), the constraint on the energy will be much more than static CWSN. Similar to static cognitive wireless sensor networks (SCWSN) in the MCWSN, a question is raised, how much energy should be spent on channel sensing? There are spectrum-sensing techniques that are quite accurate but have a very high energy budget. However, such accuracy may not be needed all the time. Because, in some cases, the interfering signal may be sporadic or may be perceived with very high power, and thus is easy to detect. Hence, a cognitive spectrum-sensing system is required to tailor the energy budget of spectrum sensing according to the signal strength.

A sensing energy budget should also be tailored according to the size of the packet to be transmitted. In WSN, there are packets of several types and sizes, and these packets have different priorities. Sensor packets can be as small as a single to few tens of bytes; therefore, selecting the right amount of energy that has to be devoted to spectrum sensing might significantly improve energy efficiency. Loss of long packets may cause retransmission of the packet, which may be even more costly. Hence, in such cases, high energy budgets can be applied to sense the spectrum. But when small packets need to be transmitted, then low-energy budget–sensing algorithms can be applied. In this way, overall lower energy consumption can be achieved [50].

There are several spectrum-sensing techniques available for CR and some of them can also be readily used in the CWSN. A good survey of spectrum-sensing techniques is given in an article by Stabellini and Zander [50]. Interested readers can refer to the same.

Spectrum sensing can be done individually or in a cooperative manner to identify the spectrum holes. In general, the cognitive transceiver devices have two important functionalities: spectrum sensing and adaptation. The spectrum-sensing hardware of cognitive transceivers keeps sensing the spectrum over a wide frequency band. This information is then passed to the SUs (in our case, the SUs were WSN nodes). When such a spectrum hole is found, the SUs adapts its transmission power, center frequency, and modulation, etc., to transmit efficiently as well as to minimize interference to the incumbents [51]. An implied assumption here is that all the nodes have reconfigurable hardware.

Also, while the transmission is in process, the cluster head or network coordinator, whichever is doing the task of spectrum sensing, should have the ability to detect the appearance of incumbents so that the SUs are able to change or give off the channel when the PUs starts transmitting in that channel.

Spectrum sensing for CWSN can be categorized as blind sensing and signal-specific sensing. Blind spectrum-sensing techniques do not rely on any special signal features. Some examples are energy detection [52] and eigenvalue-based sensing [53]. Signal-specific sensing techniques utilize specific signal features for sensing. Some examples are signature sensing for ATSC signal identification [54], FFT-based carrier sensing [55], higher-order statistics sensing (HOSs) [56], PLL-based ATSC pilot sensing [57], and wireless microphone covariance sensing [58].

Also, cooperative spectrum sensing speaks about itself when mobile scenarios come into the picture. In the mobile scenario, the hidden node problem may come often and unexpectedly.

Here, cooperative spectrum sensing can help a lot. The presence of multiple radios helps in reducing the effects of severe multipath because of the achieved diversity. It also reduces the probability that all users see deep fades at a time [60].

There are several spectrum-sensing techniques. Each one has its own pros and cons. Some are fast with less accuracy and some are slow with more accuracy. Also, the energy budgets of the different sensing techniques vary. Therefore, it needs to be optimized so that an overall lower energy consumption and reasonable accuracy is achieved.

In a previous study [61], we suggested a novel architecture for spectrum sensing specially for CWSN. We have come up with a new idea for a doubly cognitive architecture in which cognition is achieved not only with respect to time and space but also with respect to traffic.

17.9 Learning in CWSN

In this section, we will have a brief overview of the control/learning mechanisms that can be used in CWSN. Similar to spectrum sensing, there can be a centralized and a distributed approach. In the centralized approach, all the nodes are dependent on a centralized spectrum server for all of their spectrum requirements. The spectrum server determines and returns a suitable configuration to the nodes. In the distributed approach, all the nodes perform direct negotiations with their neighbors and decide on a suitable spectrum allocation. Control can also be achieved with local algorithms apart from the centralized and distributed approaches. Every node in the network follows some set of rules locally instead of having a centralized controller or distributed negotiation.

In centralized architectures, there is a demand for a control channel that is available to every node in the deployment, that is, a channel used to exchange control messages has to be known by every node and has to be unoccupied by other stations in that area. However, some situations may occur in which this control channel goes into deep fade or becomes occupied (in case it is a shared channel with the PUs). Moreover, there is an issue of the system being trapped in congestion when more nodes participate in the network. Distributed control is better suited for large environments in which the same control channel cannot be maintained throughout the large sensor field. However, this is more computationally intensive and the performance may become suboptimal. In an article by Doerr et al. [62], several CR control algorithms have been discussed.

There are several control mechanisms in the popular literature. Interested readers can refer to rule-based reasoning [63,64], responsive surface methodology (RSM)/design of experiment (DoE) [65,66], game theory [67–71], genetic algorithms [72,73], linear programming [74–76], and swarm algorithms [77].

17.10 Challenges in Designing a CWSN

17.10.1 Lifetime Maximization or Energy Efficiency

In many cases, it is impractical to recharge nodes after they are deployed because WSN nodes do not receive personal human intervention/care and usually get deployed in the field at random unknown locations. The same thing applies to CWSN. It inherits the problem of lifetime maximization and energy efficiency from the WSN. In fact, the situation deteriorates because, apart from performing data sensing, the nodes in CWSN are also involved in spectrum sensing. Also, they have to tune their RF parameters according to the requirement to achieve spectrum efficiency. Various modulation schemes, data rates, and coding schemes also directly influence power consumption at each node.

17.10.2 PU Detection and Localization

Sensor nodes are seen as the SUs in CWSN. These nodes share the licensed band with the PUs, but they must avoid interference of the PUs. To avoid interference, sensor nodes must be aware of the PUs and their locations within the region of interest. Hence, CWSN requires localizing the presence of a PU within the vicinity of the network.

17.10.3 Fusion

Information fusion deals with the combination of information from the same source or from different sources to obtain improved fused estimates with greater quality or greater relevance. In the CWSN scenario, apart from data sensing, the nodes are also involved in spectrum sensing. Most of the time, this spectrum sensing is done in a cooperative manner in which the spectrum-sensing nodes share their information with each other. This makes the fusion task even more challenging. As more sensors are deployed in harsher environments, it is important that sensor fusion techniques should be robust and fault-tolerant, so that they can handle uncertainty and faulty sensor readouts.

17.10.4 Routing

In CWSN, topology changes dynamically because sensor nodes can adjust their transmission parameters and it can turn its transceiver on or off based on the presence of PU. If a node exhausts its available energy, it ceases to function. Moreover, in the dynamic environment of CWSN, spectrum is not always available for data transfer to all the sensor nodes. This makes the routes to and from the BS to the nodes very dynamic. This adds new challenges to routing in CWSN.

17.10.5 Resource Allocation Problems

Resource allocation, that is, spectrum scheduling in CWSN should allocate spectrum fairly among all the sensor nodes and, at the same time, it should increase the spectrum utilization.

17.10.6 Power Allocation

Power allocation is another challenging problem to consider; cochannel interference is when multiple new users decide to use the same frequency. Hence, some sort of distributed/centralized control mechanism is needed in this case to manage cochannel interference.

17.10.7 Optimization of the Radio Module

By adapting the modulation type, constellation size, and channel coding rate, different data rates that could directly influence the power consumption of each node can be achieved and, in turn, will affect the lifetime of the whole network. There may be several tradeoffs between parameters to optimize the radio module. The selection of these radio parameters and changing these parameters on the fly is another challenging problem in CWSN.

17.10.8 Spectrum Sensing

Spectrum sensing should reduce the sensing duration time because, during this interval, all traffic is suspended and spectrum sensing is performed by the node. The spectrum sensing duration is also a challenge. The proper choice of quiet periods is also very important because it directly affects the throughput of the network. Apart from this, choosing the proper spectrum sensing algorithm according to the size/priority of data packets/signal strength also matters a lot in saving the energy of the spectrum-sensing nodes.

17.10.9 Representation of Network Architecture of CWSN

Network architecture helps in designing and understanding various functions of the network and nodes because CWSN has a dynamic environment and it may sometimes be mobile. Several responsibilities may be given to a network coordinator or may be given to every node in the network and it translates to different architectures. Hence, it is very difficult to find the general network architecture.

17.10.10 Futuristic Concepts

Currently, most of the research community is focused toward the application of CR in static WSNs but it has equally important applications in MWSNs. There are a very few literature available which discuss about the mobile scenarios in a CWSN. Our research group in IIIT Hyderabad is actively working on the cognitive aspects of MWSNs. We strongly feel that some of the ideas discussed below can be of extreme help while starting research in the area of cognitive MWSNs:

- The problem of energy scarcity will be more pronounced compared with the cognitive static WSN because now the same battery energy will be consumed in spectrum sensing as well as in mobility.
- Doppler will be very critical in case of moving nodes. Speed estimation may be required for proper frequency tuning in the receiver nodes.
- Channel availability parameters may frequently change as the node moves from one area to another. Frequent updates may be required, which may overload the spectrum sensing unit. This may also lead to extra consumption of limited energy.
- The location of hidden nodes may change frequently because of mobility. Cooperative spectrum sensing will be an essential requirement in MWSN and MCWSN.
- Mobility will lead to significant and frequent topological changes, which may lead to dynamic power level adjustment by the SU so that the current PU does not get interference.
- Either there has to be a mobile BS that will keep sensing with the mobile nodes on the go or the mobile nodes need to get spectrum availability data from the local spectrum-sensing units frequently.
- There will be extra load on the sensing time requirements of a mobile cognitive WSN.
- Spectrum sensing has to be done while there is quiet period to avoid false alarms. The spectrum sensing units, while sensing, may move to another zone where there is no quiet period. Or if it is taking too much time in sensing while on the move, it may get the wrong channel occupancy data.
- The scenario will become more difficult when the PUs are also moving.

- Information fusion becomes very important because of cooperating spectrum sensing as well mobility.
- PU detection as well as localization will also be difficult and cooperative sensing will be an essential requirement.

17.11 Conclusion and Summary

In this book chapter, we considered the design of WSN "one among the top 10 technologies that will change the world." We presented various applications of WSNs. The chapter focused attention on various paradigms of MWSNs and addressed various challenges in the design of such networks. Because the demand for spectrum is on the rise due to the proliferation of wireless technologies, CR-based WSN are proposed to coexist with other wireless networks such as cellular networks and vehicular networks. The design issues in mobile and cognitive WSNs are discussed. In summary, this chapter provides a concise review of the issues in the design of mobile and cognitive WSNs.

References

1. I.F. Akyildiz, W. Su, Y. Sankarasubramaniam, and E. Cayirci. A survey on sensor networks. *IEEE Communications Magazine*, 40(8):102–114, August 2002.
2. C. Intanagonwiwat, R. Govindan, and D. Estrin. Directed diffusion: A scalable and robust communication paradigm for sensor networks. In *Proceedings of the ACM MobiCom'00*, Boston, pp. 56–67, 2000.
3. C. Perkins. *Ad Hoc Networks*. Addison-Wesley, Reading, MA, 2000.
4. D. Estrin, R. Govindan, J. Heidemann, and S. Kumar. Next century challenges: Scalable coordination in sensor networks. In *ACM MobiCom'99*, Washington, DC, pp. 263–270, 1999.
5. S. Toumpis and T. Tassiulas. Optimal deployment of large wireless sensor networks. *IEEE Transactions on Information Theory*, 52:2935–2953, 2006.
6. J. Yick, G. Pasternack, B. Mukherjee, and D. Ghosal. Placement of network services in sensor networks, self-organization routing and information, integration in wireless sensor networks (special issue). *International Journal of Wireless and Mobile Computing (IJWMC)*, 1:101–112, 2006.
7. D. Pompili, T. Melodia, and I.F. Akyildiz. Deployment analysis in underwater acoustic wireless sensor networks. In *WUWNet*, Los Angeles, CA, 2006.
8. I.F. Akyildiz and E.P. Stuntebeck. Wireless underground sensor networks: Research challenges. *Ad-Hoc Networks*, 4:669–686, 2006.
9. M. Li and Y. Liu. Underground structure monitoring with wireless sensor networks. In *Proceedings of the IPSN*, Cambridge, MA, 2007.
10. I.F. Akyildiz, D. Pompili, and T. Melodia. Challenges for efficient communication in underwater acoustic sensor networks. *ACM Sigbed Review*, 1(2):3–8, 2004.
12. J. Heidemann, Y. Li, A. Syed, J. Wills, and W. Ye. Underwater sensor networking: Research challenges and potential applications. *USC/ISI Technical Report ISI-TR-2005-603* (2005).
13. http://www.coe.berkeley.edu/labnotes/0701brainybuildings.html.
14. http://bwrc.eecs.berkeley.edu/Seminars/Wright2011.22.02/PWrightSABER_SEMINAR_OCTOBER2002.ppt#1.
15. A. Rytter. Vibration based inspection of civil engineering structures. PhD dissertation, Department of Building Technology and Structural Engineering, Aalborg University, Denmark, 1993.
16. A. Mainwaring, J. Polastre, R. Szewczyk, D. Culler, and J. Anderson. Wireless sensor networks for habitat monitoring. In *ACM International Workshop on Wireless Sensor Networks and Applications (WSNA'02)*, Atlanta, GA, 2002.
17. http://www.alertsystems.org.

18. T. Hamrita, J. Durrence, and G. Vellidis. Precision farming practices. *Industry Applications Magazine, IEEE*, 15(2):34–42, March–April 2009, doi: 10.1109/MIAS.2009.931816.
19. D. Li, K.D. Wong, Y.H. Hu, and A.M. Sayeed. Detection, classification, and tracking of targets. *IEEE Signal Processing Magazine*, 19:17–29, March 2002.
20. C. Meesookho, S. Narayanan, and C. Raghavendra. Collaborative classification applications in sensor networks. In *Proceedings of the Second IEEE Multichannel and Sensor Array Signal Processing Workshop*, Arlington, VA, 2002.
21. T. He, S. Krishnamurthy, J.A.S Tankovic, T. Abdelzaher, L. Luo, R. Stoleru, T. Yan, L. Gu, J. Hui, and B. Krogh. An energy-efficient surveillance system using wireless sensor networks. In *MobiSys'04*, Boston, MA, 2004.
22. http://www.cse.ohio-state.edu/siefast/nest/nest_webpage/ALineInTheSand.html.
23. P. Juang, H. Oki, Y. Wang, M. Martonosi, L.S. Peh, and D. Rubenstein. Energy-efficient computing for wildlife tracking: Design tradeoffs and early experiences with ZebraNet. *SIGPLAN Notices*, 37(10): 96–107, October 2002, doi: 10.1145/605432.605408.
24. V.R. Jain, R. Bagree, A. Kumar, and P. Ranjan. wildCENSE: GPS based animal tracking system. In *International Conference on Intelligent Sensors, Sensor Networks and Information Processing, 2008.* ISSNIP 2008. pp. 617–622, 15–18 December 2008, doi: 10.1109/ISSNIP.2008.4762058.
25. S.-J. Kim, J.H. Seo, J. Krishna, and S.-J. Kim. Wireless sensor network based asset tracking service. In *Portland International Conference on Management of Engineering and Technology, 2008.* PICMET 2008. pp. 2643–2647, 27–31 July 2008, doi: 10.1109/PICMET.2008.4599893.
26. W. Yu. A promising pairwise key establishment scheme for wireless sensor networks in hostile environments. In *The 2010 International Conference on Multimedia Information Networking and Security (MINES)*, University of Science and Technology Beijing, China pp. 809–812, 4–6 November 2010 doi: 10.1109/MINES.2010.171.
27. M. Chen, T. Kwon, Y. Yuan, Y. Choi, and V.C.M. Leung. Mobile agent-based directed diffusion in wireless sensor networks. *EURASIP Journal on Advances in Signal Processing*, 2007.
28. Y. Xu and H. Qi. Mobile agent migration modeling and design for target tracking in wireless sensor networks. *Ad Hoc Networks*, 6(1):1–16, January 2007.
29. M. Chen, T. Kwon, Y. Yuan, Y. Choi, and V.C.M. Leung. Mobile agent based wireless sensor networks. *Journal of Computers*, 1(1):14–21, April 2006.
30. D. Konstantas and R. Herzog. Continuous monitoring of vital constants for mobile users: The MobiHealth approach. In *25th Annual International Conference of the IEEE EMBS*, Cancun, Mexico, pp. 3728–3731, 2003.
31. A.D. Wood, J.A. Stankovic, G. Virone, L. Selavo, Z. He, Q. Cao, T. Doan, Y. Wu, L. Fang, and R. Stoleru. Context-aware wireless sensor networks for assisted living and residential monitoring. *IEEE Network*, 22(4):26–33, July–August 2008.
32. http://www.tinyos.net/.
33. http://www.isi.edu/nsnam/ns/.
34. http://www.nsnam.org/.
35. R. Hajela, R.M. Garimella, and D. Sabnani. LCSD: Leveling clustering and sectoring with dissemination nodes to perform energy efficient routing in mobile cognitive wireless sensor networks. In *2010 International Conference on Computational Intelligence and Communication Networks (CICN)*, pp. 177–182, 26–28 November 2010, doi: 10.1109/CICN.2010.45.
36. D.-H. Nam and H.-K. Min. An energy-efficient clustering using a round-robin method in a wireless sensor network. In *5th ACIS International Conference on Software Engineering Research, Management and Applications, 2007.* SERA 2007. pp. 54–60, 20–22 August 2007, doi: 10.1109/SERA.2007.47.
37. Y. Gao, K. Wu, and F. Li. Analysis on the redundancy of wireless sensor networks. In *WSNA '03: Proceedings of the 2nd ACM International Conference on Wireless Sensor Networks and Applications.* ACM Press. New York, pp. 108–114, 2003.
38. Y. Yang and R. Kravets. Distributed QoS guarantees for realtime traffic in ad hoc networks. In *The Proceedings of 1st IEEE International Conference on Sensor and Ad Hoc Communications and Networks (SECON04)*, Santa Clara, CA, pp. 18–127, October 2004.

39. D. Chen and P.K. Varshney. QoS support in wireless sensor networks: a survey. In *Proceedings of the 2004 International Conference on Wireless Networks (ICWN 2004)*, Las Vegas, NV, vol. 1, pp. 227–233, June 2004.
40. I.F. Akyildiz, W.-Y. Lee, M.C. Vuran, and S. Mohanty. Next generation/dynamic spectrum access/cognitive radio wireless networks: A survey computer networks, vol. 50, no. 13, pp. 2127–2159, September 15, 2006, ISSN 13891286, 10.1016/j.comnet.2006.05.001.
41. Secondary markets initiative. http://wireless.fcc.gov/licensing/secondarymarkets/.
42. J. Mitola and G.Q. Maguire Jr., Cognitive radio: Making software radios more personal, *Personal Communications, IEEE*, vol. 6, no. 4, pp. 13–18, Aug 1999.
43. I. Howitt and J. Gutierrez. IEEE 802.15.4 low rate wireless personal area network coexistence issues. In *Proceedings of IEEE WNCN*, New Orleans, LA, 2003.
44. A.S. Zahmati, S. Hussain, X. Fernando, and A. Grami. Cognitive wireless sensor networks: emerging topics and recent challenges. In *2009 IEEE Toronto International Conference Science and Technology for Humanity (TIC-STH)*, Toronto, Ontario, Canada, pp. 593–596, 26–27 September 2009.
45. S. Shankar, C. Cordeiro, and K. Challapali. Spectrum agile radios: Utilization and sensing architecture. In *IEEE DySPAN*, Baltimore, MD, 2005.
46. D. Cavalcanti, S. Das, W. Jianfeng et al. Cognitive radio based wireless sensor networks. In *Proceedings of the 17th International Conference on Computer Communications and Network*. Virgin Islands, USA, IEEE Press. pp. 1–6, 2008.
47. IEEE 802.22 Draft Standard. IEEE P802.22TM/D0.4 draft standard for wireless regional area networks, http://www.ieee802.org/22/, November 2007.
48. D. Cavalcanti, R. Schmitt, and A. Soomro. Achieving energy efficiency and QoS for low-rate applications with 802.11e. In *The IEEE WCNC 2007*, Hong Kong, March 2007.
49. IEEE 802.15.4 Standard. Wireless medium access control (MAC) and physical layer (PHY) specifications for low-rate wireless personal area networks (LR-WPANs), 2003 Edition.
50. L. Stabellini and J. Zander. Energy-aware spectrum sensing in cognitive wireless sensor networks: A cross layer approach. In *Wireless Communications and Networking Conference (WCNC)*, Sydney, AU, 2010 IEEE, pp. 1–6, 18–21 April 2010.
51. S. Haykin. Cognitive radio brain-empowered wireless communications. *IEEE Journal on Selected Areas in Communications*, vol. 23, no. 2, pp. 201–220, February 2005.
52. R. G. Pridham and H. Urkowitz. Comment on "Energy detection of unknown deterministic signals." *Proceedings of the IEEE*, pp. 523–531, vol. 56, no. 8, pp. 1379–1380, August 1968.
53. http://www.ieee802.org/22/Meeting_documents/2006_July/22_06_0118_000000_I2R_sensing.doc.
54. Advanced Television Systems Committee. ATSC digital television standard (A/53, Part 2)—RF/Transmission System Characteristics, January 2007.
55. T. Tomioka, T. Tomizawa, and T. Kobayashi. High sensitivity carrier sensing using overlapped FFT for cognitive radio transceivers, *Vehicular Technology Conference, 2009. VTC Spring 2009. IEEE 69th*, pp.1–5, 26–29 April 2009.
56. J.M. Mendel. Tutorial on higher-order statistics (spectra) in signal processing and systems theory: Theoretical results and some applications. *Proceedings of the IEEE*, vol. 79, no. 3, p. 278–305, Mar. 1991.
57. G. Chouinard. 802.22 Presentation to the ECSG on White Space. IEEE P802.22 Wireless RANs, March 2009.
58. www.ieee802.org/22/Meeting_documents/2006_Sept/22_06_0187_00_0000_I2R_sensing2.ppt.
59. J.-K. Lee, J.-H. Yoon, and J.-U. Kim. A New Spectral Correlation Approach to Spectrum Sensing for 802.22 WRAN System, *Proceedings of the 2007 International Conference on Intelligent Pervasive Computing, IPC '07*, pp. 101–104, 11–13 October 2007.
60. S.M. Mishra, A. Sahai, R.W. Brodersen. Cooperative sensing among cognitive radios, *IEEE International Conference on Communications, 2006. ICC '06*. vol. 4, pp.1658–1663, June 2006.
61. S. Kumar, D. Singhal, and G.R. Murthy. Doubly cognitive architecture based cognitive wireless sensor networks. *International Journal of Wireless Networks and Broadband Technologies (IJWNBT)*, 1:2, 02 January 2012.

62. C. Doerr, D. Grunwald, and D.C. Sicker. Local independent control of cognitive radio networks. In *The 3rd International Conference on Cognitive Radio Oriented Wireless Networks and Communications, 2008. CrownCom 2008*. pp. 1–9, 15–17 May 2008.
63. M. Bandholz, J. Riihijärvi, and P. Mähönen. Unified layer-2 triggers and application-aware notifications. In *IWCMC '06: Proceedings of the 2006 International Conference on Communications and Mobile Computing*, New York. ACM Press. pp. 1447–1452, 2006.
64. N. Baldo and M. Zorzi. Fuzzy logic for crosslayer optimization in cognitive radio networks. In *Workshop on Digital Rights Management Impact on Consumer Communications (CCNC)*, 2008, Harrah's Las Vegas, NV, 2007.
65. T. Weingart, D.C. Sicker, and D. Grunwald. A method for dynamic configuration of a cognitive radio, networking technologies for software defined radio networks, *1st IEEE Workshop on Networking Technologies for Software Defined Radio Networks, 2006, SDR '06*. pp. 93–100, Sept. 2006.
66. J. Neel, R.M. Buehrer, J.H. Reed, and P. Gilles Robert. Game theoretic analysis of a network of cognitive radios. *The 2002 45th Midwest Symposium on Circuits and Systems* (*MWSCAS*). vol. 3, pp. III–409, III–412, August 2002.
67. J. Neel, R. Buehrer, B. Reed, and R. Gilles. Game theoretic analysis of a network of cognitive radios. In *The 2002 45th Midwest Symposium on Circuits and Systems (MWSCAS)*, Tulsa, OK, 2002.
68. J. Neel, J.H. Reed, and R.P. Gilles. Convergence of cognitive radio networks. In *IEEE Wireless Communications and Networking Conference (WCNC)*, Atlanta, GA, 2004.
69. J.O. Neel and J.H. Reed. Performance of distributed dynamic frequency selection schemes for interference reducing networks. In *Military Communications Conference MILCOM*, Washington, DC, 2006.
70. A.B. MacKenzie and S.B. Wicker. Game theory in communications: Motivation, explanation, and application to power control. In *IEEE Globecom*, Anaheim, CA, 2001.
71. R. Menon, A. MacKenzie, R. Buehrer, and J. Reed. Game theory and interference avoidance in decentralized networks. In *SDR Forum Technical Conference*, vol. 15, Phoenix, AZ, 2004.
72. T.W. Rondeau, B. Le, C.J. Rieser, and C.W. Bostian. Cognitive radios with genetic algorithms: Intelligent control of software defined radios. In *SDR Forum Technical Conference*, Phoenix, AZ, 2004.
73. C.J. Rieser, T.W. Rondeau, C.W. Bostian, and T.M. Gallagher. Cognitive radio testbed: Further details and testing of a distributed genetic algorithm based cognitive engine for programmable radios. In *Proceedings of the IEEE Military Communications Conference*, Monterey, CA, 2004.
74. O. Ileri, D. Samardzija, and N. Mandayam. Demand responsive pricing and competitive spectrum allocation via a spectrum serve. In *First IEEE International Symposium on New Frontiers in Dynamic Spectrum Access Networks*, Baltimore, MA, 2005.
75. M. Buddhikot and K. Ryan. Spectrum management in coordinated dynamic spectrum access based cellular networks. In *First IEEE International Symposium on New Frontiers in Dynamic Spectrum Access Networks*, Baltimore, MA, 2005.
76. C. Raman, R. Yates, and N. Mandayam. Scheduling variable rate links via a spectrum server. In *First IEEE International Symposium on New Frontiers in Dynamic Spectrum Access Networks*, Baltimore, MD, 2005.
77. C. Doerr, D. Grunwald, and D.C. Sicker. Local independent control of cognitive radio networks. In *3rd International Conference on Cognitive Radio Oriented Wireless Networks and Communications, 2008. CrownCom 2008*. Singapore, pp. 1–9, 15–17 May 2008.

Chapter 18

Wireless Multimedia Sensor Networks: Correlation-Based Communication

Rui Dai
North Dakota State University

Pu Wang
Georgia Institute of Technology

Contents

Wireless multimedia sensor networks (WMSNs) are networks of interconnected devices that allow the retrieval of video and audio streams, still images, and scalar data from the environment [1]. With the ability of providing enriched observation of the environment, WMSNs can enhance a wide range of applications such as environmental monitoring, industrial process control, traffic enforcement, and advanced health care delivery. Most WMSN applications require the delivery of multimedia information with a certain level of quality-of-service (QoS), in terms of delay, jitter, bandwidth, reliability, etc. To achieve this, one major factor that needs to be considered is the resource constraint in battery, memory, processing capability, and achievable data rate for sensor nodes.

A lot of energy-efficient QoS communication protocols have been proposed for different layers of the protocol stack, such as QoS MAC protocols [2,3], QoS routing algorithms [4,5], and cross-layer QoS communication modules [6]. To achieve efficient delivery of multimedia information in a WMSN, apart from properly regulating the traffic flows through networking protocols, it is essential to reduce the redundancy of traffic injected into the network. Multimedia source coding is the classic approach to reduce redundancy. For WMSNs, low-complexity image and video coding techniques have been proposed based on principles such as distributed source coding [7]. Another approach to reduce traffic redundancy is through multimedia in-network processing. According to the requirements of specific applications, sensor nodes can filter out uninteresting events locally, or coordinate with each other to aggregate correlated data [1].

It is crucial to reduce the redundancy of visual information (still images and video streams), the most bandwidth-demanding part of multimedia information in WMSNs. In a densely deployed WMSN, the observations of camera sensors with overlapped fields of view (FoVs) are correlated with each other, resulting in substantial redundancy in the traffic generated by these sensors. Studying the correlation of camera sensors can facilitate the design of multimedia source coding and in-network processing schemes for WMSNs.

In many recent research efforts for WMSNs, image processing techniques have been utilized to obtain correlation and design collaborative processing. In a study by Wagner et al. [8], images from correlated views are roughly registered using correspondence analysis. Each sensor transmits a low-resolution version of a common area, and the sink combines multiple low-resolution versions into a high-resolution image. In an article by Wu and Chen [9], spatial correlation is obtained by an image shape–matching algorithm, whereas temporal correlation is calculated via background subtraction. Based on the spatial and temporal correlation, images from correlated sensors are transmitted collaboratively. However, there are some drawbacks to implementing image processing–based methods to obtain correlation in WMSNs. On the one hand, image processing methods are very complicated and can result in considerable energy consumption for sensor nodes. On the other hand, because the performance of image processing methods is usually dependent on the specific features of images, different types of images will require different processing methods [10]. Moreover, as these methods rely on postprocessing of the images captured by camera sensors, they do not provide any way to predict correlation in advance.

Considering the above factors, we avoid using specific image processing methods; in contrast, we develop a novel analytical spatial correlation model based on the projection geometry of camera sensors [11]. This model just takes in a few parameters such as camera locations, sensing directions, and focal lengths as input, and calculates a spatial correlation coefficient that quantifies the degree of correlation between two cameras with overlapped FoVs. The model allows us to estimate the degree of correlation through low communication and computation costs in WMSNs. It also provides an application-independent and proactive solution to predicting correlation before the images are taken. We introduce the derivation of the spatial correlation model in Section 18.1.

Furthermore, we address the problem of how much information can be gained from multiple correlated cameras. The concept of entropy is used in the analysis of this problem. To estimate the amount of information gained from multiple correlated cameras, we can estimate the joint entropy of the images observed by these cameras. By definition, joint entropy should be calculated from the joint probability distribution of multiple images [12]. Because of the difficulty in image modeling and the high computation and communication costs involved, it is difficult to get an accurate estimation of the joint probability distribution [13]. Instead of calculating the joint probability distribution, we estimate joint entropy as a function of the spatial correlation coefficients between camera sensors [14]. The estimation of joint entropy can be useful to solve many problems such as predicting the coding efficiency of correlated camera sensors. Our joint entropy estimation algorithm and experimental results on the effectiveness of coding efficiency prediction are given in Section 18.2.

Multimedia source coding is infeasible to apply globally in a large-scale network. The clustering strategy has been proved to be an effective way to improve network scalability and energy efficiency for sensor networks [15,16]. This strategy uses the hierarchical concept in which the entire network is divided into regions. In many existing algorithms, the metrics for clustering are distances between nodes or residual energy [17]. However, in this work, we aim to construct clusters based on the potential coding gains so as to minimize the redundancy of network traffic. We divide the entire network into different regions. Each region corresponds to a cluster in which a group of camera sensors collaboratively perform data compression according to different coding algorithms. Because each cluster only covers a limited number of nodes, it is feasible to acquire the correlation within a cluster without incurring much extra cost. Therefore, the clustered coding strategy paves the way for the practical application of multimedia source coding in large-scale WMSNs.

With joint entropy as a prediction of joint coding efficiency, we develop a distributed multi-cluster coding protocol (DMCP) to partition the entire network into a set of coding clusters such that the global coding gain is maximized [14]. The DMCP works in a fully distributed manner. It does not involve much communication cost because each camera sensor just needs to exchange its parameters such as sensing direction and location with its one-hop neighbors, from which the joint entropy of neighboring cameras can be estimated. Because the DMCP is a heuristic solution, we also analyze its performance and derive its approximation ratio compared with an optimal coding clustering (OCC) problem. The design of the DMCP protocol is explained in Section 18.3.

Routing protocols have a significant effect on the overall performance of wireless sensor networks. Many existing QoS routing algorithms only try to provide QoS guarantee by properly distributing the network traffic [4,5], whereas the total amount of data generated by camera sensors stays the same. As a result, the existing QoS-oriented approaches are still resource-demanding for WMSNs in which large amounts of images are generated and forwarded by energy-constrained camera sensors. On the other hand, the joint compression/aggregation and routing strategy has been proved to be effective in sensor networks [18–20]. Although these algorithms work well for scalar data in sensor networks, new solutions are needed for the delivery of visual information in sensor networks. In particular, it is necessary to design algorithms that can reduce the high bandwidth demand of visual information and provide QoS support for WMSN applications.

We propose a correlation-aware QoS routing (CAQR) algorithm for the efficient delivery of visual information in sensor networks [21]. First, a correlation-aware internode differential coding scheme is developed to reduce the amount of traffic injected into the network. Then, a correlation-aware load balancing scheme is proposed to prevent network congestion by splitting the correlated flows that cannot be reduced to different paths. By integrating these correlation-aware schemes, an optimization QoS routing framework is proposed with an objective to minimize the sensors' energy consumption under delay and reliability constraints. Section 18.4 illustrates the steps of the CAQR algorithm.

Finally, Section 18.5 summarizes the contributions of this chapter, followed by some exercise questions in Section 18.6.

18.1 Spatial Correlation of Visual Information

In a WMSN, multiple camera sensors are deployed to provide multiple views, multiple resolutions, and enhanced observations of the environment [22]. Figure 18.1 gives an example of a WMSN deployed with cameras. In a typical scenario, an application specifies which area it is interested in, and the cameras that can observe this area will report their observations to the sink. For a certain area of interest, suppose there are a group of N camera sensors (C_1, C_2,..., C_N) that can observe it. We denote their observed images as $\{X_1, X_2,..., X_N\}$. There exists a correlation among the observations of these cameras. For camera C_i and camera C_j in the group, we will derive a correlation coefficient ρ_{ij} to describe the degree of correlation between X_i and X_j with respect to the area of interest.

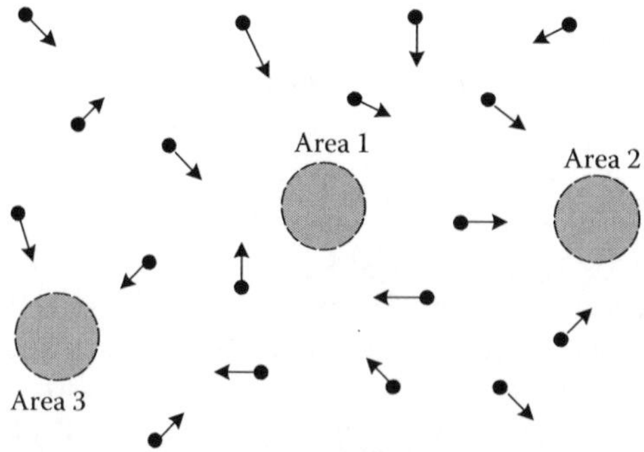

Figure 18.1 WMSN with camera sensors and areas of interest.

18.1.1 Spatial Correlation Model for Visual Information

18.1.1.1 Sensing Model

The sensing process of a camera is characterized by directional sensing and three-dimensional (3-D) to two-dimensional (2-D) projection. In computer vision, this sensing process is usually described by the pinhole camera model [23]. Figure 18.2 illustrates an example of a pinhole camera. The camera's center of projection is at the origin of a Euclidean coordinate system, and its sensing direction is along the x axis. The focal length of the camera is f, so the image plane is the plane $x = f$. A scene point P with coordinates $(x,y,z)^T$ is mapped to $P'(u,v)^T$ on the image plane, where u and v are given by

$$\begin{cases} u = fy/x \\ v = fz/x \end{cases} \tag{18.1}$$

A camera also has limited sensing range. It can only observe the objects within its FoV. If a camera sensor is deployed on a ground plane, we can use the simplified 2-D FoV model in the study by Ma and Liu [24], which models the shape of a camera's FoV as a sector. As shown in Figure 18.3, a camera's FoV is determined by four parameters $(P, R, \vec{V}, \alpha)$, where P is the location of the center of the camera, R is the sensing radius, $\vec{V}$ is the sensing direction (the center line of sight of the camera's FoV), and α is the offset angle between the sensing direction and a radius of the sector. An arbitrary point P_1 is in the FoV of the camera if it satisfies

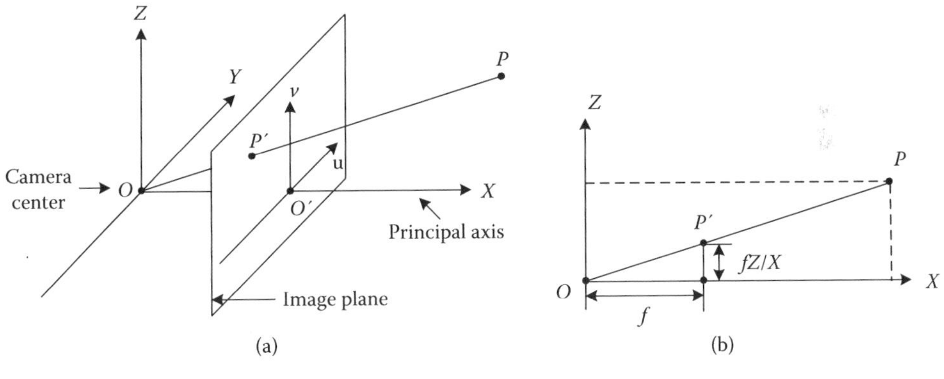

Figure 18.2 Camera projection model. (a) Image plane and principal axis. (b) Projection geometry.

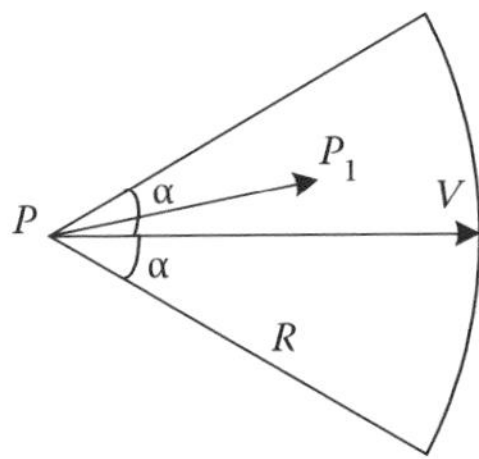

Figure 18.3 Field of view.

$$\begin{cases} \left|\overrightarrow{PP_1}\right| \le R \\ \theta \le \alpha, \end{cases} \tag{18.2}$$

where θ is the angle between $\overrightarrow{PP_1}$ and $\vec{V}$.

The focal length and the FoV parameters of a camera could be estimated through various calibration and localization methods, for example, those methods used by Forsyth and Ponce [23], Barton-Sweeney et al. [25], and Devarajan et al. [26]. In the next subsection, we derive a spatial correlation coefficient based on these parameters.

18.1.1.2 System Model

We set up a world coordinate system $(W) = (O_W, i_W, j_W, k_W)$ for the area of interest as shown in Figure 18.4a, in which the origin is the center of the area of interest, and the XOY plane is the ground plane. Seven reference points, which can also be regarded as feature points or key points in a scene, are chosen as $O(0,0,0)^T$, $A(1,0,0)^T$, $B(-1,0,0)^T$, $C(0,1,0)^T$, $D(0,-1,0)^T$, $E(0,0,1)^T$, and $F(0,0,-1)^T$. These reference points form six unit reference vectors along the orthogonal directions in the 3-D world: $\overrightarrow{OA}$, $\overrightarrow{OB}$, $\overrightarrow{OC}$, $\overrightarrow{OD}$, $\overrightarrow{OE}$, and $\overrightarrow{OF}$.

We consider the case when all the camera sensors are placed on the ground plane (XOY) and their sensing directions are also within the ground plane. For a camera sensor C_i, the coordinates of its optical center can be denoted as $(x_i,y_i)^T$. Its sensing direction can be described by a unit vector $\varphi(\varphi_x,\varphi_y)$, where $\varphi_x = \cos\theta$, $\varphi_y = \sin\theta$, and θ is the angle between the sensing direction and the x axis.

The projections of the reference points on a camera will change as the camera's location and sensing direction change. By comparing the projections of the same reference points at different cameras, we can understand the correlation characteristics among different cameras.

18.1.1.3 Projection Geometry

Figure 18.4b shows the deployment of three cameras in the world coordinate system (W), where the origin is the center of the area of interest, and the XOY plane denotes the ground plane. Camera 1 is located at $(-d,0)^T$ with its sensing direction along the x axis. Camera 2 is

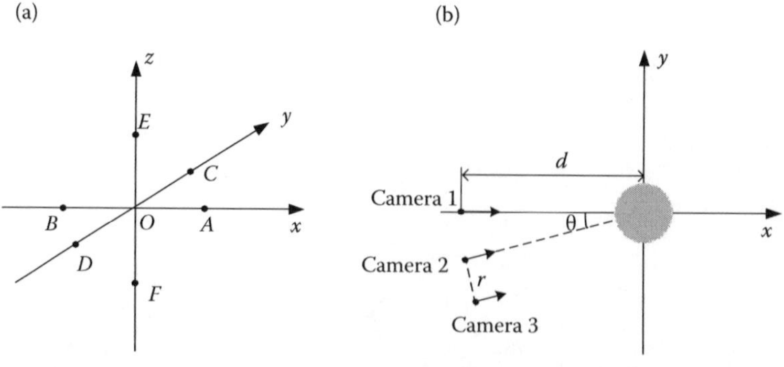

Figure 18.4 (a) Reference points in the area of interest and (b) deployment of cameras.

located at $(-d\cos\theta, -d\sin\theta)^T$, and its sensing direction rotates an angle of θ about the x axis. For both camera 1 and camera 2, their principal axes pass the center of the area of interest (the origin). Camera 3 has the same sensing direction as camera 2, but its principle axis does not pass the origin. The distance from the center to its principle axis is r, as shown in Figure 18.4b. The optical center of camera 3 is $(-d\cos\theta + r\sin\theta, -d\sin\theta, -r\cos\theta)^T$. Although the locations and sensing directions of these three cameras are different, the depths for the center of the area of interest in all three cameras have the same value d. In addition, we assume that all these cameras have the same focal length f.

To calculate the projections of the reference points in a camera, a coordinate transform is first needed to obtain the coordinates of the points in the camera's coordinate system. For example, the coordinate system of camera 1 (Figure 18.2) and the world coordinate system (W) are separated by a pure translation. For an arbitrary point P in the space, we have

$$^1P = {}^WP + {}^1O_W, \tag{18.3}$$

where ^1P is the coordinate vector of point P in the coordinate system of camera 1, whereas ^WP is the coordinate vector in the world coordinate system (W). These two vectors are related by ^1O_W, the coordinate vector of the origin in (W) seen in the coordinate system of camera 1. Here, $^1O_W = (d,0,0)^T$.

Therefore, the coordinates of the reference points in the coordinate system of camera 1 are as follows: $O_1(d,0,0)^T$, $A_1(d+1,0,0)^T$, $B_1(d-1,0,0)^T$, $C_1(d,1,0)^T$, $D_1(d,-1,0)^T$, $E_1(d,0,1)^T$, and $F_1(d,0,-1)^T$. Based on the projection model in Equation 18.1, we can obtain the projections of these reference points in camera 1: $o_1(0,0)^T$, $a_1(0,0)^T$, $b_1(0,0)^T$, $c_1\left(\frac{f}{d},0\right)^T$, $d_1\left(-\frac{f}{d},0\right)^T$, $e_1\left(0,\frac{f}{d}\right)^T$, $f_1\left(0,-\frac{f}{d}\right)^T$. The projections of reference points on camera 1 are plotted in Figure 18.5a.

As for camera 2, its coordinate system can be derived from the world coordinate system (W) as follows: rotate the world coordinate system counterclockwise for an angle of θ, and then translate the rotated system along the negative direction of x axis for a length of d, given as

$$^2P = {}^2_WR\,{}^WP + {}^2O_W, \tag{18.4}$$

where ^2O_W is the translation offset vector, $^2O_W = (d,0,0)^T$, and ^2_WR is the rotation matrix,

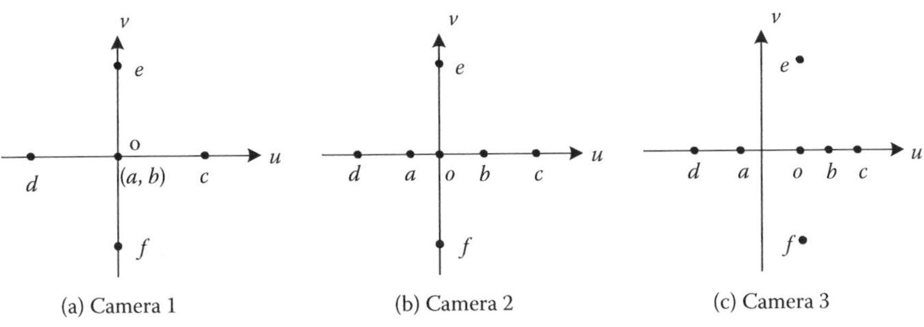

Figure 18.5 Projections of reference points and vectors on camera 1 (a), camera 2 (b), and camera 3 (c).

$$ {}^{2}_{W}R = \begin{pmatrix} \cos\theta & \sin\theta & 0 \\ -\sin\theta & \cos\theta & 0 \\ 0 & 0 & 1 \end{pmatrix}. $$

Similarly, the coordinate system of camera 3 and the world coordinate system are related as

$$ {}^{3}P = {}^{3}_{W}R\,{}^{W}P + {}^{3}O_W, \tag{18.5} $$

where the rotation matrix is the same as that of camera 2 (${}^{3}_{W}R = {}^{2}_{W}R$), whereas the translation offset vector satisfies ${}^{3}O_W = (d,r,0)^T$.

The projections of reference points on camera 2 and camera 3 can be calculated in the same way as the case of camera 1. Figure 18.5 illustrates the positions of reference points on the three cameras. Based on the coordinates of the reference points, the values of the corresponding reference vectors are calculated, which are listed in Table 18.1.

18.1.1.4 Spatial Correlation Coefficient

18.1.1.4.1 Scaling Effect

Comparing the projections of the reference vectors in Table 18.1, we find that the lengths of *OA*, *OB*, *OC*, and *OD* are different in the three cameras, but the length of *OE*/*OF* remains as a

Table 18.1 Projections of Reference Vectors

Vectors	*Projections*		
	Camera 1	*Camera 2*	*Camera 3*
$\overrightarrow{OA}$	$(0,0)^T$	$\left(\frac{-\sin\theta}{d+\cos\theta}f,0\right)^T$	$\left(\frac{r-\cos\theta}{d+\sin\theta}f-r\frac{f}{d},0\right)^T$
$\overrightarrow{OB}$	$(0,0)^T$	$\left(\frac{\sin\theta}{d-\cos\theta}f,0\right)^T$	$\left(\frac{r+\cos\theta}{d-\sin\theta}f-r\frac{f}{d},0\right)^T$
$\overrightarrow{OC}$	$\left(\frac{f}{d},0\right)^T$	$\left(\frac{\cos\theta}{d+\sin\theta}f,0\right)^T$	$\left(\frac{r+\cos\theta}{d+\sin\theta}f-r\frac{f}{d},0\right)^T$
$\overrightarrow{OD}$	$\left(-\frac{f}{d},0\right)^T$	$\left(\frac{-\cos\theta}{d-\sin\theta}f,0\right)^T$	$\left(\frac{r-\cos\theta}{d-\sin\theta}f-r\frac{f}{d},0\right)^T$
$\overrightarrow{OE}$	$\left(0,\frac{f}{d}\right)^T$	$\left(0,\frac{f}{d}\right)^T$	$\left(0,\frac{f}{d}\right)^T$
$\overrightarrow{OF}$	$\left(0,-\frac{f}{d}\right)^T$	$\left(0,-\frac{f}{d}\right)^T$	$\left(0,-\frac{f}{d}\right)^T$

constant value. The reason is that the points O, E, and F have the same depth (d) in all the three cameras, and that the cameras also have the same focal length (f). Both the depth and the focal length can influence the size of a projection. Thus, we define a scaling factor, s, as the lengths of the projections of OE and OF, given by

$$s = \frac{f}{d}. \tag{18.6}$$

18.1.1.4.2 Translation Effect

As can be seen in Figure 18.5, the projections on camera 1 and camera 2 are both in the center of the image planes, but the projections on camera 3 have an offset from the center of the image plane. The deviation from the center of the area of interest to the camera's principle axis has caused the translation of the projections. Based on the projections of reference points on camera 3, we define a translation factor as

$$t = r\frac{f}{d} = rs. \tag{18.7}$$

18.1.1.4.3 Correlation Coefficient

As shown in Table 18.1, the lengths of vectors OA, OB, OC, and OD will change as the camera's location and sensing direction change. Based on this observation, we design a disparity function to reveal the disparity between the projections of reference vectors on different cameras. Suppose that camera i and camera j are two arbitrary cameras on the ground plane that can observe the same area of interest, the disparity function is derived as below:

1. Determine the positions and sensing directions of camera i and camera j.
2. Based on the projection model in Equation 18.1, compute the projections of reference vectors in each camera.
3. Divide the projections of reference vectors by the scaling factor $s = \frac{f}{d}$ (Equation 18.6), so that we can get a set of normalized projection vectors for each camera.
4. Compute the distance for each pair of normalized vectors OA, OB, OC, and OD. For example, if the projection of OA is $o_i a_i = (u_i, v_i)^T$ on camera i, and $o_j a_j = (u_j, v_j)^T$ on camera j, the distance is calculated as

$$d_{OA} = \sqrt{(u_i - u_j)^2 + (v_i - v_j)^2}. \tag{18.8}$$

5. The disparity between the images at camera i and camera j, denoted by δ_{ij}, is defined as the average distance of the four vectors

$$\delta_{ij} = \frac{1}{4}(d_{OA} + d_{OB} + d_{OC} + d_{OD}). \tag{18.9}$$

Generally, for camera i and camera j with position parameters (d_i, r_i, θ_i) and (d_j, r_j, θ_j), the disparity between the images at the two cameras is given by

$$\delta_{ij} = \frac{1}{4}\left(\left|\frac{-d_i \sin\theta_i - r_i \cos\theta_i}{d_i + \cos\theta_i} - \frac{-d_j \sin\theta_j - r_j \cos\theta_j}{d_j + \cos\theta_j}\right| + \left|\frac{d_i \sin\theta_i + r_i \cos\theta_i}{d_i - \cos\theta_i} - \frac{d_j \sin\theta_j + r_j \cos\theta_j}{d_j - \cos\theta_j}\right| + \left|\frac{d_i \cos\theta_i - r_i \sin\theta_i}{d_i + \sin\theta_i} - \frac{d_j \cos\theta_j - r_j \sin\theta_j}{d_j + \sin\theta_j}\right| + \left|\frac{-d_i \cos\theta_i + r_i \sin\theta_i}{d_i - \sin\theta_i} - \frac{-d_j \cos\theta_j + r_j \sin\theta_j}{d_j - \sin\theta_j}\right|\right). \tag{18.10}$$

Example

For camera 1 and camera 2 in Figure 18.4b, according to the results in Table 18.1, the disparity between the images in camera 1 and camera 2 is calculated as

$$\delta_{12} = \frac{1}{4}\left(\left|\frac{d \sin\theta}{d + \cos\theta}\right| + \left|\frac{d \sin\theta}{d - \cos\theta}\right| + \left|\frac{d \cos\theta}{d + \sin\theta} - 1\right| + \left|\frac{-d \cos\theta}{d - \sin\theta} + 1\right|\right). \tag{18.11}$$

We let camera 1 stay fixed, and let the sensing direction of camera 2 (θ) change from −90° to 90°. The sensing direction difference between camera 2 and camera 1 is also θ. We set the depth d = 2.5 m for camera 1 and camera 2. The disparity between camera 1 and camera 2 in Equation 18.11 is illustrated as a function of θ (in degrees) in Figure 18.6.

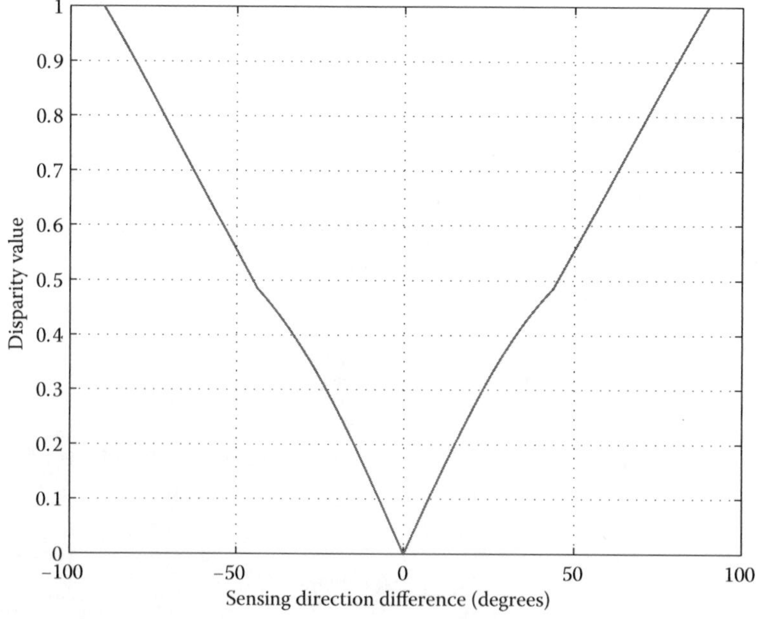

Figure 18.6 Illustration of the disparity function.

The disparity value increases as the sensing direction difference increases. The larger the disparity value, the more differences exist between the two images, that is, the images are less correlated. In the above example, the largest disparity value goes to 1 when the sensing directions of the two cameras are perpendicular, for which we can say that the two cameras are weakly correlated. For the convenience of further analysis, we bound the disparity value from 0 to 1 as follows:

$$\delta_{ij} = \min(\delta_{ij}, 1). \tag{18.12}$$

Consequently, we can define a correlation coefficient that is complementary to the disparity function:

$$\rho_{ij} = 1 - \delta_{ij}. \tag{18.13}$$

When the correlation coefficient is 0, the two images are independent of each other. If it is equivalent to 1, the two images are highly correlated. The larger the correlation coefficient, the more correlated are the two images.

18.1.1.4.4 Discussion

For WMSN applications, as long as the area of interest is specified, and the locations and sensing directions of cameras are estimated, the correlation characteristics of cameras with overlapped FoVs can be obtained as introduced above. The model depends on the selection of reference points/vectors in the area of interest. Six unit vectors along three orthogonal directions in the 3-D world are chosen in the above analysis. For a WMSN application, the reference points should be chosen properly based on specific application requirements. In addition, a camera's FoV will be reduced when it is blocked by some obstacles. To guarantee that this model works well, a camera's actual FoV needs to be estimated beforehand.

18.1.2 Performance Evaluation

We set up a scene as shown in Figure 18.4b: camera 1 and camera 2 are placed to take pictures of an area of interest. Camera 1 is placed along the x axis, and camera 2 rotates an angle of θ, so the sensing direction difference between camera 2 and camera 1 is θ. Set $d = 2.5$ m. A reference image is obtained at camera 1, and then a group of 10 images are taken for camera 2 with the following θ values: {-75°, -60°, -45°, -30°, -15°, 15°, 30°, 45°, 60°, 75°}. Figure 18.7 presents some of the images.

In Section 18.1.1, we showed that the degree of correlation is relevant to the cameras' sensing directions and their relative positions. Because the sensing directions and positions are already known, the disparity between the test images on camera 2 and camera 1 can be easily calculated by Equation 18.10. The results of the disparity values are presented in Figure 18.8a as a function

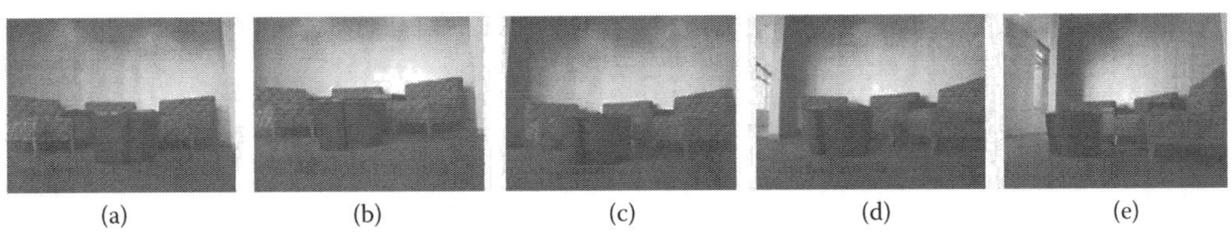

(a) (b) (c) (d) (e)

Figure 18.7 Images. (a) $\theta = 0°$, (b) $\theta = 15°$, (c) $\theta = 30°$, (d) $\theta = 45°$, (e) $\theta = 60°$.

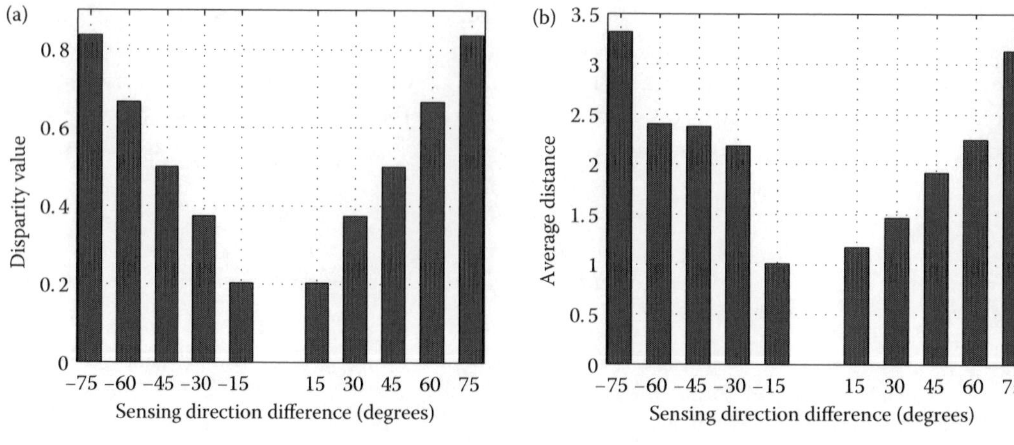

Figure 18.8 (a) Proposed disparity function versus (b) feature extraction algorithm.

of the sensing direction difference θ. The disparity increases as the sensing direction difference increases.

The correlation between images can also be obtained by applying image processing algorithms. Here, we refer to a commonly used feature extraction algorithm in an article by Manjunath and Ma [27]. In this algorithm, texture features are extracted from the images by Gabor wavelet transform, and based on the wavelet coefficients, feature vectors in multiple scales and multiple directions are constructed. Finally, an average distance is calculated by averaging all the feature distances in each feature space [27]. The average feature distances between the test images on camera 2 and camera 1 are calculated accordingly, and the results are presented in Figure 18.8b.

Comparing the results of the proposed model (Figure 18.8a) and the results of the feature extraction algorithm in the article by Manjunath and Ma [27] (Figure 18.8b), we find that in both cases, the disparity/distance value increases as the sensing direction difference increases. This is also in accordance with our common sense: if we just observe the test images in Figure 18.7 with our eyes, we can also find that the two images from camera 2 and camera 1 look more different when their sensing direction difference (θ) is larger. Therefore, the proposed spatial correlation coefficient is effective as it can reveal the correlation characteristics between images.

The slight differences between the results in Figure 18.8a and b may be explained by the intrinsic differences of the two schemes. The proposed model is derived by studying the cameras' sensing model and deployments. Therefore, the results are just dependent on a few parameters. In contrast, the feature extraction algorithm goes into detail in an image. It is sensitive to the noise in the images, and even a little change of the light condition might influence the final results.

To obtain correlation in a WMSN, the proposed correlation model just requires sensors to exchange a few sensing parameters, and the correlation coefficient can be computed in a straightforward way as shown in Equation 18.13. In comparison, image processing–based methods, such as the feature extraction algorithm, would require sensors to exchange lots of feature vectors or even entire images, and the processing load at each sensor is not negligible. A detailed evaluation of the costs for obtaining correlation can be found in an article by Dai and Akyildiz [11].

In summary, the proposed correlation coefficient can effectively model the correlation of visual information as image processing–based methods, but the computation and communication costs for the proposed method are much lower. Therefore, the proposed correlation model is suitable for use in resource-constrained WMSNs.

18.2 Joint Effect of Multiple Correlated Cameras

After we obtain the spatial correlation coefficient, we study the joint effect of multiple correlated cameras in WMSNs. In particular, we study how to measure the amount of visual information from multiple cameras in a WMSN, which can benefit from the design of multimedia in-network processing and compression schemes. We develop an entropy-based framework to estimate the amount of information from multiple correlated cameras.

18.2.1 Entropy-Based Approach

In information theory [12], the concept of entropy is used to measure the amount of information from a random source. If an image is interpreted as a sample of a "gray level source," the source's symbol probabilities can be modeled by the gray level histogram of the observed image. An estimate of the source's entropy can be generated as [10]

$$\tilde{H} = -\sum_{k=1}^{L} p(r_k)\log p(r_k), \tag{18.14}$$

where L is the number of all possible gray levels, and $p(r_k)$ is the probability of the kth gray level. It denotes the average amount of information per pixel in the image.

We consider two cameras that can observe the same area of interest. Suppose each camera has captured one image at the same time. We denote their observations about the area of interest as A and B. If we let one of the cameras report its image to the sink, the amount of information about the area of interest gained at the sink is $H(A)$ or $H(B)$. (We do not consider the information loss caused by lossy compression or packet loss during transmission.) If both of them report to the sink, the amount of information gained at the sink is the joint entropy $H(A,B)$, which is calculated by

$$H(A,B) = H(A) + H(B) - I(A;B), \tag{18.15}$$

where $I(A;B)$ is the mutual information of the two sources. $I(A;B)$ can be interpreted as the reduction in the uncertainty of one source due to the knowledge of the other source:

$$I(A;B) = H(A) - H(A|B) = H(B) - H(B|A). \tag{18.16}$$

The definition of $I(A;B)$ in probability form is given as

$$I(A;B) = \sum_a \sum_b p(a,b)\log\frac{p(a,b)}{p(a)p(b)}. \tag{18.17}$$

where $p(a)$ and $p(b)$ are the probability distributions of the pixels in image A and image B, and $p(a,b)$ is the joint probability distribution of the two sources.

Mutual information is a measure of dependence between two sources: the more A and B are correlated, the larger the mutual information $I(A;B)$. In an article by Pluim et al. [13], a normalized form of mutual information, entropy correlation coefficient (ECC), is defined as

$$\mathrm{ECC} = \frac{2I(A;B)}{H(A)+H(B)}. \tag{18.18}$$

The ECC ranges from zero to one, where zero indicates that source A and B are independent, whereas one indicates that source A is equal to source B. The larger the ECC value, the more these two sources are correlated.

Based on Equations 18.15 and 18.18, the joint entropy of A and B can be expressed as a function of $H(A)$, $H(B)$ and ECC:

$$H(A,B) = \left(1-\frac{1}{2}\mathrm{ECC}\right)[H(A)+H(B)]. \tag{18.19}$$

Because $H(A)$ and $H(B)$ can be calculated at each camera using Equation 18.14, if ECC can be estimated, the joint entropy $H(A,B)$ will be obtained. However, to calculate $I(A;B)$ and ECC by definition, a joint probability distribution of the two sources needs to be estimated. Due to the complexity of image contents and the difficulty in image modeling, it is difficult to get an accurate estimation of the joint probability distribution [13]. Besides, estimating the joint probability requires a large bulk of computation [13]. If joint probability distribution is to be estimated in a sensor network, cameras at different locations must exchange their observed images, which will introduce a lot of communication burden in the network.

It can be seen that the proposed correlation coefficient in Equation 18.13 has the same intrinsic meaning as ECC: both ranging from 0 to 1 and denoting the degree of correlation between two sources. However, if the cameras' parameters and deployment information are given, it is much easier to obtain the proposed correlation coefficient. Considering the limited processing capability of sensors, we propose to estimate ECC by the proposed correlation coefficient. If we replace ECC in Equation 18.19 with the proposed correlation coefficient ρ, we can obtain an estimation of the joint entropy of A and B as

$$H(A,B) \approx \left(1-\frac{1}{2}\rho\right)[H(A)+H(B)]. \tag{18.20}$$

Therefore, the amount of information that can be gained from A and B together depends on the degree of correlation between them. The more A and B are correlated, the less joint entropy can be gained from A and B together. That is to say, if two camera sensors transmit their images to the sink, the amount of information gained at the sink will be larger if the two sensors are less correlated.

18.2.2 Joint Entropy Estimation

The above analysis provides a way to estimate joint entropy of two cameras when an area of interest is specified. We now study a more general case: there are more than two cameras and no area of interest has been specified, such as the three cameras case in Figure 18.9. We propose a

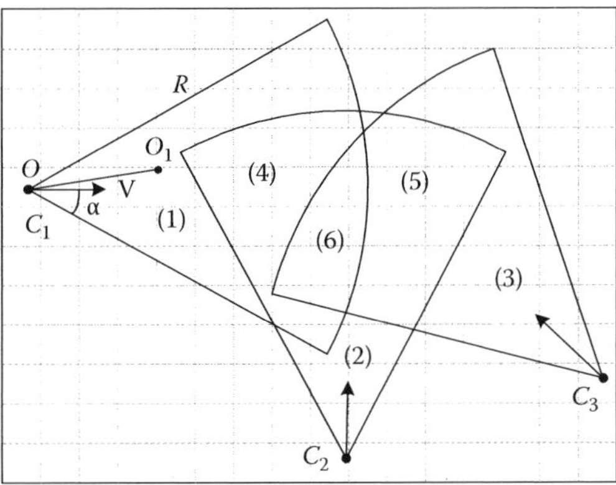

Figure 18.9 FoVs of multiple cameras.

novel entropy-based divergence measure (EDM) scheme to estimate the joint entropy of multiple cameras with no specified areas of interest. The steps of the EDM scheme are elaborated in the following paragraphs.

18.2.2.1 Area Division for Overlapped FoVs

We consider the case that N cameras (C_1, C_2,…,C_N) are deployed on the ground plane and all the cameras are homogeneous, that is, they have the same focal lengths (f), sensing radiuses (R), and offset angles (α). Denote the FoV of an individual camera C_i by $\mathcal{A}_i$, and the overall FoV for these cameras by $\mathcal{A}(\mathcal{A} = \{\mathcal{A}_1,\ldots,\mathcal{A}_N\})$. The goal of area division is to divide A into several regions (P_1, P_2,…, P_M), such that each region belongs to the FoVs of the same set of cameras. For example, the FoVs of the three cameras in Figure 18.9 are divided into six different regions.

We introduce a grid-based approach to divide the overall FoV A into regions. As shown in Figure 18.9, the overall FoV A is divided into small grids ($G(k)$, k = 1,…,K). And then we can check if a grid $G(k)$ is in a camera's FoV $(\mathcal{A}_n, n = 1,\ldots,N)$ as follows: we find the center point of the grid, and using the condition in the sensing model in Equation 18.2, we can tell if it is in the camera's FoV; if this center point is in the camera's FoV, we regard this grid to be in the camera's FoV. (This approximation is valid as long as the size of the grid is much smaller than the size of the FoV.) After traversing all the grids in the overall FoV, regions could be formed by grouping the grids that belong to the same set of cameras.

18.2.2.2 Estimating the Joint Entropy of a Region

We now introduce an algorithm to estimate the joint entropy of a region. Denote the cameras that can observe region P_i by (C_1,…,C_n). For the kth camera C_k, denote its observed visual information by X_k, and denote its observation about this region by $X_k(P_i)$. The amount of information of the region P_i is the joint entropy of the observations about this region from the cameras (C_1,…,C_n), given by

$$H(P_i) = H[X_1(P_i),\ldots,X_n(P_i)]. \tag{18.21}$$

Our solution for estimating $H(P_i)$ consists of three steps:

1. Estimate the individual entropy $H[X_k(P_i)]$ in Equation 18.21.
2. Study the correlation characteristics among the individual observations using the correlation model in the previous section.
3. With the results from the above two steps, apply a dependency graph–based algorithm to estimate $H(P_i)$ in Equation 18.21.

18.2.2.2.1 Individual Entropy Estimation

For an arbitrary camera C_k, the entropy of its observed image X_k is $H(X_k)$. The entropy of the observation X_k about the region P_i, $H[X_k(P_i)]$, can be estimated as

$$H[X_k(P_i)] \approx \frac{S(P_i)}{S(\mathcal{A}_k)} H(X_k) \tag{18.22}$$

where $S(P_i)$ is the area of P_i and $S(\mathcal{A}_k)$ is the area of the FoV. The entropy $H(X_k)$ is the total amount of information of X_k, which is provided by the projections of all the 3-D points in the FoV. Because there is no prior knowledge about where a camera is deployed or what type of scene is observed, it is assumed that when all the 3-D points in the FoV are projected on the camera's image plane, each point provides the same amount of information. Considering that the cameras are deployed on a ground plane and a 2-D FoV model is used, the amount of information that camera C_k contributes to P_i is approximately proportional to the area of P_i, so we use the ratio $\frac{S(P_i)}{S(\mathcal{A}_k)}$ to estimate $H[X_k(P_i)]$ in Equation 18.22.

This assumption works well when there are no large obstacles in a camera's FoV. A camera's FoV might be reduced in case of obstacles; therefore, when implementing the proposed algorithm in practical applications, we might need to update the camera's FoV model to reflect the effect of obstacles.

There are many different models for images, and different values of entropy may be obtained for the same image sources. In our algorithm, we avoid calculating the exact values of entropy of images. As we consider the case that all the camera sensors in a WMSN are homogeneous, without loss of generality, we assume that the entropies of the single observed images are the same, denoted by $H(X_i) = H(\cdot)(i = 1,\ldots,N)$. All the joint entropy terms in our algorithm will be expressed as relative values of $H(\cdot)$.

18.2.2.2.2 Spatial Correlation–Motivated Entropy Estimation

For cameras C_j and C_k which can observe region P_i, by taking the region P_i as the area of interest, a spatial correlation coefficient between the observations of P_i at C_j and C_k, denoted by ρ_{jk}, can be derived based on Equation 18.13. Furthermore, the spatial correlation coefficient was related to the estimation of joint entropy as shown in Equation 18.20. The joint entropy of the observations of P_i at C_j and C_k was estimated as

$$H[X_j(P_i), X_k(P_i)] \approx \left(1 - \frac{1}{2}\rho_{j,k}\right)(H[X_j(P_i)] + H[X_k(P_i)]) \tag{18.23}$$

where $X_j(P_i)$ is the observation of P_i at camera C_j, and $X_k(P_i)$ is the observation of P_i at camera C_k.

From Equation 18.23, we can obtain the conditional entropy as follows:

$$\begin{aligned} H[X_j(P_i) \mid X_k(P_i)] &= H[X_j(P_i), X_k(P_i)] - H[X_k(P_i)] \\ &\approx \left(1 - \frac{\rho_{j,k}}{2}\right) H[X_j(P_i)] - \frac{\rho_{j,k}}{2} H[X_k(P_i)] \end{aligned} \tag{18.24}$$

where $H[X_j(P_i)|X_k(P_i)]$ is the entropy of $X_j(P_i)$ with the knowledge of $X_k(P_i)$. This conditional entropy equation is used in the following algorithm to estimate the joint entropy of multiple cameras.

18.2.2.2.3 Dependency Graph–Based Joint Entropy Estimation

We propose a dependency graph–based algorithm to estimate the joint entropy of a region. To motivate our algorithm, we study a two cameras' case first. Suppose there are only two cameras (C_1 and C_2) in a region P_i. We can depict their relationship using a dependency graph: $C_2 \rightarrow C_1$. The joint entropy of the observations from C_1 and C_2 can be calculated by traversing the dependency graph. The source node C_2 contributes the entropy of its observations, $H[X_2(P_i)]$, and the node C_1 contributes the conditional entropy with respect to C_2, $H[X_1(P_i)|X_2(P_i)]$, so the joint entropy is calculated by adding these two terms: $H[X_1(P_i),X_2(P_i)] = H[X_2(P_i)] + H[X_1(P_i)|X_2(P_i)]$. The dependency graph can also be constructed as $C_1 \rightarrow C_2$, from which we can get the same result of joint entropy.

The two cameras' case can be extended to estimate the joint entropy of more than two cameras. Generally, for an arbitrary number of cameras, we construct a dependency graph to describe the dependency characteristics among them. Denote the dependency graph by $G(V,E)$, where V is a collection of cameras, and E is a collection of directed edges that stand for dependencies. Joint entropy of the region is calculated by traversing all the nodes in the graph along the directed edges. The detailed steps are described in Algorithm 1.

For a group of cameras ($C_1, C_2, \ldots, C_n$) that can observe the region P_i, we can obtain a correlation matrix $(\rho_{j,k})_{n*n}$ based on Equation 18.13. To simplify the problem, we assume limited number of dependencies: each camera is dependent on the camera that is most correlated with it. For example, if camera C_j is most correlated with camera C_k, we say that C_j is dependent on C_k, and we can construct a directed edge starting from C_k and ending at C_j: $C_k \rightarrow C_j$. C_j is said to be a direct successor of C_k, and C_k is a direct predecessor of C_j.

The dependency graph is designed to be a directed acyclic graph with the following constraints: a camera is either a source node (i.e., a node that has no predecessors), or a direct successor of one of the other cameras; a dependency graph may have several source nodes, but each node can have at most one direct predecessor; and there should be no loops in the graph, for example, $C_k \rightarrow C_j$ and $C_j \rightarrow C_k$ cannot exist in the same graph. These properties could be guaranteed through the procedure of constructing the dependency graph (lines 5–12 in Algorithm 1). For each node C_k, if another node C_j is most correlated with it, that is, neighbor $(C_j) = k$, the algorithm adds $C_k \rightarrow C_j$ into the graph only when two conditions are met: (*i*) if C_j has no predecessors and (*ii*) if C_j is not a predecessor of C_k (line 7 in Algorithm 1). The first condition guarantees that each node can have at most one direct predecessor, and the second one guarantees that there are no loops in the graph.

Given a dependency graph with the above features, the joint entropy is estimated by traversing all the nodes in the graph and adding the entropies of the nodes together, which corresponds to lines 13 to 19 in Algorithm 1. A source node contributes its individual entropy to the joint entropy, whereas a nonsource node contributes its conditional entropy with respect to its direct predecessor to the joint entropy.

Algorithm 1. Dependency graph–based entropy estimation

1: P_i: $\{C_1, C_2, \ldots, C_n\}$ with correlation matrix $(\rho_{j,k})_{n^*n}$.

2: **for** $j = 1$ to n **do**

3: $\text{neighbor}(C_j) = \arg\max_{k \neq j}(\rho_{j,k})$;

4: **end for**

5: **for** $k = 1$ to n **do**

6: **for** $j = 1$ to n and $j \neq k$ **do**

7: **if** neighbor(C_j) = k and C_j has no predecessors and C_j is not a predecessor of C_k **then**

8: Add $C_k \rightarrow C_j$ into the dependency graph;

9: Predecessor(C_j) = C_k;

10: **end if**

11: **end for**

12: **end for**

13: **for** $j = 1$ to n **do**

14: **if** C_j has no predecessor **then**

15: Add $H[X_j(P_i)]$ to $H(P_i)$;

16: **else if** Predecessor(C_j) = C_k **then**

17: Add $H[X_j(P_i)|X_k(P_i)]$ (1.24) to $H(P_i)$;

18: **end if**

19: **end for**

20: **return** $H(P_i)$.

Because the FoVs for a group of cameras are divided into several independent regions, the total joint entropy is the sum of the entropies of all the regions. For a group of cameras with observations $(X_1, \ldots, X_N)$, with their FoVs divided into regions $(P_1, \ldots, P_M)$, the total joint entropy is given by

$$H(X_1, \ldots, X_N) = H(P_1) + \ldots + H(P_M) \tag{18.25}$$

where $H(P_i)(i = 1, \ldots, M)$ is obtained by Algorithm 1.

Example

To provide an overview of the EDM algorithm, we illustrate the steps for estimating the joint entropy of the three cameras in Figure 18.9. The FoVs of the three cameras are divided into six regions. We take the sixth region (P_6) as an example. All the three cameras (C_1, C_2, and C_3) can observe P_6. Suppose we find from the geometry of the cameras' FoVs that $S(P_6) = 0.1S(\mathcal{A}_i)$. The entropy of a single image is $H(\cdot)$, by applying Equation 18.22, the individual entropies about this region are $H[X_i(P_6)] = 0.1H(\cdot)(i = 1, 2, 3)$. Furthermore, from Equation 18.13, we can obtain a correlation matrix for P_6:

$$(\rho_{jk})_{3*3} = \begin{pmatrix} 1 & 0.1 & 0 \\ 0.1 & 1 & 0.5 \\ 0 & 0.5 & 1 \end{pmatrix}.$$

By applying Algorithm 1 on the correlation matrix, we can obtain a dependency graph as $C_3 \rightarrow C_2 \rightarrow C_1$. Therefore, the joint entropy of region P_6 is $H(P_6) = H[X_3(P_6)] + H[X_2(P_6)|X_3(P_6)] + H[X_1(P_6)|X_2(P_6)]$, where the conditional entropies can be calculated from Equation 18.24. For example, $H\left[X_2(P_6) \mid X_3(P_6)\right] = \left(1 - \frac{\rho_{2,3}}{2}\right) \cdot H[X_2(P_6)] - \frac{\rho_{2,3}}{2} \cdot H[X_3(P_6)] = 0.05H(\cdot)$. In the same way, we can obtain $H[X_1(P_6)|X_2(P_6)] = 0.09H(\cdot)$. Thus, the joint entropy of P_6 is $H(P_6) = 0.24H(\cdot)$. After the entropy of each region is calculated, the joint entropy of the three cameras is calculated by $H(X_1, X_2, X_3) = H(P_1) + \ldots + H(P_6)$.

The entire EDM algorithm can be run at each sensor node. To estimate joint entropy, a node just needs to acquire the FoV parameters, locations, and sensing directions of its neighbors. Therefore, it does not require expensive communication costs in the network.

18.2.3 Coding Efficiency Prediction

One promising property of the EDM scheme is its capability of predicting the efficiency of coding among spatially correlated camera sensors. If two cameras in a WMSN are correlated, we can compress the image from one camera based on the prediction of the image from the other, which is referred to as differential coding. Multiple correlated cameras can report their observations to a multimedia processing hub and the hub can perform joint coding of the correlated images. Based on the distributed source coding principle, correlated cameras can also separately encode its own data, and the coded images are jointly decoded at the data sink. For these different coding strategies, joint entropy always serves as a lower bound of the achievable total coding rate.

Consider two correlated images (X_A and X_B) from video sensors V_A and V_B. If each sensor compresses its observed image independently, the resulting coding rates of X_A and X_B are $R(X_A)$ and $R(X_B)$, respectively. If we compress X_A using X_B as its prediction, the rate of X_A becomes $R(X_A|X_B)$ after differential coding. We define a *differential coding efficiency* as the percentage of rate saved by differential coding compared with individual coding, which is given by

$$\eta_R^d = \frac{R(X_A) - R(X_A \mid X_B)}{R(X_A)}. \tag{18.26}$$

We predict the differential coding efficiency from the entropies of the image sources. An estimated differential coding efficiency is defined as

$$\eta_H^d = \frac{H(X_A) - H(X_A \mid X_B)}{H(X_A)} = \frac{I(X_A; X_B)}{H(X_A)} \tag{18.27}$$

where $I(A;B)$ is the mutual information between X_A and X_B.

In the same way, for a group of N camera sensors with observations $X_1,\ldots,X_N$, if they are jointly coded, we introduce an *actual joint coding efficiency* as

$$\eta_R = 1 - \frac{R(X_1,\ldots,X_N)}{R(X_1) + \ldots + R(X_N)} \tag{18.28}$$

where $R(X_1, \ldots, X_N)$ is the total rate from joint coding, and $R(X_1) + \ldots + R(X_N)$ is the total rate from individual coding.

The joint entropy $H(X_1, \ldots, X_N)$ is a theoretical lower bound of the total coding rate for these cameras. To predict the percentage of rate savings of joint coding, we define an *estimated joint coding efficiency* as

$$\eta_H = 1 - \frac{H(X_1, \ldots, X_N)}{H(X_1) + \ldots + H(X_N)} \tag{18.29}$$

where $H(X_1) + \ldots + H(X_N)$ corresponds to the total coding rate needed when the cameras compress their observations individually.

18.2.4 Performance Evaluation

We investigate the effectiveness of the EDM scheme by comparing its predicted results with the actual coding performance of practical coding schemes. We evaluate the performance of joint coding for different numbers of cameras here. As for differential coding, it is a special case of joint coding among N cameras when $N = 2$.

We consider an indoor scene and an outdoor scene as representatives of various WMSN applications. We deploy 12 camera nodes at different viewpoints around each scene, and let each camera capture one image of the scene, with a resolution of 512 × 384. We record each camera's FoV parameters, so that the joint entropy and the corresponding η_H can be estimated using EDM. We also perform joint coding on the captured images, and from the resulting coding rates we can obtain η_R.

Experiments on different numbers of cameras for joint coding, coding schemes, and coding parameters are performed to evaluate the joint coding efficiency. The number of cameras for joint coding is set to three different values ($N = 2$, 3, and 4). As for coding schemes, there are many standardized solutions such as the JPEG/JPEG 2000 and the MPEG/H.26x series. For joint coding on multiple images, the redundancy among different images should be removed. The JPEG/JPEG 2000 standards can only reduce the redundancy within a single image, so they are not suitable for use here. We use two coding schemes of the H.264 standards: the Baseline profile and the recently developed Multi-View Coding (MVC) extension. And the H.264 reference software JM 8.5 [28] and JMVC 2.5 [29] are used, respectively. For both coding schemes, we obtain the coding rates under three quantization steps (QP = 28, 32, and 37). Other key parameters in the coding experiment are listed in Table 18.2.

In Equation 18.28, the rates of individual coding are obtained by performing intra coding on each image in the cluster, whereas the rate of joint coding is obtained by performing predictive coding among the images. For predictive coding, the images in the cluster are coded in a sequential order. We also use the dependency graphs in the EDM algorithm to guide the coding process. In a dependency

Table 18.2 Parameters for EDM

	H.264 Baseline	*H.264 MVC*
RD optimization	On	On
Entropy coding	UVLC	CABAC
Search range	128 pixels	128 pixels
No. reference frames	1	1

graph, each camera is connected with the camera that is most correlated with it; thus, it is beneficial to perform predictive coding between cameras that are connected in the graph. For example, if three cameras have a dependency graph as $C_1 \rightarrow C_2 \rightarrow C_3$, for joint coding of the images $\{X_1, X_2, X_3\}$, we take X_1 as the reference image and encode it first, and then encode X_2 based on the prediction of X_1, and X_3 based on the prediction of X_2. The total joint coding rate is a sum of the coding rates of the three images.

When deploying the cameras around a scene, we let their locations and sensing directions become pairwise symmetric with respect to the center of the scene. Consequently, we can have several (at least two) groups of cameras that lead to the same estimated joint coding efficiency (η_H), according to our spatial correlation model and the EDM algorithm. For each value of η_H, we perform joint coding on the corresponding groups of cameras, and take the average value of the resulting actual coding efficiencies (η_R). Comparisons of the corresponding η_H and the average η_R values for the two scenes are given in Figures 18.10 and 18.11. For both scenes, although the actual

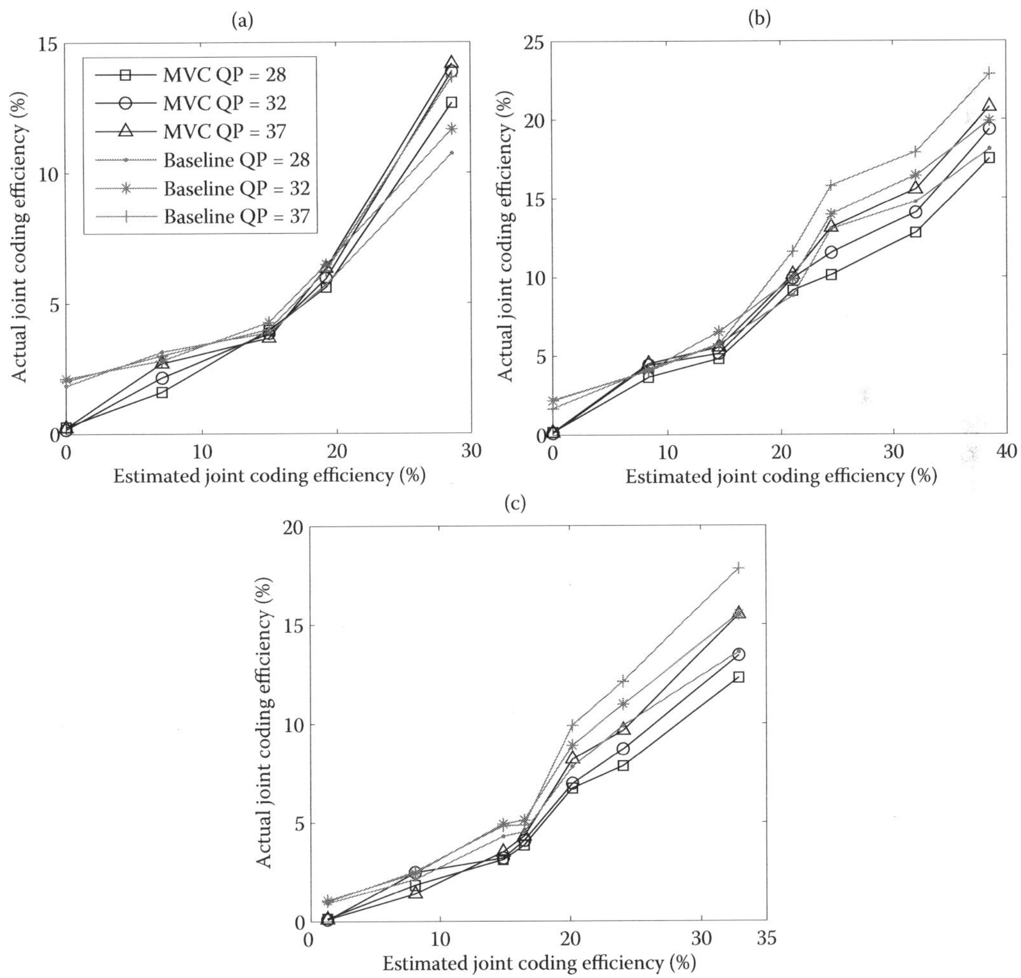

Figure 18.10 Joint coding performance of the indoor scene for different number of cameras ((a) $N = 2$, (b) $N = 3$, and (c) $N = 4$).

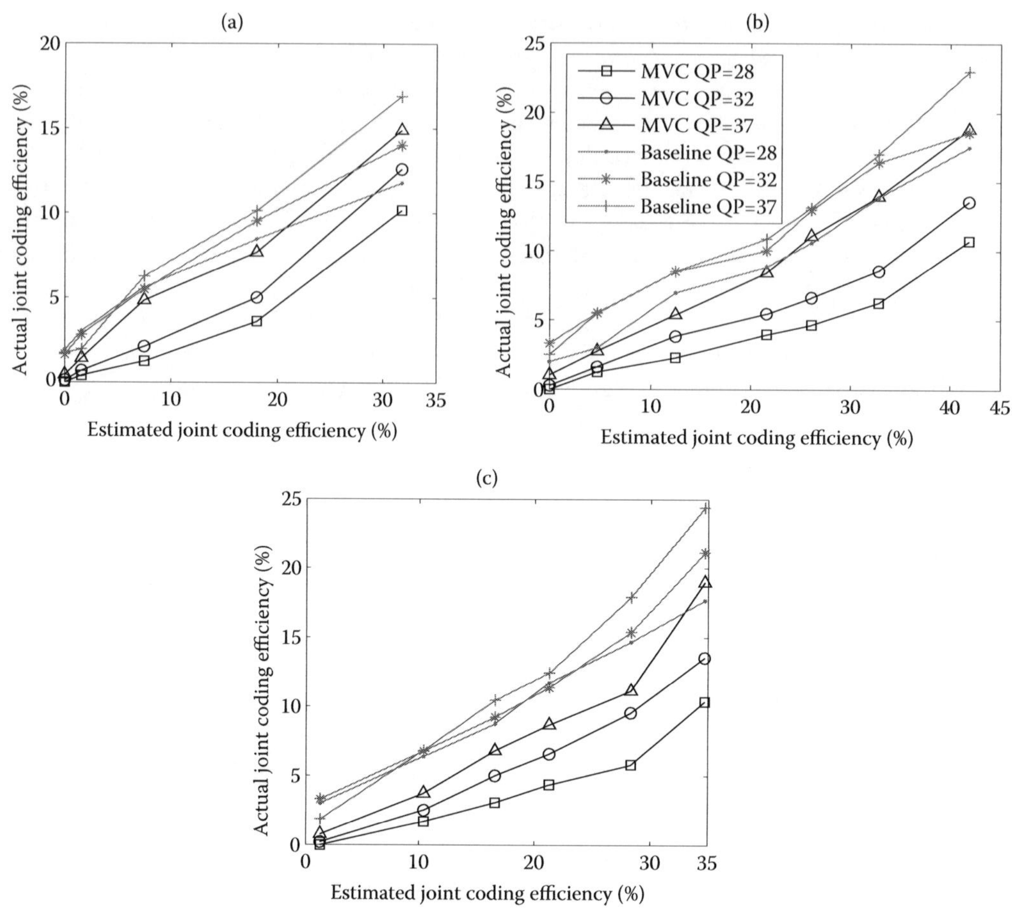

Figure 18.11 Joint coding performance of the outdoor scene for different number of cameras ((a) $N = 2$, (b) $N = 3$, and (c) $N = 4$).

joint coding efficiency might be smaller than the estimated joint coding efficiency, the actual joint coding efficiency increases as the estimated joint coding efficiency increases.

As shown in Figures 18.10 and 18.11, for the same coding scheme, the value of η_R increases as the quantization step increases: as larger quantization steps result in more distortion, they may have more potential bit savings for joint coding. The H.264 MVC extension is more advanced than the H.264 Baseline profile, and our experiments also show that the MVC extension always produce lower bit rates under the same coding parameters. However, the joint coding efficiency of the MVC extension is not necessarily larger than that of the Baseline profile. This is because the MVC extension results in smaller denominators in Equation 18.28 than in the Baseline profile.

The actual joint coding efficiency is approximately proportional to the estimated joint coding efficiency, and such a feature is independent of the number of camera sensors, coding methods, and levels of distortion. When the quantization step is fixed, the actual coding efficiency η can be predicted by a linear function of η_H, given by

$$\eta = k \cdot \eta_H \tag{18.30}$$

where k is a ratio that depends on the performance of specific encoding parameters (e.g., quantization step). We find that k ranges from 0.3 to 0.4 by performing linear regression in our data set. This gives us a prediction of how much bandwidth can be saved by coding between correlated cameras, which is very useful for the design of networking protocols for WMSNs.

18.3 Collaborative Image Compression Using Clustered Source Coding

In this section, we introduce a clustered source coding strategy to effectively reduce the redundancy of visual information from correlated cameras. The objective is to achieve a substantial compression gain and guarantee decoding reliability at the sink. First, we define an OCC problem as finding a set of coding clusters with the minimum total entropy, such that each camera node is covered by at least two different clusters. The minimization of total entropy guarantees that the global compression gain is maximized, whereas the coverage requirement ensures that the effect of cluster failures on the decoding reliability is mitigated. We prove that the OCC problem is NP-hard. As a heuristic solution, a fully distributed multi-cluster coding protocol, called DMCP, is presented to provide a $\ln\Delta$ approximation to the optimal solution, where Δ is the maximum node degree in the network.

18.3.1 Data Compression Using Clustered Source Coding

To remove the redundancy for correlated cameras, a group of camera sensors can form a cluster to collaboratively compress their data. Consider a cluster consisting of a cluster head (CH) and N camera sensors, where each sensor i produces image X_i, which is encoded with rate R_i. According to basic coding theorems, we have the following observation: The total coding rate of all nodes within a cluster is lower bounded by the joint entropy $H(X_1, X_2,\ldots, X_N)$ no matter if centralized or distributed source coding is applied. A detailed explanation of this observation is given in an article by Wang et al. [14].

18.3.1.1 OCC Problem

Because joint entropy provides a benchmark on the compression gain from joint encoding of multiple sources, we can utilize a similar entropy-based concept, called *cluster entropy*, to measure the collaborative compression gain within the scope of a single coding cluster. The target of OCC can then be correspondingly interpreted as to select a set of coding clusters according to their cluster entropies such that the total entropy of the entire network is minimized. We describe two definitions involved in the discussion above.

Definition 1

A coding cluster is a finite set comprising a camera sensor and all sensors within its transmission range. ■

Definition 2

For each coding cluster A, its cluster entropy H(A) is equal to the joint entropy of all cameras in A. ■

Now, the OCC problem can be formally stated as, given a network consisting of a finite set of camera sensors $E = e_1, e_2, \ldots e_n$ and a set of n subsets of E, $\mathcal{S} = \{S_1, S_2, \ldots S_n\}$, where each set S_i corresponds to a coding cluster with its entropy $H(S_i)$, the goal is to find a collection C from S of minimum total entropy $\sum_{S_i \in C} H(S_i)$, such that each element e_i is covered by at least two sets in C.

The minimization of total entropy guarantees that the maximum global compression gain is achieved, whereas the coverage requirement ensures that the visual information encoded by each camera has more chance to be successfully delivered to, and properly decoded at data sink.

After a WMSN is deployed in a field, we would like to select a set of coding clusters to cover the entire network with maximum compression ratio. Due to the distributed manner of WMSNs and the changing environment, a centralized algorithm is not suitable for use here. The coding cluster selection should only depend on local information. Next, we formulate the OCC problem as an integer program, and show that the OCC problem is NP-hard.

Algorithm 2. Greedy coding cluster selection algorithm

1: $C \leftarrow 0$, and $E' \leftarrow E$
2: For each $S \in \mathcal{S}$, $x(S) \leftarrow 0$.
3: **while** $E' \neq$; **do**
4: $s \leftarrow \arg\min_{X \in S} H(X)/|X \cap E|$, and $x(s) \leftarrow 1$.
5: $C \leftarrow C \cup s$, and $\mathcal{S} \leftarrow \mathcal{S} \backslash s$.
6: $E' \leftarrow E' \setminus \left\{ e \in s \cap E : \sum_{S:e \in S} x(S) \geq 2, S \in C \right\}$.
7: **end while**

18.3.1.2 Integer Program Formulation of OCC Problem

To formulate the OCC problem as an integer program, we assign a variable x_S for each set $S \in \mathcal{S}$, which is allowed 0/1 values. This variable will be set to 1 if set S is selected for the coding hierarchy. The objective function is the sum of the entropy values of all selected coding clusters. The constraint is that for each node $e \in E$, we want that at least two of the clusters containing it are selected.

$$\begin{aligned} \text{MIN} \quad & \sum_{S \in \mathcal{S}} H(S) x_S \\ s.t \quad & \sum_{S:e \in S} x_S \geq 2 \qquad e \in E \\ & x_S \in \{0,1\}, \qquad S \in \mathcal{S} \end{aligned} \tag{18.31}$$

If we treat $H(S)$ as the cost $c(S)$ associated with each coding cluster $S \in \mathcal{S}$ and let the second constraint be the coverage requirement for each node $e \in E$, the OCC problem can be reduced to the constrained set multicover (CSMC) problem. The CSMC problem is NP-hard and the greedy algorithm is essentially the best one can hope for [30]. In other words, the approximation ratio $\ln\Delta$ achieved by the greedy algorithm is the best one for CSMC problem. Therefore, the greedy strategy applies naturally to our OCC problem: let us say that the node *e* is uncovered if it occurs in fewer than two of the selected coding clusters. In each iteration, the algorithm selects, from the currently unselected clusters, the most compression-efficient cluster, in which the compression efficiency of a cluster is defined to be the average entropy of the uncovered nodes it covers. The algorithm terminates when there are no more uncovered nodes, for example, each node has been included by two different clusters. The pseudo-code of the above procedures is described in Algorithm 2.

The greedy algorithm for the OCC problem can be computed in $O(n)$ rounds if a central controller (e.g., data sink) provides the full information of the network topology along with detailed settings (e.g., sensing direction, sensing offset angle, and sensing range) for each camera. However, in a large-scale distributed network like WMSN, the centralized operations have limited flexibility and scalability. Moreover, the energy constraint of sensor nodes prohibits network-wide information exchange. Next, we introduce a distributed protocol that only needs local information exchange to achieve global compression optimization.

18.3.2 Distributed Multicluster Coding Protocol

After a WMSN is initially deployed, each camera node leads its neighbors to constitute a candidate coding cluster. Currently, each sensor node could be in one of the following four states: black, gray, half gray, and white. We call sensor nodes *black* if they are selected as the CH locaters. The CH locaters will not serve as the normal CHs but indicate the coordinates at which the future mobile or fixed CHs should be placed. We call the nodes *gray* if they are covered by at least two black nodes, and *half gray* if they are covered by exactly one black node. A node stays in the *white* state if there exists no black nodes within its one-hop range. The half gray nodes and white nodes are collectively referred to as uncovered nodes. We now describe several useful definitions.

Definition 3

The neighbor set of a node is a set consisting of the node itself and all nodes in its one-hop range. ■

Definition 4

The serving set of a node is a set comprising the uncovered nodes that are residing in its one-hop range. ■

Definition 5

The coding effectiveness of a node is the average entropy of all nodes in its serving set. ■

Definition 6

The CH counter of a node records the current number of the black nodes among its one-hop neighbors. ■

Algorithm 3. Distributed multicluster coding protocol

1: state(e) ∈ {black, gray, half gray, white, uncovered}
2: state(e) ← white, send and receive state(e) and camera settings
3: $N_e \leftarrow \{e'$: state(e') = white$\} \cup \{e\}$
4: {Discover neighbor set N_e}
5: counter(e) = 0 {Set CH counter}
6: **while** state(e) = uncovered **do**
7: $U_e \leftarrow \{e' \in N_e$: state($e'$) = uncovered$\}$
8: {Calculate serving set U_e}
9: $EC_e \leftarrow H(N_e)/|U_e|$
10: {Calculate coding effectiveness EC_e}
11: **if** $EC_e = \min_{e' \in U_e}\{EC_{e'}\}$ **then**
12: state(e) ← black, and counter = 1
13: send COVERAGE message
14: **else**
15: wait until the selection of a new black node times out
16: **if** no COVERAGE received **then**
17: state(e) remains
18: **else if** counter = 0 **then**
19: state(e) ← half gray, and counter(e) = 1
20: **else if** counter = 1 **then**
21: state(e) ← gray, and counter(e) = 2
22: send and receive ADV message-containing state(e)
23: **end if**
24: **end if**
25: **end while**
26: *Process_ Gray_ Black*()

Now, the DMCP establishes a clustered coding hierarchy. The pseudo-code of the procedures is described in Algorithm 3 and Algorithm 4. Initially (lines 1–5 in Algorithm 3), no black nodes exist in the network. Thus, every node is uncovered. Nodes in the uncovered state send out their camera settings to their neighboring nodes. After receiving the setting information, an uncovered node discovers its serving set and calculates its cluster entropy. Based on this information, an uncovered node evaluates its coding effectiveness, which is sent out along with the node state in an advertising (ADV) message to its two-hop neighbors.

A node in the uncovered (e.g., half gray or white) state collects ADV messages and extracts the coding effectiveness values from its two-hop neighbors. If the node itself is the most coding-effective node among its two-hop neighbors, it becomes a black node and sends COVERAGE messages to other uncovered nodes within its one-hop range (lines 11–13 in Algorithm 3). Otherwise, an uncovered node can encounter the following scenarios: (1) if no COVERAGE message is received within the predefined maximum duration of selecting a new black node, the node remains uncovered, recalculates its coding effectiveness, and sends out an ADV message (lines 16–17 in Algorithm 3). (2) If a COVERAGE message is received, and its CH counter is equal to zero, the node enters a half gray state and increments its CH counter by 1 (lines 18–19 in Algorithm 3). (3) If a COVERAGE message is received, and its CH counter already reaches 1, the node becomes a gray node and sets the CH counter to 2 (lines 20–21 in Algorithm 3). For the last two cases, an ADV message containing the node state is sent out to its immediate neighbors.

For a gray node, if the CH counters of all its neighbors has already reached 2, the node remains gray for the rest of cluster selection procedure and becomes a cluster member in the end. Otherwise, the node sends out an ADV message containing its coding effectiveness and collects ADV messages from all the uncovered nodes within its two-hop range. If the node itself has the highest coding effectiveness, it enters a black state and sends out COVERAGE messages to its uncovered neighbors (lines 5–6 in Algorithm 4). Otherwise, if the maximum duration of generating a new black node passes, and there still exist uncovered nodes within its one-hop range, the node remains gray (lines 8–10 in Algorithm 4). A black finally becomes a CH locater until the value of its CH counter reaches 2 upon receiving a COVERAGE message (lines 13–17 in Algorithm 4).

Algorithm 4. Process_Gray_Black()

1: $U_e \leftarrow \{e \in N_e\text{: counter}(e) < 2\}$
2: **while** $|U_e| < |N_e|$ **do**
3: **if** state(e) = gray **then**
4: recalculate coding effectiveness EC_e
5: **if** $EC_e = \min_{e' \in U_e} \{EC_{e'}\}$ **then**
6: state(e) ← black, and send COVERAGE msg
7: **else**
8: wait until the new black selection times out
9: **if** $|U_e| < |N_e|$ **then**
10: state(e) ← gray
11: **end if**
12: **end if**
13: **else if** state(e) = black, **then**
14: wait until a COVERAGE is received
15: counter(e) = 2
16: send ADV msg containing state(e) and counter(e)
17: Node e becomes a CH locator
18: **end if**
19: **end while**
20: Node e becomes a cluster member

The above procedures are performed by all nodes until each of them becomes either a CH locater or a cluster member. At the end, there is no uncovered node in the network, and the established clustered coding hierarchy covers the entire network.

Example

To provide an overview of the DMCP protocol, we illustrate the steps for constructing a clustered hierarchy for a simple WMSN with six cameras in Figure 18.12a. By this network setting, every node is within a two-hop range of each other. Initially, all cameras are uncovered and thus stay in a white state. Then, each node calculates its coding effectiveness (see Definition 5). Assuming that node 1 has

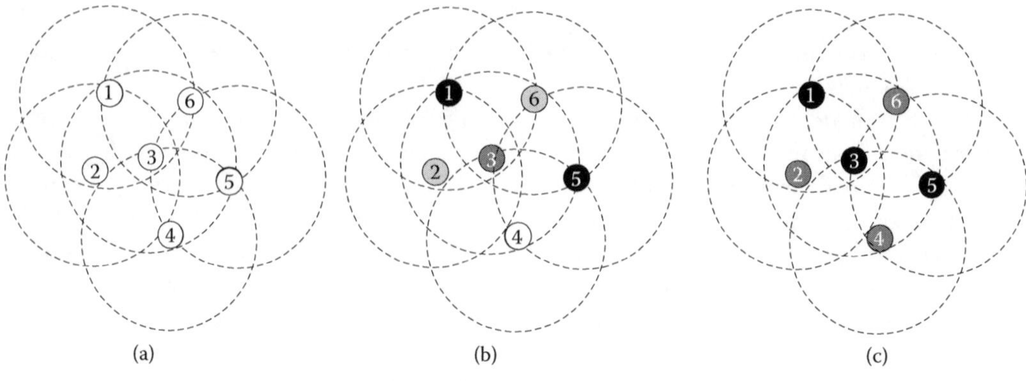

Figure 18.12 DMCP example. (a) Step 1, (b) step 2, and (c) step 3.

the minimum coding effectiveness among all its two-hop neighbors. In this case, node 1 enters a black state and its neighbors including nodes 2, 3, and 6 enter a half gray state because they are only covered by one black node. Next, every node except node 1 recalculates its coding effectiveness and exchanges the result with its two-hop neighbors. Assume that node 5 has the minimum coding effectiveness. As shown in Figure 18.12b, this indicates that node 5 enters black state and its neighbor, node 4, enters half gray state. Now, because node 3 is covered by two black nodes, node 3 enters the gray state. However, because every node except node 3 is only covered by one black node, all the half gray nodes, that is, nodes 2, 3, 4, and 6, have to recalculate their own coding effectiveness and exchange the results among two-hop neighbors. Assuming that node 3 has the minimum coding effectiveness. It implies that node 3 changes its state from the gray to the black. In this case, as shown in Figure 18.12c, every node has two black nodes in its one-hop range. This completes the DMCP protocol.

18.3.2.1 Approximation Ratio

Theorem 1

The DMCP computes a $\ln\Delta$ approximation for the OCC problem. ■

Proof

According to DMCP, the cluster entropy of a node is only related to camera settings of the nodes in its neighbor set, and the neighbor set is only determined by the local topology, the value of the cluster entropy will not change as the protocol proceeds. On the other hand, the cardinality of the serving set, which is equal to the number of its uncovered neighboring nodes, can be reduced as the protocol proceeds because some uncovered neighboring nodes could be included by some other clusters. Thus, we conclude that the coding effectiveness of a non-black node can only be reduced if the cardinality of its serving set decreases.

Based on this conclusion, we can further show that the DMCP is equivalent to the centralized greedy algorithm. According to DMCP, a non-black node v with the highest coding effectiveness within its two-hop neighborhood is eligible to become a black node. The selection of other non-black nodes outside v's two-hop range as black nodes will not affect v's eligibility to enter the black state because the status change of the nodes outside v's two-hop range cannot reduce v's serving set cardinality, and according to the conclusion above, v's coding effectiveness remains the same. Therefore, the DMCP chooses v as a black node before any nodes within its two-hop range. On the other hand, the centralized greedy algorithm always selects the most compression efficient cluster, and v leading its neighbors represents the most compression efficient cluster within its two-hop range. Therefore, the centralized approach will select the cluster led by v as a final coding cluster as the algorithm proceeds. This means that the DMCP obtains the same result as the centralized algorithm, thus achieving the same $\ln\Delta$ approximation ratio as the centralized algorithm. ■

As shown in Section 18.3.1.2, the OCC problem can be reduced to a CSMC problem, for which $\ln\Delta$ is the best approximation ratio. Thus, we can conclude that no protocols perform better than the proposed DMCP in terms of application factor.

In addition, it was shown in an article by Wang et al. [14] that DMCP terminates in $O(N)$ rounds, where N is the number of nodes in the network, with a worst case message exchange complexity of $O(1)$ per node. Because the clustered coding hierarchy only needs to be constructed when the network is initially deployed, the linear time and message complexity of the proposed protocol has a trivial effect on the network performance.

18.3.3 Performance Evaluation

We compare DMCP to the hybrid energy-efficient distributed clustering (HEED) protocol [31] and its modified version, MHEED. HEED is a well-known clustering protocol that is specially designed for wireless sensor networks that deal with scalar data. This protocol constructs a hierarchical network architecture by two phases: CH selection and cluster member assignment. In the first phase, sensor nodes are selected as CHs probabilistically. More specifically, each node is given an initial probability p (i.e., 0.05 in work by Younis and Fahm [31]) with which it becomes a CH. In the first iteration, each sensor uniformly draws a value between 0 and 1 and compares this value with the initial probability. If this value is less than p, the sensor becomes a CH and all its neighbors are covered. After this iteration, many sensors may still be uncovered because the initial probability (i.e., 0.05) is very small. Therefore, in each of the following iterations, every sensor doubles p and with this probability, the uncovered sensors become new CHs. When p reaches 1, the first phase completes. In the second phase, each sensor is assigned to the closest CH as its cluster member. Different from DMCP, HEED protocol is a compression-unaware approach. To fairly compare DMCP with HEED, we design a MHEED by incorporating the proposed EDM scheme. Specifically, MHEED uses the same procedure as HEED for the CH selection phase. In the second phase, we use the average cluster entropy, instead of node proximity to the CHs, as the metric to associate sensors with CHs. That is, each sensor joins the cluster with the minimum average entropy, a ratio of the estimated joint entropy of the cameras covered by a CH to the number of cameras it covers.

In Figures 18.13 and 18.14, we measure the coding efficiency of HEED and MHEED, respectively, and evaluate the coding efficiency enhancement of DMCP, compared with HEED and MHEED, varying the network size n and sensing radius R. Because DMCP exploits the inherent correlation structure of multiple cameras, it is expected that DMCP can achieve higher coding efficiency by finely identifying and properly selecting a set of clusters that leads to higher compression performance. Accordingly, as shown in Figures 18.13 and 18.14, DMCP achieves 28% to 50% and 20% to 40% enhancement, compared with the coding efficiency of HEED and MHEED, respectively. Meanwhile, in Figures 18.13 and 18.14, we observe that higher enhancement is achieved

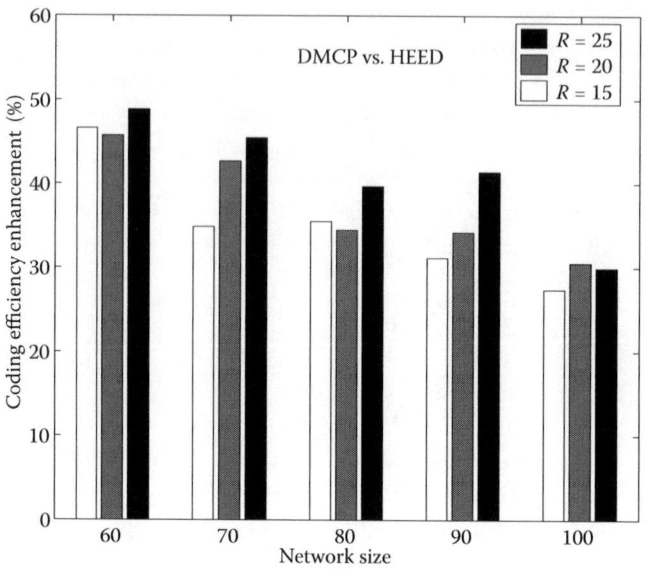

Figure 18.13 Coding efficiency of DMCP compared with HEED.

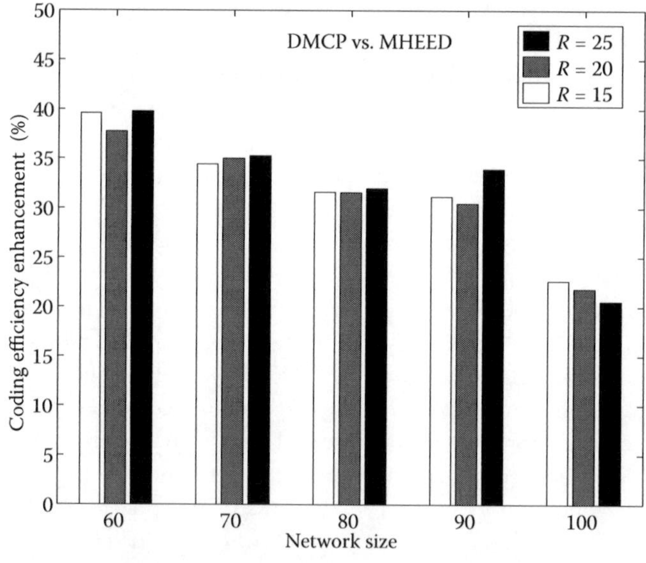

Figure 18.14 Coding efficiency of DMCP compared with MHEED.

under smaller network size (i.e., smaller camera density because of the fixed deployment area). This phenomenon is attributed to the fact that smaller camera density leads to higher variability of the joint entropy of different clusters. In this case, the correlation-based strategy, DMCP, has more evident advantage over compression-unoriented approaches like HEED and MHEED. Moreover, we observe that less enhancement is achieved in Figure 18.14 than in Figure 18.13. This implies that MHEED achieves higher coding efficiency than HEED. This is expected because MHEED uses the average entropy as the metric in selecting CHs, which is superior to just selecting the closest CH because the average entropy of a node is a measure of the expected coding performance if this node is selected as CH. It is also worth noticing that in Figure 18.14 the enhancement under different sensing radius settings is comparable, which indicates MHEED is less sensitive to the changing sensing radius compared with HEED because MHEED partially exploits the correlation of visual information in clustering procedures.

18.4 Correlation-Aware QoS Routing

We propose a correlation-aware QoS routing (CAQR) algorithm for the delivery of visual information in WMSNs. Based on the correlation characteristics of video sensors, correlation-aware operations are designed to reduce traffic redundancy and avoid congestion in the network. The correlation-aware operations are then integrated in an optimization QoS routing framework with the objective of minimizing the sensors' energy consumption under delay and reliability constraints.

18.4.1 Preliminaries

18.4.1.1 Metrics for Correlation of Visual Information

To simplify the problem, we consider the case that all the video sensors in a network are homogeneous, that is, they have the same focal lengths, sensing radiuses, and offset angles. For two arbitrary video sensors V_A and V_B with FoVs F_A and F_B, suppose at a same time, their observed images are X_A and X_B, respectively. The images X_A and X_B are correlated if F_A and F_B are overlapped with each other.

18.4.1.1.1 Overlapped Ratio of FoVs

The overlapped ratio of FoVs for sensors V_A and V_B, denoted by r_{AB}, is defined as the ratio of the overlapped area of the two FoVs to the area of the entire FoV of a sensor, given by

$$r_{AB} = \frac{S(F_{AB})}{S(F_A)} \tag{18.32}$$

where $S(F_{AB})$ ($F_{AB} = F_A \cap F_B$) is the overlapped area of F_A and F_B (Figure 18.15b), and $S(F_A)$ is the area of F_A (Figure 18.15a). If two video sensors have large overlapped ratio of FoVs, large portions of the two observed images are correlated. A large overlapped ratio also indicates that the two sensors are likely to observe the same event concurrently.

18.4.1.1.2 Differential Coding Efficiency

Consider two correlated images (X_A and X_B) from video sensors V_A and V_B. X_A can be compressed based on the prediction of X_B (differential coding). Compared with each sensor compressing their images independently, the percentage of rate saving for differential coding can be predicted by the estimated differential coding efficiency (η_H^d) in Equation 18.27. We consider the case that the individual entropies are the same for all sensor nodes [$H(X_A) = H(X_B)$]. Based on the EDM scheme in Section 18.2, η_H^d is proportional to both the overlapped FoV ratio (r_{AB}) and the correlation coefficient (ρ_{AB}). The differential coding efficiency η_H will be high when both r_{AB} and ρ_{AB} are large.

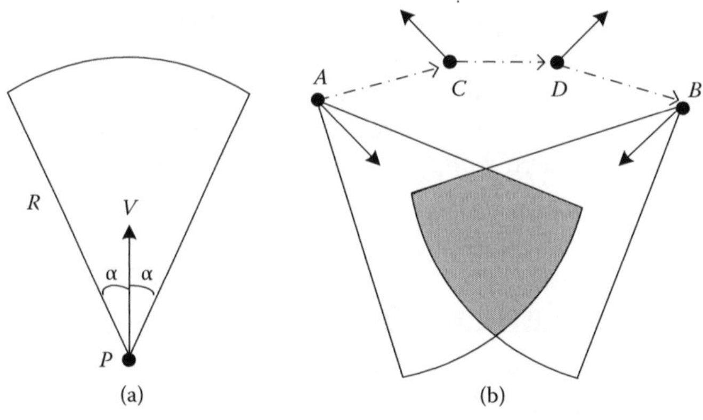

Figure 18.15 Camera sensors. (a) FoV and (b) overlapped FoVs.

18.4.1.2 Video In-Network Compression

Due to the huge size of raw visual information, images and video sequences are compressed before transmission. A lot of standardized techniques can be applied for image and video coding, such as JPEG/JPEG 2000 for coding still images and H.26x/MPEG for coding video sequences. The distributed video coding (DVC) technique [7] allows for separate encoding of correlated sources and joint decoding by the end user. It reduces the computational complexity at the encoders, but there is a lack of practical implementations of DVC in WMSNs. We consider the standardized coding techniques in this work. However, it is still possible to integrate the results of our correlation model with DVC techniques.

Standardized coding techniques, based on predictive coding concept, can be classified into intra coding and inter coding. Intra coding refers to coding techniques that reduce the redundancy within an image, whereas inter coding (also called differential or predictive coding) reduces the redundancy among multiple images. Accordingly, a compressed video sequence usually consists of periodical intra coded reference frames (*I* frames) and inter coded frames between reference frames, including forward predicted (*P*) frames, and bidirectional predicted (*B*) frames. Inter coding has much higher coding efficiency than intra coding; consequently, intra coded frames usually result in much larger sizes than inter coded frames.

In a network of correlated video sensors, nodes can cooperate with each other and remove the redundancy among their observations. Specifically, we can perform differential coding on the intra coded (*I*) frames between correlated sensors. Because video sensors that are out of the communication ranges of each other can still observe a common scene [32] (i.e., they are correlated as shown in Figure 18.15b), the differential coding of correlated sensors could be integrated in network layer operations.

Based on the discussions above, flows generated by video sensors could be classified into two categories:

1. *Intra flows*: Flows of intra coded video frames. The amount of traffic for an intra flow might be further reduced by differential coding with correlated video sensors.
2. *Inter flows*: Flows of inter coded video frames, for which the amount of traffic can hardly be further reduced.

Both types of flows need to be forwarded to the sink efficiently under QoS constraints.

18.4.1.3 Energy Consumption Models

The energy consumption for both video communication and processing are not negligible. Before introducing our proposed routing algorithm, we introduce the energy consumption models for video communication and compression here.

The energy consumption for transmitting and receiving l bits of data over a distance d is given as

$$E(l,d) = 2 \cdot E_{\text{elec}} \cdot l + \varepsilon_{\text{amp}} \cdot d_{\text{hop}}^{\alpha} \cdot l \tag{18.33}$$

where E_{elec} is the energy needed by the transceiver circuitry to transmit or receive one bit, ε_{amp} is a constant for communication energy, and α is the path loss exponent.

The energy consumption for processing can be modeled as a function of supply voltage. Suppose the execution of a task consisting of N_{cyc} clock cycles, the energy consumption for processing is estimated as

$$E_{\text{proc}}(N) = N_{\text{cyc}} C_{\text{total}} V_{dd}^2 + V_{dd}\left(I_0 e^{\frac{V_{dd}}{nV_T}}\right)\left(\frac{N_{\text{cyc}}}{f}\right). \tag{18.34}$$

The first term in Equation 18.34 is the switching energy, where C_{total} is the total capacitance switched by the computation per cycle, and V_{dd} is the supply voltage. The second term in Equation 18.34 stands for the leakage energy, where f is the clock speed, and I_0, n, K, and c are processor-dependent parameters [33].

The processing burden for video in WMSNs mainly comes from the video encoding and decoding process. Fortunately, the computational complexity of standardized video codecs has been studied a lot in the literature. From the experimental results in studies by Goto et al. [34] and Horowitz et al. [35], the number of clock cycles needed for encoding or decoding a video frame could be estimated. From these results, together with the processor-dependent parameters in Equation 18.34, we can estimate the energy consumption for encoding and decoding video frames.

18.4.2 CAQR Routing

The CAQR algorithm is a distributed algorithm to be operated on each sensor node in a WMSN. It consists of three components: correlation groups construction, intermediate node selection for correlation-aware differential coding, and QoS guaranteed next-hop selection with correlation-aware load balancing.

1. *Correlation groups construction.* When the network is initially deployed, correlated video sensors are identified by grouping together the sensors that share large overlapped FoVs.
2. *Intermediate node selection for correlation-aware differential coding.* For a video sensor V_A that generates an intra frame X_A, an intermediate node V_{B^*} is selected so that the redundancy in X_A is removed by performing correlation-aware differential coding on X_A based on the observation from V_{B^*}.

3. *QoS guaranteed next-hop selection with correlation-aware load balancing.* To forward X_A to the intermediate node V_{B*} for compression and then deliver the compressed X_A to the sink, the best next-hop node is selected, which satisfies the hop-to-hop QoS requirements while achieving minimum energy consumption. In particular, a correlation-aware load balancing scheme is applied to prevent network congestion by splitting the correlated flows that cannot be reduced to different paths.

In the following paragraphs, we explain each component and then summarize all the steps of the algorithm.

18.4.2.1 Correlation Groups Construction

Video sensors with large overlapped FoVs are likely to report the same event concurrently, and they are likely to have high differential coding gains. We introduce a centralized preprocessing step to cluster video sensors with large overlapped FoVs into correlation groups.

Let each video sensor report its focal length and FoV parameters to the sink. After receiving these parameters, the sink calculates the overlapped ratio of FoVs (r) (Equation 18.32) between any two video sensors. We apply the hierarchical clustering algorithm in the article by Jain et al. [36]. By using the overlapped ratio of FoVs (r_{ij}) as a similarity metric, the hierarchical clustering algorithm groups sensors with large overlapped ratio of FoVs together. After running the clustering algorithm, the sink broadcasts the results of clustering and assigns a group ID for each group. Each video sensor will be notified its group ID and other sensors' sensing parameters in the same correlation group. The construction of correlation groups just needs to be performed once while deploying the network. In the following steps of the routing algorithm, correlation-aware operations are performed among video sensors that belong to the same correlation groups.

18.4.2.2 Intermediate Node Selection for Correlation-Aware Differential Coding

We introduce a correlation-aware inter node differential coding scheme for the routing of intra flows.

Example

As shown in Figure 18.16, sensor V_A needs to find a route for its intra frame X_A to the sink. Suppose sensor V_B is in the same group as V_A and its distance to the sink is closer than V_A ($d_{BS} < d_{AS}$). If there

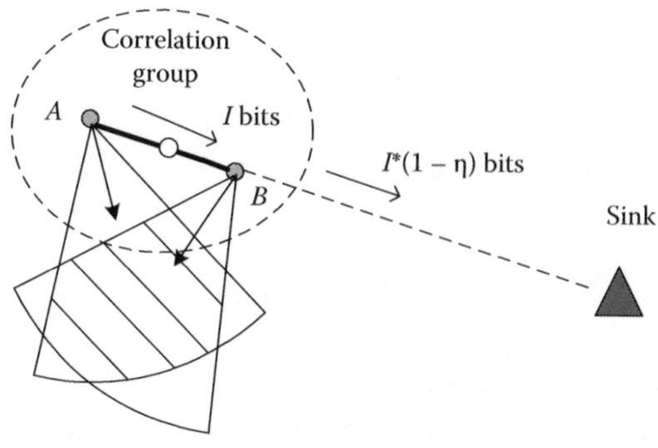

Figure 18.16 Correlation-aware differential coding.

is potential gain to compress X_A based on the prediction of X_B, the frame generated at the same time at V_B, we can send X_A to V_B to perform differential coding. From our correlation model, we can estimate the differential coding efficiency η^d_{AB} in Equation 18.27. If the size of the intra frame X_A is I_A, the saved bits from differential coding can be estimated as $I_A \cdot \eta_{AB}$. We introduce an energy gain to evaluate the potential energy efficiency of differential coding between V_A and V_B:

$$G_E(V_A,V_B) = \frac{E_{\text{com}}\{I_A \cdot \eta_{AB}\}}{E_{\text{proc}}\{I_A\}}. \qquad (18.35)$$

The numerator in the gain function is the communication energy for the bits that are saved from differential coding. It stands for the benefits brought by differential coding. This communication energy is not only related to the number of saved bits, but also related to the distance and number of hops from sensor V_B to the sink. We can estimate E_{com} using the estimated number of hops from node V_B to the sink ($\hat{N}_B^{\text{hops}}$) and the average one-hop distance (d_{hop}), given by

$$E_{\text{com}}\{I_A \cdot \eta_{AB}\} = \hat{N}_B^{\text{hops}} \cdot E(I_A \cdot \eta_{AB}, d_{\text{hop}}) \qquad (18.36)$$

where $E(I_A \cdot \eta_{AB}, d_{\text{hop}})$ is obtained from Equation 18.33 and $\hat{N}_B^{\text{hops}}$ can be estimated by

$$\hat{N}_B^{\text{hops}} = \max\left(\left\lceil \frac{d_{BS}}{d_{\text{hop}}} \right\rceil, 1\right). \qquad (18.37)$$

To estimate the average one-hop distance between nodes in the network (d_{hop}), node V_A can acquire all its one-hop neighbors' positions through a few message exchanges. And d_{hop} could be estimated as the sample arithmetic mean of the distance between V_A and all its one-hop neighbors. This estimation method involves little cost. It will be representative if the density of sensors are similar across the whole network.

The denominator in Equation 18.35 is the energy costs for performing differential coding, that is, the processing energy of differential coding at sensor V_B, including the decoding of the intra frame and the differential coding of the intra frame X_A based on the prediction of X_B at V_B. The energy for processing is related to video frame size and hardware. This term could be estimated from Equation 18.34 using the parameters in articles by Goto et al. [34] and Horowitz et al. [35].

For the routing of an intra frame generated at node V_A in correlation group $\mathcal{G}(V_A)$, we intend to find the best intermediate node in the same group so that the energy gain is maximized. This problem is formulated as follows.

18.4.2.3 Differential Coding-Based Intermediate Node Selection

$$\text{Find}: V_{B^*} = arg \max_{V_B \in \mathcal{G}(V_A)} G_E(V_A,V_B) \qquad (18.38)$$

$$\text{Subject to}: d_{BS} < d_{AS},\ G_E(V_A, V_B) > 1. \qquad (18.39)$$

A node V_B for differential coding should satisfy two conditions: (i) V_B is closer to the sink than V_A, which is given as $d_{BS} < d_{AS}$; (ii) the energy gain for differential coding is larger than 1. Among all the nodes in group $\mathcal{G}(V_A)$ that satisfy these two conditions, the node that generates the maximum energy gain (V_{B^*}) is selected as the intermediate node.

After V_A determines V_{B^*} for differential coding, it sends a request message to V_{B^*}, and V_{B^*} will send back a reply message. In this way, V_{B^*} becomes an intermediate destination for the intra frames generated by V_A. The intra frames from V_A will be forwarded to V_{B^*} first; V_{B^*} will further compress the frame and then forward it to the sink.

18.4.2.4 QoS-Guaranteed Next-Hop Selection

Each node distributively selects the optimal next-hop with the objective of minimizing energy consumption and satisfying QoS requirements in delay and reliability. Table 18.3 lists the relevant notations and variables for this algorithm.

Suppose node i needs to forward a video frame to the destination D (D can be either an intermediate node for differential coding or the sink). We define the forwarding neighbor set of node i as a set of its neighbors that are closer to the sink than itself, denoted by $\mathcal{F}_i$. The next hop node is selected from $\mathcal{F}_i$ according to the following rules.

18.4.2.5 Distributed Correlation-Aware QoS Routing

$$\text{Given}: \quad i, \mathcal{D}, j \in \mathcal{F}_i, V_{id}, \mathcal{G}(V_{id}), \{R_0^C, \ldots, R_N^C\}$$

$$\text{Find}: j^*, R_{ij^*}^C \tag{18.40}$$

Table 18.3 Notations and Variables

$\mathcal{F}_i$	Forwarding neighbor set of node i	D	Destination of a route
V_{id}	ID of a video sensor	$\mathcal{G}(V_{id})$	Correlation group of sensor V_{id}
L	Packet length	d_{ij}	Distance between node i and node j
$\{R_0^C, \ldots, R_N^C\}$	Candidate channel coding rates	R_{ij}^C	Channel coding rate for link ij
R	Transmission rate	T_{ij}	Local delay constraint
t_{ij}^q	Local queueing delay	PR_{ij}	Required local packet delivery ratio
pr_{ij}	Local packet delivery ratio	x_{v^i}	Correlation group indicator
DT	Frame decodable threshold (%)	φ	Probability that DT packets are delivered
P_D	Probability that a frame is decoded	p_b	Bit error rate

$$\text{Minimize}: E(L/R_{ij}^{C}, d_{ij}) \tag{18.41}$$

$$\text{Subject to}: \frac{L}{R \cdot R_{ij}^{C}} + \overline{t_{ij}^{q}} < T_{ij} \tag{18.42}$$

$$\frac{\gamma}{1-\gamma}(\Delta t_{ij}^{q})^{2} \le \left(T_{ij} - \frac{L}{R \cdot R_{ij}^{C}} - \overline{t_{ij}^{q}}\right)^{2} \tag{18.43}$$

$$pr_{ij} \ge PR_{ij} \tag{18.44}$$

$$\sum_{v^{j} \in \mathcal{L}\{v^{j}\}} x_{v^{j}} < w \tag{18.45}$$

The locally optimal next hop j^* is the node that results in the minimum energy consumption under local delay, local reliability, and correlation-aware load balancing requirements. A channel coding rate for the link from i to j^*, $R_{ij^*}^{C}$, is also selected from a set of predefined rates $\{R_0^C, \ldots, R_N^C\}$.

The objective is to minimize energy consumption. As shown in Equation 18.41, the minimization term is the energy consumption for transmitting a packet of L bits data and header with channel coding rate R_{ij}^{C} over a distance of d_{ij}. Equations 18.42 and 18.43 are the local delay requirements, Equation 18.44 is the local reliability requirement, and Equation 18.45 is the constraint for correlation-aware load balancing.

18.4.2.5.1 Local Delay Requirements

We use a geographic based mechanism to map end-to-end delay requirements to local delay requirements. Suppose a video flow v at node i needs to be delivered to the destination D within time T_{iD}. The local delay constraint, T_{ij}, is given as

$$T_{ij} = \left(\frac{d_{iD} - d_{jD}}{d_{iD}}\right) \cdot T_{iD} \tag{18.46}$$

where d_{iD} is the distance from node i to the destination, and d_{jD} is the distance from node j to the destination.

We consider a contention-free MAC in our context. Under this assumption, the delay of a hop mainly consists of the transmission delay and the queueing delay. The transmission delay for a packet from node i to node j can be calculated as $\frac{L}{R \cdot R_{ij}}$, where L is the length of the packet, R is the transmission rate, and R_{ij}^{C} the channel coding rate. We denote the queueing delay from node i to j by t_{ij}^{q}. Then, the total delay from node i to j is given by $\frac{L}{R \cdot R_{ij}} + t_{ij}^{q}$.

We provide probabilistic guarantee for one-hop delay, in which the probability that a packet is delivered within the deadline should not be below γ, given by $P\left(\frac{L}{R \cdot R_{ij}^{C}} + t_{ij}^{q} \le T_{ij}\right) \ge \gamma$. It can also be expressed as

$$P\left(\frac{L}{R \cdot R_{ij}^{C}} + t_{ij}^{q} \geq T_{ij}\right) \leq 1 - \gamma. \tag{18.47}$$

We let node j maintain the delays of packets in a recent period, from which we can estimate the average queueing delay $\overline{t_{ij}^{q}}$ and the variance of queueing delay $(\Delta t_{ij}^{q})^2$. As a result, the average single-hop delay is $\frac{L}{R \cdot R_{ij}^{C}} + \overline{t_{ij}^{q}}$, whereas the variance of single hop delay is $(\Delta t_{ij}^{q})^2$.

According to one-sided Chebyshev's inequality, for a random variable X with mean μ and variance σ^2, it satisfies

$$P(X - \mu \geq k) \leq \frac{\sigma^2}{\sigma^2 + k^2}, k > 0. \tag{18.48}$$

By applying Chebyshev's inequality on Equation 18.47, we find

$$P\left(\frac{L}{R \cdot R_{ij}^{C}} + t_{ij}^{q} \geq T_{ij}\right) \leq \frac{(\Delta t_{ij}^{q})^2}{(\Delta t_{ij}^{q})^2 + \left(T_{ij} - \frac{L}{R \cdot R_{ij}} - \overline{t_{ij}^{q}}\right)^2} \tag{18.49}$$

and

$$T_{ij} - \left(\frac{L}{R \cdot R_{ij}^{C}} + \overline{t_{ij}^{q}}\right) > 0. \tag{18.50}$$

Based on Equations 18.49 and 18.50, we have derived two constraints to satisfy the probabilistic delay guarantee in Equation 18.47, which are given in Equations 18.42 and 18.43. Condition 18.50 corresponds to the constraint in Equation 18.42. Comparing Equations 18.49 and 18.47, if the condition

$$\frac{(\Delta t_{ij}^{q})^2}{(\Delta t_{ij}^{q})^2 + \left(T_{ij} - \frac{L}{R \cdot R_{ij}} - \overline{t_{ij}^{q}}\right)^2} \leq 1 - \gamma \tag{18.51}$$

is met, the probabilistic delay guarantee in Equation 18.47 could be satisfied, from which Constraint 18.43 is obtained.

18.4.2.5.2 Local Reliability Requirements

We incorporate a dynamic channel coding scheme in the routing algorithm to adapt to varying wireless channel conditions. The routing algorithm selects a proper channel coding rate for link i to j, R_{ij}^{C}, from a set of predefined channel coding rates $\{R_0^C, \ldots, R_N^C\}$. A smaller channel coding rate indicates more redundancy being added to a packet and better error resilience performance.

To evaluate reliability, we use packet delivery ratio, the percentage of packets successfully delivered to the destination. If we require that each hop on a route should provide the same level of reliability, the required packet delivery ratio from node i to node j, PR_{ij}, can be estimated as

$$PR_{ij} = PR^{1/\hat{N}_{ij}} \tag{18.52}$$

where *PR* is the required packet delivery ratio given by the applications, and $\hat{N}_{ij}$ is the estimated number of hops from *i* to the destination if *j* is selected as its next hop, that is,

$$\hat{N}_{ij} = \max\left(\left\lceil \frac{d_{iD}}{\widehat{d_{ij}}} \right\rceil, 1\right) \tag{18.53}$$

where $\widehat{d_{ij}}$ is the projection of d_{ij} onto the line connecting node *i* with the sink.

An end-to-end video application usually cares if a video frame can be successfully decoded or not. We use the probability that a video frame is successfully decoded [37] as a metric to evaluate reliability. We denote this probability by P_{D}. A video frame is packed into a number of packets for transmission. The frame will be decodable only when enough packets are received correctly. We introduce frame decodable threshold [38], denoted by DT, to represent the percentage of packets needed to decode a frame. This threshold is dependent on specific video coders and their error recovering capabilities. The required packet delivery ratio *PR* in Equation 18.52 is obtained based on the required P_{D} from an application and the value of DT. The readers are referred to the article by Dai et al. [21] for details.

18.4.2.5.3 Correlation-Aware Load Balancing

For flows from the same correlation group that cannot be further compressed, the presence of traffic congestion becomes evident in that video sensors from the same correlation group tend to report the same event and generate traffic concurrently. We introduce a correlation-aware load balancing operation to split these flows to different paths, so that the probability of network congestion could be reduced.

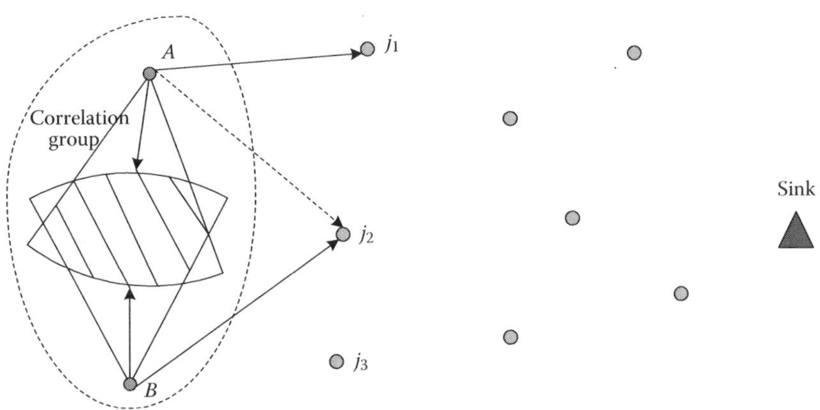

Figure 18.17 Correlation-aware load balancing.

Example

As shown in Figure 18.17, two sensors A and B are in the same correlation group. They share large overlapped FoVs, however, the differential coding gain is very low according to our correlation model. As they are likely to generate large amounts of traffic concurrently, we can try to split the flows from the two sensors to different paths.

To achieve correlation-aware load balancing, each node keeps a list of source nodes and their corresponding group IDs that it has generated or routed in a recent period. Suppose node i wants to find a next hop for a flow generated by node V_{id} at group $\mathcal{G}(V_{id})$. Its candidate neighbor, node j, has a list of source nodes of the flows that it has routed in a recent period, denoted by $(\mathcal{L}\{v^j\})$. Each source node v^j in the list is associated with its correlating group ID $\mathcal{G}(v^j)$. Node j periodically exchanges this list with its neighbors, so that node i is aware of it. For the current flow generated by V_{id}, node i can check if node j has routed flows for other nodes in the same correlation group.

We define a variable x_{v^j} to indicate if a source node v^j is in the same group as V_{id}, which is given by,

$$x_{v^j} = \begin{cases} 1, & \text{if } \mathcal{G}(v^j) = \mathcal{G}(V_{id}) \text{ and } v^j \neq V_{id} \\ 0, & \text{otherwise} \end{cases} \tag{18.54}$$

The number of nodes in list $\mathcal{L}\{v^j\}$ that are in the same group as V_{id} can be expressed as $\sum_{v^j \in \mathcal{L}\{v^j\}} x_{v^j}$. For load balancing, the algorithm should prefer to choose a next hop node with a smaller $\sum_{v^j \in \mathcal{L}\{v^j\}} x_{v^j}$, which, as indicated in Constraint 18.45, cannot exceed a threshold w. The threshold w can be set to a certain percentage of the average correlation group size of a network. In this way, for flows from the same correlation groups that cannot be further compressed, we can penalize the case that they share the same forwarding node concurrently (e.g., node j_2 in Figure 18.17), thereby reducing the possibility of congestion.

18.4.2.6 Protocol Operation

The CAQR algorithm is summarized as follows. When a WMSN is deployed, correlation groups are first constructed. After that, if a sensor V_A has a video frame X_A to transmit, it encounters two scenarios: (i) If X_A is an inter frame, V_A will send X_A to the sink. (ii) If X_A is an intra frame, V_A selects the optimal intermediate node V_{B^*} by solving the differential coding-based intermediate node selection problem in Section 18.4.2. Otherwise, if no such intermediate node can be found, V_A will send X_A to the sink node. In both scenarios, next-hop nodes are selected by solving the distributed correlation-aware QoS routing problem in Section 18.4.2 as follows. A sensor first finds out all the next hop candidate nodes that satisfy the load-balancing constraint in Equation 18.45. For each candidate node, the largest channel coding rate R_{ij}^C from $\{R_0^C, \ldots, R_N^C\}$ is found such that the reliability Constraint 18.44 can be satisfied. Using this largest channel coding rate, the sensor checks if the local delay constraints (Equations 18.42 and 18.43) are met. If so, the corresponding energy consumption (Equation 18.41) for this candidate node can be obtained. The candidate node that results in the smallest energy consumption is selected as the next hop node. In case that no candidate nodes can satisfy all four constraints, the load-balancing constraint (Equation 18.45) could be relaxed by increasing the threshold w by a small amount, so that more nodes in the neighbor set could be considered as next hop candidates.

The resource requirements and complexity for each component of the CAQR algorithm have been discussed by Dai et al. [21]. The algorithm lets each node decide the best hop locally based on local information, so the worst case message exchange complexity is proportional to the maximum degree of nodes in the network, which is natural for distributed routing protocols for sensor networks.

18.4.3 Performance Evaluation

The performance of the proposed routing algorithm is evaluated using a distributed network simulator in Java. In a field of 100 m × 100 m, 49 video sensors are deployed in a grid structure, and a sink node is placed in a corner of the field. The sensing directions of the video sensors are uniformly chosen so as to ensure full coverage of the field, and the sensing parameters of the sensors are given in Table 18.4.

The traffic for the video sensors is generated based on the features of WMSN applications. We let a target move in the field according to the random waypoint mobility model in which the pause time is set to 0. A video sensor is triggered to capture an image when it detects the target in its FoV. By launching the target from 10 different locations, we can generate 10 sequences of events representing different traffic scenarios. The captured video frames are in Quarter Common Intermediate Format (QCIF, 176 × 144 pixels), whereas the size of an encoded video frame is obtained based on the video traces provided in an article by Seeling et al. [39]. If correlation-aware coding is performed, the size of a frame is updated based on the actual coding efficiency, which is estimated from the video coding experiments in Section 18.2. The TDMA scheduling algorithm in the study by Ergen and Varaiya [38] is used as the MAC layer solution. Table 18.4 presents the TDMA settings. The parameters for estimating communication and processing energy are listed in the same table.

The proposed CAQR algorithm is compared with three other algorithms: MMSPEED [4], SPRO [18], and QoSR. MMSPEED is a representative QoS routing algorithm for sensor networks.

Table 18.4 Parameters

Offset Angle	*60°*	*Sensing Radius*	*30 m*
Image size	176 × 144	Intra period	15
Frame rate	30 fps	DT	0.8
E_{elec}	50 nJ/b	α	2
ε	10 pJ/b/m^2	C_{total}	0.67 nF
I_0	1.196 mA	V_T	26 mV
K	239.28 MHz/V	c	0.5 V
N_{cyc} (encoder)	2.3 M_{cycles}	N_{cyc} (decoder)	0.14 M
Transmission rate	1 Mbps	Transmission range	15 m
Slot length	20 ms		

SPRO is the shortest path routing with an opportunistic aggregation algorithm. QoSR is the QoS routing algorithm in Section 18.4.2 without any correlation-aware operations.

We change the amount of traffic injected in the network by varying the source coding rates of the video frames, which is achieved by varying the quantization steps (QP) in the encoder. Typical values for Peak Signal-to-Noise Ratio (PSNR) in lossy image and video compression are between 30 and 50 dB. Based on the video traces for QCIF frames in the article by Seeling et al. [39], when the PSNR ranges between 30 and 50 dB, the QP ranges between around 16 and 40, and it corresponds to average size of intra frames between 4×10^4 and 0.5×10^4 bits. We also found from the traces [39] that, on average, the size of an encoded P frame is approximately 0.2 to 0.25 times that of an encoded I frame. These statistics are used to set the coding rates in the simulation. For each rate, experiments on the aforementioned 10 different sequences of events are launched, and we measure the average performance of the 10 sequences of events.

First, we evaluate the energy efficiency of the proposed algorithm. The total energy consumption consists of the energy for sending and receiving packets and that for processing the video frames. Given the same event, the processing energy for sensing video frames and encoding local video frames will be the same for different routing algorithms, whereas differential coding along routing paths will introduce extra processing energy. Therefore, we just consider the communication energy and the processing energy for differential coding along routing paths. Figure 18.18 shows the average energy consumption for one received frame. The proposed algorithm is designed to reduce energy consumption by reducing the transmission of redundant information and selecting energy-efficient next hops. Specifically, through correlation-aware differential coding, CAQR reduces energy consumption by 15% compared with QoSR. And the energy consumption of QoSR is lower than the other two algorithms, which is brought by minimizing energy consumption in next hop selection. SPRO partially reduces the transmission of redundant information in the network, resulting in smaller energy consumption than MMSPEED, which does not consider energy consumption in routing decisions.

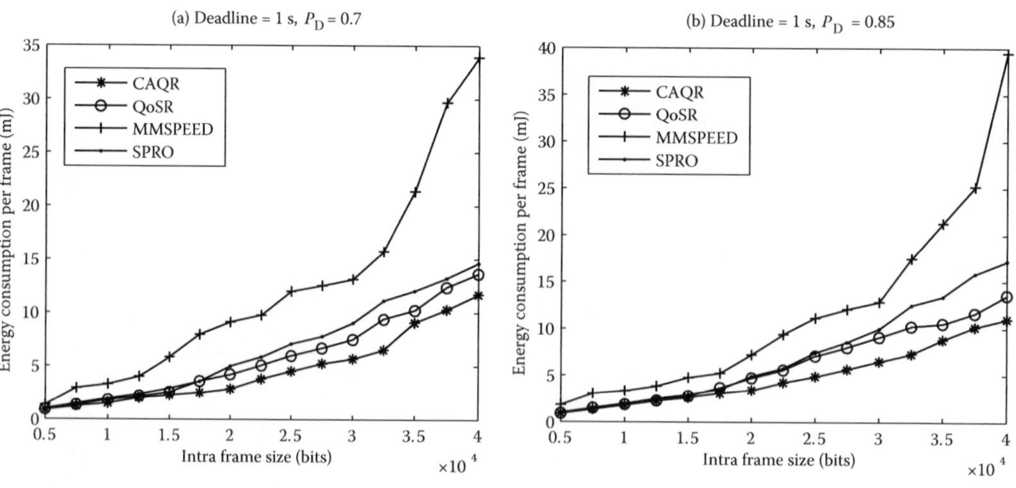

Figure 18.18 Average energy consumption per node.

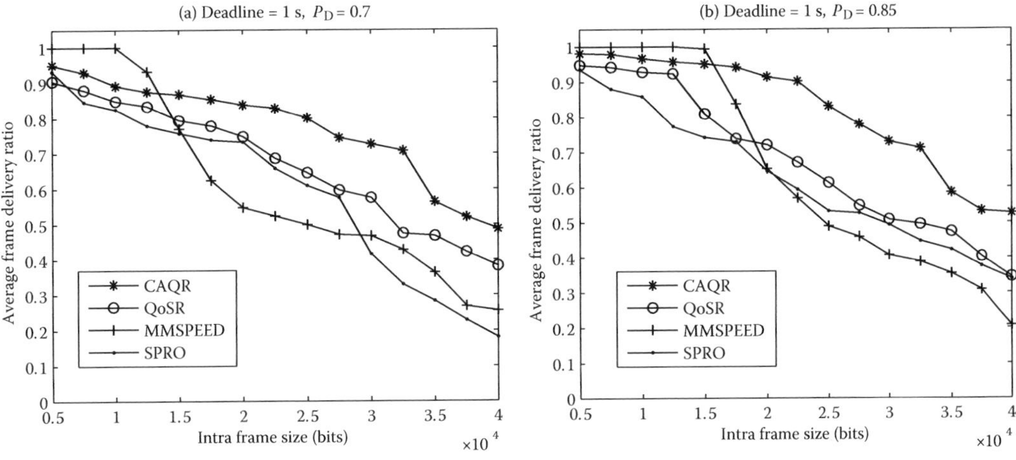

Figure 18.19 Frame delivery ratio.

We now evaluate the quality of received visual information under different reliability requirements. We set the deadline to 1 s, and vary the probability that a video frame is successfully decoded, P_D, to 0.7 and 0.85. For each reported image frame, we count the number of received packets within the deadline. If the percentage of received packets for a frame is above the frame DT, we deem that it is successfully decoded at the sink. Based on the number of decoded frames, we can obtain the percentage of successfully decoded video frames (frame delivery ratio). Figure 18.19 shows the average frame delivery ratio for the 10 sequences of events under different source coding rates. MMSPEED utilizes multipath forwarding to guarantee reliability. When there is less traffic in the network, MMSPEED produces very good reliability with 100% frame delivery ratio. However, as the traffic in the network increases, MMSPEED suffers from a lot of degradation in frame delivery ratio, in which case the traffic redundancy caused by multipath routing is prone to cause congestion in the network. The average frame delivery ratios of SPRO and QoSR are similar. By introducing correlation-aware operations in QoSR, the proposed CAQR algorithm increases the frame delivery ratio by 16% compared with QoSR.

In Figures 18.18 and 18.19, when video frame sizes become larger, that is, more traffic is injected into the network, there is more gain of the proposed algorithm over the other algorithms in terms of energy consumption and frame delivery ratio. We conclude that exploiting correlation in a video sensor network can enhance network performance, especially when the traffic load is heavy. By incorporating correlation-aware differential coding and load balancing in the routing process, the CAQR algorithm can effectively improve the quality of visual information received at the sink.

18.5 Summary

In this chapter, we have introduced a correlation-based communication framework that leverages the spatial correlation of visual information in communication protocols for WMSNs. The foundation of the framework is a novel spatial correlation coefficient, which estimates the degree of correlation of visual information observed by camera sensors with overlapped FoVs. The correlation coefficient is derived based on the sensing model and deployment information of camera sensors, providing an analytical and application-independent solution for correlation estimation. It allows WMSN applications to obtain the correlation of visual information through low communication and computation costs.

The problem of how much information can be gained from multiple cameras is also investigated. The concept of entropy (a measure of the amount of information) is used in the analysis of this problem. A joint entropy estimation method is developed by using the spatial correlation coefficient of multiple cameras. One promising property of joint entropy estimation is its capability of predicting the coding efficiency among multiple cameras. The effectiveness of the prediction of coding efficiency has been validated through practical video coding experiments.

Based on the correlation coefficient and the joint entropy estimation method, a clustered source coding protocol and a QoS routing algorithm are developed. The clustered source coding protocol divides a network of cameras into clusters and lets cameras inside a cluster perform joint coding. It maximizes the compression gain in the network and ensures the decoding reliability at the sink. The routing algorithm integrates correlation-aware operations to reduce traffic redundancy and avoid congestion in a WMSN, so that the QoS requirements of WMSN applications can be better satisfied. Evaluation results have shown that, by exploiting the correlation of visual information, the proposed communication algorithms can reduce energy consumption and improve the quality of received visual information in WMSNs.

Exercises

1. As shown in Figure 18.20, three camera sensors are deployed to observe an area of interest in a WMSN. The area of interest is in the center of the coordinate system, and the locations and sensing directions of the cameras are given in the figure (where $d = 2.5$ m). Each camera has observed one image [$X_i (i = 1,2,3)$] and the entropies of the individual images are the same [$H(X_i) = H(\cdot)$, $i = 1, 2, 3$]. Suppose an application at the sink has requested for informa-

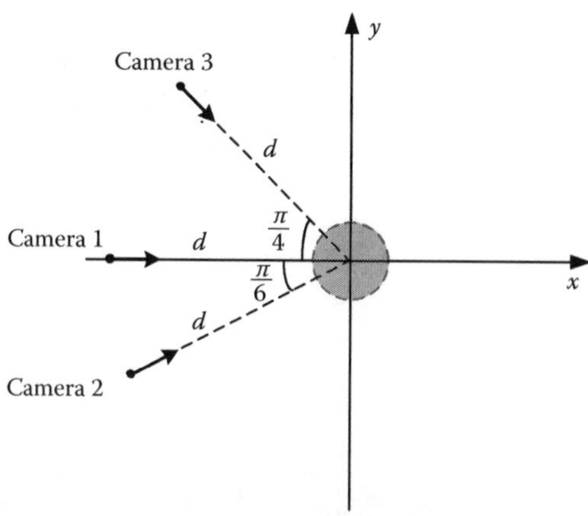

Figure 18.20 Camera deployment for exercise question 1.

tion about this area, but the network allows only two of the three cameras to report to the sink. Which two cameras should be selected to maximize the amount of information (about this area of interest) gained at the sink?

2. In the DMCP protocol, why does every node have to exchange its coding effectiveness within its two-hop range?
3. Based on the protocol operation of CAQR, what is the computational complexity for finding a next-hop node?

References

1. I. F. Akyildiz, T. Melodia, and K. R. Chowdhury. A survey on wireless multimedia sensor networks. *Computer Networks*, 51(4):921–960, March 2007.
2. Y. Liu, I. Elhanany, and H. Qi. An energy-efficient QoS-aware media access control protocol for wireless sensor networks. In *IEEE International Conference on Mobile Adhoc and Sensor Systems Conference (MASS)*, Washington, DC, November 2005.
3. N. Saxena, A. Roy, and J. Shin. A QoS-based energy-aware MAC protocol for wireless multimedia sensor networks. In *IEEE Vehicular Technology Conference*, 183–187, May 2008.
4. E. Felemban, C.-G. Lee, and E. Ekici. MMSPEED: Multipath multi-SPEED protocol for QoS guarantee of reliability and timeliness in wireless sensor networks. *IEEE Transactions on Mobile Computing*, 5(6):738–754, 2006.
5. X. Huang and Y. Fang. Multiconstrained QoS multipath routing in wireless sensor networks. *Wireless Networks*, 14(4):465–478, August 2008.
6. T. Melodia and I. F. Akyildiz. Cross-layer QoS-aware communication for ultra wide band wireless multimedia sensor networks. *IEEE Journal of Selected Areas in Communications*, 28(5):653–663, June 2010.
7. B. Girod, A. Aaron, S. Rane, and D. Rebollo-Monedero. Distributed video coding. *Proceedings of the IEEE*, 93(1):71–83, January 2005.
8. R. Wagner, R. Nowak, and R. Baraniuk. Distributed image compression for sensor networks using correspondence analysis and super-resolution. In *IEEE International Conference on Image Processing (ICIP)*, 597–600, September 2003.
9. M. Wu and C. W. Chen.Collaborative image coding and transmission over wireless sensor networks. *EURASIP Journal on Advances in Signal Processing*, 2007(1):223–231, January 2007.
10. R. C. Gonzalez, R. E. Woods, and S. L. Eddins. *Digital Image Processing Using MATLAB*. Prentice Hall, New Jersey, 2004.
11. R. Dai and I. F. Akyildiz. A spatial correlation model for visual information in wireless multimedia sensor networks. *IEEE Transactions on Multimedia*, 11(6):1148–1159, October 2009.
12. T. Cover and J. Thomas. *Elements of Information Theory*. John Wiley and Sons, New York, 1991.
13. J. P. W. Pluim, J. B. Antoine Maintz, and M. A. Viergever. Mutual-information-based registration of medical images: A survey. *IEEE Transactions on Medical Imaging*, 22(8):986–1004, August 2003.
14. P. Wang, R. Dai, and I. F. Akyildiz. A spatial correlation-based image compression framework for wireless multimedia sensor networks. *IEEE Transactions on Multimedia*, 13(2):388–401, April 2011.
15. W. R. Heinzelman, A. Chandrakasan, and H. Balakrishnan. Energy-efficient communication protocol for wireless microsensor networks. In *Proceedings of the Hawaii International Conference on System Sciences*, Hawaii, January 2000.
16. S. Bandyopadhyay and E. J. Coyle. An energy efficient hierarchical clustering algorithm for wireless sensor networks. In *IEEE Computer and Communications Conference (INFOCOM)*, San Francisco, July 2003.
17. A. A. Abbasi and M. Younis. A survey on clustering algorithms for wireless sensor networks. *Computer Communications*, 30:2826–2841, June 2007.
18. S. Pattem, B. Krishnamachari, and R. Govindan. The impact of spatial correlation on routing with compression in wireless sensor networks. In *Third International Symposium on Information Processing in Sensor Networks, IPSN 2004*, 28–35, Berkeley, CA, April 2004.

19. H. Luo and G. Pottie. A study on combined routing and source coding with explicit side information in sensor networks. In *IEEE Global Telecommunications Conference (GLOBECOM)*, St. Louis, MO, USA, December 2005.
20. H. Luo, Y. Liu, and S. K. Das. Routing correlated data with fusion cost in wireless sensor networks. *IEEE Transactions on Mobile Computing*, 5(11):1620–1632, November 2006.
21. R. Dai, P. Wang, and I. F. Akyildiz. Correlation-aware QoS routing with differential coding for wireless multimedia sensor networks. *IEEE Transactions on Multimedia*, 14(5):1469–1479, October 2012.
22. R. Cucchiara. Multimedia surveillance systems. In *Proceedings of ACM International Workshop on Video Surveillance and Sensor Networks*, 3–10, November 2005.
23. D. A. Forsyth and J. Ponce. *Computer Vision: A Modern Approach*. Prentice Hall, New Jersey, 2002.
24. H. Ma and Y. Liu. Correlation based video processing in video sensor networks. In *IEEE International Conference on Wireless Networks, Communications and Mobile Computing*, 987–992, Maui, HI, June 2005.
25. A. Barton-Sweeney, D. Lymberopoulos, and A. Savvides. Sensor localization and camera calibration in distributed camera sensor networks. In *International Conference on Broadband Communications, Networks and Systems*, 1–10, October 2006.
26. D. Devarajan, Z. Cheng, and R. J. Radke. Calibrating distributed camera networks. *Proceedings of the IEEE*, 96(10):1625–1639, October 2008.
27. B. S. Manjunath and W. Y. Ma. Texture features for browsing and retrieval of image data. *IEEE Transactions on Pattern Analysis and Machine Intelligence*, 18(8):837–842, August 1996.
28. JVT Reference Software JM 8.5. Available: http://iphome.hhi.de/suehring/tml/download/old_jm/jm8.5.zip.
29. JMVC Reference Software JMVC 2.5. Available: http://ftp3.itu.int/av-arch/jvt-site/2008_10_Busan/JVT-AC207.zip.
30. S. Rajagopalan and V. V. Vazirani. Primal-dual RNC approximation algorithms for set cover and covering integer programs. *SIAM Journal on Computing*, 28(2):525–540, 1999.
31. O. Younis and S. Fahm. HEED: A hybrid, energy-efficient, distributed clustering approach for ad hoc sensor networks. *IEEE Transactions on Mobile Computing*, 3(4):366–379, October 2004.
32. S. Soro and W. Heinzelman. A survey of visual sensor networks. *Advances in Multimedia*, 2009(640386):1–21, 2009.
33. A. Wang and A. Chandrakasan. Energy-efficient DSPs for wireless sensor networks. *IEEE Signal Processing Magazine*, 68–78, July 2002.
34. K. Goto, A. Hatabu, H. Nishizuka, K. Matsunaga, R. Nakamura, Y. Mochizuki, and T. Miyazaki. H.264 video encoder implementation on a low-power DSP with low and stable computational complexity. In *IEEE Workshop on Signal Processing Systems Design and Implementation*, 101–106, October 2006.
35. M. Horowitz, A. Joch, F. Kossentini, and A. Hallapuro. H.264/AVC baseline profile decoder complexity analysis. *IEEE Transactions on Circuits and Systems for Video Technology*, 13(7):704–716, July 2003.
36. A. K. Jain, M. N. Murty, and P. J. Flynn. Data clustering: A review. *ACM Computing Surveys*, 31(3):264–323, September 1999.
37. A. Ziviani, B. E. Wolfinger, O. M. B. Duarte J. F. de Rezende, and S. Fdida. Joint adoption of QoS schemes for MPEG streams. *Multimedia Tools and Applications*, 26(1):59–80, May 2005.
38. S. C. Ergen and P. Varaiya. TDMA scheduling algorithms for sensor networks. Technical report, University of California, Berkley, July 2005.
39. P. Seeling, M. Reisslein, and B. Kulapala. Network performance evaluation using frame size and quality traces of single-layer and two-layer video: A tutorial. *IEEE Communications Surveys and Tutorials*, 6(3):58–78, Third Quarter 2004.

DATA STORAGE AND MONITORING

Chapter 19

Distributed Data Storage and Retrieval Schemes in RPL/IPv6-Based Networks

Pietro Gonizzi and Gianluigi Ferrari
University of Parma
Jérémie Leguay and Paolo Medagliani
THALES Communications and Security

Contents

19.1 Introduction and Motivations

In contrast with conventional network data storage, storing data in wireless sensor networks (WSNs) represents a challenge because of the limited power, memory, and communication bandwidth of WSNs. Nowadays, sensors have reached higher capabilities, in terms of processing speed and local storage, than in the past years [1]. This makes them more attractive for in-network storage deployments.

This chapter deals with distributed data storage and retrieval techniques in WSNs. The focus is on WSN systems for the Internet of Things (IoT), a new vision of the Internet aiming at pushing IP connectivity into smart objects. The range of objects involved in the IoT also encompasses radiofrequency identification (RFID) and machine-to-machine (M2M) systems, to cite a few. These technologies will also be described in detail.

When WSNs are deployed for IoT observation systems, they usually consist of, on one hand, unattended nodes that sense the surrounding environment and, on the other hand, a sink node that is in charge of collecting data measurements and relaying them to a management entity. There are many reasons by which a sensor node may not be able to transmit data to the sink right after acquisition. First, the sink may not always be reachable from the sensor nodes. For instance, a mobile node can be used to periodically pull out all the collected data. Second, when applications do not require real-time collection, storing data units and sending aggregate data bursts could contribute to a reduction in the number of radio transmissions, thereby increasing the lifetime operation of the WSN. Illustrative applications include habitat monitoring, such as tracking animal migrations in remote areas [2], studying weather conditions in national parks [3], etc. Such scenarios require the collection and storage of as much data as possible between two consecutive data retrievals performed by an external agent. However, storing data on the sensor node leads to local memory overflow if data retrieval is not performed in a timely manner by the sink. To avoid data dropping or overwriting, sensor nodes can cooperate with each other by sharing acquired data.

Node failure is also a critical issue in WSNs. Periodic inactivity (e.g., for energy saving purposes), physical destruction, and (software) bugs are likely to appear in WSNs, leading to data loss. Thus, redundancy through data replication (i.e., by storing copies of the same data onto various nodes) contributes to increasing the resilience of the WSN. However, distributed data storage across nodes, with or without redundancy, is a challenge because it requires properly selected donor nodes (i.e., nodes available to store data from other nodes) and entails communication overhead to transmit data to the selected nodes. Therefore, network storage capacity and resilience have an energy cost and this limits the network lifetime.

Although distributed data storage and replication have been studied in the literature [4,5], there are still a number of challenges to tackle. First, to the best of our knowledge, most previous studies did not encompass both the distribution and the replication aspects. Second, the proposed solutions are usually not tested experimentally at a large scale, although we stress how real deployment is of primary importance when developing WSN applications.

In this work, we propose a low-complexity distributed data replication mechanism to increase the resilience and storage capacity of a WSN-based observation system against node failure and local memory shortage. We extended our previous work [6], demonstrating how this mechanism can be applied with standard low-power IP sensor networks. We evaluate our approach through experimental tests conducted on the SensLAB real platform [7] and inside the Cooja [8] environment.

The chapter is organized as follows. We first review the state of the art on distributed data storage techniques and current standard IoT protocols in Section 19.2. We then present our low-complexity greedy (LG) distributed data storage scheme in Section 19.3. We apply this approach to standard IoT networks based on RPL routing protocol in Section 19.4, and treat the data retrieval issue in Section 19.5. Finally, we provide a few conclusions and perspectives in Section 19.6.

19.2 State of the Art

This section proposes to first review ongoing activities in the IoT area, before presenting more specifically related work on distributed storage.

19.2.1 Related Work on the IoT

The IoT is a recent concept relying on the integration within the Internet of a great number of communicating objects (e.g., sensors, actuators, RFID, etc.). Even if not natively connected, the objects are provided with an interface, which lets them communicate with other interconnected objects. The interconnection between devices is mainly carried out using the IPv6 protocol [9,10] and its extensions, which we will review later in this section. On top of this, M2M architectures, such as the one developed by the European Telecommunications Standards Institute (ETSI) M2M group [11], define a set of useful Representational State Transfer (REST)-based standards for the representation of devices and their data format on the one hand and for the definition of the common interfaces on the other hand. This allowed suppliers to easily create and integrate new devices, and also allowed developers to create applications and solutions based on the abovementioned devices, abstracting from their specific characteristics and implementations. Given the continuously increasing dissemination of these devices, the perspectives, in terms of both economics and of utilization, are remarkable.

Overall, the IoT provides an information system architecture in which a large number of very heterogeneous and eventually constrained devices could interconnect and interoperate without requiring a significant effort for their adaptation [12].

19.2.2 Standard Communication Protocols

One of the greatest key enablers of the IoT are the RFID devices, especially in the version described in the ISO/IEC 18000-6 standard [13]. There are two different devices according to this standard: (i) the reader and (ii) the tags. The tags are devices used for item identification and, in this case, storage of data. The reader, on the other hand, is a device in charge of interrogating the tags, which upon request, transmit their identifier. Depending on the type of battery a device is equipped with, an RFID tag may have limited transmission range and storage capabilities. Anyway, in each case, the addressing scheme used by the RFIDs is different from the scheme adopted by the IPv6, thus a device acting as a gateway between the RFID network and the IP network is required in both IoT and M2M infrastructures.

Although RFID technology originally pushed the birth of the IoT, nowadays, we are witnessing the increasing diffusion of smarter devices, with processing and storing capabilities that could be addressed through the IPv6 protocol. The ZigBee consortium has standardized these IP-communicating devices, creating a contact point between the IoT architectures [14]. This standard is suited for a family of low-rate wireless personal area networks (LR-WPANs), allowing network creation, management, and data transmission over a wireless channel with the highest possible energy savings. This standard is based, at the first two layers of the ISO-OSI stack, on the IEEE 802.15.4 standard [15], which employs either a nonpersistent carrier sense multiple access with collision avoidance (CSMA/CA) or a slotted and beaconed medium access control (MAC) protocol. In both cases, the MAC protocol is targeted to be energy efficient, aiming to minimize collisions, and thus avoid useless packet retransmissions. Additionally, the 802.15.4 introduces the duty cycle of the RF interfaces of a transmitting device to avoid having to listen to the channel when not necessary. The ZigBee standard defines some profiles for energy efficiency and home monitoring that are totally compliant with M2M systems.

The Internet Engineering Task Force (IETF) has set the 6LowPAN, RoLL, and CoRE working groups to elaborate standard IPv6 extensions for low-power and lossy networks (LLNs) connected to the Internet. A LLN can be, for instance, a network of low-power sensors or actuators in a home or industrial automation application.

6LowPAN provides a standard adaptation layer (IETF RFC 2464, 5072, 5121) for the transport of IPv6 packets across low-power sensor networks using the IEEE 802.15.4 MAC layer for instance. The RPL (IPv6 routing protocol for LLN) [16,17] was developed to have a really limited control traffic, fit harsh and constrained environments, with limited data rate, and potentially elevated error rate. The main characteristics of this protocol will be explained in Section 19.4.1.

To efficiently collect information in LLNs, the IETF CoRE (Constrained RESTful Environments) working group has defined the constrained application protocol (CoAP) standard, supporting REST-like applications for communication between entities [18]. CoAP is a web protocol similar to HTTP, specifically developed for exploiting the functionalities of resource-constrained devices (seen as web resources). In addition, CoAP defines a mechanism for the discovery of the resources available inside a network, a multicast communication scheme, and a publish/subscribe mechanism, in which the role of initiating the communication is given to the CoAP server, which transmits the information to the subscribed CoAP clients without requiring, each time, an explicit message request by them. At the transport layer, CoAP is based on the UDP protocol, due to the reduced number of control messages introduced. To meet possible QoS requirements, because UDP is natively unreliable, CoAP has introduced the use of confirmation messages.

Especially used inside telecommunication architectures, the term M2M refers to embedded systems that are able to communicate with other systems, such as databases, applications servers, and smartphones, without requiring external human intervention. The ETSI has set a work group on this subject. At the end of 2011, the group presented the first version (release 1.0) of the ETSI M2M architecture [19]. The structure of the M2M infrastructure was split into five parts: (i) devices, (ii) area networks, (iii) gateways, (iv) core network, and (v) applications. A *device* is a node in charge of both collecting data autonomously and sending data to an application upon request. The *area network* is the element that connects the devices with a *gateway*, which in turn acts as a proxy between the *area network* and the *core network*. ETSI M2M R1 defines REST documents for representing the entities (i.e., devices, gateways, and applications), the data, and a notification mechanism. It relies on the HTTP and CoAP protocols for communications between the M2M entities.

19.2.3 Distributed Storage

Various schemes to efficiently store and process sensed data in WSNs have been proposed in the past few years [20].

In distributed WSN storage schemes, nodes cooperate to efficiently distribute data across other nodes. Previous studies focused on data-centric storage approaches [21–23], wherein data collected in some WSN regions are stored at supernodes that are responsible for these regions or for a particular type of data. Hash values are used to distinguish data types and the corresponding storage locations. Even if this approach is based on node cooperation, it is not fully distributed because supernodes store all the contents generated by the others.

In a fully distributed data storage approach, all nodes participate in sensing and storing in the same way. First, all nodes store their sensor readings locally and, once their local memories have filled up, they delegate storage to other nodes. A first significant contribution in this direction is given by Data Farms [24]. The authors propose a fully data-distributed storage mechanism with periodic data retrieval. They derive a cost model to measure energy consumption and show how a careful selection of nodes offering storage, called *donor nodes*, optimizes the system's capacity at the price of slightly higher transmission costs. They assume that the network is built on a tree topology and each sensor node knows the return path to a sink node, which periodically retrieves data. Another study was proposed in EnviroStore [25]. The authors focus on data redistribution, when the remaining storage of a sensor node exceeds a given threshold, through load balancing. They use a proactive mechanism in which each node maintains a local memory table containing the status of the memory of its neighbors. Furthermore, mobile nodes (called *mules*) are used to carry data from an overloaded area to an offloaded one and to take data to a sink node. Takahashi et al. [26] solved the data preservation problem in an isolated sensor network using graph theory, by transferring data items from low-energy nodes to high-energy ones. However, only low-energy nodes generate content. In a study by Tseng et al. [27], each data unit is assigned a priority. High-priority traffic is distributed to nodes closer to the sink. The major shortcoming of the above studies is the absence of large-scale experiments to evaluate the system's performance, which is a key contribution of our work.

Data replication consists of adding redundancy to the system by copying data at several donor nodes (within the WSN) to mitigate the risk of node failure. It has been widely studied for WSNs and several works are available in the literature [4,28]. A scoring function for choosing a suitable replicator node was proposed in a study by Neumann et al. [4]. The function is influenced by critical parameters such as the number of desired replicas, the remaining energy of a replicator node, and the energy of the neighbors of the replicator node. In TinyDSM [5], a reactive replication approach is discussed. Data bursts are broadcast by a source node and then handled independently by each neighbor, which decides whether to store a copy of the burst or not. Maia et al. [29] suggested ProFlex, a distributed data storage protocol for replicating data measurements from constrained nodes to more powerful nodes. Unlike the above approaches, we propose a replication-based distributed data storage mechanism with lower complexity because the responsibility of finding donor nodes is not centralized at the source node but is handed over to consecutive donor nodes in a recursive manner.

Overall, with respect to the related studies, our work goes further. First, we encompass, with a fully distributed mechanism, both data replication and distributed storage. Second, we conduct large-scale experiments on the SensLAB real testbed to evaluate our solution. Third, we extend our approach on top of the RPL routing protocol to be fitted within an IoT system.

19.3 Redundant Distributed Storage

19.3.1 Introduction

In the current and next sections, we propose two redundant distributed data storage mechanisms to increase the resilience and storage capacity of a WSN against node failure and local memory shortage. We evaluate our approaches through simulations and experimental tests conducted on the SensLAB real platform [7] and inside the Cooja [8] environment. Namely, the two mechanisms are LG and RPL greedy (RG). They mainly differ in the following aspects:

- The RG mechanism is built on top of the RPL routing protocol [16,17], and uses routing information to efficiently distribute and propagate data items through the WSN. On the other hand, LG does not rely on routing.
- In RG, replicas of a data item are preferably stored close to the sink, to be easily retrieved by a periodic agent. On the contrary, LG selects a neighbor node with the most available space, without considering its physical position.
- The two implementations are different. LG has been implemented in TinyOS 2.1.0 operating system [30], whereas RG has been developed in Contiki [31], on top of IPv6 and UDP. Therefore, they use different protocol stacks.

This section briefly describes the LG mechanism and its experimental validation performed on the SensLAB testbed. The reader can refer to the study by Gonizzi et al. [6] for a more accurate analysis. The RG mechanism will be presented in the next section.

19.3.2 LG Mechanism

We assume that a WSN is deployed to measure environmental data and store it until a sink node performs a periodic retrieval of the entire data stored in the WSN. Between two consecutive data retrievals, the objective is to distribute multiple copies of every data unit among nodes to avoid data losses caused by local memory shortages or by node failures. Data distribution relies on the proactive announcement, by every node, of its memory status. Each node periodically broadcasts a memory advertisement message containing its current available memory space. It also maintains an updated memory table containing the memory statuses of all its detected one-hop neighbors. Upon reception of a memory advertisement, a node updates its local memory table with the new information received. The memory table contains an entry for each neighbor. Each entry contains the address of the neighbor, the last received value of its available memory space, and the time at which this value was received.

When sending a replica of acquired data, a node looks up in its table the neighbor node with the largest available memory and most recent information. Such a neighbor is called the *donor node*. If no donor node can be found and there is no available space locally, then the acquired data is dropped. In particular, if a donor can be selected, the node sends to it a copy of the data unit, specifying how many other replicas are still to be distributed in the WSN. The number of required replicas is set to either $R - 1$ (if the node can store the original data locally) or R (if the node's local memory is full), where R denotes the maximum number of replicas. The replication process continues recursively in a hop-by-hop manner until either the last (Rth) replica is stored or stops when one donor node cannot find any suitable next donor node. In the latter case, the final number of replicas actually stored in the WSN, for that specific data item, is smaller than R.

In Figure 19.1, we show an illustrative example, where $R = 3$ copies, for two different cases. In Figure 19.1a, copies of a data unit generated by node 1 are stored (respectively, in nodes 1, 2, and 4). In Figure 19.1b, the replication process stops at node 2 and no suitable donor node can further be found. In this case, the last replica is not stored.

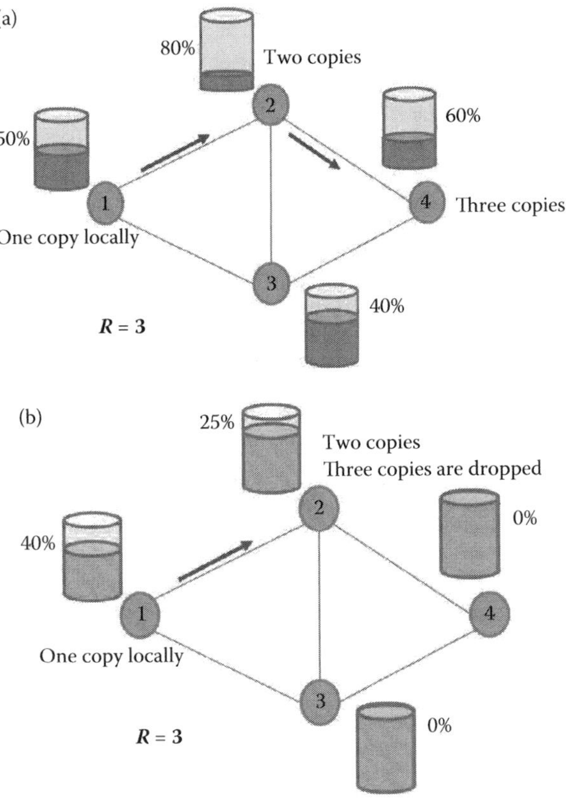

Figure 19.1 Hop-by-hop replication in the case of $R = 3$ desired replicas, for several scenarios: (a) replicas propagate up to two hops from the source node; (b) the replication process stops at an intermediate donor node and the last (Rth) copy is dropped.

At the end of the sensing period, a sink node is meant to gather all data present in the WSN by sending requests to all nodes. Upon sending the data to the sink, a node clears its local buffer and resumes sensing the environment.*

19.3.3 Performance Evaluation

The LG mechanism has been evaluated experimentally in the SensLAB testbed [7]. The SensLAB platform offers more than 1000 sensors at four sites in France (Grenoble, Strasbourg, Lille, and Rennes) where researchers can deploy their codes and run experiments. At each site, nodes are installed on an almost regular grid, as depicted in Figure 19.2. Each node's platform embeds a TI MSP430 microcontroller and operates in various frequency bands depending on the radio chip (either CC1100 or CC2420).

The LG mechanism is significantly influenced by the network topology. For instance, in a dense network, where nodes have several neighbors, data can be distributed more efficiently than in a sparse network with only a few direct neighbors per node. The network topology depends on

* We remark that data retrieval, with LG distributed storage, has not been implemented, as the protocol stack being considered does not rely on any specific multihop routing protocol.

Figure 19.2 Views from the SensLAB sites.

the transmission power of the nodes. We have run four experiments setting the transmit power P_t to 8.9, 0, –15, and –20 dBm, respectively, deploying $N = 78$ nodes. Each node has a finite local memory with size equal to 250 data units, and a sensing period chosen randomly in the interval 0.1 to 5.1 s. We have computed the number of neighbors detected in each case by counting the number of memory advertisements received by each node. We assume node x is a neighbor of node y if node y detects at least one memory advertisement from node x, within the overall experiment duration. Note that the larger the number of received memory advertisements, the higher the radio link quality.

It is of interest to evaluate the time required by the system to reach the network storage capacity, which denotes the maximum amount of data that can be stored in the WSN, in the different cases. According to a study by Gonizzi et al. [6], given the number of nodes N, the buffer sizes $\{B_i\}$, and the sensing rates $\{r_i\}$, the network storage capacity C (dimension: data units), in the absence of replications, is simply given by

$$C = \sum_{i=1}^{N} B_i \tag{19.1}$$

whereas in the case of data replication (i.e., with $R > 1$), the system has a resulting storage capacity, denoted by C_r, which is upper-bounded by C. In this case, the capacity is reduced by the maximum number R of replicas and can be lower-bounded as follows:

$$C_r \geq \frac{C}{R} \tag{19.2}$$

Obviously, $C_r = C$ when $R = 1$, that is, only the original copy is kept and no replication is performed at all.

In the case of absence of replicas, given the fixed memory size and the number of nodes equal to 250 and 78, respectively, the capacity is $C = 250 \times 78 = 19{,}500$ data units. In Figure 19.3a, the amount of data stored is shown as a function of time for the four experimental cases being considered.

As one can see, capacity is reached later when a lower transmission power is used—for instance at –20 dBm—because fewer neighbors are detected. Consequently, data cannot be distributed efficiently through the network. On the other hand, with a transmit power equal to 8.9 dBm, nodes have a larger view of the network and the capacity is reached earlier. For instance, at $t = 400$ s, the stored data at 8.9 and 0 dBm are 17,000 data units, approximately 90% of the capacity. The stored data at –15 and –20 dBm, at the same time instances, are approximately 74% and 68% of the total capacity, respectively. Moreover, two theoretical cases are considered for comparison. In the case with local storage [Local storage (anal) curve], nodes fill their own local buffer autonomously, that is, the fill-up time for the i node is $t_i = B_i/r_i$. In this case, the time interval to reach

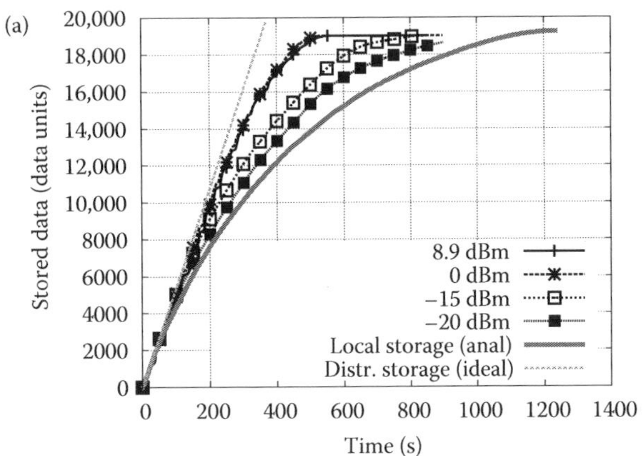

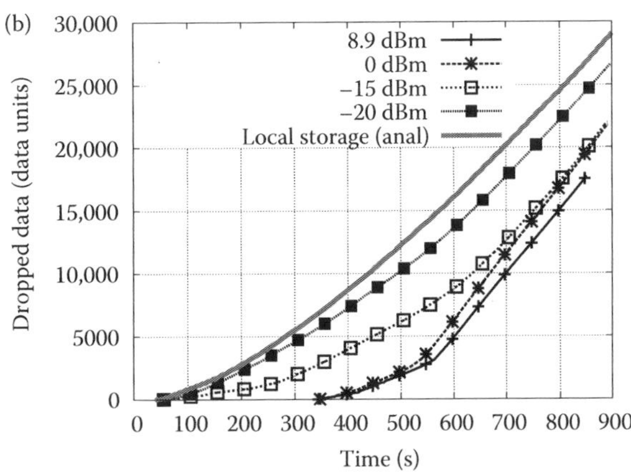

Figure 19.3 (a) Stored data and (b) dropped data in the system for various values of the transmit power, as a function of time. In all cases, we consider: no replication ($R = 1$), memory size equal to 250 data units, $N = 78$ nodes, and a memory advertisement period of 25 s.

the storage capacity corresponds to the longest storage filling time across all nodes. As expected, the analytical curve relative to local storage lower bounds the experimental curves. In the case with ideal distributed storage ["Distr. storage (ideal)" curve], a performance benchmark can be obtained considering an ideal WSN in which nodes can communicate with any other node, considering instantaneous transmission. In this case, the WSN is equivalent to a single supernode with memory size equal to the total capacity C and sensing rate equal to the sum of the individual sensing rates of all nodes.

In Figure 19.3b, the amount of dropped data is shown, as a function of time, in the four cases considered in Figure 19.3a. Nodes drop newly acquired data once their local memory has filled up and no neighbors are available to donate extra storage space. At $t = 400$ s, dropped data at 8.9 and 0 dBm equal 600 data units, approximately 3% of the stored data. In the cases with lower transmit power, for example, at −15 and −20 dBm, the amount of dropped data is significantly higher. As expected, data dropping starts in advance with lower transmission power because each node has a lower number of surrounding donor nodes to which to transfer data. In the same figure, the amount of dropped data, in the case with local storage, is also shown for comparison.

As for redundancy, replication is introduced by setting the number of copies (referred to as replicas) to a value $R > 1$. Replicas of a sensed data unit follow a hop-by-hop replication from the generator node to subsequent donor nodes. We have computed the average hop distance reached by the replicas from the generator node, for various values of R. The results show that replicas do not propagate, on average, beyond two hops from the generator node (Figure 19.4). This is in agreement with LG because donor nodes are selected on the basis of their available memory space but not their physical position.

In conclusion, LG adopts a low-complexity data distribution mechanism that performs well in a dense network with high connectivity. However, replicas of a data item are kept in the proximity of the generator node, on average within two hops: this may lead to the complete loss of some data in the case of a failure that damages all the nodes within a certain region. To reduce this risk, in the next section, an enhanced data distribution mechanism based on RPL routing protocol is presented.

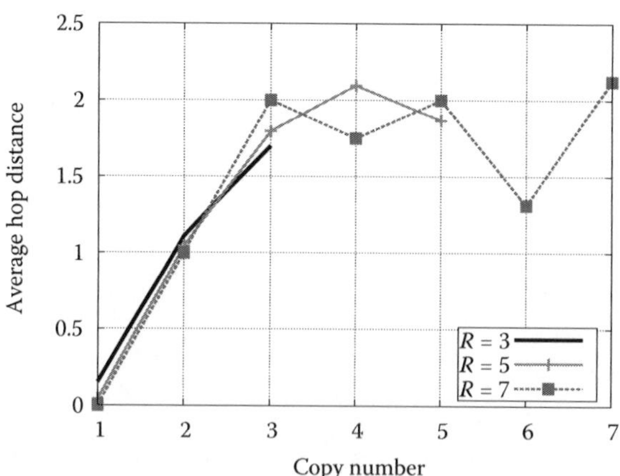

Figure 19.4 Average hop distance reached by the kth replica versus k. k varies between 1 and 3 (R = 3), 1 and 5 (R = 5), and 1 and 7 (R = 7). The average is computed on all the unique data present in the system with full redundancy, that is, with exactly R copies.

19.4 Redundant Distributed Storage with RPL

In the previous section, we have presented LG, a LG mechanism for distributed data storage. A drawback of LG is that replicas of a given data item do not spread throughout the WSN. In this section, an alternative approach, called RG, is described. RG follows a similar approach to LG, but includes information on the position of the node within the WSN. The position of a node is suggested by the RPL rank, a value provided by the RPL routing protocol, which indicates the position of a node within a routing tree, according to a routing function.

This section is organized as follows. First, an overview of RPL is given in Section 19.4.1. A detailed description of RG highlighting the main differences with LG is provided in Section 19.4.2. In Section 19.4.3, we evaluate the protocol through extensive simulations conducted in Cooja, the WSN simulator in Contiki. And finally, Section 19.4.4 concludes the discussion about data storage and introduces the data retrieval problem, which is the topic of the last part of this chapter.

19.4.1 Overview of RPL

RPL is a distance-vector protocol that creates a routing tree, referred to as destination-oriented acyclic-directed graph (DODAG), in which the cost of each path is evaluated according to the metrics defined in an objective function (OF). The goal of this protocol is the creation of a collection tree protocol (CTP), and a point-to-multipoint network from the root of the network to the devices inside the LLN, as well as a point-to-point network between any pair of devices.

To build the tree and to keep the status of the network updated, the root of the RPL tree periodically sends DODAG information object (DIO) messages. The receiving nodes may relay these messages or just consume them, if configured as leaves of the tree. The mechanism of RPL is quite simple: each node has a rank that places it in the hierarchy of the RPL tree and lets it define which nodes are its parents. When a DIO message is received for the first time, a node, before setting its rank, listens through the network discovery mechanism for the possible parents to which it can join. At the end of this discovery phase, according to the outcomes of the OF, a node sets its rank to be highest among the ranks of its possible fathers. To avoid loops or network misconfiguration, two nodes in the same network are not allowed to have the same rank; therefore, as soon as such a situation is detected, one of the conflicting nodes updates its status. According to the outcome of the OF, the node also selects the best path that it can use to transfer the data to the root of the tree. On the other side, if a DIO message has already been received at least once, the node evaluates the incoming DIO message to check whether its position in the DODAG tree can remain the same or must be updated. In the former case, the node discards the packet, whereas in the latter, it computes its new rank and it discards the list of parents to avoid creating loops due to its new position in the DODAG tree. In Figure 19.5, we show the operations that a node carries out to establish its role in the DODAG tree when it receives a DIO packet.

Because devices in LLNs are typically resource constrained, the RPL protocol also introduces a trickle mechanism to reduce the transmission frequency of DIO messages according to the stability of the network. If the network status is stable, the frequency of transmission of DIO messages decreases. As soon as an anomaly or an inconsistency within the network is detected, the frequency is kept back to the default value and a procedure of recovery is started. If the chosen procedure is a local repair, the node simply selects a new best parent. Otherwise, in the case of a global repair, the root sends a new DIO message to restart the construction of the tree.

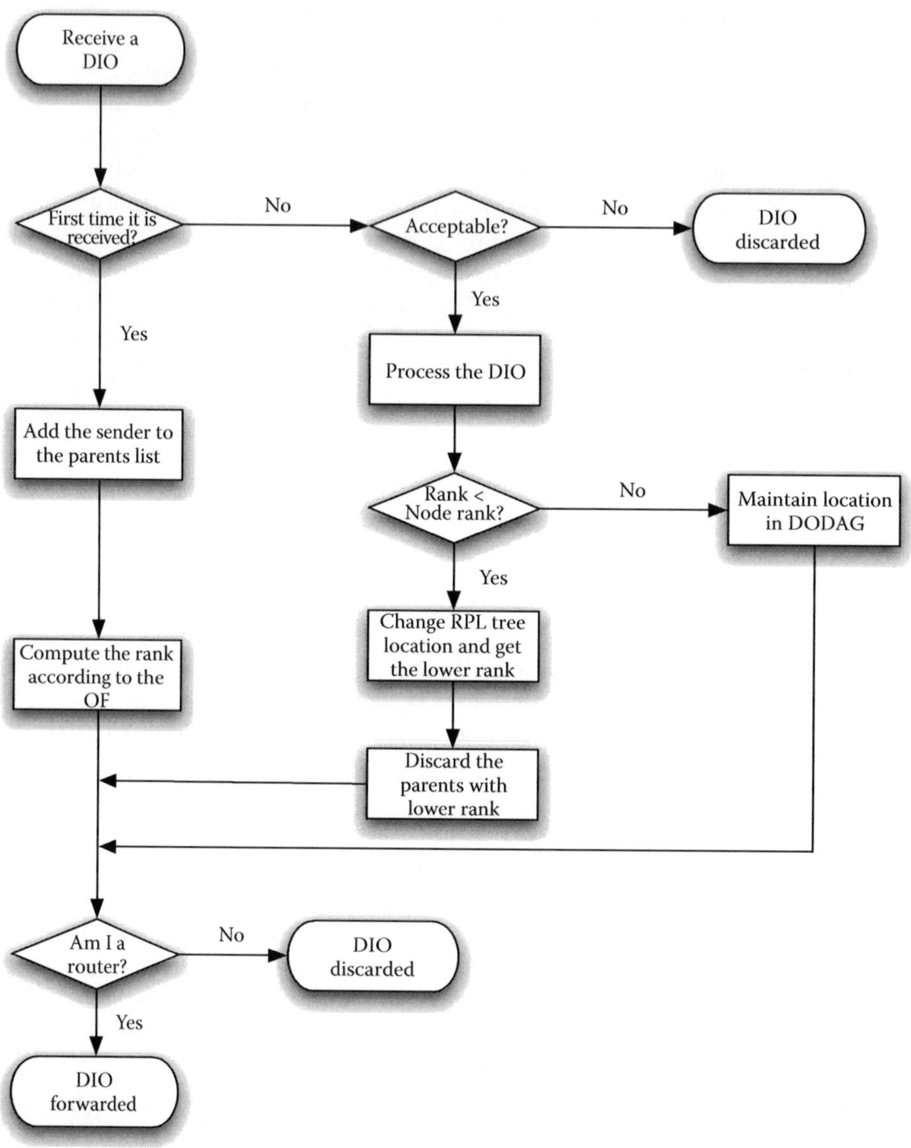

Figure 19.5 Diagram of the creation of a RPL tree in a node belonging to a DODAG tree.

Because information may also flow in the other direction, that is, from the root to the leaves, and because the communication links are asymmetric, RPL has also defined a strategy for the construction of downward routes. In this case, the remote nodes send destination advertisement object (DAO) messages to the root. While traveling back to the root, the intermediate nodes' addresses are stored in the packet, so that a complete route from the root to the node is created. There are two ways of operating while creating a downward route: (i) *storing* and (ii) *nonstoring* mode. In the former case, a message is sent from the remote node to the root and each intermediate parent node stores the addresses of the children from which it has received a DAO. In the latter, intermediate nodes do not store the addresses of their children but they are only limited to inserting their address into the DAO instead, transferring it to the root, which will be the only node with a complete topology to transfer information downward.

The control messages exchanged by RPL can also be used to convey additional information, depending on the specific application requirements. This technique, referred to as *piggybacking*, allows a significantly reduced amount of exchanged messages. For instance, it can be used to transfer information about the status of a node.

19.4.2 RG Mechanism

The RG algorithm is based on the same principles of LG: nodes of the WSN, upon joining a RPL DODAG, keep on collecting data (acquired with a given sensing rate). Data retrieval is performed by an external agent that periodically connects to the DODAG root and gathers all the data from the WSN.*

To prevent data losses, data is stored in several nodes (possibly including the generating node). This consists of copying and distributing replicas of the same data to other nodes with some available memory. As in LG, information about memory availability is periodically broadcast, by each node, to all direct neighbors.

The main parameters are listed in Table 19.1. Without loss of generality, we consider a WSN with N fixed RPL nodes deployed over an area whose surface is A (dimension: m^2). Nodes only interact with one-hop neighbors, that is, with nodes within the radio transmission range d (dimension: m). Note that an additional node acts as the RPL DODAG root, even if it does not participate in sensing and storage. Therefore, the final number of nodes is $N + 1$. The ith node has a finite local buffer of size B_i (dimension: data units) and sensing rate r_i (dimension: data units/s).

Each node broadcasts without acknowledgement and every T_{adv} (dimension: s), its memory status to all nodes within direct transmission range (i.e., one-hop neighbors). Each memory advertisement consists of six fields relative to the sending node: the RPL rank of the node, value of sensing rate, up-to-date available memory space, an aggregate that indicates the status of node memories in the down direction in the DODAG, an analogous value for the up direction, and a sequence number. Each node maintains a table that records the latest memory status received from neighbor nodes. Upon reception of a memory advertisement from a neighbor, a node updates its memory table, using the sequence number field to discard multiple receptions or out-of-date advertisements. The aggregate of the status of node memories in the down direction in the DODAG is given the minimum hop distance at which a node with some available space can be found. This distance is computed as follows: if a node detects that at least one of its children (i.e., neighbors with higher RPL rank) has some space locally, it sets this distance to 1. Otherwise, a parent increments by 1 the value of the minimum distance given by its children. Once the distance reaches a maximum value, a node assumes that there is no available memory in the down direction of the DODAG. The maximum value of the flag is given by the *MaxHopDown* parameter, listed in Table 19.1. Similarly, the status of the node memories in the up direction is computed in the same way, but in the inverse direction of the DODAG; in this case, the maximum hop distance that can be announced in a memory advertisement is given by the *MaxHopUp* parameter.

The first message exchange, depicted in Figure 19.6a, is between node 3 and node 1. Node 3 transfers a memory advertisement saying that it has no available space but one of its one-hop children does have available space, as stated by the minimum hop distance downward (mhd_{down})

* RPLC-based data retrieval is the focus of the next section.

Table 19.1 Main System Parameters

Symbol	*Description*	*Unit*
N	Number of RPL nodes	scalar
A	Surface of deployment area	m^2
d	Node transmission range	m
B_i	Node i's buffer size, $i \in \{1,\dots,N\}$	scalar
r_i	Node i's sensing rate, $i \in \{1,\dots,N\}$	s^{-1}
MaxHopDown	Maximum distance (in the down direction), that can be announced in a memory advertisement, at which some available space is present	scalar
MaxHopUp	Maximum distance (in the up direction), that can be announced in a memory advertisement, at which some available space is present	scalar
T_{adv}	Period of memory advertisement (from each node)	s
R	Maximum number of replicas per sensing data unit	scalar
T	Period of data retrieval (from the sink)	s

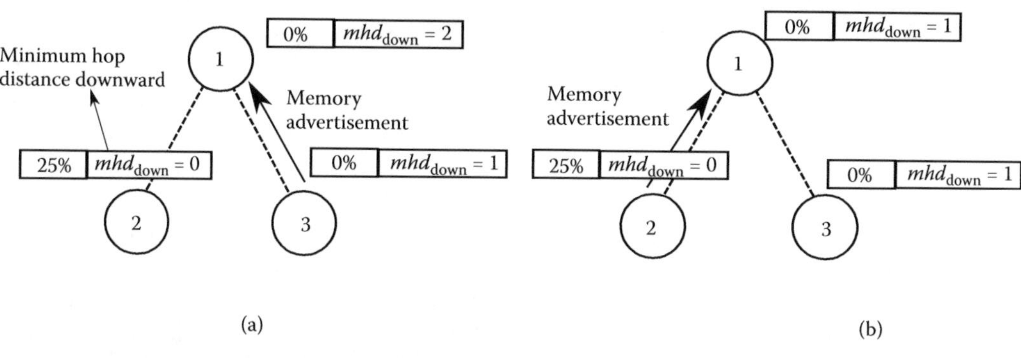

Figure 19.6 Messages exchanged for memory advertisements. In (a), node 3 transfers a memory advertisement saying that it has no available space but one of its one-hop children does have available space, as stated by the minimum hop distance downward (mhd_{down}) parameter. Node 1, which has no available memory, then sets mhd_{down} to 2, because it has received information that the closest node with available memory is at a two-hop distance. In (b), node 2 sends its memory advertisement saying that it has available space. Consequently, node 1 becomes aware that there is a closer node with available space, so it updates its mhd_{down} to 1.

parameter. Node 1, which has no available memory, then sets mhd_{down} to 2, because it has received information that the closest node with available memory is at a two-hop distance. Then, as shown in Figure 19.6b, node 2 sends its memory advertisement saying that it has available space. Consequently, node 1 becomes aware that there is a closer node with available space, so it updates its mhd_{down} to 1. An illustrative scenario is shown in Figure 19.6.

Like LG, the RG mechanism is fully decentralized, in the sense that all nodes play the same role. It consists of creating at most R copies of each data unit generated by a node and distributes them across the network, storing at most one copy per node, possibly closer to the DODAG root to later reduce the energy consumption of the retrieval phase. Each copy is referred to as a *replica*. Let us focus on node $i \in \{1,\ldots, N\}$. At time t, the node generates (upon sensing) a data unit. The memory table of node i contains one entry per direct neighbor. Node i selects from its memory table the neighbor node, called a *donor*, with the largest available memory space and the most

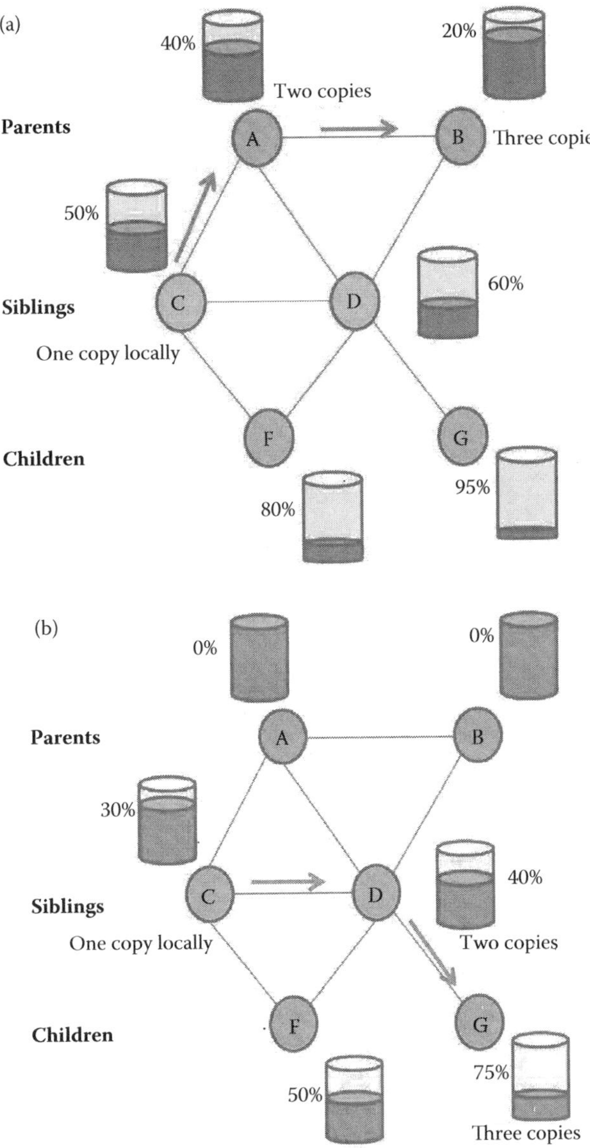

Figure 19.7 Hop-by-hop replication in the case of R = 3 desired replicas, for several scenarios: (a) replicas propagate to nodes closer to the RPL root first; (b) data is distributed to nodes in the down direction of the tree as the memories of the parents are full.

recent information. Moreover, priority is given to those donors that are parents of node i in the tree, that is, nodes with lower rank. If no parents can be selected, node i looks for a sibling in the tree, that is, a node with the same parent, providing that such a node has some available space. If no sibling can be chosen, node i searches for a child. In case all neighbors have no space locally, then node i checks if a neighbor at least knows about some available space beyond one hop, that is, in the up direction or in the down direction of the DODAG. In this case, again, priority is given to nodes in the up direction. If there is no suitable neighbor in the memory table, there is no possibility to distribute replicas of the data unit across the network. In this case, only one copy can be stored in the local memory of node i, if i has some space locally.

If a donor node can be selected, node i sends it a copy of the data unit, specifying how many other copies are still to be distributed in the WSN. As in LG, the number of required copies is set to either $R - 1$ (if node i can store the original data locally) or R (if node i's local memory is full). Upon reception of the copy, the donor node stores the copy in its memory, if it has some space locally, and selects the next donor node among its neighbors, discarding the sending node and the source node from the candidate nodes. The next donor is chosen such that its RPL rank diverges from that of the generator node. This causes replicas to spread well throughout the RPL tree. At this point, the donor sends the copy to the next chosen donor node, decrementing the number of required copies by 1. The replication process continues recursively until either the last (Rth) copy is stored or stops when one donor node cannot find any suitable next donor node. In the latter case, the final number of copies actually stored in the WSN is smaller than R. Note that, if a donor cannot find a next neighbor with a suitable RPL rank, the replica may follow a different reversed path along the DODAG, and retake the original direction later.

In Figure 19.7, we show an illustrative example with $R = 3$ desired replicas. In Figure 19.7a, nodes always try to distribute replicas to parents in the up direction of the RPL tree, for example, nodes with a lower rank. As the network is saturated, data is distributed toward nodes with the same rank or with a higher rank, that is, in the down direction of the DODAG, as shown in Figure 19.7b.

19.4.3 Performance Evaluation

RG has been implemented in Contiki 2.5 and evaluated in Cooja, a Java-based WSN simulator [8]. The scenario is depicted in Figure 19.8. A rectangular grid is composed of 60 storing nodes, shown in gray. Each node inside the grid communicates with four direct neighbors. Moreover, to simulate real conditions, the node interferes with some extra nodes: collision occurs if a node and at least one among its direct neighbors or its interfering nodes transmit a packet at the same time. For example, referring to Figure 19.8, the neighbors of node 36 are nodes 50, 51, 55, and 43. The interfering nodes, shown between the two circles, are nodes 29, 54, 56, and 57. Collisions may also be caused by the hidden terminal problem [32]. A 100% packet delivery ratio (PDR) is assumed, that is, packets are always delivered, providing that no collision has occurred. Note that nodes along the borders have fewer neighbors than the others. Finally, node 1, in the upper left part of the grid, acts as the RPL root, creates the routing topology, and performs periodic data retrieval. In Figure 19.9, a detailed view of the scenario is depicted. Nodes at the same hop distance from the RPL root are drawn with the same color. The maximum number of hops is equivalent to 13. It has been observed that RPL sometimes modifies the structure of the tree; therefore, a node may change its RPL rank.

The interval between two consecutive data retrievals is $T = 10$ min. The sensing period of the nodes is an integer number chosen randomly and independently in the range of 31 to 33 s.

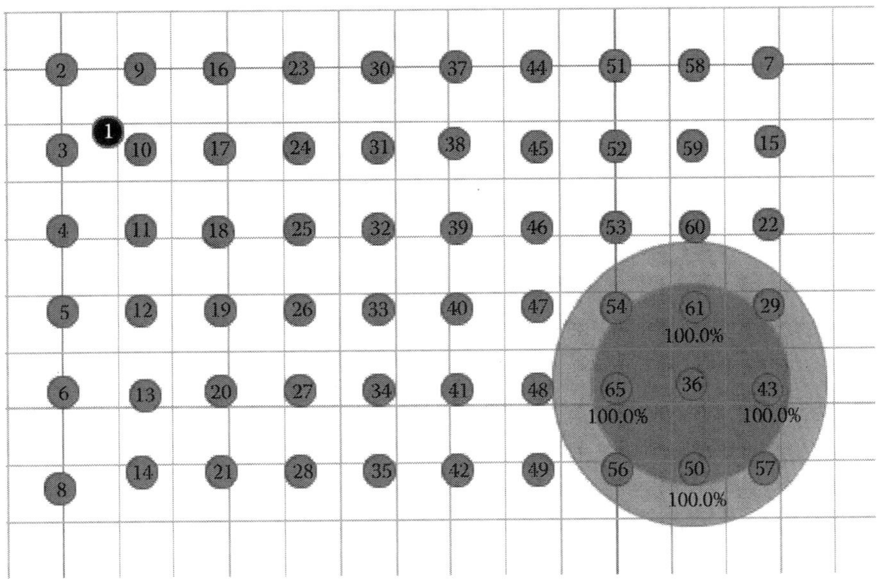

Figure 19.8 Scenario evaluated in Cooja. Sixty storage nodes (shown in gray) are deployed in a regular grid. Each node has four direct neighbors. The RPL root is node 1 of the figure (shown in black).

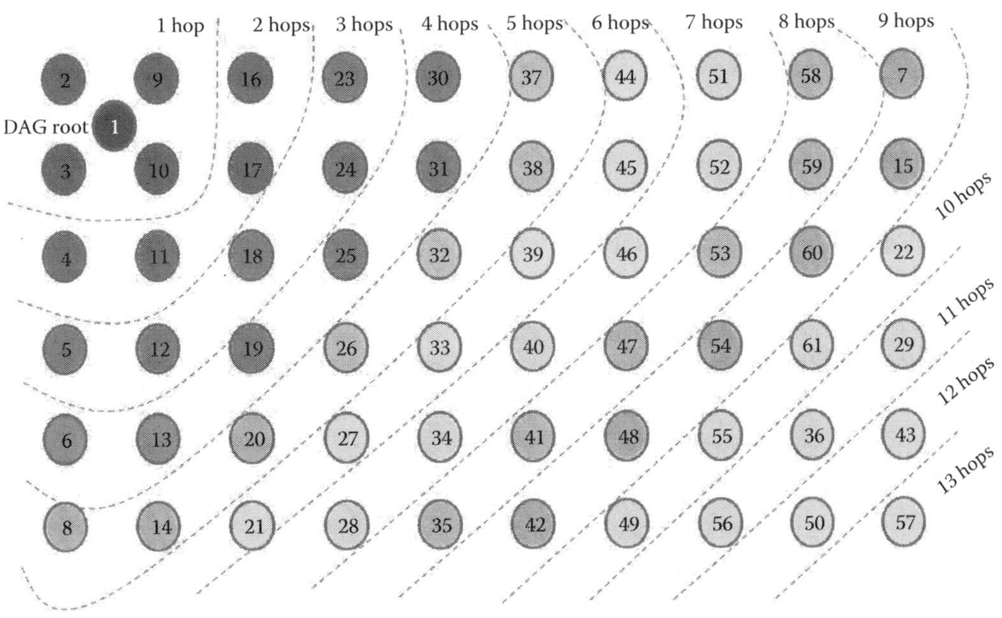

Figure 19.9 Scenario evaluated in Cooja. Sixty storage nodes are deployed in a regular grid. Each node has four direct neighbors. The RPL root is node 1 of the figure. The network has a total size equal to 13 hops.

All nodes have the same buffer size, which equivalent to $B = 100$ data units. The memory advertisement period T_{adv} is set to 30 s. We adopt the expected transmission count (ETX) as an RPL metric. According to this metric, the routing function tends to minimize the number of expected transmissions toward the RPL root. In the presence of perfectly reliable links, it is equivalent to minimizing the number of hops.

19.4.3.1 Effect of the Number of Replicas

We have run four simulations, setting the number of replicas R to 1, 3, 5, and 7, respectively. Note that the case with $R = 1$ is without replication, that is, only the original copy is stored.

The time required by the system to reach the network storage capacity has been computed in the different cases. With the setting described above, the network storage capacity C, which denotes the maximum amount of data that can be stored in the WSN, is equivalent to $C = N \times B = 60 \times 100 = 6000$ data units. In Figure 19.10a, the amount of total stored data is shown, as a function of time, for the four considered simulated cases. It can be observed that the capacity is reached later when fewer replicas are used—for instance when $R = 1$—because less data is generated. On the other hand, for higher values of R, capacity is reached earlier. However, the slope of the curve, in the case with $R = 3$, tends to approximate the ones obtained with $R = 5$ and $R = 7$. Because a higher storing rate quickly saturates the memories of the nodes at the beginning of the simulation and leads to a less efficient data distribution later.

In Figure 19.10b, the amount of dropped data is shown as a function of time. Nodes drop newly acquired data once their local memory has filled up and no neighbors are available for donating extra storage space. Remember that data is marked as dropped if no replicas at all can be stored. This explains why related curves with $R = 3$ and $R = 5$ are close to each other. This suggests that it is inconvenient to store exactly R replicas for each sensed data unit because the first replicas that are stored would quickly congest the memories and impede the next ones to be stored.

In Figure 19.10c, the amount of unique stored data is shown as a function of time. In fact, in the presence of redundancy ($R > 1$) several copies of the same data unit are distributed in the WSN, so that the amount of original node data is smaller than the total amount of actually stored data. The analysis of the unique stored data is expedient to evaluate the efficiency of data retrieval, as will be shown in Section 19.5.

At this point, it is of interest to evaluate the data placement throughout the WSN over time. As discussed previously, RG distributes replicas prioritizing donor nodes closer to the root. Figure 19.11 shows the average saturation level of the memories of the nodes at different hop distances from the root, for several observation instances. It can be noticed that, according to RG, the portion of the RPL tree closer to the root fills the memories faster. This can be of help in the data retrieval phase because data follows a shorter path to reach the sink.

As for the distribution of the replicas, Figure 19.12 shows the average hop distance reached by the redundant copies, from the owner of the original one for different cases with $R > 1$. Results show that replicas keep on moving away from each other, passing through nodes with a higher RPL rank in the tree. By comparing the results in Figure 19.12 with those, relative to the LG mechanism, shown in Figure 19.4, it can be observed that with the RG mechanism, replicas tend to spread farther from the originating node than with the LG mechanism. Therefore, the RG mechanism leads to a more resilient data preservation in the presence of a failure involving a source node and several of its neighboring nodes.

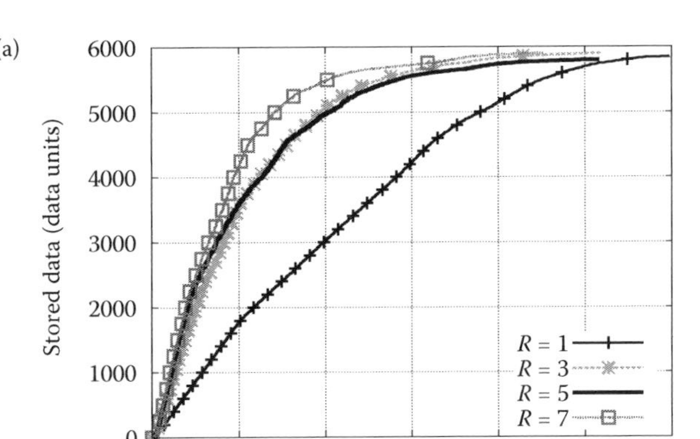

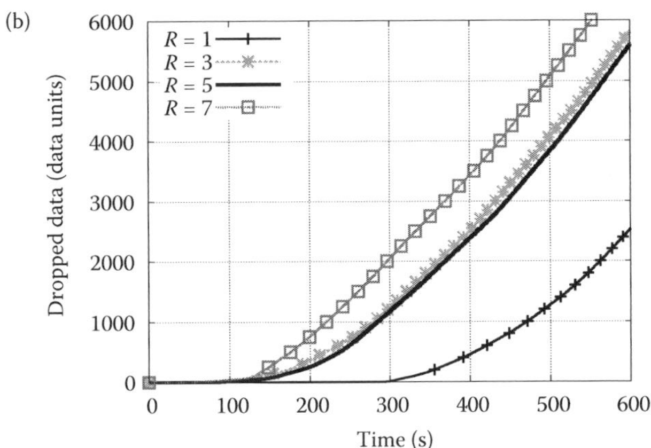

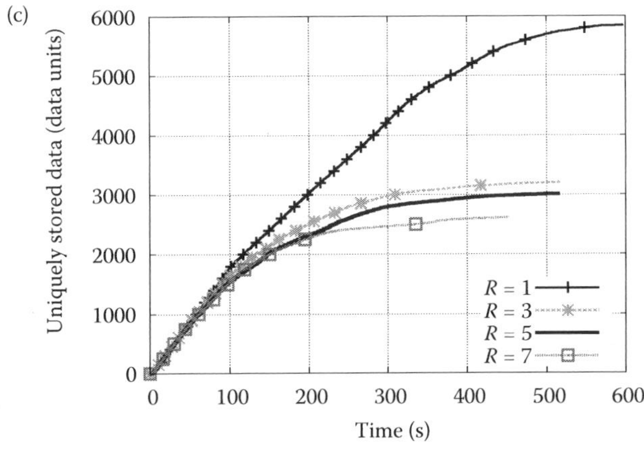

Figure 19.10 (a) Stored data, (b) dropped data, and (c) unique stored data in the system for various values of replicas *R*, as a function of time. In all cases, we consider: memory size equal to 100 data units, *N* = 60 storage nodes, and a memory advertisement period of 30 s.

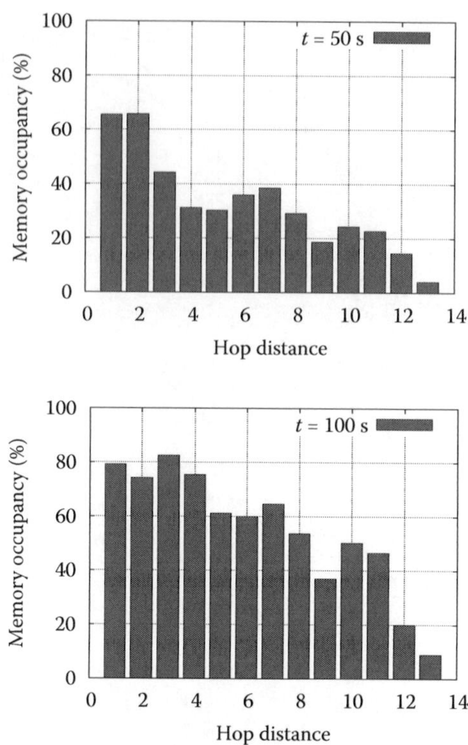

Figure 19.11 Percentage of memory occupancy varying the hop distance from the RPL root, at different time instances. Memories of nodes closer to the RPL root saturate faster. The number of replicas R is set to 7.

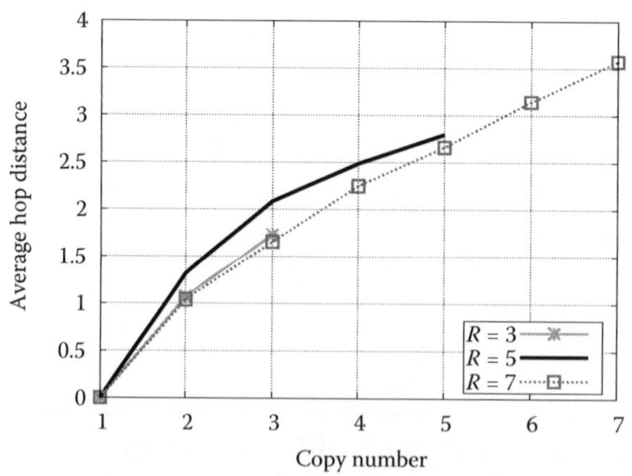

Figure 19.12 Average hop distance reached by the kth replica versus k. k varies between 1 and 3 ($R = 3$), 1 and 5 ($R = 5$), 1 and 7 ($R = 7$). The distance is calculated from the position of the first replica.

19.4.4 How to Retrieve the Stored Data?

In Section 19.3 and in the current section, two redundant data distribution mechanisms have been presented: the LG mechanism increases the network storage capacity with a slight signaling overhead; the RG mechanism, owing to a more accurate view of the network status (thanks to routing information), allows replicas to spread better across the WSN. Although the two proposed mechanisms deal with distributed storage, data retrieval is fundamental to collecting (to a sink) the stored data.

Data retrieval consists of a periodic collection of the whole sensed data, stored in the WSN, at a single node in the network, usually denoted as a sink. The sink then passes all the collected data to a monitoring center for further processing. During the retrieval phase, the major challenges are the following:

- Sending lot of data depletes the battery of the nodes, thus the lifetime of the WSN.
- Multihop communication is required for those nodes which are far away from the sink. Therefore, battery depletion is even more critical for intermediate nodes that have to forward the data of the others.
- The amount of stored data in the WSN can be extremely high, if the retrieval is not performed in a timely manner. Sending all the data instantaneously to the sink can cause many collisions, thus losing a lot of information. A solution could be letting the nodes to wait some time between two consecutive data transmissions. However, this may excessively prolong the duration of the retrieval.

Although the LG mechanism does not lead directly to a data retrieval phase, the RG does, as it builds on the RPL routing algorithm. The next section is dedicated to data retrieval with RPL and all the aspects outlined above will be covered in depth.

19.5 Data Retrieval with RPL

Data retrieval consists of forwarding the collected sensing data of the WSN to a central base station for further processing. In the literature, it often appears with various terms, for example, *data collection* and *data gathering*. This section briefly reviews some existing works in this field and presents the data retrieval mechanism that has been implemented and evaluated in RG.

19.5.1 Related Work

As stated in an article by Wang and Liu [34], data retrieval in WSNs is still in its early stage. Two main approaches can be identified: mobile data retrieval and fixed data retrieval.

Mobile data retrieval has been studied in the literature. The idea is to have a mobile entity that travels across the WSN and gathers data from every sensor node within the communication range. This should save battery at the sensors and increase the lifetime of the network. Usually, no specific routing topology is required, as the communication between the mobile sink and the sensor node is single hop. In a study by Gao and Zhang [35], the authors investigated the optimal path selection for a mobile sink in a path-constrained scenario with delay requirements. In a study by Jain et al. [36], a rich analytical framework to measure data retrieval rate, latency, and consumed power is given. Several protocols to optimize data transfer at the minimum energy cost have been proposed [37]. And in a study by Somasundara et al. [38], the problem of data collection scheduling to avoid node's buffer overflow is investigated. For a detailed survey on mobile data retrieval, please refer to the article by Francesco et al. [33]. In our work, we consider periodic data retrieval performed by a fixed sink node.

As for fixed data retrieval, a first approach, discussed in a study by Tseng et al. [27], consists of a sink node that periodically stops at a fixed point in the network. Once this point is reached, the sink advertises its presence by sending queries. Each sensor node builds its own return path to the sink and consequently sends the collected data. In such a way, a routing tree toward the sink is formed. The CTP is probably the routing mechanism most frequently used for multihop fixed data retrieval in sensor networks [39]. The strengths of CTP are its ability to quickly discover and repair path inconsistencies and its adaptive beaconing, which reduces protocol overhead and allows for low radio duty cycles. Extended versions of CTP have been proposed to deal with the nodes' mobility [40]. Dozer [41] is a data retrieval protocol aimed at achieving extremely low energy consumption. It builds a tree structure used to convey data to the sink, enriched with a time division multiple access (TDMA) scheme at the MAC layer to synchronize the nodes. In Koala [42], a data-gathering system is proposed. In contrast with Dozer, routes are built at the sink and no persistent routing state is maintained on the motes. Coupled with a low-power probing (LPP) technique at the MAC layer, Koala can achieve very low duty cycles.

With respect to the related works, we consider periodic data retrieval in an isolated RPL-based network. Instead of a cross-design among MAC layer, topology control, routing, and scheduling, we let the RPL protocol handle the routing structure autonomously. Moreover, we focus more on data availability at the sink and on latency of the retrieval than on energy efficiency.

19.5.2 Data Retrieval Mechanism with RG

We describe here the data retrieval mechanism that has been designed and implemented to work together with the RG distributed data storage mechanism proposed in Section 19.4.

Recall from Section 19.4.2 that the RG mechanism replicates and distributes copies of sensed data units toward nodes closer to the sink. The sink is supposed to periodically send retrieval requests through the RPL root. In particular, the retrieval request is broadcast by each node of the tree only once. Subsequent receptions of the same retrieval request are ignored. Moreover, a sequence number is used by the RPL root to identify each periodic retrieval request.

Data collection takes place from each sensor node toward the RPL root. Such many-to-one traffic patterns, if not carefully handled, can cause (i) many collisions and (ii) highly unbalanced and inefficient energy consumption in the whole network. To reduce these risks, the RG mechanism has been enriched with the following: A replica of a given data item is sent to the sink only if it is the closest to the RPL root among all the stored replicas of that data item. To let a node be informed that it holds a replica closer to the root, donor nodes include their IPv6 address and RPL rank within a data packet during the distribution phase. Because RG follows a hop-by-hop greedy replication scheme, the next donor node along the chain checks the rank of the closest donor node that has stored the replica before, and compares it with its own rank. In case its own rank is lower than the value announced in the data packet, that is, the node is closer to the RPL root than its ancestor donor, and providing that the node stores the replica in its local memory, then the node informs the ancestor donor about the newly closer position of the replica. In the retrieval phase, the ancestor donor will not send such a replica to the sink, as it knows that it is not the closest one. An illustrative example is depicted in Figure 19.13. In Figure 19.13a, replicas

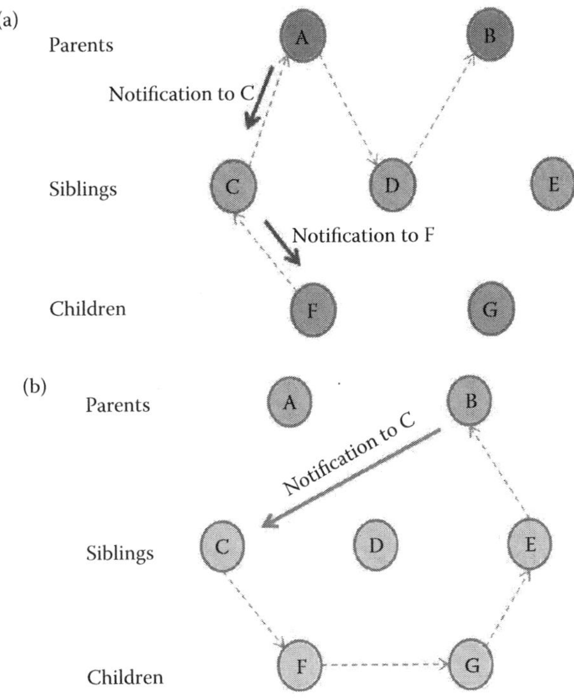

Figure 19.13 During the data retrieval phase, only the closest replica is sent to the sink. A donor node with lower rank informs the prior best donor about the new position of the replica. In (a), replicas of a data item, generated at node *F*, are stored, progressively, at nodes *C*, *A*, *D*, and *B*. Node *C* informs node *F* of its better position in the tree (*C* has rank *k*, whereas *F* has rank $k + 1$, with $k > 0$). Similarly, node *A* has to inform node *C* for the same reason. Note that *D* and *B* do not send notification to *A*, as they are not closer to the RPL root. In (b), the notification is exchanged only from *B* to *C*.

of a data item generated at node F are stored, progressively, at nodes C, A, D, and B. Node C informs node F of its better position in the tree (C has rank k, whereas F has rank $k + 1$, with $k > 0$). Similarly, node A has to inform node C for the same reason. Note that D and B do not send notification to A, as they are not closer to the RPL root. In Figure 19.13b, notification is delivered only from B to C.

19.5.3 Performance Evaluation

The data retrieval scheme developed in RG has been evaluated with the same settings of Section 19.4.3. Periodic retrieval occurs every $T = 600$ s. At this time instance, as shown in Figure 19.10a, the system has already reached the network storage capacity C, equal to 6000 data units. Therefore, the amount of data stored in the WSN to be delivered to the RPL root is considerable; if all nodes would send 100 data units contemporaneously, collisions would affect the amount of data successfully delivered to the sink. To investigate the data availability at the sink, that is, the amount of data effectively retrieved by the sink, data units have been injected in the network, from each node toward the RPL root, one at a time. The interval between the injection of two consecutive data units is denoted as I. Intuitively, higher injection rates may speed up the retrieval process but could also increase the collisions' probability, thus the percentage of data loss at the sink. On the other hand, a longer transmission interval I may increase data availability, penalizing the latency. This is critical in scenarios in which the sink cannot prolong the duration of the retrieval.

Collisions are also influenced by the amount of data that is to be retrieved. With redundancy ($R > 1$), less data is to be delivered to the sink because RG only sends one replica of a data unit, for example, only the closest replica is sent. It has already been shown in Figure 19.10c that the amount of unique stored data in the system with $R > 1$ is much less than in the case with no redundancy, that is, $R = 1$. However, a comparable volume of unique data is present in the cases with $R = 3$, $R = 5$, and $R = 7$. In Figure 19.14a, the amount of retrieved data is shown, as a function of time, for the various values of replicas R. The interval I is set to 2 s in this case. It can be observed that the amount of retrieved data decreases with R because there is more unique data in the system when no redundancy is required. With $R = 1$, all 6000 data units are sent to the sink, as there is no redundancy, but only a small fraction, that is, approximately 50% of the capacity, is successfully retrieved; the rest is lost because of collisions. For higher values of R, there are more replicas of

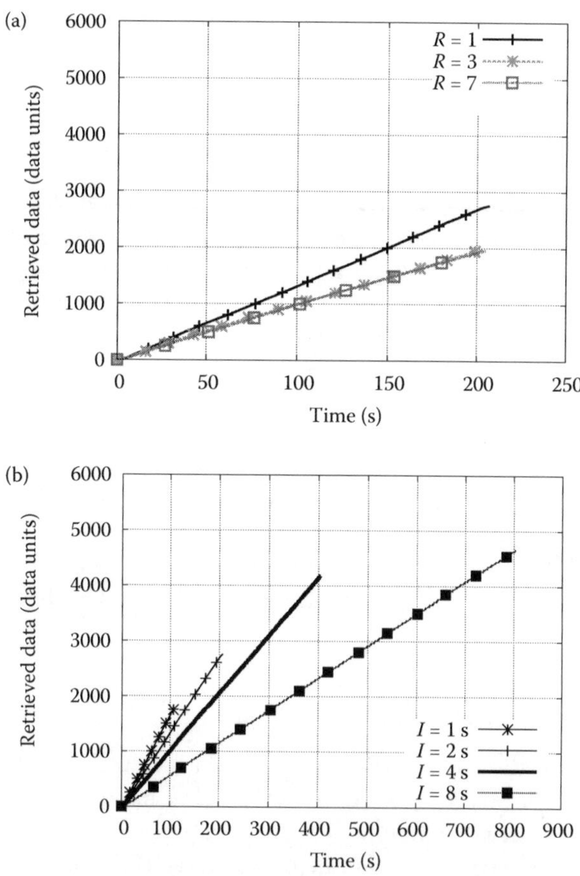

Figure 19.14 Retrieved data considering memory size equal to 100 data units, $N = 60$ storage nodes, and a memory advertisement period of 30 s. In (a), the amount of retrieved data is shown, as a function of time, for the various values of replicas R. In (b), the amount of retrieved data is shown, as a function of time, for the various values of the transmission interval I.

the same data unit in the WSN; therefore, according to the mechanism, only the closest replica is collected. However, because the amount of unique stored data is similar, the volume of retrieved data is the same in the cases with $R = 3$ and $R = 7$. The approximately 2000 retrieved data units correspond to approximately 80% of the capacity.

To reduce the effect of collisions, several simulations have been run with different values of the transmission interval I. The results depicted in Figure 19.14b for $R = 1$ show that the amount of retrieved data significantly increases for higher values of I. To summarize, the percentage of retrieved data among the total unique stored data in the system is shown in Table 19.2 for various combinations of R and I.

Table 19.2 Percentage of Retrieved Data

	$I = 1$ s	$I = 2$ s	$I = 4$ s	$I = 8$ s
$R = 1$	33%	50%	68%	82%
$R = 3$	49%	74%	95%	100%
$R = 5$	50%	74%	95%	100%
$R = 7$	55%	80%	100%	100%

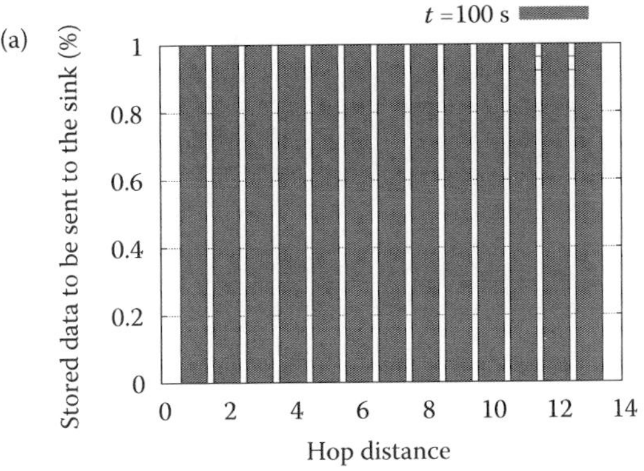

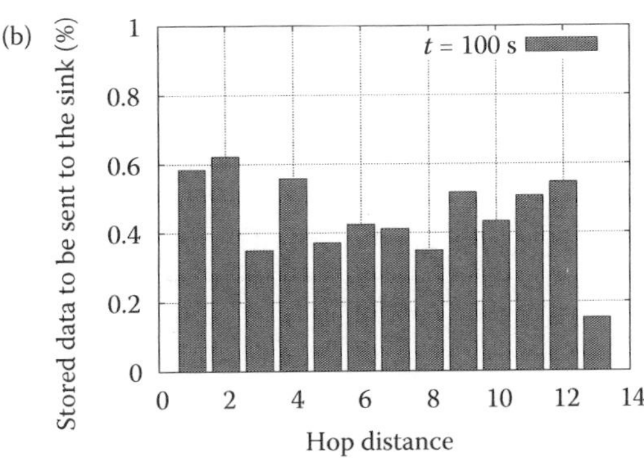

Figure 19.15 Data that is sent to the sink in the retrieval phase, for various values of R: (a) $R = 1$ (no redundancy) all stored data is to be sent. (b) $R = 7$ nodes closer to the sink are going to send more data than the others. Data retrieval period is set to 100 s.

Finally, Figure 19.15 shows the average percentage of data sent to the sink from several hop distances from the RPL root. In this case, the data retrieval period is set to T = 100 s, similar to the beginning of the simulation. Figure 19.15a corresponds to the case with R = 1: All nodes in the WSN send all their data to the sink. In Figure 19.15b, R = 7 replicas are stored in the system. In this case, nodes closer to the RPL root deliver more data than the others.

19.6 Conclusion

This chapter has addressed the problem of redundant data distribution and retrieval for IoT-based observation systems. Two redundant distributed data storage mechanisms have been proposed to increase the resilience and storage capacity of a WSN against node failure and local memory shortage. The performance of the two distributed storage mechanisms, denoted as LG and RG, has been evaluated extensively through simulations and real experiments. The RG mechanism, being built on top of RPL, lends directly to the implementation a complimentary data retrieval mechanism, whose performance has been evaluated as well. Our results clearly show a trade-off between storage redundancy (which depletes the total available storage memory) and robustness against possible node(s) failure.

Future research activities will include other parameters in the replication strategies such as the energy consumption of the nodes or the reachability of nodes, especially if they operate a low-power MAC layer with duty cycles (e.g., Contiki MAC or X-MAC). We also envision the study of dynamic reconfigurations of node behaviors (e.g., sampling) and communication layers (e.g., transmitting power and duty cycle) to meet replication demands with minimum energy cost. Finally, we would like to evaluate the performance of the RG mechanism on a real testbed such as SensLAB.

Acknowledgment

This work is funded by the European Community's Seventh Framework Programme, area "Internetconnected Objects," under grant no. 288879, CALIPSO project—Connect All IP-based Smart Objects. The work reflects only the authors' views; the European Community is not liable for any use that may be made of the information contained herein.

References

1. G. Mathur, P. Desnoyers, D. Ganesan, and P. Shenoy. Ultra-low power data storage for sensor networks. In *Proceedings of the 5th International Conference on Information Processing in Sensor Networks (IPSN'06)*, pages 374–381, Nashville, TN, USA, 2006.
2. P. Juang, H. Oki, Y. Wang, M. Martonosi, L.S. Peh, and D. Rubenstein. Energy-efficient computing for wildlife tracking: Design tradeoffs and early experiences with ZebraNet. In *Proceedings of the 10th International Conference on Architectural Support for Programming Languages and Operating Systems (ASPLOS-X)*, pages 96–107, San Jose, CA, USA, October 2002.
3. A. Beaufour, M. Leopold, and P. Bonnet. Smart-tag based data dissemination. In *Proceedings of the 1st ACM International Workshop on Wireless Sensor Networks and Applications (WSNA'02)*, pages 68–77, Atlanta, GA, USA, September 2002.

4. J. Neumann, C. Reinke, N. Hoeller, and V. Linnemann. Adaptive quality-aware replication in wireless sensor networks. In *International Workshop on Wireless Ad Hoc, Mesh and Sensor Networks (WAMSNET'09)*, pages 413–420, Jeju Island, Korea, December 2009.
5. K. Piotrowski, P. Langendoerfer, and S. Peter. TinyDSM: A highly reliable cooperative data storage for wireless sensor networks. In *International Symposium on Collaborative Technologies and Systems (CTS'09)*, pages 225–232, Baltimore, MD, USA, May 2009.
6. P. Gonizzi, G. Ferrari, V. Gay, and J. Leguay. Redundant distributed data storage: Experimentation with the SensLab testbed. In *Proceedings of the 1st International Conference On Sensor Networks (SENSORNETS)*, pages 15–23, Rome, Italy, February 2012.
7. SensLAB. A very large scale open wireless sensor network testbed, 2008. Web site: http://www.senslab.info/.
8. F. Osterlind, A. Dunkels, J. Eriksson, N. Finne, and T. Voigt. Cross-level sensor network simulation with COOJA. In *Proceedings of the 31st IEEE Conference on Local Computer Networks* (LCN), pages 641–648, Tampa, FL, USA, November 2006.
9. J.W. Hui and D.E. Culler. IP is dead, long live IP for wireless sensor networks. In *Proceedings of the 6th ACM Conference on Embedded Network Sensor Systems (SenSys'08)*, pages 15–28, Raleigh, NC, USA, November 2008.
10. J.W. Hui and D.E. Culler. IPv6 in low-power wireless networks. *Proceedings of the IEEE*, Volume 98, Issue 11, pages 1865–1878, November 2010.
11. Machine to machine (M2M) standardization-ETSI, 2012.
12. H. Wang. M2M communications. In *Proceedings of the IET International Conference on Communication Technology and Application (ICCTA 2011)*, page 1009, Beijing, China, October 2011.
13. ISO/IEC 18000-6. Information technology—Radio frequency identification for item management—Part 6: Parameters for air interface communications at 860 MHz to 960 MHz. *International Organization for Standardization (ISO)*, pages 1–134, 2004.
14. ZigBee Specification. ZigBee Standards Organization Std. r17. October 2007.
15. IEEE 802.15.4 Std. Wireless medium access control (MAC) and physical layer (PHY) specifications for low-rate wireless personal area networks (LR-WPANs), pages 1–679, October 2003.
16. O. Gaddour and A. Koubâa. RPL in a nutshell: A survey. *Computer Networks*, Volume 56, Issue 14, pages 3163–3178, July 2012.
17. T. Winter, P. Thubert, A. Brandt, J. Hui, R. Kelsey, P. Levis, K. Pister, R. Struik, J. Vasseur, and R. Alexander. RPL: IPv6 routing protocol for low-power and lossy networks. RFC 6550 (Proposed Standard), March 2012.
18. Z. Shelby, K. Hartke, C. Bormann, and B. Frank. Constrained application protocol (CoAP), draft-ietf-core-coap-11, July 2012.
19. ETSI M2M web site. http://www.etsi.org/website/technologies/m2m.aspx.
20. F. Hongping and F. Kangling. Overview of data dissemination strategy in wireless sensor networks. In *International Conference on E-Health Networking, Digital Ecosystems and Technologies (EDT)*, pages 260–263, Shenzhen, China, April 2010.
21. A. Awad, R. Germany, and F. Dressler. Data-centric cooperative storage in wireless sensor network. In *2nd International Symposium on Applied Sciences in Biomedical and Communication Technologies (ISABEL)*, pages 1–6, Bratislava, Slovak Republic, November 2009.
22. S.R. Madden, M.J. Franklin, J.M. Hellerstein, and W. Hong. TinyDB: an acquisitional query processing system for sensor networks. *ACM Transactions On Database Systems*, Volume 30, Issue 1, pages 122–173, March 2005.
23. S. Ratnasamy, B. Karp, S. Shenker, D. Estrin, and L. Yin. Data-centric storage in sensornets with GHT, a geographic hash table. *ACM Mobile Networks and Applications*, Volume 8 Issue 4, pages 427–442, August 2003.
24. A. Omotayo, M.A. Hammad, and K. Barker. A cost model for storing and retrieving data in wireless sensor networks. In *The IEEE 23rd International Conference on Data Engineering Workshop (ICDE)*, pages 29–38, Istanbul, Turkey, April 2007.
25. L. Luo, C. Huang, T. Abdelzaher, and J. Stankovic. Envirostore: A cooperative storage system for disconnected operation in sensor networks. In *The 26th IEEE International Conference on Computer Communications (INFOCOM'07)*, pages 1802–1810, Anchorage, AK, USA, May 2007.

26. M. Takahashi, Bin Tang, and N. Jaggi. Energy-efficient data preservation in intermittently connected sensor networks. In *The IEEE Conference on Computer Communications Workshops (INFOCOM WKSHPS)*, pages 590–595, Shanghai, China, April 2011.
27. Y.-C. Tseng, W.-T. Lai, C.-F. Huang, and F.-J. Wu. Using mobile mules for collecting data from an isolated wireless sensor network. In *Proceedings of the 39th International Conference on Parallel Processing (ICPP'10)*, San Diego, CA, USA, September 2010.
28. N.H. Azimi, H. Gupta, X. Hou, and J. Gao. Data preservation under spatial failures in sensor networks. In *Proceedings of the 11th ACM International Symposium on Mobile Ad Hoc Networking and Computing (MobiHoc'10)*, pages 171–180, Chicago, IL, USA, September 2010.
29. G. Maia, D.L. Guidoni, A.C. Viana, A.L.L. Aquino, R.A.F. Mini, and A.A.F. Loureiro. Proflex: A probabilistic and flexible data storage protocol for heterogeneous wireless sensor networks. Research report, INRIA, July 2011.
30. TinyOS web site. http://www.tinyos.net/.
31. Contiki OS web site. http://www.contiki-os.org/.
32. A. Tsertou and D.I. Laurenson. Revisiting the hidden terminal problem in a CSMA/CA wireless network. *IEEE Transactions on Mobile Computing*, Volume 7, Issue 7, pages 817–831, July 2008.
33. M.D. Francesco, S.K. Das, and G. Anastasi. Data collection in wireless sensor networks with mobile elements: A survey. *ACM Transactions on Sensor Networks*, Volume 8, Issue 1, pages 7:1–7:31, August 2011.
34. F. Wang and J. Liu. Networked wireless sensor data collection: Issues, challenges, and approaches. *IEEE Communications Surveys Tutorials*, Volume 13, Issue 4, pages 673–687, 4th quarter 2011.
35. S. Gao and H. Zhang. Energy efficient path-constrained sink navigation in delay-guaranteed wireless sensor networks. *Journal of Networks*, Volume 5, Issue 6, pages 658–665, June 2010.
36. S. Jain, R.C. Shah, W. Brunette, G. Borriello, and S. Roy. Exploiting mobility for energy efficient data collection in wireless sensor networks. *Journal of Mobile Networks and Applications*, Volume 11, Issue 3, pages 327–339, June 2006.
37. G. Anastasi, M. Conti, E. Monaldi, and A. Passarella. An adaptive data-transfer protocol for sensor networks with data mules. In *IEEE International Symposium on a World of Wireless, Mobile and Multimedia Networks (WoWMoM)*, Helsinki, Finland, June 2007.
38. A.A. Somasundara, A. Ramamoorthy, and M.B. Srivastava. Mobile element scheduling for efficient data collection in wireless sensor networks with dynamic deadlines. In *Proceedings of the 25th IEEE International Real-Time Systems Symposium, Lisbon, Portugal*, December 2004.
39. O. Gnawali, R. Fonseca, K. Jamieson, D. Moss, and P. Levis. Collection tree protocol. In *Proceedings of the 7th ACM Conference on Embedded Networked Sensor Systems (SenSys'09)*, pages 1–14, Berkeley, CA, USA, November 2009.
40. N. Hassanzadeh, O. Landsiedel, F. Hermans, O. Rensfelt, and T. Voigt. Efficient mobile data collection with mobile collect. In *Proceedings of the 8th IEEE International Conference on Distributed Computing in Sensor Systems (DCOSS)*, pages 25–32, Hangzhou, China, May 2012.
41. N. Burri, P. von Rickenbach, and R. Wattenhofer. Dozer: ultra-low power data gathering in sensor networks. In *6th International Symposium on Information Processing in Sensor Networks (IPSN 2007)*, pages 450–459, Cambridge, MA, USA, April 2007.
42. M.-E. Razvan, C.-J.M. Liang, and A. Terzis. Koala: Ultra-low power data retrieval in wireless sensor networks. In *Proceedings of the 7th International Conference on Information Processing in Sensor Networks, (IPSN'08)*, St. Louis, MO, USA, April 2008.

Chapter 20

Monitoring Mechanism for Wireless Sensor Networks: Challenges and Solutions

Abderrezak Rachedi
Université Paris-Est Marne-la-Vallée

Contents

20.1 Introduction

Wireless sensor networks (WSNs) attract more and more researchers and industrialists because of their potential reliability, accuracy, flexibility, cheapness, and easy deployment. In addition, WSN application is wide: natural environment monitoring (fire detection, pollution, earthquake, etc.), ecosystem tracking, health care, security (video surveillance, object tracking, etc.), and military (battlefield monitoring, object localization, border monitoring, etc.) [1–3]. One of the main constraints of these networks is the energy limitation resulting from their small size and wire independence. This constraint must be taken into account in any protocol design and sensor network deployment. The energy limitation creates vulnerabilities that are exploited by attackers. There are two kinds of attacks: passive (as in traffic analysis and selfish behavior) and active (as in false routing information injection and impersonation). The impact of passive attacks on the network is not negligible compared to the impact of active attacks. Proposing a solution to counter passive attacks is a real challenge.

In this chapter, we focus on the wireless distributed intrusion detection systems (WDIDS), particularly the monitoring mechanism to detect the misbehaving nodes such as selfish-behaving or noncooperative nodes in the context of multi-hop wireless networks, and particularly WSN [4,5]. The monitoring mechanism is an important part of the intrusion detection system (IDS). The detection of any misbehaving node by an IDS requires the running of a monitoring mechanism. The monitoring mechanism is a set of actions enabling specific sensor nodes to monitor the behavior of the other sensor nodes. This mechanism enables the evaluation of monitored nodes. The main monitoring mechanisms implemented in wireless IDS are based on the well-known mechanism called Watchdog [6]. Indeed, in network-based IDS, and particularly in wired or wireless networks with infrastructure (such as WLANs, wireless local area networks), the network's traffic can be captured and directed to specific fixed gateways with an IDS probe. Thus, the network's traffic is monitored and supervised on those networks. However, in WSNs, the nodes can only monitor the network's traffic within their radio transmission range. Except in certain situations, the nodes do not have the same observation as their neighbors. Another problem related to WSN characteristics is the hidden node problem, for example, when a collision occurs at the monitor node and prevents the observation of the monitored node during a certain period of time. The monitored node could have acted honestly or not. Therefore, the IDS for WSNs cannot efficiently monitor the activity in its neighborhood. In addition, the IDS for WSNs has to deal with another problem related to the lack of interaction between MAC and routing layers [7]. This vulnerability can be exploited by the malicious nodes in order to launch a new class of attacks called cross-layer attacks. Generally, these attacks originate from the MAC layer, but their target is the denial of service (DoS) at the routing and the upper layers. For this reason, a new approach is necessary to improve the monitoring process in WSN. That is why the design of an efficient monitoring process to detect misbehaving nodes in WSN is one of the hardest problems. For all these reasons, the already existing IDS solutions for wired or wireless networks with infrastructure (WLANs) and even for mobile ad-hoc networks (MANETs) cannot be directly applied to WSNs [6,8–10]. In this work, we focus on the cross-layer approaches, particularly MAC and routing layers in order to propose an efficient monitoring mechanism.

The chapter is organized as follows: In Section 20.2, we present some existing monitoring and energy aware mechanisms. In Section 20.3, we briefly summarize the MAC IEEE 802.15.4 standard with its both modes: beacon-enabled and non–beacon-enabled. In Section 20.4, we present the challenges and the problems in a WSN's monitoring process. The vulnerable hidden regions and their impact on the monitoring mechanism are studied. In Section 20.5, we describe and detail the analytical model for both beacon-enabled and non–beacon-enabled modes. In addition, we present the simulation results and their analysis. Finally, in Section 20.6 we give the conclusion to this chapter.

20.2 Related Work

In this section, we briefly present the existing monitoring mechanisms with a first attempt at their classification. In addition, we present energy-aware schemes for WSNs.

20.2.1 The Monitoring Mechanisms

Unlike with IDS classification [11]—host IDS, network IDS, and hybrid IDS [5]—we can classify the monitoring mechanisms according to OSI layers: physical, MAC, routing, and other upper layers.

The monitoring mechanisms at the physical layer must be able to supervise the wireless channel in order to detect anomalies or attacks sich as radio-jamming or signal-jamming modifications. The radio-jamming attacks consist of emitting signals in order to corrupt the communication by exploiting the frequency interferences [12,13].

The monitoring mechanisms at the MAC layer mainly focus on channel access and duty-cycle activities. The misbehaving nodes can cheat on the MAC parameters, such as backing off manipulation in order to get all channel resources and to prevent the other nodes from these resources [14]. This attack creates an unfair situation between well-behaving and misbehaving nodes and then negatively impacts the network performance. This attack is applicable to both IEEE 802.11 wireless networks and IEEE 802.15.4 WSNs as a result of their similar CSMA/CA–based protocols. Some solutions are proposed to face the problem of back-off misbehavior. A DOMINO mechanism is proposed to detect MAC misbehavior in the case of IEEE 802.11 [15]. However, DOMINO only supports the PCF (point coordination function) mode, and it is not applicable to the wireless multi-hop networks like WSNs. Another solution introduces a new performance metric, called order gain, to quantitatively investigate two widely used continuous misbehavior models: double-window and fixed-window back-off misbehaviors and intermittent misbehavior that performs misbehavior intermittently to evade misbehavior detection [16]. Another attack needs a monitoring mechanism at the MAC layer and is called a guaranteed time slot (GTS) attack [17]. This attack is based on the inherent properties of the IEEE 802.15.4 superframe organization in a beacon-enabled operational mode for WSNs.

At the routing layer, we quote the well-known monitoring mechanism called Watchdog [6]. It is based on packets' forwarding to detect the nonforwarding nodes. The idea is that the monitor node can listen to the traffic between its neighbors and detect if the monitored nodes forward the packets in routing operations. However, Watchdog does not take into account the physical or the MAC level's parameters. Consequently, the major wireless IDSs (WIDSs) based on the Watchdog monitoring mechanism suffer from the high ratio of false positives (false alarms). WIDS developers must consider the forwarding ratio in order to determine the rate of false positives generated during a monitoring process. Furthermore, the characteristics of wireless multi-hop networks, particularly WSNs, must be properly taken into account.

Unfortunately, the existing solutions deal with the monitoring process at physical, MAC, and routing layers separately and without taking into account the interaction between them. That is why they suffer from high false positives (false alarms) and a high cost related to energy consumption. Unlike with the existing solutions, in this work, we select the cross-layer approach in order to take into account the MAC parameters in order to improve the monitoring process at the routing layer. Moreover, we show that with this strategy we not only improve the observation of the monitor node, but we also reduce both energy consumption and false positives.

20.2.2 Energy-Aware Mechanisms

The MAC layer plays an important role in reducing the energy consumption. It is divided into four classes [18]: the consumption related to control packets, collision, idle listening, and overhearing.

Several mechanisms are proposed to tackle the energy consumption problem in WSNs, including the duty cycling [19]. This mechanism aims at saving energy by using the sleep/wakeup technique, which consists of activating the transmitter radio when the node has a packet to transmit and switching it off when there is no packet to transmit. The duty cycling needs cooperative nodes in order to coordinate the sleep/wakeup periods. This coordination is ensured by the sleep/wakeup scheduling distributed algorithm. The duty cycle in IEEE 802.15.4 is tuned by the coordinator, which selects the superframe order (SO) and beacon order (BO) parameters. However, this mechanism reduces the bandwidth and increases the delay. The study presented in [20] shows that the energy consumption decreases linearly with the size of received/transmitted packets. Unlike in IEEE 802.11, in IEEE 802.15.4 the energy consumption at packet reception is more important than at the transmission packet. That's why in our proposed model, we focus on the overhearing time to evaluate the energy consumption at the monitor node.

There are three categories of sleep/wakeup protocols: the on-demand protocol (the node wakes up only when another node wants to communicate with it), the scheduled rendezvous protocol (the nodes wake up periodically at different moments in order to avoid a collision), and the asynchronous scheme protocol (each node wakes up independently without synchronization needs).

Many MAC protocols based on CSMA and TDMA mechanisms are proposed to save energy in WSNs, such as S-MAC, TRAMA, and Z-MAC [18]. Suh et al. [21] proposed an enhancement of IEEE 802.15.4 called TEA-15.4, which consists of increasing the bandwidth and reducing the energy cost by adapting the beacon interval according to the kind of traffic. This solution is hybrid (beacon-enabled and non–beacon-enabled modes). However, its implementation is complex. In this work, we focus on IEEE 802.15.4 MAC technology with its both modes: beacon-enabled and non–beacon-enabled [22].

20.3 MAC IEEE 802.15.4

The MAC IEEE 802.15.4 has two working modes: the non–beacon-enabled mode and beacon-enabled mode [22]. The non–beacon-enabled mode is based on *nonslotted CSMA/CA*, and there is no time link between the back-off period and the beacon. In this mode, the coordinator node always stays in active, idle listening. However, the beacon-enabled mode is based on *slotted CSMA/CA*, and when the beacon starts, each node launches its back-off period. The communication between nodes is controlled by the network coordinator, which transmits beacons at regular intervals (beacon interval) in order to synchronize the sensors. The nodes use the sleep/wakeup mechanism: They have to wake up in order to receive the coordinator's beacon. The coordinator is in charge of the data routing in the network. When they receive a beacon, all nodes know the superframe duration (coordinator's activity period) and the time when they can transmit data or sleep. The advantages of this mechanism are the possibility for the coordinator to communicate with all nodes in activity periods and the reduction of energy consumption when the coordinator and nodes are inactive.

The structure of the superframe is presented in Figure 20.1. It is composed of an active and an inactive period. The active period has 16 slots divided into three parts: beacon, CAP (contention access period), and CFP (contention-free period). The beacon is transmitted at slot zero without using CSMA/CA, and then the CAP period starts.

The size of both activity and inactivity periods is calculated according to the BO and SO. The following equations illustrate how to calculate the BI (beacon interval) and the SD (superframe duration).

$$\begin{cases} BI = aBaseSlotDuration \times aNumSuperframeSlots \times 2^{BO} \\ SD = aBaseSlotDuration \times aNumSuperframeSlots \times 2^{SO}; \end{cases}$$

where aBaseSlotDuration is the symbols that form the superframe when $SO = 0$.

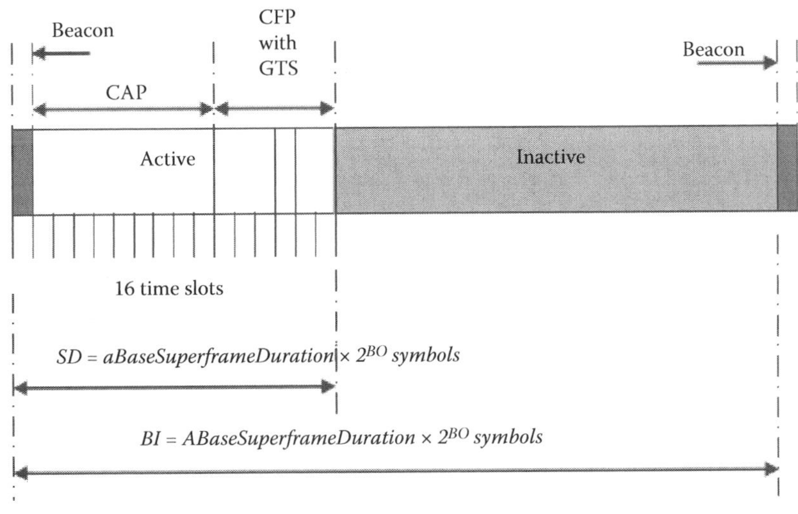

Figure 20.1 Superframe structure in IEEE 802.15.4.

aNumSuperframeSlots is the number of slots in the superframe.

The relationship between BO and SO is $0 \leq SO \leq BO \leq 14$. When the $BO = 15$, the network is in non–beacon-enabled mode.

The CFP uses a scheduled TDMA mechanism, such as a GTS (guaranteed time slot), for the network traffic with QoS requirements.

There are three ways to send data in MAC IEEE 802.15.4: direct transmission, indirect transmission, and GTS transmission. In this work, we focus on the direct transmission.

20.4 Challenges in the Monitoring Process for WSNs

20.4.1 Preliminary and Definitions

In this subsection, we give some basic and necessary definitions, such as the monitoring process, the transmission, the sensing, and the interference ranges. In addition, we present the relationship between them.

In order to differentiate between the monitoring process and the *IDS*, we give both definitions. The IDS is a complex system composed of at least two main parts: the monitoring mechanism and the detection mechanism. The goal of an IDS is *to monitor* network assets in order *to detect* misused or anomalous behavior. The complexity of the monitoring mechanism relies on the kind of network and technology of the communication link. For instance, the monitoring mechanism in multi-hop wireless networks is more complex and requires a specific design compared to the wired network. The monitoring process is a set of actions enabling specific nodes to monitor or supervise the behavior or activities of the other nodes. It is the first component located between network interfaces (NI) and the network traffic analyzer.

In the context of wireless multi-hop networks in general, and particularly WSN, some definitions are important in order to understand the challenges of the monitoring mechanism in that context.

The transmission range R_t is centered on the sender; it is defined as the area in which the power of the received signal is sufficient to correctly decode the packet. Another definition is the carrier sensing range. The carrier sensing range R_s is the range inside which nodes are able to sense the signal even though a correct packet's reception may not be possible (the receiver node may not be able to correctly decode the packet it received). Furthermore, another important aspect requires a definition that is unknown to many researchers. This aspect is the interference range. The interference range R_i is the range of the area in which each node that transmits creates a collision at the receiver node. This range is variable, and its variation depends on the distance between the transmitter and receiver nodes. On the other hand, when the receiver node is receiving the packet from the sender node, any new transmission in the interference range of the receiver node creates a collision at this node. The R_i depends on the distance between a transmitter and a receiver node (d) and on the signal-to-noise ratio (SNR) that must be above a certain threshold T_{SNR} so as to consider if the signal is valid when it reaches the receiver node. The R_i is defined by: $R_i = \sqrt{[k]T^*_{SNR}d}$ [23].

The relationship between the carrier sensing range, the interference range, and the transmission range is $R_t < R_i < R_s$ where $R_s = \beta\, R_t$, in some network simulator, like ns2, $\beta = 2.2$ [24].

Table 20.1 summarizes our notations.

Table 20.1 Table of Variables and Notations

R_t	Transmission range
R_s	Carrier sensing range
R_i	Interference range
T_{SNR}	Threshold signal-to-noise ratio
TR_A	Transmission region of a node A
IR_A	Interference region of a node A
CS_A	Carrier sense region of a node A
d_{AB}	Distance between node A and node B
$TR_{BA}(d_{AB})$	$TR_B - TR_A$
$A_{S1S2}(d_{AB})$	Monitored node's vulnerable hidden region ($IR_B - CS_A$)
$A_{S3S4}(d_{AB})$	Monitor node's vulnerable hidden region ($IR_A - CS_B$)

20.4.2 Different Hidden Regions

In this subsection, we describe the different hidden regions that have significant impacts on the monitoring process. In Figure 20.2a and b, we illustrate the hidden sensing region and the hidden transmission region of two neighbor nodes A and B. We have $CS_{AB}(d_{AB})$: The sensing region of node A does not include the sensing region of node B. $IR_{AB}(d_{AB})$ represents the interference region of node A and does not include the interference region of node B. If a node in $CS_{AB}(d_{AB})$ transmits, the signal can be sensed by node A but not by node B. The difference between $CS_{AB}(d_{AB})$ and $IR_{AB}(d_{AB})$ can be seen when a node in region $IR_{AB}(d_{AB})$ transmits: It can create a collision at receiver node A, but it is not the case for the nodes in region $CS_{AB}(d_{AB})$.

The difference between $IR_{AB}(d_{AB})$ and $TR_{AB}(d_{AB})$ is the problem of the hidden nodes, which can be resolved by the request-to-send/clear-to-send (RTS/CTS) mechanism (four-way handshaking method) in IEEE 802.11. This mechanism is designed to only take into account the transmission range but not the interference range. However, in the interference region, nodes cannot decode the packets correctly when node A transmits. The basic idea of the RTS/CTS mechanism is that when a node wants to transmit a data packet it sends the RTS packet to the receiver. Once the receiver node receives the RTS packet, it replies with the CTS packet. All nodes receive the control packet (RTS or CTS) from the transmitter nodes, and it is not the destination. They defer their transmission during *NAV* (network allocation vector), which belongs to the control packet. The NAV value is calculated by the sender and the receiver in both packets RTS and CTS. The problem is when a node senses a signal but cannot decode it, that means that it cannot calculate a NAV (network allocation vector); that is why it uses the EIFS (extended inter-frame spaces)* [25]. However, the RTS/CTS mechanism does not exist in IEEE 802.15.4, and the problem of hidden nodes in the sensing region remains.

* The EIFS is estimated to 364 μ*s* when using a 1 Mbps channel bit rate.

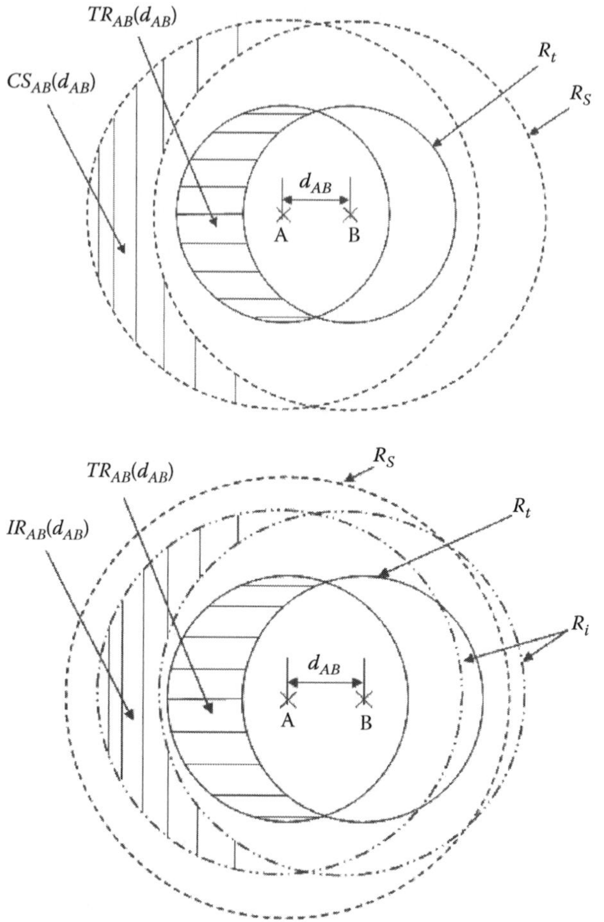

Figure 20.2 Different hidden regions, CS_{AB}, IR_{AB}, and TR_{AB}.

20.4.3 Problem with Hidden Regions

In the context of monitoring mechanisms, we distinguish two important hidden regions:

20.4.3.1 Monitored Vulnerable Hidden Region

A monitor node A wants to monitor a node B, which is its neighbor, but it has no complete knowledge about the environment of the monitored node. If the interference region of the monitored node (IR_B) is not covered by the carrier sense region of the monitor node (CS_A), we call this region a *monitored vulnerable hidden region* as shown in Figure 20.3a, and it is noted $A_{S1S2}(d_{AB})$. If a node M is located in this vulnerable region, it can reduce the forwarding rate of monitored node B by generating important traffic to another node. The attacker's goal is to prevent monitored node B from communicating and to reduce its capacity to forward the packets. However, the forwarding rate is used as a metric to determine the nodes' cooperation in the network and is also named the reputation rate [26].

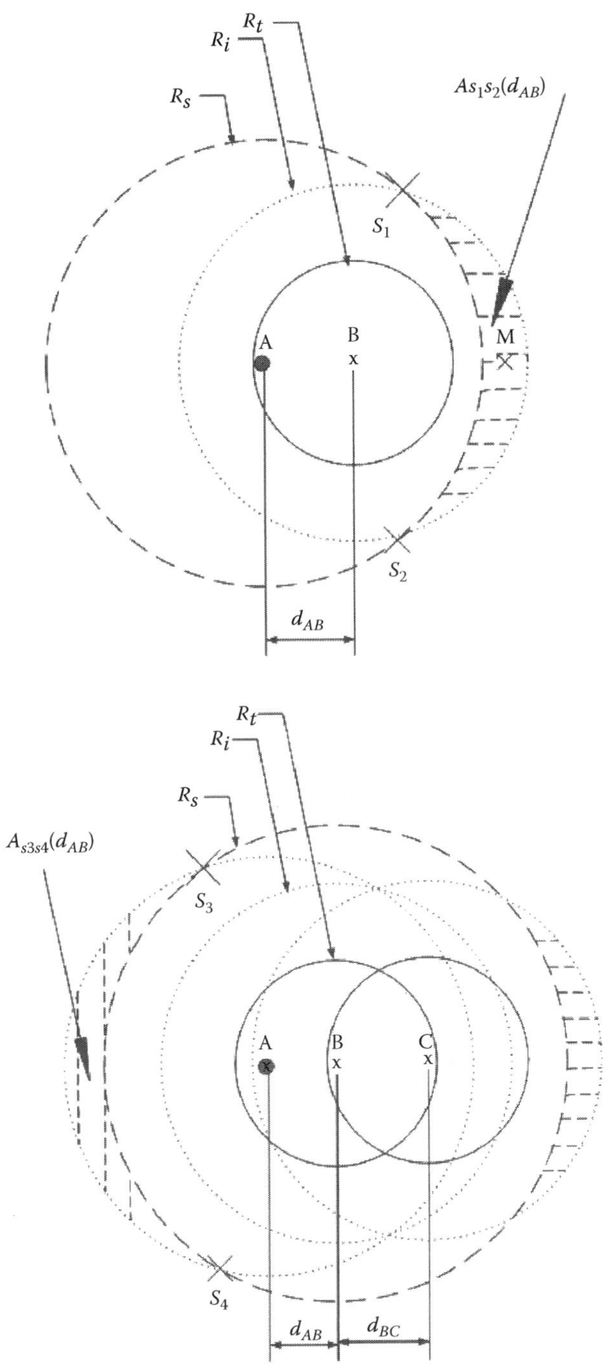

Figure 20.3 Vulnerable regions in the monitoring mechanism.

20.4.3.2 Monitor Vulnerable Hidden Region

Another vulnerable region can affect the monitoring mechanism; this region only exists if the interference region of a node A is not covered by the carrier sense region of a monitored node B. We call it the monitor vulnerable hidden region. It is illustrated in Figure 20.3b and noted $A_{S3S4}(d_{AB})$. If any node in this region starts to transmit, it disturbs the monitor node's observation. That means that if any node in region $A_{S3S4}(d_{AB})$ transmits when node A monitors node B, the observation of monitor node A is not accurate.

20.4.4 Impact of the Distance on the Hidden Areas

The distance between the monitor node and the monitored node seems to have an important impact on the monitoring mechanism. In order to show this impact, we take the example of two neighbor nodes A and B: Node B transmits to node A. Let d_{AB} be the distance between A and B. Figure 20.4a and b shows this example: In case (a), the distance between node A and node B is longer than in case (b). This means that the interference region of node A and the region

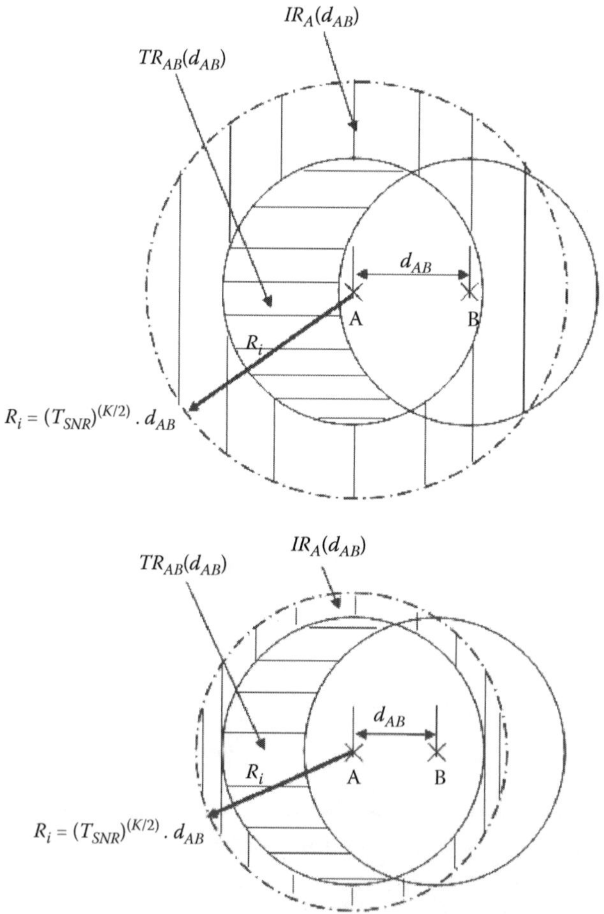

Figure 20.4 Impact of distance on interference and on hidden areas.

$TR_{BA}(d_{AB})$ are greater in case (a) than in case (b). When a node A comes closer to a node B, the region $TR_{BA}(d_{AB})$ becomes smaller, and the interference region can be covered by the transmission area when $d_{AB} \leq \frac{R_t}{\sqrt{[4]T_{SNR}}}$. The average number of nodes in region $TR_{BA}(d_{AB})$ depends on the nodes' distribution and on the mobility model. The greater a region $TR_{BA}(d_{AB})$ is, the bigger the probability of getting a great number (k) of nodes in this region is.

In order to illustrate the impact of the distance between the monitor and monitored nodes on the vulnerable region A_{S3S4}, we plot in Figure 20.5a the hidden area A_{S3S4}, according to the distance

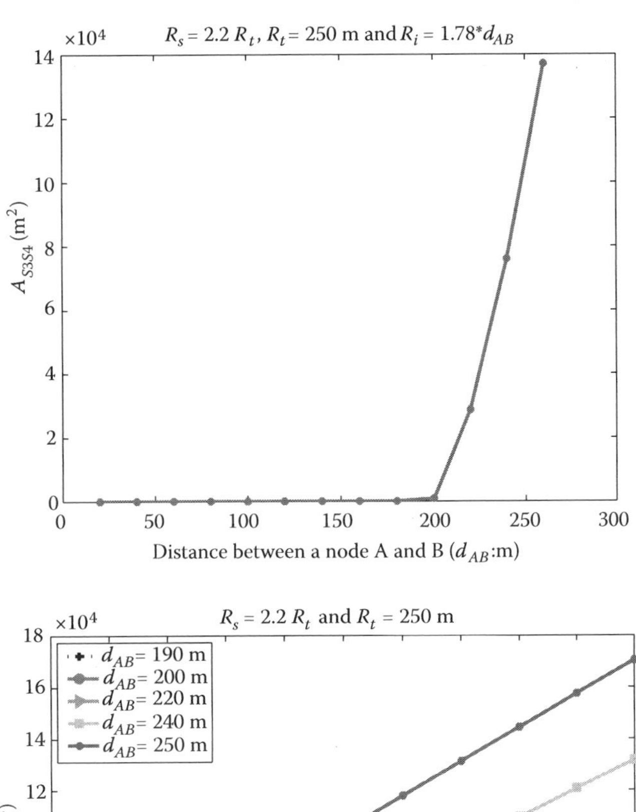

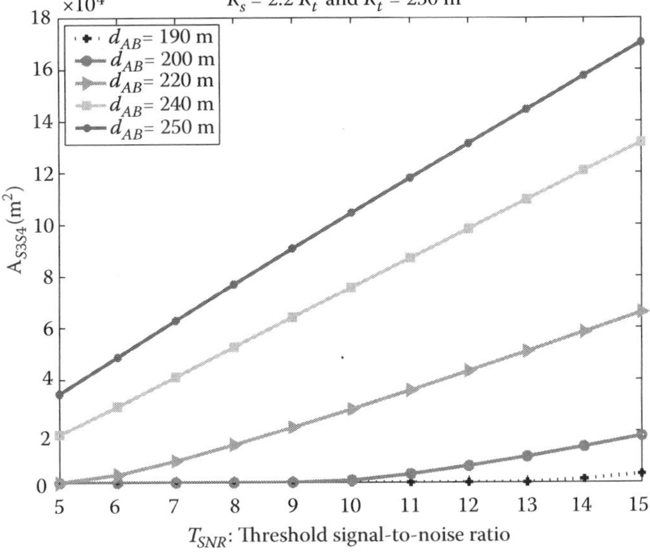

Figure 20.5 Hidden area A_{S3S4} versus T_{SNR} and distance d_{AB}.

between nodes A and B with a 550 *m* sensing range and a $R_i = \sqrt[4]{10}.d_{AB}$ interference. When the distance between both nodes is less than 200 *m*, $A_{S3S4} = 0$; that means that region A_{S3S4} is covered by the sensing region of node B. In Figure 20.5b, we plot the monitor's vulnerable hidden region versus the threshold signal-to-noise ratio in order to study the impact of the sensitivity of a signal on the monitoring process, particularly on the vulnerable region A_{S3S4}. We note that when $d_{AB} = 200$ *m* and $T_{SNR} = 10$, the vulnerable region $A_{S3S4} = 0$. However, $A_{S3S4} \neq 0$ when $T_{SNR} > 10$. Therefore, when the power of a signal increases, the interference range becomes greater, and region A_{S3S4} is less covered by the sensing region of node B ($A_{S3S4} \neq 0$). We can deduce that the power of a signal has an important impact on the monitoring process. The trade-off between T_{SNR} and the distance between the monitor and monitored nodes can significantly improve the monitoring process.

20.4.5 Energy Consumption by the Monitoring Mechanism

The monitoring mechanism is based on the overhearing process, which consists of receiving packets of neighbor nodes (even if these packets are not destined to the monitoring node). Unlike the classical wireless local area network (WLAN) or MANETs, the energy consumption of the packets reception is higher than the energy consumption of the packet transmission in WSNs [27]. This energy constraint and the vulnerabilities related to the hidden node problem (inaccurate observation of the monitor node) make the design of an energy-efficient monitoring mechanism a real challenge. In the IEEE 802.15.4 standard, they propose the beacon-enabled mode in order to reduce and optimize the energy consumption and another mode called non–beacon-enabled. In the next section, we discuss the introduction of the monitoring mechanism in both modes: beacon-enabled and non–beacon-enabled.

20.5 Solutions and Analysis

In this section, we present an analytical model for a monitoring mechanism in both modes of IEEE 802.15.4: non–beacon-enabled and beacon-enabled. We focus on the monitoring of the forwarding or the relaying activity in order to detect a network anomaly. This section is organized as follows: The network model and assumptions are presented in the first subsection. The second and third subsections are dedicated to both the non–beacon-enabled mode and beacon-enabled mode, respectively.

20.5.1 Network Model

We assume that nodes are distributed within a topology, which is a two-dimensional Poisson process with parameter λ (memory-less property of Poisson distribution). We use the two-ray ground propagation model with a threshold of the signal-to-noise ratio (T_{SNR}) set to 10 *db*. Then, the interference range is $R_i = \sqrt[k]{10} * d = 1.78 * d$, where $k = 4$ [23]. Moreover, all nodes have the same transmission range (R_t) and the same carrier sensing range (R_s). This means that the nodes within a circle of radius R_t centered at the transmitter may be able to receive packets correctly. Furthermore, the average number of nodes within a sensing range, an interference range, and a transmission range with a radius R_s, R_i, and R_t, respectively is $N_j \approx \lambda \pi R_j^2$, where $j = \{s,i,t\}$ [28].

With Poisson's parameter, the probability to get k nodes in the area $TR_{BA}(d_{AB})$ is noted by $p(k,TR_{BA}(d_{AB}))$ and obtained by

$$p(k,TR_{BA}(d_{AB})) = e^{-\lambda TR_{BA}(d_{AB})}\frac{(\lambda TR_{BA}(d_{AB}))^k}{k!} \tag{20.1}$$

The computation is the same for $CS_{BA}(d_{AB})$ and $IR_{BA}(d_{AB})$ because it is proportional to the distance d_{AB}.

20.5.2 In the Case of Non–Beacon-Enabled

In this subsection, we consider the monitoring mechanism for the case of the non–beacon-enabled mode in IEEE 802.15.4. In this mode, an unslotted CSMA/CA channel access mechanism is used [22].

According to the previous section, a disturbing event in the monitoring process occurs when a monitored node successfully transmits, and at least one node in the interference region of the monitor node and out of the sensing region of the monitored node transmits at the same time. In other words, the monitor node can correctly monitor its neighbors, if both following conditions are met:

1. The monitored node transmits a packet to another node of its neighbors successfully.
2. The monitor node can hear the monitored node's transmission correctly. No node in the interference region of the monitor node and out of the sensing region of the monitored node ($A_{S3S4}(d_{AB})$) transmits.

The monitor node needs to estimate the probability that both conditions 1 and 2 are met in order to calculate the probability that it correctly observes the monitored node when it transmits. This probability is noted P_w.

$$P_w = P\{\text{condition } 1\}\ P\{\text{condition } 2\} \tag{20.2}$$

20.5.2.1 Probability of Condition 1

The probability that condition 1 is met is defined as a successful packet's transmission of the monitored node, which is noted P_{succ} in our model. It shows the probability that a monitored node has access to the channel in order to transmit a packet coming from the monitor node. This probability may give us information about the ability of the monitored nodes to transmit packets. In order to calculate P_{succ}, we need to calculate the probability τ that a node transmits in a random time slot. τ must take into account two traffic scenarios: the saturated and unsaturated cases. For the saturated case, the traffic is intensive, which means that nodes always have a packet to transmit. For the unsaturated case, the node transmission depends on the probability that a node has a packet to transmit q. In order to calculate τ according to a probability q and a collision probability p. Much research has been carried out in order to calculate τ with the assumption that $R_t = R_i = R_s$ in the case of unslotted CSMA/CA (non beacon-enabled mode) [29–31]. This research is based on Bianchi's model for IEEE 802.11 [32] extended and improved to IEEE 802.15.4. However, most of them do not take into account the difference between the transmission, the interference, and the carrier sense ranges.

The collision probability p, necessary to compute τ, is the probability that at least one node in the interference region (R_i) transmits in the same time slot as the transmitter node. Thus, p can be expressed as follows:

$$p = 1 - (1-\tau)^{N_i - 1} \tag{20.3}$$

where N_i is the number of nodes in the interference range and, according to assumption, N_i can be calculated as follows: $N_i \simeq \lambda\pi R_i^2 = \lambda\pi\sqrt{T_{SNR}}\,(d_{AB})^2$.

The modeling of probability q is based on the traffic load, which is characterized by parameter λ^*, which represents the rate of packets that arrive at the node's buffer, and it is measured in packets per second (pkt/s). As in [30], we assume that the packet's arrival process is Poisson. So, probability q can be well estimated with a small buffer size as follows:

$$q = 1 - \exp^{\lambda^*.T_{av}} \tag{20.4}$$

where T_{av} is the expected time per slot and is calculated according to P_{tr}, P_{succ}, the successful transmission slot (T_s), and the collision slot time (T_c) as follows:

$$T_{av} = (1 - P_{tr})\,\sigma + P_{tr}\,(1 - P_{succ})\,T_c + P_{tr}P_{succ}T_s$$

where T_c and T_s are the average times a channel is sensed busy because of collisions and successful data frame transmissions.

20.5.2.2 Probability of Condition 2

The probability that condition 2 is met gives us a piece of information about the observation's disruption of a monitor node. This probability equals to one when no node in region $A_{S3S4}(d_{AB})$ transmits ($P\{cond.2\}$) in a vulnerable time. This period depends on the transmission time T_{av} of a packet: When a node B starts to transmit at t_s, the vulnerable time interval is $[t_s - T_{av} - 1, t_s + T_{av} - 1]$. The nodes in region $A_{S3S4}(d_{AB})$ must remain silent during μ slots time where $\mu = (T_{av}/\sigma)$. If any node in the region $A_{S3S4}(d_{AB})$ transmits during the vulnerable time interval, a packet cannot be received correctly by the monitor node. Region $A_{S3S4}(d_{AB})$ can be equal to zero when it is covered by the carrier sense of a node B. Otherwise, $P\{cond.2\}(d_{AB})$ (if $d_{AB} > \dfrac{R_s}{1+\sqrt{[k]T_{SNR}}}$) is given by

$$\begin{aligned} P\{cond.2\}(d_{AB}) &= \left(\sum_{k=0}^{\infty}(1-\tau)^k \frac{(N_h)^k}{k!} e^{-N_h \cdot \mu}\right) \\ &= e^{-\tau N_h \cdot \mu} \end{aligned}$$

where $N_h = \lambda\, A_{S3S4}(d_{AB})$.

The final equation of $P\{cond.2\}(d_{AB})$ is obtained by

$$P\{cond.2\}(d_{AB}) = \begin{cases} 1 & \text{if } d_{AB} \leq \varphi \\ e^{-\tau N_h \cdot \mu} & \text{Otherwise} \end{cases}$$

where

$$\phi = \frac{R_s}{1+\sqrt{[k]T_{SNR}}}$$

In order to calculate $A_{S3S4}(d_{AB})$, we calculate the intersection area between the sensing region and the interference region of two nodes X and Y, assuming that the distance between them is *d*.

$$Ar_{\{X \cap Y\}}(d) = R_s\left(arccos(\alpha) - \alpha\sqrt{1-\alpha^2}\right) + R_i\left(arccos(\beta) - \beta\sqrt{1-\beta^2}\right)$$

where

$$\alpha = \frac{R_s^2 - R_i^2 + d^2}{2dR_s} \text{ and } \beta = \frac{R_i^2 - R_s^2 + d^2}{2dR_i}$$

Therefore, $A_{S3S4}(d_{AB})$ is expressed by

$$A_{S3S4}(d_{AB}) = \begin{cases} 0 & \text{if } d_{AB} \le \varphi \\ \pi R_s^2 - Ar_{\{A \cap B\}}(d_{AB}) & \text{Otherwise} \end{cases} \quad (20.5)$$

Now, from equations 20.2 and 20.5, we can calculate $P_w(d_{AB})$ as follows:

$$P_w(d_{AB}) = \begin{cases} P_{succ} & \text{if } d_{AB} \le \varphi \\ P_{succ}.e^{-\tau N_b.\mu} & \text{Otherwise} \end{cases} \quad (20.6)$$

The main drawback of the monitoring process in non–beacon-enabled mode (or unslotted CSMA/CA) is the energy consumption because it's permanently activated.

20.5.3 In the Case of Beacon-Enabled

In this subsection, we study the case of the monitoring process with the beacon-enabled mode. In addition, we describe and analyze our recent monitoring mechanism called *muDog* [33].

Unlike the case of the non–beacon-enabled mode, the monitoring process in beacon-enabled IEEE 802.15.4 is activated only in the CAP (contention access period). That means that the monitor node wakes up to receive the beacon frame and to track the packets transmitted by the monitored node in the CAP duration in order to evaluate the cooperative metric of this node. In other words, the aim of muDog is to reduce the time of overhearing and then the energy consumption by launching targeted monitoring.

We explain the monitoring process of muDog by an example illustrated in Figure 20.6. In this scenario, we have two connections, $\{S_1,R_1\}$ and $\{S_2,R_2\}$, and a coordinator node C; its task is to route a packet from S_1 to R_1 and from S_2 to R_1. Node S_1 plays the role of monitor node, and its goal is to monitor the coordinator node C. Node C is classified as a well-behaving node if it forwards all packets to nodes R_1 and R_2; otherwise, it is classified as a selfish node. Only nodes S_1 and S_2 can act as monitor nodes, and they target the transmission of node C by the overhearing. The flowchart presented in Figure 20.7 describes the muDog mechanism run by the monitor node.

Node S_1 sends a packet Pkt_i to node R_1 as illustrated in Figure 20.6. When Pkt_i is received by node C, it transmits it to the receiver R_1 under the S_1 monitoring. Two kinds of transmission are possible with and without acknowledgement. Both transmissions are supported by our proposed monitoring mechanism. Let us suppose that we have a transmission with acknowledgement. Node S_1 is able to check the good reception of Pkt_i by coordinator C when it receives the ACK packet. This reception validates the first step of the monitoring process. Node S_1 acts as a monitor node only if both conditions of step 2 in Figure 20.7 are met: $LQI > LQI_{threshold}$ and $Pw > Pw_{threshold}$ with LQI (link quality indicator), and Pw is the probability of having a monitor node's accurate observation (see the following for more details). This step ensures the quality of the monitoring process.

We know that in the beacon-enabled mode the coordinator periodically sends a beacon frame to its neighbors in order to ensure the synchronization and to give them information related to the next transmission. The monitor node focuses on two important subfields in the beacon frame: Frame pending (it is set to one if the coordinator has a packet to transmit) and pending address fields (an address list of nodes to which the coordinator has packets to transmit). When these parameters are checked by S_1, that means that *frame pending subfield* = 1 and $@R_1 \in$ {*pending address fields*}, then the monitoring process is launched (see step 3). After this step, node S_1 starts to overhear the packets sent by node C in order to check if the coordinator correctly forwards the packets, particularly Pkt_i.

The challenge is that monitor node S_1 cannot know when the targeted packet Pkt_i will be transmitted during the overhearing period. We know that the overhearing period has an important impact on energy consumption; that is why the proposed mechanism muDog tries to optimize the overhearing period while ensuring the quality of the observation. The maximum time of overhearing does not exceed the time dedicated to CAP ($T_{overhearing\ max} = T_{CAP}$).

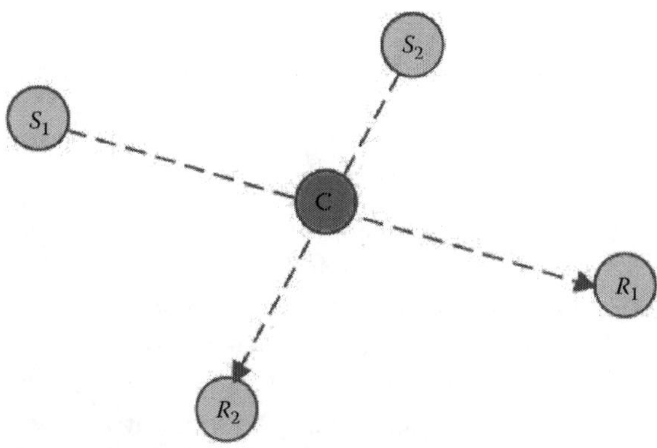

Figure 20.6 Scenario of monitoring mechanism.

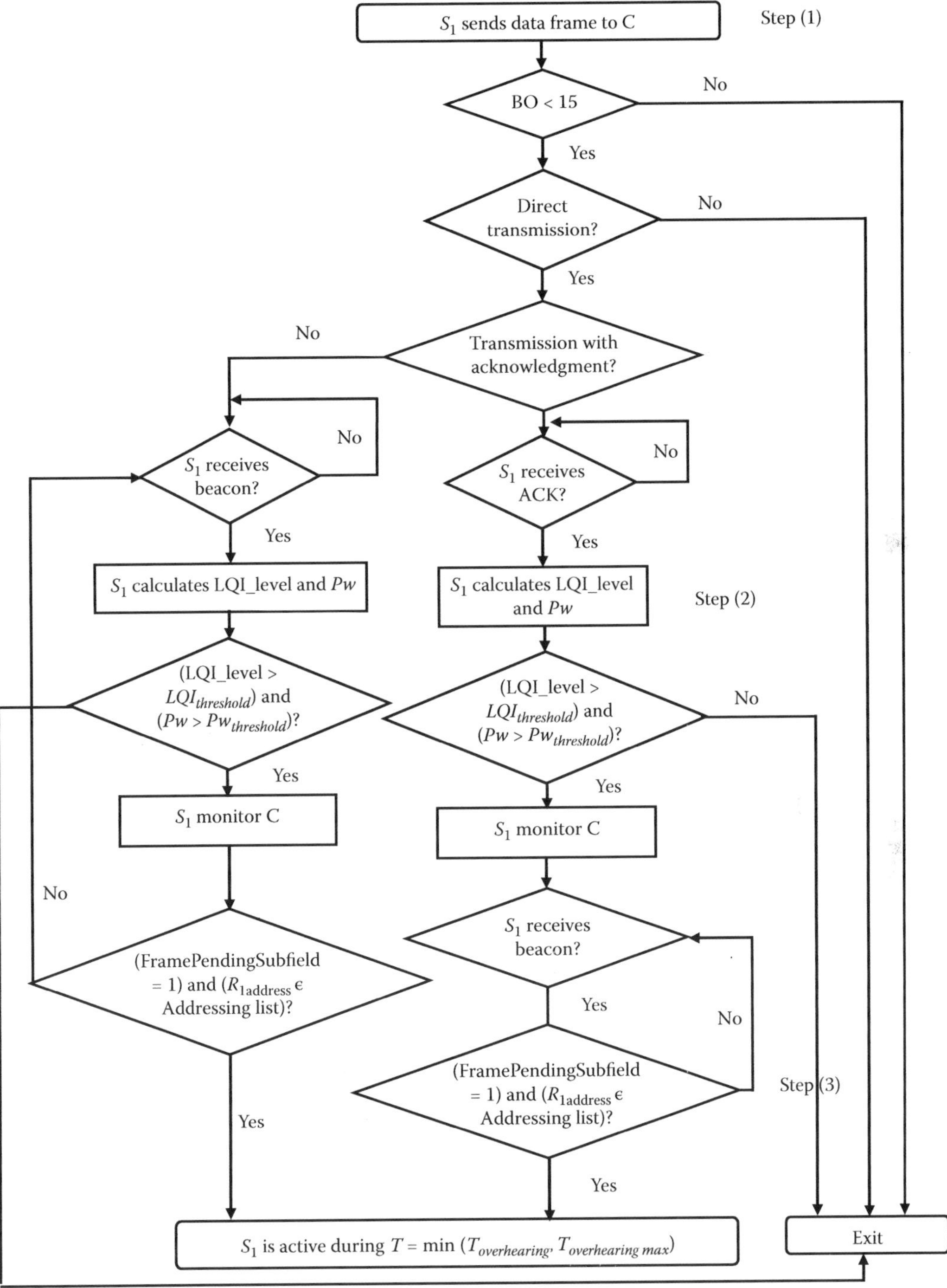

Figure 20.7 Global flowchart of muDog.

20.5.3.1 Evaluation of the Monitor's Overhearing Time

In order to evaluate the overhearing time of the monitor node, we distinguish two kinds of transmission: with and without acknowledgement. In the case of transmission with acknowledgement, we present two situations: the optimistic and the realistic. The optimistic situation is when the coordinator sends the first monitored packet Pkt_i to the R_1, which means the monitor node S_1 can switch off the monitoring process just after the overhearing of Pkt_i. In this situation, the overhearing time is equal to the one-packet transaction time $\left(T_{transac}^{1_{Pkt}}\right)$, and it is given by the following equation:

$$T_{overhearing} = T_{transac}^{1_{Pkt}} = T_{backoff}(cw_j) + T_{DataReq} + T_{Data} + 2*\gamma + 2*\omega + 2*T_{Ack} + 2*T_{IFS} \tag{20.7}$$

where

- $T_{backoff}$ is the back-off time calculated according to CSMA/CA [22]. $T_{backoff} = BOslots * TBOslots$ with *BOslots* is the random slot number selected from $[0, 2^{BE} - 1]$, and *TBOslots* is the slot time duration of the back off. *BE* is increased at each retransmission ($BE_{min} = 3$ and $BE_{max} = 5$).
- T_{Data} is the time needed to transmit the data frame (Mac payload field), and its size cannot exceed *aMaxMACPayloadSize*. T_{Data} is calculated as follows:

$$T_{Data} = \frac{8*(x + LMAC\_H + LPHY + Ladd + LMAC\_F)}{dataRate} \tag{20.8}$$

  where x is the size of the data frame. LMAC_H and LPHY are the size of the MAC and physical headers, respectively. Ladd is the size of the MAC address. LMAC_F is the size of the MAC queue (MAC footer MFR). dataRate is the binary rate in IEEE 802.15.4.
- $T_{DataReq}$ is the transmission duration of the data command frame.
- γ is the turnaround time and presents the duration between the reception of the data frame and the transmission of the ACK packet.
- ω is *macAckWaitDuration* and consists of the maximum duration necessary to receive the ACK packet after the data frame transmission. T_{Ack} is the transmission time of the ACK packet. T_{IFS} (IFS: interframe space) is the time needed to separate two consecutive data frames.

The pessimistic or realistic situation is when the monitored packet Pkt_i transmitted after a certain number of packets with different source addresses (as S_2 in the example of Figure 20.6). *NP* is the number of packets transmitted before Pkt_i, then the time of overhearing will be crossed by $NP + 1$. In addition, the packet retransmission for any reason must be taken into account, and it is limited to only three attempts. Therefore, the overhearing time in this case is calculated as follows:

$$T_{overhearing} = (NP+1) \times \left(\sum_{i=0}^{2} P_i \times \sum_{j=1}^{i+1} T_{transac}^{1_{Pkt}}(cw_j) \right) \tag{20.9}$$

where P_i is the probability to reach i^{th} attempts, and it is defined as follows:

$$P_i = \begin{cases} P_{succ}(1-P_{succ})^i & \text{if} \quad i=\{0,1\} \\ (1-P_{succ})^2 & \text{if} \quad i=2 \end{cases}$$

We based on Pollin et al. [34] a Markov model to evaluate the probability P_{succ} in IEEE 802.15.4.

$$P_{succ} = N\phi\,(1-\phi)^{N-1}(1-\alpha)(1-\beta) \tag{20.10}$$

where N is the number of nodes in the network, and ϕ is the stationary probability of the node when it attempts CCA (clear channel assessment) for the first time during one slot. α and β are the probabilities of sensing the channel busy for the first and the second CCA (in IEEE 802.15.4, the transmitter must sense the channel twice by using CCA). In order to calculate the probability ϕ, we distinguish two cases: the saturated and unsaturated network.

In the case of the saturated network, the node always has a packet to transmit. ϕ_{ACK} is calculated in the case of a transmission with acknowledgment as follows:

$$\phi_{ACK} = 1 - \left(1 - \frac{\beta_{ACK}}{(1-\beta_{ACK})(2-P_{col})}\right)^{1/N} \tag{20.11}$$

where P_{col} is the probability of collision during the transmission period, and it is calculated as follows:

$$P_{col} = 1 - \frac{N\phi_{ACK}(1-\phi_{ACK})^{N-1}}{1-(1-\phi_{ACK})^N} \tag{20.12}$$

In the case of the unsaturated network, the node does not always have a packet to transmit. We use the same model developed in [34], which consists of adding the delays X_1, X_2, and X_3. X_1 is the delay added after each transmission attempt when the channel is sensed busy. X_2 is the delay added before the next periodic transmission in the case of a transmission failure. X_3 is the delay added after the next transmission in the case of a successful first transmission. So, this model is valid only in the case of acknowledgement transmission because in the other transmission mode it is not possible for the sender to be sure that the packet is correctly received.

The probability that a collision (P_c) occurs because two nodes transmit at the same time is calculated as follows:

$$P_c = 1 - (1-\phi)^{N-1} \tag{20.13}$$

For more details you can refer to the work by Pollin et al. [34].

The number of packets sent for each pending address is not specified in the IEEE 802.15.4 standard, which makes the evaluation of NP number of packets to send before the monitored packet Pkt_i not easy. Then, we evaluate the NP_{min} and the NP_{max} in order to calculate the average number of NP (NP_{moy}). However, in the IEEE 802.15.4 standard, the number of addresses pending cannot exceed seven; then the NP_{min} = 6 (one packet to transmit for each address). In order to calculate the NP_{max}, we evaluate the maximum packet number possible to transmit during one CAP and we obtain

$$NP_{max} = \frac{T_{CAP}}{T_{transac}^{1Pkt}} \tag{20.14}$$

where $T_{transac}^{1Pkt}$ is the time transaction for one packet with a minimum back-off value.

In the case of the transmission without acknowledgement, that means that the *turnaround time* and the *macAckWaitDuration* are required. Therefore, the equation to calculate the overhearing time in the optimal case is

$$T_{overhearing} = T_{transac}^{1Pkt} = T_{backoff}(cw_j) + T_{DataReq} + T_{Data} + 2 * T_{IFS} \tag{20.15}$$

In the realistic case, the equations 20.7 and 20.11 become as follows:

$$T_{overhearing}^{noAck} = NP_{moy}\left(P_0 \times T_{transac}^{noAck}(cw_j)\right) \tag{20.16}$$

20.5.3.2 Simulation Results of Overhearing Time

Two cases have to be distinguished: the case of a saturated network and the case of a nonsaturated network. The saturated case is divided into two cases: the acknowledged and nonacknowledged cases.

In order to determine the variation interval of NP_{moy}, we have evaluated NP_{moy} variation according to the x packets size in both cases: with and without any acknowledgement. This evaluation enables us to deduce the average number of packets that may be sent during CAP.

Figure 20.8 shows that the overhearing time increases proportionally to the number of packets preceding the monitored packet, and the greatest values are observed in the saturated case with acknowledgement because the packets are more often generated and the transactions are longer than in the nonsaturated case. For each acknowledged transmission, the overhearing time remains inferior to the overhearing time in Watchdog. This is true for $NP_{moy} < 22$ packets in a saturated case and $NP_{moy} < 26$ packets in a nonsaturated case. For instance, when $NP_{moy} = 20$, the overhearing time is reduced by 9% in a saturated case and by 30% in a nonsaturated network. muDog is thus better performing than Watchdog when the queue contains less than 22 packets in a saturated network and less than 26 packets in a nonsaturated network. In the case of a transmission without any acknowledgement, muDog is much better performing. Indeed, even when the queue contains the maximum number of packets (31), the overhearing time is reduced by 63%. Figure 20.9 shows the overhearing time according to the size of the packets in six different situations. All these cases are then compared to the Watchdog. The overhearing time significantly decreases in the case of a nonsaturated network compared to a saturated network although we introduced a delay $X_1 = 100$ slots. Indeed, in the case of a nonsaturated network, the packets are less often generated, and this reduces the probability ϕ. In the case of a nonacknowledged transmission, the transaction duration is minimized, and this significantly reduces the overhearing time. muDog is quite performing in the cases of a nonsaturated network and of a nonacknowledged transmission; it prevents

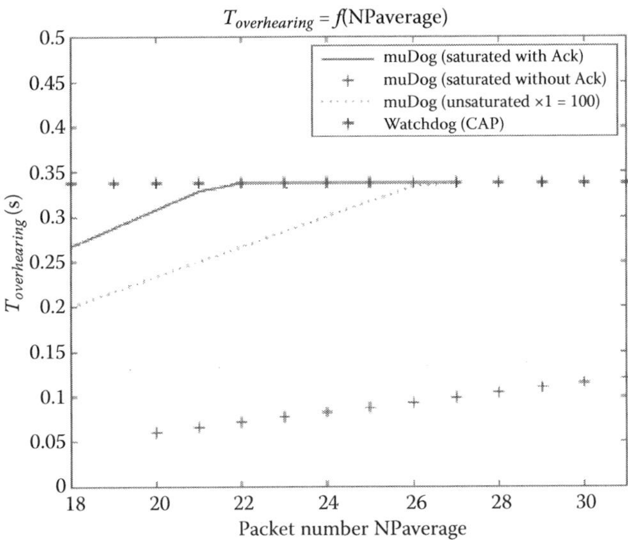

Figure 20.8 $T_{overhearing}$ **according to average number of packets (**NP_{moy}**).**

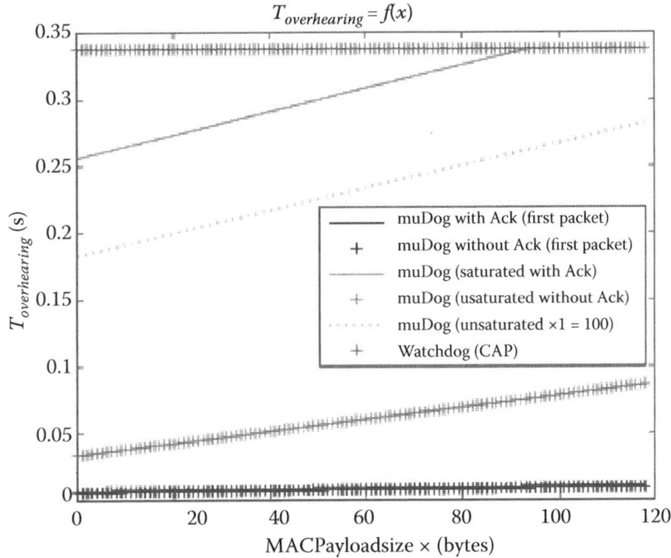

Figure 20.9 $T_{overhearing}$ **according to size of MACPayload.**

the nodes from reaching Watchdog overhearing time for any packet size. With a value of $x = 118$ (maximum value), the overhearing time is reduced by 19% in a nonsaturated network and by 72% in the case of a nonacknowledged transmission. However, in the case of a saturated network and with a great number of permanently generated packets, muDog is more efficient than Watchdog only for $x < 92$ bytes. muDog thus reduces the activity period of the monitor by 20%, which will consequently reduce the energy cost.

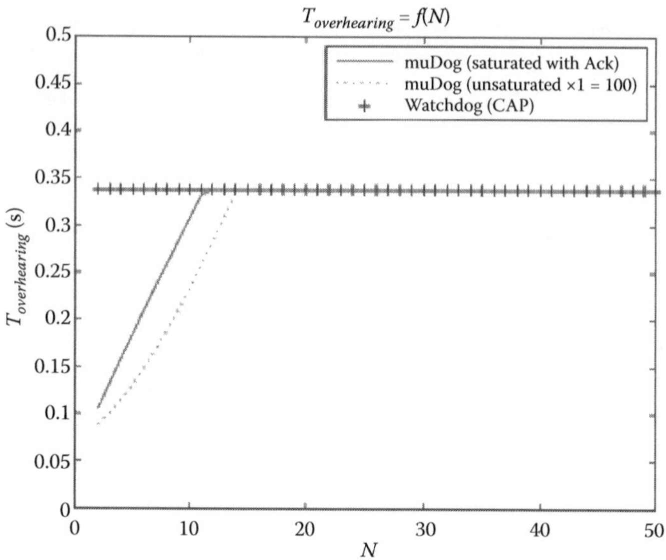

Figure 20.10 $T_{overhearing}$ **according to node density.**

In order to study the impact of node density on the overhearing time, we varied the node density N from 2 to 50 nodes in the network, and the obtained results are plotted in Figure 20.10. The overhearing time obviously increases proportionally to the node density. However, muDog is much better performing in small, saturated networks with less than 12 nodes (15 nodes in a non-saturated network) than Watchdog. For a network composed of five nodes, the overhearing time is reduced by more than 50% (0.15 s) of the time given for Watchdog. In the case of great networks, it is important to have a great overhearing time (equal to CAP) in order to monitor a great number of nodes and thus a more intense traffic.

20.5.3.3 *Evaluation of the Probability of Having a Monitor's Accurate Observation* (Pw)

As in the case of the non–beacon-enabled mode, we quote two main conditions to define the accurate observation of the monitor node: The monitored node successfully transmits the target packet ($P_{condition1}$), and no collision occurs at the monitor node during the monitored node transmission ($P_{condition2}$).

$P_{condition1}$ is calculated according to Equation 20.10 proposed in the previous subsection. However, $P_{condition2}$ is calculated by the same way as the case of non–beacon-enabled mode.

Now, we can calculate $Pw(d)$ as follows:

$$Pw(d) = \begin{cases} P_{succ} & \text{if } d \leq \varphi \\ P_{succ} . e^{-\tau N_b . \mu} & \text{Otherwise} \end{cases} \tag{20.17}$$

20.5.3.4 *Simulation Results of the Probability of Having an Accurate Observation* Pw

We evaluate the impact of the traffic charge, node density, and distance between the monitoring and monitored nodes on the probability of making an accurate observation *Pw*. In order to study the impact of the node density on the monitoring process, we have evaluated *Pw* variation according to the node density for two network traffic loads 5 *pps* and 15 *pps*. The obtained results are plotted in Figure 20.11a and b. In both cases, the probability of having an accurate

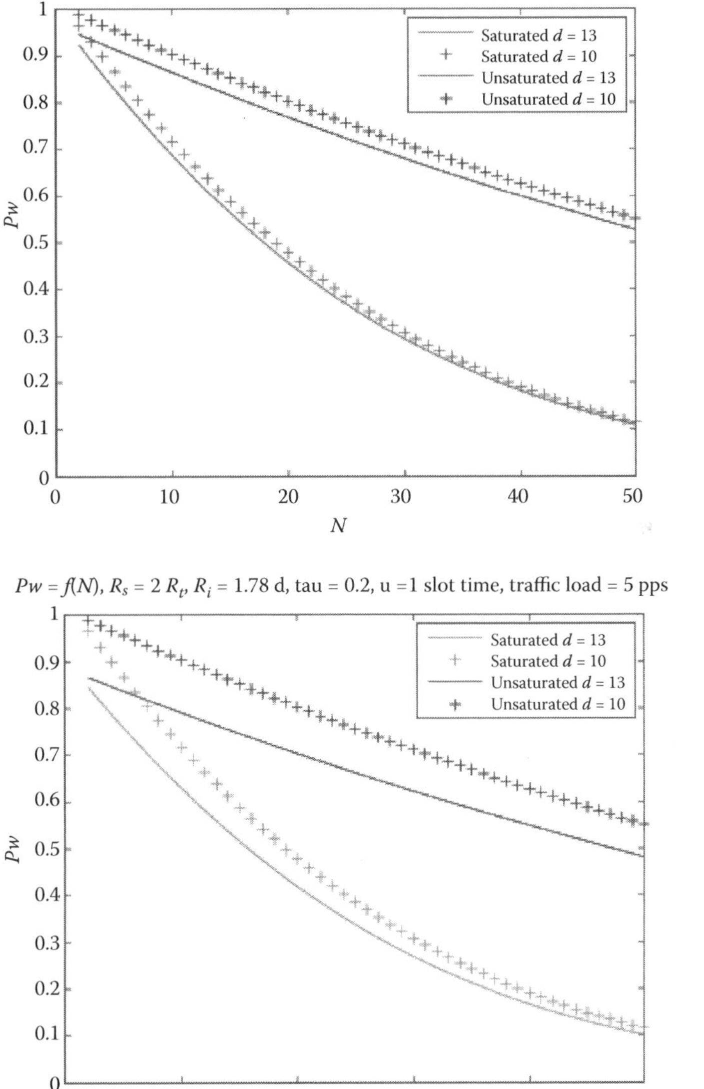

Figure 20.11 ***Pw*** **versus node density with 5 and 15 pps of traffic charge.**

observation significantly decreases when the node density increases. Indeed, when the number of nodes increases, the probability of having a successful transmission decreases (because of the collisions), and thus the probability of having an accurate observation also decreases. The decrease is even more important in the case of a saturated network. The difference between both cases may reach 80%. The best values of *Pw* are reached when the node density is small (between 2 and 10 nodes) and for short distances between the monitoring and monitored nodes (10 *m*). In such cases, they vary between 0.2 and 0.98. When the distance and the traffic charge increase, *Pw* decreases more. For instance, for $N = 10$, 15 *pps*, and $d = 13$ *m*, the probability decreases by 11% in a non-saturated network and by 13% in a saturated network. The worst situation occurs with a saturated network when the node density is high (50 nodes) and when the traffic is quite important (15 *pps*). *Pw* is then equal to 0.1.

20.5.4 Case of Reputation Metric Assessment with the Monitoring Mechanism

The reputation report of monitored node B generated by monitor node A is given by

$$R_{A,B}(d_{AB}) = \eta \cdot P_w(d_{AB}) \tag{20.18}$$

where η is the forwarding ratio assessed by a monitor node. Unlike the Watchdog mechanism, the forwarding ratio in our model is the number of packets observed by the monitor node divided by the total number of packets sent by a monitor node and well received by a monitored node (see Equation 20.19). However, in Watchdog, the denominator is the total number of packets sent to the monitored node. This difference has an important impact on the monitoring process because the total number of packets sent to the monitored node is not automatically well received when the monitor node only focuses on the routing layer. In order to avoid the vulnerability of Equation 20.19 and to correctly evaluate the total number of packets (subject of the monitoring) well received by the monitored node, we consider the cross-layer approach. This approach allows us to correctly evaluate the forwarding ratio η at the routing layer by considering the ACK packet at the MAC layer. The monitor node needs this cross-layer information to be sure that the monitored node has correctly received the packet at the routing layer. Even if there is a variation of rate, the probability that a monitored node correctly receives the packet from a monitor node enables the monitor node to evaluate the traffic in the interference range of the monitored node.

$$\eta = \left(\frac{\#\,\text{forwarded packets observed}}{\#\,\text{total of packets well received by a monitored}} \right) \tag{20.19}$$

Once monitor node A has calculated the reputation report R_{AB} of node B, it can predict the number of packets likely to be forwarded next time by node B. If the difference between the predictable number and the observed number is great, the monitor node can deduce that the monitored node has changed its behavior or that it does not have the same environment as in the first evaluation. In this situation, the monitor node needs to update the evaluation of the monitored node. The predictable number of packets forwarded by any monitored node takes the three following monitoring conditions into account: node density, node mobility, and traffic load. It can be computed by the following equation:

$$\# \text{ Forwarded packets} = \frac{(\#\text{ total sent packets}).\mathrm{R}^{*}_{(\mathrm{AB})}}{P_w} \qquad (20.20)$$

where $R^{*}_{(AB)}$ is the previous evaluation report.

The goal of Equation 20.20 is to analyze the node's behavior in the same condition as in the previous P_w evaluation. P_w depends on $P\{condition1\}$ and $P\{condition2\}$. $P\{condition1\}$ takes into account the traffic variation through the probability that a node has a packet to transmit (q) defined in Equation 20.4. However, $P\{condition2\}$ takes into account the node's density in the monitor vulnerable region called $As3s4(d)$ (illustrated in Figure 20.3b). So, this probability depends on the distance between the monitor and the monitored nodes (d). Therefore, the same condition for P_w calculation means that the estimation is done with the same probability q (traffic load) and the same distance between the monitor and the monitored nodes (d). In addition, in our proposed model, P_w is dynamically estimated for each packet sent to the monitored node (subject of monitoring). Hence, Equation 20.20 allows discussion about the node's behavior in the same condition as in the previous P_w calculation.

With our model, we can give the answer to the question: Does the monitored node intentionally refuse to cooperate, or is it unable to cooperate even if it wants to do so? That means the monitoring process is improved and the false positive is reduced. This enhancement has a positive impact on the trust model because most of the trust models used to evaluate the behavior of nodes are based on the reputation parameter.

20.6 Conclusion

In this chapter, we present and analyze the challenges to introduce an efficient monitoring mechanism for WSNs. The problem with hidden regions in the monitoring mechanism, particularly *monitored and monitor's vulnerable hidden region*, is presented and discussed. In addition, we propose two solutions based on analytical models in both cases: non–beacon-enabled and beacon-enabled modes. The constraints of WSNs, like the energy limitation, are taken into account, particularly in the case of the beacon-enabled mode with our proposed monitoring mechanism called muDog. An evaluation of our monitoring mechanism is proposed and compared to the existing Watchdog mechanism by using different parameters and metrics. The evaluations have shown that the distance between the monitor and monitored nodes and the network traffic load have a negative impact on the monitoring process. However, the nodes' density and the type of network (saturated or nonsaturated) seem to have a mostly negative impact on the observation accuracy. Moreover, the monitor node's accurate observation is thus possible in a small nonsaturated network with a short distance between the monitor and the monitored nodes. The obtained results illustrate that muDog is more efficient than Watchdog whatever the parameter is.

References

1. I. Boulanouar, A. Rachedi, S. Lohier, G. Roussel, Energy-aware object tracking algorithm using heterogeneous wireless sensor networks. In *The 4th IFIP/IEEE Wireless Days 2011 (IEEE WD2011)*, Niagara Falls, Ontario, Canada, October 10–12, 2011.
2. Z. Sun, P. Wang, M. C. Vuran, M. A. Al-Rodhaan, A. M. Al-Dhelaan, I. F. Akyildiz, BorderSense: Border patrol through advanced wireless sensor networks, *Ad Hoc Networks Journal*, vol. 9, no. 3, pp. 468–477, 2011.

3. R. Cucchiara, Multimedia surveillance systems, In *Proceedings of the Third ACM International Workshop on Video Surveillance and Sensor Networks (VSSN'05)*, 2005. New York, August 1–2, 2005.
4. R. Roman, J. Zhou, J. Lopez, Applying intrusion detection systems to wireless sensor networks. In *Proceedings of the IEEE Consumer Communications and Networking Conference (CCNC'06)*, pp. 640–644, 2006. Las Vegas, January 2006.
5. T. H. Hai, K. Hee, F. Khan, K. Hee, E.-N. Huh, K. Hee, Hybrid intrusion detection system for wireless sensor networks, In *Proceedings of the International Conference on Computational Science and Its Applications (ICCSA'07)*, pp. 383–396, 2007. San Francisco, 24–26 October 2007.
6. S. Marti, T. J. Giuli, K. Lai, M. Baker, Mitigating routing misbehavior in mobile ad hoc networks, In *Proceedings of the 6th Annual ACM/IEEE International Conference on Mobile Computing and Networking*, Boston, USA, pp. 255–265, 2000.
7. L. Guang, C. Assi, A. Benslimane, On MAC layer misbehavior in wireless networks: challenges and solutions, *IEEE Wireless Communication Magazine*, vol. 15, no. 4, pp. 6–14, 2008.
8. Y. Zhang and W. Lee, Intrusion detection in wireless ad-hoc networks, In *Proceedings of the ACM MobiCom'2000*, pp. 275–283, 2000. Boston, August 6–11, 2000.
9. C. Tseng, P. Balasubramanyam, C. Ko, R. Limprasittiporn, J. Rowe, K. Levitt, A specification-based intrusion detection system for AODV, In *Proceedings of the 1st ACM Workshop on Security of Ad Hoc and Sensor Networks*, pp. 125–134, 2003. Fairfax, Virginia.
10. Y. Huang and W. Lee, A cooperative intrusion detection system for ad hoc networks, In *Proceedings of the 1st ACM Workshop on Security of Ad Hoc and Sensor Networks*, pp. 135–147, 2003. Fairfax, Virginia.
11. A. H. Farooqi and F. A. Khan, Intrusion detection systems for wireless sensor networks: A survey, *Communications in Computer and Information Science*, vol. 56, pp. 234–241, 2009.
12. A. D. Wood, J. A. Stankovic, G. Zhou, DEEJAM: Defeating energy-efficient jamming in IEEE 802.15.4-based wireless networks, In *The 4th Annual IEEE Communications Society Conference on Sensor, Mesh and Ad Hoc Communications and Networks (SECON'07)*, pp. 60–69, 2007. San Diego, California.
13. Y. Sun, X. Wang, X. Zhou, Jamming attack in wireless sensor network: From time to space, *IEICE Transactions on Communications E95-B*, no. 6, pp. 2113–2116, 2012.
14. P. Kyasanur and N. H. Vaidya, Detection and handling of MAC layer misbehavior in wireless networks, in *Proceedings of IEEE DSN03*, pp. 173–182, Jun. 2003. San Francisco, CA.
15. M. Raya, I. Aad, J. Hubaux, and A. E. Fawal, DOMINO: Detecting MAC layer greedy behavior in IEEE 802.11 hotspots, *IEEE Transactions on Mobile Computing*, vol. 5, no. 12, Dec. 2006.
16. Z. Lu, W. Wang, C. Wang, On order gain of backoff misbehaving nodes in CSMA/CA-based wireless networks, In *Proceedings of IEEE INFOCOM*, pp. 1–9, 2010. San Diego, California.
17. R. Sokullu, O. Dagdeviren, I. Korkmaz, On the IEEE 802.15.4 MAC layer attacks: GTS attack, In *The 2nd International Conference on Sensor Technologies and Applications, (SENSORCOMM'08)*, pp. 673–678, 2008. Cap Esterel, France.
18. Y. Kwon and Y. Chae, Traffic adaptive IEEE 802.15.4 MAC for wireless sensor networks, Lecture Notes in Computer Science (LNCS), vol. 4096/2006, pp. 864–873, 2006.
19. G. Anastasi, M. Conti, M. Di Francesco, A. Passarella, Energy conservation in wireless sensor networks: A survey. *Ad Hoc Networks*, vol. 7, no. 3, pp. 537–568, 2009.
20. G. Lu, B. Krishnamachari, C. S. Raghavendra, Performance Evaluation of the IEEE802.15.4 MAC for low rate low power wireless network, In *Proceedings of the EWCN'04, Held in Conjunction With the IEEE IPCCC*, April 2004. Phoenix, AZ.
21. C. Suh, Z. Hameed Mir, Y. Ko Design and implementation of enhanced IEEE 802.15.4 for supporting multimedia service in wireless sensor networks. *Computer Networks Journal*, vol. 52, no. 13, pp. 2568–2581, 17 September 2008.
22. IEEE Standard 802.15.4, Part 15.4: Wireless medium access control and physical layer specification for low rate wireless personal area networks. IEEE Std. 802.15.4, December 2003.
23. K. Xu, M. Gerla, S. Bae, Effectiveness of RTS/CTS handshake in IEEE 802.11 based ad hoc networks, *Ad Hoc Networks*, vol. 1, no. 1, pp. 107–123, 2003.
24. UC Berkeley and USC ISI, The network simulator ns-2, part of the VINT project. Available from http://www.isi.edu/nsnam/ns, 1998.

25. The editors of IEEE 802.11, Wireless LAN medium access control (MAC) and physical layer (PHY) specification, 1997.
26. A. Rachedi and A. Benslimane, Toward a cross-layer monitoring process for mobile for mobile ad hoc networks. *Journal of Security and Communication Networks*, vol. 2, no. 4, pp. 351–368, July/August 2009.
27. Q. Wang, M. Hempstead, W. Yang, A realistic power consumption model for wireless sensor network devices, In *The 3rd Annual IEEE Communications Society on Sensor and Ad Hoc Communications and Networks*, (*SECON'06*), pp. 286–295, 2006. Reston, VA.
28. H. Takagi and L. Kleinrock, Optimal transmission ranges for randomly distributed packet radio terminals, *IEEE Transactions on Communications*, vol. Com-32, no. 3, 1984.
29. C. Burrati and R. Verdone, Performance analysis of IEEE 802.15.4 non beacon-enabled mode, *IEEE Transactions on Vehicular Technology*, vol. 58, no. 7, pp. 3480–3493, September 2009.
30. T. O. Kim, J. S. Park, H. J. Chong, K. J. Kim, B. D. Choi, Performance analysis of IEEE 802.15.4 non-beacon mode with the unslotted CSMA/CA, *IEEE Communications Letters*, vol. 12, no. 4, pp. 238–240, 2008.
31. B. Latré, P. D. Mil, I. Moerman, B. Dhoedt, P. Demeester, N. V. Dierdonck, Throughput and delay analysis of unslotted IEEE 802.15.4, *Journal of Networks*, vol. 1, no. 1, pp. 20–28, 2006.
32. G. Bianchi, Performance analysis of the IEEE 802.11 distributed coordination function, *IEEE Journal of Selected Area in Telecommunication, Wireless Series*, vol. 18, no. 3, pp. 535–547, 2000.
33. A. Rachedi, H. Baklouti, muDog: Smart monitoring mechanism for wireless sensor networks based on IEEE 802.15.4 MAC, In the *IEEE International Conference on Communications (ICC'2011)*, Kyoto, Japan, IEEE Press. 5–9 June 2011.
34. S. Pollin, M. Ergen, S. Coleri Ergen, B. Bougard, L. Van der Perre, I. Moerman, A. Bahai, P. Varaiya, F. Catthoor, Performance analysis of slotted carrier sense IEEE 802.15.4 medium access layer, *IEEE Transactions on Wireless Communications*, vol. 7, no. 9, September 2008.
35. A. Keshavarzian, H. Lee, L. Venkatraman, Wakeup scheduling in wireless sensor networks, In *Proceedings of the ACM Mobihoc 2006*, Florence, Italy, pp. 322–333, May 2006. Florence, Italy.
36. N. F. Timmons and W. G. Scanlon, Analysis of the performance of IEEE 802.15.4 for medical sensor body area networking sensor and ad hoc communications and networks, *IEEE SECON 2004*, pp. 16–24, 2004. Santa Clara, CA.

Chapter 21

Building and Orchestrating Centralized Remote Management Procedures for Wireless Sensor Networks

Michael N. Kalochristianakis
Technological Educational Institute of Crete

Evangelia Psilidou
University of Crete

Aristotelis Kretsis and Manos Varvarigos
University of Patras

Contents

21.1 Managing Wireless Sensor Networks

21.1.1 Introduction

The proliferation of wireless sensor network (WSN) technologies is challenging system administration practices because WSNs cannot be supported by standard solutions or established practices. Despite the potential value of managing remote production WSN installations, there are currently very few integrated management systems that support WSN infrastructures, besides standard workstations and smart devices. Traditional network management (NM) priorities, such as fault, configuration, security, performance, accounting management, etc., need to be complemented by additional ones, such as energy conservation, code delivery, software updates, and topology consciousness in the case of WSNs. Most commercial infrastructure management (IM) systems use remote client or agent modules that convey NM and systems management (SM) functionality. Agents may vary in terms of operation, functionality or presentation; they may run as user or system processes, support flexible or fixed types of installation and complex or simple configurations, and expose extensive or minimal graphical interfaces to users. IM agents typically exploit a family of management protocols by the distributed management task force (DMTF) and the internet engineering task force (IETF), including WBEM [1], CIM [2], WMI [3], SNMP, and others. The overall complexity and volume of such technologies is typically beyond the capabilities of WSN terminal stations, and, unless management standards are established for them, most IM solutions may not support them in the near future. The few available free, open source (FOSS) IM systems employ customizations and extensions in order to apply integrated management for WSNs.

WSNs can be categorized with respect to a variety of parameters, such as the structure of the network, the environmental conditions of the deployment, the sensing capabilities, the power source they use, etc. [4]. They can create structured or ad hoc networks deployed in terrestrial, underground, overland, or underwater environments. Sensing capabilities include temperature, sound, or even images and video. The sensing nodes may be powered by batteries, solar cells, or electrical sources. Each category of WSN exhibits unique characteristics: Structured networks are generally easier to maintain; ad hoc networks may cover more dynamic sensing scenarios than structured ones; networks deployed in harsh environments must use better packaging, protection methods, and even better sensing and communication capabilities. For instance, underwater WSNs must cope with the limited bandwidth, the high transmission delay, and the signal attenuation of the subsea environment. Such WSNs must be equipped with sufficient power sources, especially when nodes are installed in positions where they cannot be replaced or recharged. Mobile WSNs must include routing algorithms and congestion control, and WSNs for multimedia may need to include bandwidth-conservation solutions. Even if IM is one of the oldest fields in computing, managing infrastructures of such diversity under such strict conditions in terms of energy, processing, and memory resources is still a challenge.

The open source remote SM (OpenRSM) system [5] is an open, general-purpose remote management system that is capable of creating complex IM scenarios by orchestrating the execution of autonomous tasks. Its inherent IM usability and ease of customization makes it ideal for the implementation of generalized, platform-independent, task-based IM procedures and the adaptation of the already supported use cases for the management of WSN. OpenRSM is an integrated system that offers IM services namely, remote procedure call (RPC), remote desktop control (RDC), NM, assets management (AM), software delivery (SD), and reporting. The system relies on modeling tasks, managed entities, and the interactions between the two and is thus capable of supporting any IM use case within this context. OpenRSM encapsulates the pieces of management functionality offered by individual systems in tasks and offers remote execution and monitoring for them. It can thus be used to construct general, customizable use cases. This functionality has been expressed as task sequences that can be instantiated, customized, and executed, allowing any number of target stations to be managed in parallel using a central IM point. The system has been extended in various directions, such as the remote deployment and management of grid worker nodes on a large scale, the management of wireless devices, and in other directions in addition to covering the administration needs of WSNs as will be described in this chapter. The rest of the chapter describes the way WSNs can be managed via the extension of OpenRSM and the way administrators can implement usable, high-level IM use cases for WSNs. The chapter considers TinyOS as the WSN platform; however, we describe a general-purpose methodology that can be applied to any type of technology.

21.1.2 The TinyOS Platform

TinyOS [6] is a FOSS, component–based operating system for the construction of WSNs that has gained popularity during recent years. It relies on systems on chip with communication, computation, and sensing capabilities. Other similar operating systems for WSNs are MagnetOS [7], MantisOS [8], CONTIKI [9], SOS [10], PUSPIN [11], and CORMOS [12]. Recent surveys of networked sensor technologies can be found in [13] and [14]. TinyOS started as a collaboration between the University of California, Berkeley, in cooperation with Intel Research and Crossbow Technology [15] and has grown to an international consortium, the TinyOS Alliance [16]. The system offers libraries and tool chains for all the major families of embedded processors and can thus build and deploy applications for various types of boards. As any operating system, it hides any low-level details of the WSNs [17] and provides appropriate APIs for the required abstractions, that is, packet communication, routing, sensing, actuation, and storage. It is a monolithic operating system, as it uses a component model at compilation time and a static image at runtime. It is also completely nonblocking and supports a single stack. All I/O operations that last longer than a few hundred microseconds are asynchronous and have callbacks. To enable compiler optimizations, the network embedded system C (NESC) language has been created in order to statically link callbacks, called events in the TinyOS terminology. NESC has been based on C and thus supports similar syntax and also its data types that correspond to basic C types. Being nonblocking, enables the system to maintain high concurrency with a single stack but it forces programmers to write complex logic by implementing many small event handlers. TinyOS is capable of supporting complex programs with low memory requirements because elaborate applications can fit within the 16 kB of its memory; the core OS is only 400 B. To support larger computations, TinyOS provides tasks that are similar to windows deferred procedure calls and interrupt handler bottom halves. Thus, a component can post a task, which the OS will schedule to run later. Tasks are nonpreemptive and run in FIFO order. This simple concurrency model is sufficient for applications that focus on I/O, but is difficult for supporting ones that demand high CPU utilization. For that, there have been several proposals for incorporating threads into TinyOS.

Figure 21.1 MICA2 and MICA2DOT motes.

TinyOS supports WSN nodes that rely on specific hardware, abiding by its open licensing, such as MICA or TELOS. MICA as well as TELOS motes have been developed in UC Berkeley and became commercially available by Crossbow. WSN nodes may also rely on third-generation MICA hardware for WSNs, such as MICAZ, MICA2, or MICA2DOT that typically support 4 kB or more data RAM, 128 kB or more of program memory, and 512 kB or more of flash memory relying on the integrated circuit CC1000 by CHIPCON. Such devices are supported by micro-controllers, such as the ATMEL 8-bit ATmega 128L. MICA2 or MICAZ motes are illustrated in Figure 21.1. The former typically relies on a multichannel transceiver that operates at 868/916 MHz and the latter on one that can utilize the band from 2.4 to 2.48 GHz, that is, the IEEE 802.15.4 (ZigBee) spectrum. Both devices may rely on battery for a power source and typically use two batteries, AA size. The boards that host MICA2 or MICAZ motes with sensory circuits are MTS100, MTS101, MTS300, MTS310, MTS400, MTS420, and MDA300. Table 21.1 illustrates the sensors supported by each board; however, most of the motes also support expansion slots for additional sensors. MICA2DOT devices reply on multiple channel transceivers that utilize the channels 315, 433, 868, or 916 MHz. Their power source is typically lithium 3 V coin cells, and they can be installed

Table 21.1 Hardware for TinyOS

	MTS Board Model						*MDA*		
Sensor circuit	101	300	310	400	420	510	100	300	500
Light	√	√	√	√	√	√	√	√	
Temperature	√	√	√	√	√		√	√	
Microphone		√	√			√			
Buzzer		√							
Dual-axis magnetometer			√						
Dual-axis accelerometer			√	√	√	√			
Relative humidity				√	√				
Barometric pressure				√	√				
GPS					√				
Sensor extension								√	
Prototyping board									√

Figure 21.2 TELOSB motes.

on MTS510 or MDA500 boards. TELOS is another family of motes for TinyOS that includes the TELOSA, TELOSB, and TMOTE models. Figure 21.2 illustrates TELOSB hardware. These motes are packaged with the 16-bit SI MSP430 micro-controller, which is efficient in terms of energy consumption. They utilize the 2.4 to 2.48 GHz IEEE 802.15.4 band (ZigBee) and typically integrate light, temperature, humidity, and voltage sensors. The members of the family are differentiated in terms of their resources, such as their memory. TELOSA supports 2 kB of RAM, 128 kB of program memory, and 512 MB of flash memory. TELOSB supports 10 kB of RAM, 48 kB of program memory, and 1 MB of memory flash. They rely on two AA batteries for a power source or via USB ports.

Other motes that may run TinyOs are TinyNode, EyesIFX, and IntelMote2, besides others. TinyNode devices use the 16-bit SI MSP430 micro-controller and a wireless transceiver at 868 MHz. They use 8 kB of RAM, 92 kB of program memory, and 512 MB memory flash, the YE1205 integrated circuit, and lithium batteries. TinyNode boards support light, temperature, and humidity sensors and a small breadboard surface and are constructed by Shockfish. EyesIFX devices were developed by INFINEON under the energy efficient sensor networks (EYES) [18] EU research program. They employ a TDA5250 integrated circuit for their operation. They have a 16-bit SI MSP430 micro-controller, a wireless FSK and ASK transceiver that reaches up to 64 kbps of rate. Their power source comes from AA batteries, and they also come with 10 kB of RAM, 48 kB of program memory, and 512 MB of flash memory. They support high-precision temperature and light sensors. IntelMote2 or Imote2 is an advanced WSN platform that relies on the powerful PXA271 microprocessor by Intel. It uses the CC2420 integrated circuit and an IEEE 802.15.4 transceiver. It supports 256 kB SRAM, 32 kB SDRAM, and 32 MB memory of flash. The mote supports different interfaces for power, such as USB or Imote2 board batteries. The ITS400 family of Crossbow boards that use IntelMote2 support light sensors, temperature sensors, a triple axis accelerometer, and four-channel A/D converters. The IMB400 boards, also from Crossbow, feature multimedia functionality because they support camera, microphones, and sound encoding and decoding, besides passive infrared sensor radiation (PIR). IRIS devices come with the ATMEGA 1281 ATMEL 8-bit micro-controller, an IEEE 802.15.4 transceiver and also 8 kB RAM, 4 kB EEPROM, 128 kB of program memory, and 512 MB of flash memory. Such devices are powered by two AA batteries and are expandable through the use of MTS300, MTS310, MTS400, MTS420, MDA100, MDA300, and MDA320, just as MICA2 and MICAS motes.

21.1.3 Application Development in TinyOS

Applications for TinyOS consist of components called modules and configurations. They are combined to utilize the TinyOS libraries to be linked with the application code so that it is deployed in the program memory. The TinyOS includes a scheduler, typically a FIFO stack that contains events and asks, and the graph of components to be used, called the wiring. The latter is independent of the components of the application and, consequently, applications connect only to necessary components; the binary image that results from the compilation of the application is minimal in size. Modules provide the application code by implementing one or more interfaces that describe the way providers and users communicate. The former implement commands while the latter implement events; in essence, commands are responses to events, and the interaction between the two implements communication in TinyOS. Configuration objects do not contain application code; they are used for module, interface, and role declarations. TinyOS also supports the definition of logic that is not required to execute immediately with task modules. Because motes measure data in real time, they must not be interrupted, and thus tasks can be used to call other functions or events, modules that allow communication among components that support corresponding handlers. Events can be internal (hardware interrupts from timers or transceivers) or external (caused by event handlers in response to external events). The execution of events has a higher priority than tasks, and thus they may interrupt the execution of the latter.

21.1.4 The WSNs IM Scenery

The total cost of ownership (TCO) for assets such as workstations, installed software, configuration, and custom code often significantly exceeds the initial purchase or development cost. Thus, every organization using IT aims at minimizing the TCO for such assets [19]. To this end, organizations employ enterprise management systems (EMSs) [20] that are generally scalable, well organized, and cost effective. SM and NM have always been critical fields for IT organizations because they facilitate administration and provide important economies of scale [21]. The EMS market is dominated by a few well-known products, most of which are released by major enterprise system constructors [22]. Such organizations carry profound system-level knowledge and experience gained over years of producing management platforms. EMS systems offer a variety of highly specialized features and capabilities that extend their functionality. They cover areas besides standard SM and NM, such as storage management, software development, security, and business integration, and sometimes offer an ecosystem of features, services, and tools. Figure 21.3

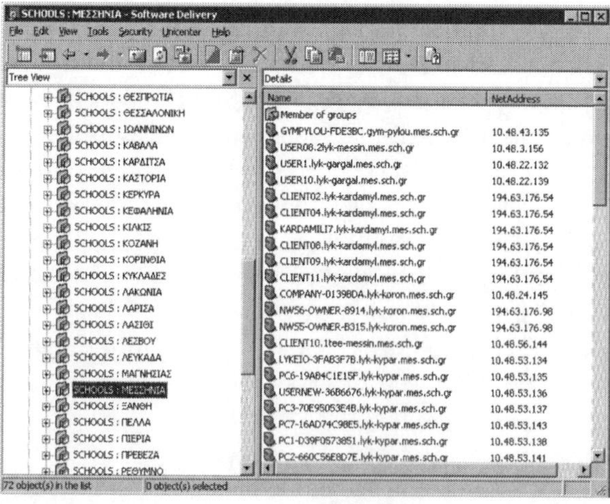

Figure 21.3 EMS integrate several technologies: This figure illustrates the SD console of the Unicenter EMS system by CA.

illustrates the SD console of the Unicenter EMS systems by CA, one of the fat consoles of the well-known EMS. Competition and marketing strategy have led EMS vendors to introduce and advertise specialized, sophisticated features, such as end-to-end root cause analysis, return-on-investment optimizations, and time-savings estimations. The EMS market has been considered relatively expensive in terms of licensing and support, at least for small-scale organizations. Today, the picture is changing [23]; high-end EMS providers respond to the changes by releasing express versions of their software, aiming to penetrate the new market of midsized corporate management systems before FOSS systems do. However, they are criticized for being overly complex; inflexible [24]; and difficult to deploy, learn, and manage. The FOSS community is, however, still rather immature as far as remote management systems are concerned. There are only a few NM projects under the SourceForge FOSS project hosting site, and only a small number of them are on the road to maturity. As far as integrated SM and NM tools are concerned, there is none at the moment. The state of the European FOSS management tool market is depicted in the catalog of available FOSS tools for the PA [25] issued in terms of the Consortium for Open Source Software in the Public Administration (COSPA), where only a couple of NM systems appear in the administration and management tools section. The OpenRSM platform [5] is a pioneering initiative aiming to open the lightweight EMS market to FOSS tools.

General-purpose IM systems, such as OpenRSM, can produce high-level, generic management functionality for WSN through customizations and integrations that exploit the facilities offered by target WSN operating systems, platforms, or toolkits. Still, problems remain even for the most traditional management cases. The case of WSN is one of the most challenging for management; for instance, not all WSN management systems support high-level management features that are autonomous and easy to exploit. Also, WSN are not designed for interoperability with informational systems, let alone IM systems. The operation of motes may not ease management; querying for the state of motes may not be straightforward because they fall frequently in sensing mode, and it might be difficult to make conclusions about their state. Motes may also be out of battery power or faulty. They may sleep if they are redundant in dense topologies or for energy-conservation purposes, or the communication channel may fail. Different WSN platforms or mote families may differ in any of the above. Thus, the most fundamental case for IM, the acquisition of identification information is a nontrivial procedure. Other cases, such as AM, RPC, RDC, and NM, also fall beyond established practices. The retrieval of AM information from motes is not directly analogous to that for workstations. While the latter relies on standards, such as WBEM/CIM or WIM, the former requires the reading of low-level information and forwarding it to the AM server. Considering the simplistic execution models supported by motes, such an application would have to be deployed for a very short period of time in order to retrieve information and then give its place to a sensing application. The AM application would additionally require the deployment of a dedicated, application, customized for each type of WSN operating system and hardware family. RPC execution for motes cannot be direct because they do not offer standard interfaces, and thus the management platform would have to rely on the functionality provided by the WSN system, provided that they offer appropriate command-line interfaces and programming APIs. RPC would be essential in order to retrieve the members of a WSN and specifically their identification address/names, so the other systems may refer to them. RDC is irrelevant for WSNs except for managing the master station through the standard IM practice of retrieving graphical interfaces. In analogy, traditional NM is not standard; most WSNs do not support even the lowest levels of the ISO architecture, and thus, SNMP cannot be exploited for them. There are systems that support queries about the state of motes, but this information cannot be considered for NM because it is, rather, discovery. SD is also irrelevant and would rely on services offered by the

WNS operating system. A general purpose, lightweight, usable platform for IM with the support of WSN would have to extend its management framework in order to include the abstractions that apply to WSNs. Management entities would have to include WSN servers or designated nodes and also motes. For each WSN, the IM system would have to support at least the configuration for the available types of motes and the interactions supported by the operating system network.

Several management tools, platforms, and architectures are currently available for WSN relying on different IM approaches. The bridge of the sensors system (BOSS) [26] implements a management platform that relies on the standard service discovery protocol UPnP. Because UPnP is difficult to run on every sensor, BOSS adopts the bridging approach and uses designated nodes as management intermediates that employ XML for the description of services and for the communication of data. The system offers basic information retrieval capabilities that regard the state of the network, the characteristics of the sensor nodes, the number of nodes in the network, and the topology. Administrators may interact with notes and configure parameters, such as the transmission power or the node state. The MANNA architecture [27] is designed to manage any WSN application by exploiting specific operation models that reflect functionality common to WSN systems. It considers three management dimensions, namely, functional areas, management levels, and WSN functionalities. Functional areas include configuration, fault, performance, security, and accounting management. WSN functionality considers configuration, maintenance, sensing, processing, and communication, and management can be applied in the level of business, service, network, and network elements. MANA is flexible and independent of the adopted WSN technology and allows all possible configurations of the managed entities. Both BOSS and MANA follow the example of established NM principles. Middleware solutions use additional logic layers within the firmware of motes in order to implement basic IM services. The MATE [28] system introduces lightweight virtualization that aims at overcoming the diversity in boards and, thus, simplifying the task of managing diverse sensors. MATE's objective is to reduce complexity by reducing the size of programs, making them capable of running on MICA, RENE, and other motes. The source code is broken into 24 instruction capsules that self-replicate through the network via ad hoc routing and data aggregation algorithms. AGILLA [29] is a mobile agent middleware that facilitates the rapid deployment of adaptive applications in WSNs. It allows users to create and inject programs, called mobile agents, which can be coordinated through local tuple spaces. Mobile agents migrate across WSNs and are capable of performing application-specific tasks. COUGAR [30] is another middleware for WSN that uses a database for scalable and flexible WSN monitoring. COUGAR issues cross-layer optimizations, such as query-layer-specific routing algorithms optimized for regular types of communication patterns. The MIRES [31] middleware adopts an asynchronous publish/subscribe model for communication with WSN nodes. NEST [32] is a real-time network coordination and control middleware that abstracts, controls, and ultimately guarantees the desired behavior of large unreliable networks. It is based on operating system tasks, called micro-cells, that provide support for migration, replication, and grouping functionality. Older middleware systems include the SCADDS and the Smart Messages project [33].

Other management systems for WSNs use a variety of implementations. SNMS [34] is an application-cooperative management system for WSNs that uses minimal resources to provide a query system that enables rapid, user-initiated acquisition of network state and performance data and also an event registration system. SNMS is based on a networking stack that runs in parallel with the applications stack. LEACH [35] and GAF [36] also fall within the category of protocol-based management platforms for WSNs. GAF exploits node redundancy and supports sleep modes for overlapping nodes while LEACH uses dynamic, efficient clustering in order to manage sensor networks. Another category of WSN tools focuses on monitoring consoles for WSNs. WSNView

[37] is a technology capable of automatically searching for and displaying network facilities, collecting and analyzing network utilization, and automatically producing notifications. Other similar visualization tools are TinyDB [38] and MoteView [39]. TinyDB uses a SQL-like syntax in order to collect data from nodes and also provides basic configuration for motes. Administrators may recover the topology of the network or create graphs of data using its versatile low-level interface that offers only minimum levels of automation because administrators need to manage network operations manually and know how to exploit its representations and operations. MoteView is a tool for NM and control for WSNs created from commercial workstations; it stores the measured data in an informational system and offers a graphical user interface that presents the network topology or the measured values. MoteView may also control mote parameters, such as transmission power, sampling frequency and node identification numbers. There are also a lot of tools that focus on battery or transmit power management by controlling the sampling frequency and transmission rate configuration. Agent-based power management (ABPM) uses intelligent mobile agents for power conservation when nodes reach critical battery levels. Other similar tools, such as SenOS and AppSleep, force WSN nodes to automatically sleep when not performing data measurements. Systems such as Siphon, DSN RM, and WinMS are capable of managing the network traffic efficiently; Siphon takes advantage of nodes that transmit in multiple directions in order to avoid congestion and heavily shared links. DSN resource management (DSNRM) evaluates traffic for incoming and outgoing links and time schedules the transmissions in order to optimize network usage.

Even if simple and small, WSNs need to be managed as any other infrastructure. One of the issues of IM for WSNs is the degree to which IM logic can be integrated at agent nodes; the diversity and simplicity of WSN execution models leaves only limited room for high-level functionality. There are cases in which lightweight threads can be exploited in order to deploy custom IM applications or in which resources allow the installation of ISO protocol stacks and thus the practice of NM or SM. The latter specifications are only met by fat systems with limited autonomy, typically positioned in fixed locations where electrical power sources are available. In the vast majority of cases or in the general case, that includes systems that rely on operating systems such as TinyOS, even lightweight IM agents cannot be designed and deployed in motes.

21.2 General Purpose, Open, IM

21.2.1 OpenRSM Overview

OpenRSM [5] extends and integrates high-value FOSS projects in order to provide an integrated management platform. The goal has been to build a remote-systems and NM platform capable of facilitating daily tasks. The system is designed to be fully functional yet simple, unlike most commercial management systems. It was designed to offer information retrieval about installed assets, deployment of software, and management of installed software, sending executable commands to stations, controlling remote desktops and also wireless access point management.

OpenRSM was developed by integrating, enhancing, configuring, and customizing FOSS tools in order to deliver general purpose IM [40]. The basic services supported by OpenRSM are AM, SD, RDC, and NM. At the time it was created, there was no FOSS system that could claim to approach functionality similar to that of commercial EMSs. The capabilities of the system extend to managing any station that can be reached through standard IP connectivity in a secure manner. OpenRSM relies on agents to convey and execute management actions, on an integration server to serve user requests, and on a graphical management console user interface. It depends on a management framework designed to model all the entities and service units necessary for IM and the description of the basic, abstract interactions between them. The design followed a layered approach; the main management entities are tasks that users can create and send to any managed workstation and the workstations where tasks are dispatched. Services correspond to the subsystems described above, that is, AM, RPC, RDC, SD, NM, and also host discovery, server tasks, access point management, task scheduling, wake on LAN, router configuration, reporting, static and dynamic entity grouping, customizable reports, and more. The hierarchy of the management entities is structured as an object-oriented tree of classes that forms the IM hierarchy. The design allows for modularity and extensibility, and thus functionality can easily be extended to include specialized tasks and complex procedures. The top-level task object holds its identification, a mnemonic name, and execution parameters, such as priority, execution method, task dependency parameters, and more. An analogous object models the abstract managed host. The framework builds on abstract classes providing entity templates that can visually be instantiated in the user console interface. For example, users can extend any task template, fill in custom parameters, and create tasks that suit their needs. For example, in order to construct a task that shuts down a remote workstation, users need to instantiate a new task object by using RPC task class. They configure it to encapsulate a shutdown command for the target platform, and they associate it with a managed station. The management framework takes care of the communication and execution details. Users can then use the same template to construct a second shutdown task for a different operating system. Both tasks will be descendants of the same framework task and inherit core functionality implemented in the task management framework.

The task-handling engine rests at the core of the IM framework because it coordinates the underlying mechanisms. Others layers include communication, task encoding/decoding, task/station verification, security. The previous levels have been dimensioned in terms of the agent, manager, and server functionality. For example, the communication between the server and the manager differs from that between the server and the agent; in the latter case, the server needs to wake the agent asynchronously and then commence state verification handshakes and controls and perform the actual communication. The manager-server communication is connection-oriented; however, both cases are implemented under the communications layer. The API hierarchy is made available via corresponding interfaces to all the components of the system.

OpenRSM has passed stress and scalability tests and has been deployed in pilot installations [5]. It is easy and quick to install by design because all the components support automatic installations created with known, open installer packagers. The agent is small in size, approximately 1.5 MB, and can be distributed easily in the case of large installations, even in remote areas connected via slow lines. After the installation, which exposes a simple graphical interface, no configuration is necessary, and the administrators can start managing their infrastructure via the management console, which is a graphical interface. The latter is designed to be simple, lightweight, and easy to learn by administrators or technical staff.

21.2.2 Fundamental IM Tasks

OpenRSM provides a controlling interface, illustrated in Figure 21.4, that can be used by the administrators to control all the subsystems and their interactions. The design has focused on synthesizing the independent functionalities provided by the subsystems in a user-comprehensive and effective manner and on the provisioning of additional supervisory functionality. The console can send any system command that is supported by the operating system of the managed stations. Commands accept parameters related to CPU priority, type, and user visibility. It can also initiate

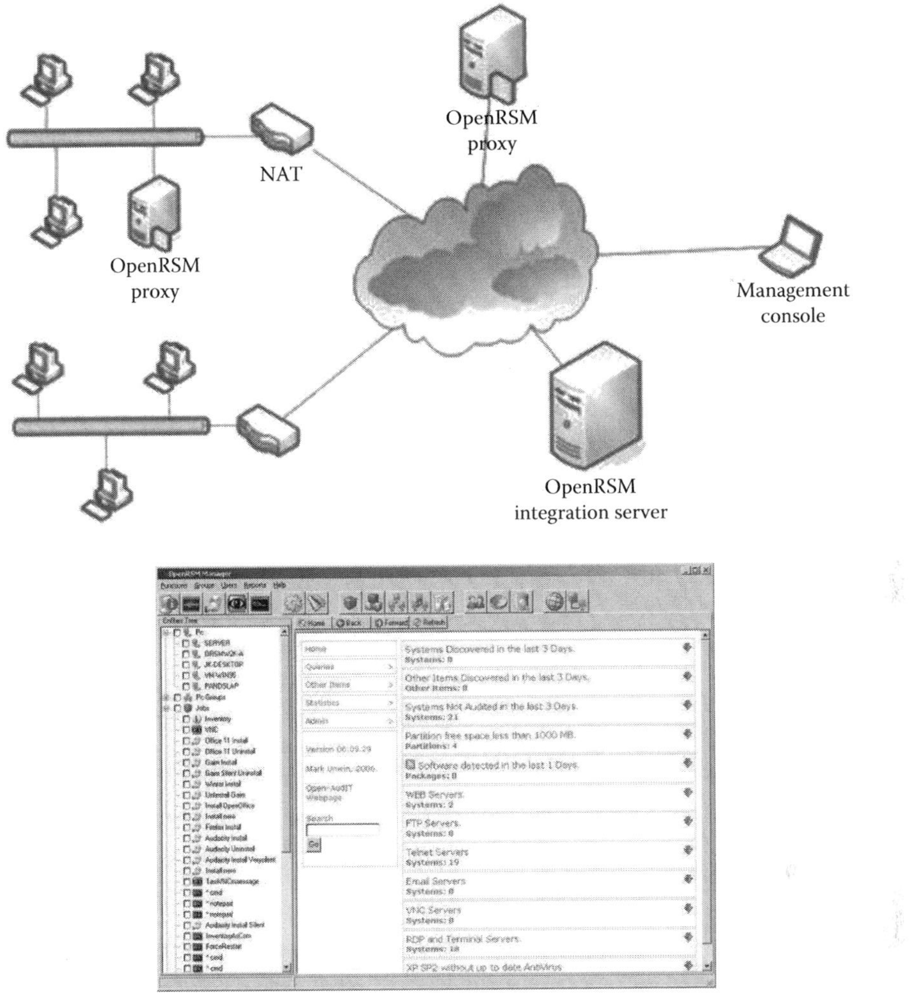

Figure 21.4 OpenRSM component architecture and graphical management console.

the discovery process to recover stations that run OpenRSM agents and are therefore manageable. This discovery can be directed toward any part of the Internet address space and results in creating active interface elements, representing corresponding managed stations. These graphical elements, along with core tasks (AM, RDC, SD, RPC), are the basic interface elements designed for interaction with the user. Workstations and tasks are fundamental entities in the management OpenRSM framework that can be combined in order to cause remote execution. They are both presented on the management tree for easy supervision. Both tasks and workstations can be grouped; groups of workstations and tasks or groups of tasks and workstations can also be combined in order to create remote execution of commands. Groups are generic: A group of tasks can contain any kind of tasks, and no dependencies are implemented. It is the administrator's responsibility to create a rational sequence of a group of tasks for execution. OpenRSM allows the user to create custom tasks. Following a clean installation, only core tasks exist. The core tasks include predefined AM, SD, RDC, and RPC tasks. As stated in the previous paragraph, a task must be visually combined with one or more agents. Thus, tasks can be considered as templates for submitted administration tasks. Each type of task is created using interface components used for that purpose only. For example, a delivery task must "know" the software that it has to install/uninstall, etc. OpenRSM provides the interface for custom task production. The tasks created are stored and made available through the interface so that they can be reused. A user can also use the workstation and task group forms to define groups of the respective entity.

The AM task rests at the core of IM, providing organizations with the ability to gather information about the hardware and software in their infrastructure. This service provides the necessary data for effective troubleshooting, infrastructure planning, and decision-making. At the technical level, AM systems use client modules to retrieve information about hosts using native interfaces. The information is submitted to the AM server where it is stored and where user interfaces are running. The technologies used, namely CIM/WBEM, are mature enough to provide vendor interoperability but, as mentioned in previous paragraphs, cannot be applied in WSNs. OpenRSM relies on OpenAudit [41], an APACHE/PHP/MYSQL application that relies on manual configuration and execution of audit software, locally on workstations. The audit software has been integrated with the OpenRSM agent module, and thus the management console of OpenRSM can trigger the remote execution of AM. The audit software has also been ported for all major operating systems. SD tasks facilitate the management of already installed software or of software that is to be installed on workstations within an administrative domain. Administrators can use the OpenRSM console to configure the software package to be delivered (executable path or URL, mode of installation, and more), registered it within the software repository, or command the server to deliver it to any selected stations. The software is conveyed to hosts, the designated installation file runs, and users may be required to complete the installation procedure. If silent or unattended modes of operation are chosen, the installation may not be interactive, and users may not be distracted in any way during the delivery procedure. Silent installations/uninstallations are very useful for routine administration and for operations in scale. The SD subsystem is complementary with the AM subsystem; administrators may use the AM system in order to supervise SD, and they may use the SD subsystem in order to install/uninstall desired modules. Figure 21.5 illustrates the SD task creation tab at the management console. For RDC, OpenRSM integrates the TightVNC package to deliver graphical remote control tasks. The management console is capable of starting a TightVNC server at a managed station or at a group of managed stations by sending an appropriate task; it then initiates a TightVNC viewer at the console. Last, but not least, OpenRSM provides full-featured NM functionality by integrating the NINO FOSS NMS tool and the OpenNMS platform.

The management console provides a means for supervising the execution of tasks by keeping a real-time list of active tasks. The list presents useful information such as task state (connecting, transferring, processing, completed), agent name, user name, related timestamps (start/stop), etc. Filters can also be applied to the list of tasks, so that it is customized according to needs. The management console relies on the combination of information from all subsystems. For instance, the grouping and reporting systems can search across the data produced by any tasks in order to recover information that answers to any criteria. Users can thus create reports or groups for workstations, tasks, software, users, etc. that share common characteristics. The matching entities can be grouped together in new groups or presented in a visual form. The retrieval of all workstations that have, for example, more physical memory than a specific value would simply require the selection of attributes and the configuration of the value for the required memory. The resulting workstation information can be presented as a report or a group of workstations, called "dynamic" because of the way it is created. Dynamic groups behave as normal groups, but they also enjoy the special feature that they are associated with the database statement that created

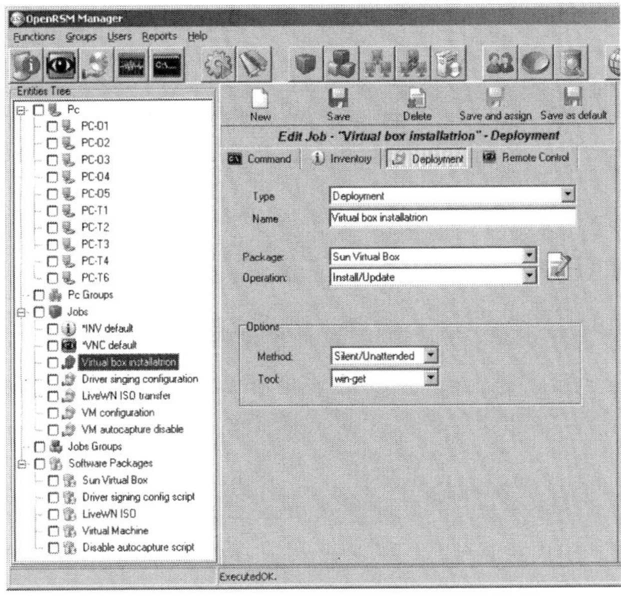

Figure 21.5 Task creation tab is used to instantiate RPC, SM, SD, and RDC tasks.

them. The query that created them may be executed at any time in which case the group is recreated based on updated workstation information. The reporting functionality is complemented by the data explorer form, created to provide complete database supervision. The user is capable of browsing database entities and combining their contents whenever internal linking is possible. Combinations are presented in the form of reports that can be exported in various formats, such as DOC, XLS, HTML, TXT, and CSV. The management console is also capable of producing cumulative and detailed statistics about system utilization. The generated statistics can record general system usage, task distribution, workstation utilization, and user actions. General information is presented in visual charts providing summary information on the task submission rates, task error rates, and task distribution with respect to task type and submitting user. Detailed information is presented in individual reports about each subsystem.

21.3 Integrating WSN Management in OpenRSM

21.3.1 Introduction

A general-purpose remote management system, such as OpenRSM, can be customized in order to exploit the management services offered by any WSN platform. Our goal has been to exploit the general-purpose nature of OpenRSM in order to offer full support for IM for TinyOS-based WSNs by constructing a finite number of IM tasks that include remote installation of TinyOS for any type of platform (FOSS/commercial), environment checks and configuration, deployment of sensing applications, recovery of mote listing, retrieval of readings and measurements in the database of the system, and WSN monitoring in real time. The above tasks were defined in terms of the OpenRSM entities, which have been extended to include motes and were made available at the management console. In the next paragraphs, we describe how custom tasks can be constructed in order to remotely create any sensing and configuration scenario and how we may retrieve measured data in the OpenRSM database. For the purposes of our case study, we used CrossBow TELOSB motes, which are typically equipped with sensors that measure battery voltage, humidity, luminosity, and temperature. The WSN uses a designated node as the intermediate between the managed station and the sensing motes.

21.3.2 Installing TinyOS Remotely

The remote installation of the TinyOS 2.1 platform presupposes the installation of a number of components, such as the JAVA programming platform; native compilers, such as the ATMEL AVR Tools [42] or the IT MSP430 tools [43]; the TinyOS tool chains; the TinyOS source code; and the GraphVIZ tool. The environment can then be configured by running appropriate shell scripts and by setting the appropriate environment variables. The installation procedure can be implemented using either SD tasks or remote procedure ones that call native package managers such as APT, YUM, ZYPPER, or RPC tasks that download and execute the necessary binaries. Whenever possible, features of the underlying operating system can be used, such as package management software. For the installation of tools and tool chains commands such as

```
"zypper— non-interactive— no-gpg-checks in -f— auto-agree-
with-licenses java-1_6_0-openjdk"
"apt-get install sun-java6-bin sun-java6-jre sun-java6-jdk
openjdk-6-jre -y -force-yes"
```

can be encapsulated in RPC tasks and be then sent to managed hosts. Commands such as the above can be massively sent by the management console. Alternatively, creating SD tasks would include creating a software package for JAVA at the management console, providing the path or URL pointing to the installation executable or archive, and then associating the package with an appropriate task. This is typical procedure for SD tasks that are created graphically and may encapsulate binaries, archive files, or binary software images. As mentioned in the previous section, software packages can be configured to support any unattended installation or uninstallation method provided by the installer of the software. Once registered with the SD system, the software is made visible on the entity tree of the management console. OpenRSM allows for any solution that can be sustained. An alternative for the installations discussed above would be to send a WGET task to agents, followed by an RPC command that utilizes the software downloaded or even a shell BASH or WSH script that would contain the aforementioned commands. Tasks that install TinyOS, NESC, deputy, TinyOS tools, and the TinyOS source tree are illustrated in Figure 21.6. The window on the left is the agent terminal window as presented by the LINUX window manager via a remote desktop connection, also opened with an OpenRSM RDC connection task. On the right, there is the management console, which has sent a "TinyOS installation" SD task. Figure 21.6 also presents tasks that install the JAVA platform and the ATMEL AVR tools, that is, avr-binutils, avr-gcc, avr-libc, avarice, avr-gdb, and avrdude. The TI MSP430 tools are installed in a similar way, namely the basic toolset, python tools, binutils, GCC, LIBC, JTAG, and GDB for MSP430. All the registered packages used by SD tasks are presented in the bottom branch of the tree. The final steps for the installation of TinyOS are the installation of the GraphVIZ tool and MAKE. Note that tasks can be grouped so that they are sent with a single click, and they can be configured to be executed sequentially in the same thread, taking advantage of task-configuration properties.

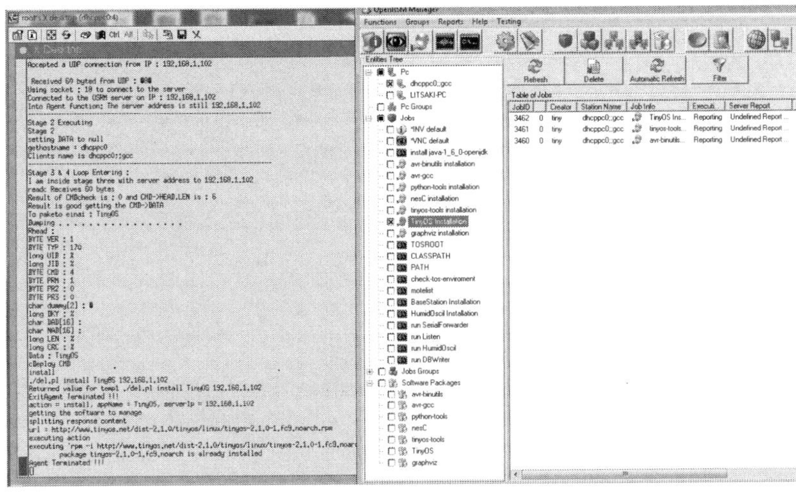

Figure 21.6 Execution of a software installation task designed to install TinyOS on remote stations.

Thus, the software is executed after the download and not in parallel with it because the two tasks are executed by the main management console thread and not by forked ones. Administrators may monitor the execution logs via the logging console of the management interface of OpenRSM, which conveys the output of the agents at the management console.

21.3.3 Running TinyOS Applications and Sensing

After installation, the environment of WSN needs to be configured so that applications can be compiled and deployed. The configuration of the environment is achieved by setting appropriate environment variables; this can be accomplished via the execution of tasks or scripts that set the TinyOS root and home directories, the CLASAPTH for JAVA and NESC, the display variable, the paths to the rules for MAKE, and any additional parameters. Additional tasks can be used to run system utilities that check and return the state of the environment, such as `tos-check-env` and `tos-install-jni`. An example set of RPC commands that can be sent, either as RPC tasks or as a SD script may include the configuration of the directory where the rules for MAKE are located, that is, "`export MAKERULES = $TOSROOT/support/make/Makerules`". After the configuration of TinyOS, RPC or SD tasks can be used to encapsulate standard deployment commands that deploy applications on motes.

TinyOS provides a toolbox of applications that offer pieces of WSN functionality. The fundamental implementation of a gateway between a serial port and the WSN is the BaseStation application. When receiving packets from serial ports, this application transmits data toward the network and, vice versa; when receiving packets from the network, it transmits them to the serial port. In order to forward traffic from the serial port toward network TCP sockets, the TinyOS platform provides the SerialForwarder. This functionality enables any tool, such as IM, NM, or data analysis, to access WSN measurements. Visualization for data is provided by Oscilloscope applications. The BaseStation application can be deployed at a designated mote in order for it to collect data from the WSN. Oscilloscope applications can also be deployed at the rest of the motes in order to take measurements and forward them to the BaseStation. Such commands can be the following:

```
make -C apps/BaseStation -f apps/BaseStation/Makefile telosb
install.0
java net.tinyos.sf.SerialForwarder -comm serial@/dev/
ttyUSB0:telosb
java -cp support/sdk/java/tinyos.jar: apps/Oscilloscope/java/
oscilloscope.jar Oscilloscope
```

The SerialForwarder application is usually executed at the managed station in order to read data from the port where the designated mote, the one that runs the base station application, is connected and to forward them over network connections. If RDC functionality is desired at the managed host, the oscilloscope graphical application can easily be executed through a corresponding task, in order to graphically present the received data. Figure 21.7 presents the configuration of an RPC command for the remote installation of the BaseStation application at the OpenRSM console. Retrieving measurements for battery voltage, temperature, luminosity, or humidity entails the deployment of the respective measuring applications, that is, VoltageOscil, TempOscil, LightOscil, or HumidOscil, and the execution of the respective client applications at the OpenRSM agent. The aforementioned applications are variations of Oscilloscope provided by TinyOS and use appropriate NESC components for data retrieval. For the measurement of luminosity, the HamamatsuS1087ParC driver is configured with SensirionSht11C for temperature

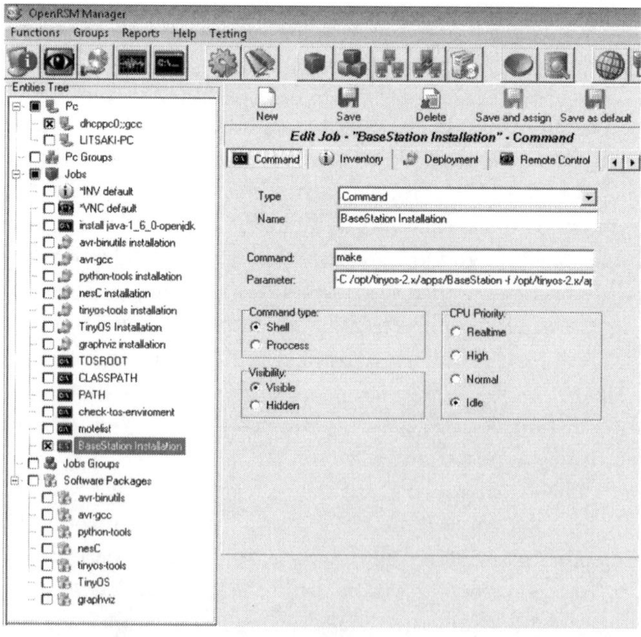

Figure 21.7 An RPC task can be configured to remotely run BaseStation and SerialForwarder applications from the OpenRSM console.

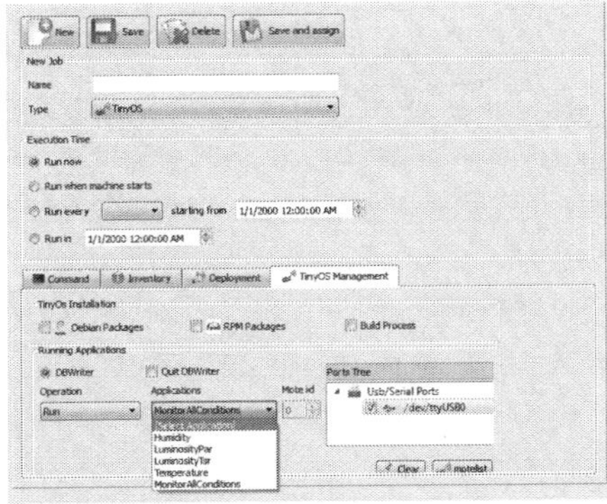

Figure 21.8 Management console supports a task-creation tab for WSN tasks.

and humidity. For each task that deploys a sensing application, an oscilloscope task presents the measurements in the managed station. Commands such as

```
"make -C apps/TempOscil -f apps/TempOscil/Makefile telosb
install"
"java -cp support/sdk/java/tinyos.jar:apps/TempOscil/java/
oscilloscope.jar TempOscil"
```

can be used to implement such tasks.

In order to concentrate the functionality described above in a single tab, a general purpose TinyOS management task instantiation panel has been developed for OpenRSM. The panel, illustrated in Figure 21.8, offers graphical WSN task creation that include tasks for the installation of TinyOS, for uninstallation, for the discovery of motes, and for the deployment of applications. Users define the name of the task, the execution time, and the type. They then instantiate it by saving it and then correlating it with managed TinyOS hosts. If the task includes the deployment of applications, users may select which of the available ones will be deployed and the target mote. This procedure can be used in order to implement any remote-sensing scenario supported by the underlying technology.

21.3.4 Concentrating Measurements

OpenRSM has been enriched with an application for TinyOS, DBWriter, that interfaces with the SerialForwarder application at the managed host in order to receive measured data as sent by motes and write them in the OpenRSM database. The application is configurable with the appropriate transformations in order to convert the data to the appropriate measurement systems. Figure 21.9 illustrates the measured data as presented in the management console. The left panel in the database view presents the navigation tree of the OpenRSM system, and the right panel illustrates the functionality for measurement presentation. Users can select a sensing dataset from each mote that corresponds to a table in the database of OpenRSM. They are then presented with the data and meta-information that includes moteIds, packet counters, sampling frequency, date, etc. They can also view the logs of the system.

As mentioned in the previous paragraphs, motes utilize sensor circuits that sample physical quantities from motes through applications such as TempMeasurement, HumidMeasurement,

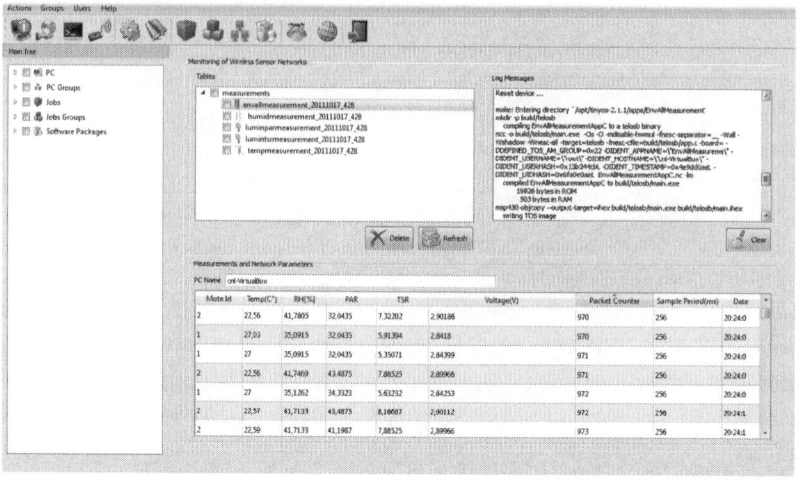

Figure 21.9 Measurements are sent by DBWriter application to OpenRSM server and can be displayed at management console in real time.

LuminParMeasurement, and LuminTsrMeasurement. The integrated EnvAllMeasurement application is used to recover data from the network besides the total number of received packets by each node, sampling frequency, and battery levels. The code of the application is presented in Table 21.2.

Each sensor uses analog to digital components that produce numeric output that can directly be converted to a metric system. The SHT11 sensor belongs to the Sensirion AG family of SHT1x temperature/humidity surface mounts. SHT11 integrates analog measuring devices and signal processing logic to provide calibrated output. Moisture is measured by a capacitive sensor and the temperature of a band gap, proportional to absolute temperature, sensor. Measurements are then converted to 14-bit digital words, which can be conveyable to TinyOS via the serial interface. Converting the digital value to relative humidity (RH) units is achieved using the following formula:

$$RH_{linear} = c_1 + c_2 {}^* SO_{RH} + c_3 {}^* SO_{RH}^2$$

where SO_{RH} represents the 12-bit measured data (Sensirion output), $c_1 = -4$, $c_2 = 0.0405$, $c_3 = -2.8*10^{-6}$. The respective conversion formula for temperature measurements is the following:

$$Temp = c_1 + c_2 * SO$$

Light sensors for TinyOS motes typically use photo-diodes manufactured by Corporation. Hamamatsu S1087 photo-diodes detect photosynthetically active radiation (PAR), and S1087-01 photo-diodes detect all the visible spectrum, including infrared (TSR). Photosynthetic active radiation can be defined as the electromagnetic spectrum of visible light, namely wavelengths from 400 to 700 nanometers that help process plant growth, photosynthesis. Both analog sensors TSR and PAR convert measurements to 12-bit digital words of length using 1.5-volt signal pulses. LEDs generate current I along a resistance 100 KOHM. The output of S1087 or S1087-01 can be converted into units for brightness that is, LUX, using the following formulas:

Table 21.2 Application EnvAllMeasurement Concentrates Data from All Nodes of the WSN

EnvAllMeasurementAppC.nc	*EnvAllMeasurementC.nc*

components new HamamatsuS1087ParC() as Sensor;
components new HamamatsuS10871TsrC() as Sensor1;
components new SensirionSht11C() as Sensor2;
components new DemoSensorC() as Sensor3;
 	
EnvAllMeasurementC.ReadPARLumin -> Sensor.Read;
EnvAllMeasurementC.ReadTSRLumin -> Sensor1.Read;
EnvAllMeasurementC.ReadExtTemp -> Sensor2.Temperature;
EnvAllMeasurementC.ReadHumid -> Sensor2.Humidity;
EnvAllMeasurementC.ReadVoltage -> Sensor3.Read;
	interface Read<uint16_t> as ReadPARLumin;
	interface Read<uint16_t> as ReadTSRLumin;
	interface Read<uint16_t> as ReadExtTemp;
	interface Read<uint16_t> as ReadHumid;
	interface Read<uint16_t> as ReadVoltage;
	
	call. ReadPARLumin.read();
	call. ReadTSRLumin.read();
	call ReadExtTemp.read();
	call ReadHumid.read();
	call ReadVoltage.read();
	
	event void ReadExtTemp.readDone(error_t result, uint16_t data){
	if (result ! = SUCCESS) {
	        data = 0xffff;
	        report_problem();
	}
	local.readingTemp = data;
	}

EnvAllMeasurement.h
typedef nx_struct envAllMeasure { nx_uint16_t version;/* Version of the interval. */ nx_uint16_t interval;/* Sampling period. */ nx_uint16_t id;/* Mote id of sending mote. */ nx_uint16_t count;/* Number of readings */ nx_uint16_t readingTemp;/*Var for temp*/ nx_uint16_t readingHumid;/*Var for humitity*/ nx_uint16_t readingPARLumin;/*Var for Par Luminosity*/ nx_uint16_t readingTSRLumin;/*Var for Tsr Luminosity */ nx_uint16_t readingVolt;/*Var for voltage*/ } envAllMeasure_t;

$$\text{LUX} = 0.625 * 1\text{e}6 * \text{I} * 1000 \text{ for photo-diode S1087}$$

$$\text{LUX} = 0.769 * 1\text{e}6 * \text{I} * 1000 \text{ for photo-diode S1087-01}$$

where I is defined as

$$\text{I} = \frac{\text{AD}_{\text{output}} * \left(\frac{1.5}{4096}\right)}{10000}$$

The micro controller MPS30 has internal sensors, such as temperature and voltage, mentioned in the description of the devices category TELOS. The analog measurements of the sensor voltage are converted into digital words of length 12 bits and conversion to physical units carried by the following formula:

$$\text{VCC} = 2 * \left(\frac{\text{AD}_{\text{output}}}{4096}\right) * V_{ref}$$

where $V_{ref} = 1{,}5$ V

21.4 Conclusions and Future Perspective

The capabilities of the system have been tested in terms of usability and functionality. We have executed scenarios that included the dynamic deployment of applications and configuration of mote parameters as described above. The tests have been executed remotely; that is, the management console, the OpenRSM server, and the managed motes were located in different domains. In cases of real-time configuration, we took advantage of the base station capabilities in order to tweak mote parameters, such as the sampling frequency and the state motes. More complex cases were based on simple or basic tasks, such as scheduling the configuration of motes and the deployment of applications. Instead of manually loading using a different configuration for each mote, we used periodic scheduling of all desired configuration or applications for motes. In order to implement a mote conservation scenario via timesharing, we organized the periodic deployment of sensing applications at each mote. For instance, we used execution windows of 2, 3, 5, and 15 min for the deployment of humidity, luminosity, acceleration, and voltage applications.

The previous sections illustrated how OpenRSM can provide remote IM for WSNs, based on the TinyOS platform. OpenRSM can create tasks for any type of WSN that offers high-level tools or utilities and, because it has been verified for scaled operation, it can support effective remote IM cases. Future work will focus on productive management of distant WSN infrastructures, on the integration of functionality for more WSN types with OpenRSM, and on the enrichment OpenRSM with new IM features. Users will be allowed to define alerts or actions that will be executed as standard tasks in case measurements reach thresholds, also defined by users. Interesting, as well as challenging, would be to include discovery and AM for WSN motes whenever the underlying WSN platform offers identification applications or data based regarding the active mote and their types. The system is also migrating to web-based technologies, such as HTML5, AJAX, and middleware.

References

1. Thompson, J.P.; Web-based enterprise management architecture, *Communications Magazine, IEEE*, vol. 36, no. 3, pp. 80–86, March 1998.
2. Tosic, V.; Dordevic-Kajan, S.; The Common Information Model (CIM) standard—An analysis of features and open issues, In *The 4th International Conference on Telecommunications in Modern Satellite, Cable and Broadcasting Services, 1999.* vol. 2, pp. 677–680, 1999, doi: 10.1109/TELSKS.1999.806301.
3. Windows management instrumentation and simple network management protocol, Microsoft technet, http://technet.microsoft.com/en-us/library/bb742612.aspx, accessed April 22, 2013.
4. Rhee, S.; Seetharam, D.; Liu, S.; Techniques for minimizing power consumption in low data-rate wireless sensor networks, In *Wireless Communications and Networking Conference*, 2004 WCNC Millennial Net, Cambridge, MA, USA. 2004 IEEE, vol. 3, pp. 1727–1731, 21–25 March 2004.
5. Karalis, Y.; Kalochristianakis, M.; Kokkinos, P.; Varvarigos, E.; OpenRSM: A lightweight open source remote management tool, *International Journal of Network Management*, May 2009, vol. 19, no. 3, pp. 237–252.
6. Levis, P.A.; TinyOS: An open operating system for wireless sensor networks, In *Proceedings of the 7th International Conference on Mobile Data Management, MDM'06*, 2006, Nara Japan (Invited Seminar).
7. The MagnetOS operating system, http://www.cs.cornell.edu/people/egs/magnetos/, October 2012.
8. Bhatii, S. et al.; Mantis OS: An embedded multithreaded operating system for wireless micro sensor platforms, *Mobile Networks and Applications*, vol. 10, no. 4, pp. 563–579, 2004.
9. Dunkels, A.; Gronvall, B.; Voigt, T.; Contiki—a lightweight and flexible operating system for tiny networked sensors, In *The 29th Annual IEEE International Conference on Local Computer Networks*, Washington, DC, pp. 455–462, 2004.
10. Han, C.; Kumar, R.; Shea, R.; Kohler, E.; Srivastava, M.; SOS—A dynamic operating system for sensor networks, In *Proceedings of the Third International Conference on Mobile Systems, Applications, and Services (Mobisys)*, Seattle, WA, 2005.
11. Lifton, J.; Seetharam, D.; Broxton, M.; Paradiso, J.; Pushpin computing system overview: A platform for distributed, embedded, ubiquitous sensor network, *Pervasive*, pp. 139–151, 2002.
12. Yannakopoulos, B.; Cormos: A communication-oriented runtime system for sensor networks, In *The 2nd European Workshop on Wireless Sensor Networks*, Februray 2005.
13. Dwivedi, A.; Tiwari, M.; Vyas, O.; Operating systems for tiny networked sensors: A survey, *International Journal of Recent Trends in Engineering*, vol. 1, no. 2, 2009.
14. Chatzigiannakis, I.; Mylonas, G.; Nikoletseas, S.; 50 ways to build your application: a survey of middleware and systems for wireless sensor networks, In *IEEE Conference on Emerging Technologies and Factory Automation, 2007*. ETFA. pp. 466–473, 25–28 September 2007.
15. Crossbow technologies, http://www.xbow.com/, October 2012.
16. The TinyOS alliance, http://www.tinyos.net/scoop/special/tinyos_alliance, October 2012.
17. Levis; Madden; Polastre; Szewczyk; Whitehouse; Woo; Gay; Hill; Wesh; Brewer; Culler; TinyOS: An operating system for sensor networks export, *Ambient Intelligence*, pp. 115–148, 2005.
18. Energy Efficient Sensor Networks (EYES), http://cordis.europa.eu/fetch?ACTION=D&CALLER=PROJ_IST&QM_EP_RCN_A=61532, accessed April 22, 2013.
19. Aziz, M.H.; Ong, C.; Yam, J.C.M.; Lee, C.W.; TCO reduction, In *The Proceedings of the 9th Asia-Pacific Conference on Communications*, vol. 3, 1147–1151, 2003.
20. Kakadia, D.; Thomas, T.; Vembu, S.; Ramasamy, J.; Enterprise management systems part I: Architectures and standards, Sun BluePrints TM, April 2002. http://www.sun.com/blueprints, October 2012.
21. Hochstein, A.; Zarnekow, R.; Brenner, W.; ITIL as common practice reference model for IT service management: Formal assessment and implications for practice. In *Proceedings of the IEEE International Conference on e-Technology, e-Commerce and e-Service*, pp. 704–710, 2005.
22. From OpenView to Open Source. GroundWork Open Source: San Francisco, CA, 2006.
23. Westerinen, A.; Bumpus, W.; The continuing evolution of distributed systems management. *IEICE Transactions on Information and Systems*, vol. 86, pp. 2256–2261, 2003.

24. Kalochristianakis, M.; Varvarigos, E.; Vassilakis, K.; Paraskeyas, M.; Considerations for successful enterprise information systems deployment: The case of the Greek School Network, In *International Conference on Service Systems and Service Management*, 2012.
25. Consortium for studying, evaluating and supporting the introduction of Open Source software and Open Data Standards in the Public Administration. Deliverable 2.1 for WP2 Catalogue of available Open Source tools for the PA. COSPA: Bozen-Bolzano, Italy, 2005.
26. Song, K.; Lee, S.; UPnP-based sensor network management architecture and implementation, In *Second International Conference on Mobile Computing and Ubiquitous Networking (ICMU 2005)*, 2005.
27. Ruiz, L.; Nogueira, J.; Loureiro, A.; MANNA: A management architecture for wireless sensor network, *IEEE Communications Magazine*, vol. 41, no. 2, pp. 116–125, 2003.
28. Levis, P.; Culler, D.; Mate: A tiny virtual machine for sensor networks, In *Proceedings of the International Conference on Architectural Support for Programming Languages and Operating Systems*, New York, pp. 100–111, 2002.
29. Fok, C.; Roman, G.; Mobile agent middleware for sensor networks: An application case study, In *Proceedings of the 4th International Conference on Information Processing in Sensor Networks*, Los Angeles, CA, pp. 382–387, 2005.
30. Cougar Project. http://www.cs.cornell.edu/database/cougar, accessed April 22, 2013.
31. Eduardo, S. et al.; A message-oriented middleware for sensor networks, In *Proceedings of the 2nd Workshop on Middleware for Pervasive and Ad-Hoc Computing*, pp. 127–134, 2004. New York.
32. A network virtual machine for real-time coordination services, http://www.cs.virginia.edu/wsn/nest.html, accessed April 22, 2013.
33. Scalable coordination architectures for deeply distributed systems, http://www.isi.edu/div7/scadds, accessed April 22, 2013.
34. Gellersen, H.W.; Schmidt, A.; Beigl, M.; Multi-sensor context-awareness in mobile devices and smart artefacts, *Journal of Mobile Networks and Applications*, vol. 7, no. 5, October 2002, pp. 342–251.
35. Heinzelman, W.; Chandrakasan A.; Balakrishan H.; Energy efficient communication protocol for wireless microsensors networks, In *Proceedings of the Hawaii International Conference on System Sc*, vol. 21, no 5, pp. 1032–1043, Maui, Hawaii, 2000.
36. Xu, Y.; Geography-informed energy conservation for ad hoc routing, *Mobicom'01*, Rome, Italy, pp. 203–212, 2001.
37. Chen, J.; Lu, H.; Lee, M.; WSNView system for wireless sensor network management, In *The 11th IASTED International Conference on Internet and Multimedia Systems and Applications*, Honolulu, Hawaii, pp. 126–131, 2007.
38. Madden, S.; Hellerstein, J.; Hong, W.; TinyDB: In network query processing in TinyOS, *ACM Transactions on Database Systems*, vol. 30, no. 1, pp. 122–173, 2003.
39. Touron; Crossbow: Moteview interface, Crossbow, 2005, http://www.xbow.com/, accessed April 22, 2013.
40. The list of available projects in the SourceForge hosting portal. http://sourceforge.net/softwaremap/index.php, accessed April 22, 2013.
41. OpenAudit. http://www.open-audit.org/, accessed April 22, 2013.
42. Atmel AVR8 microcontrollers, http://www.atmel.com/products/avr/, accessed April 22, 2013.
43. MSP430 16-bit ultra-low power MCUs, http://focus.ti.com/mcu/docs/mcuprodoverview.tsp?sectionId=95&tabId=140&familyId=342, accessed April 22, 2013

Chapter 22

Quality of Services in Wireless Sensor Networks

Rawya Yehia Rizk
Port Said University

Contents

22.1 Introduction

It is well known that quality of services (QoS) is an overused term with various meanings and perspectives. In general, QoS is a measure of the service quality that the network offers to the applications/users. Wireless sensor networks (WSNs) are an emerging technology that has gained great interest from researchers because of the widespread availability of cheap and tiny components and their countless applications. WSNs are used in numerous applications, including military, wildlife observation, fire detection, battlefield support, and environment monitoring.

Distinguished from traditional wireless networks, a sensor network has many unique characteristics, such as denser node deployment, higher unreliability of sensor nodes, asymmetric data transmission, severe power, and computation and memory constraints, which present many new challenges for the development and make QoS in sensor networks a rich area of research.

This chapter presents the challenges for QoS support in WSNs, QoS performance metrics in WSNs, QoS perspectives in WSNs, QoS protocols, types of services in WSNs, and the mechanisms to achieve QoS in WSNs.

22.2 Challenges for QoS Support in WSNs

Because WSNs have to interact with the environment, their characteristics can be expected to be very different from other conventional data networks. Thus, while WSNs inherit most of the QoS challenges from general wireless networks, their particular characteristics pose unique challenges as follows (Bhuyan et al. 2010; Karl et al. 2003; Wu et al. 2001; Chakrabarti et al. 2001; Seah 2004).

22.2.1 Autonomous

As in ad hoc networks, sensor networks can be deployed in an autonomous manner without the need for existing infrastructure. They can be set up anytime, anywhere, without the need for any central administration. As such, each node in the network has to act both as a host as well as a router to forward data packets to the sink (or other nodes in the network).

22.2.2 Scalability

A generic WSN consists of hundreds or thousands of sensor nodes densely distributed in a terrain. Therefore, QoS support designed for WSNs should be able to scale up to a large number of sensor nodes.

22.2.3 Random Deployment

Because of the nature of the applications, the network is expected to be randomly deployed, covering several hundreds or even thousands of sensor nodes in the terrain.

22.2.4 Network Dynamics

As the sensor nodes have limited power supplies and are susceptible to node failures, they may die and cause the network topology to change. Although most sensor networks are assumed to be relatively static, a sensor node may still deviate from its initial deployed location under the influence of its physical surroundings—such as winds, currents, or even wildlife—which also causes the topology to change.

22.2.5 Resource Constraints

As sensor nodes are typically small in size and battery powered, the constraints on resources involve energy, bandwidth, memory, buffer size, processing capability, and limited transmission power. Limited resources of sensor nodes, such as energy, bandwidth, and processing abilities, make these networks excellent candidates for incorporating QoS framework. Among them, energy is a primary concern because energy is severely constrained at the sensor nodes, and it may not be feasible to replace or recharge the battery for sensor nodes that are often expected to work in a remote or inhospitable environment. As a result, these constraints impose an essential requirement on any QoS support mechanisms in WSNs.

22.2.6 Unbalanced Traffic

In most applications of WSNs, traffic mainly flows from a large number of sensor nodes to a small subset of sink nodes. QoS mechanisms should be designed for unbalanced QoS-constrained traffic.

22.2.7 Data Redundancy

WSNs are characterized by high redundancy in the sensor data. However, while the redundancy in the data does help loosen the reliability/robustness requirement of data delivery, it unnecessarily spends much precious energy. Data fusion or data aggregation is a solution to maintain robustness while decreasing redundancy in the data, but this mechanism also introduces latency and complicates QoS design in WSNs.

22.2.8 Energy Balance

In order to achieve a long-lived network, energy load must be evenly distributed among all sensor nodes so that the energy at a single sensor node or a small set of sensor nodes will not be drained out very soon.

22.2.9 Data-Centric

In contrast with ad hoc networks, which are address-centric, sensor networks are usually data-centric. Instead of point-to-point communications between individual nodes in the network, the flow of data in a sensor network is always unidirectional toward a centralized sink.

22.2.10 Multiple Sinks

WSNs should be able to support different QoS levels associated with different sinks. There may exist multiple sink nodes, which impose different requirements on the network. For instance, one sink may ask sensor nodes located in the northeast of the sensor field to send a temperature report every one minute while another sink node may only be interested in an exceptionally high temperature event in the southwest area.

22.2.11 Application Specific

The possible applications of WSNs are numerous while being diverse in nature, which makes analyzing and designing QoS support for each application an important task. At the same time, these applications require different types of QoS support from the network for optimum performance. For instance, some applications may require a diverse mixture of sensors for monitoring temperature, pressure, and humidity, thereby introducing different reading rates at these sensors. Such a heterogeneous environment makes QoS support more challenging. Therefore, protocols and algorithms that are designed for WSNs are likely to be application-specific.

22.2.12 Packet Criticality

The content of data or high-level description reflects the criticality of the real physical phenomena and is thereby of different criticality or priority with respect to the quality of the applications. QoS mechanisms may be required to differentiate packet importance and set up a priority structure. As a result, QoS support for the network may have to take at least a few of the challenges described above into account when an application is specified.

22.2.13 Noisy Medium

The physical environment in which the sensor networks are deployed is usually harsh, such as on battlefields, in large warehouses, or even on the ocean floor. Sensor nodes are more susceptible to failures because they are in close contact with both the physical phenomenon to be monitored as well as the physical environment. For example, sensor nodes that are deployed in open areas may be tampered with by humans or wildlife, and sensors that are thrown in large numbers into chemically contaminated fields may corrode and become faulty.

22.2.14 Standardization

The lack of standardization in WSNs makes it hard to implement a QoS solution. There are as yet no standardizations in most WSN layers of functionality to be able to build a QoS based on them. ZigBee may consider a first attempt.

22.2.15 Variable Data Rate

Although a low data rate is sufficient for simple WSN applications, some other applications require moderate data rates. Then, the WSNs require a variable bit rate despite the required data rate for sensor networks being not as high as multimedia transmissions.

22.2.16 Heterogeneous Networking

Most sensor networks are heterogeneous, that is, there are nodes with different capabilities and requirements. Typically, the network has some full-function device (FFD) that collects data from different sensors, processes them, and forwards them to a central monitoring station. An FFD has fewer restrictions with respect to processing complexity and energy consumption. The sensor nodes themselves, on the other hand, are usually reduced function devices (RFDs) with extremely stringent limits on complexity and power consumption.

22.3 QoS Performance Metrics in WSNs

Although QoS is an overused term, there is no common or formal definition of this term. Conceptually, it can be regarded as the capability of providing assurance that the service requirements of applications can be satisfied.

Each WSN is usually deployed for a specific application, such as environmental monitoring or target tracking. In addition, each of these networks has their own unique characteristics and constraints. Consequently, the QoS performance metrics in WSNs may differ significantly from those that are used in traditional networks.

In traditional networks, most of the applications are end-to-end. These applications are often concerned with data transport between two specific autonomous computers. The QoS are perceived from a transport layer vision, which supports services for upper layers. Typical parameters for expressing QoS at the network layer are mentioned as throughput, packet delay, jitter, and packet loss. In non–end-to-end applications, communication is between a group of computers with another single computer or group of computers. Most typical applications of WSN are non–end-to-end. One end of all applications is a single sink node or multiple sink nodes, and another part is a group of deployed sensor nodes. Because of the distinct characteristics of WSN applications, traditional metrics cannot fully characterize the QoS in WSNs. Another application level QoS is needed, such as data accuracy, coverage, reliability, energy efficiency, and network lifetime, etc. In this section, the QoS metrics in WSNs are classified into two main parts: the traditional metrics and the metrics specified for the WSNs. They can be regarded as network-level QoS and application-level QoS, respectively.

22.3.1 Network-Level QoS Metrics

In any network, some QoS parameters may be used to measure the degree of satisfaction of the services, such as throughput, delay, jitter, and packet loss rate. There are many other QoS parameters worth mentioning, but these four are the most fundamental (Muthukarpagam et al. 2010).

22.3.1.1 Throughput

Throughput is the effective number of data transported within a certain period of time. In general, as the throughput of the network increases, the performance of the system increases. Throughput in a sensor network is expressed by the proportion of the information sensed in the environment, which has successfully reached the gateway. It is not usually as significant as other parameters because, in a typical use case, a sensor node sends small packets relatively seldomly. However, the use of acoustic and imaging sensors requires significant throughput as data must be streamed through the network. The nodes that generate high-speed data streams, such as a camera sensor node used to transmit images for target tracking, often require high throughput. Thus, certain WSN applications require maximizing throughput and possibly throughput guarantees. In order to improve the resource efficiency, furthermore, the throughput of WSN should often be maximized.

22.3.1.2 Delay

Delay is the time elapsed from the departure of a data packet from the source node to the arrival at the destination node, including queuing delay, switching delay, propagation delay, etc. In WSN, the delay is the time taken by the information collected by the WSN to travel from the sensing area to the gateway where the information is processed. Many of the emerging WSN deployments involve delay-sensitive applications with real-time delay constraints. Real-time does not necessarily mean fast computation or communication; a real-time system is unique in that it needs to execute at a speed that fulfills the timing requirements.

Sensor network applications are either periodic applications, such as environmental monitoring, or event-driven applications, such as target tracking. Event-driven applications tend to have stricter delay constraints; the sink must be able to receive notification that a particular event has occurred in a particular region of the network within a short time period after the occurrence so that it can react appropriately.

Meeting such delay constraints may require both hardware efficiency at the level of the clock of the WSN and software efficiency by deploying efficient routing techniques that can improve delay and on-time packet delivery.

22.3.1.3 Jitter

Jitter is generally referred to as variations in delay. It is often caused by the difference in queuing delays experienced by consecutive packets.

22.3.1.4 Packet Loss

Packet loss rate is the percentage of data packets that are lost during the process of transmission. It can be used to represent the probability of packets being lost. A packet may be lost as a result of, for example, congestion, bit error, or bad connectivity. This parameter is closely related to the reliability of the network.

22.3.2 Application-Level QoS Metrics

Some of the key performance metrics that should be considered in QoS provisioning in WSNs are node deployment, connectivity, number of sensors, data accuracy, coverage, reliability, energy efficiency, and network lifetime. These metrics cooperate to support real-time QoS for WSN applications.

22.3.2.1 Node Deployment

Node deployment in WSNs is application-dependent and affects the performance. Once deployed, sensor nodes are unattended and expected to operate for as long a time as possible. It is infeasible to replace one or more nodes in possibly harsh inaccessible environments, such as a battlefield, resulting from any kind of failure.

22.3.2.2 Network Connectivity

Network connectivity is one of the important QoS measures in WSNs. Poor connectivity causes network partitioning and disables the WSN.

22.3.2.3 Optimum Number of Sensors

An optimum number of sensors sending information toward information-collecting sinks or base stations is called sensor network resolution. This service quality, because of sensor deaths and replenishments or activation, is required to specify the optimal number of sensors for sending information at any given time.

If the WSN has N sensor nodes, the average capability of each node is scaled to $1/\sqrt{N}$; this means that the average capability of each node decreases with the increase in the number of nodes. In other words, if the number of nodes is high, the performance of the network will be poor. However, if the number is very low, there will be many areas that are not covered by the nodes, leading to the poor performance of the network. Then, tradeoffs are involved when determining the number of sensors in the network.

22.3.2.4 Reliability

Common methods for increasing reliability in communications networks are using acknowledgements and error correction. Acknowledgement mechanisms can be used to identify the packet losses, and accordingly, retransmissions can be performed in time to fix the problems (Rizk et al. 2011a).

In addition, redundancy is another method used to increase reliability as the network is able to recover from the loss. In an event-driven, object-tracking application, redundancy is used to satisfy availability and reliability. Every sensor has an affecting region. Sensors whose sensing radius crosses the affecting region of an event can sense it and report the information about this event to the sink. All sensors covered by the event will report to sink. This causes the high reliability in event delivery to the sink node but consumes more power.

22.3.2.5 Energy Efficiency

Energy efficiency is one of the most challenging aspects involved when designing protocols and considering QoS support in WSNs because of the battery-limited operation of sensor devices. The energy limitation in WSNs is directly related to the lifetime of the network. A sensor node that fails because of lack of energy is unable to sense the physical environment or communicate with its neighbors. This may lead to network partitions, which, in turn, affect the network lifetime (Kang et al. 2005).

As computation is often much less energy-consuming than transmitting, some of the communications may be traded against computation. For example, data may be processed to fit into smaller packets or by performing data aggregation. However, the aggregation has a trade-off between energy usage and reliability as a large amount of data may be lost on a missed packet. In addition, the data accuracy decreases when the aggregation is performed by combining nonredundant data.

Another method for designing a low-power WSN is to reduce the duty cycle of each node. The drawback is that as the wireless sensor node stays longer in sleep mode to save energy, there is less chance that the node can communicate with its neighbors. In addition to creating scalability challenges resulting from the need for a more complicated synchronization technique to keep more nodes in a low duty cycle, this will decrease the network responsiveness and may also lower reliability because of the lack of the exchange of control packets and delays in packet delivery.

22.3.2.6 Network Lifetime

The network lifetime is the maximum time before any node in the network drains up its energy. However, the overview of the network lifetime from the perspective of the application (or user) is different. Because of the high density in which nodes are usually deployed, the sensed data is highly correlated; therefore, even if a particular node fails after expending all its energy, a neighboring node may still be able to perform the required functionalities of sensing and communication in that spatial location. As such, network lifetime cannot be considered based on the energy consumption of the nodes alone, but must also take into account the topological location of the nodes (Kang et al. 2005).

Therefore, the more generic definition of network lifetime is the time period during which the network continuously satisfies the application requirements. These application requirements may be specified in terms of coverage or delay and may vary depending on the specifications of each application.

22.3.2.7 Coverage

In WSNs, each sensor node obtains a certain view of the environment. A given sensor's view of the environment is limited both in range and in accuracy; it can only cover a limited physical area of the environment. Figure 22.1 presents the area covered by the sensor as a disk centered at the sensor with radius R_s. Two nodes are considered neighbors if the Euclidean distance between them is less than or equal to the communication range R_c. Coverage areas may have arbitrary shape and need not be circular as shown in the figure.

One of basic problems in WSN is the calculation of sensing coverage. The coverage of a sensor network is the ratio of the space that is covered by the sensor nodes to the total space of interest. The main objective of coverage is to ensure that each physical region in the space of interest is of at least one sensor node. The coverage of the sensor network is closely correlated with the denseness of the node deployment. A sparse network results in sparse coverage, whereby the space of interest is partially covered by sensors. Dense networks result in dense coverage, whereby the space of interest is almost fully covered by the sensors. In very dense networks, which have redundant coverage, the space of interest is covered by multiple sensors, resulting in high spatial correlation in the data that is observed by nodes that have geographical proximity.

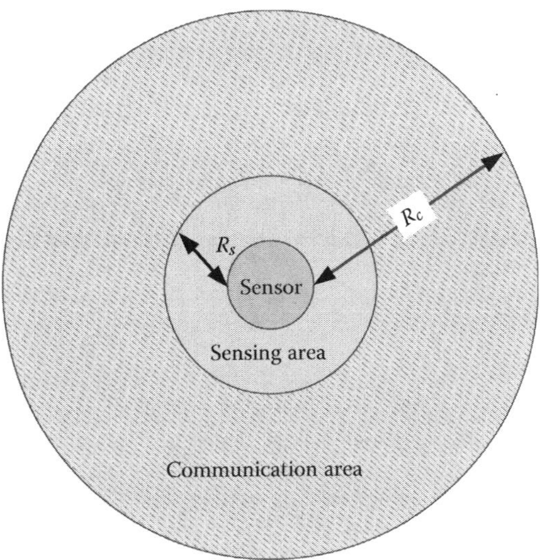

Figure 22.1 Sensor model.

This QoS metric represents the surveillance measure of WSN in a particular application. In coverage, we try to find weak points in the sensor field and activate required off-sensor nodes or suggest future deployment or reconfiguration schemes for improving the surveillance measure (Sanli et al. 2004; Cardei et al. 2005).

22.3.2.8 Spatial Accuracy

Wireless sensor nodes are usually deployed in a large terrain, and each sensor node is able to sense data from only a certain part of the space of interest. In order for the sink to collect the sensed data from the sensor nodes, it must have spatial information on all data that are collected (Adlakha et al. 2003).

Localization techniques have to be used to obtain the relative locations of the nodes and provide spatial accuracy of the sensed data. Also, the global positioning system (GPS) can be used to provide the location of each node for the WSNs that are deployed outdoors.

The type of localization technique determines the level of the spatial accuracy. It can be high spatial accuracy or low spatial accuracy. However, as high spatial accuracy usually results in higher overheads than low spatial accuracy, the appropriate level of spatial accuracy should also be based on the application requirements of the network.

22.3.2.9 Temporal Accuracy

Temporal accuracy is required in WSNs to ascertain the time period during which an event occurs. In wired and centralized networks, such as the Internet, it is possible to achieve high temporal accuracy because of the high propagation speed of the communication medium. However, WSNs typically communicate in a distributed, multi-hop manner, via a shared communication channel. In many applications of WSNs, such as underwater networks, the sensor nodes suffer from factors, such as link instability and slow propagation speed, that make it difficult to achieve time synchronization in the network, thus leading to low temporal accuracy. Some time synchronization protocols have been proposed to achieve high temporal accuracy (Adlakha et al. 2003).

22.4 QoS Perspectives in WSNs

The QoS perspectives in WSNs are classified into two categories: application-specific, which focuses on the quality of the application itself, and network-specific, which provides service quality during delivery of the data by the communication network (Yigitel et al. 2011; Min et al. 2002).

22.4.1 Application-Specific QoS

Because WSNs have diverse applications with different QoS requirements, it is impossible to analyze them individually. Also, it is unlikely that there will be a one-size-fits-all QoS support solution for each application. From this perspective, the application-specific QoS are considered the QoS parameters, which are directly related to the quality of applications. Some of the application-specific QoS parameters are lifetime, data accuracy, aggregation delay, fault tolerance, coverage, optimum number of active sensors, etc. In addition, the application demands certain requirements from the deployment of sensors, the number of active sensors, the measurement precision of sensors, and so on.

22.4.2 Network-Specific QoS

A network-specific perspective provides service quality during delivery of the data by the communication network. From this perspective, network resources are utilized efficiently in each layer of the communication protocol stack to fulfill the requirements imposed by the carried data, such as latency, packet loss, and reliability.

Although it is difficult to analyze each possible application in WSNs, it is sufficient to analyze each class of application classified by data-delivery models because most applications in each class have common requirements on the network. From the point of view of network QoS, the network is concerned with how to transmit the sensed data from the sensor field to the sink node fulfilling the corresponding required QoS.

Generally, there are three basic data-delivery models in sensor network. These are event-driven, query-driven, and continuous. The application requirements in these models are different. The factors that characterize the applications requirements are (Yigitel et al. 2011):

- Interactivity: The application may be interactive or noninteractive.
- End-to-end: The application may require end-to-end or non–end-to-end performance.
- Delay tolerance: The application may be delay tolerant or not.
- Criticality: The application may be mission critical or not.

22.4.2.1 The Event-Driven Application

In the event-driven application, sensor nodes report data only if an event of interest occurs. The sensors detect the occurrence of a certain event and take action accordingly. The application needs to detect the events and accordingly take an appropriate action as quickly as possible and as reliably as possible. It means that on one side of the application there is a sink node and on the other side is a group of sensor nodes, which are affected by these events. The data flows from these sensors are likely to be highly correlated, thereby containing much redundancy. Further, the data traffic generated by a single sensor may be of very low intensity. However, very bursty traffic may be generated by a set of sensors resulting from a common event.

The event-driven applications in WSNs are mostly non–end-to-end, i.e., one end of the application is the sink, and the other end is not a single sensor node, but a group of sensor nodes within the area that is influenced by the event. In addition they are interactive, delay tolerant (real-time), and mission critical.

22.4.2.2 The Query-Driven Application

In the query-driven application, the queries are generated by the sink node and sent to sensor nodes enquiring occurrence of certain event. To save energy, queries can be sent on demand. A query-driven data-delivery model is very similar to the event-driven model with an exception: Data is pushed to the sink without any demand by the sensor nodes in an event-driven model while data is requested by the sink and pushed by the sensor nodes in the query-driven model. Hence, contrary to the one-way traffic of the event-driven model, two-way traffic comes into the scene, which consists of requests of the sink and replies of the sensor nodes. A query may also be used to manage and reconfigure the sensor nodes. Both requests and replies must be delivered as quickly as possible and as reliably as possible.

The important points mentioned for event-driven delivery are also relevant for query-driven delivery. Most query-driven applications in WSNs are interactive, non–end-to-end, query-based, delay-tolerant, and mission-critical applications.

22.4.2.3 Continuous Application

In the continuous model, sensors send their data continuously to the sink at periodic intervals. This can be considered as the basic model for traditional monitoring applications based on data collection. The data rates can be usually low. To save energy, the radios can be turned on only during data transmissions if scalar data is collected. The data can be real-time audio, video, and image or non–real-time data. Real-time data, such as voice or image, are delay-intolerant and require a certain bandwidth requirement. Also packet losses are tolerated in a limited threshold. For periodically collected non–real-time data, latency and packet losses are tolerable.

22.4.2.4 Hybrid

In the hybrid model, two or more data-delivery models coexist in the same network. Hence, carried traffic must be classified, and requirements of these traffic classes must be satisfied.

The application requirements of the data delivery models are summarized in Table 22.1.

22.5 QoS Protocols

In WSN, as nodes cooperate to perform a single task, the wireless medium is employed to coordinate the activities of the sensors and transmit data collected to the sink. Thus, QoS in WSN implements protocols in each layer of the protocol stack. QoS protocols can be classified with respect to the layers of WSNs, starting from the top (application layer) down to the physical layer. This section presents a map for each layer with the specific functions in this layer.

Figure 22.2 shows the sensor network protocol stack, including the physical layer, data-link layer, network layer, transport layer, and application layer. These layers relative to the five protocols of the Internet protocol stack. In addition, the protocol stack also includes an energy-management platform, mobile-management platform, and task-management platform. These management platforms allow sensor nodes to work energy-efficiently, that is, forwarding data in the mobile nodes sensor network and supporting multi-task and resource sharing. The functions of the platforms and layers of protocols are as follows (Akyildiz et al. 2002):

Table 22.1 Application Requirements of the Data-Delivery Models

	Event-Driven	*Query-Driven*	*Continuous*	*Hybrid*
Interactivity	√	√	X	√
End-to-End	X	X	X	X
Delay Tolerance	X	Query-specific	√	X
Criticality	√	√	√	√

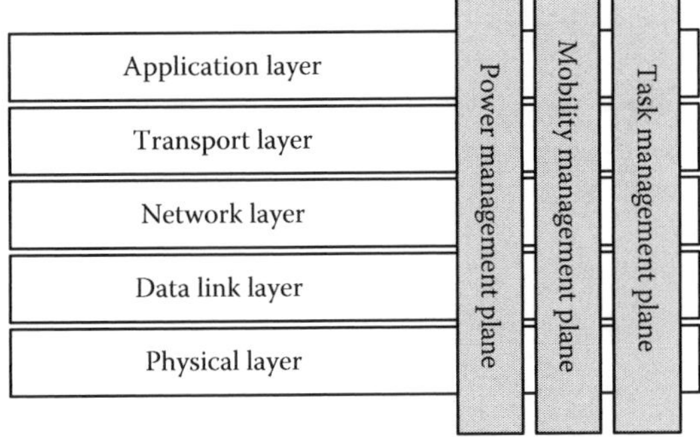

Figure 22.2 Sensor network protocol stack.

22.5.1 Management Platforms

22.5.1.1 Energy Management Platform

This platform manages the sensor nodes' energy. Saving energy is a challenge in all protocol layers.

22.5.1.2 Mobility Management Platform

It detects and registers the mobile sensor nodes to maintain routes of the sink node and allows the sensor node to dynamically track the location of its neighbors.

22.5.1.3 Task Management Platform

This platform balances and schedules the monitoring tasks in a given region.

22.5.2 Layers

22.5.2.1 Physical Layer

This layer seeks to avoid the interference with other networks or natural sources of radiation, employing techniques that modify the modulation and frequency of operation in order to enhance the signal-to-noise ratio (SNR). Providing QoS in this layer is out of scope for this chapter.

22.5.2.2 Data-Link Layer

The data-link layer ensures that data is transferred correctly between adjacent network nodes. In this layer, the scheduling of medium access and the sequence of packets to be sent are changed to supply the QoS requirements. These changes may be achieved by packet reordering and by priority control and admission policies. It is possible to adjust the amount of control packets sent to increase the QoS of a data packet.

The data-link layer is divided into two sublayers: the media access control (MAC) sublayer and the logical link control (LLC) sublayer. The MAC sublayer controls how a computer on the network gains access to the data and permission to transmit it. The LLC sublayer controls frame synchronization, flow control, and error checking. The MAC layer in WSN joins together almost all problems from traditional wired and wireless networks in additional to other new challenges, such as the lack of unique ID, power constraints, and the frequent changes in WSN topology. Current proposed MAC protocols in WSN are concerned mainly about power conserving. They don't support real QoS because of the trade offs between energy efficiency and QoS capability.

22.5.2.3 Network Layer

The network layer is responsible for determining the route from source to destination and managing traffic problems. Similar to the link layer, it is possible to use packet prioritization. In WSN, because of the variability of link quality, both the amount of copies sent and the number of distinct routes for a given packet can be adjusted according to its priority, increasing the probability of a successful delivery of a packet. In addition, routing protocols are the main concern for providing QoS in this layer.

Routing protocols in WSN can be categorized as the following:

1. Data-centric: Data are disseminated between sensors without the need for a global unique ID. It depends on the naming of desired data.
2. Hierarchical: Sensors are controlled by a sensor (cluster-head) to aggregate data. A cluster-head is either a special (more powerful) node or an elected sensor among each cluster.
3. Location-based: These protocols are location-aware; usually by utilizing a GPS. The ability to find the location makes it easier to route data to single and specific region instead of broadcasting traffic to all regions.
4. QoS based: Protocols that ensure some QoS requirements, such as optimum energy consumption, high throughput, and low delay.

22.5.2.4 Transport Layer

The transport layer is responsible for the transmission control; it is an important part of the communication service's quality. Generally, the transport layer provides main services, such as segmentation and reassembly, multiplexing, reliability, and flow and congestion control.

Normal transport protocols developed for wired or wireless communication are implemented with address-centric and end-to-end data delivery applications. Then, they do not address WSN resource constrains. Therefore, developing transport protocols specific for WSN should take the following points into consideration:

- Reliability for both ways of communication: sink-to-sensor and sensor-to-sink
- A good congestion control mechanism increases network efficiency and saves power
- Self-configuration approaches to adapt to frequent changes in network topology
- Energy-aware
- Data-centric

The need for reliable transport protocols is crucial because generating trusted data is a very important goal of any WSN. Therefore, reliability of data delivery is a main concern for providing the QoS in this layer. The congestion control mechanisms in WSNs that achieve reliable data delivery depend on the type of traffic. Traffic in WSNs is either from sensor to sink (sensed information) or from sink to sensor (control/update information):

- Sensor-to-sink: Some refer to this process as event-to-sink because it does not matter which sensor has generated the information, and the main concern is about the information itself. Thus, it is called a data-centric model of delivery.
- Sink-to-Sensor: Data sent from sink to sensor are mainly queries, updates, or operational instructions. It may include firmware or operating system updates. These need to be transferred reliably to sensors. Mostly, sink to sensor suffers less congestion than the opposite path. Traffic in this direction can be unicast or multicast. In sink to sensor, a multicast of information usually happens when there are one sender and multiple receivers. Multicasting is the process of sending a message to selected multiple recipients who have joined the appropriate multicast group. The sender has to generate only one data stream; a multicast-aware router will forward a multicast to a particular network only.

22.5.2.5 Application Layer

The application layer includes a series of tasks based on monitoring the application layer software. Also, adjusting the frequency in which data are sent and the adaptive compression techniques of audio and video applications can be applied in WSNs in order to reduce the bandwidth used. In user/application protocols, many parameters can be defined by the user to achieve some QoS in WSN:

- **Fidelity:** A user can instruct the network to send their queries back to the sink in pairs or to not accept any event that has been seen by *n* number of nodes only.
- **Update (freshness):** Sensors should send queries to the sink periodically even there are no events.
- **Mode:** User/application defines how the sink will interact to events according to the data-delivery models: event-driven, query-driven, and continuous.

From the side of the network, providing QoS to application or user defines new QoS parameters:

- **Query processing:** is the ability of the WSN to perform in-network processing instead of sending raw data to sink. For example, a sink may send a query: "What is the highest temperature in the forest?" In response to this query, each sensor will send back the temperature to the sink, which, in turn, will calculate the highest temperature or let the nodes in the network find it out themselves and then send the result only. This can be accomplished with the help of aggregation mechanisms.
- **Coverage:** High coverage is key to a robust sensor network, and it considers one of QoS measures. Having good coverage algorithms can save power and improve sensor network connectivity. In an area that is covered by multiple sensors, some sensors can be turned off in order to save power, or one or two sensors only can be instructed to sense the environment in order to provide less redundant data.

22.6 Types of Services in WSNs

This section presents the types of services in WSNs and their requirements. This facilitates understanding of the nature of these services and the required protocols for realizing QoS. We classify the services in WSNs according to the type of application into two basic classes:

22.6.1 Real-Time Services

This type of service is delay-sensitive. It corresponds to event-driven traffic or query-driven traffic that requires immediate attention in WSNs. Real-time sensor systems have many applications, especially in intruder tracking, medical care, re-monitoring, and structural health diagnosis.

22.6.2 Non–Real-Time Services

This type of service is best effort service. It corresponds to continuous applications that have periodic reporting in WSNs. Traffic in this class has no delay guarantee requirements.

Table 22.2 shows the classes of services with the associated type of applications.

Existing work on QoS of computer networks has used the following QoS metrics (Braden et al. 1994; Blake et al. 1998; Wang et al. 2005):

- Response time expected by users: This is the time elapsed between sending a request and the reception of the first response by the user.
- Delay: This is the time elapsed between the emission of the first bit of a data block by the transmitting end-system and its reception by the receiving end-system.
- Jitter: This refers to the variation of delay generated by the transmission equipment.
- Data rate: This refers to the raw data rate of encoded multimedia data before transmission, that is, the rate in which data are encoded.
- Required bandwidth: This is defined by the required data transfer rate, measured in bits per second, of each specific application in telecommunication. This metric includes raw data and overhead.
- Loss rate: This is the number of bits lost between two points in telecommunications after transmission.
- Error rate: This is the frequency of erroneous bits between two points in telecommunication after transmission.

Table 22.3 summarizes the desired QoS metrics of common categories of services, such as interactive video-on-demand, interactive audio on demand, telemetry, video broadcasting, audio broadcasting, and data represented by file transfer protocol (FTP). It is considered that most WSN applications fall into these categories. Each of these services is related to one of the two classes: either real time or non–real time (Chen et al. 2004).

Table 22.2 Classes of Services

Class of Service	*Type of Application*	*Examples*
Real-time services	Event-driven applications	Surveillance and target tracking
	Query-driven applications	Environmental control or habitat monitoring
	Hybrid applications	A surveillance application that sends both periodic temperature and event-triggered video data
Non–real-time services	Continuous applications	Surveillance or reconnaissance

Table 22.3 QoS Metrics

		QoS Metrics						
		Timelines			*Preciseness*			*Accuracy*
Class of Service	*Services*	*Response Time (s)*	*Delay*	*Jitter (ms)*	*Data Rate (bps)*	*Required Bandwidth (bps)*	*Loss Rate*	*Error Rate*
Real time	Interactive video on demand	2–5	<150 ms	<100		1.2–1.5 M	$<10^{-5}$	$<10^{-5}$
	Interactive audio on demand	2–5	<150 ms	<100		32–448 k	$<10^{-2}$	$<10^{-2}$
	Telemetry		<250 ms	<100		2k–25 M	0	0
	Video broadcasting	2–5	<10 s	<100		1.2–1.5 M	$<10^{-5}$	$<10^{-5}$
Non real time	Audio broadcasting	2–5	<400 ms	<100	56–64K	60–80 k	$<10^{-2}$	$<10^{-2}$
	FTP	2–5	Med	N/A	High	High	0	0

Interactive video-on-demand refers to a range of applications whereby users can request access to still and moving pictures. The QoS metrics for video-on-demand are given in Table 22.3. Response time should be less than 2–5 s. The typical bandwidth requirement is 1.5 Mbps.

For interactive audio on demand, the required bandwidth ranges between 32 and 448 kbps. Delay less than 150 ms is recommended for high-quality real-time traffic. A jitter rate of less than 100 ms is recommended. The loss rate and error rate should be less than 10^{-2}.

Telemetry refers to the telemetering and telecontrol of industrial processes. Telemetry is an example of a data service that requires real-time streaming performance. The QoS metrics for telemetry are given in Table 22.3. For telemetry, a delay value of 250 ms is recommended. The range of bandwidth requirements is between 2 kbps and 52 Mbps, depending on the specific telemetry system. A key differentiator from voice and video services in this category is the zero tolerance for information error and loss.

For video broadcasting, the required bandwidth is approximately 1.2–1.5 Mbps. Jitter is recommended to be less than 100 ms for a broadcast-quality video stream. The loss and error rates should be less than 10^{-5}.

For audio broadcasting, an end-to-end delay for high-quality real-time traffic is less than 400 ms. For jitter, no more than 100 ms is recommended for broadcast quality sound. A typical data stream for audio broadcasting is 56–64 kbps. The corresponding bandwidth requirement is approximately 60–80 kbps. The residual bit error rate is recommended to be lower than 10^{-2}.

FTP service requires a relatively high bandwidth, which is clearly influenced by the size of the file. Meanwhile, the size of files will also impact the time expected to finish the service. QoS metrics for FTP are given in Table 22.3. A requirement for response time is 2–5 s. Medium tolerance for delay is accepted. Jitter is not applicable (N/A), and the expected loss rate and error rate are zero.

22.7 Mechanisms to Achieve QoS in WSNs

In this section, some existing mechanisms that have been proposed in the literature to achieve QoS in WSNs are described. These mechanisms are classified according to their functions, such as sleep-wake scheduling, localization, data aggregation and/or fusion, network topology, clustering, cross-layer designs, routing, mobility, and security. Figure 22.3 summarizes the different mechanisms that achieve Qos in WSNs.

22.7.1 Sleep–Wake Scheduling

Sleep-wake scheduling can be used to achieve energy conservation in WSNs. It coordinates the sleep schedules of all the nodes such that data can still be forwarded efficiently to the sink. Sensors more or less uniformly alternate between sleeping and waking periods associated with relatively low and high power output. Most of the energy that is expended by a node is through transmission and sensing. The nodes can be put into sleep mode when they are not required to sense or transmit data to their neighboring nodes. However, this scheduling technique helps to increase energy efficiency and thus network lifetime; it causes higher latency. This is a result of nodes that are required for the data-forwarding process possibly being in sleep mode during the transmission (Kumar et al. 2005).

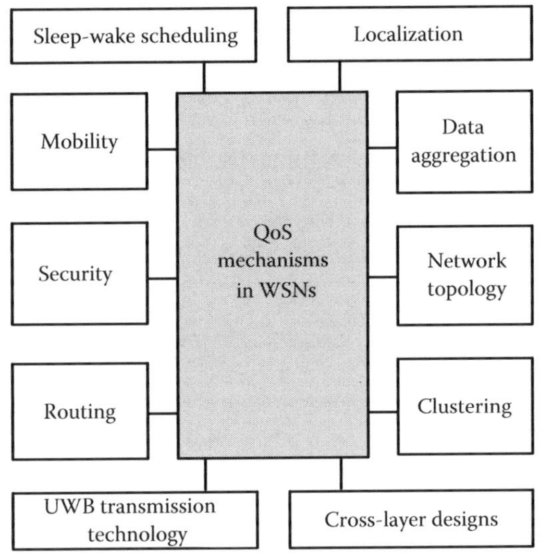

Figure 22.3 Mechanisms to achieve QoS in WSNs.

22.7.2 Localization

Localization provides an alternative mechanism of finding the physical locations of the sensor nodes in the network instead of making use of GPS, which is costly and infeasible indoors. It usually involves two phases: (i) ranging, which is the distance estimation of the node from the sink or other nodes in the network using techniques such as signal strength, angle-of-arrival (AoA), etc.; and (ii) iterative multilateration, which makes use of the range measurements from the previous phase to calculate a new location estimate. Hence, localization increases spatial accuracy at the expense of higher overheads, which reduces energy efficiency (Ganesan et al. 2003).

22.7.3 Data Aggregation

Data aggregation and/or fusion has been introduced as an in-network filtering and processing technique to help eliminate redundancy and conserve the scarce energy resources. In data aggregation, data that is coming from different sources en route is combined into a single data packet. Data fusion is similar to data aggregation in that data of different modalities, such as pressure, temperature, and salinity, are combined before data transmission (Kalpakis et al. 2002; Nath et al. 2004).

In typical sensor network scenarios, data is collected by sensor nodes throughout some area and needs to be made available at some central sink node(s), where it is processed, analyzed, and used by the application. Sometimes this data comes from more than one node located in the same area. The communication cost imposed as a result of this redundant data unnecessarily consumes energy of the nodes and bandwidth. In WSNs, data aggregation or in-network aggregation means that several information packets can be combined together and represented by same number of bits. In other words, the key idea is to combine data from different sensors to eliminate redundant transmissions and provide a rich view of the environment being monitored.

Once this is done, the energy consumption in the communication process will be reduced. This will be done if the data is cached within the aggregator, which yields to optimize the performance and reduce the number of transactions over the channels. Data aggregation techniques are tightly coupled with how data is gathered at the sensor nodes as well as how packets are routed through the network. They have a significant impact on energy consumption and overall network efficiency (e.g., by reducing the number of transmissions or the length of the packets to be transmitted).

There are two types of nodes in WSNs: normal nodes and aggregation nodes (aggregators). Normal nodes do not perform aggregation. They sense the data and send it to the sink. They also forward the data generated by other nodes. Aggregation nodes work as normal nodes and perform aggregation.

The aggregation process can be lossless or lossful. The lossless aggregation is the process of merging packets coming from different sources into the same packet without data processing. Receiving two packets carrying different physical quantities, for example, temperature and humidity, is an example of this type. These two values cannot be processed together, but they can still be transmitted in a single packet, thereby reducing overhead. At the sink, the original data can be perfectly reconstructed. This approach is also called in-network aggregation without size reduction. On the other hand, lossful aggregation is in-network aggregation with size reduction. It is the process of combining and compressing data coming from different sources in order to reduce the information to be sent over the network. As an example, assume that a node receives two packets from two different sources containing locally measured temperatures. Instead of forwarding the two packets, the sensor may compute the average of the two readings and send it in a single packet. This approach is able to reduce the amount of data to be sent over the network, but it may also reduce the accuracy with which the gathered information can be recovered at the sink. After the aggregation operation, it is usually not possible to perfectly reconstruct all of the original data.

However, stringent delay requirements can severely deteriorate the network lifetime. Henceforth, tradeoffs are involved when designing protocols for use in sensor networks.

22.7.4 Network Topology

Conventional WSNs have a single centralized sink, and all the sensor nodes have to send data to the sink. As a result, sensor nodes that are near the sink have to perform more data forwarding and packet transmissions, which lead to undesired behaviors such that increased contention and collisions near the sink and the nodes that are near the sink will drain up their energy faster. Subsequently, some proposals consider the use of more than one sink in network architecture, which provides spatially diverse routes in the network such that source nodes will avoid sending all the data in one direction and cause network deterioration. This helps to improve the load distribution of the network and increases the network lifetime at the expense of the physical deployment of more sinks (Schurgers et al. 2002; Romer et al. 2004).

Recent research considered the use of multi-input multi-output (MIMO) techniques (including MISO and SIMO). These techniques show a great improvement of energy efficiency in low-range communications for WSNs. However, physical implementation of multiple antennas at a small-size sensor node may not be feasible. As a result, a cooperative MIMO in which the multiple inputs and outputs are formed via cooperation is a solution for an energy-efficient communication.

22.7.5 Clustering

In the context of WSNs, clustering involves grouping nodes and electing a cluster head (CH) such that the non-CH nodes of a cluster can communicate with their CH directly. It is very hard to provide global synchronization in WSNs considering the large deployments and the number of sensor nodes. This challenge has led to the development of clustering mechanisms to simplify the synchronization and coordination (Heo et al. 2005).

Data aggregation is done by exploiting node clustering. CHs forward aggregated data to the sink directly or via other CHs. Thus, the collection of CHs in the network forms a connected dominating set. Clustering provides significant energy saving by improving inter-node connectivity and facilitating data aggregation and hence can be used to provide QoS support in terms of energy consumption and reliability.

Figure 22.4 shows a sensor network composed of multiple clusters of nodes. The nodes within the same cluster are closely spaced and cooperate in signal transmission and/or reception.

Clustering algorithms can be classified as static and dynamic.

- Static clustering: In static clustering, the cluster head and the members of the cluster are selected once during the deployment or initialization phase of the network. The role of the sensor nodes does not change in time. It is easy to employ and does not require any control messaging. However, the network lifetime and connectivity can be severely damaged because cluster heads consume more energy, and their batteries get depleted earlier.
- Dynamic clustering: In dynamic clustering, the clusters and the cluster head are reconstructed according to the current topology. Then, the load can be distributed evenly among cluster members. Hence, early battery exhaustion of cluster heads can be prevented. However, this method introduces significant overhead as a result of inter-cluster and intra-cluster control message exchanges.

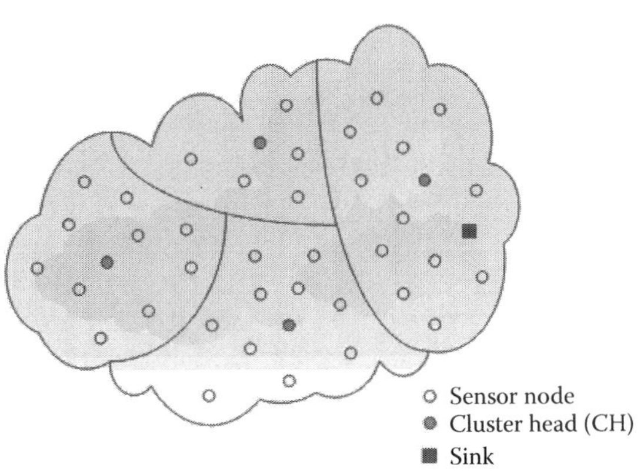

Figure 22.4 Network composed of multiple clusters of nodes.

22.7.6 Cross-Layer Designs

Although traditional networking paradigms promote the usage of a multi-layered protocol stack in which the different layers have minimal impact on each other, this does not lead to optimal performance. Cross-layered designs can help to improve network performance by providing interconnectivity across the different layers at the cost of eliminating the interdependency between adjacent layers. This architecture is based on the fact that parameters in the lower layer are reported to adjacent higher layers. This provides a mechanism in which higher layers can coordinate with lower layers (Akyildiz et al. 2006; Madan et al. 2005).

For example, if the application has QoS requirements, these requirements will be communicated and priority scheduling will be used on the data-link layer and the network layer. The whole protocol hence provides a mechanism of communicating parameters and merging of relevant protocol functions into one component.

22.7.7 Routing

Routing protocols for WSNs are used to transmit messages from sources to sinks. In WSNs, the routing protocols depend on application requirements and the network architecture, hence they need to adapt to any changes to the application requirements and network dynamics during the runtime of WSNs. Routing adaptability also includes the capability of providing fault tolerance, real-time support, energy balance, security, and handling variations in the traffic flow in WSNs. Nodes send rounds of messages (packets) and build local neighbor tables. These tables include the minimum information of each neighbor and location. This means that nodes must know their geographic location prior to neighbor discovery. Other typical information in these tables includes nodes' remaining energy, delay via that node, and an estimate of link quality. Once the tables exist, in most WSN routing algorithms, messages are directed from a source location to a destination address based on geographic coordinates (Schurgers et al. 2001).

As a result of node mobility and failures, the network topology may be continuously changing with time. As a consequence, some fundamental network operations, such as routing, localization, data exchange, etc., become much more challenging tasks compared with the stable network.

Depending on route determination, routing protocols can be classified in general into four categories as follows (Rizk et al. 2011b):

- Proactive or table-driven routing protocols: All routes are computed before they are used. This type of protocol periodically maintains all routing information. When a change of the network occurs, alternate routes are known to all nodes, and the change notification has to be propagated to maintain up-to-date routing information. Because all routes are known *a priori*, a proactive routing protocol is able to instantly respond from a sudden change of the network (low latency) as long as an alternate route exists. However, it requires large packet traffic to periodically maintain all routing information over the global network. This may cause network congestion and low effective bandwidth.
- Reactive or on-demand routing protocols: Routes are computed as they are needed. This type of protocol is initiated by only on-demand requests from a source. When a change in the network occurs, link-failure notification is propagated to reinitiate the route discovery protocol to find a new route until the message reaches the source. Reactive routing protocols are maintained by source-initiated on-demand requests with relatively lower cost over the ad

hoc network. However, it may experience increased latency to recover even a small change of the routing path on a dense network as in WSNs.

- Hybrid routing protocols: Uses both proactive and reactive routing protocols.
- Cooperative routing protocols: Nodes send data to a central node where more processing power and route information is available.

The routing protocols can be classified also as unicast, anycast, broadcast, concast, or multicast. Unicast routing is used to send a message generated by a sensor node to a single sink. Anycast routing is used to send a message from a sensor node to any of the sinks. Broadcasting is used to send a message from a sensor node to every other node in the network. Concasting is used to send a message from all the nodes to a single sink, and finally, multicasting is used to send a message from a node to a set of sinks. Figure 22.5 shows the classification of the routing protocols according to the network structure. The routing could be flat, hierarchical, or location-based routing (Rizk et al. 2011b).

In the flat network routing, each node typically plays the same role, and sensor nodes collaborate together to perform the sensing task. These protocols achieve delivery of packets from the source to the sink by means of multi-hop communication from one sensor node to another. Two main types of algorithms in flat routing are *flooding*, in which each node forwards data to all its neighbors, so too much redundant data occurs, and *data-centric routing*, in which there are no global identifiers for nodes; instead data is identified using attribute-based naming, which reduces the overhead comparing with the flooding.

The hierarchical routing protocols commonly use clustering techniques that organize the nodes into different levels, thereby giving the network a hierarchical structure. This organizes the nodes into different logical levels and defines different functionality to each sensor node based on their level in the hierarchy. The main advantage of hierarchical protocols is that they are capable of performing better in-network data aggregation, thereby improving the WSN throughput and consume less energy.

In location-based routing, sensor nodes are addressed by means of their locations. The distance between neighboring nodes can be estimated on the basis of incoming signal strengths. Relative coordinates of neighboring nodes can be obtained by exchanging such information between neighbors.

According to the protocol operation, routing protocols can be classified into negotiation-based routing where high-level data descriptors can be used in order to eliminate redundant data transmissions through negotiation. In multipath routing protocols, the data is routed through a path

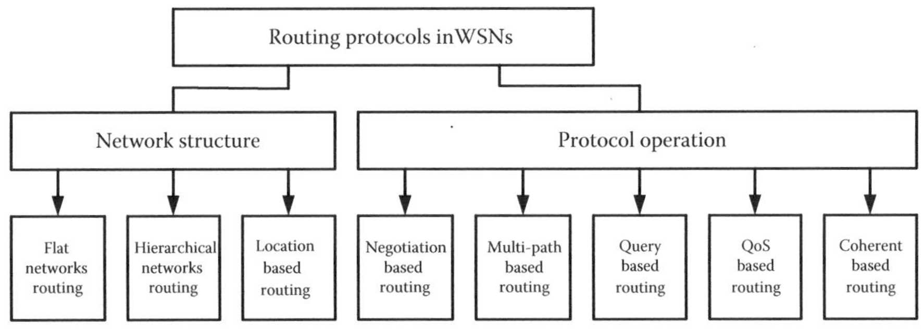

Figure 22.5 Classification of routing protocols in WSNs.

whose nodes have the largest residual energy. The path is changed whenever a better path is discovered. The primary path is used until its energy falls below the energy of the backup path, at which time the backup path is used. *Directed diffusion* is a good example of robust multipath routing and delivery.

In query-based routing, the destination nodes propagate a query for data (sensing task) from a node through the network, and the node that has the matching data sends it back to the node, which initiates the query.

In QoS-based routing, the network has to balance between energy consumption and data quality. QoS generally refers to the quality as perceived by the user while in the networking community. QoS is accepted as a measure of the service quality that the network offers to the users. In other words, QoS refers to an assurance by the network to provide a set of measurable service attributes to the end-to-end users in terms of fairness, delay, available bandwidth, and packet loss. A network has to provide the QoS while maximizing network resource utilization.

In coherent and noncoherent processing, data processing is a major component of the operation of WSNs. Hence, routing techniques employ different data-processing techniques. In general, sensor nodes will cooperate with each other in processing different data flooded in the network area.

22.7.8 Mobility

Mobility in sensor networks is highly essential for allowing communication between different connected components of the network. This also allows the operation of the sparse networks. Mobile networked systems combine the most advanced concepts in perception, communication, and control to create computational systems capable of interacting with the physical environment. Thus, it extends the individual capabilities of each network component and network user to encompass a much wider area and range of data.

Sensors often remain in a fixed location once deployed, in which case the topology of the sensor network remains static. However, if we take into consideration scenarios such that sensors may be out of service as a result of different reasons, such as nodes destroyed by high temperature or high pressure or nodes quitting the network because of energy depletion, the network cannot accomplish the task effectively, and the nodes need to be redeployed. The redeployment is good to adjust the coverage request of the network according to the changes in environment. It is desirable to deploy a minimum number of relay nodes to achieve certain connectivity requirements.

The mobility support may range from partial mobility to full mobility support. In partial mobility support, only a part of the nodes can be moved. The maximum degree of mobility may be limited to a maximum amount of mobile nodes. At the node level, mobile WSN can be categorized based on their role within the network into (Rizk et al. 2011b):

- Mobile Embedded Sensor: Mobile embedded nodes do not control their own movement, such as when attached to a shipping container or tethered to an animal.
- Mobile Actuated Sensor: Sensor nodes can also have locomotion capability, which enables them to move. With this type of controlled mobility, the deployment specification can be more exact, coverage can be maximized, and the target can be detected and followed.
- Data Mule: The sensors do not need to be mobile, but they may require a mobile device to collect their data and deliver it to the sink. These mobile devices are called data mules.

In many situations, an optimal deployment is unknown until the sensor nodes start collecting and processing data. For deployments in remote or wide areas, rearranging node positions is generally infeasible. However, when nodes are mobile, redeployment is possible.

In networks that are sparse or disjointed or when stationary nodes die, mobile nodes can maneuver to connect the lost or weak communication pathways. This is not possible with static WSNs, in which the data from dead or disconnected nodes would simply be lost. By using mobile base stations, this problem is eliminated, and the lifetime of the network is extended.

Mobility also enables greater channel capacity and maintains data integrity by creating multiple communication pathways and reducing the number of hops messages must travel before reaching their destination.

The redeployment can be done easily using mobile sensor nodes, but the network topology becomes dynamic. Topological changes in sensor networks may affect qualities of sensor coverage. Recently, there has been much focus on mobile sensor networks and the development of small-profile sensing devices that are able to control their own movement. The mobile nodes in the network enhance its capabilities. They could be used to physically collect and transport data or to recharge and repair the static nodes in the network. In other words, not only should the sensors be informative, but they should also be able to communicate efficiently.

The mobile sensor nodes must continuously obtain their position (updating the routing table) as they traverse the sensing region, and route discovery must repeatedly be performed because the topology is changed. This requires additional time, bandwidth, and energy. Fortunately, there are some researches dedicated to routing in mobile WSNs. Mobile sensor nodes require additional power for mobility and are often equipped with a much larger energy reserve or have a self-charging capability that enables them to plug into the power grid to recharge their batteries.

22.7.9 Security

Security is achieved by encrypting messages and verifying that a message is authentic. However, these may require significant processing power. In addition, encryption may widen data size, and authentication requires additional messaging, thus causing more communicational overheads (Karl et al. 2003).

22.7.10 Ultrawideband Transmission Technology

In recent years, ultrawideband (UWB) technologies have drawn great interest in WSNs. It is one of the enabling technologies for sensor network applications. UWB systems have potentially low complexity and low cost with noise-like signal properties that create little interference to other systems that are resistant to severe multi-path and jamming and have very good time-domain resolution allowing for precise location and tracking (Zhang et al. 2009).

References

S. Adlakha, S. Ganeriwal, C. Schurgers, and M. B. Srivastava, Density, accuracy, delay and lifetime tradeoffs in wireless sensor networks—A multidimensional design perspective, In *Proceedings on the 1st International Conference on Embedded Networked Sensor Systems (SenSys 2003)*, Los Angeles, November 5–7, 2003.

I. F. Akyildiz, M. C. Vuran, and O. B. Akan, A cross-layer protocol for wireless sensor networks, In *Proceedings of the 40th Annual Conference on Information Sciences and Systems*, NJ, pp. 1102–1107, 2006.

I. F. Akyildiz, W. Su, Y. Sankarasubramaniam, and E. Cayirci, Wireless sensor networks: A survey, *Computer Networks*, Vol. 38, No. 4, pp. 393–422, March 2002.

B. Bhuyan, H. Sarma, N. Sarma, A. Kar, and R. Mall, Quality of service (QoS) provisions in wireless sensor networks and related challenges, *Wireless Sensor Network*, Vol. 2, 861–868, 2010.

S. Blake, D. Black, N. Carlson, E. Davies, Z. Wang, and W. Weiss, An architecture for differentiated services, IETF RFC 2475, December 1998.

B. Braden, D. Clark, and S. Shenker, Integrated services in the Internet architecture: An overview, IETF RFC1633, June 1994.

M. Cardei and J. Wu, Energy-efficient coverage problems in wireless ad hoc sensor networks, *Journal of Computer Communications on Sensor Networks*, 2005.

S. Chakrabarti and A. Mishra, QoS issues in ad hoc wireless networks, *IEEE Communications Magazine*, Vol. 39, No. 2, pp. 142–148, February 2001.

Y. Chen, T. Farley and N. Ye, QoS Requirements of network applications on the Internet, *Information-Knowledge-Systems Management*, Vol. 4, No. 1, pp. 55–76, January 2004.

D. Ganesan, A. Cerpa, W. Ye, Y. Yu, J. Zhao, and D. Estrin, Networking issues in wireless sensor networks, *Journal of Parallel and Distributed Computing (JPDC), Special Issue on Frontiers in Distributed Sensor Networks*, Elsevier Publishers, Vol. 64, No. 7, pp. 799–814, July 2004.

N. Heo and P. K. Varshney, Energy-efficient deployment of intelligent mobile sensor networks, *IEEE Transactions on Systems, Man and Cybernetics*, Vol. 35, No. 1, pp. 78–92, January 2005.

K. Kalpakis, K. Dasgupta, and P. Namjoshi, Maximum lifetime data gathering and aggregation in wireless sensor networks, In *Proceedings of the 2002 IEEE International Conference on Networking (ICN '02)*, Atlanta, GA, August 26–29, 2002.

I. Kang and R. Poovendran, Maximizing network lifetime of broadcasting over wireless stationary ad hoc networks, mobile networks and applications, *Mobile Networks and Applications*, Vol. 10, No. 6, pp. 879–896, December 2005.

H. Karl and A. Willig, A short survey of wireless sensor networks, Technical Report TKN-03-018, Telecommunication Networks Group, Technische Universität Berlin, October 2003.

S. Kumar, A. Arora, and T. H. Lai, On the lifetime analysis of always-on wireless sensor network applications, In *Proceedings of the 2nd IEEE International Conference on Mobile and Ad Hoc Sensor Systems (MASS)*, Washington, D.C., November 7–10, 2005.

R. Madan, S. Cui, S. Lall, and A. Goldsmith, Cross-layer design for lifetime maximization in interference-limited wireless sensor networks, In *Proceedings of IEEE INFOCOM '05*, Vol. 3, pp. 1964–1975, March 2005.

R. Min, M. Bhardwaj, S. Cho, Nathan Ickes, E. Shih, A. Sinha, A. Wang, and A. P. Chandrakasan, Energy-centric enabling technologies for wireless sensor networks, *IEEE Wireless Communications*, Vol. 9, No. 4, pp. 28–39, August 2002.

S. Muthukarpagam, V. Niveditta, and S. Neduncheliyan, Design issues, topology issues, quality of service support for wireless sensor networks: survey and research challenges, *International Journal of Computer Applications*, Vol. 1, No. 6, pp. 1–4, 2010.

S. Nath, P. Gibbons, and S. Seshan, Synopsis diffusion for robust aggregation in sensor networks, In *Proceedings of the ACM Symposium on Networked Embedded Systems (SenSys'04)*, Baltimore, pp. 250–262, November 2004.

R. Rizk, A. Elmaghraby, and M. Mariee, QoS protection optimization for MPLS, *International Journal of Communication Networks and Distributed Systems (IJCNDS), Inderscience Publishers*, Vol. 6, No. 4, pp. 420–437, 2011a.

R. Rizk, H. Elhadidy, and H. Nassar, Optimized mobile radio aware routing algorithm for wireless sensor networks, *IET Wireless Sensor Systems*, Vol. 1, No. 4, pp. 206–217, December 2011b.

K. Romer and F. Mattern, The design space of wireless sensor networks, *IEEE Wireless Communications*, Vol. 11, No. 6, December 2004.

H. O. Sanli, H. Çam, and X. Cheng, EQoS: An energy efficient QoS protocol for wireless sensor networks, In *Proceedings of the 2004 Western Simulation MultiConference (WMC '04)*, San Diego, CA, January 18–21, 2004.

C. Schurgers and M. B. Srivastava, Energy efficient routing in wireless sensor networks, In *Proceedings of MILCOM '01*, Vienna, VA, October 28–31, 2001.

C. Schurgers, V. Tsiatsis, S. Ganeriwal, and M. Srivastava, Topology management for sensor networks: Exploiting latency and density, In *Proceedings of the 3rd ACM International Symposium on Mobile Ad Hoc Networking and Computing (MobiHoc '02)*, EPFL, Lausanne, Switzerland, June 9–11, 2002.

W. K. G. Seah, Quality of service in mobile ad hoc networks—Myth or reality?, In *Keynote Presentation, Australian Telecommunication Networks and Applications Conference (ATNAC 2004)*, Sydney, Australia, December 8–10, 2004.

Y. C. Wang, S. R. Ye, and Y.-C. Tseng, A fair scheduling algorithm with traffic classification in wireless networks, *Computer Communications*, Elsevier, Vol. 28, No. 10, pp. 1225–1239, June 2005.

K. Wu and J. Harms, QoS support in mobile ad hoc networks, crossing boundaries, *GSA Journal of University of Alberta*, Vol. 1, No. 1, pp. 92–106, November 2001.

M. A. Yigitel, O. D. Incel, and C. Ersoy, QoS-aware MAC protocols for wireless sensor networks: A survey, *Computer Networks*, Vol. 55, pp. 1982–2004, 2011.

J. Zhang, P. V. Orlik, Z. Sahinoglu, A. F. Molisch, and P. Kinney, UWB systems for wireless sensor networks, *Proceedings of the IEEE*, Vol. 97, No. 2, February 2009.

APPLICATIONS

Chapter 23

Artificial Eye Vision Using Wireless Sensor Networks

Shaimaa Ahmed Elsaid
Zagazig University
Aboul Ella Hassanien
Cairo University

Contents

23.1 Introduction

Recent advances in miniaturization and low-cost, low-power design have led to active research in large-scale, highly distributed systems of small, wireless, low-power, unattended sensors and actuators. Wireless sensor networks (WSNs) consist of spatially distributed wireless sensor nodes that cooperate with each other in order to monitor and collect data pertaining to physical or environmental conditions, such as temperature, pressure, motion, sound, and other phenomena. Their applications include tracking the movements of birds, small animals, and insects; monitoring environmental conditions that affect crops and livestock; macro instruments for large-scale Earth monitoring and planetary exploration; chemical or biological detection; precision agriculture; biological, Earth, and environmental monitoring in marine, soil, and atmospheric contexts; forest fire detection; meteorological or geophysical research; flood detection; bio-complexity mapping of the environment; and pollution study [1,2]. Some of the health applications for sensor networks are providing interfaces for the disabled, integrated patient monitoring, diagnostics, drug administration in hospitals, monitoring the movements and internal processes of insects or other small animals, monitoring of human physiological data, and tracking and monitoring doctors and patients inside a hospital [3]. WSNs can be effectively used in health care to enhance the quality of life provided for the patients and also the quality of health care services [4]. For example, patients equipped with a wireless body area network (WBAN) need not be physically present with the physician for their diagnostics. A body sensor network proves to be adequate for emergency cases, in which it autonomously sends data about patient health so that the physician can prepare for the treatment immediately [5–10].

Embedding smart sensors in humans has been a research challenge resulting from the limitations imposed by these sensors from computational capabilities to limited power. These WSNs carry the promise of drastically improving and expanding the quality of care across a wide variety of settings and for different segments of the population. For example, early system prototypes have demonstrated the potential of WSNs to enable early detection of clinical deterioration through real-time patient monitoring in hospitals [11,12], enhance first responders' capability of providing emergency care in large disasters through automatic electronic triage [13,14], improve the life quality of the elderly through smart environments [15], and enable large-scale field studies of human behavior and chronic diseases [16,17]. Tang et al. [18] have described their vision of a future in which one device will be able to build a WSN with a large number of nodes, both inside and outside the human body that may be either predefined or random, according to the application. This vision can only be applied through the use of common communication protocols for WSNs. The standardized hardware and software architectures can support compatible devices, which are expected to significantly affect the next generation of health care systems. Some of these devices can then be incorporated into the WBAN, providing new opportunities for technology to monitor health status [19].

The remainder of this chapter is organized as follows. Section 23.2 provides a brief description of WBANs and overviews the main differences between WSNs and WBANs. Section 23.3 gives a concise summary of some of the commercially available wireless sensor nodes for health monitoring. The benefits of using WBAN are outlined in Section 23.4. Section 23.5 presents some issues and challenges that face the usage of WSN in health applications. Section 23.6 provides a survey of WBAN application in vision rehabilitation. The architecture of the human retina, the work of artificial retinas, and the progression of disease states are described in detail. The chapter is concluded in Section 23.7. Future works are presented in Section 23.8.

23.2 Wireless Body Area Networks

Recent advances in electronics have enabled the development of small and intelligent (bio-) medical sensors, which can be worn on or implanted in the human body. The use of a wireless interface with these sensors enables an easier application and is more cost effective and makes the health care mobile; this is referred to as mHealth. In order to fully exploit the benefits of wireless technologies in telemedicine and mHealth, a new type of wireless network emerges: a wireless on-body network or a WBAN. Human health monitoring [20,21] is emerging as a prominent application of the embedded sensor networks. A number of tiny wireless sensors are strategically placed on/in a patient's body, to create a WBAN [22]. A WBAN can monitor vital signs, providing real-time feedback to allow many patient diagnostic procedures using continuous monitoring of chronic conditions or progress in recovery from an illness. An example of a medical WBAN used for patient monitoring is shown in Figure 23.1. Several sensors are placed in clothes, directly on the body, or under the skin of a person and measure the temperature, blood pressure, heart rate (HR), ECG, EEG, respiration rate, SpO_2 levels, etc. Next to sensing devices, the patient has actuators, which act as drug-delivery systems. The medicine can be delivered at predetermined moments, triggered by an external source (i.e., a doctor who analyzes the data), or immediately when a sensor notices a problem. One example is the monitoring of the glucose level in the blood of diabetics. If the sensor monitors a sudden drop of glucose, a signal can be sent to the actuator in order to start the injection of insulin. Consequently, the patient will experience fewer nuisances from his disease. Another example of an actuator is a spinal cord stimulator implanted in the body for long-term pain relief [23].

Recent technological advances in wireless networking promise a new generation of WSNs suitable for those on-body/in-body networked systems. Data acquisition across such sensors can be point-to-point or multipoint-to-point, depending on specific applications. While the distributed detection of an athlete's posture [24] would need point-to-point data sharing across various on-body sensors, applications such as monitoring vital signs as shown in Figure 23.2 will require all body mounted and/or implanted sensors [22,25] to route data multipoint-to-point to a sink node, which, in turn, can relay the information wirelessly to an out-of-body server. Figure 23.3 shows the environment of WBANs.

WBANs enable dense spatiotemporal sampling of physical, physiological, psychological, cognitive, and behavioral processes in spaces ranging from personal to buildings to even larger-scale

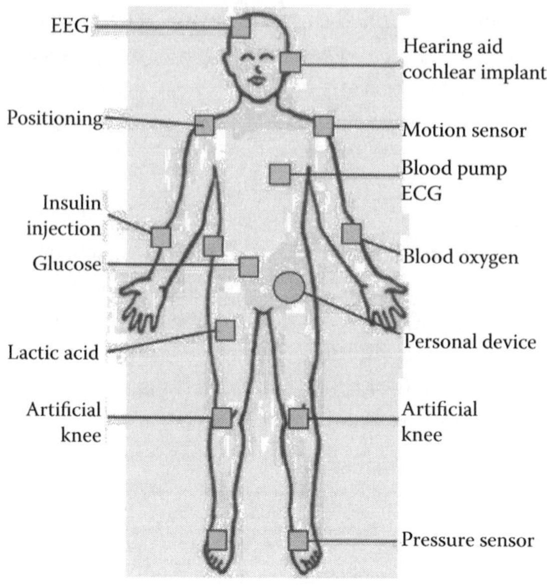

Figure 23.1 Example of patient monitoring in a WBAN. (From E. Krames, *Best Pract. Res. Clin. Anaesthesiol.*, 16, 4, 619–649, December 2002.)

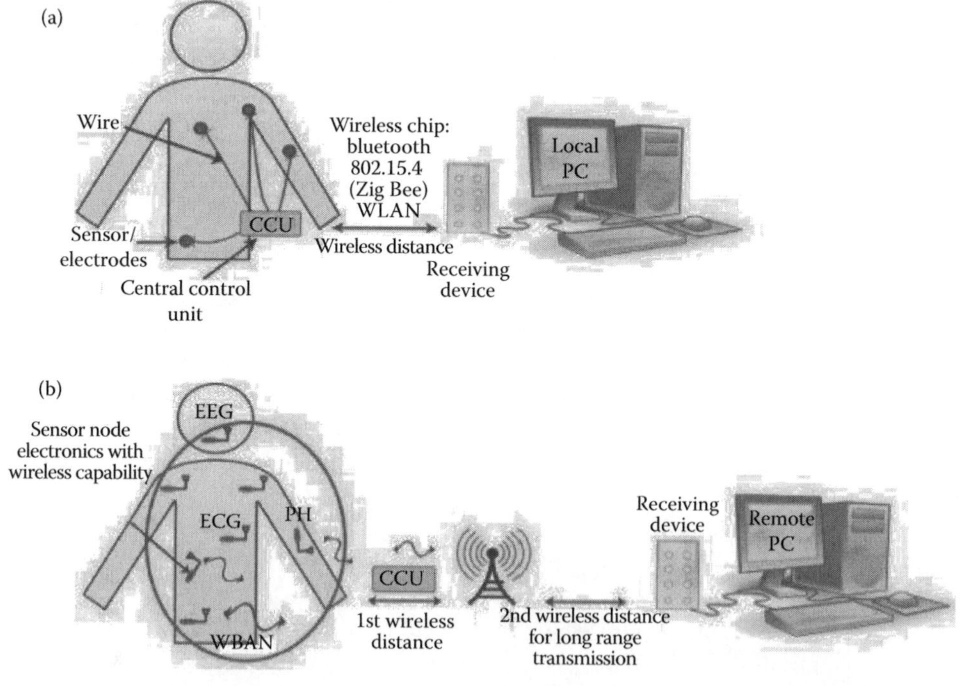

Figure 23.2 Typical WBAN system for detecting and transmitting signals from a human body: (a) current application of health care sensor network and (b) future application of health care sensor network targeted by WBAN. (From A. Darwish and A. E. Hassanien, *Sensors*, 11, 5561–5595, 2011, doi: 10.3390/s110605561.)

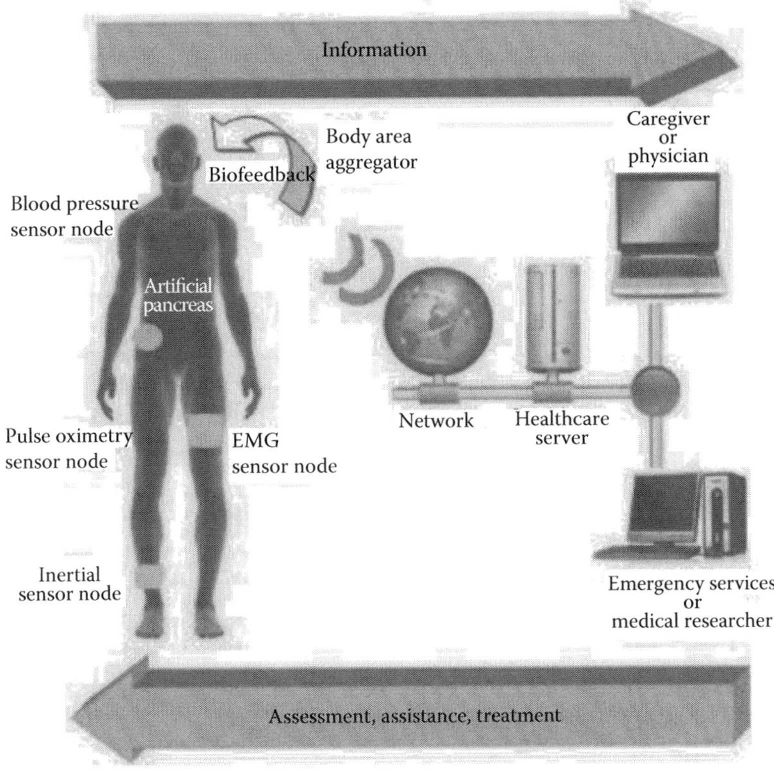

Figure 23.3 WBAN and its environment. (From A. Darwish and A. E. Hassanien, *Sensors*, 11, 5561–5595, 2011, doi: 10.3390/s110605561.)

ones. Such dense sampling across spaces of different scales results in sensory information–based health care applications. Current health care applications of WSNs target heart problems [26–28], asthma [7,29,30], emergency response [31], and stress monitoring [32]. This wireless biomedical sensor technology can also be effectively used to treat diabetes by providing a more consistent, accurate, and less invasive method for monitoring glucose levels. Currently, to monitor blood glucose levels, a lancet is used to prick a finger and a drop of blood is placed on a test strip, which is analyzed either manually or electronically. This constant pricking several times a day over a period of years can damage the tissue and blood vessels in that area. The sensor would monitor the glucose levels [33] and transmit the results to a wristwatch display. Wireless biomedical sensors can be implanted in the patient, and the glucose level can be monitored continuously [35]. Wireless biomedical sensors may play a key role in early detection of cancer [34]. Cancer cells exude nitric oxide, which affects the blood flow in the area surrounding a tumor. A sensor with the ability to detect these changes in the blood flow can be placed in suspect locations. It is likely that any abnormalities could be detected much sooner with the sensors than without. Video and various kinds of embedded sensors can be used to track and monitor the patient in their everyday activities. This information can be processed and relayed to medical personnel. A patient's routine can be assembled over the period of time, and deviations from this may be recognized and analyzed. Following is a brief review for using WBAN in medical applications.

23.2.1 Types of WBAN Devices

23.2.1.1 Wireless Sensor Node

A device that responds to and gathers data on physical stimuli processes the data if necessary and reports this information wirelessly. It consists of several components: sensor hardware, a power unit, a processor, memory, and a transmitter or transceiver [7].

23.2.1.2 Wireless Actuator Node

A device that acts according to data received from the sensors or through interaction with the user. The components of an actuator are similar to the sensor's: actuator hardware (e.g., hardware for medicine administration, including a reservoir to hold the medicine), a power unit, a processor, memory, and a receiver or transceiver.

23.2.1.3 Wireless Personal Device

A device that gathers all the information acquired by the sensors and actuators and informs the user (i.e., the patient, a nurse, a GP, etc.) via an external gateway, an actuator, or a display/LEDS on the device. The components are a power unit, a (large) processor, memory, and a transceiver. This device is also called a body control unit (BCU) [11], a body gateway, or a sink. In some implementations, a personal digital assistant (PDA) or smart phone is used.

23.2.2 WBANs Data Rates

Because of the strong heterogeneity of the applications, data rates will vary strongly, ranging from simple data at a few kilobits per second to video streams of several megabits per second. Data can also be sent in bursts, which means that it is sent at a higher rate during the bursts. The data rates for the different applications are given in Table 23.1 and are calculated by means of the sampling rate, the range, and the desired accuracy of the measurements [38,39]. Overall, it can be seen that the application data rates are not high. However, if one has a WBAN with several of these devices (i.e., a dozen motion sensors, ECG, EMG, glucose monitoring, etc.), the aggregated data rate easily reaches a few megabits per second, which is higher than the raw bit rate of most existing low-power radios.

23.2.3 Differences between WSN and WBAN

Unlike conventional WSNs, WBANs are smaller and consist of fewer nodes, less space covered, and fewer opportunities for redundancy [41]. These devices provide continuous health monitoring and real-time feedback to the user or medical personnel. Furthermore, the measurements can be recorded over a longer period of time, improving the quality of the measured data. In WBANs there are two types of devices that can be distinguished: sensors and actuators. The sensors are used to measure certain parameters of the human body, either externally or internally. Examples include measuring the heart beat or body temperature or recording a prolonged electrocardiogram (ECG). The actuators, on the other hand, take some specific actions according to the data they receive from the sensors or through interaction with the user. For example, an actuator equipped with a built-in reservoir and pump administers the correct dose of insulin to give to diabetics

Table 23.1 Examples of Medical WBAN Applications [37–40]

Application	*Date Rate*	*Bandwith*	*Accuracy*
ECG (12 leads)	288 kbps	100–1000 Hz	12 bits
ECG (6 leads)	71 kbps	100–500 Hz	12 bits
EMG	320 kbps	0–10,000 Hz	16 bits
EEG (12 leads)	43.2 kbps	0–150 Hz	12 bits
Blood saturation	16 bps	0–1 Hz	8 bits
Glucose monitoring	1600 bps	0–50 Hz	16 bits
Temperature	120 bps	0–1 Hz	8 bits
Motion sensor	35 kbps	0–500 Hz	12 bits
Cochlear implant	100 kbps	–	–
Artificial retina	50–700 kbps	–	–
Audio	1 Mbps	–	–
Voice	50–100 kbps	–	–

based on the glucose level measurements. Interaction with the user or other persons is usually handled by a personal device, for example, a PDA or a smart phone, which acts as a sink for data of the wireless devices. The following illustrates the differences between a WSN and a WBAN [42]:

- The devices used have limited energy resources available as they have a very small form factor (often less than 1 cm^3). Furthermore, for most devices, it is impossible to recharge or change the batteries although a long lifetime of the device is wanted (up to several years or even decades for implanted devices). Hence, the energy resources and, consequently, the computational power and available memory of such devices will be limited.
- All devices are equally important, and devices are only added when they are needed for an application (i.e., no redundant devices are available).
- An extremely low transmit power per node is needed to minimize interference and to cope with health concerns.
- The propagation of the waves takes place in or on a (very) lossy medium, the human body. As a result, the waves are attenuated considerably before they reach the receiver.
- The devices are located on the human body that can be in motion. WBANs should therefore be robust against frequent changes in the network topology.
- The data mostly consists of medical information. Hence, high reliability and low delay is required.
- Stringent security mechanisms are required in order to ensure the strictly private and confidential character of the medical data.

An overview of some of these differences is given in Table 23.2.

Table 23.2 Schematic Overview of Differences between WSNs and Wireless Body Area Networks

Challenges	*Wireless Sensors Network*	*Wireless Body Area Network*
Scale	Monitored environment (meters/kilometers)	Human body (centimeters/meters)
Node Number	Many redundant nodes for wide area coverage	Fewer, limited in space
Result Accuracy	Through node redundancy	Through node accuracy and robustness
Node Tasks	Node performs a dedicated task	Node performs multiple tasks
Node Size	Small is preferred, but not important	Small is essential
Network Topology	Very likely to be fixed or static	More variable because of body movement
Data Rates	Most often homogeneous	Most often heterogeneous
Node Replacement	Performed easily, nodes even disposable	Replacement of implanted nodes difficult
Node Lifetime	Several years/months	Several years/months, smaller battery capacity
Power Supply	Accessible and likely to be replaced more easily and frequently	Inaccessible and difficult to replaced in an implantable setting
Power Demand	Likely to be large, energy supply easier	Likely to be lower, energy supply more difficult
Energy-Scavenging Source	Most likely solar and wind power	Most likely motion (vibration) and thermal (body heat)
Biocompatibility	Not a consideration in most applications	A must for implants and some external sensors
Security Level	Lower	Higher, to protect patient information
Impact of Data Loss	Likely to be compensated by redundant nodes	More significant, may require additional measures to ensure QoS and real-time data delivery
Wireless Technology	Bluetooth, ZigBee, GPRS, WLAN	Low power technology required

Source: With kind permission from Springer Science+Business Media: *Body Sensor Networks*, 2006, G.-Z. Yang, Ed.

23.2.4 WBAN System Prototypes

In this subsection, we present several wireless sensing system prototypes developed and deployed to evaluate the efficacy of WBANs in some of the health care applications described before. While wireless health care systems using various wireless technologies exist, this work focuses on systems based on low-power wireless platforms for physiological and motion-monitoring studies and smart phone–based, large-scale studies.

23.2.4.1 Physiological Monitoring

In physiological-monitoring applications [70], low-power sensors measure and report a person's vital signs (e.g., pulse oximetry, respiration rate, and temperature). These applications can be developed and deployed in different contexts, ranging from disaster response to in-hospital patient monitoring and long-term remote monitoring for the elderly. Systems that automate patient monitoring have the potential to increase the quality of care, both in disaster scenes and clinical environments. Systems such as CodeBlue [44], MEDiSN [45], and the Washington University's vital sign monitoring system [46] target these application scenarios. Specifically, CodeBlue [44] aims to improve the triage process during disaster events with the help of WSNs comprising motes with IEEE 802.15.4 radios. The CodeBlue project integrated various medical sensors [e.g., EKG, SpO_2, pulse rate, electromyography (EMG)] with mote-class devices and proposed a publish/subscribe-based network architecture that also supports priorities and remote sensor control [47]. Finally, victims with CodeBlue monitors can be tracked and localized using RF-based localization techniques [48]. Figure 23.4 shows the use of WBAN to develop the CodeBlue system. This project includes pre-hospital care and in-hospital emergency care, disaster response, and stroke patient rehabilitation. Research from this project has the potential for resuscitative care, real-time triage decisions, and long-term patient observations [49]. The system integrates low-power wireless wearable vital sign sensors, handheld computers, and location-tracking tags.

Ko et al. proposed MEDiSN (see Figure 23.5) to address similar goals as CodeBlue (e.g., improve the monitoring process of hospital patients and disaster victims as well as first responders) but using a different network architecture [45]. Specifically, unlike the ad hoc network used in CodeBlue, MEDiSN employs a wireless backbone network of easily deployable relay points (RPs). RPs are positioned at fixed locations, and they self-organize into a forest rooted at one or more gateways (i.e., PC-class devices that connect to the Internet) using a variant of the collection tree protocol (CTP) [50] tailored to high data rates. Motes that collect vital signs, known as miTags, associate with RPs to send their measurements to the gateway.

Figure 23.6 shows the UbiMon system (Ubiquitous Monitoring Environment for Wearable and Implantable Sensors) that aims to provide a continuous and unobtrusive monitoring system for patients in order to capture transient events [51]. As in Figure 23.6a through c, a compact flash WBAN card is developed for PDAs, in which sensor signals can be gathered, displayed, and analyzed by the PDA as shown in Figure 23.6d.

In [52], researchers use WBAN to develop the eWatch as shown in Figure 23.7. It is a wearable sensing, notification, and computing platform built into a wristwatch. eWatch can be used for applications such as context awareness notification, elderly monitoring, and fall detection. The eWatch system can sense if the user is in trouble and then query to confirm that it is an emergency. If the user does not respond, then the eWatch can use its network abilities to call for help. The eWatch can also notify a patient when he takes certain medication.

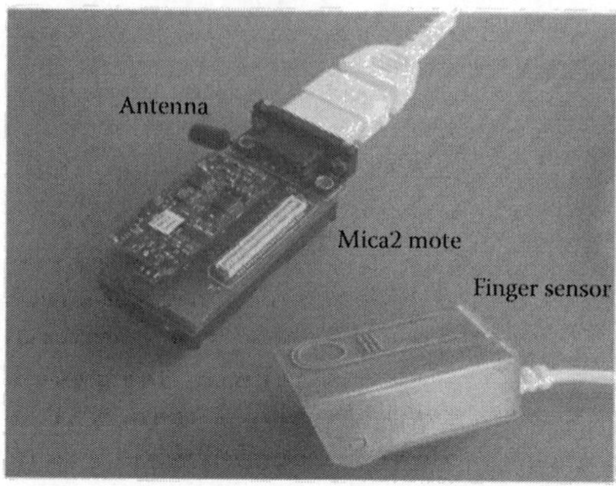

Figure 23.4 Wireless pulse oximeter sensor.

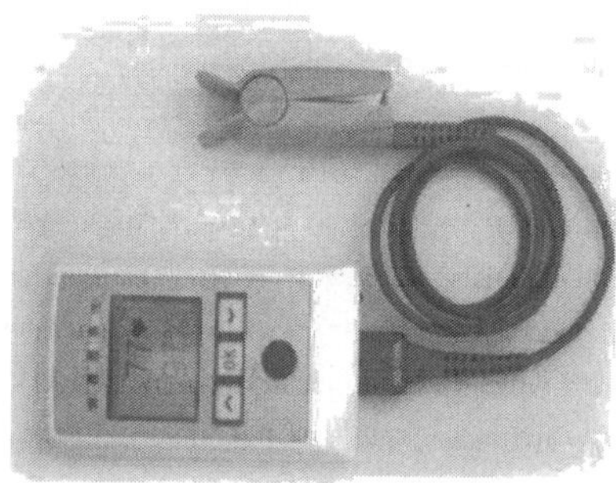

Figure 23.5 Medical information tag, or miTag for short, used in MEDiSN. (From J. Ko et al., *ACM Trans. Embedded Comput. Syst.*, 10, 1, article 11, 11:1–11:29, 2010.)

23.2.4.2 Motion and Activity Monitoring

Another application domain for WSNs in health care is high-resolution monitoring of movement and activity levels. Wearable sensors can measure limb movements, posture, and muscular activity and can be applied to a range of clinical settings, including gait analysis [53–55], activity classification [56,57], athletic performance [58,59], and neuromotor disease rehabilitation [60,61]. In a typical scenario, a patient wears up to eight sensors (one on each limb segment) equipped with MEMS accelerometers and gyroscopes. A base station, such as a PC-class device in the patient's home, collects data from the network. Data analysis can be performed to recover the patient's motor coordination and activity level, which is, in turn, used to measure the effect of treatments. In contrast to physiological monitoring, motion analysis involves multiple sensors on a single patient, each measuring high-resolution signals, typically six channels per sensor, sampled at 100 Hz each. This volume of sensor data precludes real-time transmission, especially over multi-hop paths because of both bandwidth and energy constraints. In such studies, the size and weight of the wearable sensors must be minimized to avoid encumbering the patient's movement. The SHIMMER sensor platform shown in Figure 23.8 measures 44.5 × 20 × 13 mm and weighs just 10 g, making it well suited for long-term wearable use.

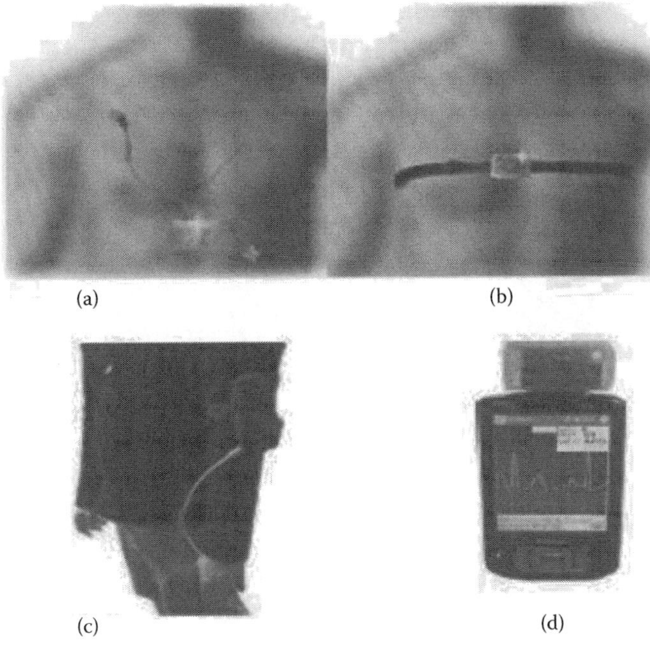

Figure 23.6 (a) Wireless 3-leads ECG sensor, (b) ECG strap (center), (c) SpO2 sensor, and (d) PDA base station.

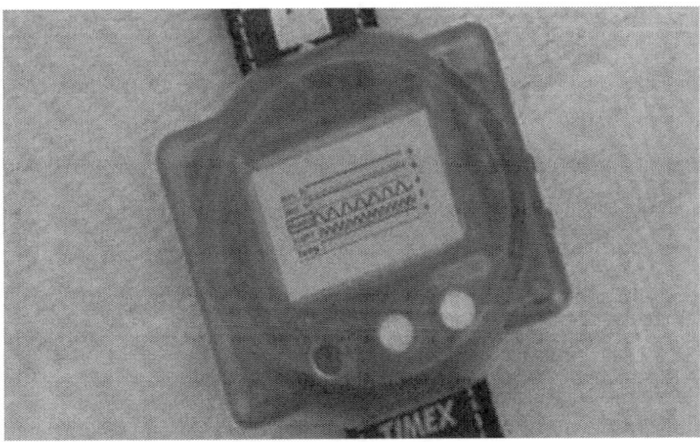

Figure 23.7 eWatch computing.

In the work of Ganti et al. [62], SATIRE is designed to identify a user's activity based on accelerometers and global positioning system (GPS) sensors integrated into Bsmart attire such as a winter jacket. SATIRE nodes, which are based on the MICAz [63] platform, measure accelerometer data and log it to local flash. These data are opportunistically transmitted using the low-power radio when the SHIMMER node is within communication range with the base station. Once the data are collected at the base station, the collected data are processed offline to characterize the user's activity patterns, such as walking, sitting, or typing. Sensor nodes perform aggressive duty

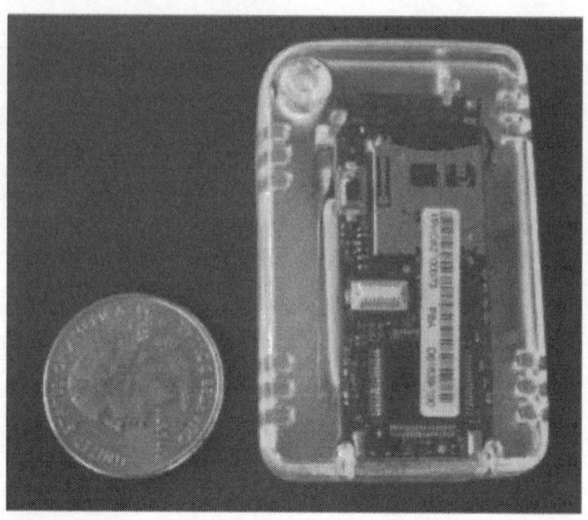

Figure 23.8 SHIMMER wearable sensor platform.

cycling to reduce power consumption, extending lifetimes from several days to several weeks. While the Mercury system [60] permits long-term studies of a patient's motor activity for neuromotor disease studies, including Parkinson's disease, stroke, and epilepsy. Energy is far more constrained in Mercury than in SATIRE because of the use of lightweight sensor nodes with small batteries. Mercury builds upon SATIRE's approach to energy management and integrates several energy-aware adaptations, including dynamic sensor duty cycling, priority-driven data transmissions, and on-board feature extraction. Mercury is being used in several studies of Parkinson's and epilepsy patients.

Figure 23.9 shows the pants for hip patient rehabilitation using WBAN. The HipGuard system is developed for patients who are recovering from hip surgery. This system monitors the patient's leg and hip position and rotation with embedded wireless sensors. Alarm signals can be sent to patient's wrist unit if hip or leg positions or rotations are false, and hence the HipGuard system can provide useful real-time information for the patient rehabilitation process [64].

Researchers use WBAN and wireless telephony technology in the MobiHealth project that uses GPRS/UMTS wireless communication technology for transferring data to create a generic platform for home health care. MobiHealth aims to provide continuous monitoring to patients outside the hospital environment [65]. It targets improving the quality of life of patients by enabling new value-added services in the areas of disease prevention, disease diagnosis, remote assistance, physical state monitoring, and even in clinical research [66]. Figure 23.10 shows a MobiHealth system.

Figure 23.11 shows using WBAN in the LifeShirt project. It is a comfortable and completely noninvasive "smart garment" that gathers data during a patient's daily routine, providing the most complete remote picture of a patient's health status. It enables health care professionals and researchers to accurately monitor more than 30 vital life-sign functions in the real-world settings where patients live and work [42].

WBAN can also be applied in the Vital Sign Monitoring System as shown in Figure 23.12. In [67] a real-time patient monitoring system was designed and developed to integrate vital sign sensors, location sensors, ad hoc networking, electronic patient records, and web portal technology to allow remote monitoring of patient status.

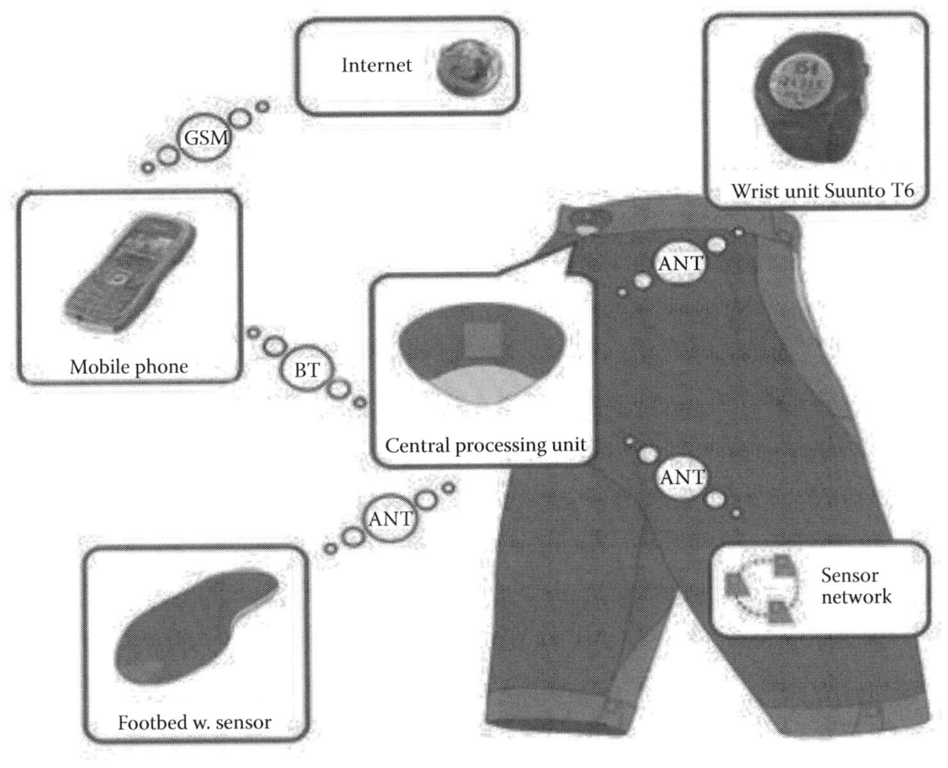

Figure 23.9 HipGuard Pants for hip patient rehabilitation.

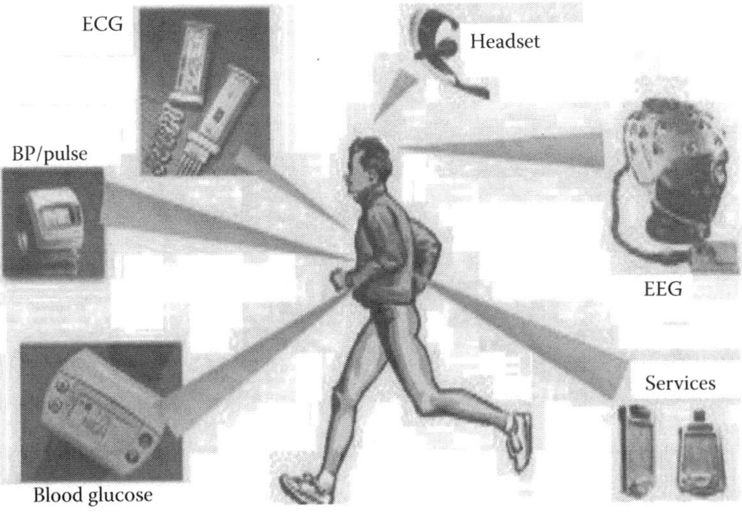

Figure 23.10 MobiHealth system, monitoring a patient outside the hospital environment.

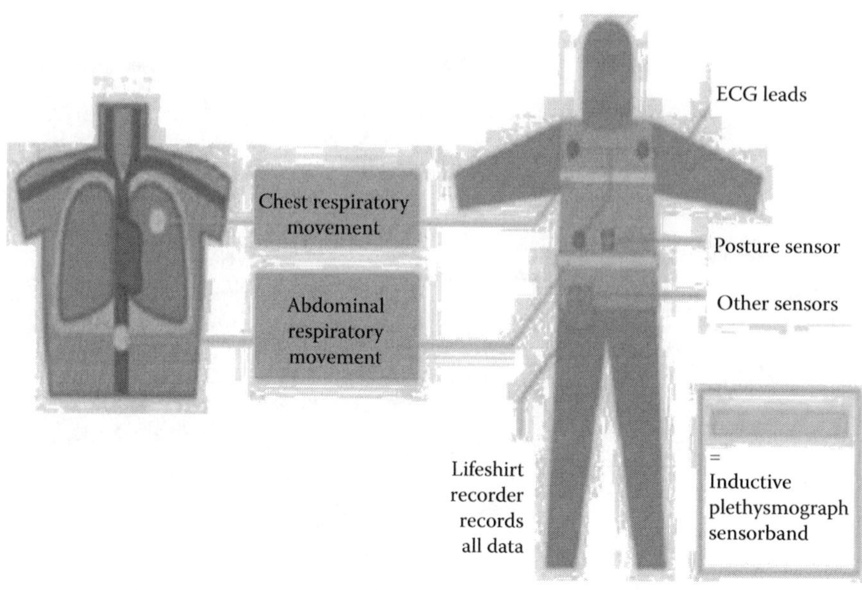

Figure 23.11 Smart LifeShirt.

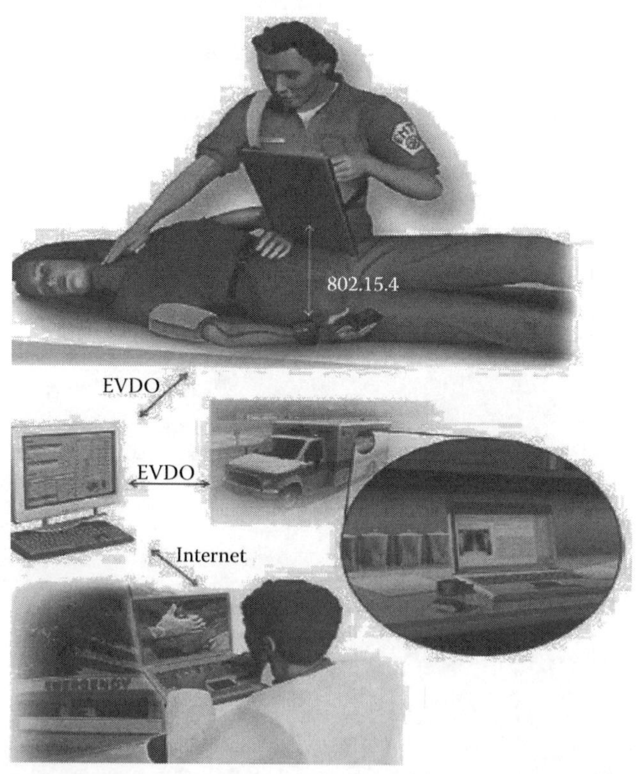

Figure 23.12 Vital Sign Monitoring System. (From T. Gao et al., Vital signs monitoring and patient tracking over a wireless network, In *Proceedings of the 27th Annual International Conference of the IEEE EMBS*, Shanghai, China, pp. 102–105, 1–4 September 2005.)

23.2.4.3 Large-Scale Physiological and Behavioral Studies

The third WBAN applications category in health care that was discussed is their use in conducting large-scale physiological and behavioral studies [70]. The confluence of body-area networks of miniature wireless sensors, always-connected sensor equipped smart phones, and cloud-based data storage and processing services is leading to a new paradigm in population-scale medical research studies, particularly with ailments whose causes and manifestations relate to human behavior and living environments. The combination of body-area WSNs, smart phones, and cloud services permits physical, physiological, behavioral, social, and environmental data to be collected from human subjects in their natural environments continually, in real time, unattended, and in an unobtrusive fashion over long periods. Typically, data are collected from wireless sensors worn by subjects, wireless medical instruments, and sensors embedded in devices such as smart phones.

AutoSense systems [68] collects objective measurements of personal exposure to psychosocial stress and alcohol in the study participants' natural environments. A field-deployable suite of wireless sensors form a body-area wireless network and measure HR, HR variability, respiration rate, skin conductance, skin temperature, arterial blood pressure, and blood alcohol concentration. From these sensor readings, which, after initial validation and cleansing at the sensor, are sent to a smart phone, features of interest indicating onset of psychosocial stress and occurrence of alcoholism are computed in real time. The collected information is then disseminated to researchers answering behavioral research questions about stress, addiction, and the relationship between the two. Moreover, by also capturing time-synchronized information about a subject's physical activity, social context, and location, factors that lead to stress can also be inferred, and this information can potentially be used to provide personalized guidance about stress reduction. This system for population-scale medical studies is still in its early stages, and several technical and algorithmic challenges remain to be addressed. Energy is certainly one challenge. While some on-body sensors have high sampling rates leading to significant energy consumption (i.e., low battery life), the desire to facilitate easy compliance with the study protocols precludes a frequent charging schedule. However, a bigger challenge with this technology is the issue of information privacy and its tension with the quality and value of information [69].

23.3 Available Sensors

The research community has been very active in developing a plethora of wearable and wireless systems that seamlessly monitor HR, oxygen level, blood flow, respiratory rate, muscle activities, movement patterns, body inclination, and oxygen uptake (VO_2). Given below is a concise summary of some of the commercially available wireless sensor nodes for health monitoring:

- Pulse oxygen saturation sensors: They measure the percentage of hemoglobin (Hb) saturated with oxygen (SpO_2) and HR
- Blood pressure sensors
- ECG
- Electromyogram (EMG) for measuring muscle activities
- Temperature sensors, both for core body temperature and skin temperature
- Respiration sensors
- Blood flow sensors
- Blood oxygen level sensor (oximeter) for measuring cardiovascular exertion (distress)

23.4 Benefits of Using WBAN

Whenever a WBAN is introduced to public use, the first question is how will it help me or what can it do for me? In this section, we discuss who benefits from WBAN applications in the health care field.

Everybody hears in the news every day that there is a shortage of specialized doctors or nurses around the country. One of the benefits of these applications is that they increase doctors' and nurses' efficiency. Now they can care for more patients than before. Their work is also improved and made easier by having access to accurate data on patients in real time. For patients whose health is online, the benefits are even greater. They have increased access to specialized doctors. They do not have to stick around the hospital any longer. This ease in mobility allows them to do their own work while still under the doctor's care. Safety is another issue that is helped here because the rate of mistakes can significantly be decreased. Also, patients can be picky when making changes in their daily lives when signing up for a treatment. With wireless technologies, their health care can be less intrusive, for example, in the case of wearable sensors. They do not have to show up to the hospital for a blood-pressure check. It can be done while they are working through wearing wireless sensors that transmit this information in real time to their doctors.

As mentioned earlier, negligent mistakes by doctors and nurses cost hospitals, insurance companies, and the government a large sum of money each year. These costs can be reduced by reducing the number of mistakes. Efficient and secure data handling and resource management is another area where wireless networks can help. Deploying huge machines around the hospital can be very expensive and time consuming. With wireless technology, interfaces can be designed such that access to the machines can be provided from anywhere in the hospital. This allows for rapid and flexible deployment. By increasing the doctors' and nurses' efficiency, hospitals can provide care for more patients and increase their profits.

23.5 Challenges in a WBAN

The human body consists of a complicated internal environment that responds to and interacts with its external surroundings but is, in a way, separate and self-contained. The human body environment not only has a smaller scale, but also requires a different type and frequency of monitoring with different challenges than those faced by WSNs. The monitoring of medical data results in an increased demand for reliability. The ease of use of sensors placed on the body leads to a small form factor that includes the battery and antenna part, resulting in a higher need for energy efficiency. In this section, some of the core challenges in designing WSNs for health care applications are described. While not exhaustive, the challenges in this list span a wide range of topics, from core computer system themes, such as scalability, reliability, and efficiency, to large-scale data mining and data association problems and even legal issues. The challenge of developing a tiny device that can perform complicated tasks and be tough enough to survive in the eye's salty environment demands a diversity of talents and capabilities. Generally, health monitoring is performed on a periodic check basis, in which the patient must remember its symptoms; the doctor performs some tests and plans a diagnostic, then monitors patient progress along the treatment [71]. Health care applications of WSNs allow in-home assistance, smart nursing homes, and clinical trial and research augmentation [72]. Before describing usage of WBANs in vision repair, the following subsections focus on several challenges and general aspects that describe this kind of technology. Challenges in health care application include low power, limited computation, security and interference, material constraints, robustness, continuous operation, and regulatory requirements [73].

23.5.1 Power

As most wireless network–based devices are battery operated, power challenge is present in almost every area of application of WSNs, but limitation of a smart sensor implanted on a person still poses even further challenge. In a full active mode, a node cannot operate more than a month because a typical alkaline battery provides about 50 W h of energy. In practice, for many applications, they have to guarantee that the device will work for a year or two without any replacement. This could include devices such as heart pacemakers. To deal with these power issues, the developers have to design better scheduling algorithms and power management schemes [74]. Although there is a lot of effort in designing low-power sensors to solve this problem [75], there is still a need for robust energy techniques, which is related to the new term "green technology." The solar cells that can provide up to 15 m under direct sun cannot be used with body-worn wireless sensors because sensors are preferably to be placed under the clothing. Thus, motion [76] and body heat [77,78] based energy techniques should be explored for health care systems.

23.5.2 Computation

Because of both limited power as well as memory, computation should also be limited. The biosensors cannot perform large-bit computations because of lack of enough memory. Unlike conventional WSN nodes, biosensors do not have that much computational power. Because communication is vital, and memory is low, little power remains for computation. A solution is that some sensors may have varying capabilities that communicate with each other and send out one collaborative data message [73].

23.5.3 Security and Interference

One of the very important issues that could be considered, especially for medical systems, is security and interference. Physiological data collected by the sensor network is health information, which is of personal nature. It is critical and in the interest of the individual to keep this information from being accessed by unauthorized entities. This is referred to as confidentiality, which can be achieved by encrypting the data by a key during transmission. Data authenticity is also one of the security requirements. This property is very important for the biosensor network because absence of this property may lead to situations in which an illegal entity is disguised as a legal one and reports false data to the control node or gives the wrong instructions to the other biosensors possibly causing significant harm to the host [79]. In [80], elliptic curve cryptography for key distribution to decrease energy consumption is proposed. However, there is still a need for more efficient cryptography methods to meet the security requirements. Another challenging issue with the security requirements for WBANs arises when the patients are unconscious as in a case when obtaining passwords may not be possible. In this case, biometric methods can be used for accountability as in [81] as well as the physiological signal-based authentication method proposed in [82]. Security attacks are especially problematic with low-power WSN platforms because of several reasons, including the strict resource constraints of the devices, minimal accessibility to the sensors and actuators, and the unreliable nature of low-power wireless communications. The security problem is further exacerbated by the observation that transient and permanent random failures are common in WBANs, and such failures are vulnerabilities that can be exploited by attackers. For example, with these vulnerabilities, it is possible for an attacker to falsify context, modify access rights, create denial of service, and, in general, disrupt the operation of the system. This could result in a patient being denied treatment, or worse, receiving the wrong treatment.

23.5.4 Material Constraints

Another issue for WSN application to health care is material constraints. A biosensor should be implanted within the human body; therefore the shape, size, and materials must be harmless to the body tissue. For example, a smart sensor designed to support the retina prosthesis might be small enough to fit within an eye. Also chemical reactions with body tissue and the disposal of the sensor are of extreme importance.

23.5.5 Mobility

The purpose of WBANs and health-monitoring systems is to let people go about their life while accessing high-quality medical services. The use of sensor networks for this purpose is not new [83, 84]. However, the emergence of WBANs has enabled the development of applications to ensure and promote the mobility of users. Thus, the WBAN technologies enable ubiquitous health care systems. Mobility requires the development of multi-hop, multi-modal, ad hoc sensor networks, which bring the problems mentioned in the previous subsections along with the problem of location awareness.

23.5.6 Robustness

Whenever the sensor devices are deployed in harsh or hostile environments, robustness rates of device failure becomes high. Protocol designs must therefore have built-in mechanisms so that the failure of one node should not cause the entire network to cease operation. A possible solution is a distributed network in which each sensor node operates autonomously though still cooperates when necessary. For instance, if the sensor part is not working, the communication part should be used if it benefits the network and communication is operating as expected. One way to achieve this would be that a node might be comprised of a sensing block, a communication block, a scheduling block, and a data block. This would be a good way to isolate the malfunctioning block from the rest of the components in the node as well as reducing power consumption among the various components. In order to ensure that the proper data is being sent and received, there are a few alternatives that can be used, such as checksums, parity check, and cyclic redundancy check [73].

23.5.7 Continuous Operation

Continuous operation must be ensured along the lifecycle of a biosensor as it is expected to operate for days, sometimes weeks without operator intervention. Hence, it is important to keep the amount of communications to the minimum. It is necessary that those communications that occur for purposes other than the actual data communication should be minimized if it is not possible to eliminate them [79].

23.5.8 Regulatory Requirements

Regulatory requirements must always be met; there must be some testimony that these devices will not harm human body. The wireless transmission of data must not harm the surrounding tissues, and the chronic functioning and power utilization of these devices must also be nonmalignant. Design for safety must be a fundamental feature of biomedical sensor development even at the earliest stages [73]. It is conceivable that some immoral researchers could perform tests and trials with devices that are dangerous to the volunteers. Therefore, it is imperative to have diligent oversight of these testing operations.

23.5.9 Sensitivity of Sensors

The sensitivity of the sensor devices is especially important when users wear these sensors in harsh environments such as in fire situations. Sweat can affect the transducers of the sensor devices negatively, causing a reduction in the sensitivity of the body-worn sensors or requiring recalibration of the sensors. Gietzelt et al. [85] proposed an automatic self-calibration algorithm for triaxial accelerometers. Yet, the self-calibration and sensitivity enhancement algorithms are still needed for sensor devices different than accelerometers. Low-maintenance and highly sensitive vital-sign monitoring sensors will attain importance as pervasive health care systems evolve.

23.6 Artificial Retina

Pervasive computing will change the computing landscape, enabling the implementation of new applications that were never imagined. As the population of the United States ages, vision loss is becoming an increasingly serious public health problem. Already, approximately 3.3 million Americans over the age of 40, or one person in 28, are either blind or have low vision [86] (vision so poor that it significantly interferes with everyday life and cannot be corrected, even with eyeglasses). The main goal of applying WBAN in vision rehabilitation is to build a chronically implanted artificial retina for visually impaired people, addressing two retinal diseases: age-related macular degeneration (AMD) and retinitis pigmentosa (RP).

23.6.1 Retinal Diseases: AMD and RP

Two major causes of blindness, age-related macular degeneration (AMD) (severe vision loss at the center of the retina in over 60) and retinitis pigmentosa (RP) (photoreceptor dysfunction → loss of peripheral vision), damage the photoreceptor cells in the light-sensitive membrane in the eye (retina) but leave the nerve connections to the brain intact; see Figure 23.13. Patients eventually lose their vision. A healthy photoreceptor stimulates the brain through electric impulses when light is illuminated from the external world. When damaged, vision is blocked at the locations of the photoreceptors.

AMD is an eye disease primarily affecting the central vision regions in people age 60 and older [87]. According to the Macular Degeneration Research Fund, a case of AMD is diagnosed in the United States every 3 min. Each year, 1.2 million of the estimated 12 million people with AMD will suffer severe vision loss. Patients with AMD have dark areas in their vision caused by fluid leakage or bleeding in the macula, the center of the retina that produces the sharpest vision. The brain initially compensates for these dark patches. Early cellular dysfunction or spotting in the macula may go undetected until the disease is in advanced stages.

RP is the most common inherited cause of blindness in people between the ages of 20 and 60 worldwide. Approximately 500,000 people in the United States suffer some level of visual impairment from RP, and, of those, 20,000 are totally blind. RP is a degenerative disease of the retina that affects the photoreceptor, or rod cells, which control a person's ability to see in dimly lit surroundings. Vision loss is gradual and may result in diminished or lost peripheral vision or blindness.

A person with normal vision has frontal and peripheral vision as shown in Figure 23.14. The eye can see an image by interpreting light hitting the back light-sensing cells, the rods and cones that line the retina. If something happens to these cells, then vision distortion or even blindness occurs [88].

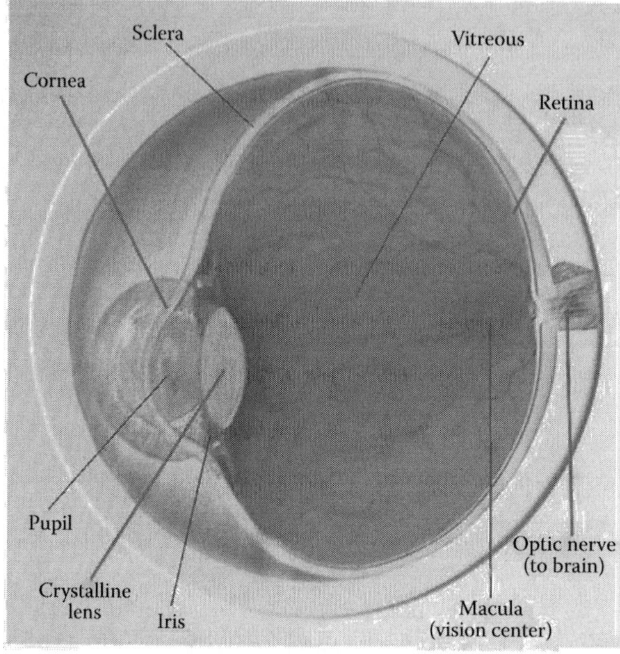

Figure 23.13 Human eye viewed from the side.

Figure 23.14 Normal vision.

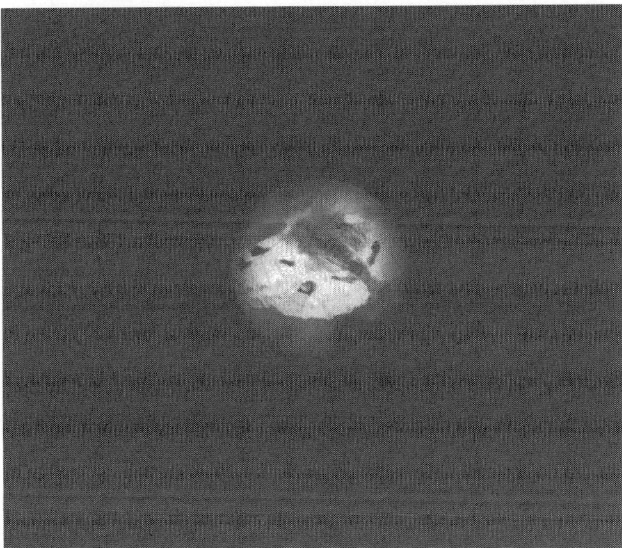

Figure 23.15 How RP and AMD affect vision.

A person with RP loses much of his or her peripheral vision and sees an effect sometimes called "tunnel vision." A person with AMD loses some vision usually in the center of their field of vision as shown in Figure 23.15.

23.6.2 Architecture of the Human Retina

For the layman, it is necessary to understand the unique architecture of the human retina. There are five layers to the retina (as determined by looking at slices through a light microscope). In a strange way, the light coming into the eye must pass through four of the layers before it reaches the section of the retina where the photocells receive and react to the light. When the photocells fire, they send signals back toward the surface of the retina through the four layers above. Various levels of image processing take place as the signals leave the photocells and transverse the other layers. The ganglion cells are the final layer for retinal processing. Their axons (nerve fibers leaving the cell body) are actually the nerves (about a million of them) that leave the retina and travel back into the brain. There is debate among scientists about the best approach to creating artificial vision. The first epi-retinal team believes that resting the chip on the ganglion cells (and their axons) is easier to do surgically and avoids the damaged (diseased) tissue below. The second sub-retinal group feels that it is best to surgically "cut" beneath the four layers and place the chip directly in the (damaged) photoreceptor layer. The eye is only about one inch in diameter. Retinal implants have to be very small and be surgically placed precisely into a thin strip of tissue (the retina is sometimes compared to a wet slice of Kleenex). To get to the retina, surgeons have to remove much of the internal part of the eye. The lens may have to be removed and replaced and the vitreous is removed and later a salient solution is injected as a replacement. The skill of the surgeon is very important in these delicate operations.

23.6.3 How an Artificial Retina Works

Normal vision begins when light enters and moves through the eye to strike specialized photoreceptor (light-receiving) cells in the retina called rods and cones, as shown in Figure 23.16 [97]. These cells convert light signals to electric impulses that are sent to the optic nerve and the brain. Retinal diseases such as AMD and RP destroy the photoreceptor cells. The artificial retina device bypasses these cells to transmit signals directly to the optic nerve. The device consists of a tiny

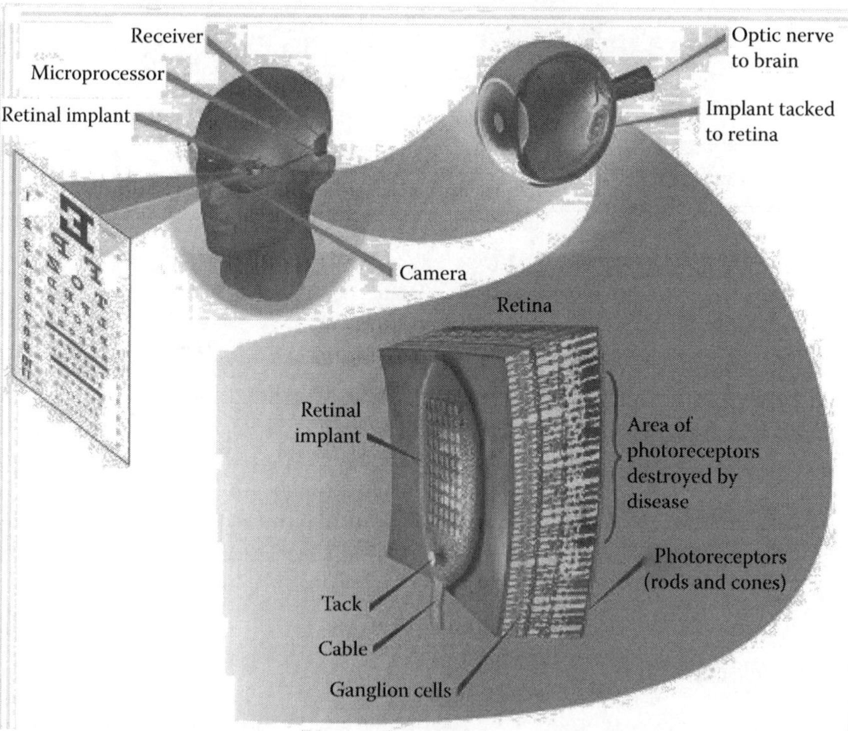

Figure 23.16 Implantable artificial retina for blind people (From Wise, K.D., *IEEE Eng. Med. Bio. Mag.*, 24, 22–29, 2005.)

camera and microprocessor mounted in eyeglasses, a receiver implanted behind the ear, and an electrode-studded array that is tacked to the retina. A wireless battery pack worn on the belt powers the entire device. The camera captures an image and sends the information to the microprocessor, which converts the data to an electronic signal and transmits it to the receiver. The receiver sends the signals through a tiny cable to the electrode array, stimulating it to emit pulses. The pulses travel through the optic nerve to the brain, which perceives patterns of light and dark spots corresponding to the electrodes stimulated. Patients learn to interpret the visual patterns produced.

23.6.4 Progression of Disease States

The design and success of an implant depends on the situation in the eye or cortex at the time of surgery [87]. Most of the retinal implants are (currently) targeted at RP patients. There are many kinds of RP depending on the location and extent of genetic damage; some appear sooner in life, some later. Some forms are severe and acute; other forms are more chronic and play out over a longer time frame.

The success of a retinal implant depends on how much retina is left (that is healthy enough to receive a chip). There are stages of disease progression in RP. A remodeling (of the five-stage retinal cell layer architecture) and a rewiring of the neurons occur as the disease progresses. Blood vessels also reposition and relocate as the PR unfolds. The astrocytes and meuller cells (especially) proliferate as the disease advances. No two eyes are ever alike. The success or failure of an implant depends on individual circumstances.

A very chemically powerful neurotransmitter called glutamate is associated with retinal photocells. Glutamate works fine as long as it is contained inside its normal cycle of operation. When disease processes begin in the retina, glutamate is released into tissues that it was not designed to come in contact with. In effect glutamate is "dynamite;" it becomes a cellular toxin, poison to everything around it. Glutamate is somehow involved in the breakdown of the photocell layer in RP. But what makes RP patients eligible for retinal implants is that, although the rods and cones deteriorate (from glutamate poisoning), 50% to 75% of the ganglion cells survive, and approximately 80% of the bipolar and amacrine cells survive. In other words, RP is a disease that targets one layer of the retina severely (only about 5% of the photoreceptor survives) but more or less spares the other (processing and transmission) layers; that is, only the sensors are (initially) damaged severely in PR. Other complicating factors with RP include a marked decrease in blood flow (circulation drops 80%), a breakdown in the layers (so that it is difficult to determine them), and a progression, in some cases, to a retinal soup that leaves no viable retina for implantation (end stage of some varieties of the disease). For an implant to work, there must be preservation of some retinal layers.

23.6.5 Advancements in Using WBAN in Artificial Retina

In the 1960s, Giles S. Brindley of the University of Cambridge in England attached 80 electrodes to miniature radio receivers and implanted them into the brain of a blind patient. The patient reported seeing phosphenes (flashes of light) [89]. William Dobelle began experiments with cortical stimulation in human volunteers during the 1970s. His early implant system was (understandably) bulky by today's standards. Large wires protruded from a hole drilled in the subject's skull and ran to a large computer processing system. The patient saw light blobs (phosphenes) in black and white. Some of Dr. Dobelle's first implanted patients have used the system (without infection or severe complication) for more than 20 years. In 1992, a blind volunteer at the US National Institute of Neurological Disorders and Stroke learned to recognized phosphene letters (using an implant).

In the early 1990s, scientists led by Mark Humayun (then at Johns Hopkins University, now at Doheny) made an important discovery that led to an opportunity for intervention. They had discovered that, in the cases of certain retinal diseases RP and AMD, the parts of the eye that were damaged were the photoreceptor cells of the retina, the rods and cones that normally convert light into electrical impulses. But the diseases leave intact the neural paths to the brain, paths that transport the electrical signals. Thus, while the input from the photoreceptors eventually ceases, up to 70% to 90% of the nerve structures set up to receive those inputs remained intact.

The general thought at this time was that once you lose the photoreceptors you lose the rest of the circuitry as well. They found that the nerve cells in the eye are left fairly intact, so the task they had was how to take images and convert them into tiny electrical pulses. That is exactly what the photoreceptors do; they capture light and convert it into chemical pulses, and this information is sent to the brain, and it allows us to see.

Schwiebert et al. [73] developed a microsensor array that can be implanted in the eye as an artificial retina to assist people with visual impairments. In the Smart Sensors and Integrated Microsystems (SSIM) project, a retina prosthesis chip consisting of 100 microsensors is built and implanted within human eye as shown in Figure 23.17. This allows patients with no vision or limited vision to see at an acceptable level. The wireless communication is required to suit the need for feedback control, image identification, and validation. The communication pattern is deterministic and periodic, so TDMA fits best in this application to serve the purpose of energy conservation. Two group communication schemes are investigated: a LEACH-like cluster head–based approach and a tree-based approach. The system consists of an integrated circuit and an array of sensors. The integrated circuit is a multiplexer with on-chip switches and pads to support a 10 × 10 grid of connections; it operates at 40 kHz. Moreover, it has an embedded transceiver for wired and wireless communications. Each connection in the chip interfaces a sensor through an aluminum probe surface. Before the bonding is done, the entire integrated circuit, except the probe areas, is coated with a biologically inert substance. Each sensor is a microbump. It starts with a rectangular shape but, near the end, tapers to a point and rests gently on the retina tissue. The sensors are sufficiently small and light to be held in place with relatively little force. The distance between adjacent microbumps is approximately 70 μm. The sensors produce electrical signals proportional to the light reflected from an object being perceived. The ganglia and additional tissues transform the electrical energy into chemical energy, which, in turn, is transformed into optical signals and

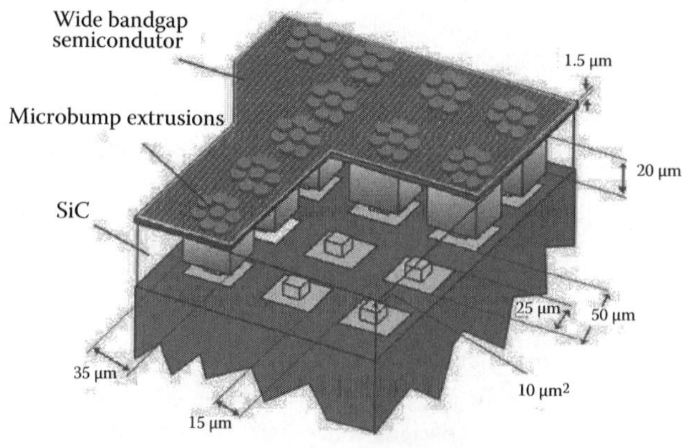

Figure 23.17 Illustration of the microsensor array.

communicated to the brain through the optic nerves. The magnitude and wave shape of the transformed energy corresponds to the response of a normal retina to light stimulation.

The system is a full duplex system, allowing communication in a reverse direction. In addition to the transformation of electrical signals into optical signals, neurological signals from the ganglia can be picked up by the microsensors and transmitted out of the sensing system to an external signal processor. In this way, the sensor array is used as a reception and transmission system in a feedback loop. Two types of wireless communications are foreseen. First, the eventual mapping of an input electrical signal to brain patterns cannot be realized internally with the sensing system alone because signal processing is a computationally intensive process. Second, diagnostic and maintenance operations require the extraction of data from the sensing system. For these reasons, interconnecting the sensing system with an external system is essential. In addition to these requirements, communication is required to transfer data from a charge-coupled device (CCD) camera (embedded in a pair of spectacles) to the sensor array. Figure 23.18 illustrates the signal-processing steps of the artificial retina. A camera embedded in a pair of spectacles directs its output to a real-time digital signal processor (DSP) for data reduction and processing. The camera can be combined with a laser pointer for automatic focusing. The output of the DSP is compressed and transmitted through a wireless link to the implanted sensor array, which decodes the image and produces a corresponding electrical signal.

A recent research project at Wayne State University and the Kresge Eye Institute developed an artificial retina using wireless biomedical sensors [89]. The project aimed to build a chronically implanted artificial retina with sufficient visual functionality to allow persons without vision or with limited vision to "see" at an acceptable level. In June 2002, at the 48th meeting of the American Society for Artificial Organs, Dr. William Dobelle announced that his team had created a commercially available artificial eye. The cost for the system is $98,000. The Dobelle artificial eye uses a miniature digital video camera mounted on the lens of sunglasses. Images from the camera are sent to a computer-processing unit on the belt. The cortical chip implant is embedded in the brain through a hole drilled in the skull. Cables run out of the head down to a stimulating control system (also on the belt). Inside the head, the chip contains electrodes that penetrate the brain in the area of the primary visual cortex. Images from the camera go to the microcomputer on the belt where the processor determines which electrodes are to be stimulated. Signals are sent to the stimulator unit, which activates selected electrodes in the implant. The implants hold 16 electrode arrays (placed on the mesial surface of the occipital lobe). It is possible that, eventually,

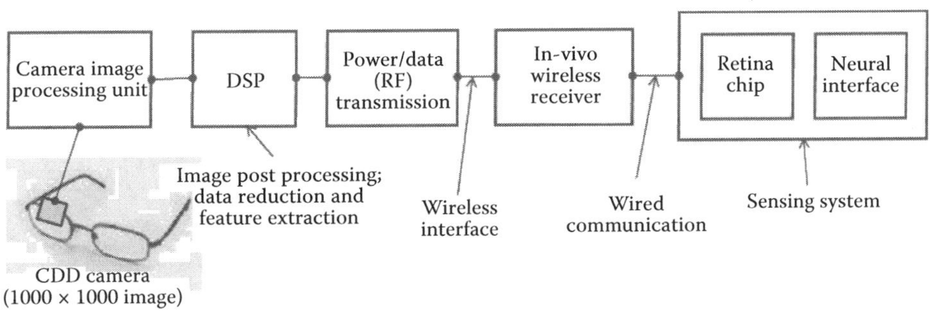

Figure 23.18 Processing components of the artificial retina. (From L. Schwiebert et al., Research challenges in wireless networks of biomedical sensors, In *The 7th Annual International Conference on Mobile Computing and Networking*, Rome, Italy, pp. 151–165, 2001.)

combinations of cortical and retinal chips networked together might be used to provide even higher degrees of visual processing than can be attained using either approach alone. The system can be disconnected from the camera and plugged directly into a television or a computer, allowing direct access to these technologies. Low resolution does not allow for "normal" viewing of these screens, but it does demonstrate how future chip technologies could be directly linked to output devices. A video monitor can also be patched into the system so that observers can monitor what the patient is viewing.

The DOE Artificial Retina Project [87] is a multi-institutional collaborative effort to develop and implant a device containing an array of microelectrodes into the eyes of people blinded by retinal disease. Ultimately, the goal is to restore useful vision for patients blinded by AMD and RP. This work builds upon a first-generation device containing a 16-electrode array on a miniature disc that can be implanted in the back of the eye to replace a damaged retina. After years of research and laboratory prototypes, artificial vision is being tested in humans for the first time. Artificial vision researchers take inspiration from another device, the cochlear implant, which has successfully restored hearing to thousands of deaf people. But the human vision system is far more complicated than the hearing system. The eye perceives millions of distinct points of light. Light entering through the pupil is converted to electrical signals, anatomical rods and cones, the light sensitive cells in the retina. Those electrical pulses are carried through the optic nerve to the brain, where they yield images to the world.

The Model 1 device [developed by Second Sight Medical Products, Inc. (SSMP)] was implanted in six blind patients between 2002 and 2004, whose ages ranged from 56 to 77 at the time of implant and all of whom have RP; see Figure 23.19. The device consists of a 16-electrode array in a one-inch package that allows the implanted electronics to wirelessly communicate with a camera mounted on a pair of glasses. It is powered by a battery pack worn on a belt. This implant enables patients to detect when lights are on or off, describe an object's motion, count individual items,

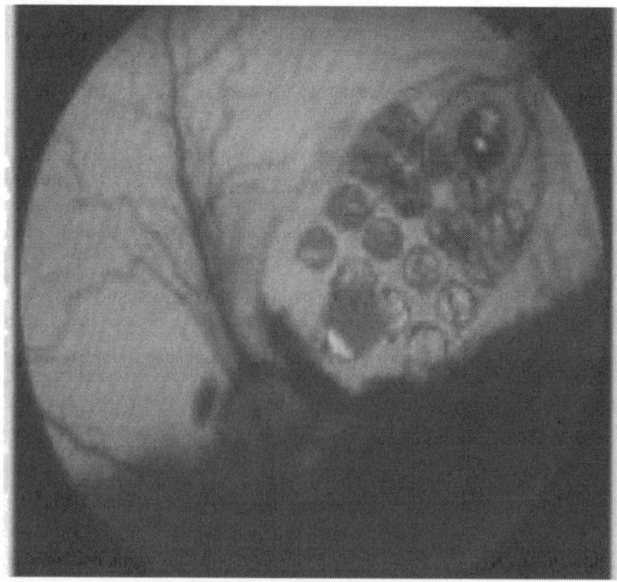

Figure 23.19 Argus I is a retinal prosthesis that contains a small implantable chip with electrodes. These electrodes stimulate the retina and help people regain limited vision.

and locate objects in their environment. To evaluate the long-term effects of the retinal implant, five devices have been approved for home use. Figure 23.20 shows the use of a small external camera to transmit images to an implanted 4 × 5 mm retina chip with 16 electrodes, which is positioned near the ganglion cell layer of the eye.

The smaller, more compact Model 2 retinal prosthesis (developed by SSMP with DOE contributions) is currently undergoing clinical trials to evaluate its safety and utility. This model is much smaller, containing 60 electrodes, and surgical implant time has been reduced from the 6 h required for Model 1 to 2 h.

The Model 3 device, which will have more than 200 electrodes, has undergone extensive design and fabrication studies at the DOE national laboratories and is ready for preclinical testing. The new design uses more advanced materials than the two previous models and has a highly compact array. This array is four times more densely packed with metal contact electrodes and required wiring connecting to a microelectronic stimulator. Simulations and calculations indicated that the 200+ electrode device should provide improved vision for patients.

Figure 23.21 approximates what patients with retinal devices see. Increasing the number of electrodes will result in more visual perceptions and higher-resolution vision. Scientists worked with patients who received the Model 1 implant (a 4 × 4 array of 16 electrodes) for about one month after surgery to help them interpret what they saw.

In [90], a practical WSN application that will enable blind people to see is proposed. The challenge is that the actual mapping between the input images to the brain pattern is too complex and

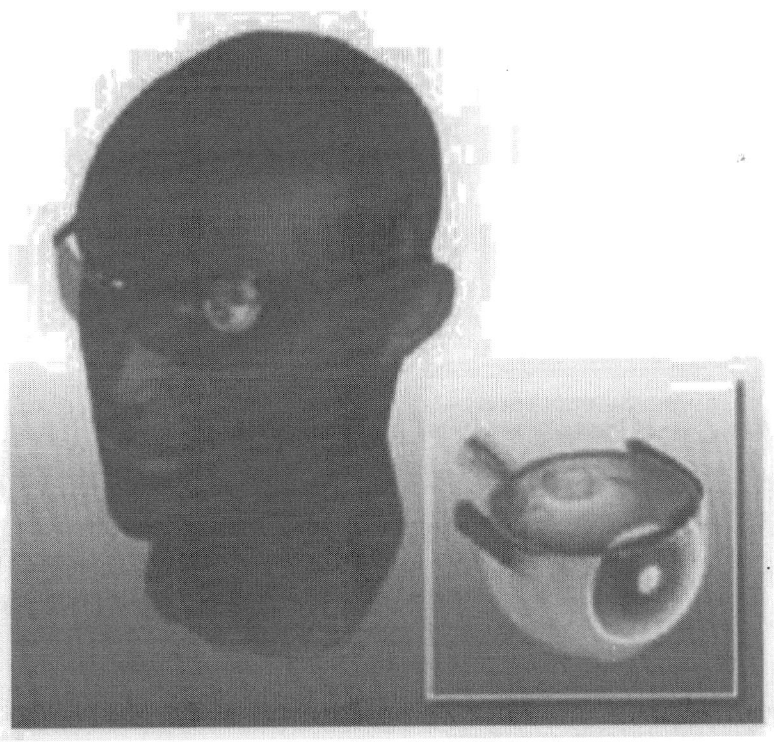

Figure 23.20 Using a small external camera to transmit images to an implanted 4 mm × 5 mm retina chip with 16 electrodes, which is positioned near the ganglion cell layer of the eye.

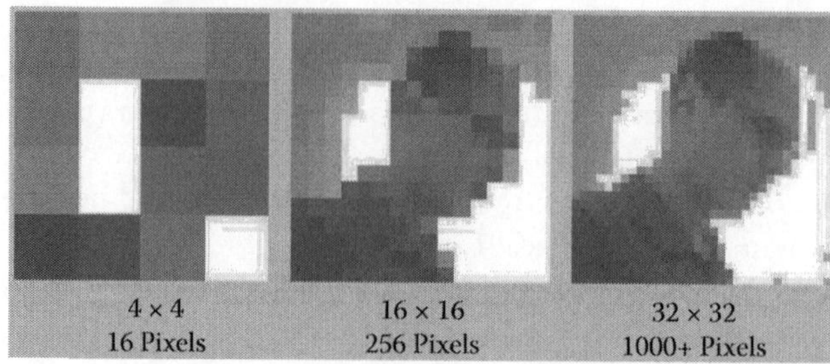

Figure 23.21 Images generated by the Artificial Retinal Implant Vision Simulator devised and developed by Wolfgang Fink at the Visual and Autonomous Exploration Systems Research Laboratory, California Institute of Technology.

not well understood. The connectivity problem in 3D/2D WSNs is studied, and the distribution of efficient algorithms to accomplish the required connectivity of the system are proposed. A new connectivity algorithm CDCA to connect disconnected parts of a network using cooperative diversity is provided.

The goal of this proposed approach is to allow people with no vision or with low vision to see what their neighboring partners can see. Unlike some other artificial vision proposals, this approach is applicable to virtually all causes of blindness. This device may also help some legally blind low vision patients because the cortex of sighted people responds to stimulation similarly to the cortex of blind people. A lot of work has been done to try to imitate the functionality of the human eye, and artificial vision techniques have been proposed. The challenge is that the actual mapping between the input images to a brain pattern is too complex and not well understood. In our case, we do not need to understand that complex mapping pattern because human brains function in a similar way, and the input to one brain (sent from another human with normal vision) could well be analyzed in the same way as if it was originated from the blind man's brain. Because the sensor devices are placed in the human body, the research problems differ from traditional WSNs. Most work concerning artificial vision has included a digital camera transmitting wireless data to a microchip in the eye [91]. The challenge is that the actual mapping between the input images to a brain pattern is too complex and not well understood. Because the sensors are irreplaceable and in case of emergency, we need to report to the central computer, and we require that our network is connected at all times if some sensor nodes were deployed on nearby objects to form a connected network.

The sensor implants are inside both the eye of the blind man's eye and a normal man's (or dog's) eye (Figure 23.22). Vision is established using the following steps:

A. Light entering through the pupil is converted to electrical signals. Those electrical pulses are converted to data and are stored in the retinal implant of the normal eye.
B. The data is then transmitted using the wireless transceiver to the blind man's eye.
C. The signal travels along a tiny wire to the retinal implant.
D. The back side of the retina is stimulated electrically by the sensors on the chip.
E. These electrical signals are converted to chemical signals by tissue structures, and the response is carried via the optic nerve to the brain.

The unique difference between this approach and any other is using the data from one man's eye. There is no need to understand the complex mapping of an image from the retina to the brain

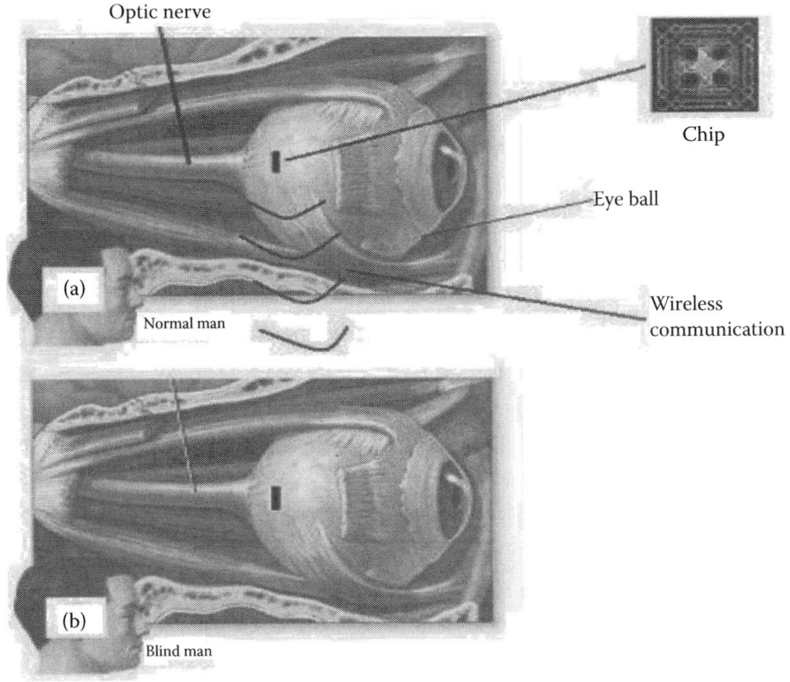

Figure 23.22 Proposed approach for eye vision. (From M. K. Watfa, *Int. J. Inform. Commun. Eng.* 4, 5, 2008.)

via the optic nerve. This application might be very useful in scenarios where the blind person and his partner are focusing their attention on a static phenomenon (e.g., watching a movie, attending a lecture, etc.). Cooperative diversity utilizes intermediate transmitters as repeaters to extend the range of transmission of a source. The advantage of cooperation is the accumulation of power from all relays, which is a source of diversity, and also of signal-to-noise ratio improvement [92–95].

Other retinal prostheses projects are underway in the United States and worldwide, including in Germany, Japan, Ireland, Australia, Korea, China, and Belgium. These programs pursue many different designs and surgical approaches. Some show great promise for the future, but have yet to demonstrate practicality in terms of adapting to and lasting long-term in a human eye. Thus far, the projects that have progressed to clinical (human) trials are the collaborative DOE effort, a project at the now-defunct Optobionics (Chicago), and two efforts in Germany at Intelligent Medical Implants AG and Retinal Implant AG. (For more information on worldwide projects, see *Science* 312, 1124–1126, 2006).

23.7 Conclusion

This survey shows that WSN technology offers significant potential in numerous application domains. Given the diverse nature of these domains, it is essential that WSNs perform in a reliable and robust fashion. WSNs have proven their usage in the future distributed computing environment. However, there are significant amounts of technical challenges and design issues, and those needs to be addressed.

This survey paper shows that WBANs can be widely used in medical applications and how these new applications can be used to potentially reduce costs for hospitals, government, and insurance companies. With wireless network–based medical technologies, applications can be designed to be less intrusive in patients' daily lives. It has surveyed a number of systems and applications for health care, mainly wearable and implantable sensors. As nanotechnology provides smaller and multiple-functional sensor nodes, the further these new small body networks may evolve, making their use as natural as wearing clothes in the near future. Given the importance of addressing ways to provide smart health care for the elderly, the chronically ill, and children, researchers have started to explore technological solutions to enhance the provision of health and social care in a way that complements existing services. The common types of wireless sensors are mentioned, and the benefits of using wireless networks for medical applications are discussed. The major challenges and open research problems of WBANs are discussed briefly. This paper concentrates on applying WBAN in vision rehabilitation and how to build a chronically implanted artificial retina for visually impaired people. This approach addresses two retinal diseases: AMD and RP. The architecture of the human retina, the work of artificial retinas, and the progression of disease states are described in detail. Advancements in using WBAN in artificial retinas are discussed to illustrate the enormous development this approach and its extent of technical efficiency.

23.8 Future Directions

With their potential uses in the medical and home health care fields, wireless networks have an important contribution to improving lives of patients. Besides bringing comfort to patients, there are large commercial benefits in the area of reducing costs and improving equipment and patient management. In the near future, the use of WBANs in vision repair will increase because it allows patients to be more self-sufficient, and the implant could turn out to be cheaper than governments paying for higher levels of health care or in-home care for patients. The system can therefore reach ubiquity, when each individual would have a computational module able to seamlessly interact with the smart space's system to overcome blindness.

As stated, using WBAN in artificial retinas is certainly a step in the right direction but, unfortunately, does not help those with traumatic retinal detachments or with tears that may lead to retinal detachments. Nonetheless, it is progress, and hopefully that progress will continue so that one day a retinal detachment does not result in a complete loss of vision. Also improving operations is the providing of early warning of system failure and a predictive maintenance tool, improving energy management, providing automatic record keeping for regulatory compliance, eliminating personnel training costs, or reducing insurance costs. The collaboration and synergy of sensing, processing, communication, and actuation is the next step to exploit the potential of these technologies.

These are the general considerations or challenges that will face the future of retinal prosthesis. All these considerations are important and should be taken into consideration in future works.

1. The artificial retina must deliver visual information in real time, just like the human retina. Having extra delay would mean that the patient would see the scene later than a person with a normal vision. This could lead the patient to make wrong and dangerous decisions such as crossing a street at the wrong time.
2. Artificial retinal power must not burn or injure the remaining part of the retina or the eye. The human retina consumes a 10th of a watt [96]. An artificial retina should consume no more than that.

3. The artificial retina must be small just like human retina, that is, half a millimeter thick and less than a cm^2 in area.
4. The artificial retina must be biocompatible.
5. The prosthesic device interface that will stimulate the remaining functional part of the human retina must be convex to conform to the convex shape of the retina.
6. The signal amplitude should be variable from patient to patient. Retinal tissue resistance will vary from person to person, and this implies that the device should accommodate increasing or decreasing the electrical signal amplitude.
7. The human retina processes the visual scene information to enhance the image and possibly to extract useful and variant information from the scene. In various layers, the human retina performs spatial averaging, contrast detection, adaptation, and more. Creating a retinal prosthesic device should process the visual scene in a similar way.
8. A chip that can be implanted with a smaller surgical cut is highly preferred. Engineering a chip that requires the surgeons to create a large surgical cut might be considered risky and may not implantable.

References

1. N. Bulusu, D. Estrin, L. Girod, and J. Heidemann. Scalable coordination for wireless sensor networks: Self-configuring localization systems. *Proc. Sixth International Symposium on Communication Theory and Applications (ISCTA '01)*, Ambleside, Lake District, UK, pp. 1797–1800, 2001.
2. A. Cerpa and D. Estrin, ASCENT: Adaptive self-configuring sensor networks topologies, *IEEE Transactions on Mobile Computing (TMC)*, 3(3), pp. 272–285, September, 2004.
3. Y. H. Nam et al. Development of remote diagnosis system integrating digital telemetry for medicine, In *Proceedings of International Conference IEEE-EMBS*, Hong Kong, pp. 1170–1173, 1998.
4. U. Varshney, Pervasive healthcare and wireless health monitoring, *Mobile Netw. Appl.*, 12, pp. 113–127, March 2007.
5. K. W. Goh, J. Lavanya, Y. Kim, E. K. Tan, and C. B. Soh, A PDA-based ECG beat detector for home cardiac care, In *IEEE Engineering in Medicine and Biology Society*, Shanghai, China, pp. 375–378, 2005.
6. H. Zhou, K. M. Hou, J. Ponsonnaille, L. Gineste, and C. D. Vaulx, A real-time continuous cardiac arrhythmias detection system: RECAD, In *IEEE Engineering in Medicine and Biology Society*, Shanghai, China, pp. 875–881, 2005.
7. H. Chu, C. Huang, Z. Lian, and T. Tsai, A ubiquitous warning system for asthma inducement, In *IEEE International Conference on Sensor Networks, Ubiquitous and Trustworthy Computing*, Taichung, Taiwan, pp. 186–191, 2006.
8. L. Huaming and T. Jindong, Body sensor network based context aware QRS detection, In *Pervasive Health Conference and Workshops*, Innsbruck, Austria, pp. 1–8, 2006.
9. J. Luprano, J. Sola, S. Dasen, J. M. Koller, and O. Chelelat, Combination of body sensor networks and on-body signal processing algorithms: The practical case of MyHeart project, In *International Workshop on Wearable and Implantable Body Sensor Networks (BSN 2006)*, Cambridge, MA, USA, 2006.
10. R. Jafari, A. Encarnação, A. Zahoory, F. Dabiri, H. Noshadi, and M. Sarrafzadeh, Wireless sensor networks for health monitoring, In *Second Annual International Conference on Mobile and Ubiquitous Systems: Networking and Services (MobiQuitous 2005)*, pp. 479–481, 2005.
11. O. Chipara, C. Lu, T. C. Bailey, and G. C. Roman, Reliable patient monitoring: A clinical study in a step-down hospital unit, Department of Computer Science and Engineering, Washington University. St. Louis, St. Louis, MO, Technical Report WUCSE-2009-82, December 2009.
12. J. Ko, J. Lim, Y. Chen, R. Musaloiu-E., A. Terzis, G. Masson, T. Gao, W. Destler, L. Selavo, and R. Dutton, MEDiSN: Medical emergency detection in sensor networks, *ACM Trans. Embedded Comput. Syst.*, 10, 1, 11:1–11:29, 2010.

13. T. Gao, C. Pesto, L. Selavo, Y. Chen, J. Ko, J. Lim, A. Terzis, A. Watt, J. Jeng, B. Chen, K. Lorincz, and M. Welsh, Wireless medical sensor networks in emergency response: Implementation and pilot results, In *IEEE Int. Conf. Technol. Homeland Security*, pp. 187–192, 2008.
14. D. Malan, T. Fulford-Jones, M. Welsh, and S. Moulton, Code Blue: an ad hoc sensor network infrastructure for emergency medical care, In *Proceedings of the MobiSys/Workshop on Applications for Mobile Embedded Systems*, pp. 12–14, June 2004.
15. G. Virone, A. Wood, L. Selavo, Q. Cao, L. Fang, T. Doan, Z. He, and J. A. Stankovic, An advanced wireless sensor network for health monitoring, In *Proceedings of the Transdisciplinary Conference on Distributed Diagnosis and Home Healthcare*, pp. 95–100, April 2006.
16. S. Kumar, Autosense, NIHGEI project at The University of Memphis, 2007. [Online]. Available: http://www.cs.memphis.edu/~santosh/AutoSense.html. Accessed on September 8, 2011.
17. K. Patrick, A tool for geospatial analysis of physical activity: Physical activity location measurement system (PALMS). NIHGEI project at the University of California at San Diego, 2007. [Online]. Available: http://www.gei.nih.gov/exposurebiology/program/docs/KevinPatrick.pdf. Accessed on October 10, 2011.
18. Q. Tang, N. Tummala, S. Gupta, and L. Schwiebert. Communication scheduling to minimize thermal effects of implanted biosensor networks in homogeneous tissue, *IEEE Trans. Biomed. Eng.*, 52, pp. 1285–1294, 2005.
19. E. Jovanov, A. Milenkovic, C. Otto, and P. de Groen, A wireless body area network of intelligent motion sensors for computer assisted physical rehabilitation, *J. Neuroeng. Rehabil.*, 1, pp. 2–6, 2005.
20. J. Coosemans and R. Puers, An autonomous bladder pressure monitoring system, *Sens. Actuat. A*, 155–161, pp. 123–124, 2005.
21. P. Troyk and M. Schwan, Closed-loop class E transcutaneous power and data link for micro-implants, *IEEE Trans. Biomed. Eng.*, 39, pp. 589–599, 1992.
22. M. Quwaider and S. Biswas, Body posture identification using hidden Markov model with a wearable sensor network, In *Proceedings of the ICST 3rd International Conference on Body Area Networks*, Tempe, AZ, USA, pp. 1–8, March 2008.
23. E. Krames, Implantable devices for pain control: Spinal cord stimulation and intrathecal therapies, *Best Pract. Res. Clin. Anaesthesiol.*, 16, 4, pp. 619–649, December 2002.
24. K. Chen and D. J. Bassett, The technology of accelerometry-based activity monitors: Current and future, *Med. Sci. Sports Exerc.*, 37, pp. 490–500, 2005.
25. M. Quwaider and S. Biswas, Physical context detection using multimodal sensing using wearable wireless networks, *J. Commun. Softw. Syst.*, 4, 191–202, 2008.
26. K. W. Goh, J. Lavanya, Y. Kim, E. K. Tan, and C. B. Soh, A PDA-based ECG beat detector for home cardiac care, In *IEEE Engineering in Medicine and Biology Society*, Shanghai, China, pp. 375–378, 2005.
27. J.-L. Lin, H. C. Liu, Y.-T. Tai, H.-S. Wu, S.-J. Hsu, F.-S. Jaw, and Y.-Y. Chen, The development of wireless sensor network for ECG monitoring, In *The 28th Annual International Conference of the IEEE, Engineering in Medicine and Biology Society*, New York, USA, pp. 3513–3516, 2006.
28. S. A. Taylor and H. Sharif, Wearable patient monitoring application (ECG) using wireless sensor networks, In *28th Annual International Conference on the IEEE Engineering in Medicine and Biology Society*, New York, USA, pp. 5977–5980, 2006.
29. J. Kolbe, W. Fergursson, and J. Garret, Rapid onset asthma: A severe but uncommon manifestation, *Thorax*, 53, pp. 241–247, April 1998.
30. S. Sur, T. Crotty, G. Kephart, B. Hyma, T. Colby, C. Reed, L. Hunt, and G. Gleich, Sudden-onset fatal asthma: A distinct entity with few eosinophils and relatively more neutrophils in the airway submucosa?, *Am. Rev. Respir Dis.*, 148, 3, pp. 713–719, 1993, PMID: 8368644 [PubMed – indexed for MEDLINE].
31. K. Lorincz, D. J. Malan, T. R. F. Fulford-Jones, A. Nawoj, A. Clavel, V. Shnayder, G. Mainland, M. Welsh, and S. Moulton, Sensor networks for emergency response, *IEEE Pervasive Comput.*, 3, pp. 16–23, October–December 2004.
32. E. Jovanov, A. Lords, D. Raskovic, P. Cox, R. Adhami, F. Andrasik, Stress monitoring using a distributed wireless intelligent sensor system, *IEEE Engineering in Medicine and Biology Magazine*, 22, 3, pp. 49–55, May/June 2003.
33. Y. J. Zhao, A. Davidson, J. Bain, S. Q. Li, Q. Wang, and Q. Lin, A MEMS viscometric glucose monitoring device, In *The 13th International Conference on Solid-State Sensors, Actuators and Microsystems*, TRANSDUCERS '05, pp. 1816–1819, 2005.

34. I. F. Akyildiz, W. Su, Y. Sankarasubramaniam, and E. Cayirci, A survey on sensor networks, *IEEE Commun. Mag.*, 40, 8, pp. 102–114, August 2002.
35. National Center for Health Statistics, URL: http://www.cdc.gov/nchs/Default.htm, Accessed in December 2011.
36. A. Darwish and A. E. Hassanien, Wearable and implantable wireless sensor network solutions for healthcare monitoring, *Sensors*, 11, pp. 5561–5595, 2011, doi: 10.3390/s110605561.
37. L. Theogarajan, J. Wyatt, J. Rizzo, B. Drohan, M. Markova, S. Kelly, G. Swider, M. Raj, D. Shire, M. Gingerich, J. Lowenstein, and B. Yomtov, Minimally invasive retinal prosthesis, *IEEE Solid-State Circuits Conference*, San Francisco, pp. 15–17, February 2006.
38. T. Penzel, B. Kemp, G. Klosch, A. Schlogl, J. Hasan, A. Varri, and I. Korhonen, Acquisition of biomedical signals databases, In *IEEE Engineering in Medicine and Biology Magazine*, 20, 3, pp. 25–32, May/June 2001.
39. S. Arnon, D. Bhastekar, D. Kedar, and A. Tauber, A comparative study of wireless communication network configurations for medical applications, *IEEE Wirel. Commun.* [see also *IEEE Pers. Commun.*], 10, 1, pp. 56–61, February 2003.
40. B. Gyselinckx, J. Penders, and R. Vullers, Potential and challenges of body area networks for cardiac monitoring, In *ISCE 32nd Annual Conference, J. Electrocardiol.*, 40, 6, S165–S168, November–December 2006.
41. A. H. Mark, C. P. Harry, T. B. Adam, R. Kyle, H. C. Benton, H. A. James, and L. John, Body area sensor networks: Challenges and opportunities, In *IEEE Computer Society*, Atlantic City, NJ, USA, pp. 58–65, 2009.
42. IEEE standard for safety levels with respect to human exposure to radio frequency electromagnetic Fields, 3 kHz to 300 GHz, IEEE C95.1-1991 (1999 ed.).
43. G.-Z. Yang, Ed., *Body Sensor Networks*, Springer-Verlag London Limited, 2006.
44. D. Malan, T. Fulford-Jones, M. Welsh, and S. Moulton, CodeBlue: An ad hoc sensor network infrastructure for emergency medical care, In *Proceedings of the MobiSys/Workshop on Applications for Mobile Embedded Systems*, pp. 12–14, June 2004.
45. J. Ko, J. Lim, Y. Chen, R. Musaloiu-E., A. Terzis, G. Masson, T. Gao, W. Destler, L. Selavo, and R. Dutton, MEDiSN: Medical emergency detection in sensor networks, *ACM Trans. Embedded Comput. Syst.*, 10, 1, 11:1–11:29, 2010.
46. O. Chipara, C. Lu, T. C. Bailey, and G.-C. Roman, Reliable patient monitoring: A clinical study in a step-down hospital unit, Department of Computer Science and Engineering, Washington University St. Louis, St. Louis, MO, Technical Report WUCSE-2009-82, December 2009.
47. B. Chen, K. K. Muniswamy-Reddy, and M. Welsh, Ad-hoc multicast routing on resource-limited sensor nodes, In *Proceedings of the 2nd International Workshop on Multi-Hop Ad Hoc Networks: From Theory to Reality*, Florence, Italy, pp. 87–94, 2006.
48. K. Lorincz and M. Welsh, MoteTrack: A robust, decentralized approach to RF-based location tracking, *Pers. Ubiquit. Comput.*, 11, 6, pp. 489–503, 2007.
49. R. Jain, Survey paper: Medical applications of wireless networks, This paper is available on-line at http://www.cse.wustl.edu/~jain/, Accessed in December 2011.
50. O. Gnawali, R. Fonseca, K. Jamieson, D. Moss, and P. Levis, Collection tree protocol, In *Proceedings of the 7th ACM Conference On Embedded Network Sensor Systems*, Berkeley, CA, pp. 1–14, 2009.
51. K. Laerhoven, B. Lo, J. Ng, S. Thiemjarus, R. King, S. Kwan, H. Gellersen, M. Sloman, O. Wells, P. Needham, N. Peters, A. Darzi, C. Toumazou, and G. Yang, Medical healthcare monitoring with wearable and implantable sensors, In *UbiHealth 2004: 3rd International Workshop on Ubiquitous Computing for Pervasive Healthcare Applications*, Nottingham, England, 2004.
52. U. Maurer, A. Rowe, A. Smailagic, and D. Siewiorek, eWatch: A wearable sensor and notification platform, In *The International Workshop on Wearable and Implantable Body Sensor Networks (BSN 2006)*, Cambridge, MA, USA, 2006.
53. A. Pentland, Healthwear: Medical technology becomes wearable, *Computer*, 37, 5, 42–49, May 2004.
54. A. Salarian, H. Russmann, F. J. G. Vingerhoets, C. Dehollain, Y. Blanc, P. R. Burkhard, and K. Aminian, Gait assessment in Parkinson's disease: Toward an ambulatory system for long-term monitoring, *IEEE Trans. Biomed. Eng.*, 51, 8, pp. 1434–1443, August 2004.

55. M. Visintin, H. Barbeau, N. Korner-Bitensky, and N. E. Mayo, A new approach to retrain gait in stroke patients through body weight support and treadmill stimulation, *Stroke*, 29, 6, 1122–1128, June 1998.
56. J. He, H. Li, and J. Tan, Real-time daily activity classification with wireless sensor networks using hidden Markov model, In *Proceedings of the 29th Annual International Conference of the IEEE Engineering in Medicine and Biology Society*, pp. 3192–3195, August 2007.
57. E. Miluzzo, N. D. Lane, K. Fodor, R. Peterson, H. Lu, M. Musolesi, S. B. Eisenman, X. Zheng, and A. T. Campbell, Sensing meets mobile social networks: The design, implementation and evaluation of the CenceMe application, In *Proceedings of the 6th ACM Conference on Embedded Network Sensor Systems*, Raleigh, NC, pp. 337–350, 2008.
58. A. Ahmadi, D. D. Rowlands, and D. A. James, Investigating the translational and rotational motion of the swing using accelerometers for athlete skill assessment, In *Proceedings of the 5th IEEE Conference on Sensors*, pp. 980–983, October 2006.
59. F. Michahelles and B. Schiele, Sensing and monitoring professional skiers, *IEEE Pervasive Comput.*, 4, 3, pp. 40–46, July–September 2005.
60. K. Lorincz, B. Chen, G. W. Challen, A. R. Chowdhury, S. Patel, P. Bonato, and M. Welsh, Mercury: A wearable sensor network platform for high-fidelity motion analysis, In *Proceedings of the 7th ACM Conference on Embedded Network Sensor Systems*, pp. 353–366, 2009.
61. S. Patel, K. Lorincz, R. Hughes, N. Huggins, J. H. Growdon, M. Welsh, and P. Bonato, Analysis of feature space for monitoring persons with Parkinson's disease with application to a wireless wearable sensor system, In *Proceedings of the 29th IEEE Engineering in Medicine and Biology Society Annual International Conference*, pp. 6290–6293, August 2007.
62. R. Ganti, P. Jayachandran, T. Abdelzaher, and J. Stankovic, SATIRE: A software architecture for smart AtTIRE, In *Proceedings of the 4th International Conference on Mobile Systems. Applications, and Services*, Uppsala, Sweden, pp. 110–123, June 2006.
63. Crossbow Corporation. MICAz specifications. [Online]. Available: http://www.xbow.com/Support/Support_pdf_files/MPR-MIB_Series_Users_Manual.pdf. Accessed in November 2011.
64. M. Soini, J. Nummela, P. Oksa, L. Ukkonen, and L. Sydänheimo, Wireless body area network for hip rehabilitation system, *UbiCC J*, 3(5), pp. 42–48, 2008.
65. Mobihealth project, URL: http://www.mobihealth.org. Accessed in December 2011.
66. P. Neves, M. Stachyra, and J. Rodrigues, Application of wireless sensor networks to healthcare promotion *J. Commun. Softw. Syst.*, 4, 3, 1845–6421/08/8143, September 2008.
67. G. Mulligan and L. W. Group, The LoWPAN architecture, In *The 4th Workshop on Embedded Networked Sensor*, Cork, Ireland, 2007.
68. J.G. Ko, C. Lu, M. B. Srivastava, J. A. Stankovic, A. Terzis, and M. Welsh, Wireless sensor networks for healthcare, *Proc. IEEE*, 98, 11, pp. 1947–1960, November 2010.
69. S. Kumar, Autosense, NIH GEI project at The University of Memphis, 2007. [Online]. Available: http://www.cs.memphis.edu/~santosh/AutoSense.html. Accessed in August 16, 2011.
70. T. Das, P. Mohan, V. N. Padmanabhan, R. Ramjee, and A. Sharma, PRISM: Platform for remote sensing using mobile smartphones, In *Proceedings of the 8th International Conference on Mobile Systems, Applications, and Services*, San Francisco, pp. 63–76, June 2010.
71. P. Khan, M. A. Hussain, and K. S. Kwak, Medical applications of wireless body area networks, *JDCTA* 3, pp. 185–193, 2009.
72. J. A. Stankovic, Q. Cao, T. Doan, L. Fang, Z. He, R. Kiran, S. Lin, S. Son, R. Stoleru, and A. Wood, Wireless sensor networks for in-home healthcare: Potential and challenges, In *High Confidence Medical Device Software and Systems Workshop*, Pennsylvania, PA, USA, 2005.
73. L. Schwiebert, S. K. S. Gupta, and J. Weinmann, Research challenges in wireless networks of biomedical sensors, In *The 7th Annual International Conference on Mobile Computing and Networking*, Rome, Italy, pp. 151–165, 2001.
74. B. Krishnamachari *Networking Wireless Sensors*, Cambridge University Press 2005.
75. J. Yoo, L. Yan, S. Lee, Y. Kim, and H. Yoo, A 5.2 mw self-configured wearable body sensor network controller and a 12 w wirelessly powered sensor for a continuous health monitoring system, *IEEE J. Solid-State Circ.* 45, pp. 178–188, 2010.

76. T. Gao, D. Greenspan, M. Welsh, R. Juang, and A. Alm, Vital signs monitoring and patient tracking over a wireless network, In *Proceedings of the 27th Annual International Conference of the IEEE EMBS*, Shanghai, China, pp. 102–105, 1–4 September 2005.
77. M. Renaud, K. Karakaya, T. Sterken, P. Fiorini, C. Hoof, and R. Puers, Fabrication, modelling and characterization of MEMS piezoelectric vibration harvesters, *Sens. Actuat. A*, 145–146, pp. 380–386, 2008.
78. C. Lauterbach, M. Strasser, S. Jung, and W. Weber, Smart clothes self-powered by body heat, In *Proceedings of Avantex Symposium*, Frankfurt, Germany, pp. 5259–5263, May 2002.
79. S. Cherukuri, K. Venkatasubramanian, and S. Gupta, BioSec: a biometric based approach for securing communication in wireless networks of biosensors implanted in the human body, In *Proceedings of the 2003 International Conference on Parallel Processing Workshops (ICPPW'03)*, 1530-2016/03 2003 IEEE.
80. J. Misic, Enforcing patient privacy in healthcare WSNs using ECC implemented on 802.15.4 beacon enabled clusters, In *Proceedings of the Sixth Annual IEEE International Conference on Pervasive Computing and Communications*, Hong Kong, pp. 686–691, March 2008.
81. G. Zhang, C. Poon, Y. Li, and Y. Zhang, A biometric method to secure telemedicine systems, In *Proceedings of the 31st Annual International Conference of the IEEE Engineering in Medicine and Biology Society*, Minneapolis, MN, USA, pp. 701–704, September 2009.
82. S. Bao, Y. Zhang, and L. Shen, Physiological signal based entity authentication for body area sensor networks and mobile healthcare systems, In *Proceedings of the 27th Annual International Conference of the IEEE EMBS*, Shanghai, China, pp. 2455–2458, 1–4 September 2005.
83. T. Hori, Y. Nishida, T. Suehiro, and S. Hirai, SELF-Network: Design and implementation of network for distributed embedded sensors, In *Proceedings of IEEE/RSJ International Conference on Intelligent Robots and Systems*, Takamatsu, Japan, pp. 1373–1378, 30 October–5 November 2000.
84. Y. Mori, M. Yamauchi, and K. Kaneko, Design and implementation of the vital sign box for home healthcare, In *Proceedings of IEEE EMBS International Conference on Information Technology Applications in Biomedicine*, Arlington, VA, USA, pp. 104–109, November 2000.
85. M. Gietzelt, K. Wolf, M. Marschollek, and R. Haux, Automatic self-calibration of body worn triaxial-accelerometers for application in healthcare, In *Proceedings of the Second International Conference on Pervasive Computing Technologies for Healthcare*, Tampere, Finland, pp. 177–180, January 2008.
86. E. Biagioni and G. Sasaki, Wireless sensor placement for reliable and efficient data collection, In *Proceedings of the Hawaii International Conference on Systems Sciences (HICSS 2003)*, IEEE, New York, pp. 127–136, Hawaii, January 2003.
87. Base URL: http://artificialretina.energy.gov, Accessed January 2012.
88. National Alliance for Eye and Vision Research, www.eyeresearch.org. Accessed at November 10, 2012.
89. W. Dobelle, Retinal and cortical chip implants: Artificial vision, In *Eye and Chip World Conference on Artificial Vision held at Lawrence Tech University in Southfield Michigan*, June 2002. http://www.wayfinding.net/iibnNECtextchip.htm.
90. M. K. Watfa, Practical applications and connectivity algorithms in future wireless sensor networks, *Int. J. Inform. Commun. Eng.* 4, 5, 2008.
91. C. Veraart, F. Duret, M. Brelen, and J. Delbeke, Vision rehabilitation with the optic nerve visual prosthesis, In *Proceedings of the 26th Annual International Conference of the IEEE EMBS*, San Francisco, USA, pp. 4163–4165, September 2004.
92. I. Maric and R. Yates, Efficient multihop broadcast for wideband systems, In *Proceedings of DIMACS Workshop on Signal Processing for Wireless Transmission*, Piscataway, pp. 285–299, October 2002.
93. A. Sendonaris, E. Erkip, and B. Aazhang, User cooperation diversity—Part I: System description, *IEEE Trans. Commun.*, vol. 51, pp. 1927–1938, November 2003.
94. A. Sendonaris, E. Erkip, and B. Aazhang, User cooperation—Part I: System description and user cooperation—Part II: Implementation aspects and performance analysis, *IEEE Trans. Commun.*, 51, 11, pp. 1927–1938, November 2003.
95. J. Laneman, D. Tse, and G. Wornell, Cooperative diversity in wireless networks: Efficient protocols and outage behavior, *IEEE Trans. Inform. Theory*, 50, 12, pp. 3062–3080, December 2004.
96. https://www.stanford.edu/group/brainsinsilicon/vision.html, Accessed December 17, 2011.
97. K.D. Wise. Silicon microsystems for neuroscience and neural prostheses: Interfacing with the central nervous system at the cellular level. *IEEE Eng. Med. Bio. Mag.*, 24, 22–29, 2005.

Related Web Sites for More Information

AMD Alliance www.amdalliance.org.
American Council of the Blind www.acb.org.
Biomimetic MicroElectronic Systems bmes-erc.usc.edu.
DOE Artificial Retina Project ArtifcialRetina.energy.gov.
Doheny Eye Institute www.usc.edu/hsc/doheny/.
Foundation Fighting Blindness www.blindness.org.
National Alliance for Eye and Vision Research www.eyeresearch.org.
Prevent Blindness America www.preventblindness.org.
Second Sight Medical Products www.2-sight.com.

Chapter 24

Wireless Sensor Networks: A Medical Perspective

Mohamed Mostafa M. Fouad and Nashwa El-Bendary
Arab Academy for Science, Technology, and Maritime Transport
Rabie A. Ramadan and Aboul Ella Hassanien
Cairo University

Contents

24.1 Introduction

Wireless sensor networks (WSNs) consist of many tiny nodes that cooperate to form a wireless network. These sensors collect different phenomena from the monitored field. Collected data is usually sent to a centralized node, called a sink node, for data analysis and decision-making. A common architecture of WSNs is presented in Figure 24.1, in which the sink node disseminates its data to the Internet for user usage. However, sensor nodes suffer from many limitations, including scarce energy sources, small memory footprint, and limited processing capabilities. In addition, in many of the WSN applications, sensors are deployed in large numbers. Consequently, there are many other issues, such as scalability, data reliability, security and key management, data analysis, and efficient multi-hop routing. Nevertheless, WSNs have many applications including the battlefield [1], health care [2], critical infrastructure [3,4], smart environment [5], and transportation network monitoring [6].

So, motivated by user demand and driven by recent advances in hardware and software along with the emergence of micro-electro-mechanical systems (MEMS) technology, the usage of WSNs for health care has shown a good potential to alter the practice of medicine. Accordingly, WSNs can be found in various health care applications, providing an active tool for continuous monitoring and analysis of patients' and elderly people's physiological parameters. In addition to that, the recently proposed wireless body area networks (WBANs), incorporating context-aware sensing for increased sensitivity and specificity, has been widely used in health care.

Figure 24.1 WSN architecture.

This chapter presents an overview of the impact of WSNs on medical health care as well as highlighting the state of the art in the applications of WSNs for tele-health care. Also, many challenges introduced by health care applications of WSNs result from the required level of trustworthiness and the need to ensure the privacy and security of medical data. So this chapter highlights in more detail the major challenges and design issues that such wireless sensing systems face and have to consider or overcome when implemented in medical health care applications via providing insights on the techniques and features that researchers and system designers should consider for successful deployments along with emphasizing some ongoing research directions.

24.2 WSNs and Medical Sensing

In recent times, WSNs carried the promise of intensely improving and mounting the quality of care for different segments of the population. Hence, WSNs presented a new paradigm for early system prototypes in order to demonstrate the potential of WSNs for enabling early detection of clinical deterioration through real-time patient monitoring in hospitals, enhancing first responders' capability of providing emergency care in large disasters through automatic electronic triage, improving the life quality of the elderly through smart environments, and enabling large-scale field studies of human behavior and chronic diseases [7].

Thus, before moving forward in this chapter, it is worthwhile to assume the scale of deployment of health care applications using medical WSNs via giving a brief idea about the research groups and projects that have started to develop health monitoring using WSNs. So far, several projects have been introduced in this regard. For example, CodeBlue [8], LiveNet [9], MobiHealth [10], Alarm-Net [11], UbiMon [12], ReMoteCare [13], MobiCare [14,15], Lifeguard [16], AID-N [17], CareNet [18], ASNET [19], WiMoCA [20], SAPHIRE [21], THE-MUSS [22], MEDiSN [23], etc.

The following briefly illustrates the most popular health care projects applying medical WSNs, namely, CodeBlue, Alarm-Net, UbiMon, MobiCare, and MEDiSN.

24.2.1 CodeBlue

CodeBlue [8,24] is a popular health care research project, based on a medical sensor network developed at the Harvard Sensor Network Lab. Its architecture contains several medical sensors, board onto the Mica2 motes [25], which are placed on the patient's body. These medical sensors sense the patient body data and transmit it wirelessly to the end-user devices (personal digital assistants [PDAs], laptops, and personal computers) for further analysis.

As shown in Figure 24.2, presenting the general architecture of the CodeBlue project for emergency response, the basic idea of CodeBlue is that a doctor or medical professional issues a query for patient health data using a PDA, which is based on a publish-and-subscribe architecture. The medical sensors publish their relevant data to a specific channel, and the end-user subscribes the channel by using a hand-held devices (e.g., PDA and laptop). A TinyADMR routing component, based on an adaptive demand-driven multicast routing (ADMR) protocol, is used for facilitating node mobility, multicast routing, and minimal path losses. Further, the CodeBlue architecture facilitates RF-based localization, called MoteTrack [26], in order to locate a patient's or medical professional's position. CodeBlue is anticipated for deployment in pre-hospital and in-hospital emergency care, stroke patient rehabilitation, and disaster response.

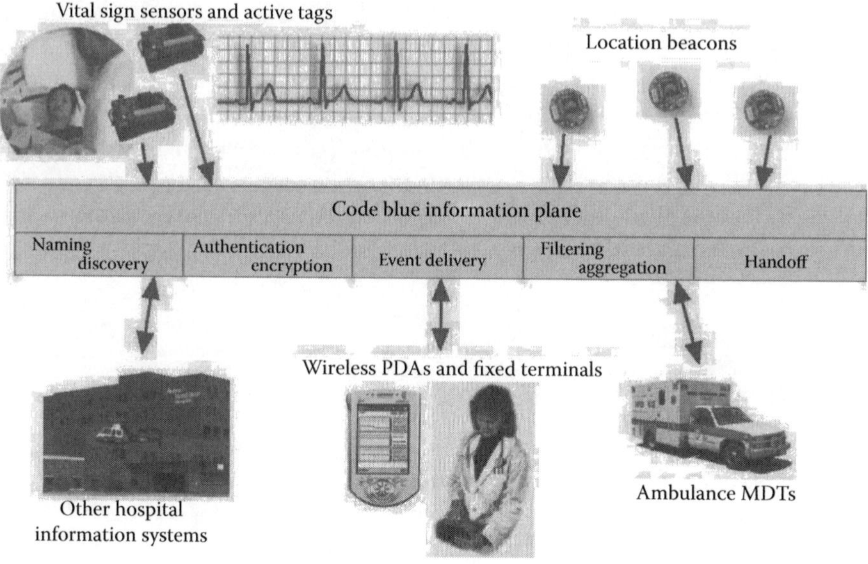

Figure 24.2 CodeBlue architecture for emergency response. (From CodeBlue healthcare project. Available online at: http://fiji.eecs.harvard.edu/CodeBlue/; Malan D. et al., CodeBlue: an ad hoc sensor network infrastructure for emergency medical care, 2004.)

24.2.2 Alarm-Net

Alarm-Net [11] was specifically designed at the University of Virginia as a heterogeneous network for patient health monitoring in the assisted-living and home environment. Alarm-Net consists of body sensor networks and environmental sensor networks. Three network tiers are applied to the proposed assisted-living and home environment as shown in Figure 24.3. In the first tier, a resident wears body sensor devices, which sense individual physiological data, and in the second tier, environmental sensors, such as temperature, dust, motion, light (i.e., MicaZ boards) are deployed in the living space to sense the environmental conditions. In the third tier, an Internet protocol (IP)-based network is used, which is comprised of Stargate gateways called AlarmGate.

The idea of Alarm-Net is that body sensors broadcast individual physiological data using a single hop to the nearest stationary sensor (i.e., second tier). Thereafter, the stationary emplaced sensor nodes forward the body data using multi-hop communication to the AlarmGate, which is a gateway between the wireless sensor and IP networks and is also connected to a back-end server.

24.2.3 UbiMon

UbiMon (ubiquitous monitoring environment for wearable and implantable sensors) [12] is a WBAN architecture composed of wearable and implantable sensors (e.g., electrocardiography [ECG], saturation of peripheral oxygen [SpO2], and blood oxygen) using an ad hoc network. The aim of the project is to provide continuous monitoring of an individual's physiological states and to capture transient as well as life-threatening abnormalities that can be detected and predicted.

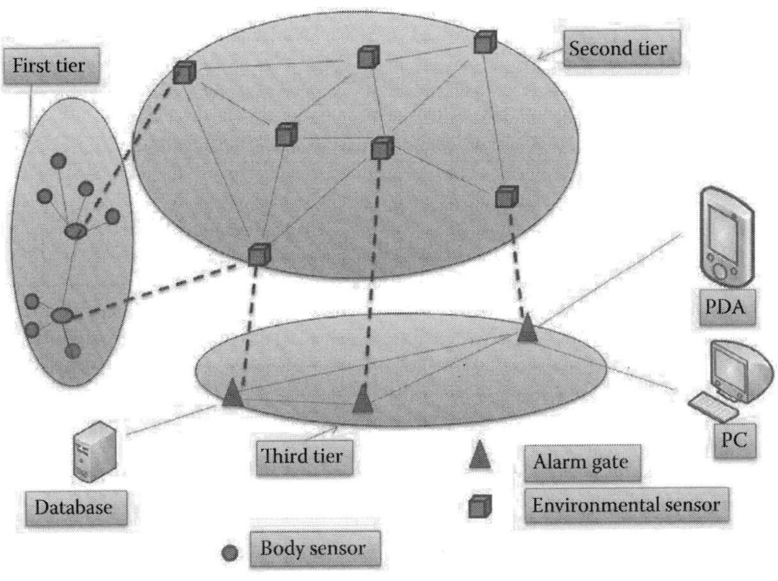

Figure 24.3 ALARM-NET architecture. (From Kumar, P., Lee, H.J., *Sensors*, 12, pp. 55–91, 2012.)

24.2.4 MobiCare

MobiCare [14] provides a wide-area mobile patient monitoring system that facilitates continuous and timely monitoring of a patient's physiological status. Similar to UbiMon, the MobiCare system comprises of a WBAN having wearable sensors (e.g., ECG, SpO2, and blood oxygen), a WBAN manager called "MobiCare client that is an IBM wristwatch," and a back-end infrastructure (MobiCare server). In a timely manner, the medical sensors sense the patient's body data and broadcast it to the MobiCare client. The MobiCare client aggregates the body data and sends them using a cellular link to the MobiCare server. The MobiCare server supports the medical staff for offline physiological analysis and for patient care [14].

24.2.5 MEDiSN

MEDiSN, a health care system developed at Johns Hopkins University, was especially designed for patients' monitoring in hospitals and during disaster events [23]. It comprises multiple physiological monitors (called PMs), which are battery-powered motes (Telos boards [28]) equipped with medical sensors for collecting patients' physiological health information (e.g., blood oxygenation, pulse rate, electrocardiogram signals, etc.). The PMs are mobile, temporarily storing sensed data and transmitting it (after encrypting and signing the sensed data) to the relay points (RPs). MEDiSN incorporates different stationary RPs that are self-organized into a bidirectional routing tree and forwards PM data to the gateways and vice versa. The RPs use a collection tree routing protocol (CTP) to forward their measurements to the gateway. Moreover, MEDiSN is connected with a back-end database that constantly stores medical data and presents them to authenticated GUI clients. Specifically, this research focused on reliable communication, data rate, routing, and QoS.

24.3 Tele-Health Care: Different Perspectives

A broad definition of tele-health care is the use of information technology to provide health care services at a distance, which, hence, typically can include anything from medical services at the inpatient or at the outpatient stage. Tele-health care could even include a situation in which a doctor in one hospital supports surgery with a doctor in another hospital somewhere on another continent, for example. What one is looking at with tele-health care is much more focused health care where it is available to monitor patients in their homes or when they are mobile for capturing vital parameters.

Moreover, recent advances in wireless communication technology simplify the traditional paradigm used for patient's monitoring and treatment. This paradigm consists of three intersected perspectives: patient, physician, and health care center. The use of biomedical wireless sensors adds many benefits for each perspective. The following details these additional benefits from patient, physician, and health care center perspectives.

24.3.1 From the Patient's Perspective

Traditional health care systems use wired biomedical sensors, which attach patients to machines in order to read different values of vital data. In contrast, modern health care systems use small-sized portable biomedical wireless sensors that allow many patients to perform their daily work while they are still under their physician's supervision. There have been significant research efforts for minimizing the size of these biomedical wireless sensor devices to be invisible in order to be accepted and wearable by patients. Most of the existing solutions integrate these sensor devices into fabric or other substances or implant them in human body [29].

The main benefit of providing health care services and consultations where some patients exist at a certain time (e.g., in the patient's own home), granted by using these wireless sensors, is to reduce clinic visit intervals and provide cheaper health care services accordingly. A patient may also configure a wireless sensor for his own purposes. For instance, Bell's MyLifeBits project [30], which includes text search, text and audio annotations, and hyperlinks, can be used as a personal reminder to recall everything a patient reads, writes, and hears.

24.3.2 From the Physician's Perspective

Sensor devices provide great services to physicians as they are always on to afford real-time continuous monitoring and online communication with patients. These tiny wireless sensors reduce language barriers between physicians and patients by collecting and storing the vital signs of each patient directly into his health record, such as an electronic medical record (EMR), which results in enhancing the physician's decision-making for diagnostics and accurate patient treatment. Also, using a patient's EMR reduces paper trails and time for both physicians and patients to go through the patient's whole medical life story every time visiting a new physician. Furthermore, that technology may also assist as a new approach to monitoring infectious disease outbreaks.

24.3.3 From the Health Care Center's Perspective

For any organization, quality and process improvements are primary drivers. The reliance on wireless sensors reduces the 24/7 labor cost and minimizes potential human errors. Accordingly, health care centers can provide high-quality medical services with competitive prices. Health care centers store biomedical signals collected periodically or continuously from remote patients in a database server. This data is used for real-time monitoring of patients in case of emergencies and is also useful in periodic checkups. The health care center uploads this data directly to the responsible physicians. As a result, this provides more efficient utilization of physicians, decreases scheduled clinical visits, and supports long-term care while shortening hospital stays and, consequently, contributing to reduced health care costs. For some educational health care centers, the stored data is used to support the training of medical interns and students [31]. In some cases, such as emergencies or disasters occurring for a remote patient, the health care center alerts other in-home relatives in order to serve that patient until the arrival of an ambulance.

24.4 Health Care Information Technology: Challenges and Design Issues

Lately, many research communities and many application domains focus on the use of wearable sensor devices. The resource scarcity that is inherent in these wireless sensor devices, in terms of limited computational capabilities, memory constraints, limited network capacity, and limited energy sources, imposes technical constraints on their network's functionalities.

Furthermore, the new health care pattern will raise many primary and biomedical security, as well as privacy, concerns, which are considered to be the nontechnical side of health care information technology design issues. Figure 24.4 illustrates the main technical and nontechnical design issues for health care information technology.

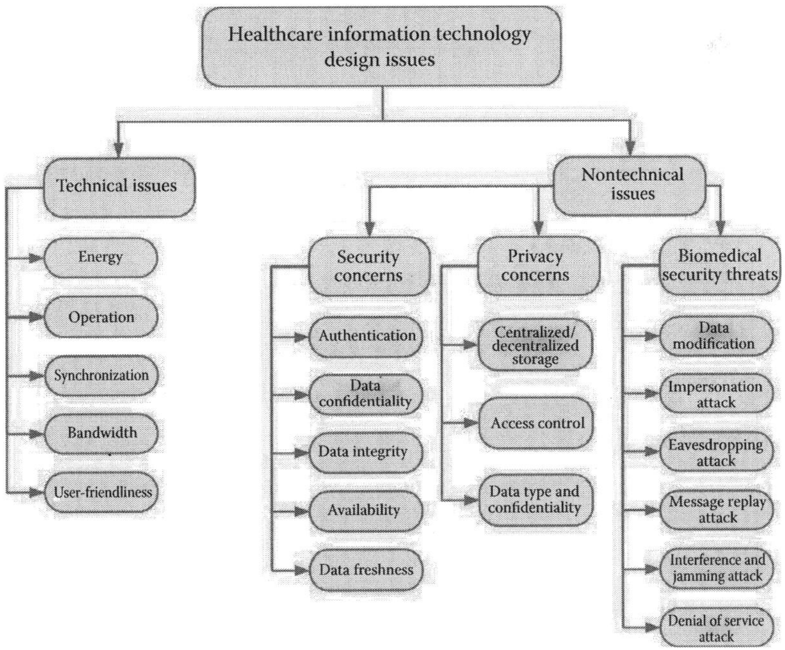

Figure 24.4 Technical and nontechnical design issues for health care information technology.

24.4.1 Technical Issues

For the health care domain, technical challenges reached above and beyond the constrained resources of the WSN, such as (a) how to make these sensor devices easy to operate and manage and reliable under different operating conditions, and (b) how to ensure that the synchronization of the data-delivery model is affordable and friendly for end users. The following gives a description of these technical challenges.

24.4.1.1 Energy

Although there are many proposed techniques for sensor networks that are energy efficient, still there is a dire need for more energy saving, especially for the health care dedicated sensors, in which recharging the batteries may be forgotten or difficult, especially for elderly and handicapped people. On the other hand, most of the wireless technologies based on Bluetooth and Wi-Fi failed to provide energy-efficient systems because they are manufactured to be run by a small energy source, such as a coin battery [29,31]. For example, even with the Shimmer sensor [32], which is one of the best-designed wearable sensor device supporting multiple sensing solutions, such as kinematics, ambient biophysical sensing, and supported with MicroSD interface for a Flash memory up to 2 GB, the energy is still constrained because it is manufactured by a rechargeable Li-polymer battery of 450 mAh. Figure 24.5 shows the Shimmer wireless sensor platform that measures 53 × 32 × 19 mm and weighs just 22 g [32].

24.4.1.2 Operation

For most remote patient monitoring situations, a patient's data needs to be collected continuously during day and night; however, transmitting these huge amounts of data may be useless and quickly drain the medical node's energy. Thus, the operation of the health care sensor should perform the self-care principle [33], in which data is collected, processed, and analyzed by the patient's wearable device. So, this wearable device prompts the patient regarding the best solution for his current situation. Otherwise, the device automatically reports to the health care center in

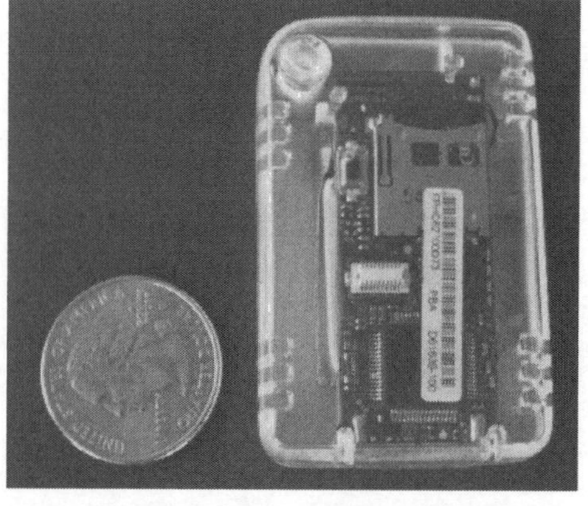

Figure 24.5 The Shimmer wearable sensor platform incorporates a TI MSP430 processor, a CC2420 IEEE 802.15.4 radio, a triaxial accelerometer, and a rechargeable Li-polymer battery.

critical and unmanageable situations. Another example is applying the self-care principle for helping diabetic patients to manage their own conditions via having the wearable sensor device alerting the patient to take an action regarding any change in glucose level in his blood or correlating glucose level with an insulin bump that provides (injects) the patient with the required amount of insulin.

24.4.1.3 Synchronization

For pervasive health care systems, the number of sensor nodes may be high; thus, the data collection rate will increase accordingly. So, in order to measure the accurate delivery delay between the health care sensor devices, a real-time synchronization in addition to a calibration between their clocks has to be performed. Moreover, time synchronization plays an important role in allocating the patient's place through estimating the patient's distance using time difference techniques, such as time-of-flight and time difference of arrival (TDoA)–based ranging techniques [34]. Finally, synchronization may be considered as a basis for energy saving; it coordinates the sleep schedules of existing devices, so that they can communicate with each other efficiently [35]. Although there are several schemes that tackle nodes' synchronization and calibration [36,37], it is still an open research problem.

24.4.1.4 Bandwidth

The quality of a health care wireless sensor device not only depends on the ability to sense and to analyze data locally but also relies on the available bandwidth. However, in order to tune the lifetime of a sensor, the device is forced to decrease its power consumption. Consequently, this decreases the used transmission bandwidth. This bandwidth limitation may cause a problem for transmitting a huge amount of data, such as medical diagnostic imaging data [35]. Therefore, the health care sensor device should be supported by an efficient lightweight compression technique [38,39]. Moreover, the bandwidth is subject to the mobility of sensor devices that need to be taken in consideration for designing professional health care systems.

24.4.1.5 User-Friendliness

For continuous indoor and outdoor activity monitoring, the health care sensor should have specific characteristics, such as to be portable and wearable, lightweight and small sized with standalone systems in order to perform medical analysis. Also, a health care sensor should be easily plug-and-play by the patient. Finally, the health care sensor must provide feedback through a professional user-friendly interface [40]. User interaction with the system interface should be identified clearly as it may be varied from one patient to another; for instance, handicapped or elderly patients should have different interaction mechanisms with the system (e.g., voice, gesture, and visual animation) than capable patients.

24.4.2 Nontechnical Issues

Many security and privacy concerns, facing pervasive health care systems and applications, such as who has the right to access the data, how and when data is stored, the trustworthiness in the security mechanism selected for data transfer, data analysis rights, and other related issues.

Generally, threats that biomedical sensors face can be classified into categories of passive versus active attacks. Passive attacks occur while routing the patient data. They are in the nature of eavesdropping on, or monitoring of, packets exchanged within a health care WSN. On the other hand, active attacks involve modifications of the patient's data stream by creating a false stream or attempting to gain unauthorized access to the health care network.

Another classification for attacks facing biomedical sensors is external versus internal attacks. In external attacks (intruder node attack), the attacker uses external nodes that do not belong to the health care wireless network. The internal attacks, on the other hand, occur when a legitimate sensor of a wireless network behaves in unintended or unauthorized ways. An external attacker has no access to most cryptographic materials in sensor networks while an internal attacker may have a partial security key; thus, it gains the trust of other sensor nodes. Thus, internal attacks are much harder to be detected. The majority of external attacks aim to consume the network resources by means of performing passive eavesdropping on data transmissions, injecting false data, or by extending these attacks to perform denial of service (DoS) attacks [41].

The following highlights security and privacy concerns, facing pervasive health care systems and applications.

24.4.2.1 Security Concerns

Because of the wireless nature of the new health care sensors, they are vulnerable to a variety of potential attacks, e.g., false data injections, eavesdropping, and denial of service attacks. There are primary security requirements that need to be achieved in order to ensure the safety of health care systems.

24.4.2.2 Authentication

Authentication in sensor networks is essential for each sensor node and health care base station. They must have the ability to verify whether the data received was really sent by a trusted sender or by an attacker. The main concern in WSNs is to find the security scheme that enables authentication in an energy-efficient way.

24.4.2.3 Data Confidentiality

Confidentiality is the ability to protect packets that travel between biomedical sensor nodes from a passive attacker, i.e., an outsider should not be able to access the information that transits between two nodes. If authentication is taken care of correctly, then confidentiality is a relatively simple process.

24.4.2.4 Data Integrity

Data integrity in sensor networks is needed to ensure the reliability of the data and guarantee that the message or packet being delivered has not been tampered with, altered, or changed.

24.4.2.5 Availability

Availability requirement ensures that the services provided by the biomedical sensor devices are always available even with presence of an internal or external attack.

24.4.2.6 Data Freshness

The data confidentiality and data integrity services should be followed by a data freshness mechanism to ensure the freshness of each packet, in other words, to insure that no old packet has been replayed.

24.4.2.7 Biomedical Security Threats

Table 24.1 illustrates some common attack behaviors, targeting the health care sensor networks and their suggested countermeasures [42,43].

24.4.2.8 Privacy Concerns

Privacy is frequently highlighted as a basic concern in WSNs with regard to health care applications. Health care systems based on wireless technology continuously collect and process personal information using wireless media. Therefore, they can pose serious threats to the privacy of an individual. There are many privacy implications concerning the use of biomedical sensors in

Table 24.1 Biomedical Security Attacks and Countermeasures

Attack	*Description*	*Countermeasures*
Data modification	The attacker can delete or generate false error information and send this modified information back to the health care representative. This attack may threaten the patient's life.	Data integrity
Impersonation attack	The attacker eavesdrops a wireless sensor node's identity information or compromises a wireless sensor node to present multiple identities, so it can be used to cheat other nodes and reduce the effectiveness of the medical data storage process.	Authentication
Eavesdropping attack	The attacker is secretly listening to transmitted messages of the sensor nodes and then used stolen data in performing various malicious acts. This eavesdropping could either be by a passive or active eavesdropper.	Encryption
Message replay attack	The malicious node is resending previously sent packets (i.e., logging in without username and password) to trick the receiver into unauthorized operations or to duplicate transactions.	Data freshness
Interference and jamming attack	By using a powerful transmitter, an attacker can interfere with and jam sensor radio signals. A jamming attack causes the message to be corrupted or lost.	Electromagnetic interference filters
Denial of service attack	The attempt to make a network resource unavailable to its intended users, so it may have serious consequences to the health and well-being of patients.	Authentication

health care, such as [42,44]. One privacy issue is where should the health data be stored, which defines the medical database server type, whether it is centralized or it is decentralized and consisted of many connected local databases.

Another privacy issue is the access control for viewing a patient's medical record as normally there are two categories that have access to a patient's medical record. The first category includes users having the rights to read and write in the patient's medical record, such as the physicians, nurses, and some other clinical staff members. The second category comprises users having read-only rights, such as the third party insurance provider. In some cases, such as emergencies or disasters happening to a remote patient, it is a necessity to compromise the patient's privacy to a certain degree in order to disclose some information to other people, such as relatives in order to serve the patient in need.

A third health care privacy concern is the type of data to be collected and its confidentiality. The medical data is of varied formats, such as doctor notes, x-ray images, lab tests, etc. The open problem is which of these differently formatted data needs to be electronically saved within a patient's medical record. Some of this data are sensitive pieces of information and saving them might pose risks, such as patient's location information.

In order to protect sensitive data and protect the patient's privacy, an encryption mechanism should be applied to all communications over wireless networks and the Internet. Also, it is necessary to in advance inform the patient that his medical information will be revealed and to whom it may be revealed.

24.5 Wireless Body Area Network

A WBAN is a special type of WSN, in which health care is the major contribution of this network [45–50]. In fact, with an aging population problem in many developed countries and the cost of health care, WBANs are considered to be one of the solutions to such problems. Different tiny biosensors have been introduced to the community; these tiny sensors can easily be implanted in the human body without problems. The health data measured by these sensors is required to be transferred to a centralized computer for analysis. However, wired communication might not be a suitable solution for patients who experience high physical mobility or no longer need to stay at hospital.

Devices in WBAN can be classified into sensors and actuators. Sensors are used to measure certain parameters from the human body, internally or externally, such as body temperature, blood pressure, or recording a prolonged electrocardiogram (ECG). sSmart sensors are also utilized in WBAN that can make a decision by itself. The actuator, on the other hand, is used to execute the decision of the network. For example, a biosensor that is used for measuring glucose levels might be correlated with an insulin bump to satisfy the patient's required amount of insulin.

Although a WBAN is a type of WSN, there are more restrictions that are posed for it when it deals with the human body and might be implanted in it. Therefore, sensors need to be very tiny, and they are required to live for a long time because it is not practical to change the sensor's battery while the sensor is implanted in the body. In addition, health care data, in general, is very sensitive and requires being secured. Therefore, some new protocols need to be investigated and developed specifically for WBAN.

24.5.1 Structure, Methodology, and Components

For WBANs, sensors are attached and/or implanted in the human body as shown in Figure 24.6. In addition, as can be seen in the figure, different types of sensors and actuators are used, such as motion sensors, blood oxygen, and insulin injection. These sensors send their data to a local processing unit (LPU), such as a PDA, as shown in Figure 24.7, presenting the general architecture of the WBAN. The reason behind that is the complexity of analyzing the data on the sensor itself.

The LPU is hooked up to an access point for Internet connection. Then, the data is transmitted via the Internet to the patient's hospital and/or ambulance. In some other architectures, the PDA might be used to directly send a message to the patient's doctor reporting the patient's status. Based on this architecture, a patient is no longer obligated to stay in bed, and he or she is allowed to move around freely. However, in some of the WBANs, it is assumed that there are two types of networks, namely intra-network and inter-network. The latter is concerned with communication among sensors attached to the human body; whereas the intra-network refers to the communication between the body sensors and the LPU and consequently to the Internet. In addition, there are many new standards for WBANs; the most-utilized communication technologies are IEEE 802.15.1 (Bluetooth) or IEEE 802.15.4 (ZigBee); a WLAN uses IEEE 802.11 (WiFi) and a WMAN IEEE 802.16 (WiMax). At the same time, the communication in a WAN can be established via satellite links.

Furthermore, a conceptual framework for a WBAN is introduced in [52] and depicted in Figure 24.8. The framework considers a smart sensor that has on board a sensing device, single

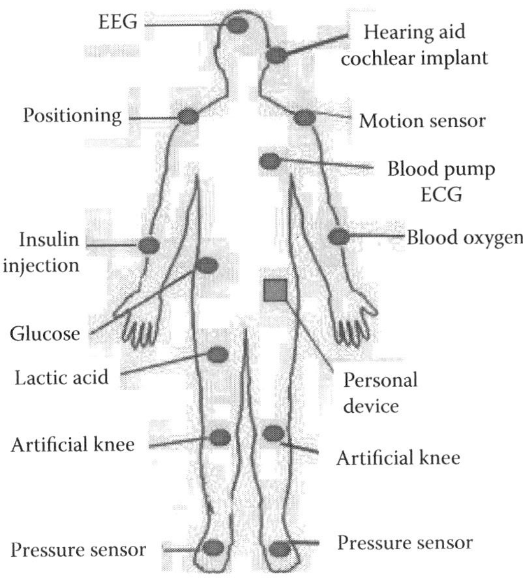

Figure 24.6 WBAN sensors. (From Latré, B. et al., *Wirel. Netw.*, 17, 1, pp. 1–18, 2011.)

Figure 24.7 WBAN architecture.

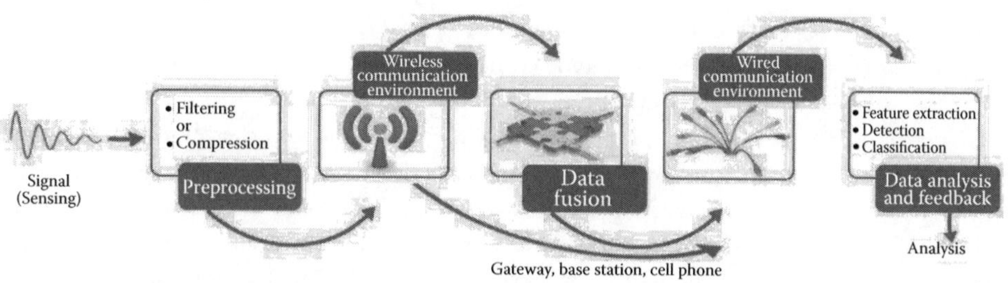

Figure 24.8 WBAN conceptual framework. (From Abu Al-sand, K. et al., *J. Comput. Sci.*, 8, 1, pp. 121–132, 2012.)

filter(s), and/or a certain compression technique. The filtering and the compression modules are used as preprocessing to the signals received from the patient's body (e.g., ECG signals). Through a wireless media (ZigBee, Wi-Fi, or Bluetooth), the received data is transferred to a data fusion module (usually a cell phone) or bypassed through a wired network to a centralized computer (e.g., at the hospital) for data analysis and feedback. Of course, the data analysis and feedback module may contain feature extraction, detection, and classification sub-modules. These sub-modules are very important because the received ECG signal is a very weak signal, and it is required to guarantee the runtime detection to the type of received signal.

24.5.2 Applications

As can be recognized from the previous discussion, there are many components of WBAN that they can be connected together to support a variety of applications. Some of these applications are related to medical, lifestyle, and sports, etc.

24.5.2.1 Medical Applications

A WBAN is used in many medical applications in which sensors are utilized to report human physiological data, for administration of drugs in hospitals and as an aid to rehabilitation, especially for continuous monitoring and logging vital parameters of patients suffering from chronic diseases, such as diabetes, asthma, and heart attacks. For example, the authors in [53] described WBANs for health monitoring at home. The prototype is used to measure the user's heart rate and locomotive activity then periodically upload time-stamped information to the home server. The system consists of two tiers; in the first tier, sensors are connected to the human body as shown in Figure 24.9. These sensors report their data to a home server. The second tier is responsible for sending the data via Internet where a gateway is used for this purpose; the gateway is either connected to the home server or collects the data wirelessly from the human body sensors. The data is sent through the Internet to the medical center for analysis and decision-making.

Another prototype is reported in [54] where a ubiquitous health care system is developed. As shown in Figure 24.10, the sensor consists of different medical sensing devices measuring different phenomena, such as ECG pulse rate, respiration for respiration rate, SpO2 for the saturation of peripheral oxygen (SpO2) value that estimates the oxygen saturation level, and so on. Other environmental sensors are also used, such as temperature and humidity sensors. Based on the collected

Figure 24.9 WBAN for home health monitoring. (From Otto, C. A. et al., A WBAN-based system for health monitoring at home, In *3rd IEEE/EMBS International Summer School, Medical Devices and Biosensors (ISSS-MDBS)*, Boston, 2006.)

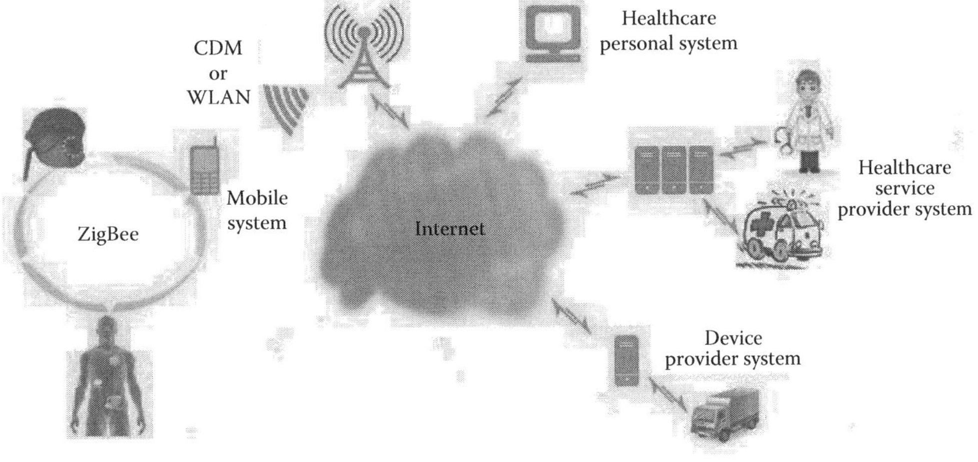

Figure 24.10 WBAN for ubiquitous health care system.

information from sensors, the context-aware ubiquitous health care system may know the location of the patient by using the sensor device's information.

24.5.2.2 Lifestyle, Sports, and Other Applications

Nowadays, people carry their WBAN while they are walking, running, or doing their daily exercises. Applications in areas such as lifestyle, assisted living, sports, entertainment functionalities, toxic levels in the lungs of a firefighter, radiation exposure, navigation support in the car or while walking, a museum or city guide, and infant monitoring, are now possible through WBAN. For example, a blood-pressure sensor is introduced by MIT in [55]; a MEMS accelerometer is used, as shown in Figure 24.11, in order to adjust the hydrostatic pressure offset for the pulse wave transit time (PTT) of the blood pressure sensor. Therefore, a patient may move his hand freely.

Another prototype is introduced in [56] where a WBAN is used for the measurement of kinematic parameters as shown in Figure 24.12. Three ProMove nodes [57] are placed along the left

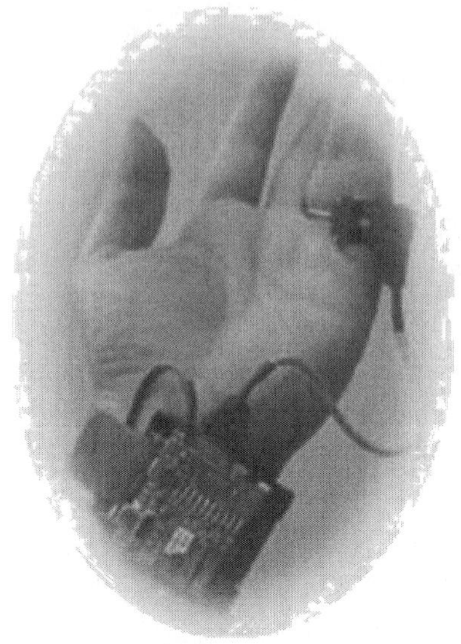

Figure 24.11 Blood pressure sensor. (From Shaltis, P. et al., Wearable, cuffless PPG-based blood pressure monitor with novel height sensor, In *Proceedings of 28th Annual International Conference of the IEEE Engineering in Medicine and Biology Society (EMBS),* New York, pp. 908–911, 2006.)

Figure 24.12 WBAN for sports. (With kind permission from Springer Science+Business Media: *Pers. Ubiquit. Comput.,* A performance analysis of a wireless body-area network monitoring system for professional cycling, 2011, Marin-Perianu, R.S., Marin-Perianu, M., Havinga, P.J.M., Taylor, S., Begg, R., Palaniswami, M., Rouffet, D.)

leg of the cyclist, on the thigh, and for shank and foot. These sensors are used to identify the joint motions of the three segments of the lower limb during biking sports.

Moreover, for military applications, WBAN technology will also help protect those exposed to potentially life-threatening environments, such as soldiers, first responders, and deep-sea and space explorers. For example, soldiers may have different sensors attached to their bodies, including biomedical and GPS sensors for location identification. WBAN might be used to detect and identify chemicals to prevent casualties from friendly fire and monitoring of a soldier's physiological condition. Furthermore, WBAN researchers are already working to improve deep brain stimulation, heart regulation, drug delivery, and prosthetic actuation [58].

24.6 Conclusion and Future Directions

The ongoing improvements in wireless sensor technology have been resulting in various applications, which are making life easier and more comfortable. Similarly, a variety of applications of WSN can be seen in modern health care systems. WSN has made an enormous impact on medical health care in terms of reducing the patient's risks of severity in emergency situations and providing them with a friendly environment at home. Also, decreasing electronics prices and their growing power, coupled with sensing technologies, promise to make health monitoring in one's home, rather than frequent trips to the hospital, a reality. This chapter highlighted the impact of using WSNs on medical health care. There are various applications that can monitor patients and alert physicians and emergency care providers if the patient is at risk. The small-sized wearable wireless medical devices were designed in order not to affect patient's normal activities. Correspondingly, their applications face several challenges.

Further efforts are necessary to manage major challenges that primarily attached with WSNs including improving quality of service (QoS) of wireless communication, reliability of sensor nodes, security, privacy, and fault tolerance, and standardization of interfaces and interoperability. Moreover, some other research challenges can be considered for WSN-based health care systems, such as developing scalable and flexible routing infrastructure; introducing rapid, robust, and ad hoc deployment mechanisms; allowing fully distributed decision-making techniques; studying communication methodologies that adapt to widely varying network conditions; and designing a network that copes with enormous ranges of density and node volatility. In addition, further studies of different medical conditions in hospitals plus clinical and ambulatory settings are necessary to determine specific limitations and possible new applications of this technology.

References

1. Lee, S.H., Lee, S., Song, H., Lee, H.S. (2009). Wireless sensor network design for tactical military applications: Remote large-scale environments, In *Proceedings of the 28th IEEE Conference on Military Communications (MILCOM)*, IEEE Press, Piscataway, NJ, USA, pp. 911–917.
2. Stankovic, J.A., Cao, Q., Doan, T., Fang, L., He, Z., Kiran, R., Lin, S., Son, S., Stoleru, R., Wood, A. (2005). Wireless sensor networks for in-home healthcare: Potential and challenges, In *High Confidence Medical Device Software and Systems (HCMDSS) Workshop*, Philadelphia, PA.
3. Ramadan, R.A., Hesham, E. (2008). Deployment of sensor networks on critical infrastructures: A survey, In *The 4th International Computer Engineering Conference Information Society Applications in the Next Decade (ICENCO)*, Cairo, Egypt.
4. Ramadan, R.A. (2009). *Deployment of Mobile Heterogeneous Sensors on Critical Infrastructure: Sensor Deployment Problem and Its Solutions*, VDM Verlag Publisher, Saarbrücken, Germany.
5. Ramadan, R.A., Hagras, H., Nawito, M., Eldesouky, A.E.B. (2010). The intelligent classroom: Towards an educational ambient intelligence testbed, In *The 6th International Conference on Intelligent Environments (IE)*, pp. 344–349.
6. Haoui, A., Kavaler, R., Varaiya, P. (2008). Wireless magnetic sensors for traffic surveillance, *Transportation Research, Part C: Emerging Technologies*, pp. 294–306.

7. Ko, J.G., Lu, C., Srivastava, M., Stankovic, J., Terzis, A., Welsh, M. (2010). Wireless sensor networks for healthcare, *Proceedings of the IEEE*, vol. 98, 11, pp. 1947–1960.
8. CodeBlue healthcare project. Available online at: http://fiji.eecs.harvard.edu/CodeBlue/
9. Chen, B.R., Peterson, G., Mainland, G., Welsh, M. (2008). LiveNet: Using passive monitoring to reconstruct sensor network dynamics. In *Proceedings of the 4th IEEE International Conference on Distributed Computing in Sensor System (DCOSS'08)*, Santorini Island, Greece, 11–14 June 2008.
10. Halteren, A.V., Bults, R., Wac, K., Konstantas, D., Widya, I., Dokovsky, N., Koprinkon, G., Jones, V., Jerzog, R. (2004). Mobile patient monitoring: the MobiHealth system. *J. Inform. Tech. Healthcare*, 2, pp. 365–373.
11. Wood, A., Virone, G., Doan, T., Cao, Q., Selavo, L., Wu, Y., Fang, L., He, Z., Lin, S., Stankovic, J., (2006). ALARM-NET: wireless sensor networks for assisted-living and residential monitoring; Technical Report CS-2006-01; Department of Computer Science, University of Virginia: Charlottesville, VA, USA.
12. Ng, J.W.P., Lo, B.P.L., Wells, O., Sloman, M., Peters, N., Darzi, A., Toumazou, C., Yang, G.-Z., (2004). Ubiquitous monitoring environment for wearable and implantable sensors (UbiMon). In *Proceedings of 6th International Conference on Ubiquitous Computing (UbiComp'04)*, Nottingham, UK, 7–14 September 2004.
13. Fischer, M., Lim, Y.Y., Lawernce, E., Ganguli, L.K. (2008). ReMoteCare: Health monitoring with streaming video. In *Proceedings of 7th International Conference on Mobile Business*, Barcelona, Spain, pp. 280–286, 7–8 July 2008.
14. Chakravorty, R. (2006). A programmable service architecture for mobile medical care. In *Proceedings of 4th Annual IEEE International Conference on Pervasive Computing and Communication Workshop (PERSOMW'06)*, Pisa, Italy, 13–17 March 2006.
15. Rajasekaran, M.P., Radhakrishnan, S., Subbaraj, P. (2010). Sensor grid applications in patient monitoring. *Future Generat. Comput. Syst.*, 26, pp. 569–575.
16. Montgomery, K., Mundt, C., Thonier, G., Tellier, A., Udoh, U., Barker, V., Ricks, R., Giovangrandi, L., Davies, P., Cagle, Y., Swain, J., Hines, J., Kovacs, G. (2004). Lifeguard—A personal physiological monitor for extreme environments. In *Proceedings of the 26th Annual International Conference of the IEEE EMBS*, San Francisco, CA, USA, pp. 2192–2195, 1–5 September 2004.
17. Gao, T., Massey, T., Selavo, L., Crawford, D., Chen, B.R., Lorincz, K., Shnayder, V., Hauenstein, L., Dabiri, F., Jeng, J., Chanmugam, A., White, D., Sarrafzadeh, M., Welsh, M. (2007). The advanced health and disaster aid network: A light-weight wireless medical system for triage. *IEEE Trans. Biomed. Circ. Syst.*, 1, pp. 203–216.
18. Jiang, S., Cao, Y., Iyengar, S., Kuryloski, P., Jafari, R., Xue, Y., Bajcsy, R., Wicker, S. (2008). CareNet: An integrated wireless sensor networking environment for remote healthcare. In *Proceedings of 8th International Conference on Body Area Networks*, Brussels, Belgium, 13–15 March 2008.
19. Mahmoud, A.S., Sheltami, T.R., Amara, M.H.A. (2006). Wireless sensor network implementation for mobile patient. In *Proceedings of International Conference on Grid and Cooperative Computing*, Manama, Bahrain, 20–22 March 2006.
20. Farella, E., Pieracci, A., Brunelli, D., Benini, L., Ricco, B., Acquaviva, A. (2005). Design and implementation of WiMoCA node for a body area wireless sensor network. In *Proceedings of System Communication*, Montreal, QC, Canada, pp. 342–347, 14–17 August 2005.
21. Nee, O., Hein, A., Gorath, T., Hulsmann, N., Laleci, G.B., Yuksel, M., Olduz, M., Tasyurt, I., Orhan, U., Dogac, A., Fruntelata, A., Ghiorghe, S., Ludwig, R. (2008). SAPHIRE: Intelligent healthcare monitoring based on semantic interoperability platform: pilot applications. *IET Commun.*, 2, pp. 192–201.
22. Han, D., Lee, M., Park, S. (2009). THE-MUSS: Mobile u-health service system. *Biomed. Eng. Syst. Tech.*, 25, pp. 377–389.
23. Ko, J., Lim, J.H., Chen, Y., Musaloiu-E. R., Terzis, A., Masson, G.M. MEDiSN: Medical emergency detection in sensor networks. *ACM Trans. Embed. Comput. Syst.*, 10, 1, pp. 1–29, August 2010, New York.
24. Malan, D., Jones, T.F., Welsh, M., Moulton, S. (2004). CodeBlue: An ad hoc sensor network infrastructure for emergency medical care, In *Proceedings of the MobiSys Workshop on Applications of Mobile Embedded Systems (WAMES)*, Boston.
25. Mica2datasheet. Available online: https://www.eol.ucar.edu/rtf/facilities/isa/internal/CrossBow/Data Sheets/mica2.pdf.

26. Lorincz, K., Welsh, M. (2006). MoteTrack: A robust, decentralized approach to RF-based location tracking. *Pers. Ubiquit. Comput.*, 11, pp. 489–503.
27. Kumar, P., Lee, H.J. (2012). Security issues in healthcare applications using wireless medical sensor networks: A survey, *Sensors*, 12, pp. 55–91.
28. Telos datasheet. Available online: http://www.eecs.harvard.edu/~konrad/References/TinyOSDocs/telos-reva-datasheet-r.pdf
29. Alemdar, H., Ersoy, C. (2010). Wireless sensor networks for healthcare: A survey, In *Proceedings of the Computer Networks.*, vol. 54, 15, pp. 2688–2710.
30. Bell, G. and Gemmell, J. (2009). *Total Recall: How the E-Memory Revolution Will Change Everything.* Dutton, New York.
31. Omary, Z., Mtenzi, F., Wu, B., O'Driscoll, C. (2011). Ubiquitous healthcare information system: Assessment of its impacts to patient's information, *IJISR*, 1, 1, pp. 71–77.
32. Shimmer research, the Shimmer sensor platform data sheet (2012). Available online at http://www.shimmer-research.com/download/documentation
33. Kim, J., Beresford, A.R., Stajano, F. (2006). Towards a security policy for ubiquitous healthcare systems, In *Proceedings of the 1st International Conference on Ubiquitous Convergence Technology*, pp. 263–272.
34. Krishnamachari, B. (2005). *Networking Wireless Sensors*, Cambridge University Press. ISBN 0521838479.
35. Darwish, A., Hassanien, A.E. (2011). Wearable and implantable wireless sensor network solutions for healthcare monitoring, *Sensors*, 11, pp. 5561–5595.
36. Sachan, V.K., Imam, S.A., Beg, M.T. (2012). Energy-efficient communication methods in wireless sensor networks: A critical review, *Int. J. Comput. Appl. (0975–8887)*, 39, 17, pp. 35–48.
37. Mottola, L., Picco, G. (2011). Programming wireless sensor networks: Fundamental concepts and state of the art, *ACM CSUR*, 43, p. 3.
38. Marcelloni, F., Vecchio, M. (2010). Enabling energy-efficient and lossy-aware data compression in wireless sensor networks by multi-objective evolutionary optimization, *Inform. Sci.*, 180, 10, pp. 1924–1941.
39. Walder, J., Krátk, M., Baca, R., Platos, J., Snasel, V. (2012). Fast decoding A for variable-lengths codes. *Inform. Sci.*, 183, 1, pp. 66–91.
40. Jasemian, Y. (2008). Elderly comfort and compliance to modern telemedicine system at home, In *Proceedings of Second International Conference on Pervasive Computing Technologies for Healthcare*, Pervasive Health, Tampere, Finland, pp. 60–63.
41. Sharifnejad, M., Shari, M., Ghiasabadi, M., Beheshti, S. (2007). A survey on wireless sensor networks security, in *The 4th International Conference: Sciences of Electronic Technologies of Information and Telecommunication (SETIT)*, Tunisia.
42. Al-Ameen, M., Liu, J., Kwak, K. (2010). Security and privacy issues in wireless sensor networks for healthcare applications, *J. Med. Syst.*, pp. 1–9.
43. Lupu, T.G. (2009). Main types of attacks in wireless sensor networks. In *Proceedings of the 9th WSEAS International Conference on Signal, Speech and Image Processing*, and *9th WSEAS International Conference on Multimedia, Internet and Video Technologies (SSIP '09/MIV'09)*. World Scientific and Engineering Academy and Society (WSEAS), Stevens Point, Wisconsin, USA, pp. 180–185.
44. Meingast, M., Roosta, T., Sastry, S. (2006). Security and privacy issues with health care information technology, In *Proceedings of the 28th IEEE EMBS Annual International Conference*, New York City, USA.
45. Istepanian, R.S.H., Jovanov, E., Zhang, Y.T. (2004) Guest editorial introduction to the special section on m-health: Beyond seamless mobility and global wireless health-care connectivity, *IEEE Trans. Inform. Technol. Biomed.*, 8, 4, pp. 405–414.
46. Van Dam, K., Pitchers, S., Barnard, M. (2001). Body area networks: Towards a wearable future, In *Proceedings of WWRF Kick Off Meeting*, Munich, Germany.
47. Schmidt, R., Norgall, T., Mörsdorf, J., Bernhard, J., von der Grün, T. (2002). Body area network ban—A key infrastructure element for patient-centered medical applications, *Biomed. Technol. Berlin*, 47, 1, pp. 365–368.
48. Gyselinckx, B., Van Hoof, C., Ryckaert, J., Yazicioglu, R.F., Fiorini, P., Leonov, V. (2005). Human++: autonomous wireless sensors for body area networks, in *The Proceedings of the IEEE in Custom Integrated Circuits Conference*, pp. 13–19.

49. Otto, C., Milenkovic, A., Sanders, C., Jovanov, E. (2006). System architecture of a wireless body area sensor network for ubiquitous health monitoring, *J. Mobile Multimedia*, vol. 1, 4, pp. 307–326.
50. Lo, B., Yang, G.Z. (2006). Body sensor networks: Infrastructure for life science sensing research, In *Life Science Systems and Applications Workshop*, IEEE/NLM, Bethesda, MD, pp. 1–2.
51. Latré, B., Braem,B., Moerman, I., Blondia, C., Demeester, P. (2011). A survey on wireless body area networks, *Wirel. Netw.*, 17, 1, pp. 1–18.
52. Abu Al-sand, K., Mahmuddin, M., Mohamed, A. (2012). Wireless body area sensor networks signal processing and communication framework: Survey on sensing, communication technologies, delivery and feedback, *J. Comput. Sci.*, 8, 1, pp. 121–132.
53. Otto, C.A., Jovanov, E., Milenkovic, E. (2006). A WBAN-based system for health monitoring at home, In *3rd IEEE/EMBS International Summer School, Medical Devices and Biosensors (ISSS-MDBS)*, Boston.
54. Jung, J., Ha, K., Lee, J., Kim, Y., Kim, D. (2006). Wireless body area network in a ubiquitous health-care system for physiological signal monitoring and health consulting, *International Journal of Signal Processing (IJSIP), Image Processing and Pattern Recognition*, 1, pp. 47–54.
55. Shaltis, P., Reisner, A., Asada, H. (2006). Wearable, cuffless PPG-based blood pressure monitor with novel height sensor, In *Proceedings of 28th Annual International Conference of the IEEE Engineering in Medicine and Biology Society (EMBS)*, New York, pp. 908–911.
56. Marin-Perianu, R.S., Marin-Perianu, M., Havinga, P.J.M., Taylor, S., Begg, R., Palaniswami, M., Rouffet, D. (2011). A performance analysis of a wireless body-area network monitoring system for professional cycling, *Pers. Ubiquit. Comput.*, Springer, ISSN 1617–4909.
57. ProMove wireless inertial sensing platform. http://www.inertiatechnology.com
58. Khan, I.M., Jabeur, N., Khan, M.Z., Mokhtar, H. (2012). An overview of the impact of wireless sensor networks in medical health care, In *The 1st International Conference on Computing and Information Technology (ICCT)*, Al-Madinah Al-Munawwarah, Saudi Arabia, pp. 576–580.

Chapter 25

Smart Environmental Monitoring Using Wireless Sensor Networks

Nashwa El-Bendary, Mohamed Mostafa M. Fouad, Rabie A. Ramadan, Soumya Banerjee, and Aboul Ella Hassanien

Cairo University

Contents

25.1 Introduction

The heavy industry that has grown around the world and the diversity of chemicals along with contamination resulting from humans' daily life, released into the environment have raised serious concerns on the effects of such materials on the ecosystem and human health. Therefore, countries around the world have increased their regulations in terms of monitoring as well as controlling and treating pollution. Such regulations raised the issue of finding a suitable, cost-effective, and reliable technology to encounter the pollution effects.

Efficient environmental monitoring and protection, based on precautionary or timely actions, could be realized only if there are appropriate environmental diagnostic systems or, moreover, real-time environmental monitoring systems that are capable of collecting and analyzing all information relevant to the environmental status.

The environmental monitoring system becomes a smart monitoring system when the environment itself becomes a self-monitoring and self-protecting environment that is aware of its current status with the possibility of an automatic alarm rising if some event occurred. So smart environmental monitoring not only includes environmental pollution monitoring, but also controls the effect of environmental changes on humans, animals, plants, and even on the shape of the Earth. Accordingly, for providing innovative environmental monitoring methodologies and procedures, wireless sensor networks (WSNs) are the key that enables more flexible real-time environmental monitoring, diagnostics, and finally protection of further serious degradations.

Via integrating wireless sensor devices in the environment itself, the level of environmental protection could be substantially raised, giving the environment a lot of new intelligent features. These features are primarily self-monitoring and self-protecting with the possibility of not only reactive, but also proactive, reactions for different phenomena taking place in this environment [1].

Through utilizing the self-configuration ability of the sensor nodes, they can then form a network that can provide access to information anytime and anywhere by collecting, processing, analyzing, and disseminating data. Thus, WSN actively participates in creating a smart environment. Currently, via having WSNs as the backbone for smart environments, there is a great and enormous challenge in the hierarchy of detecting certain phenomena, monitoring and collecting relevant data, assessing and evaluating the resulting information, conveying meaningful user displays, and carrying out decision-making and alarm functions.

This chapter discusses the concept of smart environmental monitoring along with presenting its different architectures, applications, and related design issues. Also, it highlights how smart environments embody the trend toward increasingly connected and automated environmental monitoring through linking wireless sensing devices to environmental phenomena and events. Moreover, this chapter addresses some challenges introduced by environmental monitoring applications to WSNs resulting from the need to ensure the privacy and security of collected data in addition to considering the localization and deployment of sensor nodes. Furthermore, the chapter provides insights on the techniques and features researchers and system designers should consider for successful deployments and operation of real smart environmental monitoring systems. Finally, the chapter concludes by outlining ongoing research directions for smart environmental monitoring.

25.2 Environmental Monitoring: A Background

Currently, the world faces unprecedented challenges in environmental monitoring. Therefore, the target is to collect and analyze environmental data in order to avoid any potential risks. In fact, this is an essential aspect of decision makers; however, environmental data is scarce. At the same time, increasing population, urbanization, energy, transportation, and agricultural developments are the main sources of environmental pollution. In addition, natural disasters, such as earthquakes, floods, tsunamis, and landslides, are sources of environmental phenomena that might affect a large number of people. Moreover, global warming, ocean acidification, and biodiversity loss can also lead to large-scale impacts on the environment. Furthermore, air and water pollution are considered to be the most serious environmental problems. So, the more the relationship between air and water pollution and human health is understood, the more risk is mitigated. As an individual usually breathes once every three to four seconds, air pollution is considered to be one of the environmental parameters that most directly affects human health. In fact, toxic gases, such as sulfur dioxide (SO_2), nitrogen oxides (NOx), ozone, etc., are the pollutants produced by industrial activities and/or road transportation. In addition to atmospheric air pollution, indoor air pollution from burning traditional fuels in small spaces poses serious health risks. It has been discovered that the most effected humans are the women and girls involved in cooking over wood and charcoal. This problem appears especially in undeveloped countries where modern sources of energy are not available and the common fuels are wood, agricultural residues, animal dung, charcoal, and coal.

On the other hand, water is the main source of life for humans, plants, and all living creatures. Therefore, large agencies, including government agencies, industry, academic researchers, and a wide variety of private organizations, dedicate significant resources to monitoring, protecting, and restoring water resources and their watersheds. The main sources of water pollution involve untreated sewage, chemical discharge, petroleum leaks and spills, dumping in old mines and pits, and agricultural chemicals that are washed off or seep into the ground from farms. However, water-quality monitoring for large water surfaces is not an easy task. Each one of the important parameters reflecting water quality has a measuring sensitivity level, such as temperature, pH, conductivity, dissolved oxygen, nitrogen, and phosphorous. Water security is another challenge in environmental monitoring in which the increasing population and growth of urbanization, especially in developing countries, increases water demands. In fact, it is expected that by 2030 developing countries will suffer from a shortage of water resulting from the increasing number of the world's population from one billion to 3.9 billion [2]. In addition to the water shortage problem, degradation of the land might be an environmental problem as well when soil erosion risk is increasing because of the use of soil to support food production. It is also projected that by the year 2030, the land affected by soil will increase from 20 million km^2 to 30 million km^2 [3].

As well, noise is another serious problem in which different sources might produce the noise, including vehicles, trains, music, factories, and many other sources. The traditional method for detecting noise is the manual method; however, such a method does not scale to the high demand in time and space. Therefore, sensor networks could be the solution to the collection of noise data over time for large areas. However, special types of sensors are required.

25.3 Smart Environmental Monitoring: An Overview

Several years ago, digital data loggers replaced the old mechanical mechanisms for environmental monitoring. The digital data loggers are more easy to operate and maintain and cheaper than the old mechanisms used to record data at specific intervals and which required human intervention. However, digital data loggers usually provide monitoring at one point only, and in many cases, multiple points need to be monitored. So several different solutions are used. Recent advances in low-power wireless network technology have created the technical conditions to build multifunctional tiny sensor devices with several types of sensors, such as chemical, optical, thermal, and biological, available to be attached to these wireless sensor devices. WSNs can be used to observe and react according to the physical phenomena of the surrounding environment without the need for human supervision or intervention [4]. So widespread networks of inexpensive wireless sensor devices offer a substantial opportunity for smart environmental monitoring in which the surrounding physical phenomena in certain environments is being monitored more accurately compared to traditional sensing methods [5].

Accordingly, the word "smart" in "smart environmental monitoring" does not mean only automatic collection without supervision. It means involving ambient intelligence (AmI) in data analysis and decision making. Ambient intelligence (AmI) refers to electronic environments, which are sensitive and responsive to the presence of people, where devices work in concert to support people in carrying out their daily life activities and tasks easily and naturally using information and intelligence that is hidden in the network connecting these devices. As these devices grow smaller and more integrated into the environment, the technology disappears into the surroundings until only the user interface remains perceivable by users [6]. The achievement of AmI largely depends on the technology deployed (e.g., sensors and devices interconnected through wireless networks) as well as on the intelligence of the software used for decision making [7]. The science of AmI has been advanced to become the science covering different fields, such as robotics, WSNs, human–computer interfaces (HCIs), pervasive computing, and artificial intelligence (AI) as shown in Figure 25.1. We can say that ambient intelligence is an extension of AI, in which sensing is added as well as interactivity with the used system. In addition, AmI is closely related to what is called "service science" [8], in which smart services are introduced to the user.

The science of robotics is no longer related to mechanical fields only. Different and smarter robots are designed and introduced to the world. Most current robots look like humans, and new embedded technology makes them walk and think like humans. Some of these robots are mobile,

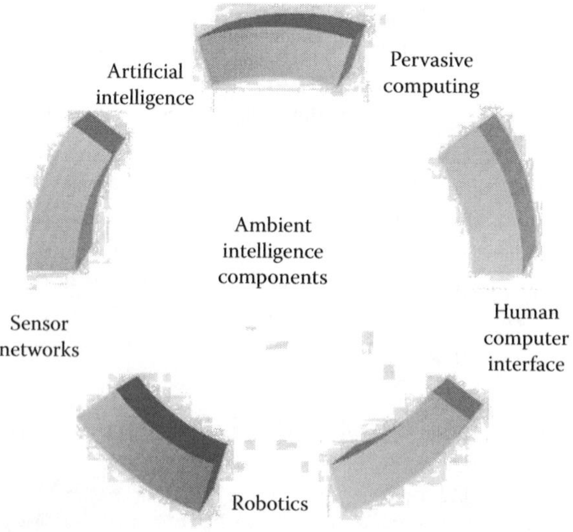

Figure 25.1 Ambient intelligence components.

and they are able to move from one place to another. Such robots are usually responsible for searching areas and measuring different environmental parameters. An example of such robots are the ones that explored Mars and were designed to roam its surface. Another type of robots is the rolling robots that have wheels and usually look like cars. These robots are used to move around and explore flat or rocky areas. A third type of robots is walking robots. These robots have legs to walk on; some of them have four or more legs and are used to scan rocky terrains that are difficult to navigate with wheels. A fourth type is the humanoid robotics as shown in Figure 25.2 showing Sony humanoid robots that emulate the movement of humans and can also dance like humans and interact with music as well. Although the capabilities of these robots are currently used in very limited activities, it is expected in the near future they will be utilized in many of the smart environmental monitoring applications.

The science of AI is the science of building computers or other machines that have the ability to perform those activities that are normally thought to require intelligence. It is related to the usage of computers to understand human intelligence, however, without the need for confining itself to methods that are biologically observable. The main problems of AI include reasoning, knowledge, learning, planning, communication, and perception [7].

Pervasive computing is also called ubiquitous computing, which is related to embedding microprocessors in everyday usable objects to communicate information. This science tries to allow objects to wirelessly communicate unobtrusively with each other. In addition, this science investigates the communication between smart devices and the Internet as well as making the data sent by these devices easily available for Internet users [10].

The science of HCI deals with the study, planning, and design of the interaction between the computer and users. Many fields are involved in this science, such as operating systems (OSs), programming languages, and computer graphics. Other fields are involved, such as social science, cognitive psychology, linguistics, and human factors [11].

Figure 25.2 Sony humanoid robots. (From Adaptive robotics, Available online at: https://inlportal.inl.gov/portal/server.pt/community/robotics_and_intelligence_systems/455, August 2012.)

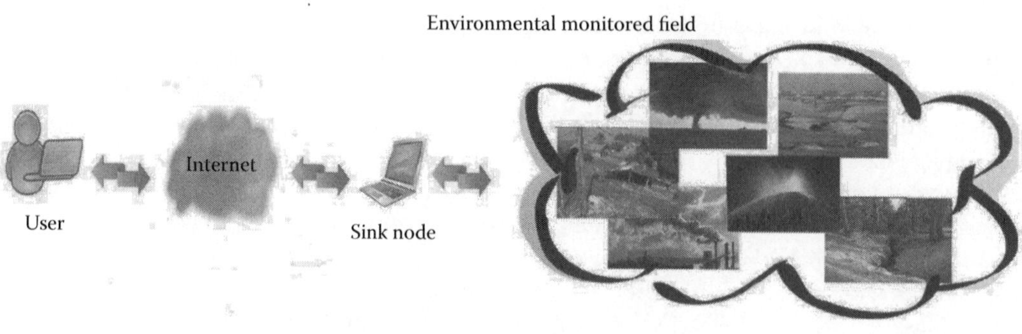

Figure 25.3 Monitoring environmental phenomena using WSNs.

Finally, like any responsive creature, the smart environment relies first and foremost on sensory data from the real world. Sensory data comes from multiple sensors of different modalities in distributed locations. The smart environment needs information about its surroundings as well as about its internal workings [12]. The concept of WSNs is a new science that started with the first prototype sensor node in the year of 2000. WSNs, which deal with tiny sensors that capture environmental parameters, such as temperature, humidity, daylight, pressure, etc., consist of many inexpensive, low-power communication tiny wireless devices that collaborate together to form a wireless network. These wireless sensors can be deployed into a physical space, providing dense sensing close to physical phenomena, and collect data about different phenomena from the monitored field. Collected data is usually sent to a centralized node called a sink node for data analysis and decision making. A common architecture of WSNs is presented in Figure 25.3, in which the sink node disseminates its data to the Internet for user usage. However, sensor nodes suffer from many limitations, including scarce energy sources, small memory footprint, and limited processing capabilities. In addition, in many of the WSN applications, sensors are deployed in large numbers. Consequently, there are many other issues with WSNs, such as scalability, data reliability, security and key management, data analysis, and efficient multi-hop routing. Nevertheless, WSNs have many applications, including on the battlefield [13], in health care [14], with critical infrastructure [15,16], smart environment [17], and transportation network monitoring [18]. As shown in Figure 25.3, WSNs are considered to be the basic infrastructure of many smart environmental monitoring applications (e.g., air pollution, water pollution, volcanoes, twisters, floods, fires, etc.).

25.3.1 Architecture

There are many architectures for smart environments; in fact, the architecture, in most of the cases, is based on the application. However, in this section, we present the most common architectures for smart environmental monitoring applications. As can be seen in Figure 25.4, there are four basic layers of the architecture, which are the physical layer, OS abstraction layer, middleware layer, and application components layer. The physical layer consists of different components, including the smart devices and the communication interface. Smart devices could be smart sensors, such as pH, oxygen, SO_2, etc., as well as humidity and temperature sensors. The communication component could be wired as in indoor applications or wireless as in outdoor applications. Smart environmental networks are usually utilizing the ZigBee [19] and Bluetooth [20] communication standards in addition to satellite communication.

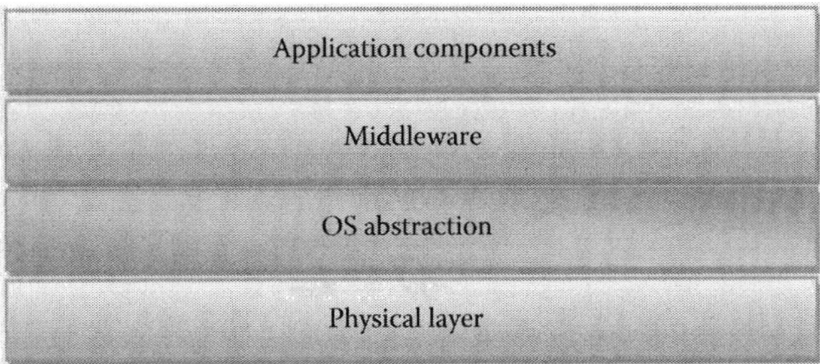

Figure 25.4 Basic components of smart environmental monitoring system.

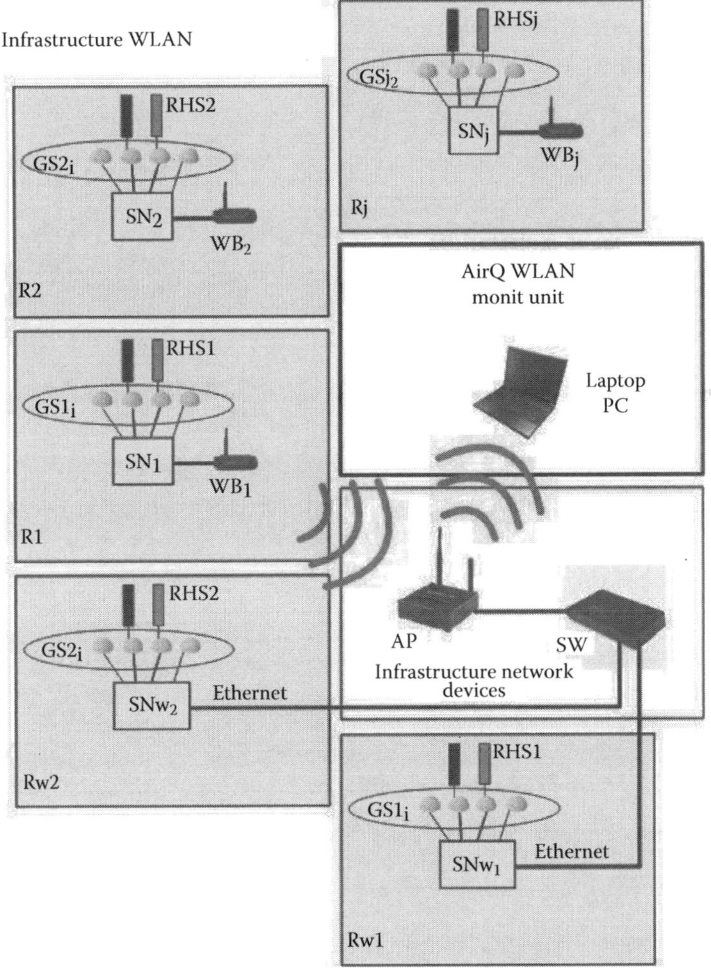

Figure 25.5 Example of indoor sensor network for air quality monitoring. (From Postolache, O. et al., *IEEET Instrumentation and Measurement*, vol. 58, 9, pp. 3253–3262, 2009.)

The OS abstraction layer is the kernel of any smart environmental monitoring in terms of including the basic required instructions for environment operations. Above the OS abstraction layer sits the middleware layer, which includes the message controllers and communication channel identifier. In addition, it is the interface between the application components and the OS abstraction layer in which different applications might require different requirements. The final layer is the application components layer, in which it is assumed that the application consists of many components that might be implemented separately. At the same time, these components can communicate to each other through the middleware layer.

An example of the architecture of smart environmental monitoring is the system proposed by Postolache et al. [21], in which they proposed a WSN for air pollution monitoring. The authors presented a network for indoor air quality monitoring and another network for outdoor air quality monitoring. As shown in Figure 25.5, sensors are wired or wirelessly connected to a centralized node for data analysis. As can be seen, sensors are distributed in different rooms. The data analysis phase is based on single-output neural networks taking the environmental sensors as inputs. In this application, sensors and wireless devices represent the physical layer, according to Figure 25.4, while the other three layers are implemented on the centralized node.

Another architecture, named MoDisNet, is proposed in [22] for air pollution monitoring as well; however, it adopts a grid-based architecture. Figure 25.6 shows the basic blocks of the MoDisNet architecture, in which four components are formed. Sensors represent the first

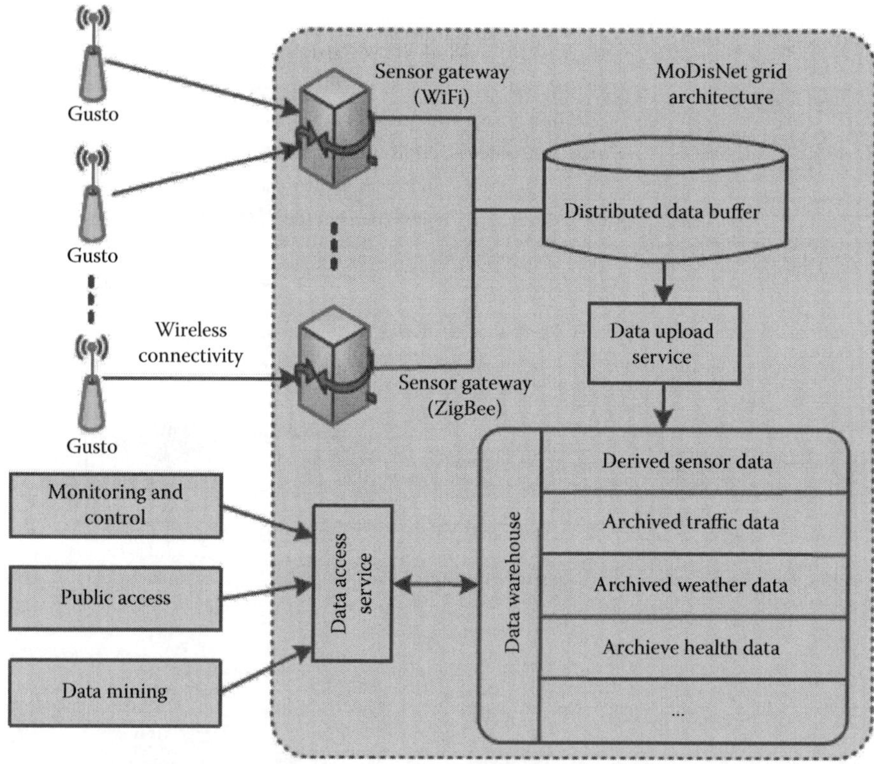

Figure 25.6 Grid-based MoDisNet air pollution monitoring architecture. (With kind permission from Springer Science+Business Media: *Lecture Notes in Computer Science*, The disappearing computer, vol. 4500, 2007, Streitz, N. A., Kameas A., Mavrommati, I., eds.)

component while the communication either by ZigBee or Wi-Fi is the second component. The third component involves the data warehouse where the received data is stored. The final component is the different services involved, such as data upload service and data access services. More architectures for indoor applications are presented in [23].

25.3.2 Challenges and Design Issues

Recent advances in communications, location and sensing technologies, and data processing have led to the emergence of smart environments as one important side of research in this domain. Nevertheless, they pose serious problems that should be resolved in order to gain the full benefit of these smart systems. These problems include deployment, localization, security, and privacy.

25.3.2.1 Deployment

Whether the deployment of the sensor nodes is being done dynamically (nodes are arbitrarily located) or sensor nodes are fixed in specific locations, it is a major concern in designing any environmental monitoring application because the functionality and the desired quality of service of such applications are subjective to the deployment behavior. Furthermore, the scarce resources of sensor nodes (communication bandwidth, processing capability, energy source, sensing range) place extra physical constraints on selecting the best deployment strategy. In this context, there are a number of deployment issues that should be considered, such as

- Considering the maintenance plan of WSN because the sensor nodes may be placed in dangerous or inaccessible areas.
- Investigating the deployment site for any phenomena that could affect the functionality of the environmental WSN, such as radio interference, bad weather, etc.
- Checking the energy level of the environmental WSN nodes prior to deployment in order to reduce the early failure of the whole network. Even the energy depletion rate of each of the network's nodes should be predicted in advance for avoiding incorrect sensor readings [25].
- Supporting the availability of a strong peer-to-peer communication link for each node with at least one of its neighboring nodes. The communication links should be well managed in order to prevent a network's congestion and to avoid the missing short links problem [26].
- Using a suitable communication protocol that is responsible for message routing between nodes. In the real world, if this route is not established properly, this may cause a reporting latency problem [26].

25.3.2.2 Localization

Currently, many of the new emerging applications are targeting node localization strategies in order to accurately determine the node position. Defining the coordinates of the reported data is critical for a number of reasons, such as [27]

- Facilitating the monitoring and tracking of people or entities within their environments.
- Quick reporting for the location of an occurred phenomenon.

- The assurance of a specific quality of coverage provided by the network's nodes at any time.
- Achieving a load-balancing routing that utilizes the information location to periodically switch the routing between selected groups of nodes. This switching process conserves the nodes' energy along with achieving the desired load balance.
- Conserving the network's resources; whereas location information can be used to broadcast a query for a specific geographic area instead of flooding the whole network.

Although the node's localization has many benefits, it raises two important problems [28]. First is the problem of defining the node's coordinates, which can be absolute physical (x, y, z) coordinates obtained by geographical locating systems, such as a global positioning system (GPS), or might be relative coordinates to other predefined global coordinates. Attaching a GPS receiver with each node is very expensive as it raises the sensor node's price. GPS adds extra overhead on the energy dissipation rate; therefore, it is not suitable for energy-constrained sensor nodes.

The second problem is calculating the distance between sensor nodes using their radio communication range. Radio signals can be used by sensor nodes to assist localization through using the strength of the received signal to calculate its distance from the transmitter. This approach is known as the received signal strength indication (RSSI). However, the accuracy of the RSSI approach could be affected by the unstable and dynamic environmental conditions [29].

25.3.2.3 Security

In the real world, smart environments are equipped with a large number of small and heterogeneous sensor devices. Therefore, providing information security protection and guaranteeing privacy within such environments are challenging tasks. The security and privacy of smart environments need a well-defined security policy, including authentication, data encryption, and access rights mechanisms. The first concern with such environments is to make sure that services are provided to an authenticated user. The main requirement of such authentication mechanisms is to provide their service in a calm and transparent way [30]. For that reason, there are two calm authentication mechanisms for smart environments [31], namely, biometric or carried token authentication and usage of point of authentication.

The first mechanism of authenticating an individual is via using biometric measurements or through a carried authenticate token, e.g., radio frequency identification (RFID). These biometric measurements include voice recognition, facial recognition, fingerprints, retina or iris scan, etc. While biometric technologies are more efficient than using common passwords, research has shown that there is the potential of false negatives and false positives with individual users [32]. Alternatively, though the authenticate token is an easy authentication technique, it has a serious problem in that it may be misused by an intrusion if it is lost or stolen from the authenticated user.

The second mechanism of the point of authentication is a technique in which the authentication machines are located within the smart environment. Sensors within that environment are responsible for tracking and authenticating the user once passing from one point of authentication to another.

Because of the sensitive nature of some smart environmental applications (e.g., fire alarm systems, smart surveillance systems, intelligent control systems, etc.), information security mechanisms, such as data encryption or access control, constitute the bottom line for individual data protection.

Although authentication mechanisms can be used to ensure that data is coming from an authenticated person or entity, it cannot prevent attackers from packet eavesdropping or injecting incorrect data into the network. A data encryption service provides confidentiality against the previously mentioned attacks. It can be accomplished in hardware as well as in software or in both of them to ensure the highest level of security. Several symmetric and asymmetric key cryptography algorithms can be used to provide such service [33]. The main issue that should be addressed by these data encryption algorithms is the balance against the computational, memory, and power constraints of sensor nodes.

25.3.2.4 Privacy

Privacy has been always a concern when using WSNs for many applications. Smart environments collect information from both sensors and users. Once this information is available electronically and, moreover, online, it opens the door for hackers and other malicious attackers as well as those who are originally authorized to access that information. For that reason, smart environment applications have raised a few issues regarding protecting the privacy of individuals. These issues include data access rights, how and when data is stored, data transfer security mechanisms, data analysis rights, and governing policies.

Recently, smart environments stated a new privacy term, privacy control [34], that is not only setting rules and enforcing them, but also defining a way of adaptively managing that privacy. Adaptive management for user privacy is a main concern because of gradual changes in the degree of disclosure of personal/entity information or as a result of the user mobility from one smart space to another.

25.4 Smart Environmental Monitoring Applications

As a result of the ease of their deployment and their relatively low cost, WSNs continue to be a very popular technology for monitoring and acting upon events in dangerous or harsh environments for humans. That necessity of monitoring depends on the type of data that must be gathered by the network devices. Almost any application could be classified into two categories: event detection (ED) and spatial process estimation (SPE) [35]. For the first category, sensors are deployed to detect an event (e.g., a fire location in a forest, an earthquake, etc.) while the second category aims at estimating a given physical phenomenon (e.g., the humidity forecast in a wide area, the temperature variations in a greenhouse, etc.). The estimated behavior of the spatial process is based on samples taken by sensing devices that are typically placed in random positions. Typical sensor measurement parameters include the following [36]:

- *Physical measurements:* such as two-axis magnetometers, light and ultraviolet intensity, radiation levels, radio waves and microwaves, humidity, temperature, atmospheric pressure, fog and dust, sound and acoustics, two-axis accelerometers, shock waves, seismicity, physical pressure, video and image, and location (GPS) and locomotion measurements.
- *Chemical and biological measurements:* such as the presence or concentration of a substance or agent at specified concentration levels.
- *Event measurements:* such as determination of the occurrence of human-made or natural events (e.g., cyber-level events, tracking of internal and external events, etc.).

Although there are different application domains that use WSNs, the structure of these networks for each of these applications has more or less similar components. Also, the pattern arrangement of each component depicts the same functionality applied in different environmental fields. The following subsections discuss some of WSN's environmental applications.

25.4.1 Agriculture

The integration of WSNs in agriculture is a recent concept, which leads to what is called precision agriculture [37]. The agricultural WSNs support measurement of the following key parameters: air temperature, air humidity, soil temperature, soil moisture, leaf wetness, atmospheric pressure, solar radiation, trunk/stem/fruit diameter, wind speed, wind direction, and rainfall. A number of researches worked on a large-scale vineyard on very large-scale deployment (65 motes with a maximum of eight hops) in order to collect the data of pH values, titratable acidity (TA), and grape berry weight [38]. The Lofar_Agro project [39] is another application of precision agriculture that is fighting phytophthora using microclimates in potato crops. Lofar_Agro sensors were employed to map out a temperature and soil humidity distribution that gives an effective strategy for controlling the phytophthora disease.

Precision agriculture is now becoming popular in greenhouse techniques. Using such WSNs allows the farmer to easily control the desired crop's environment conditions, which opened a new chance for the farmer to produce different crops in different climates and various seasons. Chaudhary et al. [40] estimate that a greenhouse (70 m × 150 m) designed for a typical crop, will require approximately 40 to 50 wireless nodes (sensors and actuates). Figure 25.7 shows some of the crops' monitoring wireless sensors that may be deployed in a typical greenhouse.

25.4.2 Air and Water Quality

Despite air pollution (air quality) monitoring using traditional periodic data collecting techniques being a very complex and expensive task; it is a vital class of environmental monitoring. However, using WSNs decreases the complexity of air pollution and dangerous gas monitoring via obtaining more immediate readings [22]. The WSN air pollution monitoring system (WAPMS) proposed in [41] gives an example of an air quality monitoring application. WAPMS partitions the region

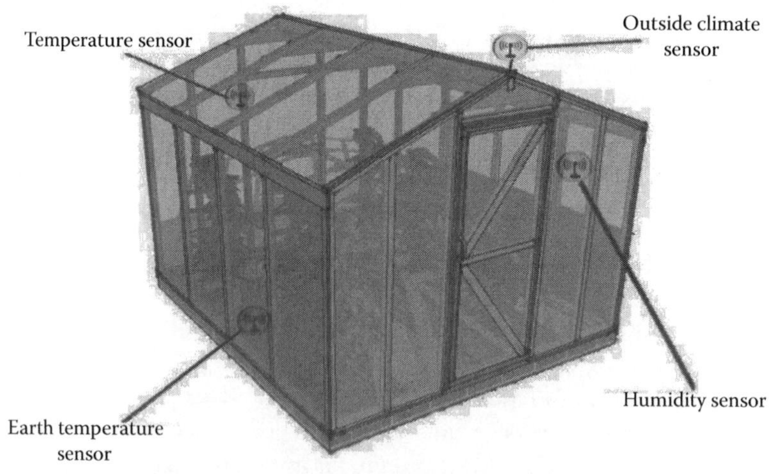

Figure 25.7 Typical greenhouse based on WSNs. (From Chaudhary, D. D. et al., *International Journal of Wireless and Mobile Networks (IJWMN)*, vol. 3, 1, pp. 140–149, 2011.)

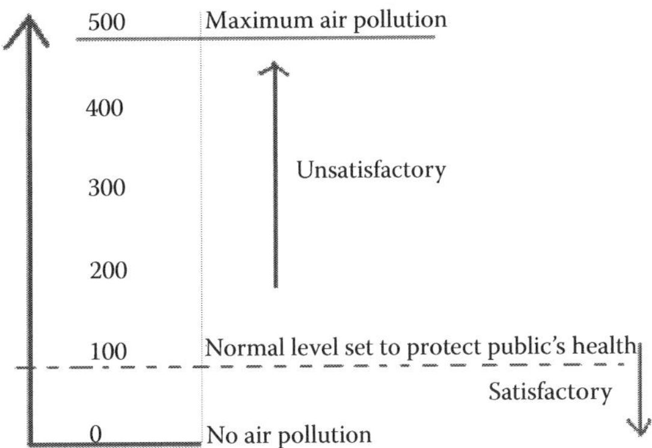

Figure 25.8 Range of AQI values. (From Khedo, K. K. et al., *International Journal of Wireless and Mobile Network (IJWMN)*, vol. 2, 2, pp. 31–45, 2010.)

of interest to be monitored into small zones with a cluster head assigned within each zone. For each zone, sensor nodes sense the data and send them to the cluster head in their respective zone through multi-hop routing. Each set of cluster heads sends its data to a specific sink node that retransmits them to the gateway to be stored within a database to be evaluated. That evaluation is based on a predefined air quality indicator (AQI) that evaluates air quality based on air pollutants that have adverse effects on both human health and the environment. The pollutants are ozone, fine particulate matter, nitrogen dioxide, carbon monoxide, sulfur dioxide, and total reduced sulfur compounds. Figure 25.8 illustrates the ranges of AQI values used for air quality data valuation.

On the other hand, WSN applications for water quality monitoring aim to keep water resources within a standard determined for domestic usage because these resources are prone to a threat of

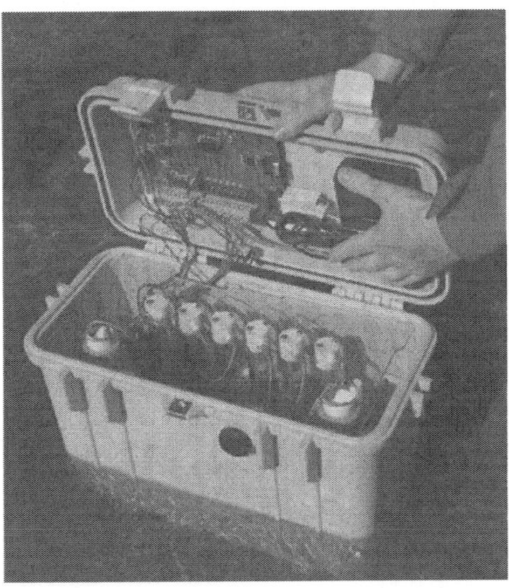

Figure 25.9 SmartCoast multisensor system. (From O'Flynn, B. et al., SmartCoast: A wireless sensor network for water quality monitoring, In *Proceedings of the 32nd IEEE Conference on Local Computer Networks*, LCN 2007, Dublin, Ireland, pp. 815–816, 2007.)

pollution; especially the deliberate contamination of industrial activities. Determination parameters include pH level, turbidity, phosphates, dissolved oxygen, conductivity, and temperature. The usage of WSN techniques for water quality monitoring are not only providing real-time data acquisition, but also reducing the cost of the whole monitoring system as well as the time needed for such a system to be deployed.

The SmartCoast project is an example of applying WSN for water quality monitoring that was designed by Irish researchers [42] in order to follow the EU Water Framework Directive (WFD). The WFD aims at improving the European water environment by 2015. It requires governments to ensure the good chemical status of both surface water and groundwater bodies across Europe [43]. That project is designed to provide six plug-and-play interfaces for measuring temperature, pH, conductivity, depth, phosphate, and turbidity as illustrated in the multisensor system unit shown in Figure 25.9. Also, other real-time water quality monitoring systems based on WSNs are available, such as EMNET [44], Fleck [45], and LakeNet [46].

25.4.3 Noise Pollution

Many environments, such as airports, road works, factories, construction sites, and other environments producing loud noises, require effective noise pollution monitoring systems. Noise pollution is a common environmental problem that affects people's health by increasing the risk of hypertension, ischemic heart disease, hearing loss, and sleep disorders, which also influence human productivity and behavior [47].

Harmful noise is defined as any sound pressure above a certain volume threshold. Sound pressure level (SPL) is a logarithmic measure of the effective pressure of a sound, measured in decibels (dB), relative to a reference value above a standard reference level. The commonly used "zero" reference sound pressure in air is 20μPascal, which is usually considered the threshold of human hearing (at 1 kHz) [48]. On top of the corrective measures applicable to some sources of noise, in 2002, the European Union adopted a directive setting out a community approach to the management and evaluation of ambient noise in order to protect public health [49]. This approach enforces member states to regularly provide accurate mappings of noise levels and exposing that information to citizens, such as the London road traffic noise map website [50].

WSNs can provide a cheap and flexible infrastructure to support the detection of noise pollution sources. A specific noise sensor is used to detect the SPL from the surrounding environments. The collected noise levels are essential for the preparation of noise maps.

A recent example of using WSNs for noise monitoring is the NoiseTube project [51]. The main goal of that project is to turn the ubiquitous mobile phones into noise sensors with the purpose of enabling individuals to measure the level of their daily exposure to the surrounding noise. Moreover, each user can also participate in creating a collective map of noise pollution by sharing geo-localized measurement data with the NoiseTube community. The CitySense project [52] is another example that provides a fixed network of 100 stationary wireless sensors on buildings and streetlights throughout a city (the current target is Cambridge, Massachusetts) and allows collecting noise and air pollution data as well as deliver it in real time to the users. Furthermore, the project supports diverse research communities by providing the essence of open participation on shared resources.

25.4.4 Climate Change Monitoring and Weather Forecasting

As a fact, climate change is changing us; the severity of natural disasters is increasing as an impact of climate change, for example, typhoons, river floods, tropical storms, droughts, landslides, forest fires, ice quakes, and occasional earthquakes. Climate change monitoring is one of the promising areas of WSNs. Wireless sensor nodes trace different climate parameters (e.g., temperature, CO_2 concentration, wind speed, wind direction, soil temperature, light intensity, air humidity, etc.).

Climate monitoring sensor networks are an early stage in terms of disaster reduction, which use historical and contemporary remote sensing data in combination with other geospatial data sets as input to compute predictive models and early warning systems [53]. For example, the GlacsWeb project [54,55] was designed to monitor glacier behavior using a single-hop sensor network of eight glacier probes placed in, on, and under glaciers in Norway. The project aims to understand glacier dynamics in response to climate change (global warming). Each probe periodically sends collected data (e.g., changes in temperature, strain, glacier stress, water pressure, orientation, subglacial sediment, etc.) to a surface base station. While Figure 25.10 demonstrates the glacier probe case, Figure 25.11 shows the general view of the GlacsWeb's communication architecture where glacier probes in Norway communicate via the base and the reference stations back to the server in the UK.

The climate change monitoring process can be extended to find a local/global weather forecast by analyzing historical monitored data. Weather plays a key role in plants' growth as well as plants' health and quantity; that is, weather-related disease, along with heavy rain and high relative

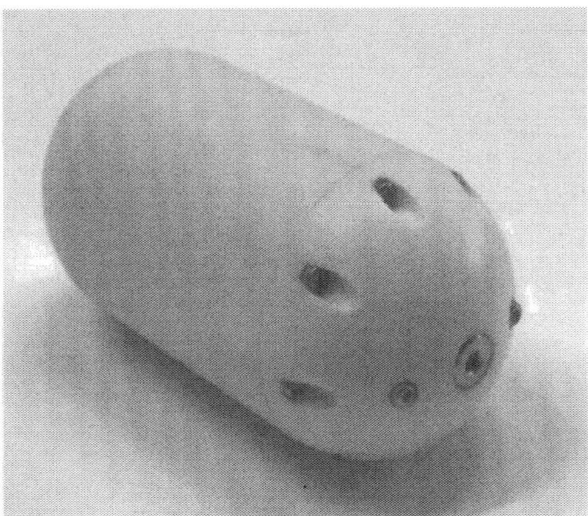

Figure 25.10 Glacier probe case. (From The GlacsWeb project, University of Southampton, Available online at: http://glacsweb.org/technology/, August 2012.)

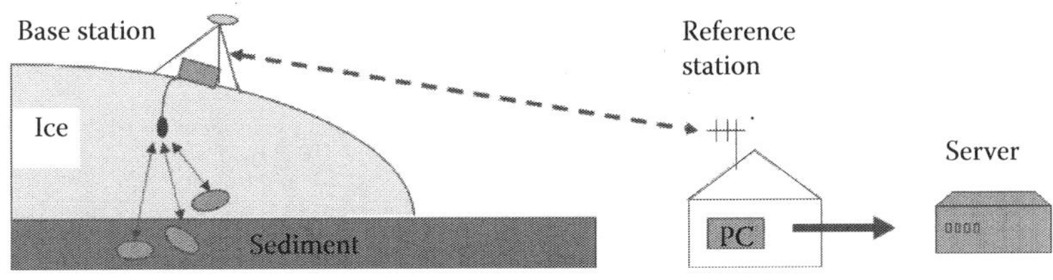

Figure 25.11 General view of the GlacsWeb's communication architecture. (From The GlacsWeb project, University of Southampton, Available online at: http://glacsweb.org/technology/, August 2012.)

humidity, may cause heavy losses in production each year. The high costs of planting materials and labor-motivated efforts to develop weather forecasting systems. For that purpose, WSNs technologies are used for continuous monitoring for some agricultural parameters, such as air temperature, air relative humidity, soil temperature, soil humidity, etc. For example, Ghobakhlou et al. [57] proposed a model that uses WSNs for vineyard farm management in order to prevent freeze damages to the crops. Ghobakhlou's experiments were conducted in nine vineyards chosen in different countries (Chile, Uruguay, Japan, and New Zealand). The main feature of that model is that it enables users to access the WSN collected data via the Internet for further analysis and to support future decision-making.

25.4.5 Structural Health Monitoring

Structural health monitoring (SHM) [58] is an emerging technology, dealing with the development and implementation of continuous and reliable monitoring systems for civil infrastructure using a dense network of smart sensors. Figure 25.12 depicts a SHM system with Smart Sensors [59]. SHM applications detect conditions of bridges, buildings, aircraft, wind turbines, dams, oil pipelines, and naval ships, etc. SHM system development is not a trivial approach because it encompasses many disciplines, such as structural dynamics, materials and structures, fatigue and fracture, nondestructive testing and evaluation, sensors and actuators, microelectronics, signal processing, and much more. Without this global view of the SHM system, it will be difficult for engineers to holistically manage the operation of an engineering structure through its life cycle. Even with old structures that do not have built-in sensors, WSNs provide a reliable and low-cost solution to meet SHM requirements. For the Golden Gate Bridge in San Francisco, as an example, 64 nodes are distributed over the main span and the tower collecting ambient vibrations synchronously at a 1 kHz rate. Sensors capture important modes of dynamic bridge behavior to determine its health status [60]. Anastasi et al. [61] proposed a project to monitor historical heritage buildings. The main aim behind such a project is to preserve these ancient buildings from deterioration because they need periodic restoration interventions.

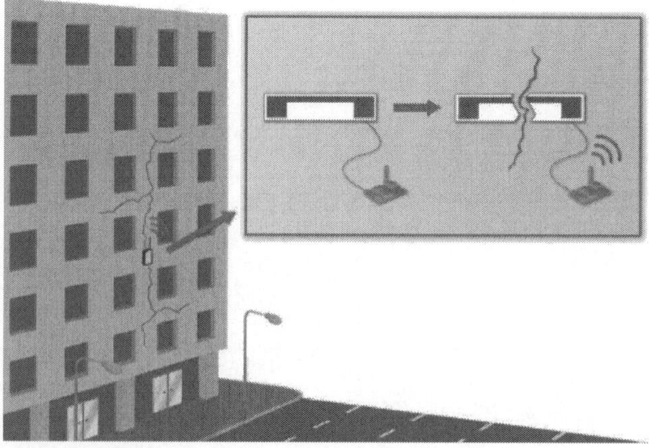

Figure 25.12 SHM system with Smart Sensors. (From Libelium, A. A. Smart cities platform from Libelium allows system integrators to monitor noise, pollution, structural health and waste management, Available online at: http://www.libelium.com/smart_cities, August 2012.)

25.4.6 Natural Disaster Detection

Natural disaster prediction and detection, such as volcano, fire, twister, flood, tsunami, and earthquake detection, is of great interest to public safety and scientific exploration. WSNs could be used to provide predictive early-warning systems that would effectively help to mitigate the damages caused by these natural disasters.

As an example, millions of acres are lost as a result of forest fires every year. By deploying specialized wireless sensor nodes in strategically selected high-risk areas within forests [62], the early detection of fires is increased (as it is illustrated in Figure 25.13). Consequently, this increases the likelihood of success in early extinguishing efforts.

In other situations, where the region of interest is inaccessible and hostile to human beings, such as near the crater of a volcano, deploying WSNs is the only solution to such situations. WSNs are used to measure the dynamic magnitude of the volcano (seismic signals) in order to predict volcano eruption time [63]. Figure 25.14 illustrates an example of using WSNs in volcano monitoring

Figure 25.13 Example of forest early warning system architecture.

Figure 25.14 WSN-based volcano monitoring architecture. (Werner-Allen, G., Lorincz, K., Ruiz, M., Marcillo, O., Johnson, J., Lees, J., Welsh, M., Deploying a wireless sensor network on an active volcano, *IEEE Internet Computing, Special Issue on Data-Driven Applications in Sensor Networks*, vol. 10, no. 22, pp. 18–25. © 2006 IEEE.)

architecture, which uses a network consisting of 16 sensor nodes, each with a microphone and seismometer. They collect seismic and acoustic data on volcanic activity. Nodes relay data via a multi-hop network to a gateway node connected to a long-distance modem, providing radio connectivity with a laptop at the observatory.

On the other hand, earthquakes can be considered to be one of the most devastating natural disasters. Earthquakes occur when the rocks break and move as a result of stresses caused by plate movements. The wireless acceleration sensor devices are deployed as the sensing element in order to recognize the earthquake [65]. The time for spreading information about the probability of the occurrence of an earthquake is critical. Therefore, in Japan, some smart phones offer an early warning service in the event of earthquakes. These phones have a built-in notification system that is tied to Japan's sophisticated early-earthquake-detection service, which can provide between a few seconds and a couple of minutes of advance notice prior to an earthquake's actually occurring [66,67].

Furthermore, global warming is the rise in the average temperature of Earth's atmosphere and oceans since the mid-20th century. Such influences of Earth's climate cause a number of the climate's significant problems. Water flooding is the most considerable problem of such climate change. For that reason, there is a need for an event-driven flood monitoring system. However, monitoring rivers that are several thousand kilometers long is not an easy task and is even expensive to apply using traditional techniques. Therefore, utilizing WSNs plays a main role for lowering the monitoring cost and gaining efficient coverage for the river of interest from flood event monitoring to information dissemination. Figure 25.15 presents a simple deployment for a riverbank monitoring system using WSNs [68]. Sensor nodes are deployed in groups within predefined locations along the riverbank and each group of these sensors transmits water level and water flow data to a local base station. Base stations, in turn, retransmit these data to the user or a data processing center for further analysis.

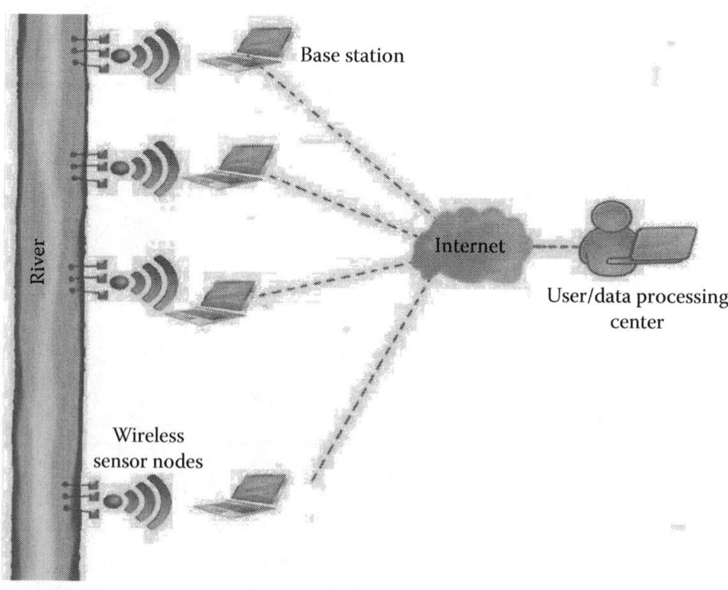

Figure 25.15 Riverbank WSN-based monitoring.

25.5 Conclusions and Future Directions

For smart environmental monitoring, integrating WSNs in the environment itself will considerably raise the degree of environmental protection, enabling a lot of new intelligent features for that environment, such as self-monitoring and self-protection and the possibility of both reactive and proactive reactions to different events. To implement these features, it is necessary to work not only on the development of new sensor units and communication between them, but also on the development of new procedures and algorithms for environmental data collection, analysis, and verification [1].

Smart environmental monitoring systems adapt to changing conditions and intelligently communicate with phenomena in order to obtain certain observations and take certain actions accordingly. Their design and implementation require multidisciplinary knowledge of human-machine interfaces, decision-making, databases, wireless sensor networking, multimedia, and pervasive computing.

Research on smart environmental monitoring has recently made great strides, and for the first time, data of different types and places can be merged together and accessed from anywhere. However, a number of ongoing challenges remain. This chapter addressed the concept of smart environmental monitoring along with presenting its different architectures, applications, and related design issues. Also, it highlighted how smart environments embody the trend toward increasingly connected and automated environmental monitoring through associating wireless sensing devices to environmental phenomena and events. Some recent relevant environmental monitoring projects with real deployments were presented and analyzed. Moreover, the chapter emphasized the techniques and features that should be considered by researchers and system designers for successful deployments and operation of smart environmental monitoring systems. Furthermore, the following outlines some of the issues and directions for future work.

Recent progress in indoor-networking technologies and the rapid explosion of smart handheld devices (such as palmtops and PDAs) have already set the stage for the required networking support for smart indoor environmental monitoring. However, the real challenge is to enable achieving smart environmental monitoring systems with seamless connectivity between heterogeneous networking components and energy-efficient, secured routing in resource-poor wireless mobile networks, including smart sensor devices.

Moreover, as many of the current projects view a smart environmental monitoring system as a stationary entity that is deployed and responds to events and actions of mobile creatures, a broader definition can encompass mobile robotic agents that support extendible monitoring for the surrounding environment.

Also, database design and support for smart environmental monitoring applications presents exceptional challenges as a large amount of sensor data is constantly collected and must be integrated in real time to support prediction, decision-making, and related tasks. Research work on both active and streaming databases for data collected via WSNs can be used for responding quickly to environmental conditions and reduce the burden on the decision maker as well as for efficient processing of multimedia information acquisition and dissemination by sensors. In addition, for monitoring remote or hostile environments using environmental monitoring systems installed in isolated locations that cannot be visited regularly, an intelligent remote and a secured access standard protocol is necessary for operating, managing, reprogramming, and configuring the wireless sensor devices, regardless of manufacturer. Additionally, as WSNs continue to emerge as a technology that will transform the way we measure, understand, and manage the natural environment, producing cheaper and disposable sensor platforms is also a challenge.

Environmental monitoring usually uses a limited type of sensing device, such as temperature, light, humidity, and atmospheric pressure. As previously presented in this chapter, for some recent applications, new environmental monitoring applications will be developed and new sensing units will be necessary to measure new physical phenomena, such as image and video. Transmitting image and video data on resources and power-constrained WSNs are a challenge that has to be concurrently addressed by both scientific researchers and by industry in order to create successful commercial solutions [69]. Also, power management schemes are essential for long-term operation via applying harvesting schemes, cross-layer protocols, and new power storage devices as possible solutions to increase the sensors' lifetime [70].

Finally, other issues and directions for future work still have to be considered, such as the fact that although current real WSNs for environmental monitoring proposes use tens to hundreds of nodes, considering scalability, it is necessary to prove that the available theoretical solutions are suited to large real WSNs. Also, another basic issue is that the IEEE 802.15.4 represents a milestone in standardization efforts. So it is important to specify standard interfaces to allow interoperability between different module vendors in order to reduce the costs and to increase the available options. That's besides reducing the size of sensor devices, which is essential for many applications as battery size and radio power requirements play an important role in size reduction. So the production of platforms compatible with the smart dust can be a determinant in WSN environmental monitoring.

References

1. Stipanicev, D., Bodrozic, Lj., Stula, M. (2007), Environmental intelligence based on advanced sensor networks, In *Proceedings of the 14th International Workshop on 2007 IWSSIP and ECSIPMCS*, Maribor, Slovenia.
2. OECD (2009), Conference proceedings: ICTs, the environment and climate change, In *Proceedings of the High-Level OECD Conference* Helsingør, Denmark.
3. OECD (2008), OECD environmental outlook to 2030. Organization for Economic Cooperation and Development, Paris.
4. Yick, J., Mukherjee, B., Ghosal, D. (2008), Wireless sensor network survey, *Computer Networks*, vol. 52, 12, pp. 2292–2330.
5. Cardell-Oliver, R., Kranz, M., Smettem, K., Mayer, K. (2005), A reactive soil moisture sensor network: Design and field evaluation, *International Journal of Distributed Sensor Networks*, vol. 1, pp. 149–162.
6. David, B., Zhou, Y., Xu, T., Chalon, R. (2011), Mobile user interfaces and their utilization in a smart city, In *Proceedings of the International Conference on Internet Computing (ICOMP'11) as part of WorldComp'2011 Conference*, Las Vegas, Nevada, USA. pp. 383–388.
7. Augusto, J. C., McCullagh, J. P. (2007), Ambient intelligence: Concepts and applications, *Computer Science and Information Systems*, vol. 4, 1, pp. 1–27.
8. Spohrer, J., Maglio, P., Bailey, J., Gruhl, D. (2007), Steps toward a science of service systems, *IEEE Computer*, vol. 40, 1, pp. 71–77.
9. Adaptive robotics, Available online at: [https://inlportal.inl.gov/portal/server.pt/community/robotics_and_intelligence_systems/455], (August 2012).
10. Saha, D., Mukherjee, A. (2003), Pervasive computing: A paradigm for the 21st century, *IEEE Computer*, vol. 36, 3, pp. 25–31.
11. Dix, A., Finlay, J., Abowd, G., Beale, R. (2003), *Human–Computer Interaction*, 3rd Edition, Prentice Hall, New Jersey, USA. 2003.
12. Lewis, F. L. (2004), Smart environments: Technologies, protocols, and applications, In Cook, D. J. Das, S. K. (Eds.), *Wireless Sensor Networks*, Wiley, New York.

13. Lee, S. H., Lee, S., Song, H., Lee, H. S. (2009), Wireless sensor network design for tactical military applications: Remote large-scale environments, In *Proceedings of the 28th IEEE Conference on Military communications (MILCOM)*, IEEE Press, Piscataway, NJ, USA, pp. 911–917.
14. Stankovic, J. A., Cao, Q., Doan, T., Fang, L., He, Z., Kiran, R., Lin, S., Son, S., Stoleru, R., Wood, A. (2005), Wireless sensor networks for in-home healthcare: Potential and challenges, In *Proceedings of the High Confidence Medical Device Software and Systems (HCMDSS 2005) Workshop*, Philadelphia, PA.
15. Ramadan, R. A., Hesham, E. (2008), Deployment of sensor networks on critical infrastructures: A survey, In *Proceedings or the 4th International Computer Engineering Conference Information Society Applications in the Next Decade (ICENCO 2008)*, Cairo, Egypt.
16. Ramadan, R. A. (2009), *Deployment of Mobile Heterogeneous Sensors on Critical Infrastructure: Sensor Deployment Problem and Its Solutions*, VDM Verlag Publisher, Saarbrücken, Germany.
17. Ramadan, R. A., Hagras, H., Nawito, M., Eldesouky A. E. B. (2010), The intelligent classroom: Towards an educational ambient intelligence testbed, In *Proceedings of the 6th International Conference on Intelligent Environments (IE)*, Kuala Lumpur, Malaysia, pp. 344–349.
18. Haoui, A., Kavaler, R., Varaiya, P. (2008), Wireless magnetic sensors for traffic surveillance, *Transportation Research, Part C: Emerging Technologies*, vol. 16, 3, pp. 294–306.
19. ZigBee, Available online at: [www.zigbee.org], (August 2012).
20. Bluetooth, Available online at: [www.bluetooth.org], (August 2012).
21. Postolache, O., Pereira, J. M. D., Silva-Girao, P. M. B. (2009), Smart sensors network for air quality monitoring applications, *IEEET Instrumentation and Measurement*, vol. 58, 9, pp. 3253–3262.
22. Ma, Y., Richards, M., Ghanem, M., Guo, Y., Hassard, J. (2008), Air pollution monitoring and mining based on sensor grid in London, *Sensors*, vol. 8, 6, p. 3601.
23. Ramadan, R. A. (2009), Leveraging RFID technology for intelligent classroom, In *Proceedings of the 4th IEEE International Design and Test Workshop (IDT09)*, Riyadh, Saudi Arabia.
24. Streitz, N. A., Kameas A., Mavrommati, I. (Eds.) (2007), The disappearing computer, In *Lecture Notes in Computer Science*, Berlin, Heidelberg, vol. 4500, Springer.
25. Tolle, G., Polastre, J., Szewczyk, R., Culler, D., Turner, N., Tu, K., Burgess, S., Dawson, T., Buonadonna, P., Gay, D., Hong, W. (2005), A macroscope in the redwoods, In *Proceedings of SenSys'05*, San Diego, California, USA.
26. Ringwald, M., Romer, K. (2007), Deployment of sensor networks: problems and passive inspection, In *Proceedings of the Fifth International Workshop on Intelligent Solutions in Embedded Systems*, Madrid, Spain.
27. Krishnamachari, B. (2005), *Networking Wireless Sensors*, Cambridge University Press, Cambridge, England.
28. Bachrach, J., Taylor, C. (2005), Localization in sensor networks, In *Handbook of Sensor Networks*, John Wiley & Sons, Inc., New York, USA.
29. Bahl, P., Padmanabhan, V. (2000), Radar: An in-building RF-based user location and tracking system, In *Proceedings of the INFOCOMM 2000*, vol. 2, pp. 775–784.
30. Weiser, M., Brown, J. S. (1997), The coming age of calm technology, In Denning, P., Metcalfe, R. (Eds.), *Beyond Calculation: The Next Fifty Years of Computing*, Springer-Verlag, New York.
31. Hansen, M., Kirshmeyer, M., Jensen, C. (2008), Persistent authentication in smart environments, In *Proceedings of the 2nd International Workshop on Combining Context with Trust, Security, and Privacy (CAT08)*, Trondheim, Norway, pp. 31–44.
32. Bolle, R., Connell, J., Pankanti, S., Ratha, N. (2003), Guide to biometrics, In *Springer Professional Computing*, Springer-Verlag New York Inc., USA.
33. Liu, D., Ning, P. (2006), *Security for Wireless Sensor Networks, Advances in Information Security*, vol. 28, Springer, Heidelberg.
34. Nixon, P., Wagealla, W., English, C., Terzis, S. (2004), Security, privacy and trust issues in smart environments, In D.J. Cook, S.K. Das (Eds.), *Smart Environments: Technology, Protocols, and Applications*, John Wiley & Sons, Inc., New York, USA.
35. Buratti, C., Conti, A., Dardari, D., Verdone, R. (2009), An overview on wireless sensor networks technology and evolution, *Sensors*, vol. 9, pp. 6869–6896.

36. Sohraby, K., Minoli, D., Znati, T. (2007), *Wireless Sensor Networks, Technology, Protocols, and Applications*, John Wiley & Sons, Inc., New York, USA.
37 Blackmore, S. (1994), Precision farming: An introduction, *Outlook on Agriculture*, vol. 23, 4, pp. 275–280.
38. Beckwith, R., Teibel, D., Bowen, P. (2004), Unwired wine: Sensor networks in vineyards, In *Proceedings of the 3rd IEEE Conference on Sensors*, Vienna, Austria vol. 2, pp. 561–564.
39. Baggio, A. (2005), Wireless sensor networks in precision agriculture, *Computer Networks*, vol. 49, 6, pp. 743–765.
40. Chaudhary, D. D., Nayse, S. P., Waghmare, L. M. (2011), Application of wireless sensor networks for greenhouse parameter control in precision agriculture, *International Journal of Wireless and Mobile Networks (IJWMN)*, vol. 3, 1, pp. 140–149.
41. Khedo, K. K., Perseedoss, R., Mungu, A. (2010), A wireless sensor network air pollution monitoring system, *International Journal of Wireless and Mobile Network (IJWMN)*, vol. 2, 2, pp. 31–45.
42. O'Flynn, B., Martínez-Català, F., Harte, S., O'Mathuna, C., Cleary, J., Slater, C., Regan, F., Diamond, D., Murphy, H. (2007), SmartCoast: A wireless sensor network for water quality monitoring, In *Proceedings of the 32nd IEEE Conference on Local Computer Networks*, LCN 2007, Dublin, Ireland, pp. 815–816.
43. Water Note 8: Pollution: Reducing dangerous chemicals in Europe's waters, Available online at: [http://ec.europa.eu/environment/water/water-dangersub/index.htm], (August 2012).
44. EmNetLLC.Technology, Available online at: [http://www.heliosware.com/], (August 2012).
45. The CSIRO ICT Centre, Wireless sensor network devices, Available online at: [http://www.ict.csiro.au/page.php?did=80], (August 2012).
46. Seders, L. A., Shea, C. A., Lemmon, M. D., Maurice, P. A., Talley, J. W., LakeNet: An integrated sensor network for environmental sensing in lakes, *Environmental Engineering Science*, vol. 24, pp. 183–191.
47. European Commission (1996), Green paper on future noise policy, COM(96) 540, Brussels, Belgium.
48. European Commission Working Group Assessment of Exposure to Noise (WG-AEN) (2006), Good practice guide for strategic noise mapping and the production of associated data on noise exposure, Brussels, Belgium.
49. European Commission (2002), Directive 2002/49/EC of the European Parliament and of the Council of 25 June 2002 relating to the assessment and management of environmental noise. *Official Journal of the European Communities*, Luxemburg, vol. L189, pp. 12–25.
50. London noise mapping. The London road traffic noise map, [Available online at: www.londonnoisemap.com], (August 2012).
51. NoiseTube, website: http://www.noisetube.net (August 2012).
52. Murty, R. N., Mainland, G., Rose, I., Chowdhury, A. R., Gosain, A., Bers, J., Welsh, M. (2008), CitySense: An urban-scale wireless sensor network and testbed, In *Proceedings of the IEEE International Conference on Technologies for Homeland Security*, Waltham, Massachusetts, USA, pp. 583–588.
53. Bahrepour, M., Meratnia, N., Poel, M., Taghikhaki, Z., Havinga, P. (2010), Distributed event detection in wireless sensor networks for disaster management, In *Proceedings of the 2nd International Conference on Intelligent Networking and Collaborative Systems (INCOS2010)*, Thessaloniki, Greece, pp. 507–512.
54. Padhy, P., Martinez, K., Riddoch, A., Ong, H. L. R., Hart, J. K. (2005), Glacial environment monitoring using sensor networks. In *Proceedings of the Real-World Wireless Sensor Networks Workshop (REALWSN'05)*.
55. Martinez, K., Basford, P. J., De Jager, D., Hart, J. K. (2012), A wireless sensor network system deployment for detecting stick slip motion in glaciers. In *IET International Conference on Wireless Sensor Systems 2012*, London.
56. The GlacsWeb project, University of Southampton, Available online at: [http://glacsweb.org/technology/], (August 2012).
57. Ghobakhlou, A., Shanmuganthan, S., Sallis P. (2009), Wireless sensor networks for climate data management systems, In *Proceedings of the 18th IMACS World Congress (MODSIM09)—International Congress on Modeling and Simulation*, Australia, pp. 959–965.
58. Chang, F.-K. (1997), *Structural Health Monitoring: Current Status and Perspectives*, United States Air Force, Office of Scientific Research, United States Army Research Office, National Science Foundation (U.S.).

59. Libelium, A. A. (2011), Smart cities platform from Libelium allows system integrators to monitor noise, pollution, structural health and waste management, Available online at: [http://www.libelium.com/smart_cities], (August 2012).
60. Kim, S., Pakzad, S., Culler, D., Demmel, J., Fenves, G., Glaser, S., Turon, M. (2007), Health monitoring of civil infrastructures using wireless sensor networks, In *Proceedings of the 6th International Conference on Information Processing in Sensor Networks (IPSN '07)*, Cambridge, MA, ACM Press, pp. 254–263.
61. Anastasi, G., Lo Re, G., Ortolani, M. (2009), WSNs for structural health monitoring of historical buildings, In *Proceedings of the 2nd Conference on Human System Interactions*, pp. 571–576.
62. Hefeeda, M., Bagheri, M. (2009), Forest fire modeling and early detection using wireless sensor networks, *Ad Hoc and Sensor Wireless Networks*, vol. 7, pp. 169–224.
63. Werner-Allen, G., Lorincz, K., Johnson, J., Lees, J., Welsh, M. (2006), Fidelity and yield in a volcano monitoring sensor network, In *Proceedings of the 7th Symposium on Operating Systems Design and Implementation (OSDI '06)*, ACM Press, New York, pp. 381–396.
64. Werner-Allen, G., Lorincz, K., Ruiz, M., Marcillo, O., Johnson, J., Lees, J., Welsh, M. (2006), Deploying a wireless sensor network on an active volcano, *IEEE Internet Computing, Special Issue on Data-Driven Applications in Sensor Networks*, vol. 10, 22, pp. 18–25.
65. Suroso, D. J., Cherntanomwong, P., Sunarno. (2012), Earthquake mitigation: An application of wireless sensor networks, *TEKNOFISIKA*, vol. 1, 1, pp. 27–33.
66. CNNMoney, Elmer-DeWitt, P., Earthquake coming? There's an app for that, August 22, 2011, Available online at: [http://tech.fortune.cnn.com/2011/08/22/earthquake-coming-theres-an-app-for-that] (accessed 27 August 2012)
67. Gigaom, D. E., Early earthquake warning pops up in iOS 5 beta, Available online at: [http://gigaom.com/apple/early-earthquake-warning-pops-up-in-ios-5-beta/] (accessed 27 August 2012)
68. Karanasios, S. (2011), New and emergent ICTs and climate change in developing countries, in the climate change, Innovation and ICTs project in coordination with the University of Manchester's Centre for Development Informatics, 2011, Available online at: [http://www.niccd.org] (accessed 28 August 2012)
69. Zhou, L., Xiong, N., Shu, L., Vasilakos, A. V., Park, J. H. (2010), Context-aware middleware for multimedia services in heterogeneous networks, *IEEE Intelligent Systems*, vol. 25, 2, pp. 40–47.
70. Oliveira, L. M. L., Rodrigues, J. J. P. C. (2011), Wireless sensor networks: A survey on environmental monitoring, *Journal of Communications*, vol. 6, 2, pp. 143–151.

Index

Page numbers followed by f and t indicate figures and tables, respectively.

A

B

C

E

F

I

J

K

L

M

N

Q

R

T

U

V

W

X

Z

Printed and bound by CPI Group (UK) Ltd, Croydon, CR0 4YY
25/10/2024
01779408-0005